Modern Crystal & Mechanical Filters

OTHER IEEE PRESS BOOKS

Electrical Noise: Fundamentals and Sources, *Edited by M. S. Gupta*
Computer Methods in Image Analysis, *Edited by J. K. Aggarwal, R. O. Duda, and A. Rosenfeld*
Microprocessors: Fundamentals and Applications, *Edited by W. C. Lin*
Machine Recognition of Patterns, *Edited by A. K. Agrawala*
Turning Points in American Electrical History, *Edited by J. E. Brittain*
Charge-Coupled Devices: Technology and Applications, *Edited by R. Melen and D. Buss*
Spread Spectrum Techniques, *Edited by R. C. Dixon*
Electronic Switching: Central Office Systems of the World, *Edited by A. E. Joel, Jr.*
Electromagnetic Horn Antennas, *Edited by A. W. Love*
Waveform Quantization and Coding, *Edited by N. S. Jayant*
Communication Satellite Systems: An Overview of the Technology, *Edited by R. G. Gould and Y. F. Lum*
Literature Survey of Communication Satellite Systems and Technology, *Edited by J. H. W. Unger*
Solar Cells, *Edited by C. E. Backus*
Computer Networking, *Edited by R. P. Blanc and I. W. Cotton*
Communications Channels: Characterization and Behavior, *Edited by B. Goldberg*
Large-Scale Networks: Theory and Design, *Edited by F. T. Boesch*
Optical Fiber Technology, *Edited by D. Gloge*
Selected Papers in Digital Signal Processing, II, *Edited by the Digital Signal Processing Committee*
A Guide for Better Technical Presentations, *Edited by R. M. Woelfle*
Career Management: A Guide to Combating Obsolescence, *Edited by H. G. Kaufman*
Energy and Man: Technical and Social Aspects of Energy, *Edited by M. G. Morgan*
Magnetic Bubble Technology: Integrated-Circuit Magnetics for Digital Storage and Processing, *Edited by H. Chang*
Frequency Synthesis: Techniques and Applications, *Edited by J. Gorski-Popiel*
Literature in Digital Processing: Author and Permuted Title Index (Revised and Expanded Edition), *Edited by H. D. Helms, J. F. Kaiser, and L. R. Rabiner*
Data Communications via Fading Channels, *Edited by K. Brayer*
Nonlinear Networks: Theory and Analysis, *Edited by A. N. Willson, Jr.*
Computer Communications, *Edited by P. E. Green, Jr. and R. W. Lucky*
Stability of Large Electric Power Systems, *Edited by R. T. Byerly and E. W. Kimbark*
Automatic Test Equipment: Hardware, Software, and Management, *Edited by F. Liguori*
Key Papers in the Development of Coding Theory, *Edited by E. R. Berlekamp*
Technology and Social Institutions, *Edited by K. Chen*
Key Papers in the Development of Information Theory, *Edited by D. Slepian*
Computer-Aided Filter Design, *Edited by G. Szentirmai*
Laser Devices and Applications, *Edited by I. P. Kaminow and A. E. Siegman*
Integrated Optics, *Edited by D. Marcuse*
Laser Theory, *Edited by F. S. Barnes*
Digital Signal Processing, *Edited by L. R. Rabiner and C. M. Rader*
Minicomputers: Hardware, Software, and Applications, *Edited by J. D. Schoeffler and R. H. Temple*
Semiconductor Memories, *Edited by D. A. Hodges*
Power Semiconductor Applications, Volume II: Equipment and Systems, *Edited by J. D. Harnden, Jr. and F. B. Golden*
Power Semiconductor Applications, Volume I: General Considerations, *Edited by J. D. Harnden, Jr. and F. B. Golden*
A Practical Guide to Minicomputer Applications, *Edited by F. F. Coury*
Active Inductorless Filters, *Edited by S. K. Mitra*
Clearing the Air: The Impact of the Clean Air Act on Technology, *Edited by J. C. Redmond, J. C. Cook, and A. A. J. Hoffman*

Modern Crystal & Mechanical Filters

Edited by

Desmond F. Sheahan

Chief Engineer, Components and Processes
GTE Lenkurt

Robert A. Johnson

Senior Design Engineer
Collins Commercial Telecommunications Division
of Rockwell International

A volume in the IEEE PRESS Selected Reprint Series,
prepared under the sponsorship of the
IEEE Circuits and Systems Society.

IEEE PRESS

The Institute of Electrical and Electronics Engineers, Inc. New York

PRINTED IN THE UNITED STATES OF AMERICA

IEEE International Standard Book Numbers: Clothbound: 0-87942-095-2
Paperbound: 0-87942-096-0

Library of Congress Catalog Card Number 76-57822

Sole Worldwide Distributor (Exclusive of the IEEE):

JOHN WILEY & SONS, INC.
605 Third Ave.
New York, NY 10016

Wiley Order Numbers: Clothbound: 0-471-03237-9
Paperbound: 0-471-03236-0

Contents

Introduction

The world continues to move in the direction of more extensive and more complex electronic communication links between its people. Individual messages have either a frequency or a time spectrum, and some form of filtering process is needed to separate them. New methods of filtering have been developed in response to demands for lower system costs and reduced equipment size. Some of the concepts being explored are in new technologies such as digital filtering or surface acoustic wave devices. However, there has also been a significant upsurge of interest in the older crystal and mechanical filter technologies, and modern production facilities have been set up to exploit them. This collection of papers addresses itself to the significant developments that have occurred in these older technologies during the last few years.

The planar acoustically coupled crystal filter of today bears little resemblance to the crystal lattice filter manufactured over the past four decades. The performance of present day voice-channel mechanical filters used in telephone systems is an order of magnitude better than that of their predecessors. Factories building telecommunication filters are now highly automated in terms of fabrication, tuning, and testing. The engineering and manufacturing of mechanical and crystal filters have become sophisticated technologies, as evidenced by the papers that comprise this book.

The crystal filter technologies included in this book concentrate on devices that employ acoustically coupled resonators. We have not emphasized papers on lattice-type crystal filters or electrically coupled piezoelectric ceramic filters. In addition, we have not included surface acoustic wave (SAW) filters because of the uniqueness of their realization in comparison to ladder-type mechanical and crystal filters. Therefore, this is a book on the subject of electromechanical bandpass filters which are ladder networks employing input and output electromechanical transducers, resonators, and acoustic coupling elements. The "crystal" filters described in this book are generally "monolithic" or "polylithic" filters utilizing deposited electrodes on quartz-crystal substrates, the resonators being trapped-energy regions between electrodes, and the coupling elements being the nonplated regions between the electrode pairs. The term "mechanical" filter is used to describe discrete resonator (rod, disk, bar), wire-coupled designs.

The first paper, "Crystal and Mechanical Filters," serves as a good introduction. In that paper filters are treated from a physical and acoustic perspective rather than from an electrical equivalent circuit standpoint. The paper is tutorial in nature and was written for the purpose of introducing the network theorist to the subject of acoustics as applied to electromechanical bandpass filters.

Crystal and Mechanical Filters

DESMOND F. SHEAHAN, MEMBER, IEEE, AND ROBERT A. JOHNSON, MEMBER, IEEE

***Abstract*—As the design of crystal and mechanical filters requires a knowledge of techniques not generally known to the designer of *LC* filters, these techniques are emphasized in this paper. The paper discusses the commonality of mechanical resonance possessed by these filters and how this leads to design restrictions. The basics of resonator theory are discussed and also the coupling techniques used to make multi-resonator filters. A listing is supplied of the types of filters that are generally made with mechanical or crystal resonators. A flow chart shows the normal design process, and the electrical transformations that are commonly used to develop the *LC* designs into useful form are also described. The subjects of tuning techniques and manufacturing methods, in general, are discussed as they can have an important bearing on these types of filters. The properties of the materials commonly used for the resonators and the transducers are described, as well as the restrictions that they in turn place on the design process. The paper concludes with a discussion of some topics where more work can and should be done. The aim of the paper is to impart to the nonspecialist some idea of how this important class of filters is designed and manufactured, as well as giving an idea of the performance that can reasonably be expected from them.**

I. Crystal and Mechanical Filters

CRYSTAL and mechanical filters utilize mechanical resonances to accomplish the filtering of electrical signals. In that respect, they differ substantially from other filters where purely electrical resonances are involved.

Because these devices mechanically filter electrical signals, they must utilize transducers to translate electrical energy into mechanical energy and vice versa. Just as we can readily identify crystal and mechanical filters by the presence of a transducer, the way in which the transducer is realized is the main distinguishing feature between the crystal and the mechanical types of filter. In crystal filters, the transducer is integral with the filter resonators and the transducer action comes from the piezoelectric properties of the crystal material. Mechanical filters, on the other hand, have transducers that are a part of the end resonators and, therefore, they can take a number of different forms. As a general rule, however, we can categorize the three most common types of crystal and mechanical filters in the manner shown in Fig. 1.

In the past, the distinction between crystal and mechanical filters was based on the method used to couple the various resonators. Crystal filters were electrically coupled single-resonator crystals in a lattice or hybrid lattice form. Mechanical filters were mechanically coupled resonators in ladder or bridged ladder form. Confusion arose with the development of monolithic crystal filters that utilized mechanical coupling but were manufactured from quartz crystals. This mechanical coupling in a crystal is commonly referred to as acoustic coupling. The monolithic filters are usually classified as crystal filters, which is justified on the basis of manufacturing processes that are common to those of single-resonator quartz crystals, but the design techniques have more in common with mechanical ladder filters.

This paper will discuss the many-sided nature of crystal and mechanical filter design. It will emphasize the many considerations that enter into this design process without putting a major emphasis on synthesis techniques which are used extensively. In order to keep the paper to a manageable size, we will discuss bandpass filters only, because by far the vast majority of crystal and mechanical filters are used for bandpass applications.

The photographs in Fig. 2 show some of the filter types that will be discussed. It is worth noting that all of these filters are in large volume production at various companies [1]–[3].

II. Resonator Types

Mechanical and crystal filters utilize the bulk resonances of the resonators. This means that the total volume of the resonator is involved in the resonance as distinguished from other types such as surface-wave resonators, where only a thin surface layer is involved. The boundaries of the resonator should be noted carefully as they are not always obvious. An example of this occurs in the thickness mode and also in the thickness-shear mode of vibration where a phenomenon called energy trapping occurs. As we will see later, the resonator is defined by the crystal area where the electrodes on opposite surfaces overlap and not by the actual boundaries of the crystal itself.

In designing a resonator, we will generally wish to obtain one dominant mode of vibration. Spurious modes will tend to be present in any resonator and usually we will want to suppress them or, failing that, their presence will have to be accounted for in the filter design. There are some special cases, however, where resonators will be deliberately designed to have more than one mode of vibration [4]. Additional modes are sometimes used in mechanical filters to shape the stopband response or to provide an additional natural resonant frequency [5].

Fig. 4 shows various resonators along with their useful frequency ranges and their dominant frequency determining dimensions. Of these various resonators, the ones that have found the greatest usage in recent years are the flexure, extensional, and torsional modes for mechanical filters and the thickness-shear mode for crystal filters.

Examination of Fig. 2 shows the dominant frequency determining dimensions for each of the resonator modes, and it can be seen that these dimensions are common in many cases. Consequently, care must be taken to mount

Manuscript received September 13, 1974.
D. F. Sheahan is with GTE Lenkurt, Inc., San Carlos, Calif. 94070.
R. A. Johnson is with Collins Radio/Group-RI, Newport Beach, Calif. 92663.

Reprinted from *IEEE Trans. Circuits and Syst.*, vol. CAS-22, pp. 69–89, Feb. 1975.

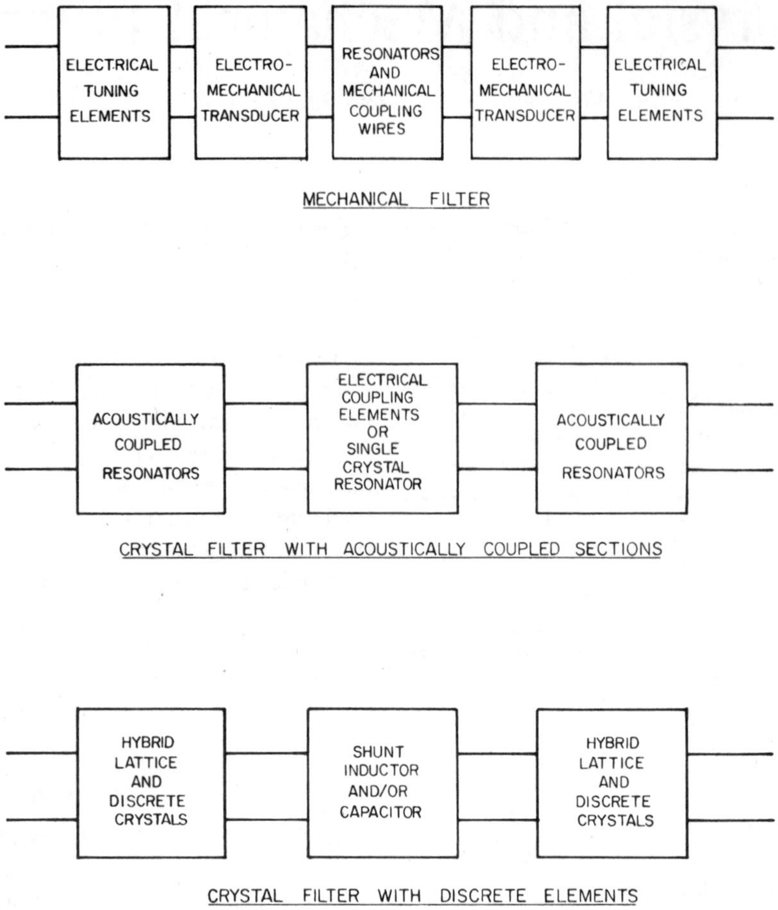

Fig. 1. The 3 main categories of crystal and mechanical filters.

or drive the resonator so that only the desired mode is excited. For example, the flexure-mode resonator could be suspended at two points, as shown, without hindering the flexure mode, while this mounting would offer resistance to the extension mode. The dotted lines or points in the diagrams correspond to nodal lines, or points of no motion in specific directions. For example, there is no linear motion at the nodal points of the bar flexure-mode resonators but there is torsion at these two points. Low-frequency mechanical filters in the 3–30-kHz frequency range are designed to couple torsional energy from resonator to resonator with small diameter wires [6]. These wires are also used as supports, but the fact that torsional energy is also transmitted to the mounting structure must be taken into account in the design.

In thickness-shear-mode crystal resonators, the mechanical energy is trapped underneath the electrodes. The crystal is therefore supported from points away from the area of electrode overlap, and since there will be no mechanical motion at these points, the supporting structure will, to a first approximation, have negligible influence on the resonator's frequency. The nodal lines also indicate the velocity distribution in various directions along the surfaces of the resonators. In the mechanical filter case, the coupling wire position, its size, and, ultimately, the filter bandwidth is a function of the velocity distribution across the resonator surfaces. The question of support and coupling wire position may seem like a trivial problem from an academic point of view, but a proper solution often means the difference between the success or failure of a product.

Basic Mechanical Resonator Concepts

Of greatest importance to the filter designer are the parameters governing the frequency of the resonator. Also of importance, in particular to the mechanical filter designer, is the velocity distribution or equivalent mass of the resonator at points on the resonator surfaces [1], [7]. The equivalent mass corresponds to the mass of a comparable spring-mass resonator tuned to the same frequency as the actual distributed-mass resonator. The equivalent mass is proportional to the inverse of the velocity squared or more exactly, equal to the total kinetic energy in the resonator divided by one-half the velocity (in a specified direction) squared at a point on the resonator. The importance of the equivalent mass is related to both the design of the transducer and the mechanical coupling means, or more specifically, the physical size of the transducer and coupling wires is proportional to the equivalent mass of the resonator at the points of attachment.

Regardless of the particular mode of vibration of isotropic (properties the same in all directions) materials such as iron–nickel alloys, the natural frequency of a resonator will be of the form

$$\omega = \frac{K}{X}\sqrt{\frac{E}{\rho}} \tag{1}$$

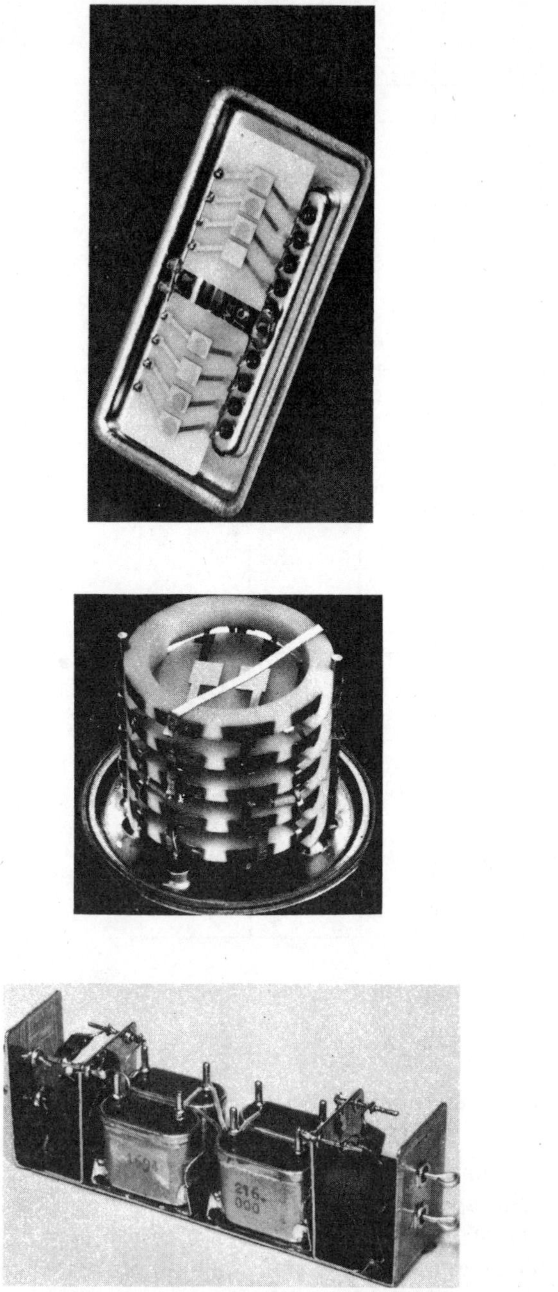

Fig. 2. (a) Bilithic (also called monolithic) crystal filter. (b) Polylithic crystal filter. (c) Conventional crystal filter.

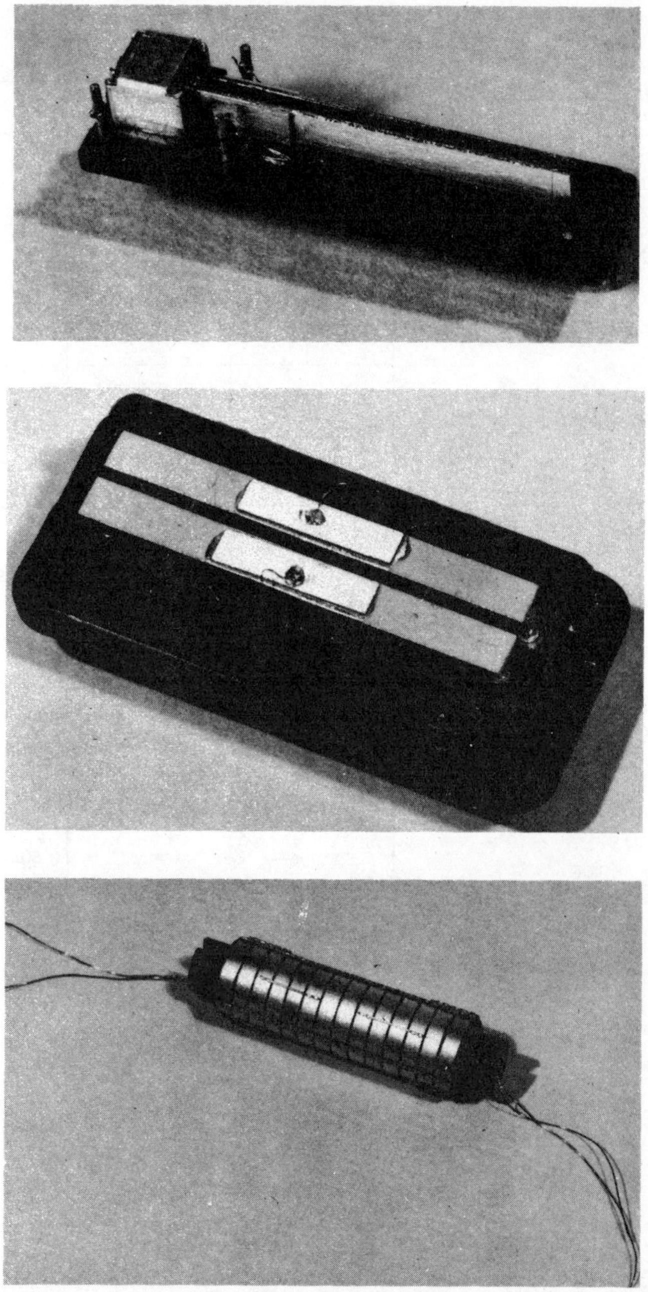

Fig. 3. (a) Tuning fork mechanical filter. (b) Bar flexure mechanical filter. (c) Disk-wire mechanical filter.

where K may be some complex function or simply a constant, as is the case in the extensional-mode resonator equation. X is a function having the dimension of length, E is Young's modulus (or the shear modulus), and ρ is the density of the material. In the simple case of an extensional-mode resonator,

$$\omega_n = \frac{n\pi}{L}\sqrt{\frac{E}{\rho}} \tag{2}$$

where n is the number of half-wavelengths along the length L, and E is Young's modulus. In the torsion case, the shear modulus G is substituted for Young's modulus.

In both of the preceding cases, the velocity distribution is sinusoidal along the length of the bar. The equivalent mass of the extensional bar at the end of the bar and in the direction of motion of the bar is

$$M_{eq} = \frac{K \cdot E \cdot}{\frac{1}{2} v_L^2} \tag{3}$$

$$M_{eq} = \frac{\int_{x=0}^{x=L} \frac{1}{2} m v_x^2}{\frac{1}{2} v_L^2} \tag{4}$$

since

$$v_x = v_L \cos\left(\pi \frac{x}{L}\right)$$

$$M_{eq} = \frac{\dfrac{v_L^2 \rho A}{2} \displaystyle\int_{x=0}^{x=L} \cos^2\left(\pi \frac{x}{L}\right) dx}{\frac{1}{2} v_L^2} \tag{5}$$

	NAME	MECHANICAL	CRYSTAL	FREQ. DETERMINING DIMENSION	USEFUL FREQ. RANGE
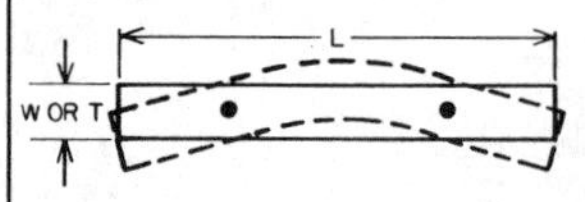	BAR FLEXURE	√	√	L, T L, W	3 kHz → 60 kHz
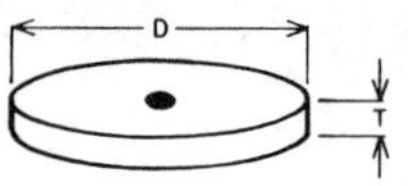	DISC FLEXURE	√	√	D, T	60 kHz → 600 kHz
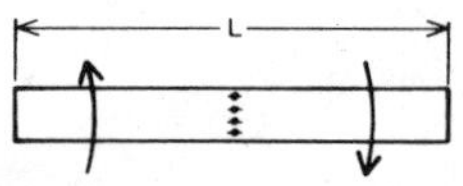	TORSION	√		L	60 kHz → 300 kHz
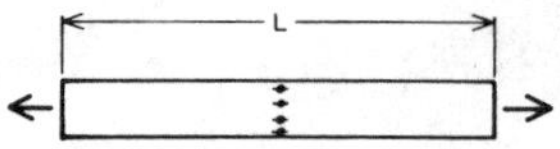	EXTENSION	√	√	L	60 kHz → 500 kHz
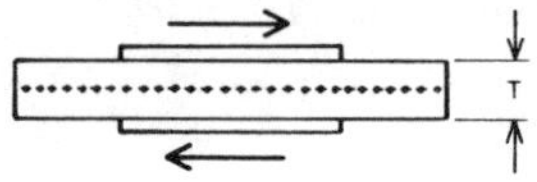	THICKNESS SHEAR		√	T	1 MHz → 100 MHz
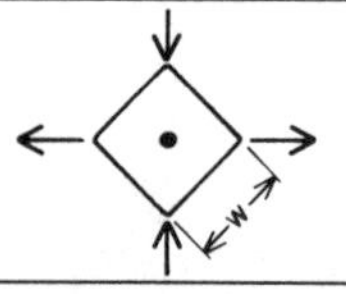	FACE SHEAR		√	W	300 kHz → 1 MHz

••••• = NODAL LINES OR POINTS
L = LENGTH
T = THICKNESS
D = DIAMETER
W = WIDTH

Fig. 4. The most common types of resonators.

and after integrating,

$$M_{eq} = \frac{\rho AL}{2} \tag{6}$$

which means simply that the equivalent mass is equal to one-half the static mass of the bar where A is the cross-sectional area of the bar.

The equation for the bar vibrating in flexure has the same form as (1), but the K/X term is found by solving an equation involving cos and cosh terms [8], [9]. The displacement at any point x on the bar is found from

$$\delta = A(\cosh kx - \cos kx) + B(\sinh kx - \sin kx) \tag{7}$$

where

$$k^4 = \rho A\omega^2/EI \tag{8}$$

and I is the moment of inertia of the section.

Flexure-mode disk resonators are more complicated as they have a velocity distribution that is in the form of a Bessel function along the radius and sinusoidal along the circumference (or constant around the circumference when vibrating in a circular nodal point mode) [10].

Mechanical-filter resonators are made from so-called constant-modulus alloys with trade names like Ni-Span C, Sumi-Span, and Thermelast. Since the dimensions and density of most iron alloys change very little with changing temperature, the major factor in temperature drift is the change in Young's modulus, and so these alloys are blended to produce a temperature-stable modulus. The addition of nickel to produce an alloy can decrease the temperature shift to a relatively low value. The addition of chromium (or molybdenum) reduces the sensitivity of the temperature coefficient to nickel–iron composition changes and an additional element, titanium (or beryllium), makes it possible to adjust the temperature coefficient after fabricating the resonator [11], [12].

Thickness-Shear Crystal Resonators

The design of crystal resonators is more complicated than that of mechanical resonators because we are dealing with an anisotropic material, which is one whose properties vary with the direction in which they are measured [13]. Density is an isotropic property, i.e., not direction dependent but elasticity is anisotropic. Consequently, mechanical resonance characteristics are direction dependent in a crystal. Piezoelectricity is another direction-dependent property, and since in a crystal we are relying on the piezoelectricity for the transducer action, there is no point in even considering the mechanical resonance properties for a particular crystalline direction unless the desired resonance can be excited by the integral transducer, i.e., the piezoelectric effect. In a crystal, therefore, the transducer and the mechanical resonance properties cannot be considered in isolation whereas with mechanical filters, there is some freedom to consider them separately.

The thickness-shear-mode crystal resonator has attracted a great deal of attention because it can be mounted in such a manner as to decouple the dominant mode of vibration from the supports. This is due to a phenomenon called energy trapping [14]–[17], which means that when electrodes are deposited on a crystal, the mechanical energy is confined to the area underneath the plated electrodes thus leaving room outside of this area for supporting the crystal and making the necessary external electrical connections.

The thickness-shear mode of vibration is generally used in the temperature stable *AT*-cut of quartz. This is a plate of quartz cut at a particular angle from the basic crystal and having the property that electrodes on the top and bottom surfaces of the plate can be used to exite a shear mode of vibration in the plate. In order to understand energy trapping, let us consider the solution of the wave equation for a thickness-shear vibration in an infinite plate.

A cross section of the plate is shown in Fig. 5 where we are assuming a plate that is infinite in the z direction. The wave equation is

$$\nabla^2 u = \frac{1}{v^2}\frac{\delta^2 u}{\delta t^2} \tag{9}$$

where $u(y,x,t)$ is the acoustic disturbance and the displacement due to this disturbance is in the x direction, as shown in Fig. 5. v equals $(c/\rho)^{1/2}$ where v is the acoustic velocity, c is the crystal elasticity, and ρ is its density. The solution of the wave equation has the form

$$u = (A \sin \alpha x + B \cos \alpha x) \sin \beta y \exp(-j\omega t). \tag{10}$$

When the expressions for the boundary conditions have been satisfied, we obtain the equation for α as follows:

$$\alpha^2 = \frac{\rho}{c}(\omega^2 - \omega_p{}^2) \tag{11}$$

where

$$\omega_p = (2p - 1)\frac{\pi}{h}\sqrt{\frac{c}{\rho}}. \tag{12}$$

This tells us that a nondecaying standing-wave solution of the wave equation only exists when ω is greater than ω_p, but we can have many values of ω_p, depending on the integer p. When p is equal to unity, we have what is called the fundamental mode of vibration, whereas we have overtone modes when p is greater than one.

Let us now consider the situation shown in Fig. 6 where electrodes have been plated on a section of the crystal. We can consider the effect of this plating to be one of either increasing the crystals effective density or of increasing its effective thickness. In any case it will result in a lowering

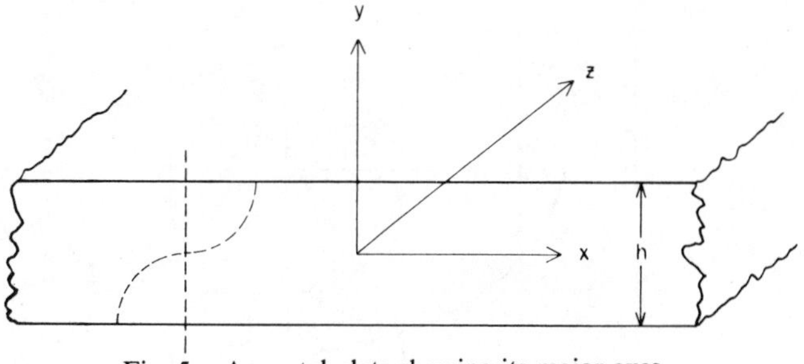

Fig. 5. A crystal plate showing its major axes.

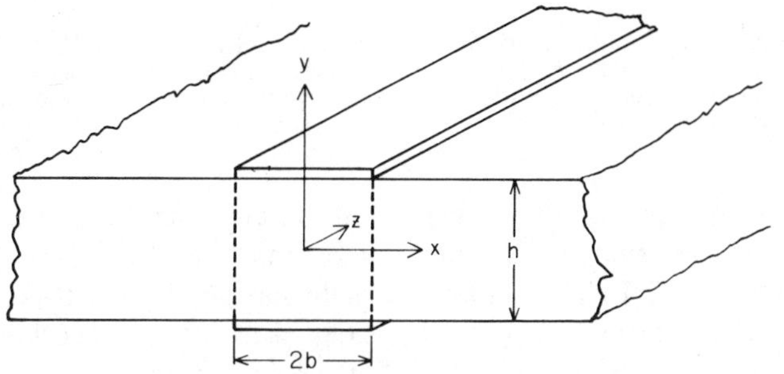

Fig. 6. A crystal plate with electrodes having infinite length in the z direction.

of the frequency in the plated area, i.e.,

$$\rho \text{ (plated area)} > \rho \text{ (unplated area)}$$

therefore,

$$\omega_p \text{ (plated area)} < \omega_p \text{ (unplated area)}.$$

We will therefore have a range of frequencies in which α^2 is positive in the plated area of the crystal and negative outside, i.e., when

$$\omega_p \text{ (plated area)} < \omega < \omega_p \text{ (unplated area)}. \tag{13}$$

There will be standing waves in the electroded region and the waves will decay exponentially outside of this region. We therefore say that the wave energy is trapped in the electroded region and the trapping is caused by the mass loading of the electrodes.

In the simplified analysis just described an infinite z dimension was assumed. This was because the analysis is very complex with a finite dimension. However, we will now consider in general terms what happens when all the dimensions are finite. The solution for the wave equation in the simplified analysis for one finite lateral dimension has standing waves in the x direction. These are caused by reflections off the discontinuities at the transitions between the electroded and the unelectroded regions. When we have a finite dimension in the z direction, we will have corresponding reflections and standing waves in that direction also. However, because of the anisotropy, the propagation constant for the z direction will be different from that in the x direction and this leads to the complexity of an exact analysis.

The effects of restricting lateral dimensions have been computed by Spencer [18] for two cases; 1) where the z dimension is infinite, x finite, and 2) where the x dimension is infinite and the z dimension is finite. These calculations

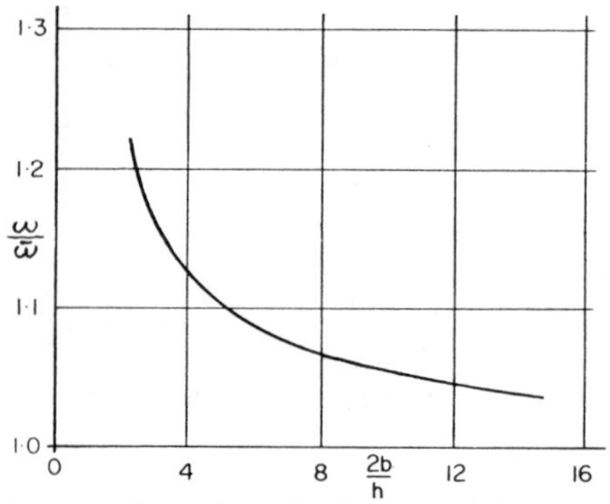

Fig. 7. This curve shows how the frequency of a resonator rises as its lateral electrode dimensions are restricted; ω is the frequency of an infinite plate. ϖ is the frequency of a semi-infinite plate having a restricted lateral dimension $2b$.

take the form shown in Fig. 7. On this curve, ϖ is the resonant frequency of an infinite plate, and ω is the resonant frequency of the semi-infinite plate having the length-to-thickness ratio $2b/h$. This tells us that as the finite dimension of a resonator plate is restricted, the resonant frequency will rise above that of an infinite plate [19]. This is true whether the restriction is in the x direction, the z direction, or both. The rise becomes greater as the finite dimension gets closer to the thickness.

The significance of this becomes clearer when we discuss inter-resonator coupling. There we will see that coupling is dependent on the frequency difference between the plated resonator and the surrounding unplated quartz. The resonator frequency is below that of the surrounding quartz because of the mass loading of the electrodes, but due to the finite length effect described above, we can see that the amount of mass loading required will increase as the lateral resonator dimensions get smaller. When designing a resonator, we must consider its length-to-thickness ratios to make sure that the desired plateback, or frequency lowering due to the plating process, can be achieved with a moderate amount of mass loading. If the amount of mass loading required is too large, the resonator Q will be adversely affected.

One other subject that must be discussed when considering the resonator design is that of anharmonic modes [20], [21]. These are weak spurious modes of vibration that occur at frequencies slightly above each overtone mode. They occur at frequencies that correspond to the cases when the lateral dimensions are an integral number of half wavelengths for the propagation in the lateral direction. In spite of the fact that optimum electrode length-to-width ratios can be chosen for rectangular resonators to keep these spurious modes away from the main resonances, anharmonic modes can still cause problems in a multi-resonator monolithic filter. If however, no more than two coupled-resonator crystals are used in the filter, any electrical energy that results from a spurious mode of vibration at a frequency in the filters stopband will see an electrical mismatch outside of the resonator and, therefore, not cause a major problem in the filter.

III. Coupling Elements

Up to now, we have considered only separate resonators and the factors that must be considered when they are being designed for use in filters. The present widespread interest in crystal and mechanical filters is due to the fact that separate resonators can be coupled mechanically (i.e., acoustically). In the case of crystals where electrical access can be made, electrical coupling can also be used, and this points out one of the main differences between mechanical and crystal filters. In the conventional ladder-type mechanical filter, electrical access is available to the filter from the extreme ends of the filter on the electrical sides of the transducers. Consequently, only mechanical coupling can be used, and the allowable electrical elements are located at the terminations or else connecting the two terminations in a bridging sense. Exceptions to this are cascaded mechanical filters [22], and single-resonator tuning forks that are coupled like discrete-resonator crystal filters [23]. In these cases, each filter or tuning fork has an input and an output transducer allowing a wide variety of interconnections.

Coupling in Mechanical Filters

In most instances, mechanical coupling between resonators is realized through the use of small diameter wires welded to the resonator surfaces. The size of the coupling wire is a function of the equivalent mass of the resonator at the point of attachment and the fractional bandwidth of the filter (BW/f_0). The wire material is an iron–nickel alloy with constant modulus characteristics, and it is similar to the material used in the resonators. The coupling may be flexural, torsional, extensional, or some combination of these depending both on the mode of vibration of the resonator and on the position of attachment. The three possible modes of vibration can be described in two-port matrix form, in much the same way as an electrical transmission line, as seen by

$$\begin{bmatrix} v_1 \\ f_1 \end{bmatrix} = \begin{bmatrix} \cos B_e x & jZ_e \sin B_e x \\ j/Z_e \sin B_e x & \cos B_e x \end{bmatrix} \begin{bmatrix} v_2 \\ f_2 \end{bmatrix} \text{(Extension)} \quad (14)$$

$$\begin{bmatrix} \dot{\theta}_1 \\ m_1 \end{bmatrix} = \begin{bmatrix} \cos B_t x & jZ_t \sin B_t x \\ j/Z_t \sin B_t x & \cos B_t x \end{bmatrix} \begin{bmatrix} \dot{\theta}_2 \\ m_2 \end{bmatrix} \text{(Torsion)} \quad (15)$$

and

$$\begin{bmatrix} f_1 \\ m_1 \\ v_1 \\ \dot{\theta}_1 \end{bmatrix} = \begin{bmatrix} \alpha_{11} & \alpha_{12} & \alpha_{13} & \alpha_{14} \\ \alpha_{21} & \alpha_{22} & \alpha_{23} & \alpha_{24} \\ \alpha_{31} & \alpha_{32} & \alpha_{33} & \alpha_{34} \\ \alpha_{41} & \alpha_{42} & \alpha_{43} & \alpha_{44} \end{bmatrix} \begin{bmatrix} f_2 \\ m_2 \\ v_2 \\ \dot{\theta}_2 \end{bmatrix} \text{(Flexure)} \quad (16)$$

where f is force, v is velocity, m is the bending moment, $\dot{\theta}$ is the angular velocity, and x is the distance from a free end of the wire.

In addition,

$$B_e = \omega\sqrt{\rho/E}$$

$$B_t = \omega\sqrt{\rho/G}$$

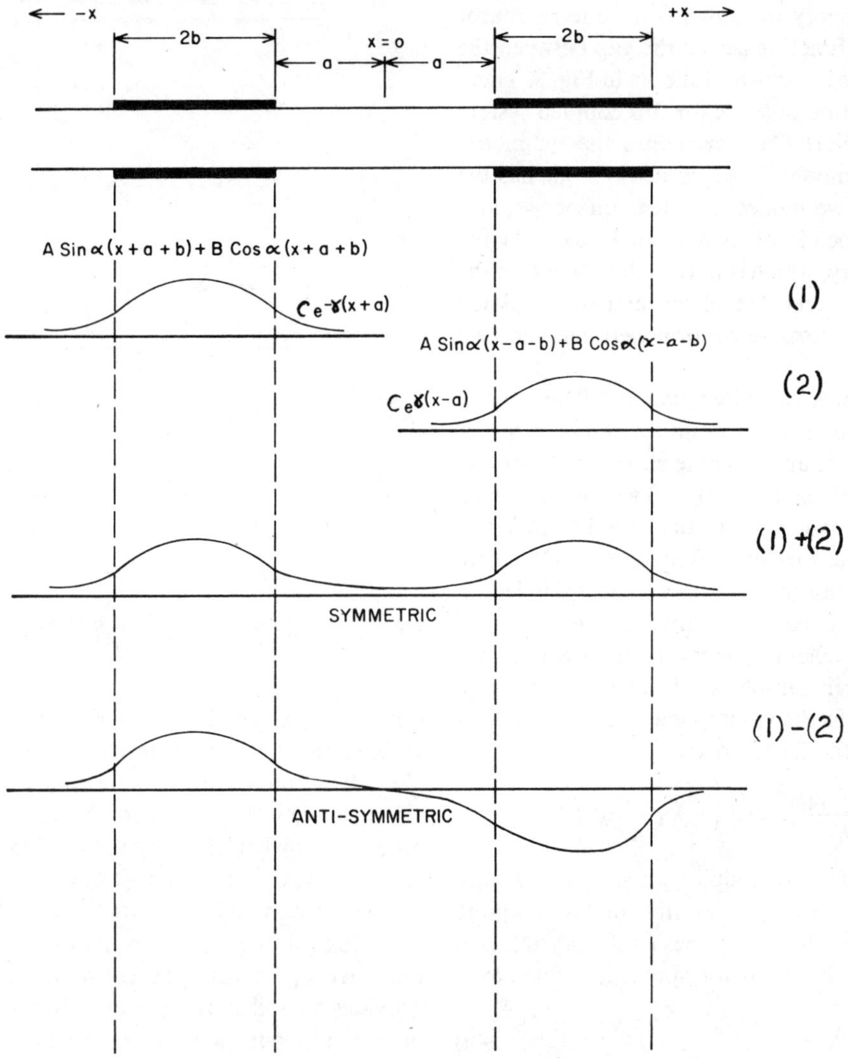

Fig. 8. Mode patterns for two acoustically coupled resonators.

$$Z_e = A\sqrt{\rho E}$$
$$Z_t = A\sqrt{\rho G}$$

and the α's are trigonometric and hyperbolic sine and cosine functions. As an example,

$$\alpha_{11} = \cos \alpha + \cosh \alpha \tag{17}$$

where

$$\alpha^4 = (\rho A/K)\omega^2 x. \tag{18}$$

K is the product of Young's modulus and the moment of inertia of the cross section of the bar.

When the coupling wire is used in the extensional mode and it is less than one-eighth of a wavelength long, it acts like a π network with a series arm spring and shunt masses; each mass being equal to one-half of the static mass of the wire. When the coupling wire is one-quarter wavelength long, it acts as an impedance inverter [5].

Although we are able to write equations describing first-order coupling effects, the distortion of a coupling wire due to welding the wire to the resonators makes it necessary to build and measure resonator-wire pairs or quads in the lab before constructing the entire filter. In the case of disk-wire filters, the analytical problem is compounded by the fact that the coupling is a combination of both flexure and extension.

Acoustically Coupled Crystal Resonators

To understand acoustic coupling in crystal filters, let us go back to the simplified analysis that was used to explain energy trapping. This demonstrated that in a region where we have $\omega > \omega_p$, the propagation constant α^2 is negative. This means that the wave function will decay exponentially with distance away from the resonator. The energy itself does not decay but only the wave function.

Two similar resonators (electrode pairs) placed close together as shown in Fig. 8 will have standing-wave solutions in the resonators themselves and exponentially decaying wave functions outside of the resonators [7], [16], [24], [25]. These wave functions will overlap and they will provide a means for energy to tunnel from one resonator to the other. The wave function across the gap between the resonators can therefore be drawn as shown in Fig. 8. There are two modes of vibration possible for this coupled system since it is of second order. These are called the symmetric and the antisymmetric modes. The symmetric mode has the lower frequency of the two modes and the frequency separation between them depends on two main factors: 1) the rate of exponential decay, which is in turn dependent on the frequency difference between the plated and the unplated regions, and 2) the physical separation between the two resonators.

Coupling equations developed by Sykes and Beaver [24], [26], [27], provide a good starting point for design purposes when we wish to know the approximate coupling. To obtain these equations, the total length ($4b + 2a$) was first assumed to be exactly one-half wavelength for the lateral direction of propagation, thus giving the lower frequency mode where both resonators are vibrating in phase. The total length ($4b + 2a$) was next assumed to be equal to one full wavelength, thus giving the higher frequency mode where the two resonators are vibrating in anti-phase. The difference between these two frequencies was then multiplied by an empirically-derived exponential decay factor to give

$$K = k_1 \left(\frac{4b + 2a}{h}\right)^2 \exp\left(-k_2 \frac{2a}{h} \Delta^{1/2}\right) \tag{19}$$

where K is the inter-resonator coupling, and k_1 and k_2 are constants that depend on the orientation of the coupling direction with respect to the major axes of the crystal. Δ is called the plateback of the resonator, and it is defined as

$$\Delta = \frac{f_u - f_p}{f_p} \tag{20}$$

where f_u and f_p are, respectively, the frequencies of the unplated and the plated resonator.

Since the particle motion due to a thickness-shear disturbance is along the so-called X-axis of the AT-cut quartz crystal, there will be greater coupling along this axis than along the Z-axis. However, when choosing a coupling axis for a filter, we must also keep in mind the method to be used for making the external electrical connections. If the coupling is along the Z-axis and the electrical connections are made at the extremities of the X-axis, the resonator will be more susceptible to external stresses than would be the case if the axes were interchanged. Coupling along the X-axis is referred to as thickness-shear coupling and coupling along the Z-axis is called thickness-twist coupling.

So far, the discussion has focused only on coupling between a pair of resonators. The coupling phenomenon portrayed in Fig. 8 can be extended to more than two resonators if they are deposited in a row. Acoustic energy will couple from one to the other in sequence and the n-resonator system will have n modes of vibration where, in the lowest order mode, all of the resonators are vibrating in phase and for the highest order mode they are alternately out of phase, as shown in Fig. 9. Beaver experimentally verified these mode patterns by means of X-ray pictures [28]. It is interesting to note that this same displacement effect shown in Fig. 9 is also typical of disk-wire and flexure mode filters when the phase of the input and output of each resonator is the same. This can easily be demonstrated with simple spring-mass resonators loosely coupled with springs.

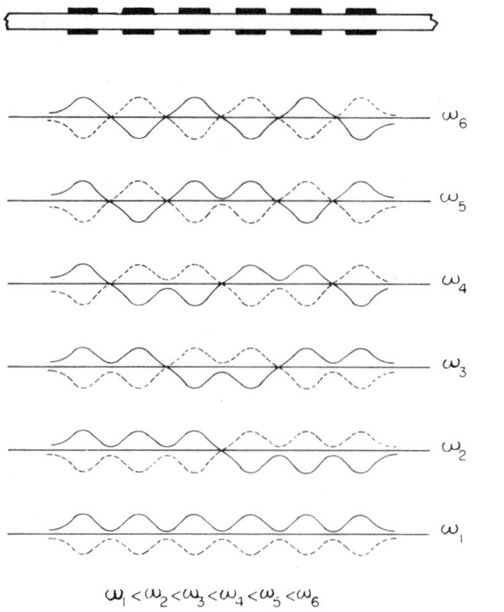

Fig. 9. Wavefunctions of the coupled modes of vibration in a monolithic crystal filter or in a mechanical filter.

In designing coupled-resonator crystals that have more than two resonators, the coupling equation just described provides a good starting point. More complex analyses have been made but they have limited usefulness, as will be discussed more fully later. We will see that, for practical reasons, multi-resonator filters have an additional adjustable electrode placed between each pair of resonators, i.e., at $x = 0$ in Fig. 8. Most lengthy analyses of multi-resonator systems are not able to take this important element into account simply because, in an "exact" analysis that takes into account anisotropy and piezoelectric effects, the mathematics are already very complex.

IV. Electrical Equivalent Circuits and Transducers

The essential operating features of crystal and mechanical filters can be described by electrical equivalent circuits. This facilitates the filter design process as it makes available the vast amount of theory that has been accumulated on inductor and capacitor filters. The physical basis for these equivalent circuits will now be explained. Wherever possible, we will emphasize the so-called mobility analogy [7], which relates "through" variables such as current and force and "across" variables such as voltage and velocity. With the mobility analogy the electrical and mechanical networks have the same topology which is a useful conceptual aid.

Let us consider a mechanical resonance system and its electrical mobility analogy as shown in Fig. 10. The nodal

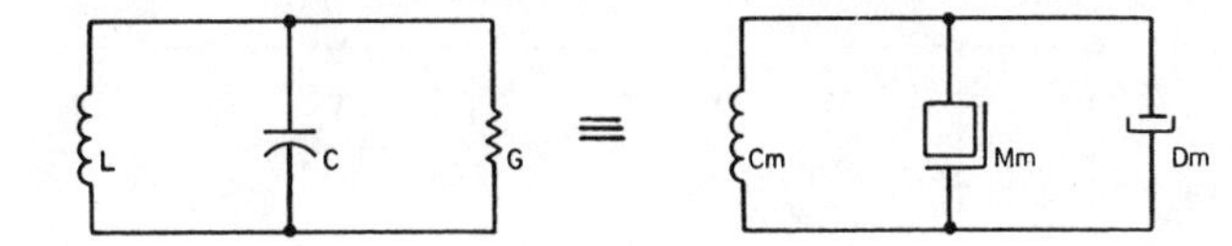

	ELECTRICAL CIRCUIT			MECHANICAL CIRCUIT	
	L	INDUCTANCE	⟷	Cm	COMPLIANCE
	C	CAPACITANCE	⟷	Mm	MASS
	G	CONDUCTANCE	⟷	Dm	DAMPING
ACROSS VARIABLE	V	VOLTAGE	⟷	Vm	VELOCITY
THROUGH VARIABLE	I	CURRENT	⟷	fm	FORCE

Fig. 10. General equivalences of mechanical and electrical resonant circuits.

differential equations for the mechanical and electrical circuits can be written as

$$M_m \frac{dv_m}{dt} + D_m v_m + \frac{1}{C_m}\int_{t_0}^{t} v_m \, dt = F \sin \omega t \quad (21)$$

$$C \frac{dV}{dt} + GV + \frac{1}{L}\int_{t_0}^{t} V \, dt = I \sin \omega t. \quad (22)$$

Both equations have the same form which shows that there is an analog relationship between them. We can construct a table of these electrical-to-mechanical analogies as shown in Fig. 10.

Since either of these differential equations by itself adequately describes the resonance properties of the system, a desired resonator could be designed either as a mass-compliance system or as an inductance–capacitance system. In either case, the pertinent parameters such as inductance and capacitance in the electrical case will have to be related to the actual physical dimensions and material properties of the resonator. If we deal strictly in terms of mechanical quantities such as mass and compliance, we would require a means for measuring velocity and force in order to ensure that we were actually achieving the desired mechanical performance. Electrical measurements, on the other hand, are relatively straightforward. Crystal resonators are directly measurable in electrical terms because of their integral transducer action due to the piezoelectric effect. On the other hand, external transducers are required in order to make electrical measurements on mechanical resonators, and a detailed knowledge of these transducers would be required in order to relate grams to farads, etc. Because of these difficulties and also because of the fact that the desired end result is an electrical device, the design and measurement of crystal and mechanical resonators is almost always done in electrical terms. The electrical measurements are generally simplified to the point where they are mainly dependent on the measurement of resonant frequencies. In making electrical measurements on mechanical resonators, the magnetostrictive properties of the resonator alloys can be used to achieve the necessary

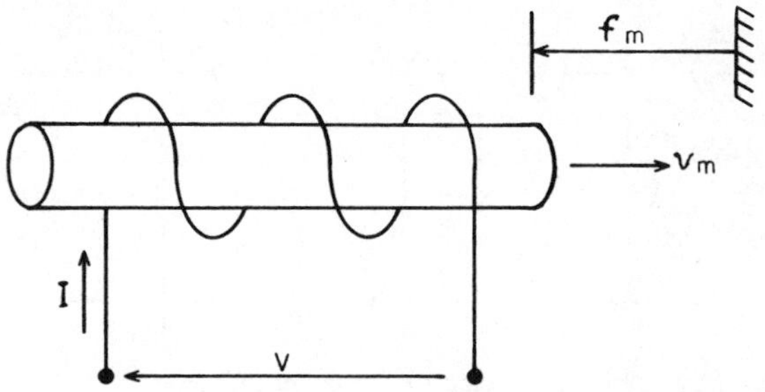

Fig. 11. Pictorial diagram of a magnetostrictive transducer.

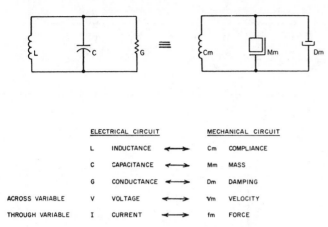

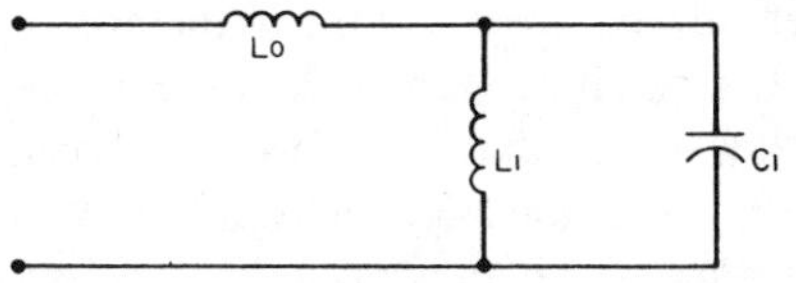

Fig. 12. Electrical equivalent circuit of a magnetostrictive transducer.

transducer action. This is done by placing a coil near the resonator and using it to couple magnetic energy into the magnetostrictive alloy. The electrical measurements can then be made on this coil.

In this section, we will discuss equivalent circuits and their mechanical analogs. As already discussed however, resonator design is almost always done in strictly electrical terms. The value of the mechanical analogs resides in the fact that they provide a good conceptual aid. It is generally easier to visualize a mechanical resonator in strictly mechanical terms. We can then construct its mechanical equivalent circuit and obtain its electrical equivalent circuit by using the mechanical-to-electrical analogs just described.

Let us now turn our attention to a comparison between magnetostrictive transducers and piezoelectric transducers such as quartz, ceramic, etc. As shown in Fig. 11, a magnetostrictive transducer has a coil wound around a bar of magnetostrictive material, which can be either iron–nickel alloy or ferrite. A current through the coil will produce a magnetic field which will, in turn, produce a force via the magnetostrictive effect. We therefore have current which is a "through" variable producing force which is also a "through" variable. If we make electrical measurements at the electrical terminals, we observe a driving-point impedance pole (f_1) and a zero (f_2), as well as a positive reactance outside of the region of resonance. From this data (and a knowledge of the driving-point synthesis of a reactance function), we can construct the simple three reactance-element network of Fig. 12. The electrical inductance of the coil L_0 and the equivalent or so-called motional inductance of the mechanical network are related by the electromechanical coupling coefficient as seen by

$$k_{em}^2 = \frac{2(f_2 - f_1)}{f_1} \approx \frac{L_1}{L_0}. \quad (23)$$

If the position of the transducer rod in the coil or the distance between the coil and the rod is changed, the coupling coefficient will change and so will the ratio of L_1 to L_0.

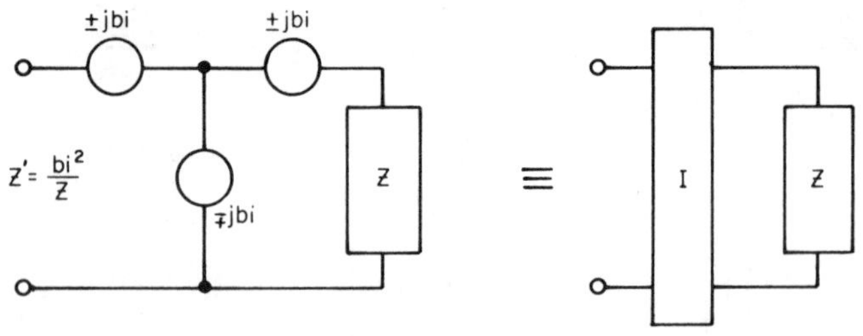

Fig. 13. Reactive impedance inverter.

Therefore, although the mass and compliance of the transducer bar remain the same and the inductance L_0 is essentially constant, since it is mainly the inductance of the coil itself, the value of L_1 can therefore be varied. From this we see that the relationship between compliance and inductance, or mass and capacitance, is based on the electromechanical coupling.

In a piezoelectric transducer, on the other hand, we have a voltage input producing a force via the piezoelectric effect. We therefore have voltage which is an "across" variable producing force which is a "through" variable. Consequently, whenever we are using piezoelectric transducers we need to have a gyrator or impedance inverter in the circuit to effect the transition from "through" variables to "across" variables and vice versa. The impedance inverter can be a reactive T as shown in Fig. 13, where the reactances are assumed to be constant and capacitors are a close enough approximation for narrow-band filters at high frequencies. We should keep in mind that this kind of impedance inverter introduces a 90° phase shift which becomes important when we bridge reactive components across the filter in order to produce finite-frequency attenuation poles. Alternatively, we can introduce an electro-mechanical gyrator to accomplish the impedance inverter function. This is done in the example of a single crystal resonator, in the next section, where the gyrator is placed exactly at the electrical-to-mechanical interface. The main function of the impedance inverters however is to change shunt branches into series branches and vice versa.

Equivalent Circuit of a Single Crystal Resonator

To illustrate the method of finding electrical equivalent circuits, let us start with a single quartz crystal or piezoelectric ceramic resonator. This is shown in Fig. 14(a) which is a composite electrical and mechanical circuit where the electrical-to-mechanical transition is represented by an electro-mechanical gyrator G. The capacitance C_0 is the static capacitance of the crystal which is the capacitance across the crystal electrodes at a frequency away from the mechanical resonance, and the other terms represent the mechanical motional parameters. Assuming that the mechanical resonator is not clamped ($v_m \neq 0$), we can move the mechanical network through the gyrator and obtain the circuit of Fig. 14(b), which is the conventional representation of a crystal resonator's equivalent circuit. Finally the circuit of Fig. 14(c) can be obtained simply by

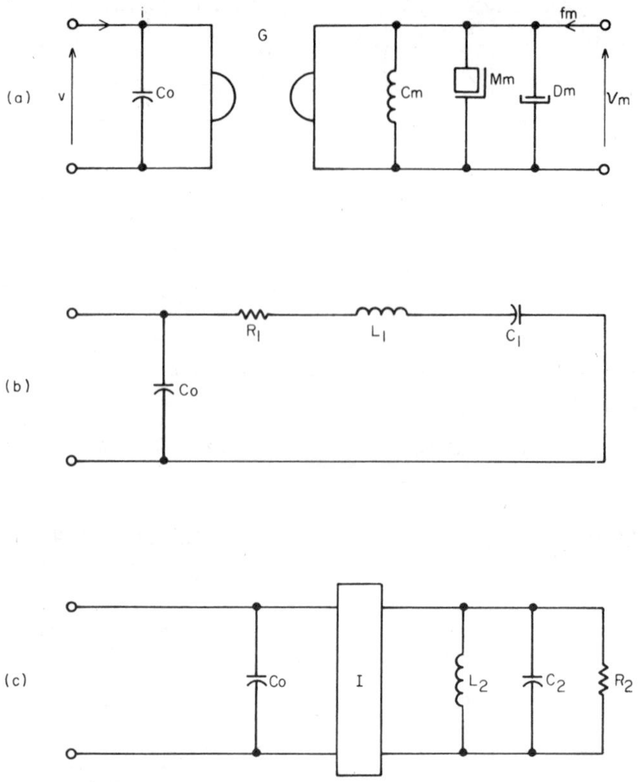

Fig. 14. Equivalent circuits for a single crystal resonator.

inserting an electrical impedance inverter I into Fig. 14(b). This last equivalent circuit has the advantage of retaining the mechanical circuit topology. Measurements of pole and zero frequencies will determine the coupling coefficient which is the ratio of the two capacitances in Fig. 14(b). We need to know the exact value of one of the components in order to be able to use the pole and zero information to determine the others. This can most easily be done by measuring C_0 which is simply a static capacitance. These equivalent circuit elements can also be computed directly by relating the mechanical and electrical elements via the piezoelectric coefficients [29], [30], but this is a cumbersome practice and it is not generally done. The elements of the equivalent circuit in Fig. 14(b) are related by the electromechanical coupling coefficient, but unlike the magnetostrictive case, the piezoelectric electro-mechanical coupling coefficient is a fixed property of the piezoelectric material. With reference to Fig. 14(b), the coupling coefficient can be written as

$$k_{em}^2 = \frac{2(f_2 - f_1)}{f_1} \approx \frac{C_1}{C_0} \tag{24}$$

where, in this case, f_2 is a driving-point impedance pole and f_1 is an impedance zero. In discussing mechanical resonators it was mentioned that the coupling coefficient can be varied by adjusting the transducer rod. We do not have any similar freedom with crystal resonators as the coupling is a fixed quantity that is determined both by the properties of the piezoelectric material and its crystalline orientation.

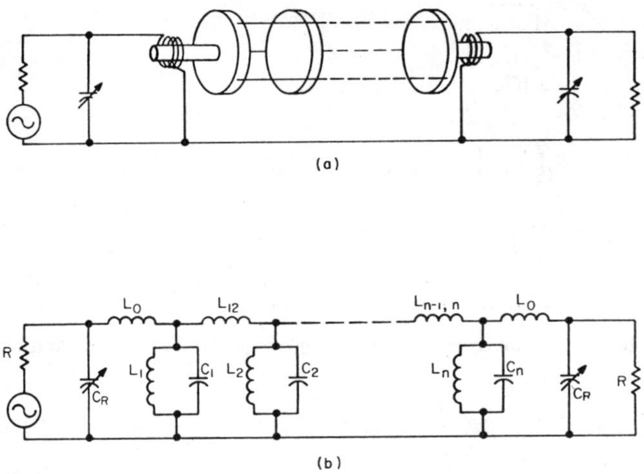

Fig. 15. Magnetostrictively driven disk-resonator mechanical filter and its electrical equivalent circuit.

This in turn leads to one of the limitations of quartz filters which is their relatively narrow bandwidth as determined by the low coupling coefficient of quartz.

Mechanical Filter Equivalent Circuit

Fig. 15 shows a pictorial diagram and an electrical equivalent circuit of an extensional mode magnetostrictively driven mechanical filter.

L_i and C_i correspond to the desired flexure modes of the disk resonators and the inductances $L_{i,i+1}$ correspond to the compliances of coupling wires. The transducer coil inductance is represented by L_0 and the capacitance C_R is used to resonate with it. This capacitance C_R could alternatively have been placed in series with L_0 if a low value of terminating resistance R were desired. The motional parameters of the magnetostrictive transducers are included in L_1C_1 and in L_nC_n. As can be seen, this equivalent circuit has the same topology as the mechanical filter and it is therefore a good conceptual aid.

An alternate configuration would be the piezoelectric-ceramic transducer driven filter where the network of Fig. 14(c) is substituted for L_0, L_1, and C_1 in Fig. 15 and inductive tuning is used.

Equivalent Circuit of Acoustically Coupled Crystal Resonators

In order to obtain an electrical equivalent circuit for the case of acoustically coupled crystal resonators, we just extend the reasoning used to derive the circuit for a single resonator [31]. The basic difference will be in the representation of the acoustic coupling. Coupling in the mechanical circuit is represented by a spring or compliance since the medium acts like a short transmission line, so this will be represented by an inductance in the topologically equivalent electrical circuit [7].

Considering three identical acoustically coupled resonators, as shown in Fig. 16(a), the electrical equivalent circuit in Fig. 16(b) is simply three of the single resonator equivalent circuits in Fig. 14(c) with inductive coupling between them. This circuit gives the best conceptual view of what is happening in the filter. Normally the intermediate resonators will be short circuited, and when this is done, the short circuit reflects through the impedance inverter I as an open circuit, thereby giving the circuit in Fig. 16(c). If we now redraw the circuit as it appears through the inverters, or if we move one of the inverters through the circuit and combine it with the second inverter, Fig. 16(d) is obtained. Note that the ideal transformer is used to account for the 180° of phase change which occurs in the two 90° inverters. In actual practice however, this 180° could be accounted for just as well by making one of the coupling capacitors C_c negative, and this is commonly done when we are dealing with just two coupled resonators. The value of the coupling for this circuit is expressed as a capacitance ratio, i.e., $K = C_1/C_c$ etc., where K is obtained from the coupling relation of (19).

In all of the equivalent circuits discussed so far, we have only considered the dominant modes of vibration. There may be spurious modes and they can best be handled by additional resonant circuits in series with the dominant mode parallel resonant circuits in Fig. 16(c) or in parallel with the series resonant circuits in Fig. 16(d) [32].

V. Broad Classification of Filter Types

Mechanical and crystal filters can be divided into broad categories of narrow band, intermediate band, and wide band [33]–[35]. The wide-band category is really a special case of the intermediate band, covering filters with percentage bandwidths in the range of 5–10 percent. Mechanical and crystal bandpass filters are not made with percentage bandwidths greater than 10 percent because they are limited by transducer bandwidths and spurious modes of vibration in the resonators.

The main distinguishing feature about the narrow-band category is the fact that no coils are used. In this category, we can include the following:

1) two-resonator low-frequency mechanical filters used for the detection of single tones or cascades of two-resonator filters with capacitive coupling between sections [36];
2) crystal ladders in which a ladder structure is built from discrete one-port resonators and discrete capacitors [37], [38];
3) a cascade of two or more resonators-per-section crystal lattice sections [38]–[40];
4) tuning-fork, one-pole mechanical filters in hybrid form [23];
5) monolithic crystal filters in which the entire filtering function is performed on a single piece of quartz with acoustic coupling between the resonators [2]; and
6) polylithic crystal filters using several pieces of quartz and some capacitors. Each piece of quartz may contain a single resonator or two resonators that are acoustically coupled [3].

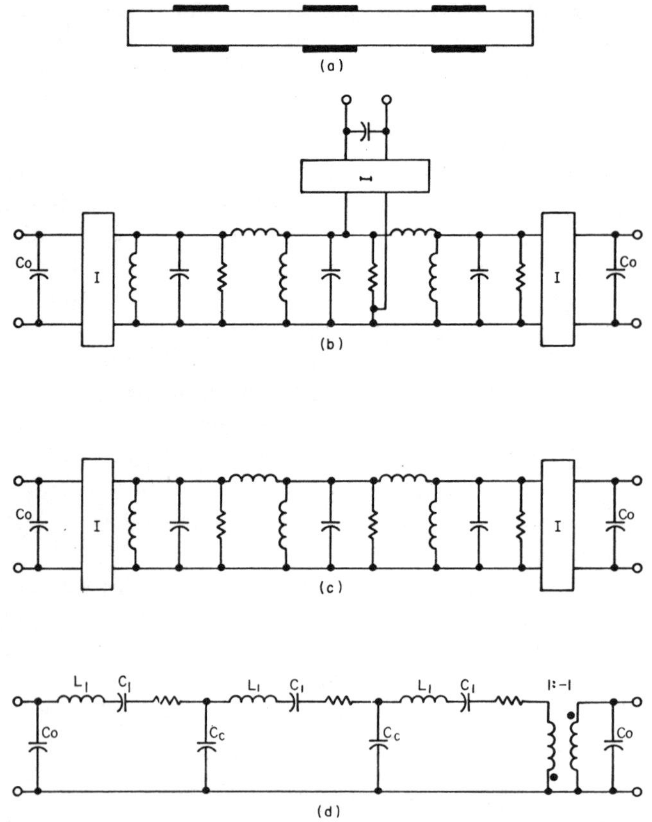

Fig. 16. Three-resonator monolithic crystal filter and its electrical equivalent circuits.

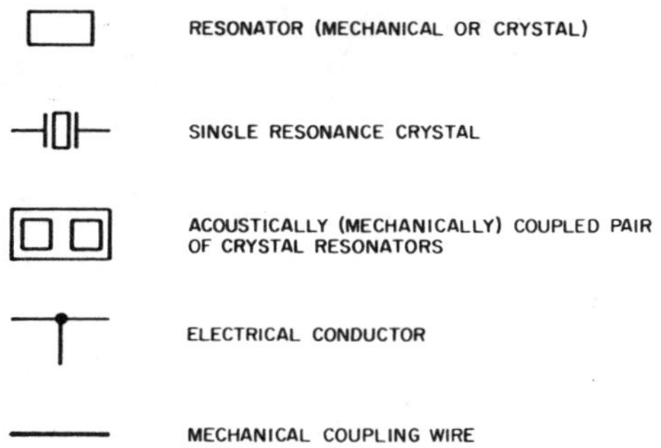

Fig. 17. Symbols used in the topology diagrams. Figs. 18, 19, and 20.

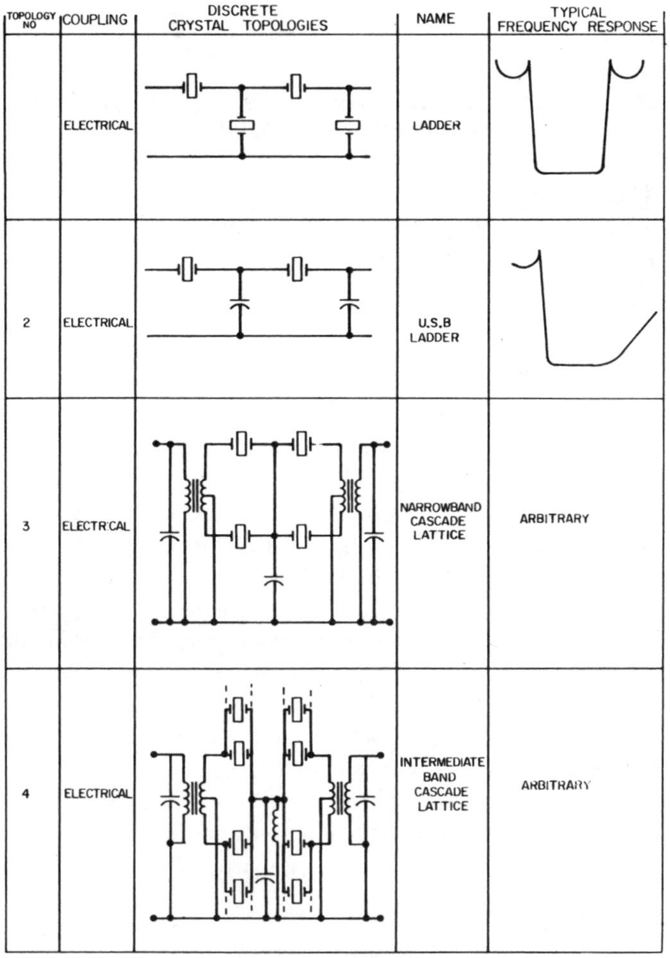

Fig. 18. Topologies of discrete crystal filters and their typical frequency responses.

In the intermediate-bandwidth category, resonating coils or capacitors are used to help obtain additional bandwidth. We can further classify intermediate bandwidth filters as follows:

1) disk or bar mechanical filters that are composed of a number of resonators with wire coupling between them and also bridging across them [1];
2) crystal lattices in cascade using coils and hybrid transformers at the input and output of each section [34], [40]; and
3) *LC*-crystal ladders that use the crystals to realize attenuation poles with reduced degradation of the passband performance compared to using lower *Q* *LC* elements [41]–[45].

Wide-band crystal and mechanical filters also make use of coils or capacitors for the purpose of realizing a wide bandwidth, but the coils also act as the input and output resonant elements in a section, thus increasing the filter selectivity but at the expense of less stable end resonators.

VI. Filter Topologies

In this section, commonly used crystal and mechanical filter topologies are tabulated along with their typical frequency responses. The symbols in Fig. 17 are used on the topology diagrams in Figs. 18, 19, and 20. The crystal filter topologies are divided into two main categories depending on whether or not acoustically coupled resonator pairs are used. The topology to actually be used in a particular application will be dependent very much on the available technology in terms of design aids and factory processes as well as filter specifications such as center frequency, bandwidth, passband ripple, required stopband behavior, etc.

The topologies 8, 9, and 11–14 all make use of bridging elements such as capacitors or mechanical paths in order to realize attenuation poles. Topologies 1 and 2 use poles and zeroes of series and shunt-arm impedances, whereas networks 3 and 4 realize attenuation poles by balancing lattice arms.

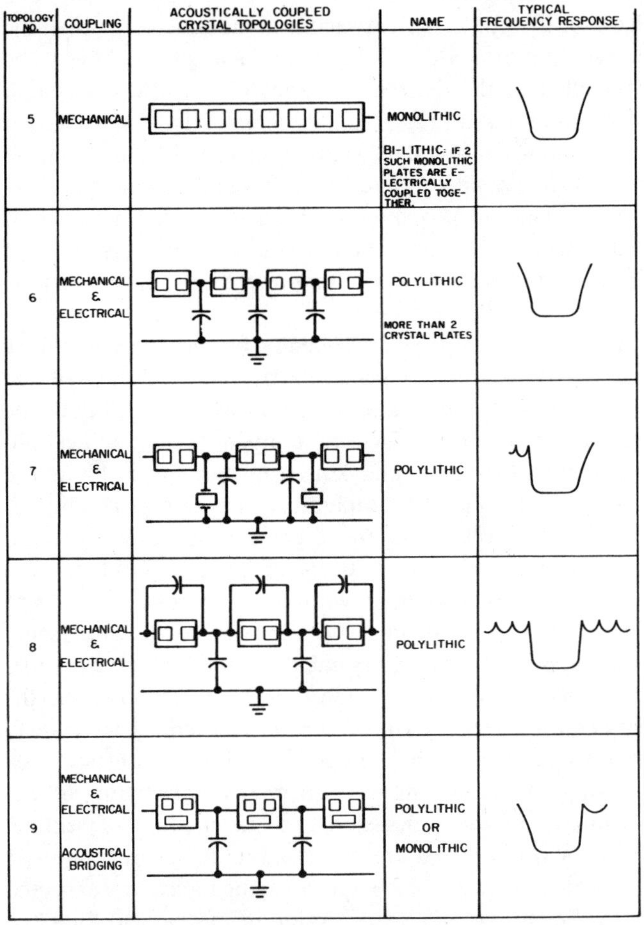

Fig. 19. Topologies of acoustically coupled crystal filters and their typical frequency responses.

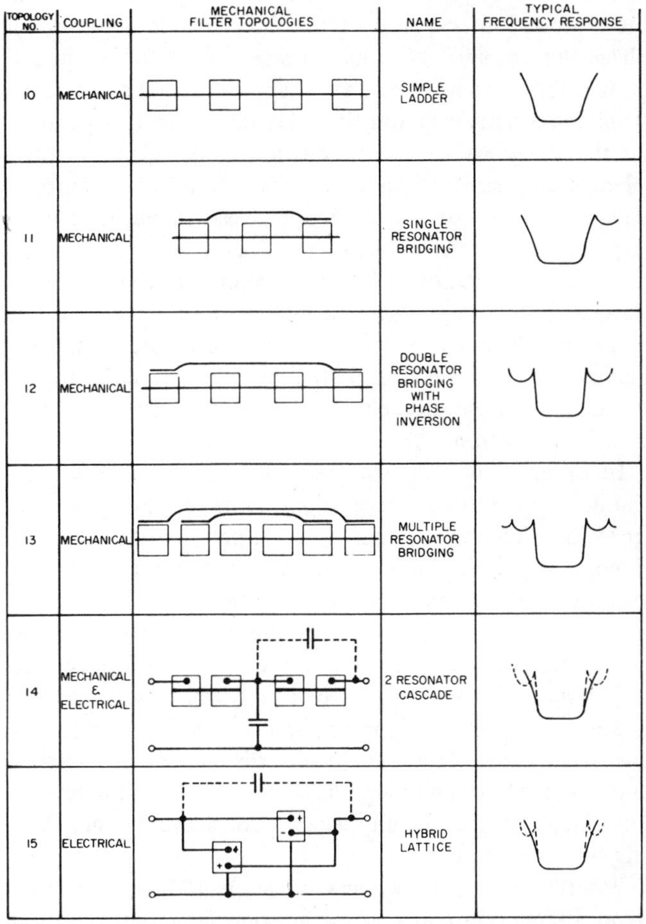

Fig. 20. Topologies of mechanical filters and their typical frequency responses.

VII. Filter Design Criteria

In order to fully characterize any filter, we must know its frequency response, insertion loss, delay distortion, absolute delay, spurious response, temperature stability, and aging stability. We will now discuss the limitations on these parameters as they apply to crystal [46] and mechanical filters [47].

Mechanical filters have been built in the frequency ranges of 100 Hz–600 kHz with percentage bandwidths ranging from 0.1 to 10 percent. The physical size of the filter is the main limitation at the lower frequencies because the filter becomes too large, especially when compared to active filters. At the higher frequencies, the filter components are small and the manufacturing tolerances are difficult to control, but we must also contend with the fact that the frequency shifts due to temperature and aging effects become a significant fraction of the filter's bandwidth. The resonator Q values will determine the minimum fractional bandwidth that can be achieved and this minimum bandwidth will not be less than $1/Q$ and most likely it will be greater than $10/Q$. Even with a fractional bandwidth of $10/Q$, the insertion loss will still be relatively high. Q values for mechanical filter resonators range from 1000, in the case of composite metal–ceramic resonators, to 20 000 for metal resonators.

For crystal filters, the frequency range of fundamental-mode AT-cut resonators is 2–15 MHz; whereas frequencies up to 200 MHz have been achieved with overtone modes. AT-cut resonators have been built at frequencies lower than 2 MHz, but it becomes more difficult to prevent spurious modes at the lower frequencies and the cost is also high. Fundamental mode resonators have also been built at frequencies greater than 15 MHz, but the cost is high due to the difficulty of handling the very thin pieces of quartz.

Crystal-filter fractional bandwidths are as low as 0.01 percent, and the resonator Q values range from 10 000 to 300 000. The resonator Q actually obtained will be determined by both the resonator design and by the manufacturing processes. The method of supporting the resonator most often has the greatest effect on the Q. For a fixed-fractional-bandwidth narrow-band crystal filter, insertion loss is a quantity that is almost entirely dependent on the resonator Q values and it is typically less than 2 dB for a good telephone channel filter design. With intermediate-bandwidth crystal filters, however, insertion loss is also influenced by the coil Q values. For narrow-bandwidth mechanical filters, on the other hand, insertion loss is most affected by mechanical resonator and transducer Q values and in wide-bandwidth designs coil losses play the major role in determining insertion loss.

The delay distortion and absolute delay are related to the attenuation response of the filter if it is a minimum phase design, and the same rules apply as for all filters, namely that the absolute delay is dependent on the integrated stopband attenuation and the delay distortion is dependent on the steepness of the attenuation slope. Consequently, when it is desired to have low absolute delay, we should have attenuation only at the frequencies where it is actually required. This means using topologies that have finite frequency attenuation poles. An exception to the preceding statement regarding delay distortion occurs when lattice or bridged ladder sections are used to generate right half-plane attenuation poles to flatten the delay response. In this instance, the absolute delay is increased by the amount of delay equalization.

In order to prevent spurious stopband responses, care should be taken to avoid spurious paths for energy to flow from the filter input to the filter output. As already mentioned, we could have spurious longitudinal modes of vibration in a monolithic filter where energy can couple from the input to the output resonators directly, without going through the intermediate resonators. This effect can be nullified if a bilithic [48] or a polylithic filter topology is used. When this is done, the spurious modes can couple energy only part of the way through the filter to the electrical coupling network and there the signal will see an impedance mismatch; the net result being a considerably improved stopband response.

Whereas in monolithic crystal filters, the overall stopband attenuation is decreased by spurious input-to-output modes and this is usually considered to be an annoyance. In bar flexure mode mechanical filters, on the other hand, the spurious modes can be controlled to a certain extent, and they act like input-to-output bridging paths that give various combinations of finite frequency attenuation poles near the filter's passband. Spurious modes of vibration in flexure mode disk-wire filters act like additional tuned circuits in the shunt arms of the ladder equivalent circuits, resulting in finite attenuation poles between the desired and unwanted passbands [5]. The attenuation poles can be controlled by the coupling wire placement so as to shape the stopband response for single-sideband applications. Spurious modes in conventional single-resonator crystal configurations usually result in unwanted stopband responses and reduced transition-band steepness.

Temperature and aging stability [49] are ultimately tied to the properties of materials used in the filter. For crystal filters, the temperature stability is governed by the well-known cubic frequency drift versus temperature curve for *AT*-cut crystals. The temperature performance actually obtained will be determined by the tolerance on the *AT* angle. This can generally be done to an accuracy of ± 1 min of arc which means theoretical temperature drifts of ± 3 ppm, for temperature changes of $\pm 25°C$, about a nominal temperature of $+25°C$. There are other secondary factors such as mechanical stresses on the quartz that influence the temperature performance, but they can be controlled in the resonator design. Once again however, for a given filter design, temperature stability must be considered if a decision is to be made between a monolithic or a polylithic design. In a monolithic design, the *AT* angle will have to be controlled to the desired tolerance over the whole length of the monolithic quartz plate, whereas it is much easier to maintain the angular accuracy in just the small central area of the polylithic plate. In addition, an averaging effect will be achieved if a filter is made up from a number of plates with the same nominal tolerance, thus stabilizing the passband edges but resulting in greater passband ripple at the temperature extremes.

For mechanical filters, the center frequency temperature stability is determined primarily by the properties of the resonator alloy and to a lesser extent by the stability of the transducer materials. Typical temperature stabilities obtained for iron–nickel alloys used in mechanical filters are 1–10 ppm/°C [1]. The temperature variation of ceramic or ferrite transducers tends to be greater than this value and it will generally be of the order of 20 ppm/°C [11], [50]. By temperature variations we have only been referring to the variation of frequency as Q variations with temperature are insignificant for both crystal filters and for the resonators in mechanical filters. For mechanical filters however, the temperature dependence of the transducers Q value may have to be included in the determination of insertion loss.

Aging of crystal filters is primarily a function of the cleanliness of the processing and the method of sealing. For crystal filters that are processed under clean conditions and sealed in metal covers, typical aging figures are expected to be 3 ppm over 20 years. Aging in mechanical filters is, for the most part, due to relief, with time, of internal strains within the resonators [51]. The frequency aging of multiple resonator filters over a 10-year period will be approximately 50 ppm. In the case of two-resonator filters composed of composite ceramic/metal alloy resonators, the aging will be less than 5 percent of the filter bandwidth.

VIII. Design Flow Chart

The design process, for crystal and mechanical filters, starts with a classical *LC* design that can be produced by insertion-loss [38], [52], or image-parameter techniques [53]–[55]. The design flow chart is shown in Fig. 21, and apart from the final transformation step, it is seen to be a conventional one. During this process, the final configuration has to be kept constantly in mind to ensure that it will be possible to achieve a realizable low manufacturing cost topology and also to ensure that the range of element values will be realizable. In crystal filters, for example, the range of motional inductances achievable may be very narrow and less than 3:1. This will generally mean that many trial designs may have to be made before a realizable filter is obtained. In order to minimize mechanical filter manufacturing costs, equal-size resonator designs are realized by varying the coupling between resonators. This is done by varying either the diameter or the length of the coupling wire.

Some filters may not be amenable to classical synthesis procedures and optimization techniques may be required;

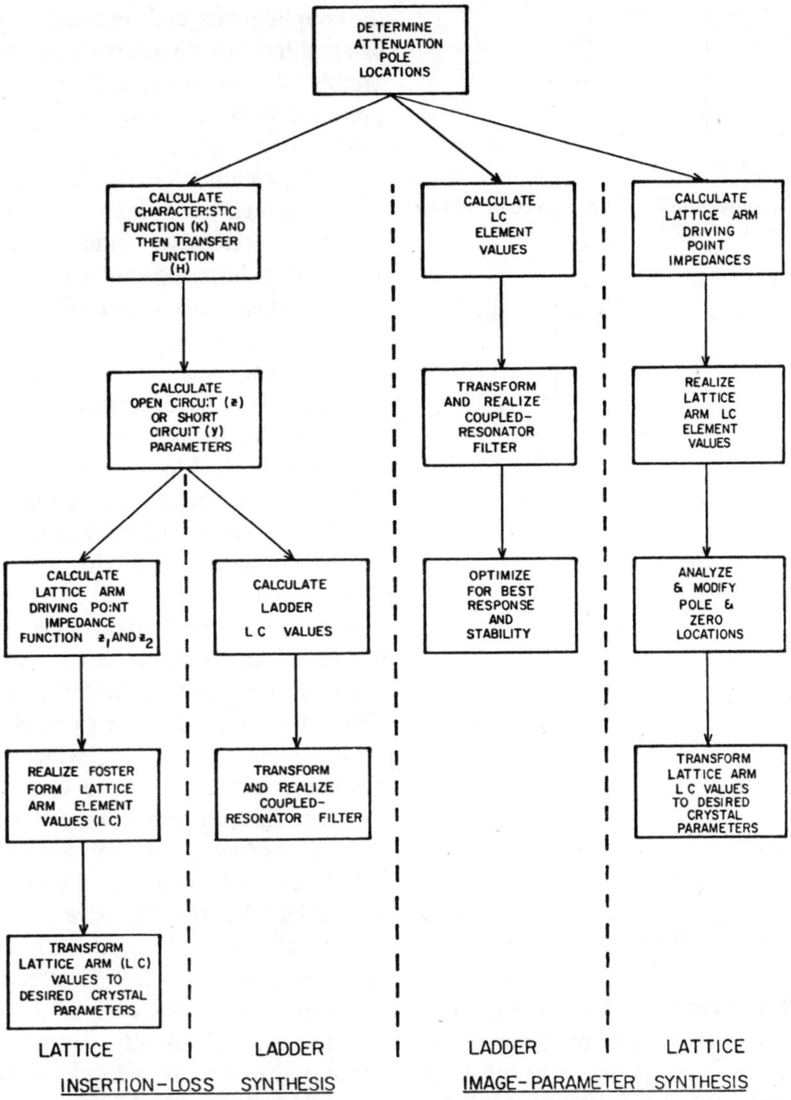

Fig. 21. Design flow chart for crystal and mechanical filters.

an example of this may be some bridged capacitor topologies as in Fig. 19. Also, special techniques are needed to incorporate coils at the input and output of mechanical ladder networks and intermediate-bandwidth crystal filters.

Although insertion-loss techniques may be more exact, image-parameter methods are still used and they have the advantage of excellent flexibility and visibility in the design process. Coupled with optimization programs, image-parameter methods provide a relatively fast design tool that does not need computer programming for special cases.

IX. TRANSFORMATIONS

Once a basic *LC* prototype filter has been designed, the *LC* configuration must be developed into a form that is realizable with crystal or mechanical resonators. Some of the transformations that are widely used [38], [56], [57] for this purpose are listed in this section.

Since all crystal and mechanical filters are basically narrow-band devices, the reactive impedance inverter in Fig. 22 is the most useful of these transformations. A good example of its utility occurs when designing a symmetrical all-pole bandpass filter from a lowpass prototype. This procedure is used to design the filters of topologies 5, 6, 10, and 14 in Figs. 19 and 20.

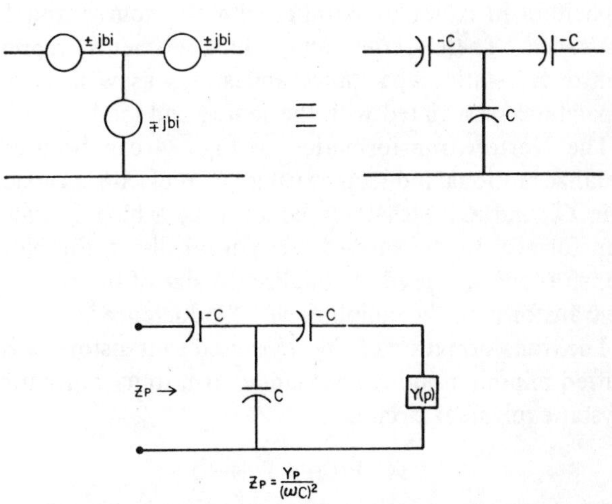

Fig. 22. Reactive impedance inverter and its narrow-band approximation.

$Z \text{ in}'(\text{p}) = \frac{R}{K} \longrightarrow$

$L = \frac{R^2 C}{K}$

$K = 1 + (\omega CR)^2$

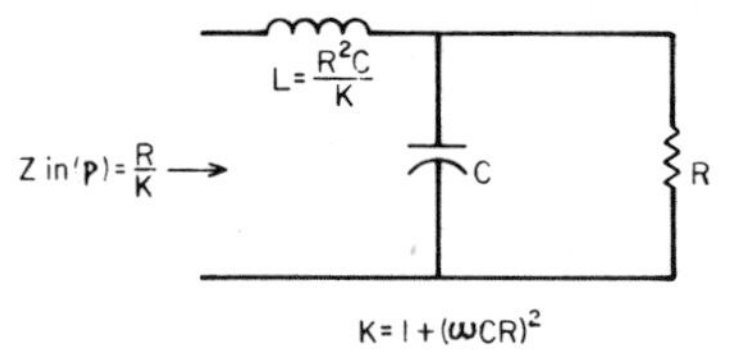

Fig. 23. Method of absorbing source and load capacitance by adding a low-pass section.

$n > 1$

$n = 1 + \frac{Cb}{Ca}$ $Cc = \frac{Cb}{n}$ $Cd = \frac{Ca}{n}$

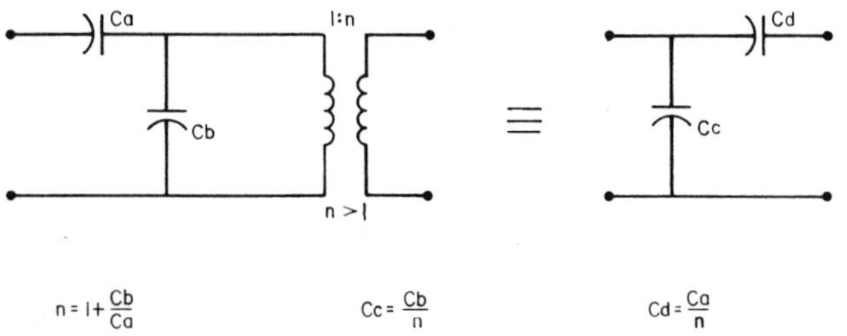

Fig. 24. Norton transformation.

$C_2 = \frac{C_3}{1 + \frac{C_1}{C_3}}$ $L_2 = L_1 \left(1 + \frac{C_1}{C_3}\right)^2$ $C_4 = \frac{C_1}{1 + \frac{C_1}{C_3}}$

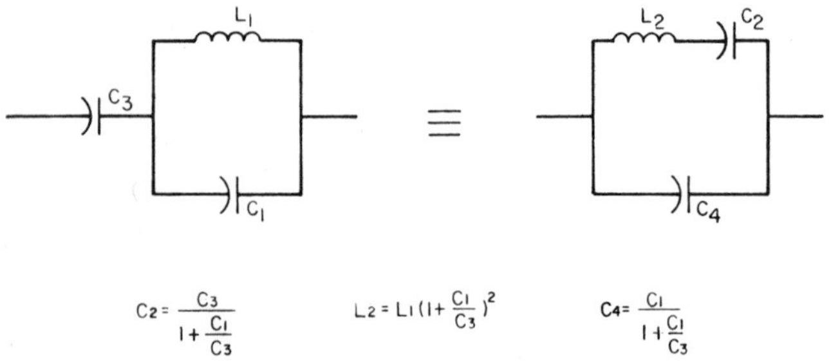

Fig. 25. This transformation is used to transform a zero shifting section into a form that is realizable with a crystal.

The transformation of Fig. 23 can be viewed as a parallel to series RC transformation, at the center frequency of the filter. This transformation is used in the case of both mechanical and crystal narrow-band filters to realize a parametric type of design, i.e., a design with nontuned capacitors in series or parallel with the source and load resistance. These series and shunt elements represent transducer static capacitance and strays as well as other capacitance associated with the source and load.

The Norton transformation of Fig. 24 can be used to equalize motional inductance values by replacing a capacitor pair C_c and C_d with their equivalence which includes a transformer. In the case of mechanical filters, the Norton transformation is used to equalize the size of the resonators by transforming the coupling wire "inductance."

The transformation of Fig. 25 is used to transform a zero-shifted minimum-inductance ladder arm to its conventional crystal equivalent circuit.

X. FILTER TUNING

Mechanical and crystal filters can be tuned by either an additive or a subtractive process, but generally not by both simultaneously. In discussing tuning, a distinction must be made between the tuning of resonant elements and the adjusting of coupling elements as different techniques are used on each.

Tuning of mechanical filter resonators takes place before the coupling-wire and resonator assembly process, although some filters also require frequency adjustment after assembly. In all cases, the resonators themselves can be pretuned to a sufficiently tight tolerance as to result in a satisfactory response, but the effects of resonator or filter supports, welding variations, and resonator spacings often cause enough mistuning to require readjustment after assembly. Resonator tuning and adjustments in tuning make up a large percentage of the cost of a mechanical filter, so this subject must be treated with great care in the development of new filter types. Preassembly resonator tuning is made feasible by the fact that the constant Young's modulus alloys used as resonators are also magnetostrictive, and therefore, they can be vibration excited by an alternating magnetic field. Most resonator tuning methods remove material by means of drill tuning, grinding, sanding, sandblasting, or laser machining, although the additive methods of solder loading and wire mass welding are often used after assembly.

The method of supporting a mechanical filter and the tuning of the resonators are closely related. Filters such as disk-wire types are supported at each end by the coupling wires themselves. For this to be possible, the coupling wires must be relatively large, which in turn tends to cause resonator mistuning due to variations of disk spacing and welding. This problem is avoided in the case of extensionally-coupled flexure mode mechanical filters where a small coupling wire is used at a low mechanical impedance point on the bar. Because the wire is small, the resonators must individually be supported with wire supports connected to the nodal points [32]. This requires additional assembly operations and extra care in processing resonators. Wire coupling is usually tightly controlled but, because of welding variations, either on-line welding voltage adjustments must be made before assembly, or filing, sanding, or additional wires must be added after assembly to realize the desired bandwidth and passband performance.

AT-cut crystals are generally tuned by placing the crystal in a vacuum and condensing a metal vapor on to the electrodes thereby increasing the mass loading and lowering the frequency. The resonant frequency can be monitored continuously by a bridge circuit and arranged to shut off the process when the frequency reaches its desired value. YAG lasers [58] are also used to tune crystals by cutting holes in the electrodes to decrease the mass loading and cause the frequency to rise. Quartz is transparent to the YAG laser energy and the power is therefore only absorbed by the electrodes; but nevertheless, the power has to be carefully controlled to avoid creating small cracks in the crystal as these can decrease the Q. The Q will also be decreased if the metal is not evenly removed from the entire electrode surface. An even removal can often be difficult to achieve in practice.

The coupling between a pair of acoustically coupled resonators can also be varied either by additive or subtractive techniques, but the most popular method is the subtractive technique with the YAG laser. The laser lends itself better to the coupling adjustment rather than the resonator adjustment because, with the coupling adjustment, metal can be completely removed from the quartz in any pattern without affecting the resonator Q. This is because the resonator itself is not being altered by the laser but only the area in between the resonators. To understand this better, let us consider Fig. 8. This tells us that as mass is added to the midpoint between the two resonators, i.e., at $x = 0$, the frequency of the symmetric mode will be lowered but the antisymmetric mode will not be affected since it has zero amplitude at $x = 0$. The addition of this inner-resonator mass, therefore, has the effect of increasing coupling and decreasing the value of the coupling capacitor C_c, as shown in the electrical equivalent circuit of Fig. 16(d).

Fig. 16(d) also helps us to understand one of the difficulties encountered when making coupling and tuning adjustments on monolithic crystal filters. We will normally wish to tune the resonant circuits L_1C_1 or the coupling element C_c but they are not independently accessible. For example, if we look at the center resonator with the other two open circuited, we see approximately L_1C_1 in series with $C_c/2$. Therefore, if a coupling capacitor C_c is adjusted, we will be required to readjust the adjacent resonator frequencies. The adjustment of coupling, therefore, requires two extra processing steps although in some cases it can be avoided entirely if tight process control is exercised. The preceding remarks on resonator frequency and coupling adjustment do not of course apply to the low-frequency crystal resonators that are not mass sensitive.

XI. Manufacturing Considerations

While manufacturing considerations cannot be ignored in any filter technology, they are very important for crystal and mechanical filters because the dimensional and material related process steps do influence the mechanical resonances and thereby the filter's performance.

An example of these manufacturing considerations can be obtained by considering the topologies for acoustically coupled resonators as shown in Fig. 19. From this we see that the same type of all-pole response can be obtained with either a monolithic or a polylithic filter design. The monolithic approach should permit a low-cost filter to be built at high production volumes as it uses planar processing techniques that can be automated [59]. The cost of a monolithic crystal blank will, however, be more than the cost of four smaller polylithic blanks, and if the monolithic blank should break all is lost. The difficulty of maintaining parallelism of the crystal plate faces means that a coupling adjustment is definitely required in critical filter designs, if a monolithic approach is used, and as already discussed, the need to perform a coupling adjustment adds two extra steps to the process. It may also happen that there is a very tight specification on spurious modes of vibration in the filter's stopband. If such spurious modes other than the desired thickness-shear mode are excited in a monolithic filter, they can directly couple the input signal to the output resonator and give rise to a spurious output in the filters stopband. A polylithic approach, on the other hand, would provide an electrical mismatch to such spurious modes at the junctions between crystals [60]. The design process must therefore make a choice between these two approaches, keeping in mind also the much larger capital equipment cost required for the monolithic approach [65], [66].

A comparison of topologies 8 and 9 in Fig. 19 will show another example where the same type of response is achieved by two different manufacturing techniques. In these cases, the cost of the processing complexity required to achieve accurate acoustical bridging [61], [62] has to be balanced against that of straightforward electrical bridging, as shown in topology 8.

The metal used for the electrode of any quartz resonator filter must be chosen carefully. It must have high conductivity and it must be stable because any change in the material such as oxidation would cause a mass change with a consequent change in the resonator frequency. Because of its good stability and high conductivity, gold is generally used as the final coating material on crystal electrodes. It is used in conjunction with some other metal such as chromium, molybdenum, or titanium that acts as an adhesive to the quartz.

Probably the most basic manufacturing considerations when designing a mechanical filter are those of frequency tolerances and coupling or bridging tolerances. The basic network topology of a mechanical filter, namely whether or not bridging is to be used, determines, to a great extent, these tolerances. The use of bridging to produce finite attenuation poles, in order to reduce both differential and absolute delay, results in a less sensitive passband and a more sensitive stopband response [63]. System requirements often determine the construction of the filter but, as often as not, the filter topology or construction sets the system parameters. This is true in the design of radios where the IF stage frequency is most often set by the available mechanical filter design.

The mechanical filter designer must keep in mind frequency drift due to both temperature and aging as well as the ability to build filters in a production environment within certain passband cutoff frequency "windows." The temperature coefficient of resonant frequency can be held to 1 ppm/°C over a 100°C temperature range if proper resonator heat treatment procedures and sorting methods are used. The heat treatment procedures are dependent on the particular batch (lot or heat) of material used as is the variation of the temperature coefficient from part-to-part (i.e., some batches have a very narrow scattering of individual resonator temperature coefficients).

The designer must concern himself with the stability of the mechanical resonant frequency of the composite transducer-metal end sections as well as the stabilities of the metal resonators. He has to consider the stability of the electro-mechanical coupling coefficient and also the stability of the transducer inductance if he is using a magnetostrictive transducer or alternatively the stability of the static capacitance if he is using a piezoelectric transducer. These considerations force the designer to make decisions with regard to transducer size and means of attachment, the transducer material, tuning inductor size and material (for piezoelectric

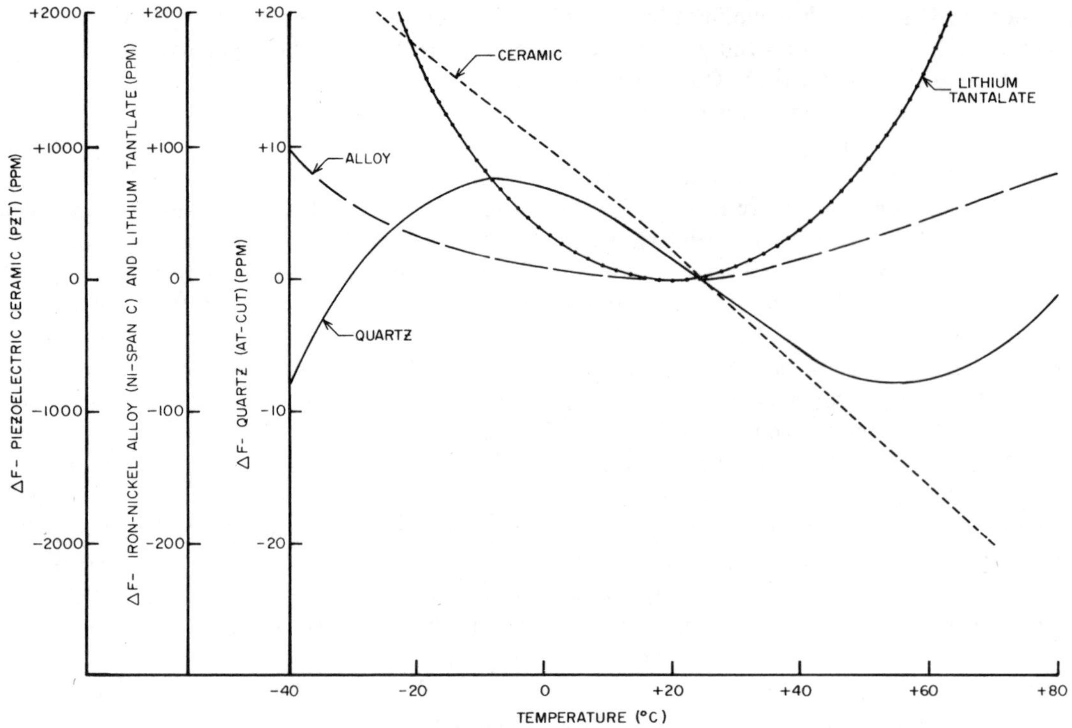

Fig. 26. Temperature characteristics of 4 common transducer materials.

intermediate-bandwidth designs), and tuning inductor temperature compensation.

XII. Transducer and Resonator Material Considerations

Crystal filters have transducers that are integral with the resonators. In mechanical filters, on the other hand, the end resonators are usually resonating transducers that consist of metal alloy parts bonded to the main transducer components, which are made from either ceramic or ferrite materials. The internal resonators in mechanical filters are made from iron–nickel alloy only. Because the end resonators are composite components, they will be less stable than the internal resonators. This does not present a serious problem to the filter designer since the frequency response of multiple-resonator filters is less sensitive to changes in the end resonator frequencies as compared to the frequencies of the internal resonators. This difference in sensitivity is typically four to one.

Fig. 26 shows the temperature characteristics of four common transducer materials: quartz, thermelast iron–nickel alloy, lithium tantalate, and a piezoelectric ceramic. It should be noted that there is an order of magnitude difference between each frequency change axis. Although the *AT*-cut quartz has the best temperature coefficient, it has the lowest electro-mechanical coupling coefficient of the materials listed. This coupling coefficient (k_{em}) has previously been defined by (23) and (24).

The coupling coefficient of *AT*-cut quartz is 7.2 percent, lithium tantalate is 21.6 percent, and commercially available piezoelectric ceramics (driven in what is called a k_{31} mode) is 22.6 percent. The bandwidth of a filter is limited by the value of coupling coefficient according to the approximate equations

$$\text{BW max (narrow band)} = k_{em}^{2} \tag{25}$$

$$\text{BW max (wide band)} = \sqrt{2}\, k_{em}. \tag{26}$$

The coupling coefficient of a composite ceramic–metal transducer can, in the flexure mode, be varied to a maximum of approximately 15 percent depending on the amount of ceramic attached to the alloy bar. In the case of narrow-band mechanical filters, only enough piezoelectric ceramic material is used on a composite resonator to satisfy (25) thus minimizing both temperature shift and aging as well as increasing the Q of the resonator.

Aging of a mechanical resonator is closely related to its temperature stability. For constant modulus alloys, lifetime aging is roughly one-half of the frequency shift of the resonator over a 100°C temperature range. Both the aging rate and the temperature shift of mechanical resonators are greatly influenced by the support means, and by the amount of temperature cycling. They are also influenced by the stress relief, the alloy composition, the heat treatment, and the particular mode of vibration. The amounts by which the frequency of a crystal resonator shifts with temperature variations and with age are influenced by the particular crystal cut, by the presence of contaminating substances on the crystal, and also by the environment in which the crystal has been sealed. The mode of vibration and the drive level also play a very important role in determining the amount of aging. As a general rule, the drive level should be kept low if a low aging rate is required.

Crystal and mechanical resonators are not generally

designed to have specific Q values. Instead, the designer follows certain rules and he sees that tight process control is exercised so as to get the best possible Q from the particular material being used. As an example of these design rules, consider AT-cut quartz crystal resonators. The amount of plating used to form the resonator area is a design variable and it should be thick enough to achieve the amount of energy trapping desired. However, if the plating is too heavy, the Q will be adversely affected. On the other hand, if the plating is too light, not only will the energy trapping be ineffective, but the Q will also be poor due to the low conductivity of the thin metal film.

Typical Q values for quartz, constant modulus alloys, and piezoelectric ceramic are 200 000, 20 000, and 2000, respectively. Higher values have been achieved with carefully designed and processed resonators. The Q of lithium tantalate is almost 100 000, which combined with its excellent coupling coefficient, makes it a good material for wider bandwidth crystal filters and also as a transducer for mechanical filters.

XIII. Advantages and Disadvantages of Crystal and Mechanical Filters

By far, the major advantage of crystal and mechanical filters is their low cost, when compared to LC filters. They also occupy substantially less volume and they weigh a great deal less than the LC filters. When compared to active filters, the major advantage of cost still applies and there is the added bonus that crystal and mechanical filters are passive and they do not generate any internal noise. The preceding comparisons only apply to those cases where an alternative filtering technology can be used. However, this cannot always be done and in those applications where the filter is required to have narrow bandwidth, high Q, good aging, and temperature stability, a crystal or mechanical filter is often the only type that is suitable.

One major drawback with the technologies of mechanical and crystal filters is the fact that they are not suitable for lowpass or highpass filters and the optimum frequency ranges are determined by the material properties. In some applications, this can be circumvented by the technique of two-step modulation which translates the energy to be filtered to a frequency which is optimum for the filter technology to be used. This is the technique that has been used successfully in the design of telephone channel filters [64]. The telephone channel filter is one of the most demanding types of filters, and yet a new generation of channel filters has emerged that uses either the crystal or the mechanical technologies to replace older LC filter designs.

Crystal filters and some mechanical filters offer an additional advantage of permitting flexible modulation plans for frequency division multiplex systems since the filtering is actually performed at a frequency that is well removed from the frequencies of both the input and the output of the multiplexer.

The design and manufacture of crystal and mechanical filters is more complex than that of LC filters; this is one of the reasons why they are not more widely used. The reasons for this added complexity are the inter-disciplinary nature of the devices. They involve, to varying degrees, technologies that are quite foreign to passive LC filters, such as ceramic-to-metal bonding, crystallography, vacuum deposition, to mention just a few. Because of this complexity, it will probably continue to be true that crystal and mechanical filters will be used only in those applications where there is a large enough production volume to justify the initial investment required.

XIV. Future Developments

Despite the increasing trend towards digital techniques, there will always be a major role for analog systems. As long as this is so, mechanical and crystal filters will continue to have an important place in electronics. In most cases, they are by far the least expensive way of achieving high-quality narrow-band filtering. Manufacturing costs will continue to decrease as the volumes increase and production methods become more automated. This should be most noticeable in the area of acoustically coupled crystal filters since their processes lend themselves to planar technology.

Materials technology will play a major part in future developments. This will lead to a wider variety of applications of the crystal and mechanical filter techniques that have been described. In the crystal area, we have really only spoken of quartz as the piezoelectric material. There are other possibilities that are now starting to emerge. Lithium tantalate is one, and it offers the possibility of realizing crystal filters with percentage bandwidths of approximately 6–8 percent. Properties such as coupling mechanisms are similar in lithium tantalate to those that have been described for quartz. In mechanical filters, materials technology offers the possibility of resonator materials with higher Q and better aging and temperature stability. So far, work has concentrated almost entirely on iron–nickel alloys. Transducer materials is another area that will probably benefit from future research, and in this case materials like lithium tantalate and new piezoelectric ceramics may have definite advantages for mechanical filters.

We will probably see more complex filtering functions increasing in popularity. Methods of achieving finite frequency attenuation poles in both stopbands are known now but they are not in widespread use on highly selective filters. These use techniques of mechanical bridging either by wires in mechanical filters or by parallel acoustical paths in crystal filters. Although it is difficult to control the bridging tolerances with today's techniques, we should expect an improvement in the future. Flatter and improved passbands is another area for future development. There has been little pressure for this up to now but we will probably see stiffer requirements in the future. This will mean improved materials technology as well as improved resonator design and better processing methods. As processing techniques continue to improve, there should also be an increase in the Q values that are routinely obtained.

In summary, mechanical and crystal filters owe their increase in usage to their stability, passivity, low cost, and small size. We expect them to be around for a long time.

References

[1] R. A. Johnson, M. Börner, and M. Konno, "Mechanical filters—a review of progress," *IEEE Trans. Sonics Ultrason.*, vol. SU-18, pp. 155–170, July 1971.

[2] R. A. Sykes, W. L. Smith, and W. J. Spencer, "Monolithic crystal filters," *IEEE Int. Convention Record*, vol. 15, part II, pp. 78–93, 1967.

[3] D. F. Sheahan, "Channel bank filtering at GTE Lenkurt," *Proc. 1974 IEEE Int. Symp. Circuits Systems*, pp. 115–120, Apr. 1974.

[4] M. Konno, Y. Tomikawa, T. Takano, and H. Izumi, "Electromechanical filters using degeneration modes of a disk or ring," *Electron. Commun. Jap.*, vol. 52A, pp. 19–28, May 1969.

[5] R. A. Johnson, "Electrical circuit models of disk-wire mechanical filters," *IEEE Trans. Sonics Ultrason.*, vol. SU-15, pp. 41–50, Jan. 1968.

[6] K. Yakuwa, "Characteristics of an electro-mechanical filter using piezoelectric composite bending vibrators," *Electron. Commun. Jap.*, vol. 48, pp. 206–215, Nov. 1965.

[7] R. A. Johnson, "Mechanical bandpass filters," *Modern Filter Theory and Design*, G. C. Temes and S. K. Mitra, Eds. New York: Wiley, 1973.

[8] M. Börner, "Biegeschwingungen in mechanischen filtern," *Telefunken J.*, vol. 31, pp. 115–123, June 1958; also, pp. 188–196, Sept. 1958.

[9] M. Konno and H. Nakamura, "Equivalent electrical network for the transversely vibrating uniform bar," *J. Acoust. Soc. Amer.*, vol. 38, pp. 614–622, Oct. 1965.

[10] R. L. Sharma, "Equivalent circuit of a resonant, finite, isotropic, elastic circular disk," *J. Acoust. Soc. Amer.*, vol. 28, pp. 1153–1158, Nov. 1956.

[11] M. Börner, "Magnetische werkstoffe in electromechanischen resonatoren und filtern," *IEEE Trans. Magnetics*, vol. MAG-2, pp. 613–620, Sept. 1966.

[12] H. Albert and I. Pfeiffer, "Uber die Temperaturabhangigkeit des Elastizstatsmoduls von Niob-Titan-Legierungen," *Z. Metallkde*, vol. 63, no. 3, pp. 126–131, 1972.

[13] J. F. Nye, *Physical Properties of Crystals.* New York: Oxford, 1957.

[14] W. S. Mortley, "Frequency-modulated quartz oscillators for broadcasting equipment," *Proc. IEE* (London), vol. 104B, pp. 239–253, Dec. 1956.

[15] R. Bechmann, "Quartz AT-type filter crystals for the frequency range 0.7 to 60 MHz," *Proc. IRE*, vol. 49, pp. 523–524, 1961.

[16] D. R. Curran and D. J. Koneval, "Energy trapping and the design of single- and multi-electrode filter crystals," *Proc. 18th Ann. Symp. Frequency Control*, pp. 93–119, May 1964.

[17] W. H. Horton and R. C. Smythe, "On the trapped-wave criterion for AT-cut quartz resonators with coated electrodes," *Proc. IEEE*, pp. 598–599, Apr. 1967.

[18] W. J. Spencer, "Transverse thickness modes in BT-cut quartz plates," *J. Acoust. Soc. Amer.*, vol. 41, pp. 994–1001, Apr. 1967.

[19] R. D. Mindlin and P. C. Y. Lee, "Thickness-shear and flexural vibrations of partially plated, crystal plates," *Int. J. Solids Structures*, vol. 2, pp. 125–139, 1966.

[20] D. F. Sheahan, "An improved resonance equation for AT-cut quartz crystals," *Proc. IEEE*, vol. 58, pp. 260–261, Feb. 1970.

[21] A. Glowinski, R. Lancon, and R. Lefevre, "Effects of asymmetry in trapped energy piezoelectric resonators," *Proc. 27th Ann. Symp. Frequency Control*, pp. 233–242, Apr. 1973.

[22] R. A. Johnson and W. D. Peterson, "Build stable compact narrow-band circuits," *Electron. Design*, pp. 60–64, Feb 1, 1973.

[23] M. Konno, *Electro-Mechanical Resonators and Applications.* Tokyo: Corona, 1973.

[24] R. A. Sykes and W. D. Beaver, "High-frequency monolithic crystal filters with possible application to single frequency and single sideband use," *Proc. 20th Ann. Symp. Frequency Control*, pp. 288–308, Apr. 1966.

[25] G. Sauerbrey, "Amplituden verteilung und elektrische ersatzdaten von schwing quarz platten," (AT-Schnitt), *A.E.U.*, vol. 18, no. 10, pp. 624–628, 1964.

[26] W. D. Beaver, "Theory and design principles of the monolithic crystal filter," Ph.D. Thesis, Lehigh University, Pa., 1967.

[27] W. D. Beaver, "Analysis of elastically coupled piezoelectric resonators," *J. Acoust. Soc. Amer.*, vol. 43, pp. 972–981, May 1968.

[28] ——, "Theory and design of the monolithic crystal filter," *Proc. 21st Ann. Symp. Frequency Control*, pp. 179–199, Apr. 1967.

[29] R. A. Heising, *Quartz Crystals for Electrical Circuits.* New York: Van Nostrand, 1946.

[30] W. G. Cady, *Piezoelectricity.* New York: McGraw-Hill, 1946.

[31] R. C. Rennick, "An equivalent circuit approach to the design and analysis of monolithic crystal filters," *IEEE Trans. Sonics. Ultrason.*, vol. SU-20, no. 4, pp. 347–356, Oct. 1973.

[32] A. E. Guenther, "High-quality wide-band mechanical filters, theory and design," *IEEE Trans. Sonics Ultrason.*, vol. SU-20, no. 4, pp. 294–301, Oct. 1973.

[33] E. Christian and E. Eisenmann, "Consideration for the design of crystal filters," *Proc. 3rd Allerton Conf. Circuit System Theory*, pp. 806–816, 1965.

[34] H. Betzl, "Ein Beitrag zur Berechnung von eingliedrigen Quarz-Bruckenbandpässen mittlerer Bandbreite nach der Betriebsparametertheorie," *Frequenz*, vol. 19, pp. 206–209, June 1975.

[35] A. E. Guenther, "Electromechanical filters: satisfying additional demands," *Proc. IEEE Int. Symp. Circuit Theory*, pp. 142–145, Apr. 1973.

[36] R. A. Johnson, "Mechanical filters," *Proc. 1973 Int. Symp. Circuit Theory*, pp. 402–405, Apr. 1973.

[37] A. I. Zverev, *Handbook of Filter Synthesis.* New York: Wiley, 1967.

[38] G. Szentirmai, "Crystal and ceramic filters," *Modern Filter Theory and Design*, G. C. Temes and S. K. Mitra, Eds. New York: Wiley, 1973.

[39] E. Christian and G. C. Temes, "On the Szentirmai transformation," *IEEE Trans. Circuit Theory*, vol. CT-13, pp. 450–452, Dec. 1966.

[40] Y. Rainsard, "Sur la suppression du transformateur differentiel utilise dans les structures de jaumann," *Cables and Transmission*, vol. 21, no. 4, pp. 226–237, 1967.

[41] J. W. Poschenrieder, "Die Wellenparametertheorie als einfaches Hilfsmittel zur Realisierung von Quarzbandfiltern in Abzweigschaltung," *NTZ*, pp. 132–138, Mar. 1959.

[42] H. Yoda, "Quartz crystal mechanical filters," *Proc. 13th Ann. Symp. Frequency Control*, May 1959.

[43] W. Poschenrieder and F. Schöfer, "Das Elektromechanische Quarzfilter-ein neues Bauelement für die nachrichtentechnik," *Frequenz*, vol. 17, pp. 88–94, Mar. 1963.

[44] M. Börner, "Mechanische filter mit dampfungspolen," *Arch. Elek. Übertragung*, vol. 17, pp. 103–107, Mar. 1963.

[45] R. A. Johnson, "A single-sideband disk-wire type mechanical filter," *IEEE Trans. Component Parts*, vol. CP-11, pp. 3–7, Dec. 1964.

[46] J. L. Hokanson, "The monolithic crystal filter: the device, its operation and choice of piezoelectric materials," *6th Ann. Integrated Circuits Seminary IEEE*, Hoboken, N.J., pp. 32–43, Apr. 1969.

[47] E. M. Frymoyer, R. A. Johnson, and F. H. Schindelbeck, "Passive filters: today's offerings and tomorrow's promises," *EDN*, vol. 18, pp. 22–30, Oct. 5, 1973.

[48] J. F. Werner, A. J. Dyer, and J. Birch, "The development of high performance filters using acoustically coupled resonators on AT-cut quartz crystals," *Proc. 25th Ann. Symp. Frequency Control*, pp. 65–75, Apr. 1971.

[49] R. J. Byrne and J. L. Hokanson, "Effect of high-temperature processing on the aging behavior of precision 5 MHz quartz crystal units," *IEEE Trans. Instrum. Meas.*, vol. IM-17, pp. 76–79, Mar. 1968.

[50] C. M. van der Burgt, "Performance of ceramic ferrite resonators as transducers and filter elements," *J. Acoust. Soc. Amer.*, vol. 28, pp. 1020–1032, Nov. 1956.

[51] M. Börner, "Progress in electromechanical filters," *Radio Electro. Eng.*, vol. 29, pp. 173–184, Mar. 1965.

[52] H. J. Orchard and G. C. Temes, "Filter design using transformed variables," *IEEE Trans. Circuit Theory*, vol. CT-15, no. 4, pp. 385–408, Dec. 1968.

[53] W. Herzog, *Siebschaltungen mit Schwingkristallen.* Braunschweig, Germany: F. Vieweg, 1962.

[54] D. Indjoudjian and P. Andrieux, *Les Filtres a Cristaux Piezoelectriques.* Paris: Gauthier-Villars, 1953.

[55] D. I. Kosowski, "Synthesis and realization of crystal filters," Tech. Rep. 298, Research Lab of Electronics, MIT, Cambridge, June 1955.

[56] J. Lang and C. E. Schmidt, "Crystal filter transformations," *IEEE Trans. Circuit Theory*, vol. CT-12, pp. 454–457, Sept. 1965.

[57] D. S. Humphreys, *The Analysis Design and Synthesis of Electrical Filters.* Englewood Cliffs, N.J.: Prentice-Hall, 1970.

[58] P. Lloyd, "Monolithic crystal filter," *Proc. 25th Ann. Symp. Frequency Control*, pp. 65–75, Apr. 1971.

[59] R. J. Byrne, "Monolithic crystal filters," *Proc. 24th Ann. Symp. Frequency Control*, pp. 84–92, Apr. 1970.

[60] D. F. Sheahan and C. E. Schmidt, "Coupled resonator quartz crystal filters," WESCON, Session 8, Aug. 1971.

[61] H. Yoda, Y. Nakazawa, and N. Kobori, "High-frequency crystal monolithic (HCM) filters," *Proc. 26th Ann. Symp. Frequency Control*, pp. 76–93, Apr. 1972.

[62] Y. Masuda, I. Kawakami, and M. Kobayashi, "Monolithic crystal filter with attenuation poles utilizing 2-dimensional arrangement of electrodes," *Proc. 27th Ann. Symp. Frequency Control*, pp. 227–232, Apr. 1973.

[63] H. Schüssler, "Filters for channel bank filtering," *Proc. of the 1974 IEEE Int. Symp. Circuits Systems*, pp. 106–110, Apr. 1974.

[64] "Session on Channel Bank Filtering," *Proc. of the 1974 IEEE Int. Symp. Circuits Systems*, Apr. 1974.

[65] P. Lloyd, "Monolithic crystal filters for frequency divison multiplex," *Proc. 25th Ann. Symp. Frequency Control*, pp. 280–286, Apr. 1971.

[66] A. J. Miller, "Preparation of quartz crystal plates for monolithic crystal filters," *Proc. 24th Ann. Symp. Frequency Control*, pp. 93–103, Apr. 1970.

Part I
Mechanical Filters

Section I-A
Historical and Review Papers

Although patents were taken out on mechanical filters in the early 1920's and excellent work was done in the area of sound reproduction using bandpass filter concepts, it was not until 1946 that a practical electromechanical filter was introduced to the public by Robert Adler. Adler's plate-type filter was quickly followed by the development of rod-neck and disk-wire filters. These filters were first used in high performance receivers because of their high resonator Q's and good temperature and aging stability. In fact, much of the credit for the rapid development and widespread use of single-sideband equipment must be given to these early mechanical filters.

Although the first paper in this section "Compact Electromechanical Filters," was written three decades ago, it is timeless in its explanation of how a mechanical filter works. In fact, Adler takes the reader through the thinking process of a person developing his first mechanical filter by discussing equivalences between electrical and mechanical elements, as well as the dual role of transducers in termination and energy conversion. He also describes an actual filter and its use in a communication receiver. Adler gives credit to Warren Mason and others at Bell Laboratories for their early work with spring-mass systems [1], but he could also have included Mason's early work with transmission-line mechanical filters [2].

Hathaway and Babcock's paper was the first survey paper written on the subject of mechanical filters. In this paper, the authors describe, in considerable detail, the three filters developed in the previous decade, namely, the mechanical transmission-line filter of Roberts, the disk-wire filter of Doelz, and a further description of the extensional-mode plate-wire filter first developed by Adler. Included in this paper is a description of various types of magnetostrictive transducers and the use of mechanical filters as channel filters in FDM telephone systems.

In the mid-to-late 1950's, work on mechanical filters was started in Europe and Japan. The European designs employed magnetostrictive ferrite transducers, whereas the Japanese designs used piezoelectric ceramics. Börner's paper, "Progress in Electromechanical Filters," describes the work done in Europe on torsional-mode and longitudinal-mode filters to the year 1965. This is still a helpful paper because the author deals with the tough real-world subjects of spurious responses, material characteristics, parts tolerances, and production methods. Börner also discusses methods of realizing attenuation poles; some of the methods work well and some, because of spurious modes, do not.

The paper "Mechanical Filters—a Review of Progress" reviews work done prior to 1971 by industry and universities throughout the world. Rather than being a tutorial paper, it is a compilation of various mechanical configurations, resonator and coupling element models, and transducer types. Also, there is a discussion on designing very small mechanical filters and their comparative advantages and disadvantages with respect to planar process devices. In addition, this paper discusses tuning fork and flexure-mode bar filters designed to operate at frequencies from a few hundred Hz to 50 kHz that are used in narrow bandwidth signaling and tone selection applications such as rapid transit, FDM, FSK, paging, and navigation systems.

The final paper in this section, "Mechanical Filters and Resonators," acts as an update to the previous paper in terms of filter configurations as well as references. The most important of these configurations described is the 12-resonator, bar-flexure mode filter designed in West Germany for use as a channel filter in FDM telephone systems. Also included in this paper are a large number of equivalent circuits and network transformations, all of which are based on the force-current mobility analogy.

References

[1] W. P. Mason, *Electromechanical Transducers and Wave Filters*. New York: Van Nostrand, 1948.

[2] —, "Wave transmission network," U.S. Patent 2 345 491, Mar. 1944.

Compact ELECTROMECHANICAL FILTER

By ROBERT ADLER

Engineering Department
Zenith Radio Corp.
Chicago, Ill.

Interconnected metal plates that transmit vibrations act as transmission-line type filter. Plates are coupled to electrical circuit by magnetostriction. Filter for 455-kc i-f channel of broadcast receiver has very sharp cutoffs, is small, cheap, easily constructed and efficient

THIS PAPER presents principles of operation and practical design of a novel and rather unconventional wave filter.

In its present form the filter is best suited for the 455-kc intermediate-frequency channel of broadcast and communication receivers; its frequency response is characterized by a flat pass band of well defined width, and by extremely sharp attenuation outside the band limits. Its adjacent-channel selectivity easily surpasses that obtained with much more expensive conventional filters.

Principle of Operation

The new filter is of the electromechanical type. Intermediate frequency currents, upon entering the filter, are converted into mechanical vibrations of the same frequency. These vibrations then pass through a structure resembling a ladder, consisting of several mechanically resonant metal plates coupled to each other by wires which act as springs.

This structure forms a bandpass filter for mechanical vibrations. Width of the pass band, as will later be shown, is determined by the design of each individual section; and, because the several sections are all alike, bandwidth does not depend upon the number of sections. Attenuation outside the band limits, however, increases with the number of sections.

To understand the operation of the filter most easily, let us first consider a familiar electrical filter, composed of inductors and capacitors; the type shown in Fig. 1A is rather fundamental and can be found in many books on electrical wave filters.

Such a filter will transmit a band of frequencies starting at the resonant frequency F of the two series elements L_1, C_1 and ending at a somewhat higher frequency $F + W$, where W stands for the bandwidth of the filter. If shunt capacitances C_2 are large compared to the series ones C_1, the fraction C_1/C_2 becomes small compared to unity; the bandwidth ratio W/F is then equal to 2 C_1/C_2. For an i-f filter in which $F = 455$ kc and $W = 9$ kc, the bandwidth ratio W/F becomes 0.02 and C_1/C_2 becomes 0.01.

Series elements in the terminating half sections should have only half the reactance of the other series elements. Also, no resistances at all are needed inside the filter, so that, strictly speaking, all inductors

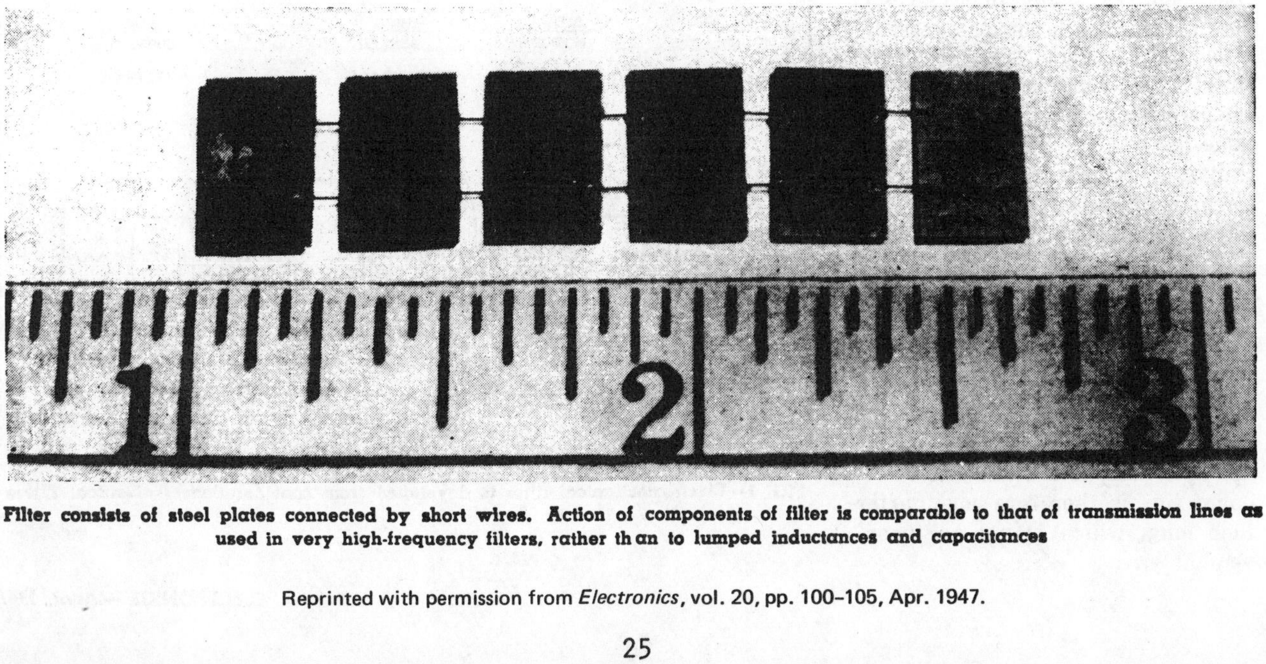

Filter consists of steel plates connected by short wires. Action of components of filter is comparable to that of transmission lines as used in very high-frequency filters, rather than to lumped inductances and capacitances

Reprinted with permission from *Electronics*, vol. 20, pp. 100–105, Apr. 1947.

and capacitors should have infinite Q's; the generator and the load are resistive, however. Their resistance R should be so chosen that the filter will be correctly terminated; for the circuit of Fig. 1A, R becomes numerically equal to the reactance of the shunt capacitances C_2.

The analogy between electrical and mechanical network elements, with masses substituted for inductors and springs for capacitors, is well known; and it appears quite feasible to build a mechanical filter structure, which is fully equivalent to the electrical filter shown in Fig. 1A, by combining masses and springs in analogous fashion. Such filters have been built, especially at the Bell Telephone Laboratories, by Mason, Blackman and Lakatos, and probably by others (See, for instance, Electromechanical Transducers and Wave Filters, by Warren P. Mason, D. Van Nostrand Co., Inc., New York, 1942, page 86).

Distributed Constants

For operation at 455 kc, the masses and springs of such a filter would become inconveniently small, and it is questionable whether a practical design could be developed on this basis.

The way around this obstacle is once more suggested by an electrical analogy. Radio engineers have long been familiar with the transition from lumped circuits to lines with distributed constants. (Line lengths are proportioned to give required input impedances.) This transaction becomes necessary when the frequency of electrical circuits is increased to a point where inductors and capacitors become too small to handle. A similar transition exists in the field of mechanical vibrations; only, because sound travels so much more slowly than do electromagnetic fields, the transition must be made at much lower frequencies. Velocity of sound in steel and many other solids is of the order of three miles per second, making the length of a longitudinal wave at 455 kc about one-half inch, and a half-wave resonant line is therefore one-quarter inch long, which is not an inconvenient size.

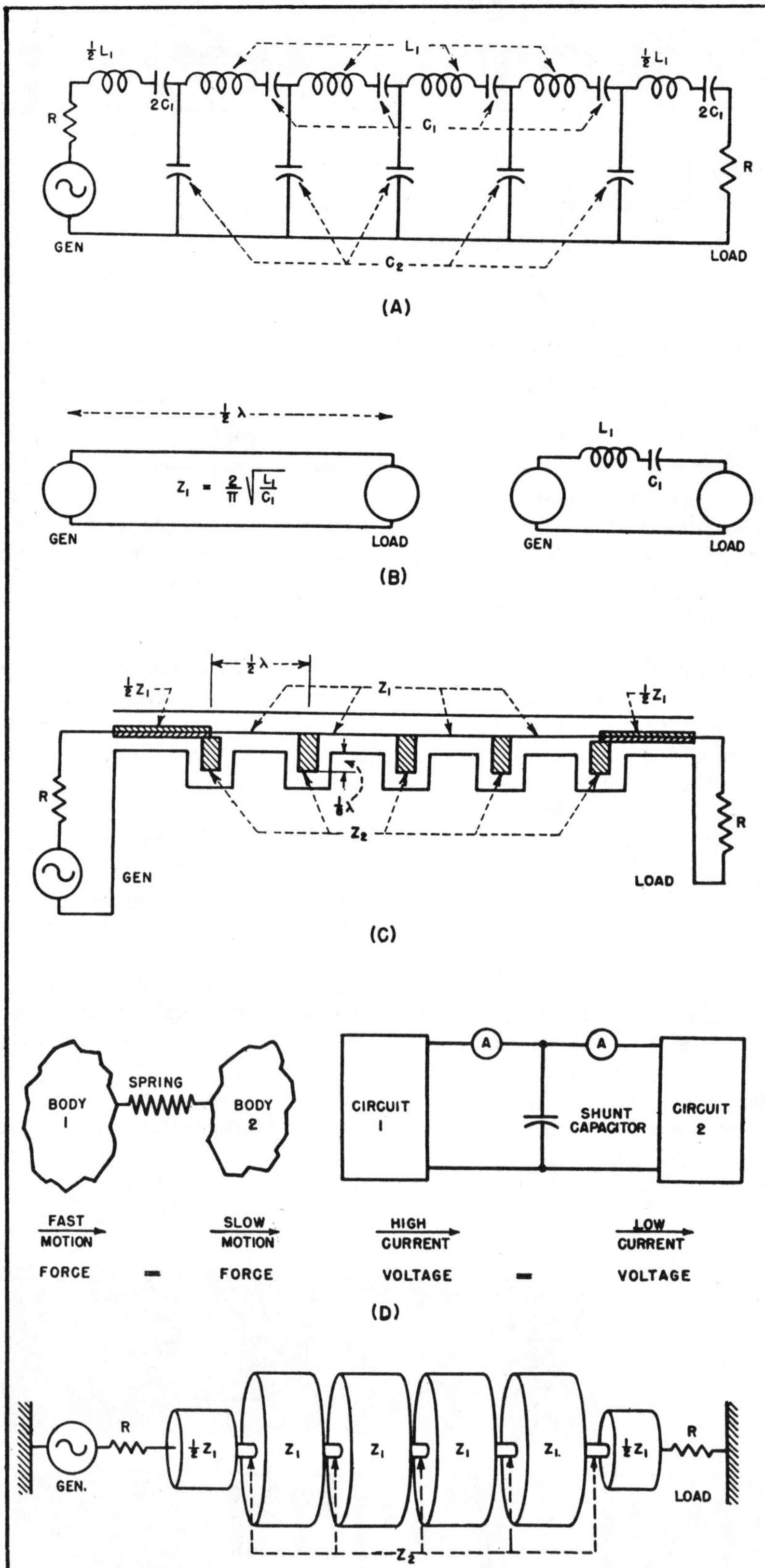

FIG. 1—Electromechanical filter is developed from consideration of electrical filters

To learn how to arrange mechanical lines to obtain a filter, let us look first at electrical lines which are more familiar. In the left portion of Fig. 1B, a half-wave line is shown which connects a generator to a load; to the right there appears its lumped-circuit equivalent, an inductance and capacitance in series. The following conditions must be met: Resonant frequency left and right should be the same, and characteristic impedance Z_1 of the line should be equal to $2/\pi$ times (about 64 percent) the reactance of L_1 or C_1 at resonance. Then the two networks become equivalent at all frequencies not too far from the resonant frequency F.

Using the equivalence just stated, series inductors and capacitors of the original filter can be replaced by half-wave lines of the proper impedance. There remain the shunt capacitors. These can best be taken care of by means of open-circuited line stubs less than a quarter wave long. For instance, if the stubs are an eighth wavelength long, their reactance becomes numerically equal to their characteristic impedance, which we may call Z_2.

With these two kinds of line elements, we can now draw the line equivalent of the original filter. Figure 1C shows several half-wave sections of impedance Z_1, shunted at their junctions by open eighth-wave lines of a much lower impedance Z_2. Bandwidth ratio W/F, which was previously equal to $2\ C_1/C_2$, becomes now $(4/\pi)\ (Z_2/Z_1)$. With these relations we design a line.

To obtain $W = 9$ kc at $F = 455$ kc, Z_2 must be made 1.57 percent of Z_1. It is apparent that the shunting stubs must be cut from lines of very low impedance; also, the impedances of the terminating half-wave lines should be only $0.5Z_1$.

Equivalent Mechanical Filter

We are now almost ready to apply the knowledge gained with the aid of our electrical model to an equivalent device in which mechanical lines are used. Half-wave and eighth-wave mechanical lines can easily be constructed for a given frequency; all we need to know is the velocity of sound propagation in the medium we select. There remains, however, the need for defining the ratio between the impedances of two mechanical lines, for upon such a ratio depends the bandwidth ratio W/F of the filter.

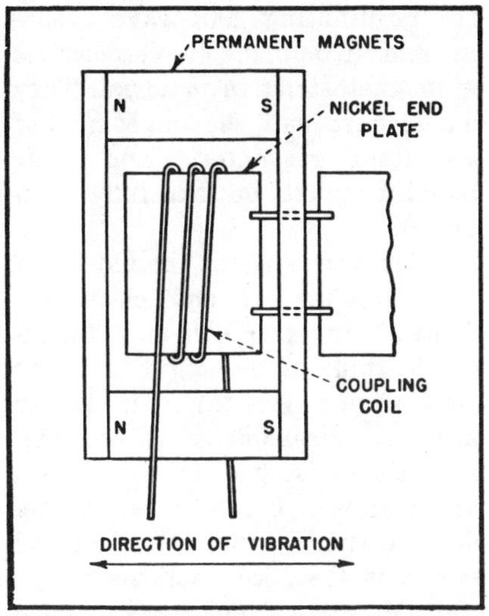

FIG. 2—Magnetostrictive end plates are magnetized and coupled by coils to the electrical circuit

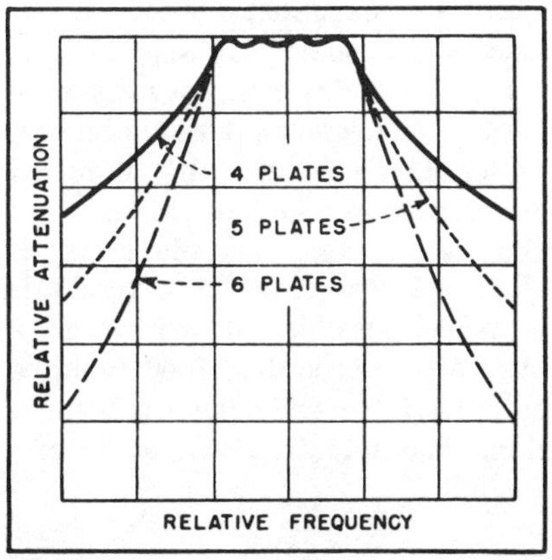

FIG. 3—Cutoff at edges of pass band is sharper for mechanical filters having more plates

Let us consider a section of mechanical line made of a given solid material in which vibration is propagated longitudinally. For a fibre of line having one unit of cross-sectional area, isolated from the surrounding body, it takes a certain force to produce a desired motion; then, if total cross-section of the line is A units, the total force required to impart the same motion to the entire area will be A times this certain force. In other words: Force for a given motion is proportional to the cross-sectional area. If we think of mechanical force as equivalent to electrical voltage, it follows that mechanical impedance of the line is proportional to its cross-sectional area.

The rules just derived enable us to design mechanical line elements which, when combined, will produce the desired pass band, but we still need to know how these elements should be interconnected. Let us again refer to the electrical analogy, shown in detail in Fig. 1D. A coupling spring between two mechanical objects exerts equal forces upon both, while taking up the difference between their velocities. Similarly, an electrical shunt connected between a generator and a load maintains equal voltages across both, but takes up the difference between generator current and load current. To act in a manner equivalent to electrical shunts, our mechanical eighth-wave stubs should therefore be connected like coupling springs between the half-wave resonating lines.

Figure 1E shows a highly schematized view of the structure which is obtained by joining mechanical lines together in the manner just derived. Any mechanical structure built in accordance with this schematic design will act as a bandpass filter.

Electromechanical Terminations

The figure calls for a mechanical generator on the left and for a mechanical load on the right, both having the proper resistance but no reactance. Because this filter is to be used between electrical circuits, generator and load should both be electromechanical converters, somewhat like a speaker and a microphone. For the frequency range near 455 kc, the only suitable converters are piezoelectric or magnetostrictive bars. Such bars always have mass and elasticity, however, and so it seems impossible to construct the desired converters.

To find a way out of this difficulty, let us look once more at the lumped electrical circuit (Fig. 1A) from which our filter was first derived. It is the resistive elements R which cannot be built without introducing reactances. But instead of attempting the impossible, consider the terminating elements R, 0.5 L_1, and 2 C_1 as a unit; together they form a tuned circuit with $Q =$ of $0.5[(L_1/C_1)/R]^{1/2}$. This Q, incidentally, is equal to the reciprocal of the bandwidth ratio W/F, becoming 50 for the 9-kc filter.

If it is permissible to consider the resistive elements as damping tuned circuits which terminate this electrical filter at both sides, then, in analogous fashion, the required resistance can be incorporated into the terminating half-wave lines of the mechanical filter by giving them the proper damping.

Finally, the resistances in the mechanical filter should simultaneously serve as electromechanical converters, which is possible only if the terminating half-wave lines themselves are piezoelectric or magnetostrictive.

Electromechanical Damping

The electromechanical filter must therefore consist of several half-wave resonators with the least possible mechanical damping, coupled to each other and to the terminating elements by eighth-wave lines of much smaller cross section. Electrical properties of all these elements are of no consequence. But the terminating half-wave resonators must be made of piezoelectric or magnetostrictive material. They should have half the impedance of the other resonators, and their damping must be carefully controlled.

A few words about the nature of this damping: If any electromechanical converter—we may, for instance, think of the oldest kind, the rotating d-c generator—is loaded on its electrical side, a reflected mechanical resistance is established which causes expenditure of mechanical energy whenever electrical energy is absorbed from the generator. Mechanical energy consumed by this reflected resistance is not lost but reappears in electrical form.

Evidently, then, reflected mechanical resistance produced by loading the electrical side of converters provides damping. From an efficiency viewpoint it would be best to produce exactly the required amount of damping in this manner and use no frictional damping at all. In piezoelectric converters this is indeed possible, and with them a filter of the type described could be built having nearly ideal transmission. Magnetostrictive converters, at frequencies near one-half megacycle, are not quite as efficient. Only part of the required mechanical resistance can be produced by electrical loading; the remaining portion is inherent in the mechanical damping—internal and external friction—acting upon the two end pieces. Accordingly, a fraction of the incoming electrical energy is lost in the filter input, and an equivalent fraction of the mechanical energy arriving at the filter output is lost there.

The question may therefore be asked why, in spite of these losses, the magnetostrictive type should have been chosen for a practical design. The answer lies in its simplicity, economy and stability.

A typical filter consists of flat nickel end plates 0.005 inch thick, stainless steel plates twice as thick (to maintain the required impedance ratio) and pairs of parallel steel wires 0.006 inch in diameter, which connect adjacent plates. It is put together by spotwelding the steel wires to the rectangular plates. To put it into operation, each of the nickel end plates is premagnetized in the direction of vibration, by means of small permanent magnets as shown in Fig. 2, and coils are

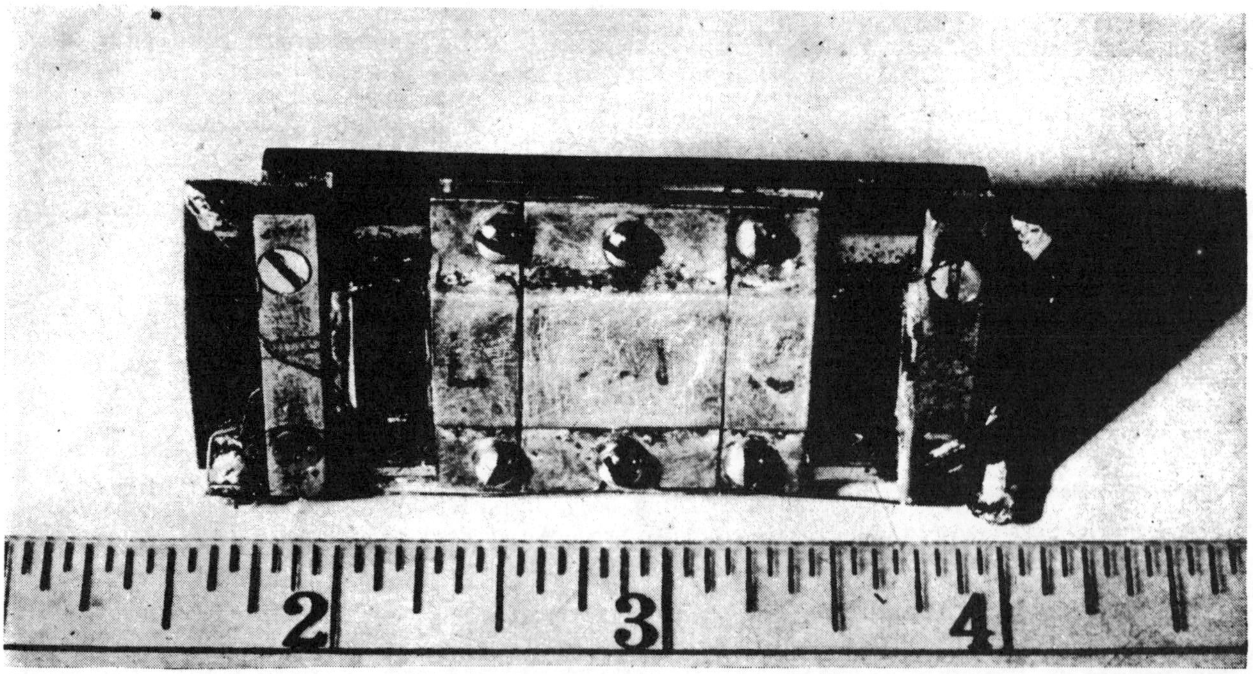

For use in intermediate-frequency amplifier of broadcast receivers, the electromechanical filter is housed in a protective case. Coupling coils are wound around end plates, which are located in the field of permanent magnets

arranged around each nickel end plate to make it vibrate as a magnetostrictive half-wave bar.

The filter structure is held loosely in place between linings of soft cloth or neoprene inside a flat metal cover. As long as no undue pressure is exerted upon the stainless steel plates, their mechanical *Q* is quite high, between 2,000 and 4,000, and they come fairly close to pure reactances. Consequently, the losses inside the filter, aside from the conversion loss at the terminations which was discussed before, are very small, and additional stainless steel sections cause no noticeable increase in the attenuation inside the pass band. Every additional section, however, causes sharper cutoff at the band limits. The curves of Fig. 3 show how frequency response varies with the number of sections.

For the six-piece filter shown in the pictures, the combined cross-section of both coupling wires is equal to about 1.5 percent of the cross-section of the stainless steel plates, measured in the direction of vibration. Because they are made of very similar materials, their impedance ratio is also about 1.5 to 100, leading to a theoretical bandwidth of 8.5 kc at 455-kc center frequency. Measured response of this filter is shown in Fig. 4. For purposes of comparison, the i-f response of a conventional receiver of good quality is plotted under the same zero line. It is interesting to note that the mechanical filter transmits the higher audio frequencies, up to about 4 kc, somewhat better than do conventional i-f transformers; but the adjacent carrier (10 kc away) is attenuated 1,000 times by the mechanical filter. The conventional filter needs 20-kc spacing to do the same.

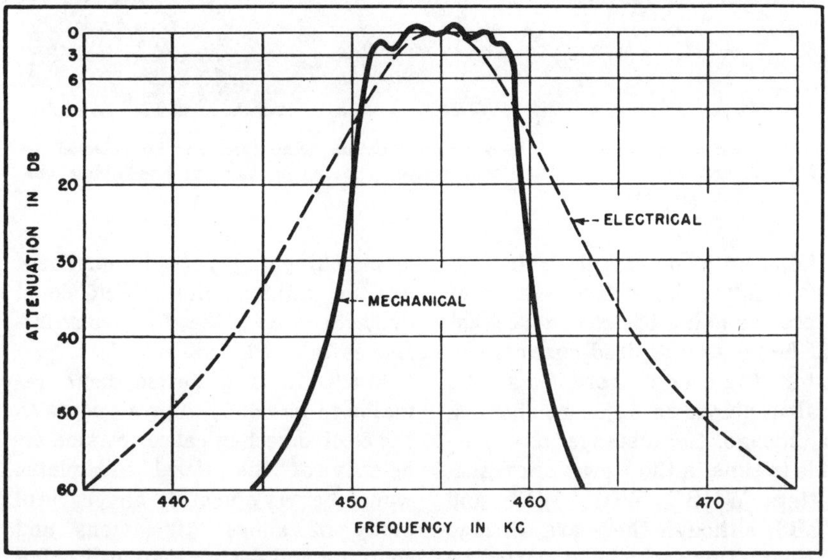

FIG. 4—Comparison of electrical and electromechanical filter bandpass characteristics

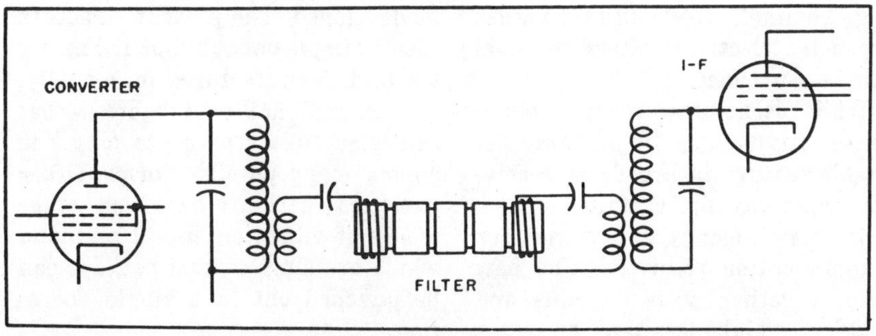

FIG. 5—Electromechanical filter is most effective if used between converter and first i-f amplifier

By using coupling wires with other diameters, pass bands up to 14 kc and down to 4 kc have been obtained in experimental filters. For the wider bands, the low Q required in the nickel end plates becomes more difficult to realize.

Use in Receivers

In a radio receiver, the best place for the electromechanical filter is right after the converter, as shown in Fig. 5. The coupling coils which surround the nickel end plates of the filter structure can most easily be wound for impedances of the order of 100 ohms; transformers are used to match this low impedance to the plate circuit of the converter and to the grid circuit of the following i-f tube.

The insertion loss of the filter in its present form is about 14 db. In designing a practical receiver, however, it has been found that some of this loss can be recovered. The matching transformers around the filter can be built with higher impedances than are normally used in i-f transformers, because the mechanical filter greatly relieves the stability requirements for the electrical circuits, permitting the use of a higher *L/C* ratio. Furthermore, the insertion loss in the filter reduces the overall i-f regeneration. It seems fair to say that, in a balanced i-f design, the net loss caused by introducing the filter in its present form is between 6 and 10 db, corresponding to a reduction in gain by a factor of two or three for a given tube combination.

Temperature variations affect the mechanical resonant frequency of the half-wave plates, causing the pass band as a whole to shift by a small amount. If the plates were made of plain steel, a temperature rise of 50 degrees F would shift the band by somewhat more than one kilocycle. Stainless steels are available which show much smaller frequency shift; plates are preferably made from such materials.

Receiver Performance

The performance of a broadcast receiver equipped with the six-piece filter described (two times down

In experimental receiver, four filters with different bandwidths are assembled in a single structure. Any filter can be selected by four-position switch (switching is simple at the 100-ohm impedance of the filter terminations). Merits of different bandwidths can thus be determined

4.5 kc off center, 1,000 times down 10 kc off center) is interesting in many respects. It was first suspected that such a set would be hard to tune, but tests with a number of lay listeners did not bear this out. Change in tone quality caused by incorrect tuning sets in at two clearly defined points much more abruptly than in conventional receivers; listeners seem to find it quite easy to tune between these two points.

With correct tuning, tone quality appears to be quite similar to that of conventional sets. One might expect to hear unfamiliar transients around 4,000 cps caused by the unusually sharp cutoff of the pass band, but nothing unusual could be detected. It appears that at 4,000 cps, where the sensitivity and discrimination of the human ear are not as good as in the medium range, an even greater rate of cutoff would be required to produce noticeable transients.

Adjacent-channel selectivity, of course, is the distinctive feature of a receiver using the electromechanical filter. Any desired channel can be tuned in even if there is a strong local signal on an adjacent channel. In Chicago, for instance, it is possible to tune in the New York transmitters WNBC, WOR, WJZ, and WCBS, although there are Chicago stations (three of 50,000 watts and one of 10,000 watts) separated only 10 kc from each of these four New York channels. It would take rather expensive electrical filters to equal this performance.

While economical adjacent-channel selectivity may be a highly desirable feature in broadcast receivers, especially for certain regions, it is of paramount importance for communication receivers. The new filter, together with its coils and magnets, weighs less than an ounce and takes less than one cubic inch of space. Its essential parts are stainless steel, easily manufactured by punching and spotwelding. There are no adjustments which could drift or vary; the frequency response is fixed.

Much further development remains to be done; improvements in the electromechanical conversion efficiency of the nickel end plates would be very useful; and careful study of alloys, dimensions and coupling coil design may well bring such improvements about. Economical production methods must be developed. The present structure seems simple enough, but it has not yet been manufactured in quantity.

It should not be forgotten that this structure represents only one among many possible forms. Other varieties, perhaps based on other modes of vibration, might be found which are simpler, and perhaps can be punched out in a single operation.

Survey of Mechanical Filters and Their Applications*

J. C. HATHAWAY†, MEMBER, IRE, AND D. F. BABCOCK†, SENIOR MEMBER, IRE

The following paper is one of a planned series of invited papers, in which men of recognized standing will review recent developments in, and the present status of, various fields in which noteworthy progress has been made.—*The Editor*

Summary—**Since the inception of mechanical filters several years ago, their many desirable features have resulted in many applications. As filtering requirements have become more stringent, the improved performance of mechanical filters has become more pronounced. Q's in the order of a hundred times better than those of comparable electrical circuits are possible. These high Q's allow the use of lossless filter design for narrow bandwidth flat-top filters with more than one section. The mechanical filter allows compact design which is consistent with the miniaturization of modern equipment.**

Although numerous mechanical filters structures have been built, three types have found the most application. These are the ladder type with resonant plates interconnected by fine wires, the cylindrical rod structure machined to produce alternate necks and slugs, and a cylindrical arrangement with disk resonators interconnected by coupling wires. The center frequency, bandwidth, and filter skirt selectivity are a function of element sizes, spacing, and number of elements used. Proper selection of resonator size and shape, and proper arrangement of driving and coupling elements will suppress spurious responses, which are a major problem in mechanical filter design. Transducers used with mechanical filters provide for the converting of electrical to mechanical energy or mechanical to electrical energy and impedance matching of the filter. Of the four kinds of transducers that have been employed—electromagnetic, electrostatic, magnetostrictive, and piezoelectric—the magnetostrictive have been the most promising in regard to high frequency of operation, stability, efficiency, and economy.

Mechanical filters are especially applicable to carrier systems and single-sideband equipment, and are equally useful in band-pass filtering in high-performance communications receivers. Filters used in carrier systems have proved beneficial in providing improved performance in selecting channels and carriers with dependable, compact design. In single-sideband transmission and reception, mechanical filters are very effective in rejecting the carrier frequency and the undesired sideband. The use of mechanical filters will often eliminate an extra stage of conversion. Simplicity of design is afforded in receivers when mechanical filters are employed by eliminating the need for many IF stages to obtain the desired selectivity. A selective bandwidth arrangement is possible which occupies no more room than one conventional IF shield can. Mechanical filters may be employed in nearly any environmental condition where other filters are used. Filter designs are sufficiently rugged for application in portable and mobile equipment. Also, they are not subject to appreciable change with temperature variations. Mechanical filters have revolutionized electrical equipment design and with new design and fabrication techniques, they will continue to exert an accelerated influence in the field.

Introduction

THE APPLICATION of mechanical elements to electrical filtering problems has been receiving an increasing amount of study in recent years. As filtering requirements have become more and more stringent and the shortcomings of electrical elements have become a limiting factor in filter designs, mechanical elements have shown desirable properties. The losses and instability of electrical resonant elements have placed a definite lower limit on the fractional bandwidth of intermediate frequency electrical filters. The development of metal alloys with exceedingly low loss and very good temperature stability has added impetus to the development of mechanical filtering elements. Moreover, the increased demand for smaller size in components associated with electronic circuitry has made the small size of the mechanical filters look very desirable.

The development of constant modulus alloys suitable for use in mechanical filters can be traced to early work done by Guillaume[1] and Chevenard.[2] The alloys which have been of greatest importance have been those of iron and nickel. The range of investigation has been primarily between 30 and 50 per cent nickel with smaller amounts of other alloying elements. The alloy which has found frequent use in recent filter designs is Ni-Span C.[3] This is a heat treatable alloy with a nickel content of 42 per cent and small concentrations of chromium and titanium. With proper heat treatment this alloy has realized temperature coefficients of frequency less than three parts per million per degree over the range of −50°F to 150°F.

Early developments in the field of mechanical filtering dealt primarily with the use of isolated resonators as replacements for electrical resonant elements. This includes work by W. P. Mason,[4] R. V. L. Hartley,[5] H. A. Burgess,[6] and G. W. Pierce.[7] Although this work dealt with isolated resonators, it formed a background for later work in multiple section filter design. Since World War II, several organizations have pursued development of electromechanical filters for use in electronic circuitry. Among those who have developed multiple sec-

* Original manuscript received by the IRE, July 30, 1956.
† Collins Radio Co., Burbank, Calif.

[1] G. E. Guillaume, "Action des additions metallurgiques sur l'anomaliede dilatabilite desaciers au nickel," *Compt. Rend.*, vol. 170, pp. 1433–1435; June, 1920.
[2] M. P. Chevenard, "Etude de l'elastice de torsion des aciers au nickel a'haute teneur en chrome," *Compt. Rend.*, vol. 171, pp. 93–96; July, 1920.
[3] *Eng. Data Bull.*, H. A. Wilson Co., New York, N. Y.
[4] U. S. Patents 2,342,813; 2,345,491.
[5] U. S. Patent 1,654,123.
[6] U. S. Patent 1,666,681.
[7] U. S. Patents 1,750,124; 1,882,394; 1,882,396; 1,882,397.

Reprinted from *Proc. IRE*, vol. 45, pp. 5-16, Jan. 1957.

tion electromechanical filters are R. Adler,[8] W. Van B. Roberts,[9] M. L. Doelz,[10] L. L. Burns Jr.,[11] M. L. Anthony,[12] R. M. Virkus,[12] and V. D. Landon.[13] Nearly all of this late work has been in the frequency range of 50 to 1000 kc using mechanical couplers, magnetostriction transducers, and alloy resonators.

The loss in resonant elements in the band-pass filter must be such that the fractional loss, or $1/Q$, be small with respect to the fractional bandwidth (bandwidth/center frequency) of the filter. It can be seen that, as the filter fractional bandwidth becomes 1 per cent or less, the requirements on the Q of the electrical elements become restrictive. Mechanical resonant sections commonly have Q's in the order of 10,000 or approximately a hundred times that realizable with electrical elements. This allows lossless filter theory to be used in designing narrow bandwidth filters with flat tops and with many sections having low insertion loss. An example of the comparison between a mechanical filter and a multiple section LC filter is shown in Fig. 1. These filters are compared on the basis of similar skirt characteristics. The 60-db bandwidth of the mechanical filter is 16 kc. The rounding of the LC filter response is obvious with the low Q obtainable from electrical elements. In narrower bandwidths this effect is even more pronounced. Typical mechanical filter selectivity curves are shown in Fig. 2 for bandwidths of 0.5 kc, 6 kc, and 35 kc at a center frequency of 455 kc.

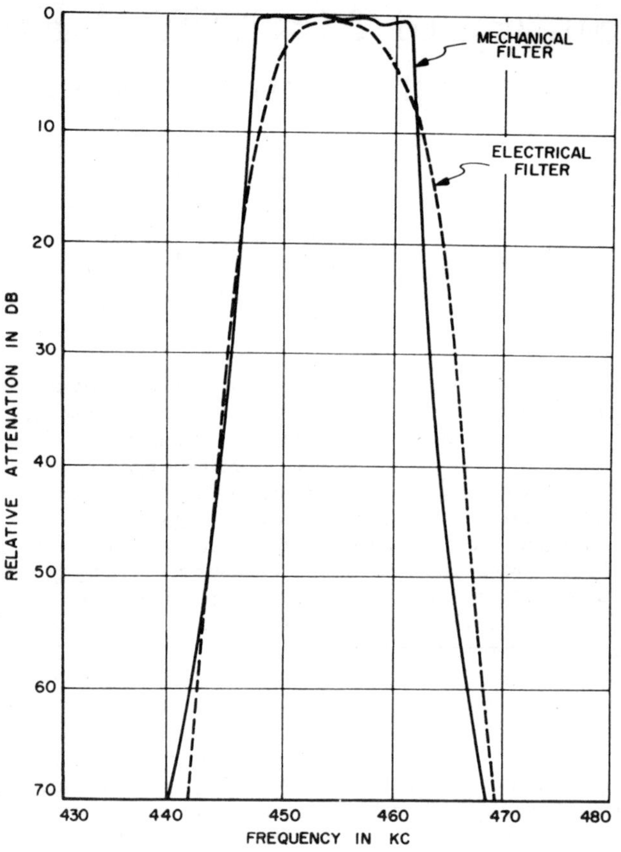

Fig. 1—Frequency response curves for mechanical and electrical LC filters.

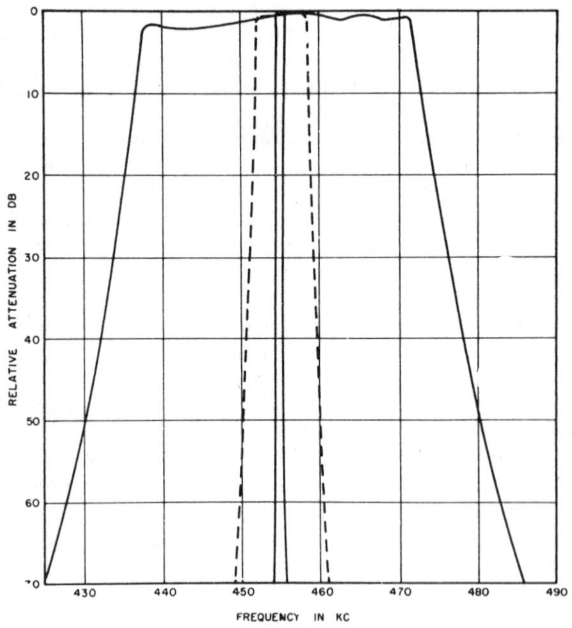

Fig. 2—Selectivity curves of 455-kc filters with nominal 6-db bandwidths of 0.5 kc, 6 kc, and 35 kc.

Filter Structures

This survey is concerned primarily with IF filters in which the selectivity characteristic is determined entirely by mechanical elements. Filters for low frequencies and those which include both electrical and mechanical elements, such as crystal lattices, are beyond the scope of this paper. Each mechanical filter has a number of resonators and couplers which form a central structure. The design of this structure is determined by center frequency, bandwidth, and selectivity requirements. Transducers located at each end of the filter provide coupling between the mechanical elements and the electrical circuit. They also provide the correct impedance for filter termination. This loading impedance is obtained from mechanical damping in the transducer and from resistance in the electrical circuit to which it is coupled.

A large number of structures have been proposed for use in mechanical IF filters. The arrangement of elements in the structure is determined largely by the choice of mechanical resonators. Of the many possible designs, three types are the most common. One employs a ladder type structure with resonant plates interconnected by fine wires. Another employs a cylindrical rod machined to produce alternate necks and slugs. In this case, either the necks or the slugs may be utilized as resonators. A third type of mechanical filter has a cylindrical arrangement with disk resonators interconnected by coupling wires. These three structures will be discussed in the following paragraphs.

[8] U. S. Patent 2,501,488.
[9] U. S. Patents 2,578,452; 2,617,882; 2,696,590; 2,647,948.
[10] U. S. Patent 2,615,981.
[11] U. S. Patent 2,647,949; 2,605,354.
[12] U. S. Patents 2,652,542; 2,652,543.
[13] U. S. Patent 2,660,712.

In 1947, Adler presented a practical design for a mechanical filter with half-wave longitudinal resonant plates.[14] A more recent model of this type has been discussed in an article by Lapin.[15] The filter structure consists of a series of flat plates interconnected by a pair of fine metal wires, as shown in Fig. 3. Each of the plates vibrates in an extensional mode with motion parallel to the long axis of the filter as shown. Resonant plates have Q's ranging from 2000 to 10,000. The two thin wires between each pair of plates expand and contract to provide elastic coupling. The center frequency of a filter is determined by the length of the plates, which must be a half-wave long at the center of the pass band. Bandwidth is a function of coupling wire diameter and length, as well as the position of wires on the plates. With a given length and position, bandwidth is approximately proportional to the cross sectional area or to the square of the diameter. The filter skirt selectivity on each side of the pass band is determined by the number of resonant plates in the structure.

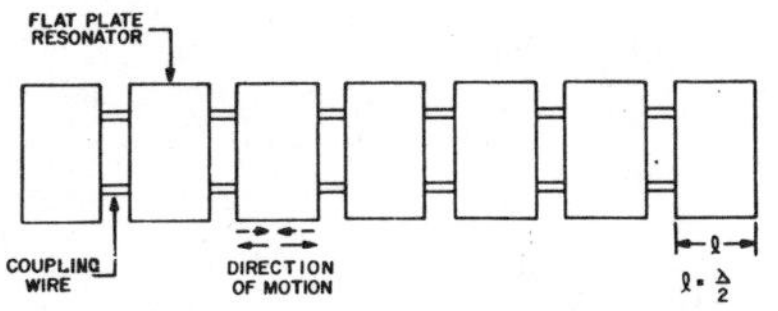

Fig. 3—Diagram of central structure—plate type filter.

Since resonators in mechanical filters are distributed elements, they have a number of natural vibration frequencies in addition to that utilized for operation of the filter. These undesired resonances give rise to spurious responses outside the pass band. Thus, the suppression of spurious responses is a major design problem. For all of the filters discussed here, spurious peaks have been minimized by proper selection of resonator size and shape, and by proper arrangement of driving and coupling elements. In recent plate type filters, spurious responses do not occur within 50 kc of the center frequency, and are more than 70 db down for all other frequencies.

A photograph of the plate type mechanical filter structure is shown in Fig. 4. This filter is designed for operation at 455 kc with a 6-db bandwidth of 10 kc. Each of the seven resonant plates is 0.250 inch by 0.398 inch by 0.010 inch. The two coupling wires between each pair of plates are $\frac{1}{8}$ wavelength, or about 0.055 inch long, with a diameter of 0.0055 inch. Filters have been constructed with up to 13 plates. The width of the central structure is approximately $\frac{3}{8}$ inch and the length slightly less than $1\frac{7}{8}$ inch. A photograph of the filter assembly is shown in Fig. 5. With this mounting, the central ladder structure is sandwiched between a bakelite retainer plate on the bottom and a cellulose retainer sponge on top. In this manner, it is protected against damage due to shock and vibration. The over-all case size is $2\frac{5}{8}$ inches by $\frac{3}{4}$ inch by $\frac{5}{8}$ inch.

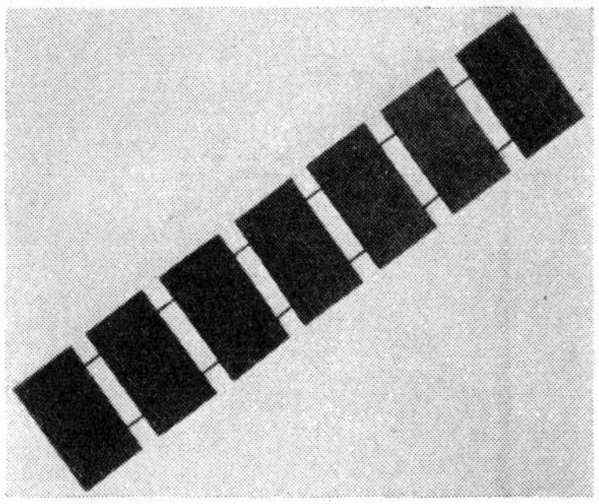

Fig. 4—Plate type filter structure. (Courtesy of Motorola, Inc.)

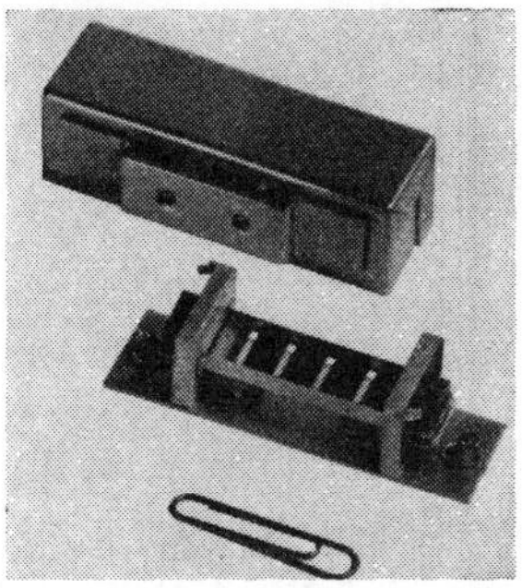

Fig. 5—Plate type filter assembly. (Courtesy of Motorola, Inc.)

In fabricating these filters, strips the width of the plates are first cut from sheets of Ni-Span C in a precision shear. These are then cut to the correct length in a micrometer shear. Wires are attached to plates by a spot welding technique. Special fixtures are employed to position and hold these elements during the welding operation. This operation is sufficiently uniform to preclude misadjustment of the elements during assembly.

Another type of mechanical filter, the neck-and-slug type, has been discussed in articles by Roberts, Burns, and George.[16–18] It is illustrated in Fig. 6. The neck-and-slug type filter may vibrate in either the longitudinal mode or the torsional mode. The longitudinal mode is illustrated in Fig. 6(a). This filter has a number of small diameter neck resonators, each one-half wavelength long. Resonators vibrate with motion parallel to the long axis of the filter. Slugs are provided between resonators for coupling. This filter differs from the others discussed here in that it employs mass coupling rather than spring coupling. The center of the pass band is determined by the length of each neck resonator. Also, for a given resonator diameter, bandwidth is determined by the length and diameter of the slugs. As on the other

[14] R. Adler, "Compact electromechanical filters," *Electronics*, vol. 20, pp. 100–105; April, 1947.

[15] S. P. Lapin, "Electromechanical filters," *Radio and Television News, Radio-Electronic Eng. Sec.*, vol. 50, pp. 9–11+; December, 1953. Also, *Proc. NEC*, vol. 9, pp. 353–362; February, 1954.

[16] R. W. George, "Electromechanical filters for 100-kc carrier and sideband selection," Proc. IRE, vol. 44, pp. 31–35; January, 1956. (See p. 34.)

[17] L. L. Burns and W. Van B. Roberts, "Mechanical filters for radio frequencies," *RCA Rev.*, vol. 10, pp. 348–365; September, 1949.

[18] L. L. Burns, "A band-pass mechanical filter for 100 kilocycles," *RCA Rev.*, vol. 13, pp. 34–36; March, 1952.

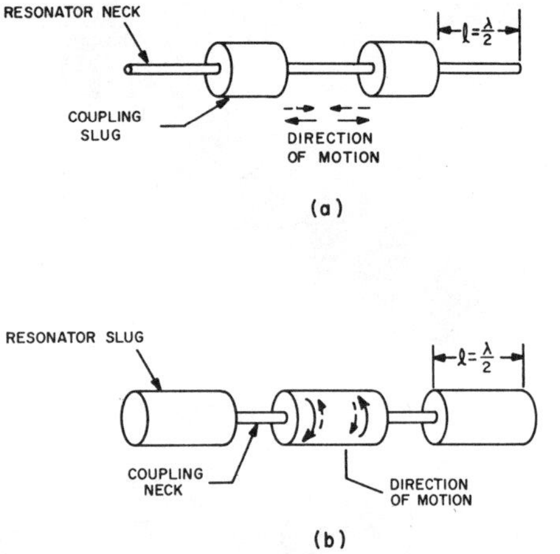

Fig. 6—Diagram of central structure, neck-and-slug type filter. (a) Longitudinal mode, slug coupled. (b) Torsional mode, neck coupled.

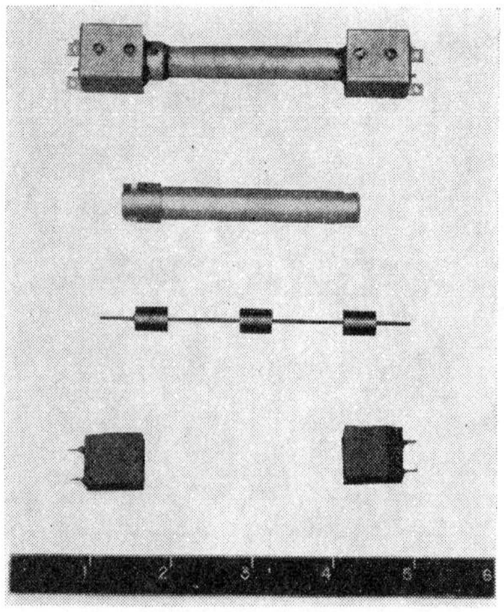

Fig. 7—Longitudinal mode, slug coupled filter. (Courtesy of Radio Corp. of America.)

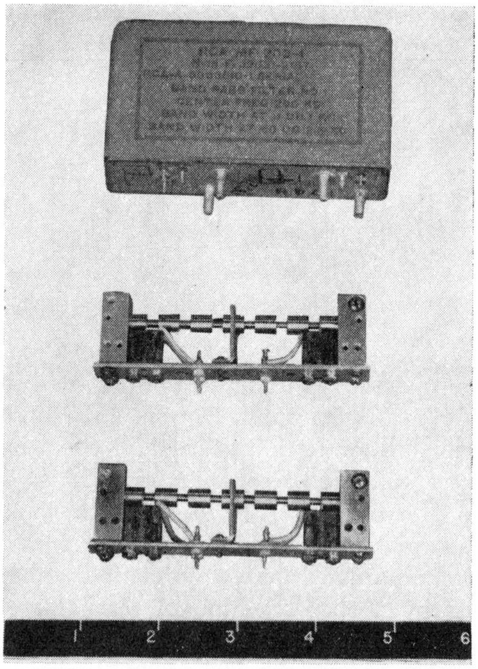

Fig. 8—Torsional mode, neck-coupled filter. (Courtesy of Radio Corp. of America.)

filters, skirt selectivity is determined by the number of resonant sections. The neck-and-slug type filter may also vibrate in the torsional mode, as illustrated in Fig. 6(b). Each large cylinder is a half-wave resonator tuned to the center of the pass band. At resonance, one end vibrates in rotation about the axis in the opposite direction to the motion of the other end. The small diameter quarter-wave sections provide spring coupling between resonators. Coupler length and the ratio of coupler neck diameter to resonator diameter determines the fractional bandwidth of the filter. For most applications, the torsional mode offers advantages over the longitudinal mode. Since the velocity of propagation for the torsional mode is about 60 per cent of the velocity for the extensional mode, the filter is more compact. Also, since the mechanical impedance of couplers is proportional to the fourth power of the diameter rather than to the square, a narrow-band torsional filter can be constructed without very great difference in diameter between the resonator and coupler portions.

An example of the longitudinal slug coupled filter is illustrated in Fig. 7. This has two half-wave resonators in the central structure as shown in the center of the figure. The three large cylinders are coupling slugs. The filter structure, consisting of neck resonators and coupling slugs, is mounted in the tubular case shown above it. Transducers at each end are inserted in coil mounting blocks at the bottom. The complete assembly is illustrated at the top of the photograph. This has over-all dimensions of approximately $\frac{5}{8}$ inch by $\frac{5}{8}$ inch by $4\frac{1}{8}$ inches. The filter is designed to have a center frequency of 105 kc and a 6-db bandwidth of 0.4 kc.

An example of the torsional neck coupled filter is illustrated in Fig. 8. This assembly employs two filters in a single case. Both have center frequencies of 200 kc. The unit at the bottom of the photograph has a 6-db bandwidth of 1 kc, and the unit in the center, a bandwidth of 3 kc. Each filter has seven half-wave torsional resonators coupled by six small diameter necks. The central structure of each filter is supported between two mounting blocks at the ends. Angle brackets are provided in the center to increase shock resistance. These two filters are mounted side by side in a case shown at the top of the picture. This has over-all dimensions of $1\frac{1}{4}$ inch by $2\frac{1}{16}$ inches by $3\frac{13}{16}$ inches.

Very narrow bandwidths may be obtained by employing multiple couplers between resonant sections. This is illustrated by the filter in Fig. 9. This model employs five half-wave resonant cylinders. These are located on each side of multiple couplers consisting of two quarter-wave necks separated by a quarter-wave slug, Fig. 10. With this arrangement, a very small coefficient of coupling can be achieved without employing extremely small diameter necks. The filter described here has a 6-db bandwidth of about 50 cycles at a center frequency of 100 kc.

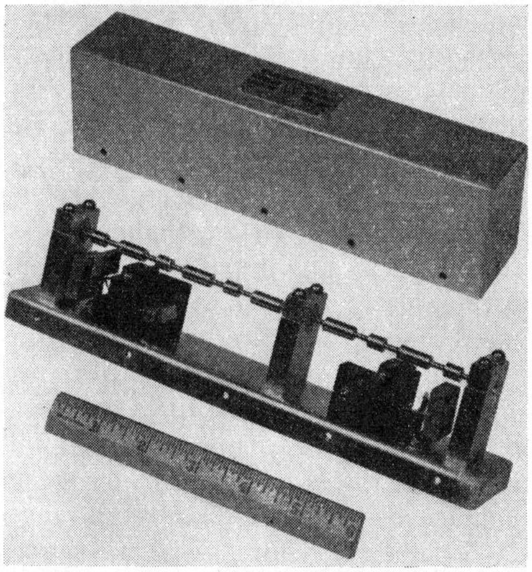

Fig. 9—Torsional neck-coupled filter with multiple couplers. (Courtesy of Radio Corp. of America.)

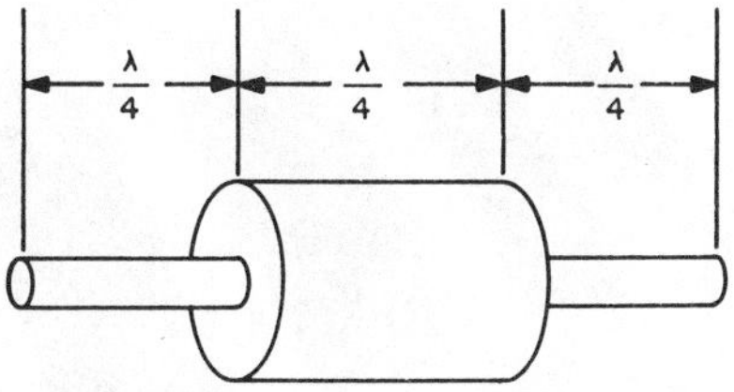

Fig. 10—Multiple coupler—neck-coupled filter.

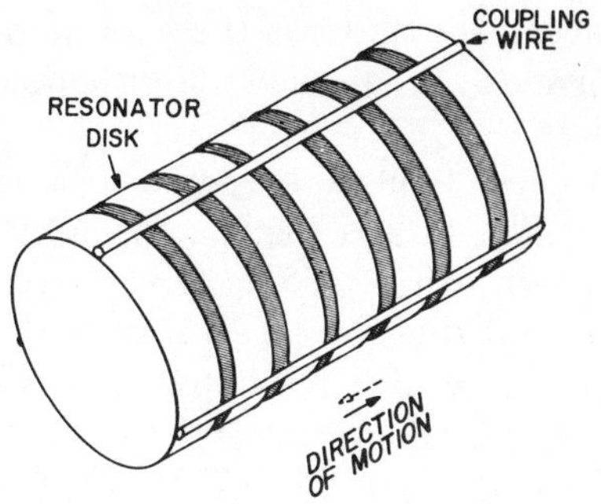

Fig. 11—Diagram of central structure—disk type filter.

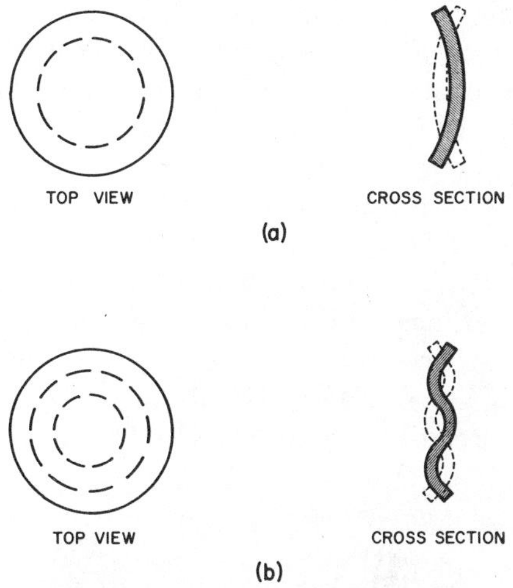

Fig. 12—Two normal modes of vibration for circular disks. (a) One nodal circle. (b) Two nodal circles.

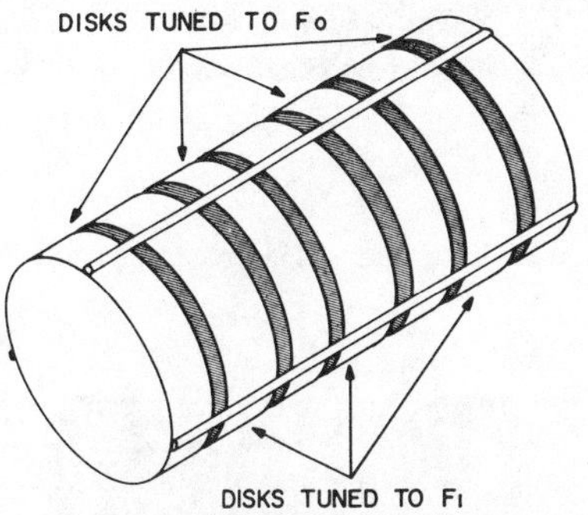

Fig. 13—Disk filter with multiple couplers.

The neck-and-slug type filter is fabricated by grinding necks in a solid rod of Ni-Span C. After grinding, resonators are individually tuned to the center frequency. During this operation, adjacent slugs are clamped, and the frequency of the resonator is determined by coupling into a coil and external measuring circuit. Frequency can be raised by removing material from a narrow band around the center or lowered by removing material from the ends. After each resonator has been tuned, the filter is securely clamped in the end supports.

A third type of mechanical filter employs Ni-Span C disks as high Q resonant elements. It has been discussed in articles by Doelz, Hathaway, and Brown.[19–21] This disk type filter is illustrated in Fig. 11. Resonators vibrate with motion perpendicular to the flat surface as shown by the two patterns in Fig. 12. Here, deflections are illustrated by cross sectional views and nodal circles by dashed lines on top views. In designing mechanical filters, a mode is selected to obtain the most convenient disk size. In most cases, the pattern with one circle is employed below 200 kc and the pattern with two circles from 200 kc to 600 kc. The resonant frequency of each disk may be varied by adjusting thickness and diameter. Disk resonators have Q's of about 10,000. Coupling in the filter is provided by wires welded to the peripheries of disk resonators. Bandwidth varies approximately as the total area of the coupling wires, while selectivity is determined by the number of disks. With the disk type filter, spurious responses have been reduced 60 to 100 db below the pass band level. Here, as on the slug filter, it is possible to design ruggedized narrow-band models by employing multiple couplers between resonant sections as illustrated in Fig. 13. Four disks are tuned to the center frequency, F_0, while three alternate impedance matching disks are tuned to F_1, either above or below the pass band. Since detuned disks have a high mechani-

[19] M. L. Doelz and J. C. Hathaway, "How to use mechanical I-F filters," *Electronics*, vol. 26, pp. 138–142; March, 1953.
[20] E. M. Brown, "Using the Collins F455A-31 filters," *CQ*, vol. 9, pp. 13–19; March, 1953.
[21] E. M. Brown, "More on mechanical filters," *CQ*, vol. 9, pp. 34–36; October, 1953.

cal impedance at the center of the pass band, coupling is reduced between F_0 disks without employing small coupling wires.

Two examples of the disk type mechanical filter are illustrated in Fig. 14 and Fig. 15. Fig. 14 shows a filter with 3-kc bandwidth at 455-kc center frequency. This filter has a rigid tubular type arrangement of disks and wires. Each of the disk resonators has a diameter of 0.350 inch, and a thickness of 0.071 inch. They are separated by a space of 0.040 inch. Each of the three coupling wires attached to the disk peripheries has a diameter of 0.010 inch. The small wires attached between the end disk and adjacent disk are for the purpose of impedance matching. The central structure of the filter is supported by means of two brass tube assemblies, which also contain transducer coils. Since they do not resonate in the pass band, they have negligible effect on filter characteristics. The over-all length of the filter structure with its supporting tubes is 2 inches. Fig. 15 shows a filter with 8.5-kc bandwidth at a center frequency of 250 kc. Here, each disk resonator has a diameter of 0.630 inch and a thickness of 0.124 inch. Resonators are separated by a space of 0.050 inch. A total of 12 coupling wires 0.017 inch in diameter are required to obtain this relatively wide fractional bandwidth.

Fig. 14—Disk filter—455-kc center frequency.

Fig. 15—Disk filter—250-kc center frequency.

Four different mechanical filter mounting cases are illustrated in Fig. 16. The filter on the right is designed for single-sideband operation at 250 kc. Its over-all dimensions are 1 inch by 1 inch by $3\frac{1}{2}$ inches. The two filters in the center are designed for AM or cw application at 455 kc. The bathtub has over-all dimensions of 1 inch by $\frac{15}{16}$ inch by $2\frac{13}{16}$ inches, while the tubular arrangement has a diameter of $\frac{3}{4}$ inch and a length of $2\frac{5}{8}$ inches. The can on the left is a new tubular mounting with a diameter of $\frac{7}{16}$ inch and a length of $1\frac{3}{4}$ inch. It is especially desirable where miniaturization is important.

In fabricating disk type filters, resonators are cut from rods of Ni-Span C. They are then ground and lapped to a very fine tolerance. After grinding, disks are heat treated to relieve strains. Final tuning is accomplished by measuring the resonance of each disk and by grinding the required amount of material from the flat surface. Automatic equipment has been designed to perform this operation rapidly on large quantities of disks. Specially designed jigs and fixtures are employed for welding coupling wires to disks. By maintaining precise control of this operation, it is possible to obtain uniform characteristics in assembled filters without subsequent adjustment.

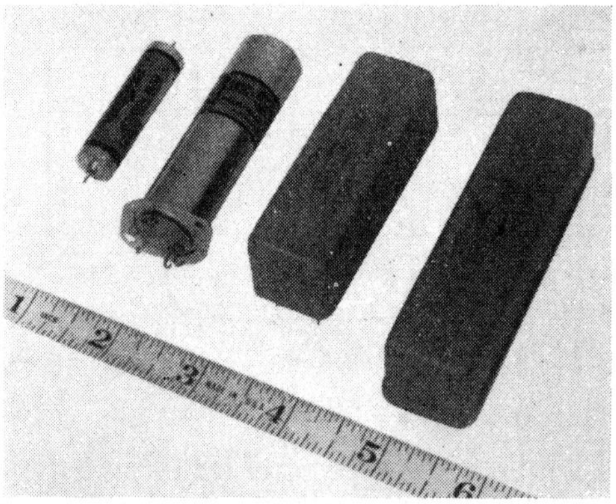

Fig. 16—Disk type filter mounting cases.

Electromechanical Transducers

In general there are four methods of electromechanical transduction available to the filter designer. These are electromagnetic, electrostatic, magnetostrictive, and piezoelectric. Electromagnetic and electrostatic transducers were used in several early filtering devices. Since these are lumped constant systems, their use has been restricted to the audio and ultrasonic range by parasitic resonances which limit their high-frequency performance. Magnetostrictive and piezoelectric transducers, however, are basically distributed constant systems, and may be employed at radio frequencies when proper dimensions are chosen. Transducers in mechanical filters perform a double function. They terminate the filter with the correct impedance, as well as converting electrical energy to mechanical (and vice versa). For minimum loss, it is desirable to terminate a filter with the reflected load which the transducer receives by coupling into the electrical circuit. Since electromechanical coupling is not always sufficient to provide this loading, it is frequently necessary to employ a terminating element with inherent mechanical dissipation, in addition to utilizing reflected resistance. A transducer may be employed as the end resonator in a filter or it may be attached to the end resonator. Magnetostrictive transducers have been employed most frequently in mechanical filters because of their efficiency, economy, and stability.

Mechanical filters have been designed with magnetostrictive transducers employing either nickel-iron alloys[22] or ferrites. The principal requirements for these transducers are good magnetostrictive coupling and low temperature coefficient of frequency. A magnetic bias field is required for optimum magnetostrictive coupling. This is normally obtained from a permanent magnet located near the transducer. The signal input current is fed to a coil around the magnetostrictive element. The resulting ac field causes the transducer to expand and contract. This vibration is coupled directly to mechanical elements in the filter. Alloys have been employed as magnetostrictive transducers for many years. The principal advantage of alloy transducers is their ease of fabrication. Since they normally have high eddy current loss, it is not possible to obtain sufficient electromechanical coupling for termination of wide-band filters. Additional loading may be obtained by employing a dissipative metal, such as annealed nickel, or by coating the transducer with a viscous material, such as petroleum jelly or silicone oil. Magnetostrictive ferrites have very low eddy current loss. Filters constructed with transducers of the equimolar iron-nickel type have very small transmission loss values.[15,23] These filters have nearly constant coupling coefficients over a wide temperature range, as well as low temperature coefficients of frequency. Ferrites are particularly advantageous in providing low loss termination for wide-band filters.

Magnetostrictive drives have been employed in the three types of filter structures discussed above. The driving arrangement for a plate type filter is illustrated in Fig. 17. Here the transducer and first resonator plate are shown. The transducer plate is made of nickel or a magnetostrictive alloy. The coil provides an alternating magnetic field, causing the plate to vibrate as shown, while a ferrite permanent magnet furnishes the necessary bias. This method of coupling gives insertion loss values of 8 to 15 db per filter. Comparable inductance-capacitance filters have an insertion loss of 15 to 17 db. In a mechanical filter with high Q resonant sections, the insertion loss is concentrated almost entirely in the transducers. Thus, narrow-band multisection mechanical filters offer less loss than the corresponding electrical designs. The range of electrical termination impedance for the plate type filter discussed here is 1500 ohms to 8000 ohms.

[22] E. M. Wise, "Design of Nickel Magnetostrictive Transducers," The International Nickel Co.

[23] R. L. Harvey, "Ferrites and their properties at radio frequencies," *Proc. NEC*, vol. 9, pp. 287–299; February, 1954.

The torsional neck-and-slug type filter employs a composite end resonator for transduction, Fig. 18. A pair of longitudinal mode ferrite transducers are attached tangentially through connecting wires to a torsional resonator at one end of the filter. Two coils wound around this pair of resonators are connected to produce opposing fields. The ac field, together with the dc bias from the Alnico magnet, causes one transducer to become longer while the other becomes shorter, thereby exciting the torsional resonator. With this pair of balanced transducers, a minimum bending moment is applied to the filter. Thus objectional spurious modes are minimized. Magnetostrictive transducers have been fabricated from either a magnetostrictive alloy or a ferrite. With the ferrite, transmission loss is less than 2 db. Typical values of termination impedance range from 1500 ohms to 30,000 ohms.

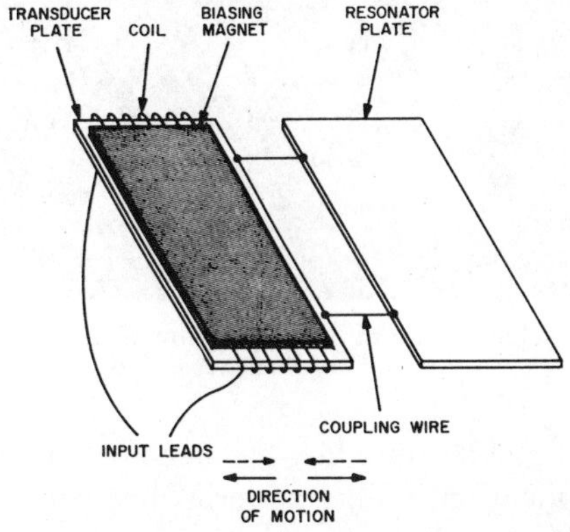

Fig. 17—Transducer for plate type filter.

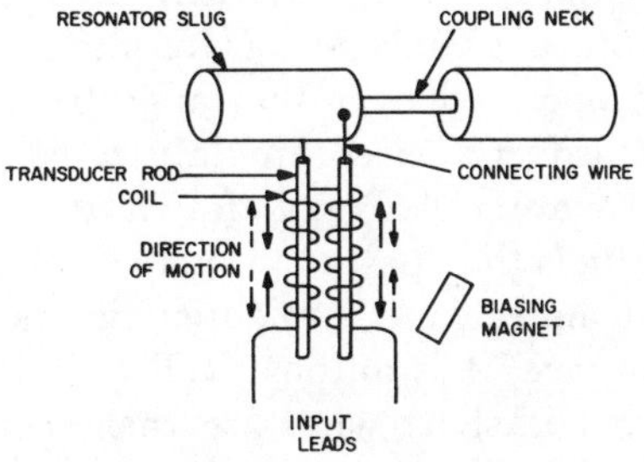

Fig. 18—Transducer for slug type filter.

Fig. 19 illustrates the transducer normally employed with disk type filters. A nickel-iron alloy wire is attached to the center of the first disk resonator. A coil is placed around the wire and a bias magnet located adjacent to the coil. Alternating fields cause the alloy wire to expand and contract with a longitudinal motion to drive the first disk. The wire is an odd multiple of a quarter wavelength long. Transducer elements may be seen in the exploded view in Fig. 20. In most filters, the transducer wire is attached to the center of the first active disk. This symmetrical drive minimizes objectionable spurious responses. A typical value of transmission loss with high impedance load is 10 db. Here, transmission loss is equal to 20 $\log_{10} E_i/E_o$, where E_i and E_o are input and output voltages respectively. Filter input impedance may be obtained from a few hundred ohms to 50,000 ohms.

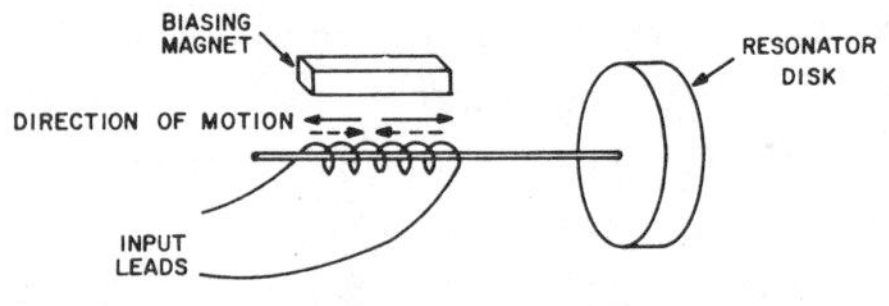

Fig. 19—Transducer for disk type filter.

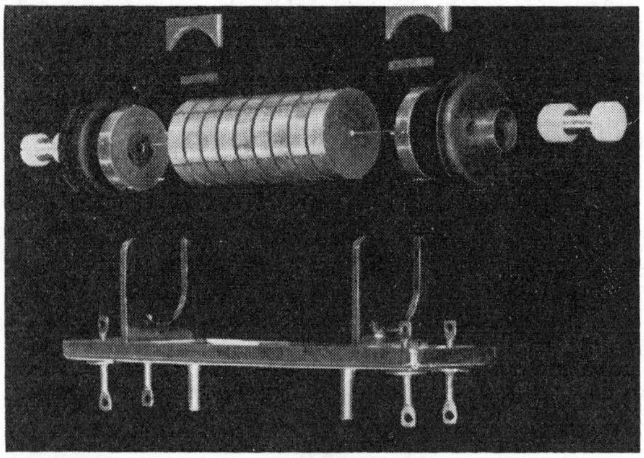

Fig. 20—Exploded view of disk filter showing transducer wire, coil, and magnet.

A special nickel-iron alloy has been developed for use in this application.[24] This material has good magnetostrictive characteristics together with a low value of Q. Thus it provides adequate impedance match for relatively wide-band filters. In addition, this material has a low temperature coefficient of frequency permitting filters to be operated over the range from −40°C. to +85°C. Transducer wires are fabricated from alloy ingots. These are rolled and drawn to the required diameter. Wire is then passed through a vertical furnace for the final anneal and straightening. As it emerges from this furnace, it is automatically sliced to the desired length. The short wires are then secured to the center of the first disk.

Most mechanical filter designs employ magnetostrictive transducers. However, it is also feasible to utilize piezoelectric elements for this application. A quartz transducer was employed to drive the plate type filter discussed above. This had a −18.5° X-cut quartz element with plating on the X-faces. Electromechanical conversion was good, but Q and impedance level were too high for satisfactory termination. Moreover, fabrication problems were more difficult than with the magnetostrictive plates. Polarized ceramic drivers, such as barium titanate, have more favorable impedance and Q values, but these have not been employed in standard designs because of difficulties with temperature coefficient and aging.

[24] U. S. Patent 2,719,084.

Filter Characteristics

The application of mechanical filters involves most of the same considerations observed in the use of electrical filters. The basic parameters of the filter are the selectivity and transfer characteristics. One essential difference in the case of mechanical filters is that the termination is only partly determined by the external electrical circuit, as some of the loading may be mechanical in the transducer. Various circuit parameters and environmental conditions which affect mechanical filter operation must be considered in order that the filter may operate in an optimum fashion.

Typical selectivity curves are shown in Fig. 21. The

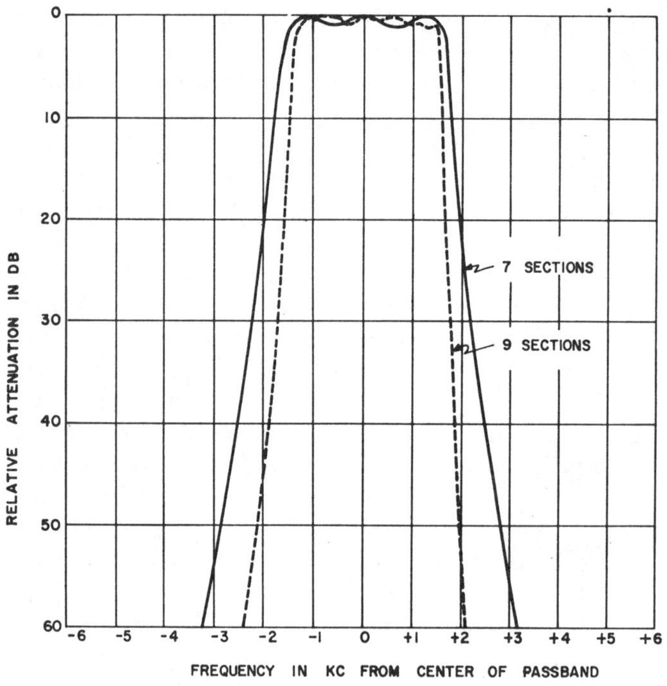

Fig. 21—Selectivity curves for filters with seven sections and nine sections.

steepness of the skirts is determined by the number of sections in each filter. The seven-section design has a 6-db bandwidth of 3.4 kc and a 60-db bandwidth of 6.3 kc. The ratio of these bandwidths, or shape factor, is 1.85 to 1. For the nine-section model, the 6-db bandwidth is 3.0 kc, and the 60-db bandwidth, 4.5 kc. This gives a shape factor of 1.5 to 1. A greater number of resonators would produce even steeper skirts. The ratio of maximum to minimum output level in the pass band is defined as ripple or peak-to-valley ratio. Here, pass band is the useful frequency range. For most applications, ripple amplitude must be 3 db or less. To produce mechanical filter designs having a minimum peak-to-valley ratio, it is essential that elements be accurately tuned and mechanical impedances be held within close limits. This requires precise control of materials and fabrication techniques. The problem is particularly critical for designs with a large number of sections. By accurate adjustment of filter elements, it is possible to fabricate models with peak-to-valley ratio of less than 1 db.

In designing circuits with mechanical filters, it is important to consider the rated impedance and loss values. A circuit diagram of a mechanical filter network is shown in Fig. 22. In this diagram, R_s and R_0 are source and load

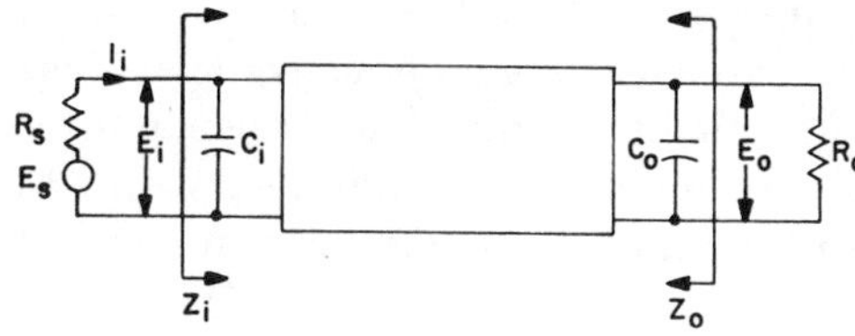

Fig. 22—Network parameters.

resistance, respectively. Since electrical loading affects the pass band ripple, it is important to employ rated values of R_s and R_0, especially with low-loss filters. The capacitors, C_i and C_o, are required to resonate input and output transducer coils. Since the coils reflect reactance back into the filter, it is necessary that the capacitance be maintained within a given tolerance to maintain the specified values of transmission loss and peak-to-valley ratio. Filter losses in the pass band have been defined in terms of insertion loss or transmission loss with open circuit load. Insertion loss values are useful for filters operated with matched impedances, while transmission loss values with open circuit load are useful in pentode IF amplifiers where both source and load impedance are much greater than filter impedances. Filter insertion loss values range from less than 2 db for ferrite transducers to 16 db for alloy transducers. Another useful parameter is the transfer impedance. The over-all gain of a pentode amplifier stage with a mechanical filter may be determined by multiplying the g_m of the tube by the transfer impedance of the filter. The input and output impedances, Z_i and Z_o, are measured across the parallel resonant capacitor and transducer coil. These impedances are largely resistive and have typical values ranging from 1000 ohms to 50,000 ohms. Filter characteristics are adversely affected by excessive dc, I_i and by excessive ac input voltage, E_i. Direct current affects the magnetic biasing field of the transducer. For minimum transmission loss and peak-to-valley ratio, this current must be maintained below a rated value. An excessive input voltage applied across the transducer coil causes nonlinear operation. Thus for minimum distortion, filter input voltage should be less than the rated value.

Mechanical filters are suitable for application in environments generally considered suitable for electrical filters. The elements in mechanical filters are fabricated from special alloys to minimize temperature changes. With Ni-Span C resonators it is possible to have frequency drift below 10 parts per million per degree centigrade. Also, with low temperature coefficient transducer alloys or ferrites, it is possible to minimize variations in peak-to-valley ratio. A typical operating range for a mechanical filter is $-40°$C. to $+85°$C. With mechanical structures fabricated from metal alloys, they are not damaged by extreme variations in temperature during storage. Also, with strain-relieved alloys, there is no observable drift with time.

The ability of mechanical filters to resist mechanical shock and vibration is a function of electrical design parameters. In general, wider bandwidth filters are more rugged. Filters with 3 kc and wider bandwidths have proven suitable for portable and mobile application in civilian and military equipment.

Applications

Mechanical filters may be employed for most band-pass filtering requirements in the frequency range from 50 to 500 kc. They are especially applicable to carrier systems and single-sideband equipment, as well as communications receivers with high performance requirements.

The characteristics of available filters greatly influence the design of carrier systems. The mechanical filter has been applied to these systems because of its improved performance and smaller size compared with electrical filters. Filters are used here for two purposes—channel separation and carrier selection.

In carrier system applications, it is often required that a family of separation filters be provided every 4 kc through the frequency range of the system. Fig. 23

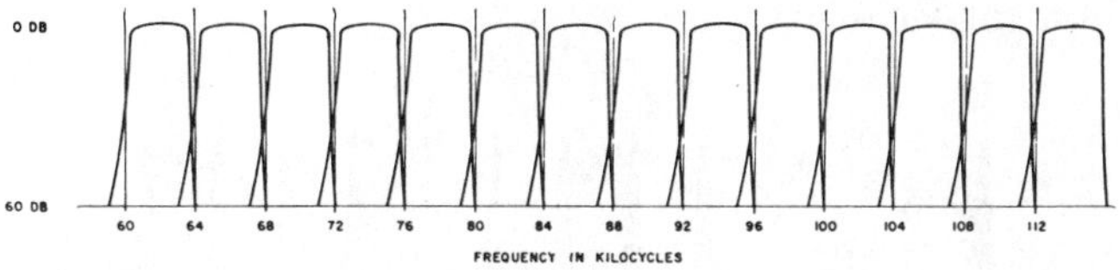

Fig. 23—Composite curves of family of mechanical channel filters designed for carrier systems.

shows a combined curve of a group of channel separation filters. In general this requirement is similar to the single-sideband requirement which will be mentioned later. However, since the carrier system signal may pass through many filters in tandem, the requirements on peak-to-valley ratio are more stringent. For optimum system performance, it is desirable that the pass band response variation be minimized.

To obtain satisfactory transmission of digital type signals through a carrier system, it is necessary to have minimum phase distortion. For no distortion, variation in phase shift β must be linear with respect to frequency, or the envelope delay, represented by $d\beta/d\omega$, must be constant. An envelope delay curve for a filter with ten sections is shown in Fig. 24. Sharp delay peaks at the edges of the pass band are associated with the very rapid cutoff on the attenuation characteristics. From theoretical analysis, it may be shown that envelope delay is related to the variation in amplitude response.[25]

[25] H. W. Bode, "Network Analysis and Feedback Amplifier Design," D. Van Nostrand Co., Inc., New York, N. Y., 1945.

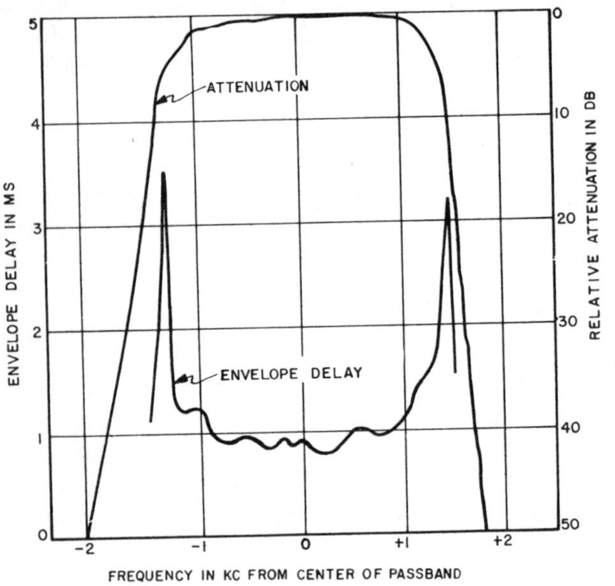

Fig. 24—Envelope delay and attenuation for ten-section filter.

In a frequency stabilized carrier system, it is also necessary to provide carrier selection filters whose function it is to select the injection carrier out of a spectrum of frequencies. A curve showing a family of these carrier selection filters is shown in Fig. 25. The compactness of mechanical filters simplifies the design of carrier frequency equipment, as shown in Fig. 26, a photograph of a carrier selector.

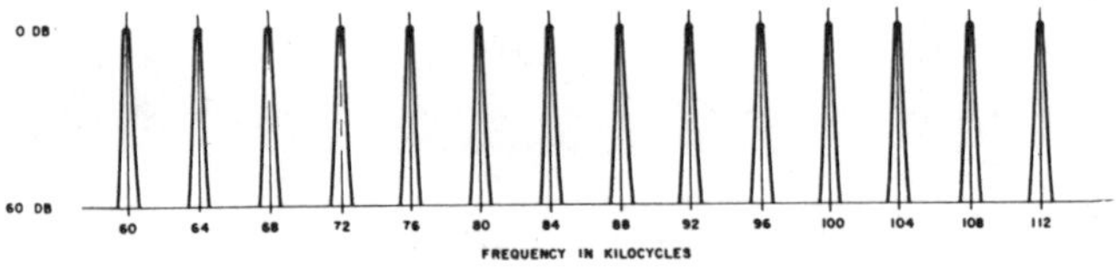

Fig. 25—Composite curves of mechanical carrier selector filters designed for carrier systems.

Fig. 26—Carrier system module with mechanical filter.

Mechanical filters find frequent application in single-sideband transmitters and receivers. In the transmitter, the single-sideband signal is generated at a low radio frequency where a filter of appropriate characteristics is available. This frequency should in general be as high as is practical considering the requirements of the filter. Electrical filters for this purpose have been used in the frequency range of 20 to 50 kc. Crystal lattice filters have in general been used at 100 kc. In the succeeding stages, the sideband signal is heterodyned to the final operating frequency. The low frequencies of electrical filters require one extra step of conversion. This is true because the ratio of the injection frequency to the signal frequency should always be maintained within the range of 5/1 to 10/1. This compromise is required in order to operate between the limits of spurious frequency generation and practical selectivity requirements.

A specification placed upon a filter in single-sideband applications requires a specific amount of carrier rejection. Since the filter must have very rapid cutoff from the lowest desired audio frequency to the carrier frequency, it is necessary that very high performance filters be used. In Fig. 27 are shown curves for a pair

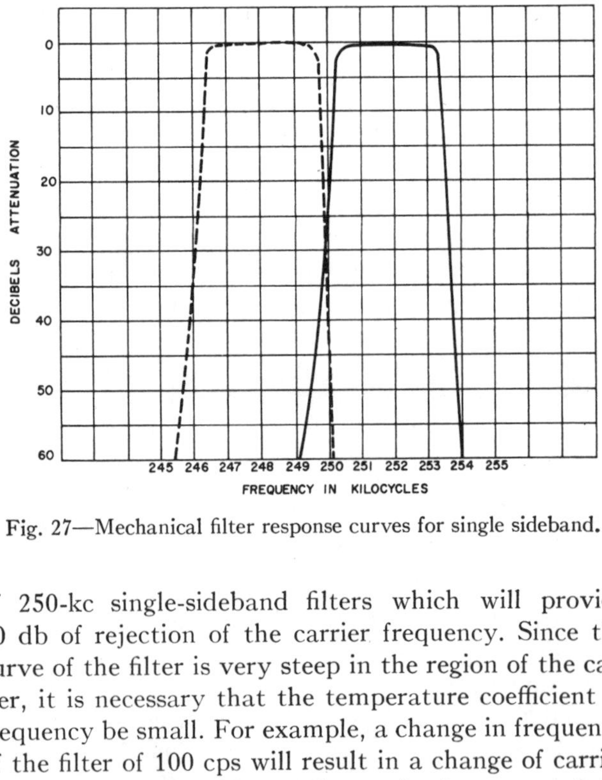

Fig. 27—Mechanical filter response curves for single sideband.

of 250-kc single-sideband filters which will provide 20 db of rejection of the carrier frequency. Since the curve of the filter is very steep in the region of the carrier, it is necessary that the temperature coefficient of frequency be small. For example, a change in frequency of the filter of 100 cps will result in a change of carrier rejection of 10 db. Also, a change in the corner of the filter will be directly translated as a change in the audio frequency response of the single-sideband transmitter. Another function of the mechanical filter in the single-sideband generator is to reject the undesired sideband. As is shown in Fig. 27, the filter will provide excellent sideband rejection.

In radio receiver applications, a mechanical filter fulfills the need for selectivity in a fashion superior to that usually provided by a multiplicity of IF transformers. Not only is the mechanical filter capable of providing a flatter topped pass band and steeper skirts on the selectivity curve, but also its availability as a lumped selective network allows the designer considerably more freedom in providing an optimum receiver design. The problem of cross modulation or desensitization has always plagued the radio designer. The accepted ideal is to provide only enough gain ahead of the selectivity to overcome mixer noise. However, with distributed selectivity it is necessary to pick up some gain with each IF stage in order to keep the number of tubes from becoming too large. Comparison of the desensitization performance between the mechanical filter IF strip and a well-designed conventional IF transformer strip is shown in Fig. 28. By preventing overloading of the first and

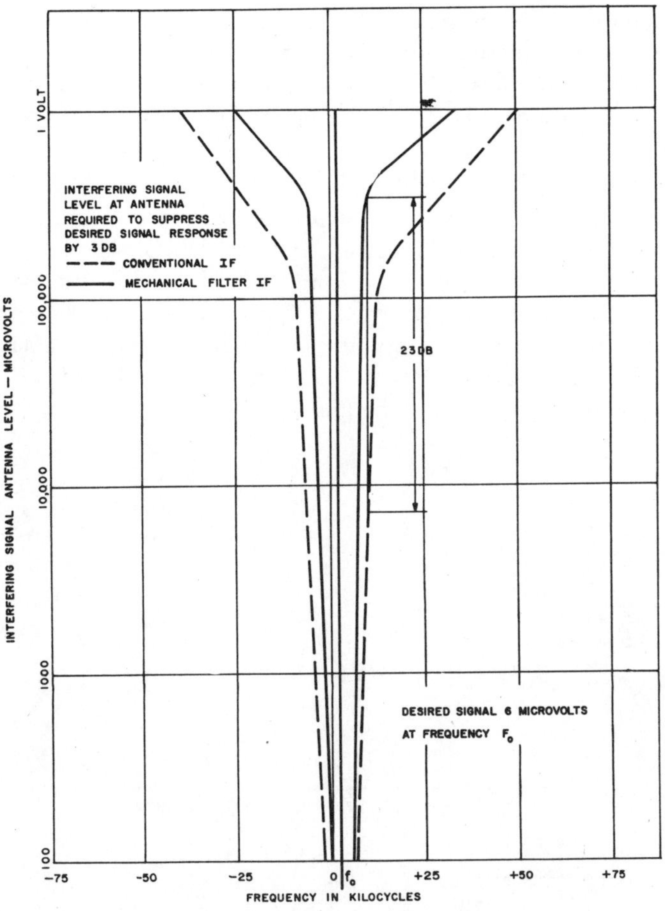

Fig. 28—Desensitization performance for mechanical filter IF strip and conventional IF strip.

second IF tubes, 23 db of improved protection is given by the mechanical filter IF strip. The use of selectivity in such concentrated form brings with it certain problems as well as certain advantages. If a high degree of selectivity is to be realized across the filter, it is necessary that the coupling be through the filter and not around it. In Fig. 29 is shown a schematic diagram of a mechanical filter stage indicating the paths by which unwanted signals may find their way around the filter.

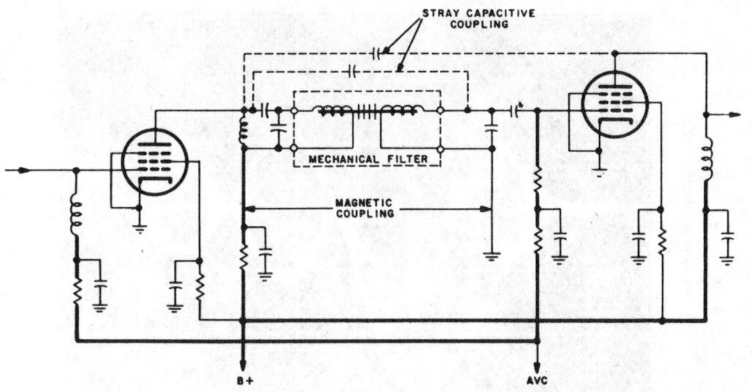

Fig. 29—Undesired coupling paths which must be reduced below 100 db for proper filter operation.

Since the filter will provide rejection band losses greater than 100 db, it is necessary that the shielding and decoupling be good to this value also.

The small size of the mechanical filter makes possible a selection of various bandwidths in very compact form. As an example of what can be done, Fig. 30 shows a four bandwidth turret which is not much larger than one conventional IF transformer. With bandwidths available from 500 cycles to 35 kc, the radio designer may easily provide whatever degree of selectivity he wants in very convenient fashion.

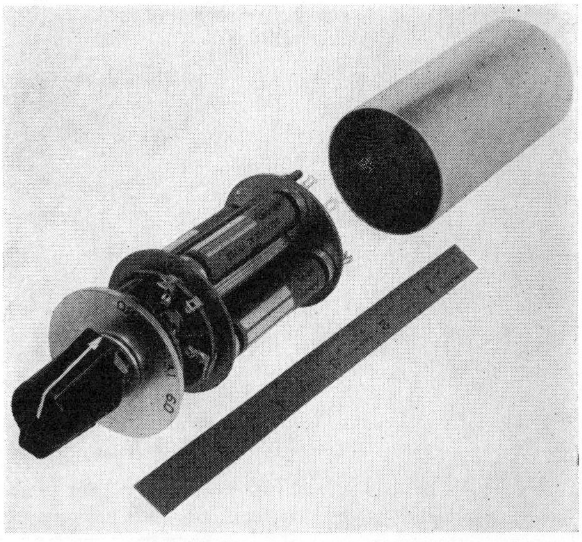

Fig. 30—Mechanical filter turret to provide selection of various bandwidths.

In the design of receivers which may be subjected to severe impulse noise interference, the response of the filters to impulses is an important consideration. Impulse response measurements are shown in Fig. 31 for filters with seven sections and ten sections. Both designs have bandwidths of 3 kc. The envelope of the response may be expressed approximately by the relation $\sin x/x$, where x is a direct function of time, for times greater than that of the main response. In the seven-section case, the impulse response rises to the first maximum 0.8

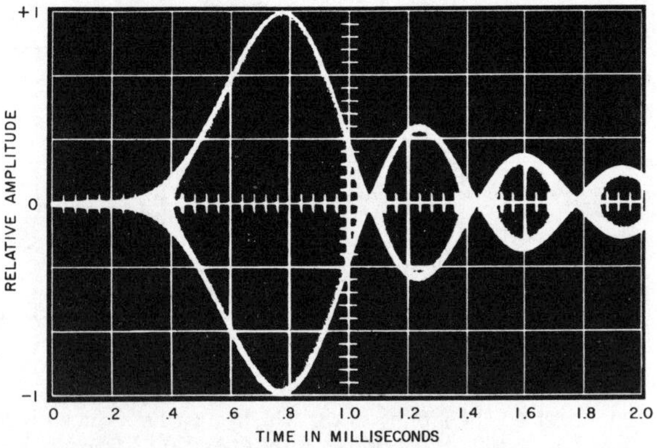

(a) Seven-section Filter.

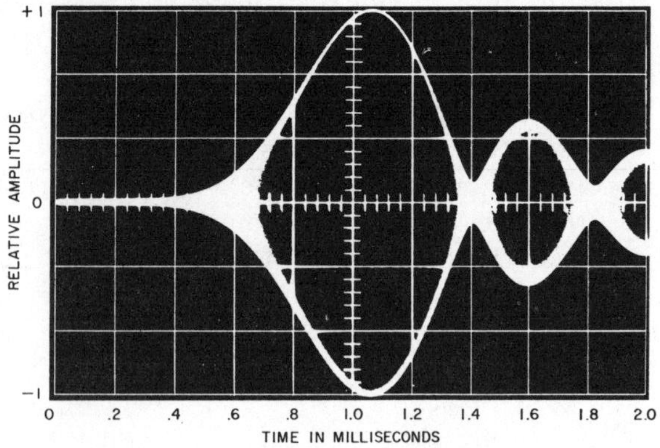

(b) Ten-section Filter.

Fig. 31—Impulse response for 3-kc bandwidth filters.

millisecond after the input pulse at $t=0$. In the ten-section case, the response is shifted with respect to time so that the maximum value occurs 1.05 milliseconds after $t=0$. The impulse response may be obtained for other bandwidths by changing the time scale. Thus, for a filter with 6-kc bandwidth, the horizontal scale in Fig. 31 would be halved.

Conclusion

Mechanical IF filters were first presented as a practical radio component nine years ago. Since that time, they have found wide application in equipment designs to provide improved performance over other types of filters. In addition to providing improved performance, mechanical filters have introduced new concepts in communication system design. It is reasonable to expect that further advancements in design and fabrication techniques will result in superior mechanical filters with an even greater range of applications.

Acknowledgment

The authors wish to express their appreciation to B. Niederman of Motorola, Inc., R. W. George of the RCA Laboratories Division, and P. D. Gerber and D. L. Lundgren of the RCA Victor Division for their kind assistance in supplying information for this paper. Thanks are also due to M. L. Doelz of the Collins Radio Company for his advice and suggestions.

Progress in Electromechanical Filters

By

M. BÖRNER, Dr.rer.nat.†

Presented at a meeting of the Electro-Acoustics Group in London on 11th November 1964.

Summary: The development of techniques for manufacturing electromechanical filters is discussed and the change from single rods with turned-in necks to built-up construction described. Single-sideband filters and miniature types are then described and methods of eliminating spurious modes are shown.

1. Introduction

Fifteen years ago, first in the United States, later on also in Europe, research and development work began in the field of new types of electromechanical filters for frequencies up to 500 kc/s. Quite a number of research institutes in industry and in universities participated, but of the many variations of possible filter constructions which were at first proposed, only a small number have proved to be practicable. Initially, these were those manufactured by the Collins Company, who have been active in this field since its inception. An early filter, still in use, had bending-plate resonators and coupling rods, attached by means of spot-welding. Later on further practicable filter types were developed which will be discussed in this paper.

The filters used in the early stages of development, and to some extent until now, were manufactured out of a single metal rod by turning in necks (RCA, Marconi, Telefunken). In newer types, however, a technique is used which consists of first turning the individual resonators piece by piece, then tuning them to the right frequency, and afterwards assembling the complete structure by spot-welding the coupling members on to the resonator surfaces. This spot-welding process can now be performed so precisely that no post-tuning is necessary. In perfecting this technique, however, it has been found that all possible requirements for mechanical filters cannot be met by one type only. On the one hand filters must be made smaller and smaller, following the present trend of miniaturization in electronics. On the other hand some types of filters require ever-increasing accuracy with respect to their tolerances for frequency and attenuation behaviour. This is in keeping with the demand for more and more transmission channels to be established within very limited frequency ranges.

Such requirements led to the development of special single-sideband filters (15–20 circuits, carrier frequency approximately 200 kc/s) and to the development of miniaturized filters (6–10 circuits, passband 455 kc/s). These two trends are dealt with in this paper. Filters are described with regard to their electrical and mechanical properties as well as their production techniques.

Finally, possible future developments are discussed. All the known electromechanical bandpass filters are minimal-phase-shift networks, in which the transmission properties depend strongly on the phase properties. In the case of fast data-transmission, filters are needed with a frequency-independent delay-time for transmitted signals. This is achieved with additional all-pass elements. It seems possible to achieve integrated mechanical bandpass-all-pass structures for electromechanical filters in a very simple way, using a construction similar to that employed in the single-sideband filters. Not very different from such phase-compensated filters are filters in which attenuation poles have been achieved in the neighbourhood of the passband.

2. Single Sideband Filters

The development of electromechanical filters will be discussed in chronological order, starting with the single-sideband filter. It will be seen how the knowledge gathered with respect to tuning and production techniques for this type of filter assisted in further development. In a technique described by Roberts and Burns,[1] an attempt was made to produce single-sideband filters,[2] by turning in necks into a single metal rod (Fig. 1). Using longitudinally vibrating magnetostrictive transducers, the first cylindrical metal resonator is brought to torsional resonance by means of tangential forces acting on its surface. The torsional vibrations are coupled to the second resonator by means of a thin torsional $\lambda/4$ line. This process is then continued along a series of resonators and coupling elements until at the output end the mechanical energy is again converted to electrical energy by two longitudinal transducers vibrating in antiphase, as at the input end. Figure 2 shows one reason why such filters have not been successful. In addition to the proper transmission range for torsional vibrations, further transmission ranges exist, which are caused by

† Telefunken Research Institute, Ulm, Western Germany.

Reprinted with permission from *Radio Electron. Eng.*, pp. 173–182, Mar. 1965.

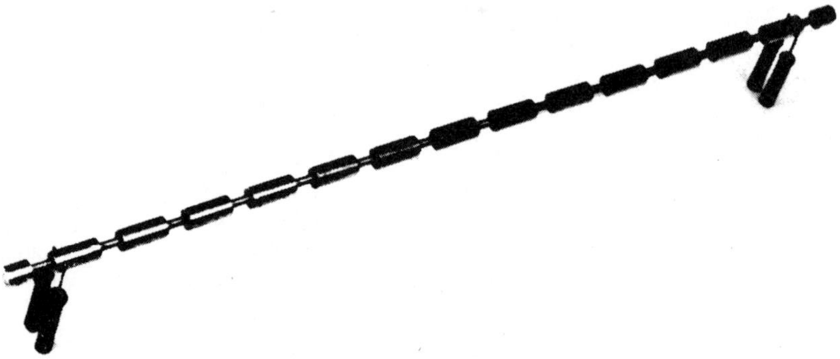

Fig. 1. Torsional-mode single-sideband filter made by turning necks in a rod (with ferrite transducers).

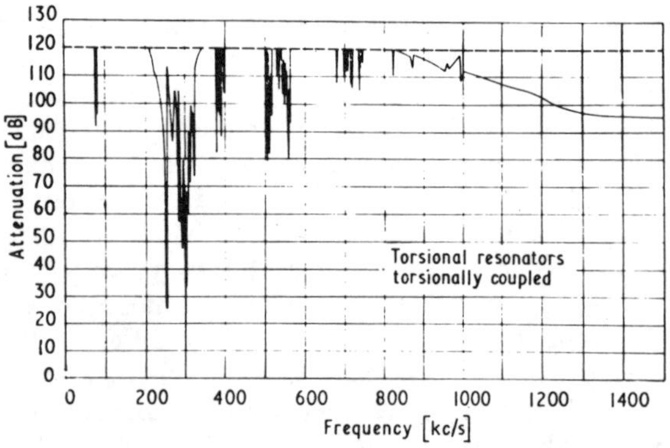

Fig. 2. Spurious responses in a filter of the type shown in Fig. 1.

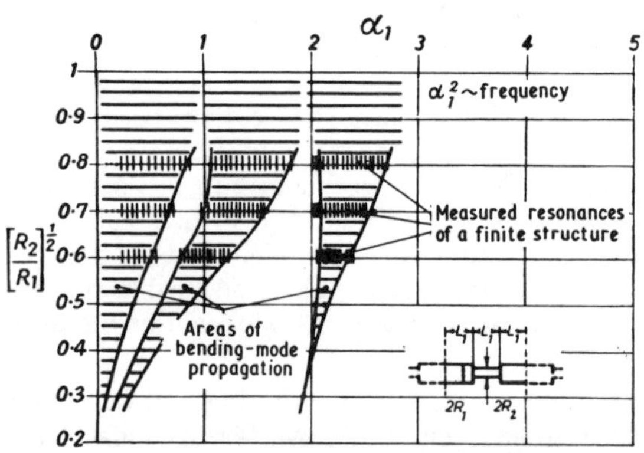

Fig. 3. Calculated bending-mode propagation in an infinite periodic structure.

unwanted resonances. The bending modes are particularly disturbing due to their large bandwidth. They cannot be suppressed by changing the cross-sectional dimensions, due to the following fact:

For torsional vibrations, the coupling between neighbouring resonators is given by $(R_c/R_r)^4$ where R_c is the radius of the coupling elements and R_r that of the resonators. This fourth-power dependence of coupling factor upon diameter ratio calls for relatively small diameter differences at the usual bandwidth for single-sideband filters. In complete contrast to this is the case where longitudinal vibrations are used. There the coupling only depends on the square of the diameter ratios, and for the same relative bandwidth greater diameter differences are needed. The coupling for bending modes in particular is then also much weaker.

Figure 3 shows the theoretical bandwidth of periodic bending-mode-lines as a function of diameter ratio.[6] With larger diameter differences the bending-mode propagation disappears, and the few remaining bending-mode resonances could be suppressed for example by using different diameters for the filter resonators.[7, 8]

Another difficulty with such one-piece filters still remains: Each individual resonator has to be tuned after turning, to a few cycles per second. This can be done for a particular resonator only when the neighbouring resonators are clamped (or alternatively very strongly detuned). This clamping method, however, no longer works at ultrasonic wavelengths of the order of 1 cm, because the required mechanical short circuits can no longer be realized.

Figure 4 shows a single-sideband filter, assembled with individual resonators.[7, 9] The resonators oscillate torsionally, while the coupling wires, which are spot-welded on the surfaces of the resonators and are a quarter of a wavelength long, vibrate longitudinally. Using this construction it is possible to manu-

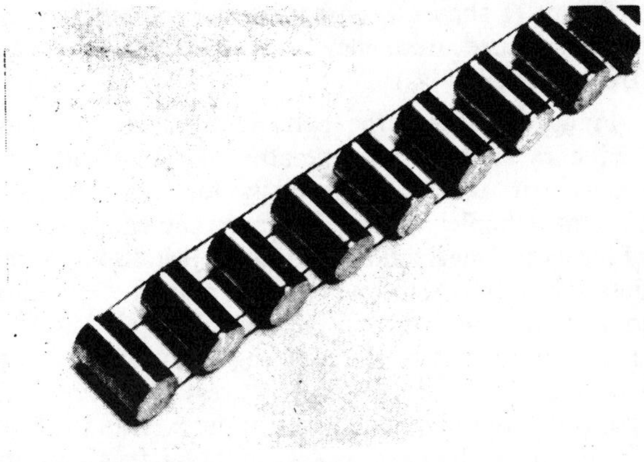

Fig. 4. Single-sideband filter structure using torsional $\lambda/2$-mode resonators coupled with longitudinally vibrating $\lambda/4$-wires.

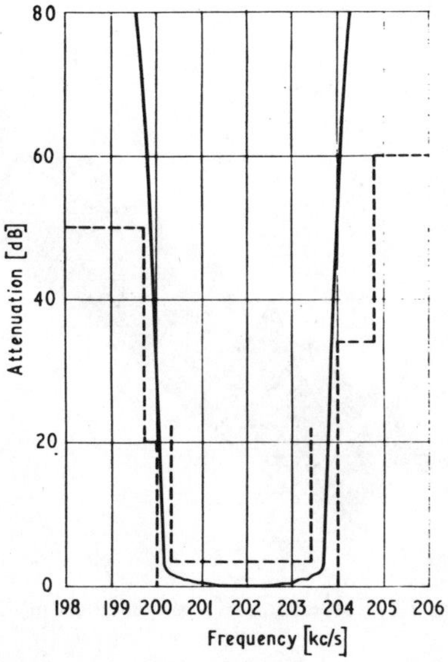

Fig. 5. Transmission curve of a FE 21 type single-sideband filter.

facture single-sideband filters with high precision and very few subsidiary modes. Figures 5 and 6 show transmission curves for two different types, first a filter for one single-sideband channel and then one for two single-sideband channels. Figure 7 shows the subsidiary mode characteristic. Because the driving forces for torsional vibrations are applied to the resonator surfaces, and not as moments in the middle of the end surface—as in the filter shown in Fig. 1—only small wire diameters are needed for the coupling elements. Bending modes are therefore practically not propagated. The remaining subsidiary modes come from λ-vibration of the coupling wires and harmonic overtones of the $\lambda/2$-torsional resonators. It is sufficient to drive this filter with an asymmetrically

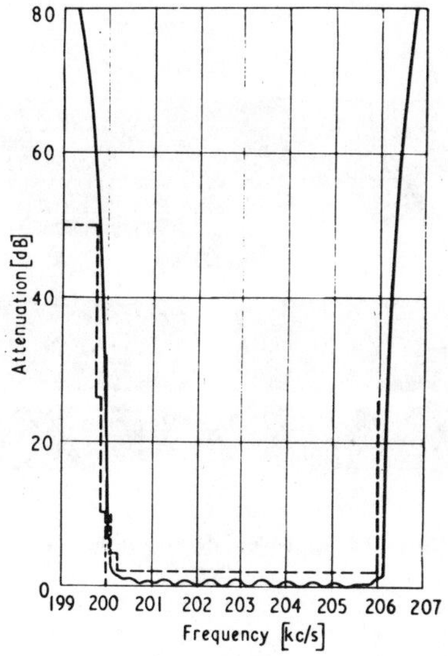

Fig. 6. Transmission curve of a FE 25 type single-sideband filter.

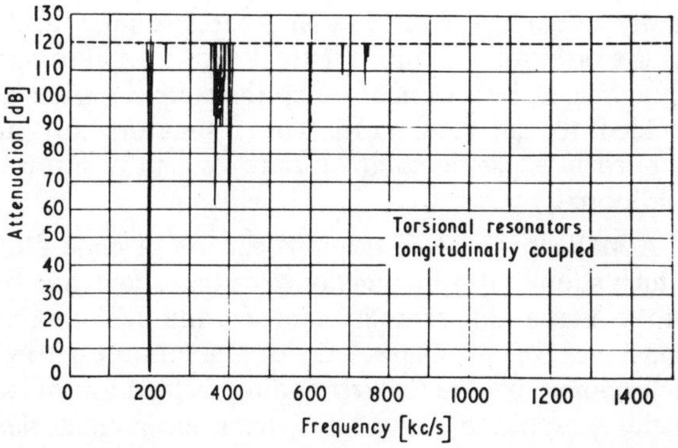

Fig. 7. Spurious responses in a FE 22 type single-sideband filter.

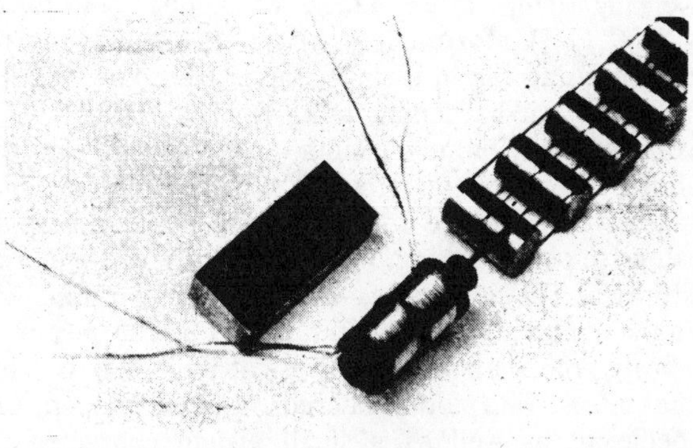

Fig. 8. The electromechanical transducer of a FE 21 type single-sideband filter.

Fig. 9. Automatic spot-welding machine for single-sideband filters.

attached, longitudinally vibrating electromechanical transducer which consists of a ferrite rod, $\lambda/2$ long, biased by a permanent magnet (Fig. 8). The resonators are tuned in a largely automatized apparatus. The tuning accuracy is of the order of ± 3 c/s. The diameter tolerance of the coupling wires is small and is of the order of $\pm$ 1 μm. Figure 9 shows an automatic spot-welder, by means of which the coupling wires are welded to the resonators. The resonators are suspended in a plastic comb and are kept in position by additional wires.

A major problem in the construction of good filters which should also be mentioned is that the behaviour of the filter at different temperatures and the long-term constancy of its characteristics are almost entirely determined by the corresponding behaviour of the metal resonators and the transducer materials. Nickel-iron alloys are considered as resonator materials and it is possible, by making use of an adjustable interaction between magnetostrictive characteristics and internal stresses, to hold the resonance frequency constant to $\pm 1 \times 10^{-6}$ deg C in the working temperature range of about -40°C to $+80$°C. Unfortunately these materials mostly exhibit an ageing in their characteristics, and only by means of age-hardening using further additives (for example, beryllium, titanium, molybdenum) in the alloys, can usable materials be achieved. The internal stresses are adjusted by means of a correspondingly extensive cold-working, when the rod-shaped material is drawn. The hardening, which also has an influence on the temperature coefficient, takes place by means of a fairly precise annealing process at about 600° C. This guarantees, with good materials, a good temperature coefficient as well as a small ageing characteristic ($< 1 \times 10^{-7}$/week) and a high quality factor (between 15 000 and 25 000).

Figure 11 shows the behaviour with temperature of the resonance frequency of a torsional resonator (frequency 200 kc/s).

Figure 12 shows the behaviour of the resonance frequency shift with temperature for a cobalt-substituted ferrite which was specially developed for use in electromechanical filters.[10, 11, 12] The characteristics of the electromechanical transducer just discussed may deteriorate by about one order of magnitude, without influencing the filter characteristics. In addition to these characteristics, the coupling factor of the electromechanical transducer is of interest. It is important that this should be constant, as it determines the transformation of the electrical terminating resistance into a mechanical termination which has to be matched to the mechanical impedance of the filter.

Figure 13 shows the behaviour of the electromechanical coupling factor with temperature.

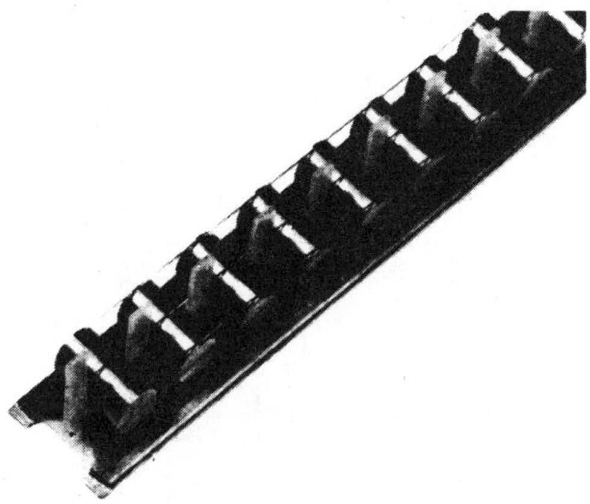

Fig. 10. Suspension of a single-sideband filter.

3. Miniaturized Filters

When the first electromechanical band filters were developed, no one foresaw the technique of miniaturization which has since become more and more successful in the field of electronics. Filter sizes were then in keeping with the dimensions of miniature tubes. It will be shown in this section how it has proved possible to miniaturize electromechanical filters. Some cautionary remarks with respect to the possibilities of miniaturization will also be made.

It may be seen that the filter quality depends materially on the quality of the resonators, which is in turn strongly dependent on the volume to surface ratio of these resonators. This is apparent from the consideration that disturbances in the resonator quality (ageing, temperature coefficient of the resonance frequency, Q-factor) emanate principally from

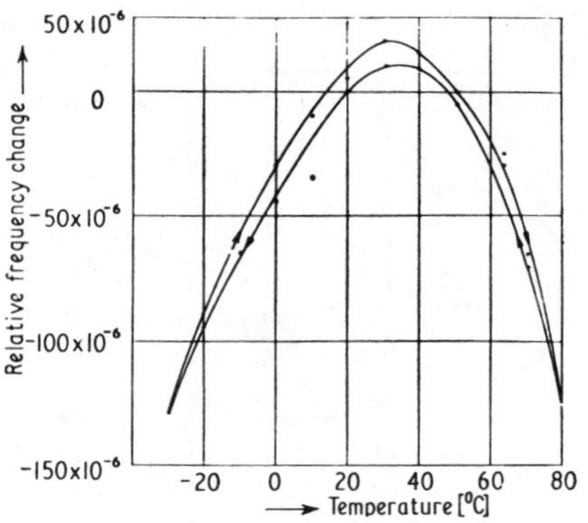

Fig. 11. Relative change of the resonance frequency of 200 kc/s torsional resonators with temperature.

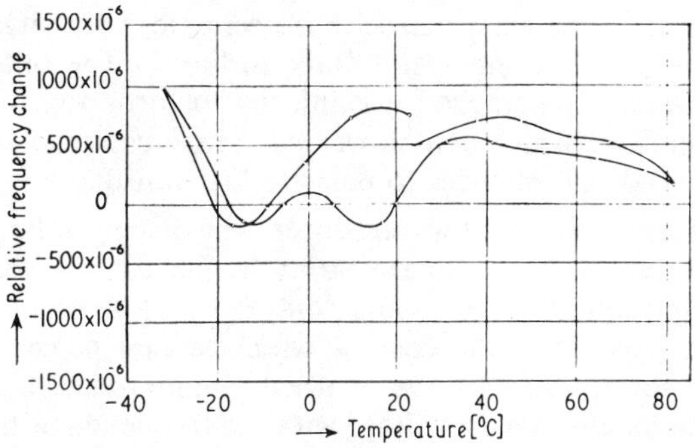

Fig. 12. Relative change of the resonance frequency of a 200 kc/s longitudinally vibrating ferrite transducer with temperature.

the surface in the form of changes in the crystal lattice and wandering of dislocations. Here one has to consider primarily the surface damage during the essential machining and tuning processes.

This then has two consequences:

(1) When geometrically similar resonators are compared, the lower the frequency the better the resonator.

(2) At the same frequency, the resonator whose shape comes closest to the spherical is best.

On account of spurious mode suppression and manufacturing considerations, as well as requirements for the filter structure as a whole (of which the resonator is always only a part), it is necessary to deviate from these somewhat oversimplified geometrical standards. Yet the fact remains that, for example, a resonator for a good single-sideband filter at 200 kc/s has better characteristics than one for 300 kc/s. The fact that, for equal *relative* quality (equal $T_k(f)$, equal relative ageing rate, equal Q-factor) the filter with the better *absolute* characteristics will always be achieved at the lower frequency has not even been taken into consideration here. With the resonator materials available at present, single-sideband filters with commercial quality can only be achieved at frequencies up to about 200 to 250 kc/s. A further substantial miniaturization of this type of filter is not contemplated at the moment.

The position is somewhat different in the case of normal i.f. filters for a.m. and f.m. receivers, which are also needed in large quantities, as well as filters for single-sideband transmission with simple requirements.

For these filters a midband frequency of about 455 kc/s is prescribed. At this frequency the greatest reduction in size can be achieved by selecting a filter structure in which both resonators and coupling elements are driven in flexure. However, apart from the quality deterioration already mentioned which takes place when the resonators become too small, and the fact that very small bending-mode transducers are hard to make, considerations with respect to spurious resonances in particular call for a different approach.

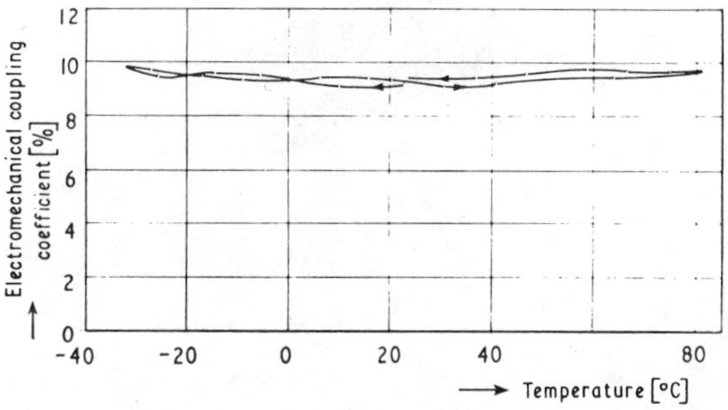

Fig. 13. Electromechanical coupling coefficient of a longitudinally vibrating ferrite transducer vs. temperature.

We therefore prepared to use $\lambda/2$-longitudinal resonators, which are easy to make and tune. Such resonators, coupled by means of short bending-mode leads, were used to build the miniature filter structure shown in Fig. 14. Several coupling wires fixed to the cylindrical resonator and transducer surfaces by spot-welding may be seen.[13] Once again the resonators are made out of a stabilized nickel-iron alloy and this offers the additional advantage that, using a coil and slight magnetic bias, the resonators can be checked by means of the magneto-strictive effect, so that for production tuning the resonance frequency can be determined accurately. Instead of the magneto-strictive ferrite transducers used previously, piezo-electrical transducers were used for this filter with obvious advantages.

Since the transducer now needs no driving coil, it fits very snugly into the filter structure. It consists of a small tube made of piezoelectric ceramic, with silver coatings on the inner and outer surfaces. The outer coating is strengthened galvanically to form a silver-nickel layer which can be welded. These coatings are also used as electrodes to polarize the transducer.

Figure 15 shows this transducer. The driving voltage is connected to the inner coating by means of a wire making contact at the nodal point for $\lambda/2$ longitudinal vibrations. The outer coating, which carries the coupling wires, is also the earthed counter-electrode. When an alternating voltage with the frequency of the $\lambda/2$ longitudinal resonance is applied, this resonance is excited in a roundabout way via the cross-contraction.

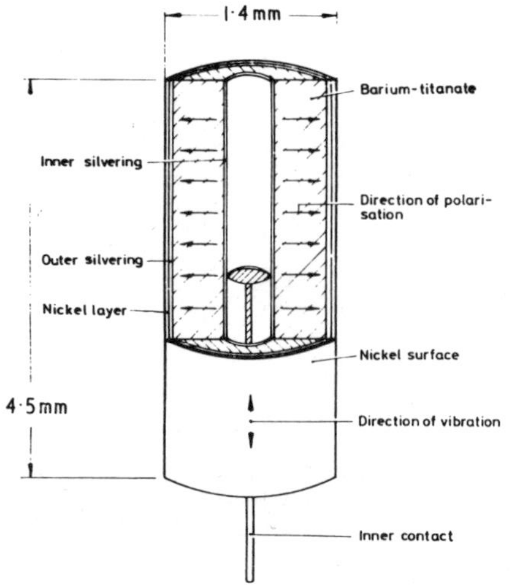

Fig. 15. The electromechanical transducer of a miniaturized filter.

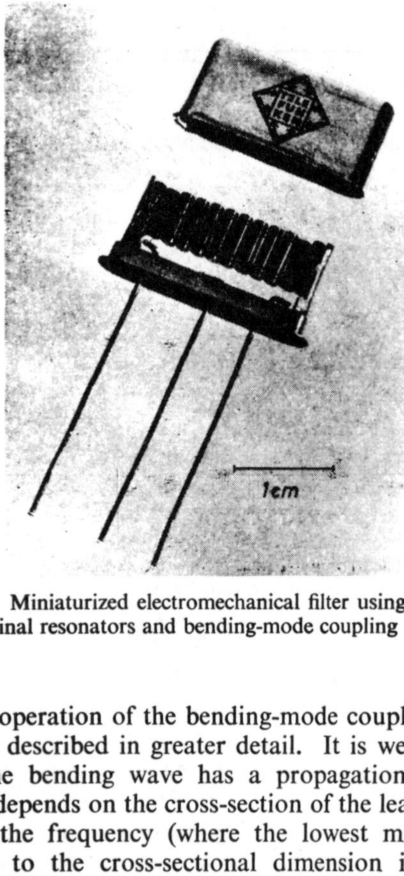

Fig. 14. Miniaturized electromechanical filter using $\lambda/2$ longitudinal resonators and bending-mode coupling wires.

The operation of the bending-mode coupling wires will be described in greater detail. It is well known that the bending wave has a propagation velocity which depends on the cross-section of the lead as well as on the frequency (where the lowest mode with respect to the cross-sectional dimension is always understood). Thus it is possible to make a bending-mode vibrator as short as desired at a given frequency. The same applies to a coupling wire which nevertheless should not show any resonances in the relevant frequency range (here approximately 455 kc/s).

The diameter of the coupling leads may be chosen in such a way that the resonators are packed together as tightly as possible. The length of the coupling leads is judiciously chosen as a quarter wavelength. This has the advantage that all the resonators can be pre-tuned to the midband frequency, even when the coupling between neighbouring resonators varies, corresponding to optimal dimensioning of the filter. The reason why the dimensioning becomes complicated is because the propagation of waves in a bending-mode vibrator is characterized by four components instead of two, as with longitudinal or torsional vibrators. These components vary along the length of the lead corresponding to the wave propagation.[14]

Figure 16 shows the basic element of a miniature filter. At the connecting point between bending lead and resonator the following bending values apply:

Displacement y

Displacement slope $\dfrac{\partial y}{\partial x} = y'$.

Total moment of the normal forces

$$N = EJ\frac{\partial^2 y}{\partial x^2}$$

(E = Young's modulus, J = second moment of area of the rod cross-section around an axis through the neutral bending plane and perpendicular to the vibration plane).

Shear force $S = EJ\dfrac{\partial^3 y}{\partial x^3}$ on the rod.

If the diameters of the longitudinal resonators and of the bending rods were optional, then all four values would have to be considered with respect to their influence on the vibrational behaviour. However, for relatively narrow-band filters we try to achieve very weak coupling only, so that we can assume that the bending leads must have a much smaller diameter than that of the resonators. Apart from this we wish to attach the thin coupling wires by electrical spot-welding. Both stipulations immediately lead to the simplification $y' = 0$, and this simplification further leads to the possibility of mathematically treating the bending coupling lead in complete analogy to the well-known torsional or longitudinal coupling leads. The bending coupling lead, which furnishes $\lambda/4$ coupling in particular, is then described by a chain-matrix with vanishing elements in the main diagonal.

For the length l_{K1} of the $\lambda/4$ coupling lead (the corresponding 'free' wave propagation of course does not exist!) one obtains[14]

$$l_{K1} = 0{\cdot}667\sqrt{R_B . \lambda_L} \qquad \text{......(1)}$$

and for the $3\lambda/4$ lead

$$l_{K2} = 1{\cdot}552\sqrt{R_B . \lambda_L} \qquad \text{......(2)}$$

R_B is the radius of the coupling lead and λ_L the wavelength for longitudinal vibrations in the bending lead at the filter's midband frequency.

The coupling provided between two $\lambda/2$ longitudinal resonators with radius R_L, denoted as $(_{K12})_1$ for $\lambda/4$ coupling or $(K_{12})_2$ for $3\lambda/4$ coupling, is given by[14]

$$(K_{12})_1 = 1{\cdot}396\frac{R_B^{\frac{5}{2}}}{R_L^2 \lambda_L^{\frac{1}{2}}} \qquad \text{......(3)}$$

or

$$(K_{12})_2 = 1{\cdot}605\frac{R_B^{\frac{5}{2}}}{R_L^2 \lambda_L^{\frac{1}{2}}} \qquad \text{......(4)}$$

The method of calculation has proved successful. A filter with midband frequency of 455 kc/s was built, for example, with a calculated bandwidth of 5·2 kc/s. The measured bandwidth was 4·9 kc/s. The validity of these relationships, however, reaches a limit when plate-shaped resonators are considered.

The transmission characteristics of this filter are of course not only determined by the vibration types used for the resonators and coupling leads, which only come into play in the actual transmission range. This already emerged clearly from the remarks about single-sideband techniques. The bending-mode resonances of the longitudinal resonators, as well as bending- and longitudinal coupling of these bending vibrations to one another, have also proved to be important for the behaviour of miniature filters in the neighbourhood of the passband.[15]

A most important way of suppressing these spurious resonances, which could make a filter completely useless, consists in making the electromechanical transducer sensitive to the longitudinal mode only. However, it is not possible to achieve more than 20 to 30 dB spurious mode suppression with this method due to small residual unsymmetry, which cannot be completely avoided in miniature filters. The only remaining way out is to look for arrangements whereby the bending modes of the resonators are only coupled very weakly. If, for example, one attaches the bending coupling wires to those positions on the longitudinal resonators where the bending modes of the resonators have their nodes, then the coupling approaches zero. In Fig. 17, R_s denotes a value which is proportional to the coupling of bending resonances (with $s+1$ nodes) for the longitudinal vibrator. For the position of attachment of the coupling wires given by $l_a/_bl = 0{\cdot}13$, optimum decoupling is achieved for the second ($s = 2$) and the third bending harmonic ($s = 3$).[13] This is also the case which is of most interest in practice. With arbitrary increase of the diameter of the longitudinal resonators (at constant longitudinal resonance and therefore constant length except for the Rayleigh correction factor), the first bending mode always lies far below the $\lambda/2$ longitudinal resonance, because, according to Timoshenko's theory, the bending wave velocity soon becomes independent of the diameter. The second and third bending harmonics ($s = 2$ and $s = 3$) however, can already lie in

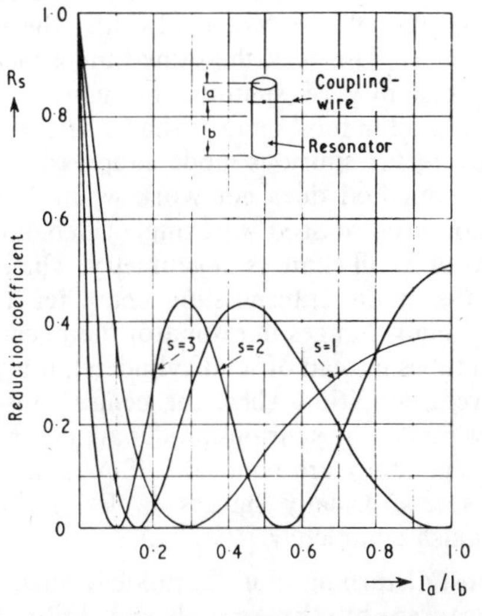

Fig. 17. Reduction of the coupling of undesirable bending-modes by changing the point of connection between coupling wires and longitudinal resonators.

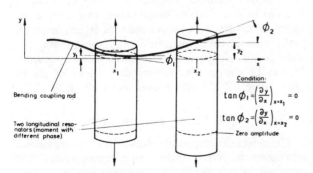

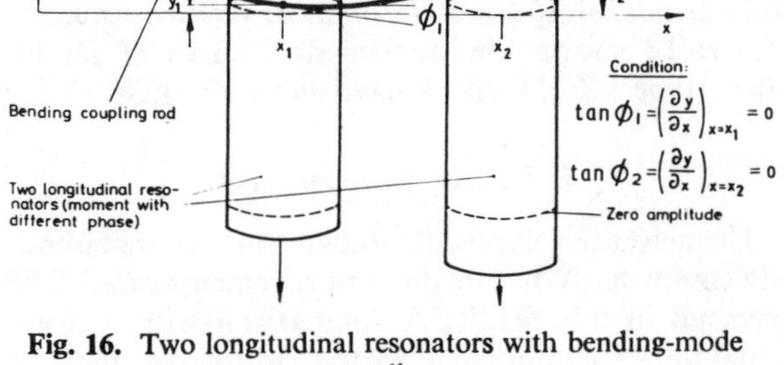

Fig. 16. Two longitudinal resonators with bending-mode coupling.

the immediate vicinity of the actual transmission band, so that the adjustment of $l_a/l_b = 0{\cdot}13$ becomes technically very interesting. The use of so thin a longitudinal resonator such that very high bending modes ($s > 3$) occur in the neighbourhood of the transmission band, is in any case not to be recommended, as with the higher orders the relative frequency separation of bending resonances, and therefore also the spurious mode separation, becomes ever smaller.

To sum up, one can say that by this method, neighbouring resonators can be decoupled, with regard to the second and third bending mode, to the order of 1 : 40.

The question of how great the coupling of bending modes actually is, and how great this is in comparison with the coupling of the main mode in the actual transmission band of the filter, is also interesting. For the sake of comparing the bandwidth for spurious resonances to the main transmission bandwidth, we define, as ratio of the two bandwidths the spurious mode ratio N. This is at the same time a measure of the suppressibility of spurious resonances. If this ratio N is small in relation to 1, then a further method of improving the spurious mode suppression may be used. This method does not work when $N > 1$, (as has already been noticed with single-sideband filters). By means of small changes in geometry, which do not disturb the main transmission characteristics (for example, small changes in resonator diameters, or using small tubes instead of solid cylinders), it is possible to achieve a condition where the bending resonances of the few remaining spurious modes are detuned more strongly than they are coupled. If $N \ll 1$, the production spread usually suffices to let the spurious modes vanish completely.

The most disturbing spurious mode spectrum in our case is caused by the second and third bending modes of the longitudinal resonators, where the bending vibrations are polarized in the plane of the filter, and the energy is transmitted by the thin coupling leads through longitudinal drive. In this case the spurious mode ratio N is calculated as

$$N = VF_r \frac{1{\cdot}41}{m} \sqrt{\frac{\lambda_L}{R_L}} \sqrt{\frac{\lambda'_L}{R_B}} \qquad \ldots\ldots(5)$$

In this equation m has the values $m = 7{\cdot}85$ (for $s = 2$) and $m = 11{\cdot}00$ (for $s = 3$) respectively. λ_L is the wavelength for longitudinal waves in the middle of the main transmission band and λ'_L the wavelength for longitudinal vibrations in the middle of the relevant spurious mode spectrum (for example at $s = 2$ or $s = 3$). The quantity V is found from

$$V = \frac{1}{\sin(2\pi l_{kr}/\lambda'_L)} \qquad \ldots\ldots(6)$$

and F_r equals 1·14 or 0·99 according to whether respectively $\lambda/4$ or $3\lambda/4$ bending coupling is used for the main transmission band.

For the filter shown in the following example, we have

$$l_k = 1{\cdot}7\,\text{mm}\ (3\lambda/4\ \text{coupling}),$$
$$\lambda_L \simeq \lambda'_L = 10\,\text{mm}$$
$$2R_L = 1{\cdot}5\,\text{mm}, \qquad 2R_B = 0{\cdot}1\,\text{mm},$$
$$V \simeq 1{\cdot}25 \quad \text{and} \quad F_r \simeq 1.$$

For $m = 7{\cdot}85$ N becomes 6 and for $m = 11{\cdot}00$, N becomes 4. The spurious mode ratio is therefore > 1 in both cases, and only by reducing the coupling by choosing the point of attachment of the coupling wires to the resonator so that $l_a/l_b = 0{\cdot}13$ (optimum for

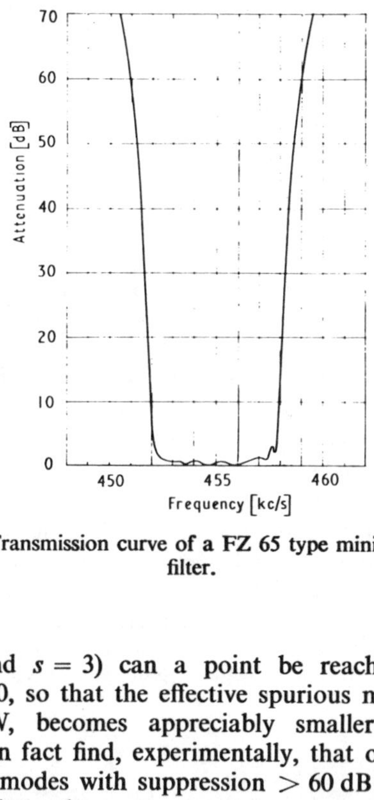

Fig. 18. Transmission curve of a FZ 65 type miniaturized i.f. filter.

$s = 2$ and $s = 3$) can a point be reached where $R_s = 1/40$, so that the effective spurious mode ratio $N' = R_s N$, becomes appreciably smaller than 1. We did in fact find, experimentally, that only a few spurious modes with suppression > 60 dB occurred, while without these measures to suppress them, very wide bands of spurious resonances readily occurred. Figure 18 shows the transmission curve of an I.f. filter, (type FZ 65) with a bandwidth of 6 kc/s.

4. Further Developments

The newer developments shown thus far are essentially complete. What further improvements can still be expected in this field? A comparison with conventional filter techniques shows that a broad category of

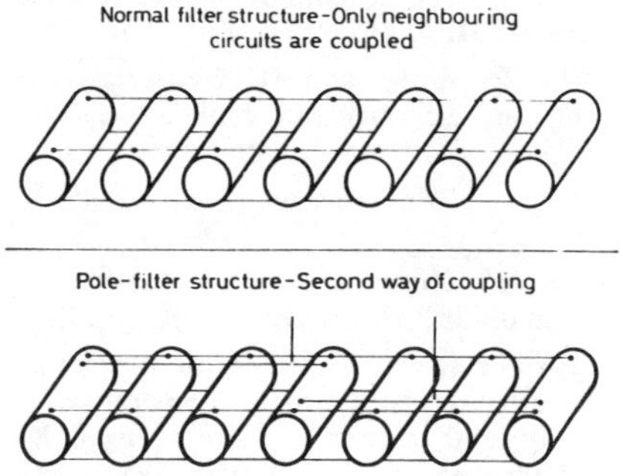

Fig. 19. Realization of pole-filters by two-way coupling.

filters have up to the present not yet been realized mechanically. The mechanical filter types realized up to now correspond to simple coupling filters. How can one realize filters with attenuation poles and non-minimal-phase filters? There has been a whole series of interesting experiments in this field, dating back several years. Mason[16, 17] was the first to try such experiments. Unfortunately all these approaches failed in their practical realization, principally on account of a large amount of uncontrollable spurious modes. Lukas[18] describes how a second resonance of the same resonator may be used to achieve definite attenuation poles. We found a simple way to build filters with attenuation poles by not only coupling neighbouring circuits to one another, but also to non-neighbouring circuits by secondary coupling leads.[19] (Fig. 19.)

It is surprising that the relatively long coupling leads needed for several apparently possible structures (Fig. 19 shows only one example of many) are completely free of spurious modes, although one would expect them to have a multitude of bending resonances and therefore closely spaced spurious modes. It is true that with the excitation of the longitudinal mode in long thin wires (as in the example of Fig. 19) a large number of bending resonances exist. The characteristic impedance, however, that can be assigned to these resonances, differs so strongly from that of the substantially more massive resonators that have to be coupled that these resonances are practically not coupled at all to the actual energy exchange in the filter. Figure 20 shows a picture of a realized pole filter, and Fig. 21 shows the measured attenuation curve of such a filter. Trceva treats filters with secondary coupling as pure electrical coupling filters.[20] Fairly concrete concepts about the realization of all-pass filters, with which the delay curve may be adjusted independently of the attenuation curve,

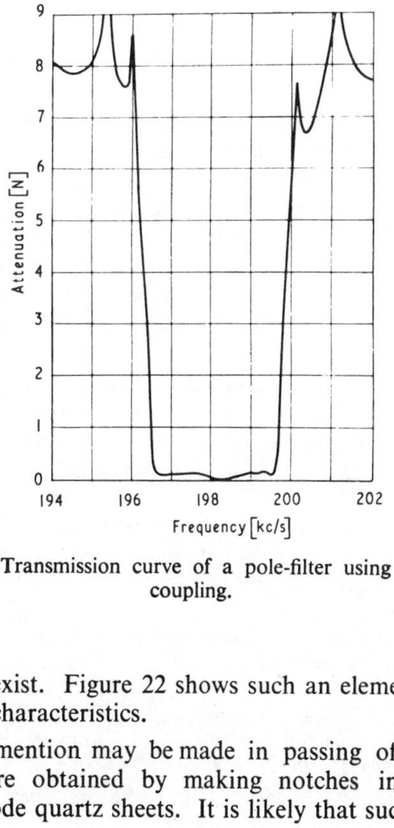

Fig. 21. Transmission curve of a pole-filter using two-way coupling.

already exist. Figure 22 shows such an element with all-pass characteristics.

Brief mention may be made in passing of filters, which are obtained by making notches in plate-shear mode quartz sheets. It is likely that such filters

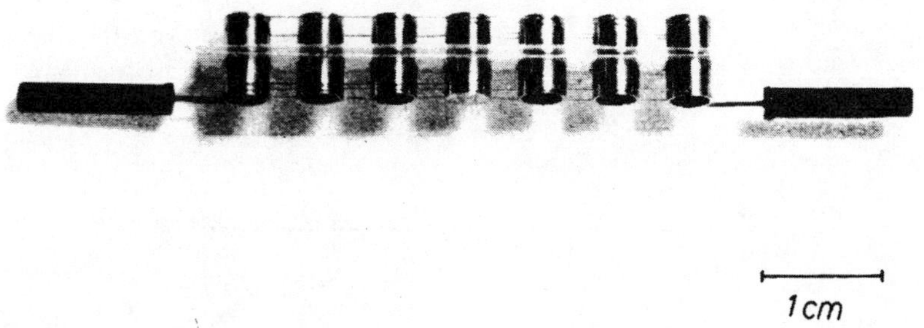

Fig. 20. A pole-filter using two-way coupling.

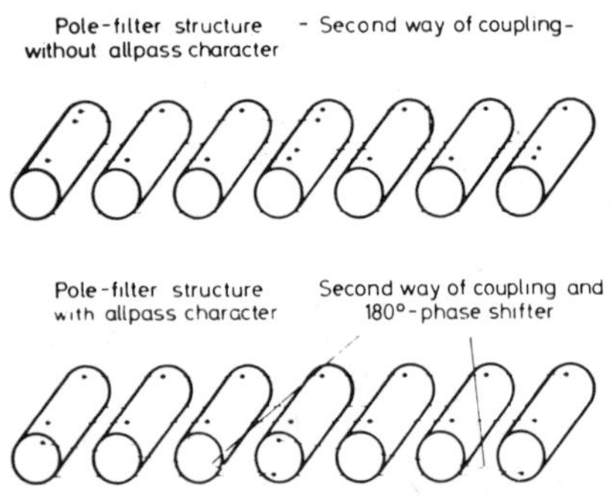

Fig. 22. Realization of non-minimal-phase mechanical filters with all-pass portion.

can only be manufactured with a small number of circuits.

A further demand is for filters of the present quality, but for frequencies above one megacycle. The requirements set, however, regarding their behaviour with temperature and above all their Q-factors cannot be fulfilled with present materials. A few interesting trial experiments which have been undertaken with a view to mastering the possible vibrational structures at such high frequencies[22] make little reference to these requirements in fact and therefore do not point the way towards realizing the present quality at still higher frequencies. Nevertheless it is possible that new applications will be found for such mechanical resonators in electrical filters.

5. References

1. W. van B. Roberts and L. L. Burns, "Mechanical filters for radio frequencies", *R.C.A. Review*, **10**, p. 348, September 1949.
2. R. W. George, "Electro-mechanical filters for 100 kc carrier and sideband selection", *Proc. Inst. Radio Engrs*, **44**, p. 14, January 1956.
3. G. L. Grisdale, "Electro-mechanical filters for use in telecommunication equipment", *Brit. Commun. Electronics*, **6**, p. 768, November 1959.
4. H. Bache, "A practical electromechanical filter", *Marconi Review*, **22**, No. 134, p. 144, 1959.
5. W. Struszynski, "A theoretical analysis of the torsional electromechanical filters", *Marconi Review*, **22**, No. 134, p. 119, 1959.
6. M. Börner, "Biegeschwingungen in mechanischen Filtern", *Telefunken Z.*, **31**, p. 115 and p. 188, 1958.
7. M. Börner, E. Kettel and H. Ohnsorge, "Mechanische Filter für die Nachrichtentechnik", *Telefunken Z.*, **31**, p. 105, 1958.
8. H. D. Pieper, "Mechanische filter", *Telefunken Z.*, **32**, p. 279, 1959.
9. M. Börner, E. Dürre and H. Schüssler, "Mechanische Einseitenbandfilter", *Telefunken Z.*, **36**, p. 272, 1963.
10. C. M. van der Burgt, "Ferrites for magnetic and piezomagnetic filter elements with temperature independent permeability and elasticity", *Proc. Instn Elect. Engrs*, **104B**, Supplement No. 7, p. 550, 1957. (I.E.E. Paper No. 2287R, 1957.)
11. C. M. van der Burgt, "Controlled crystal anisotropy and controlled temperature dependence of the permeability and elasticity of various cobalt-substituted ferrites," *Philips Res. Rep.*, **12**, p. 97, 1957.
12. S. Schweitzerhof, "Über Ferrite für magnetostriktive Schwinger in Filterkreisen", *Nachrichtentechn. Z.*, **11**, p. 179, 1958.
13. M. Börner and H. Schüssler, "Miniaturisierung mechanischer Filter", *Telefunken Z.* (To be published.)
14. M. Börner, "Mechanische Filter mit Biegekopplung", *Archiv Elektr. Übertrag.*, **15**, p. 175, 1961.
15. M. Börner, "Zum Nebenwellenverhalten von Mikromodulfiltern (Biegenebenwellen in biegegekoppelten Longitudinalfiltern)", *Archiv Electr. Übertrag.*, **16**, p. 532, 1962.
16. W. P. Mason, U.S. Patent 2,345,491 (1944).
17. W. P. Mason, "Physical Acoustics", I Part A, (Academic Press, New York, 1964).
18. J. Lukas, "Plattenförmige elektromechanische Filter mit Dämpfungspolen", *Archiv Elektr. Übertrag.*, **17**, p. 230, 1963.
19. M. Börner, "Mechanische Filter mit Dämpfungspolen", *Archiv Elektr. Übertrag.*, **17**, p. 103, 1963.
20. E. Trceva, "Hochfrequenzbandfilter mit Dämpfungspolen bei endlichen Frequenzen und geebnetem Betrag im Durchlassbereich", *Nachrichtentechnik*, **12**, No. 12, p. 450, 1962 and **13**, No. 1, p. 36, 1963.
21. W. Poschenreider and F. Schöfer, "Das elektromechanische Quarzfilter—ein neues Bauelement für die Nachrichtentechnik", *Frequenz*, **17**, No. 3, p. 88, 1963.
22. E. Gikow, A. Rand and J. Gianotto, "Functional circuits through acoustic devices", *I.R.E. Trans. on Military Electronics*, **MIL-4**, No. 4, p. 469, 1960.

Manuscript received by the Institution on 12th October 1964. (Paper No. 968/EA18.)

Mechanical Filters—A Review of Progress

ROBERT A. JOHNSON, MEMBER, IEEE, MANFRED BÖRNER, AND MASASHI KONNO, MEMBER, IEEE

Abstract—**This paper is a review of electromechanical bandpass filter, resonator, and transducer development. Filter types discussed include intermediate and low-frequency configurations composed of rod, disk, and flexure-bar resonators and magnetostrictive ferrite and piezoelectric ceramic transducers. The resonators and transducers are analyzed in terms of their dynamic and material characteristics. The paper also includes methods of realizing attenuation poles at real and complex frequencies. The last section is a look at future developments.**

Introduction

MORE THAN 20 years have passed since the first practical mechanical filters were introduced by Adler [1], Roberts [2], and Doelz [3]. These filters met an existing need for greater selectivity in the IF stages of AM and SSB receivers designed for voice communication. Modern receivers and telephone communication equipments carry not only voice but data messages as well, thus requiring lower passband ripple and well-defined and stable passband limits. In addition, greater selectivity and lower loss are often required.

Mechanical filter development has kept pace with the demands of communication engineers. For instance, filters having passband-ripple requirements of 0.5 dB or less or 60/3-dB bandwidth ratios of 1.3/1 are not unusual. Package size has also been reduced; a 10-pole filter can be designed into a 1-cm^3 package. In addition, the lower frequency limits of mechanical filters are now in the audio range; the upper limit is that of an AT-cut crystal. Although the mechanical filter was conceived in the U.S., a great amount of developmental work is now being conducted in Europe and Japan.

Early interest in the use of mechanical elements to provide passband characteristics resulted from the development of the electrical bandpass filter by Campbell and Wagner in 1917 [4]. A number of patents relating to spring-mass systems [5] were filed soon after, but the most significant work in terms of practical devices was done by Maxfield and Harrison on phonograph recording and playing equipment [6]. Fig. 1 shows a comparison of the frequency response of a phonograph designed as an electromechanical filter (solid curve) and the best of phonographs available at the time. As dramatic as these results were, very little work was done in the next two decades on mechanical filters designed to act as frequency selective devices.

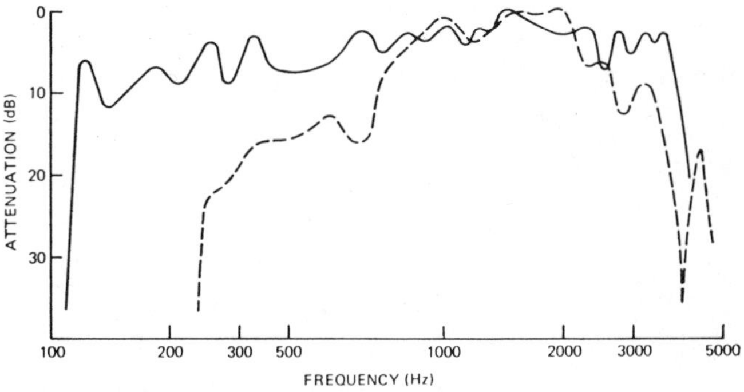

Fig. 1. Improvement in phonograph design through the use of filter design methods [6].

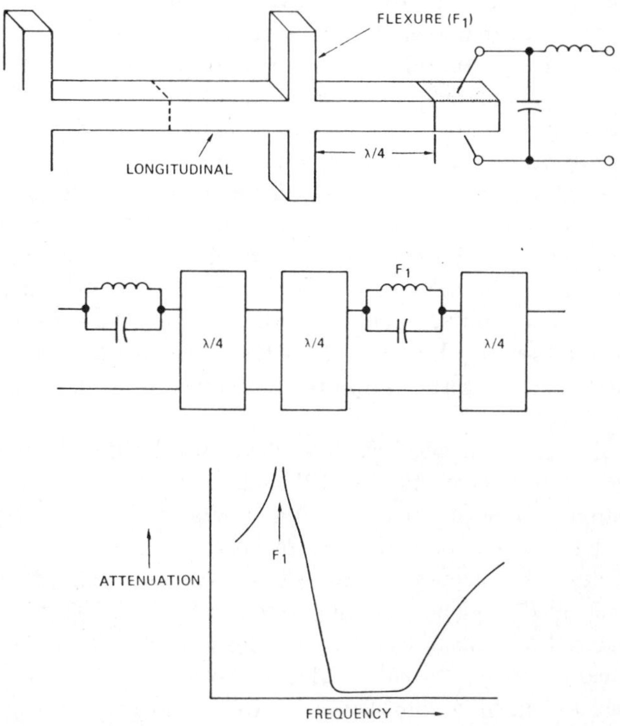

Fig. 2. Distributed element mechanical filter (Mason, 1941).

In the early 1940's, Mason followed up his excellent work on crystal filter design with the development of the very interesting, but not widely used, mechanical filter shown in Fig. 2 [7]. Some of the significant features of this design included the use of piezoelectric transducers, distributed-element half-wavelength resonators, and an attenuation-pole-producing coupling element.

Just a few years later, the first practical IF mechanical filter, shown in Fig. 3(a), was developed by Adler.

Manuscript received January 6, 1971.
R. A. Johnson is with the Collins Radio Company, Newport Beach, Calif.
M. Börner is with AEG—Telefunken Research Institute, Ulm, Germany.
M. Konno is with the Department of Electrical Engineering, Yamagata University, Yonezawa, Japan.

Reprinted from *IEEE Trans. Sonics Ultrason.*, vol. SU-18, pp. 155–170, July 1971.

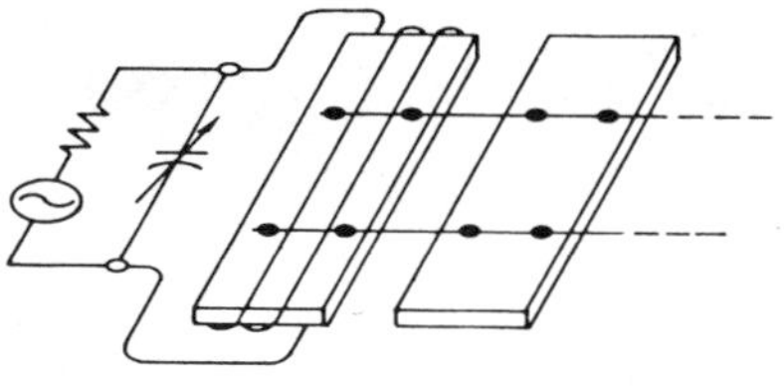

(a)

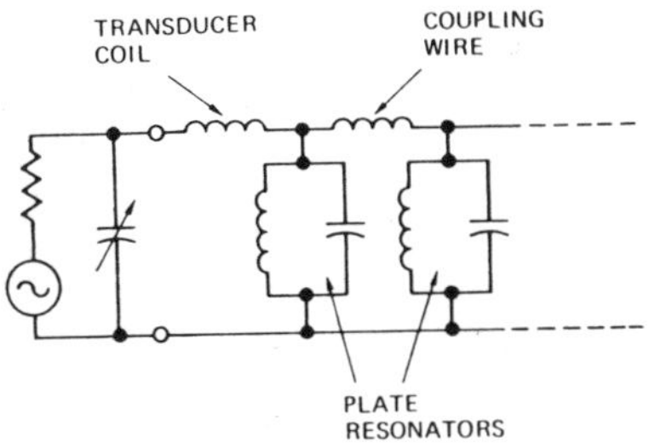

(b)

Fig. 3. Resonator-coupling wire mechanical filter (Adler, 1946). (a) Pictorial diagram. (b) Electrical equivalent circuit.

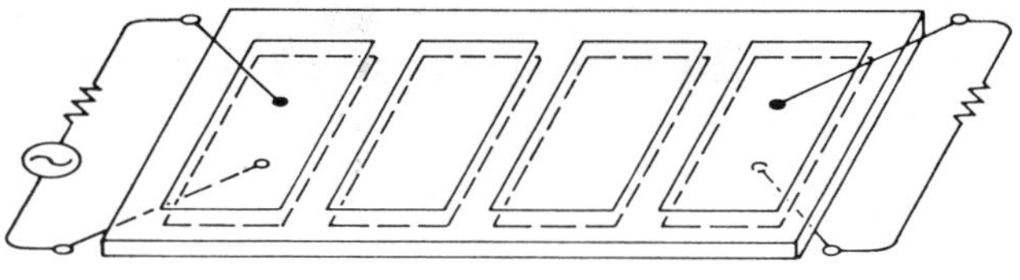

(a)

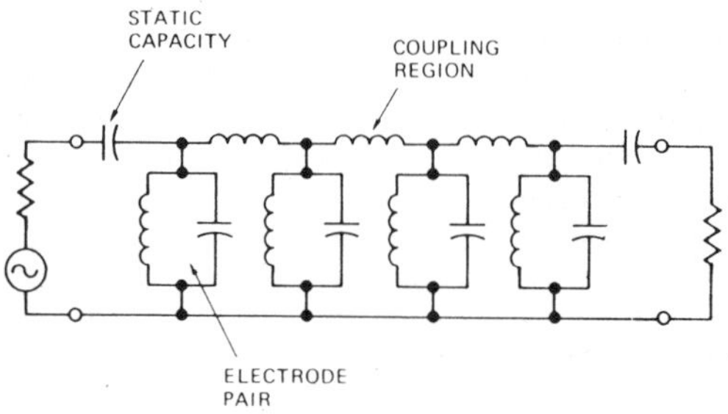

(b)

Fig. 4. Quartz monolithic mechanical filter. (a) Pictorial diagram. (b) Electrical equivalent circuit.

This filter made use of half-wavelength plates coupled by short small-diameter wires. The plate resonators can be represented by parallel tuned circuits, the coupling wires by series inductors, as shown in the electrical equivalent circuit of Fig. 3(b). The nickel end plates act as magnetostrictive input and output transducers.

The frequency response characteristics are monotonic, that is there are no finite attenuation poles. Adler's filter is typical of most modern mechanical filters in that distributed parameter resonators are coupled through nonresonant lines. There is a striking similarity between the plate-wire filter and the latest type of mechanical filter, the monolithic or quartz mechanical filter shown in Fig. 4(a).

The quartz mechanical filter developed by Beaver and Sykes [8] and Nakazawa [9] makes use of the energy-trapping concept where standing waves are set up under each electrode pair. In the regions between the electrodes, the acoustic wave decays exponentially, the electroded regions act like resonators, and the non-electroded regions act like coupling elements. The electrical equivalent circuit of the plate-wire and the quartz mechanical filter shown in Fig. 4(b) are identical except that in the plate-wire case the input and output capacitors are replaced by coils. The monolithic filter also has monotonic frequency response characteristics. A construction similar to that of the monolithic filters described above has been used by Börner and Schüssler [10].

Basic Filter Structures

Monotonic Designs

Adler's development of the plate-wire mechanical filter was soon followed by Doelz's disk-wire filter [11], shown in Fig. 5(a). The disk-wire filter makes use of flexural modes of vibration of the disk resonators. The two most commonly used modes are the one-nodal circle (50–200 kHz) and the two-circle mode (200–600 kHz). The resonators are coupled by small-diameter wires that are spot welded around the circumference of the disks. As a result of the very complex displacement at the edge of the disk, the coupling mode is a combination of extension and flexure and is, therefore, very difficult to analyze. The earliest designs made use of small-diameter odd-number quarter-wavelength iron-nickel alloy transducers. Later designs, such as that shown in Fig. 5(a), use one-half and full-wavelength ferrite transducers.

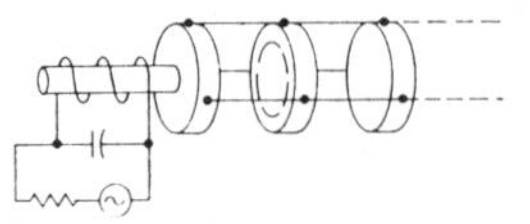

(a)

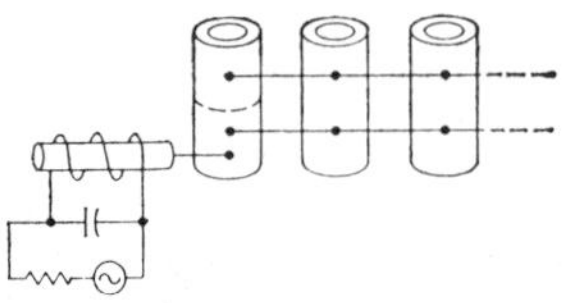

(b)

Fig. 5. (a) Disk-wire mechanical filter, (Doelz, 1949). (b) Rod-wire mechanical filter (Tanaka and Börner, 1957).

Ferrite transducers are also used on the rod-wire filter shown in Fig. 5(b). The torsional rod-wire filter was developed independently by Börner in Germany [12], [13], and Tanaka in Japan [14]; the Tanaka design uses Langevin metal alloy/ceramic transducers in place of the ferrites. The cylindrical rod resonators are designed to vibrate in a half-wavelength torsional mode at frequencies up to 250 kHz and in a half-wavelength longitudinal mode when the filters are designed to operate at 455 kHz. Use of the torsional mode results in longitudinal coupling between resonators, whereas at 455 kHz the coupling involves bending or flexure.

Use of a wire to connect the transducer to the end resonator, as shown in Fig. 5(b), results in a reduction of spurious or unwanted responses, particularly if it is attached at a nodal point of the strongest unwanted modes. This technique, which is often used on disk-wire filters, results in spurious modes being suppressed more than 60 dB, as shown in Fig. 6(a). In contrast to the excellent spurious response rejection of the torsional rod-wire filter is the response [Fig. 6(b)] of the rod-neck filter of Fig. 7(a), which is driven in a similar manner, but is subject to broad bandwidth bending modes.

The rod-neck filter was developed by Roberts [2] at the same time the disk-wire filter was being developed. The basic design concept is that of coupling half-wavelength torsional or longitudinal resonators with quarter-wavelength necks. The two major problems of the early designs were the easily excited spurious bending modes [15], and the difficulty in construction and tuning due to being turned out of a single rod. In addition, at lower frequencies like 100 kHz, the filters become extremely long. In order to reduce the length, Tanaka developed the folded line filter shown in Fig. 7(b). The resonators vibrate in a half-wavelength longitudinal mode and are coupled by relatively large-diameter short wires. The use of longitudinal-mode Langevin transducers plus the reduced overall length results in a greater rejection of unwanted modes. Like the rod-wire and folded designs, the disk-wire filter has a relatively low spurious response level.

All four filters discussed so far operate at frequencies above 50 kHz. With the exception of some work reported by Mason and Konno, there was little activity at frequencies below 50 kHz before 1960. One of the first practical designs was that of Mason and Thurston [16], who used antisymmetric mode flexural resonators coupled in torsion as shown in Fig. 8(a). The antisymmetric mode makes this design less susceptible to microphonic excitation. A more widely used design is the symmetrical mode filter shown in Fig. 8(b). Work by Konno [17], [18], Yakuwa [19], and Albsmeier [20], have made this a very practical device in the frequency range of 300 Hz to 30 kHz. The symmetric-mode filter is driven by piezoelectric-ceramic transducers. The coupling wires are attached to the resonators at the nodal points, which results in torsional coupling. This type of filter is very sensitive to changes in the position of the coupling wires so considerable care is taken in manufacturing to ensure that bending modes are not propagated. The microphonic problem has been solved, in part, by supporting the filter with high-damping silicon–rubber supports.

In addition to the flexural-bar/wire low-frequency mechanical filters, a considerable amount of work has been done in Japan on the development of tuning-fork mechanical filters [21], such as those shown in Fig. 9(a) and (b). The three-prong filter is interesting in that it acts like a coupled two-resonator filter, thus providing a

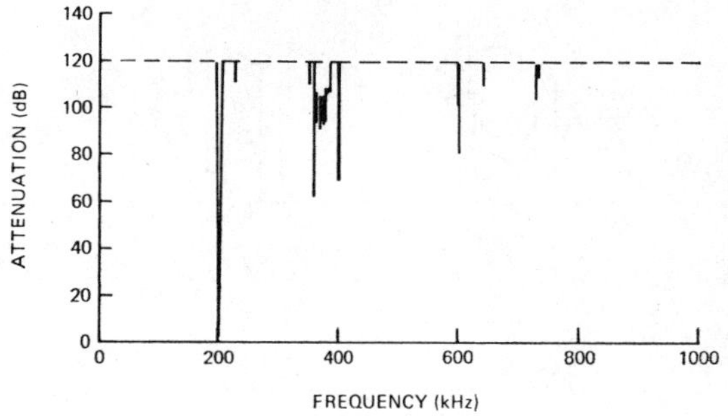

(a)

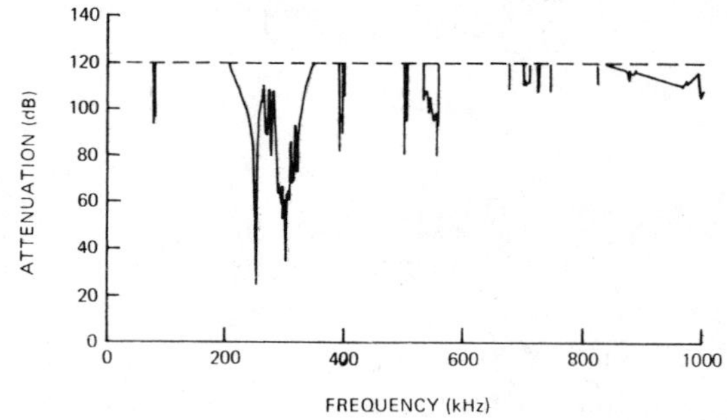

(b)

Fig. 6. Spurious responses of (a) torsional rod-wire, and (b) rod-neck mechanical filters.

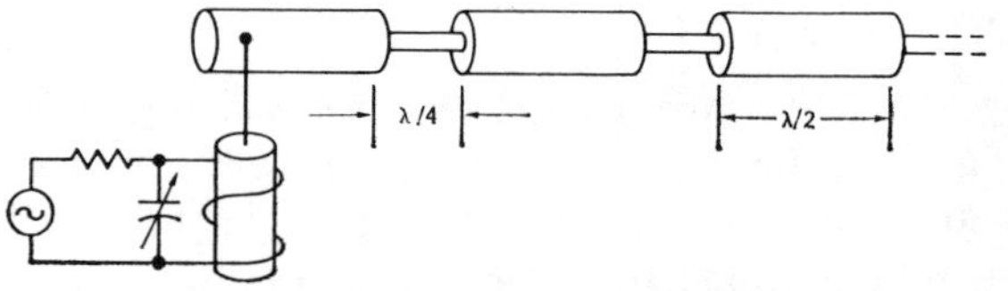

(a)

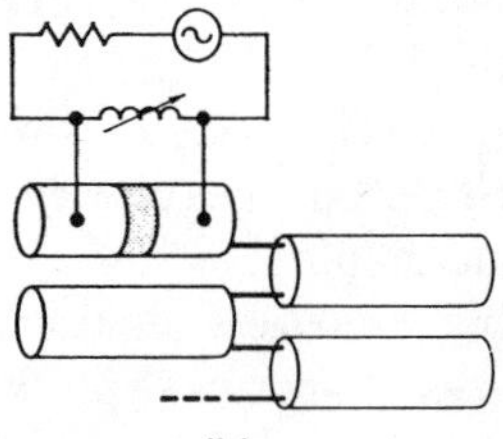

(b)

Fig. 7. Distributed element mechanical filters. (a) Rod-neck designs [2]. (b) Folded line designs (Tanaka, 1958).

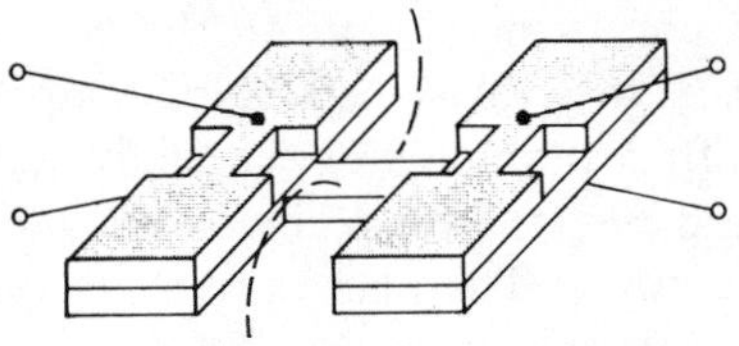

(a)

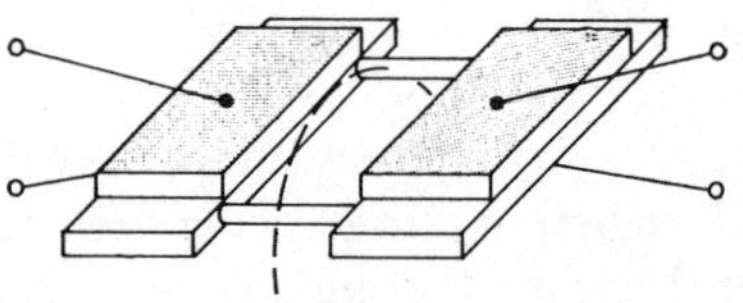

(b)

Fig. 8. Low-frequency mechanical filters. (a) Antisymmetric mode [16]. (b) Fundamental mode [18].

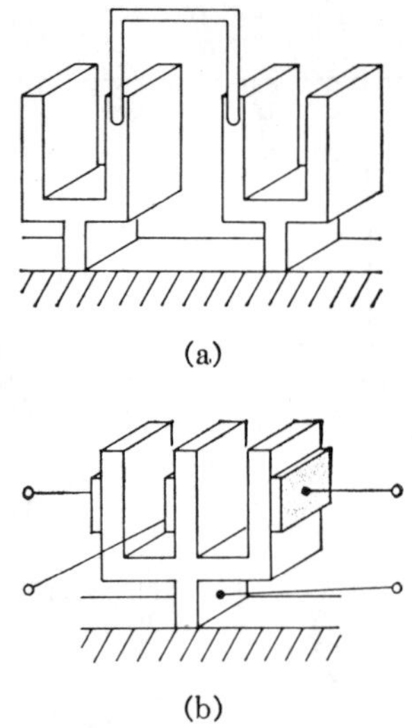

Fig. 9. Tuning fork mechanical filters. (a) U-shaped coupling type. (b) Two-pole three-prong type.

greater amount of selectivity than a simple two-prong device. The tuning fork and flexural-bar filters are widely used in Japan for remote and automatic control, selective calling and paging, telemetry, measuring instruments, and pilot and carrier signal pickup.

Finite Attenuation Poles.

By using as many as 15 resonators, highly selective mechanical filters of the bar-wire, disk-wire, and folded-resonator types are realizable. Although the resultant passband amplitude and delay responses are satisfactory for voice communication, the ripple amplitude and differential delay variations may be excessive when data are to be transmitted. By making use of finite-frequency attenuation poles, fewer resonators are needed, thereby reducing the differential delay, size, cost, and, most often, the passband ripple.

One of the first practical devices to realize finite attenuation poles was the crystal-plate filter with capacitive bridging [22], [23], shown in Fig. 10(a). Although the plate configuration was described in Adler's basic patent [24], the idea of using quartz and capacitive bridging was new. Connecting the capacitor from the top electrode of the input plate to the bottom electrode of the output plate is the same as adding a phase inverter across the output terminals of the equivalent circuit of Fig. 10(b) and results in a pair of attenuation poles as shown in Fig. 10(c). The phase inverter adjacent to the coupling element represents the 180-deg phase shift between resonators at the lowest natural mode of the mechanical system. The resonators vibrate in a length-extension mode ($5° - X$ or $-18.5° - X$) or face shear mode (CT) and are coupled through so-called Poisson coupling. Although these filters have been designed and built with more than two resonators, most currently manufactured filters are designed as 455-kHz 2 poles in cascade for use in SSB and FM receivers.

The use of wires to couple mechanically alternate

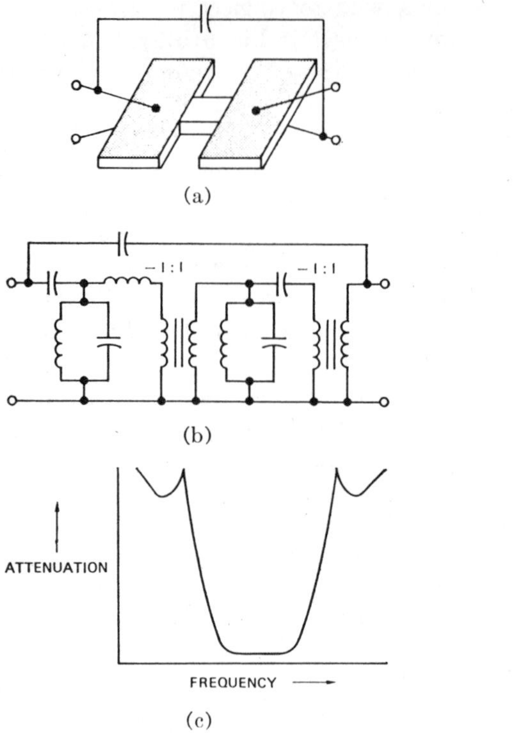

Fig. 10. Crystal mechanical filter with capacitive bridging [22]. (a) Pictorial diagram. (b) Electrical equivalent circuit. (c) Frequency response.

resonators was independently considered by Börner [25] and Johnson [26]. Börner's filter shown in Fig. 11(a) makes use of half-wavelength torsional-mode resonators and extensional mode-coupling elements. The disk-wire filter of Fig. 11(b) is composed of flexural mode disks and basically extensional mode-coupling elements. When the spacing between the resonators is such that the bridging wire is less than one-half of an acoustic wavelength long, the ideal transformer of Fig. 11(c) acts like a simple one-to-one transformer and an attenuation pole is realized on the high-frequency side of the filter passband. This can be understood by converting the π network, composed of the two adjacent resonator coupling wires (inductors) and the bridging wire, to its T equivalent circuit. The inductor in series with the center resonator produces an impedance zero, which results in an attenuation pole above the filter passband, as shown in Fig. 11(d).

When the bridging wire is between one-half and a full-wavelength long, the transformer in the electrical equivalent circuit acts like a phase inverter and an attenuation pole is realized below the filter passband. An alternate method of realizing the phase inversion in the rod-wire filter is shown in Fig. 11(a) by the dashed lines. In this case, out-of-phase regions of the alternate resonators are coupled. A similar method is used with disk-wire filters where one of the alternate disks vibrates in a diameter mode (rather than a nodal-circle mode), the three adjacent disk wires being connected to in-phase sectors, and the bridging wire to an out-of-phase portion of the resonator [27].

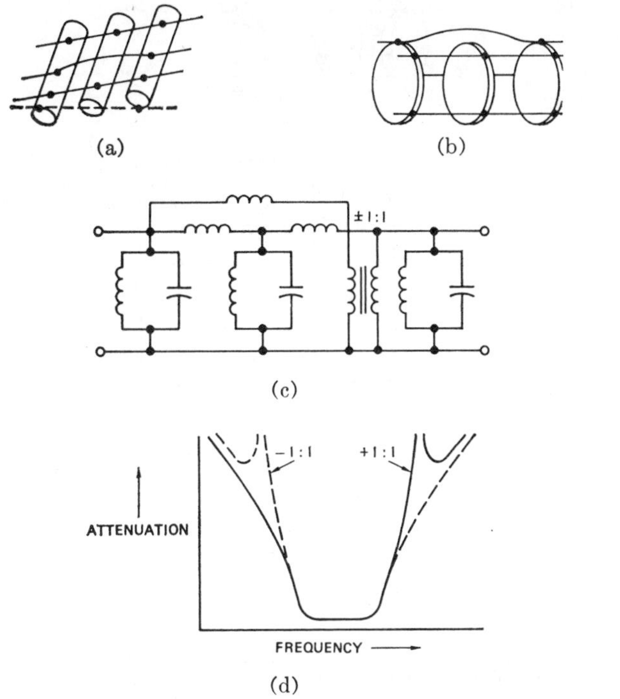

Fig. 11. Single-resonator acoustic bridging. (a) Rod-wire, [25]. (b) Disk-wire (Johnson, 1964). (c) Electrical equivalent circuit. (d) Frequency response.

If a high degree of selectivity is needed both above and below the filter passband, a coupling wire can be used to bridge two resonators producing a symmetrical pair of attenuation poles. This type of design is subject to being able to realize a phase inverter with each bridging wire. When no phase inversion takes place, the resultant frequency response is less selective than the monotonic case but delay compensation due to right-half plane attenuation poles is possible (see Table I). The most common method used to realize the phase inverter, in the case of the rod-wire filter, is to space adjacent resonators a quarter-wavelength apart, which results in the bridging wire being three-quarters of a wavelength long. The advantages of the quarter-wavelength coupling are that the resonators are all tuned to the center frequency of the filter and, in addition, the coupling is relatively insensitive to variations in wire length.

In order to maintain a small package size, the coupling wire length between disk resonators is usually less than one-eighth of a wavelength. At 455 kHz, due to the thickness of each disk being on the order of a quarter-wavelength, simple two-disk bridging results in symmetrical finite-attenuation poles. Fig. 12 shows the frequency response of the 12-disk filter shown in Fig. 13. Note that the passband ripple is quite low. This is in part due to the fact that the disk-wire as well as the rod-wire finite-pole filters are designed with modern insertion-loss techniques. Included in the design method are transformations such as that shown in Fig. 14(a) where a low-pass or bandpass ladder network is converted to an equivalent bridged form [28]. Because of the narrow-band nature of mechanical filters, an inverter I can be

TABLE I
ATTENUATION POLE LOCATIONS FOR VARIOUS BRIDGING CONFIGURATIONS

	Direct (1 : 1)	Inverted (−1 : 1)
One resonator	Upper stopband pole	Lower stopband pole
Two resonators	Delay correction	Upper and lower stopband poles

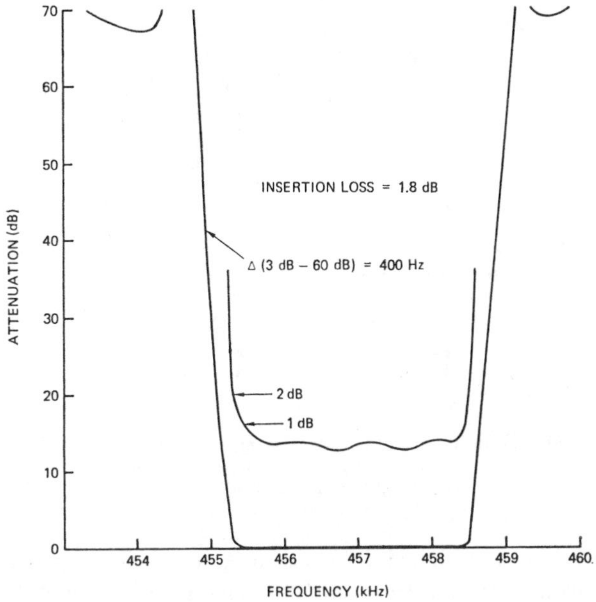

Fig. 12. Frequency response of a double-disk bridging mechanical filter.

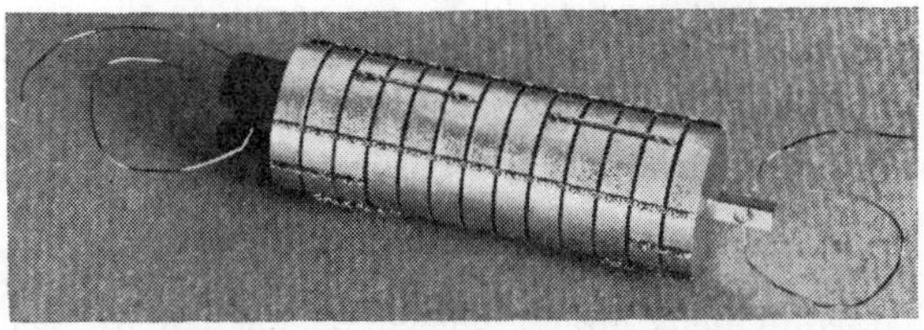

Fig. 13. Mechanical filter with wire bridging across two resonators (Collins Radio).

converted to a π network composed of two negative inductors and one positive inductor, all having the same absolute value [29]. As an example, the low-pass network of Fig. 14(b), which may be a Cauer-type filter [30], can be converted to the bandpass double-resonator bridging topology shown in Fig. 14(c). The use of double (coincident) poles results in a physically symmetrical mechanical configuration, which is helpful in reducing manufacturing costs. The short coupling wire length between resonators, as shown in Fig. 13, not only decreases the package size but increases the strength of the filter. The ability of the structure to withstand high shock and vibration levels is also due to the coupling wires being located away from the centroid of the structure.

For the past 15 years, disk-wire mechanical filters

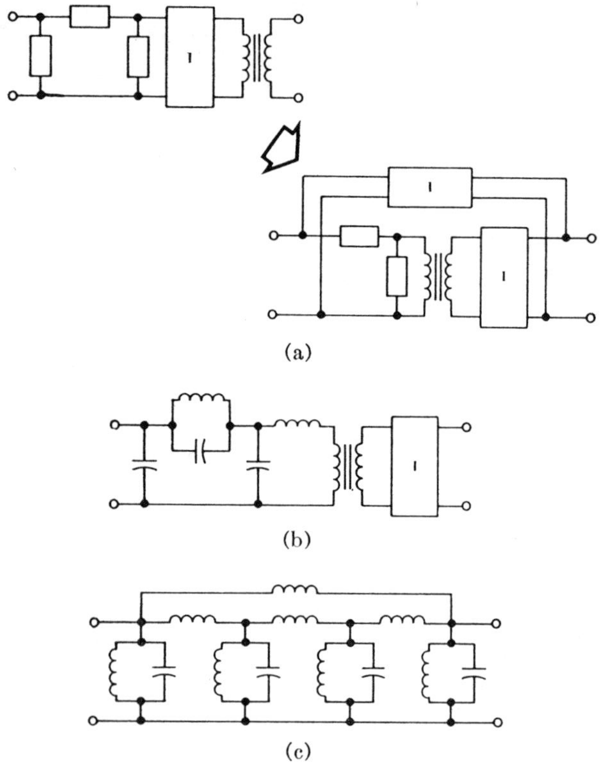

Fig. 14. Double bridging transformations. (a) General transformation. (b) Low-pass prototype. (c) Final bridging-wire bandpass electrical equivalent circuit.

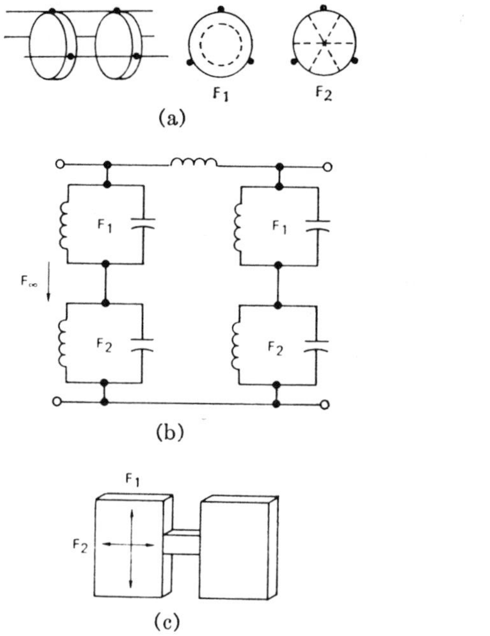

Fig. 15. Multiple-mode finite pole configuration. (a) Disk-wire. (b) Simplified electrical equivalent circuit. (c) Plate-type design.

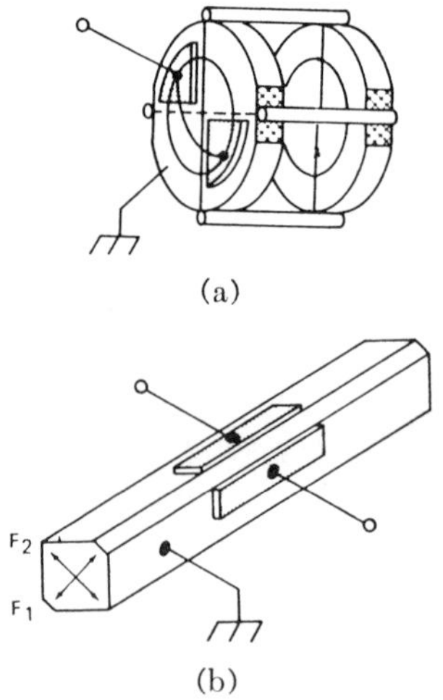

Fig. 16. Multiple resonant mode filters. (a) Disk-wire design. (b) Low-frequency flexural bar.

have been designed to make use of spurious modes of vibration to control stopband selectivity. Although the theory was not understood until recently [31], it was found that by varying the coupling-wire orientation around the circumference of the disks the slope of one side of the response could be increased at the expense of the other side. For instance, in Fig. 15(a) each disk has a natural resonance near-frequency F_1 in the passband. Each disk also has a natural resonance (actually two as well as many others at different frequencies) near F_2 above the filter passband. Making use of the simplified equivalent circuit shown in Fig. 15(b), we see that an attenuation pole is produced at F_∞ between F_1 and F_2. This results in a steeper response above the filter passband, which can be controlled by varying the coupling wire orientation, which in turn controls the effect of resonator F_2. The same technique has been used in the design of plate filters [32] where the length and width dimensions control the frequencies F_1 and F_2 as shown in Fig. 15(c).

By removing a segment from the edge of a disk resonator, two controllable diameter modes corresponding to each pair of degenerate modes can be produced [33]. This technique has been used to design a variety of multiple mode filters [34]–[36], most of which are still only laboratory models. An example of this type of design is shown in Fig. 16(a). This particular filter employs one-circle one-diameter modes of vibration, as well as piezoelectric ceramic transducers and has a frequency response equivalent to a four-resonator design. The solid nodal lines correspond to the highest natural mode.

In the case of low-frequency filters, a similar technique can be used where the corner of a flexural bar can be removed to produce two natural modes by destroying symmetry of the moments of inertia. An example of this type of resonator is shown in Fig. 16(b) where the arrows show the displacement directions of the two modes [37]. Fig. 17 shows an early filter that makes use of this technique to obtain a total of six natural resonances as well as additional attenuation poles [38]. A large variety of devices have been designed using this method, including a three-resonance two-attenuation pole filter constructed from a single bar [39].

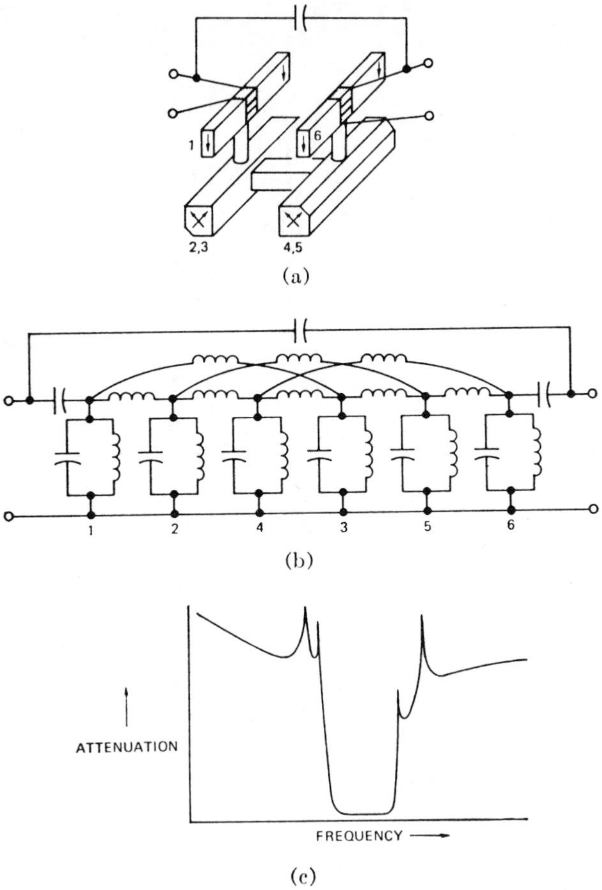

Fig. 17. Six-pole-segmented flexural bar design.

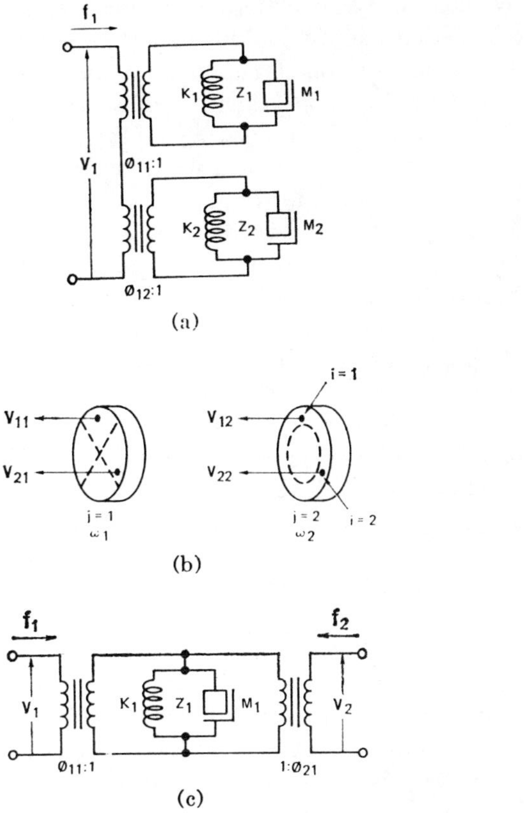

Fig. 18. Two-port two-mode disk resonator. (a) Driving point impedance model. (b) Disk nodal patterns. (c) Single-mode two-port equivalent circuit.

Analysis of Resonators and Coupling Elements

A considerable amount of work has been done in the past few years on the problem of analyzing and modeling mechanical resonators and coupling elements. This work has resulted in a better understanding of the effects of unwanted modes found in both resonators and coupling wires on mechanical filter response characteristics. In addition, the improved network models have been most helpful in the development of new mechanical filter types.

Lumped Element Equivalent Circuits

The driving point mobility (ratio of velocity to force) of a linear mechanical resonator at a point i can be expressed as [40]

$$v_i/f_i = (\omega/j) \sum_{i=1}^{\infty} [M_{ij}(\omega^2 - \omega_j^2)]^{-1} \tag{1}$$

where v_i and f_i are the velocity and the applied force at i and M_{ij} is the equivalent mass of mode j corresponding to the natural frequency ω_j. By making use of a velocity transformer of turns ratio ϕ_{ij} and recognizing (1) as the partial fraction expansion of a reactance function, we can construct the schematic diagram shown in Fig. 18(a).

This network describes the characteristics of a two-mode resonator at a point (or port in electrical terminology) $i = 1$. The resonator could be a rod or bar vibrating in a flexural and a longitudinal mode or, as shown in Fig. 18(b), a disk vibrating in a two-diameter flexure mode and a single-circle flexure mode. If only one of the modes, but two points on the disk are considered, the force-velocity relationships can be found from the schematic diagram of Fig. 18(c). In general, a resonator having M ports and N natural modes can be described by the matrix equation

$$v] = [Z]\, f] \qquad z_{kl} = \sum_{i=1}^{N} \phi_{ki}\phi_{li} z_i . \tag{2}$$

In (2), $v]$ and $f]$ are column matrices of order M and z_{kl} is an element in the $M \times M$ Z matrix. A generalized equivalent circuit based on (2) is shown in Fig. 19.

Disk Resonators

To be able to make use of any resonator model, it is necessary to be able to calculate or measure natural resonant frequencies and equivalent mass values at specific points on the resonator. In the case of thick disk resonators, excellent agreement between calculated and experimental values of frequency were obtained by Deresiewicz and Mindlin [41] and Sharma [42] for symmetrical modes; that is, flexural modes having circular nodal patterns. More recently, Onoe solved the problem of nonaxisymmetrical modes (nodal diameters) by making use of a linear combination of independent waves that satisfy the differential equations of motion and boundary conditions on the major surfaces of the disk as well as approximately satisfying boundary conditions on the lateral surfaces [43]. Unlike the exact solutions that are expected when analyzing an electrical network, solutions for frequency and equivalent mass of a mechanical resonator are only as good as the number of higher order waves that are taken into account. The greater the number of waves, the more accurately the boundary conditions can be satisfied.

The earliest work on finding the equivalent mass of a thick disk resonator was performed by Sharma for the case of axisymmetric modes [44]. This analysis was based on the method of finding the equivalent mass by dividing the total kinetic energy in the system (disk) by one-half of the square of the velocity in a specified direction at a point on the disk [45]. This same technique coupled with that of Onoe's [43] has been used to calculate the equivalent mass of disks vibrating in both symmetric and nonsymmetric modes [46]. Examples of equivalent mass versus position on the surface of the disk are shown in Fig. 20 for two adjacent vibration modes. Note that the equivalent mass at the disk edge ($r = 1.0$) is lower in the case of the adjacent two-diameter one-circle mode than the two-circle mode. As a rule, the lower the impedance of a mode, the broader is its coupled response and the more difficult it is to suppress.

Within the frequency range of interest (50–600 kHz) in the case of disk-wire filters, there are various other unwanted modes present such as radial and concentric shear modes. These can often be troublesome in the case of wideband filter designs where the modes sometimes fall in the range of the passband. These particular modes, called contour modes because they involve no transverse (flexural) vibration but only a change in the shape or contour of a disk, have studied by Onoe [47].

Bar Resonators and Coupling Elements

Flexural-mode resonators are also used at low frequencies. In the 500-Hz-to-50-kHz frequency range, the resonators are in the form of bars such as those shown in Fig. 8. Analysis of the resonant frequencies of thick flexural-bar resonators has been performed by Mason [45] and Näser [48]. If the resonator is treated as a thin bar, equations similar to those describing a transmission line can be written as [15], [49]

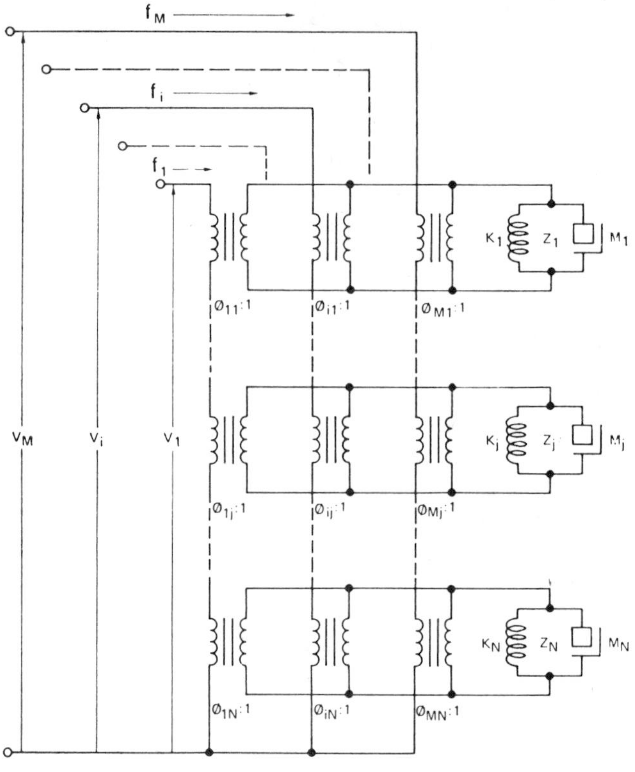

Fig. 19. Generalized schematic of an M-port N-node resonator.

$$
\begin{aligned}
H_1 &= S\cdot s & H_6 &= S\cdot c + C\cdot s & S &= \sinh\alpha\\
H_2 &= C\cdot c & H_7 &= s + S & s &= \sin\alpha\\
H_3 &= C\cdot c - 1 & H_8 &= s - S & C &= \cosh\alpha\\
H_4 &= C\cdot c + 1 & H_9 &= c + C & c &= \cos\alpha\\
H_5 &= S\cdot c - C\cdot s & H_{10} &= c - C
\end{aligned}
$$

$$\alpha^4 = (\rho A/K)\omega^2 l^4 \qquad K\alpha^2/(j\omega l^2) = (\rho AK)^{1/2}/j.$$

$$
\begin{bmatrix} F_1\\ M_1\\ V_1\\ \dot\theta_1 \end{bmatrix}
=
\begin{bmatrix}
H_9 & -H_8(\alpha/l) & -H_7(K\alpha^3/j\omega l^3) & -H_{10}(K\alpha^2/j\omega l^2)\\
H_7(\alpha/l) & H_9 & H_{10}(K\alpha^2/j\omega l^2) & -H_8(K\alpha/j\omega l)\\
H_8(j\omega l^3/K\alpha^3) & H_{10}(j\omega l^2/K\alpha^2) & H_9 & -H_7(l/\alpha)\\
-H_{10}(j\omega l^2/K\alpha^2) & H_7(j\omega l/K\alpha) & H_8(\alpha/l) & H_9
\end{bmatrix}
\begin{bmatrix} F_2\\ M_2\\ V_2\\ \dot\theta_2 \end{bmatrix}
\tag{3}
$$

In (3), which relates to Fig. 21(a), K represents the product of Young's modulus times the moment of inertia of the cross section of the bar, and ρ, A, and l are, respectively, the density, cross sectional area, and length of the bar.

Using (3), we can represent a flexural-mode resonator by the equivalent circuit shown in Fig. 21. This equivalent circuit represents a one-resonator two-port system similar to that shown in Fig. 18(c). Note that in this case the representation (where force is an across variable) is used resulting in a dual formulation where the resonator is represented by a series-tuned circuit. In addi-

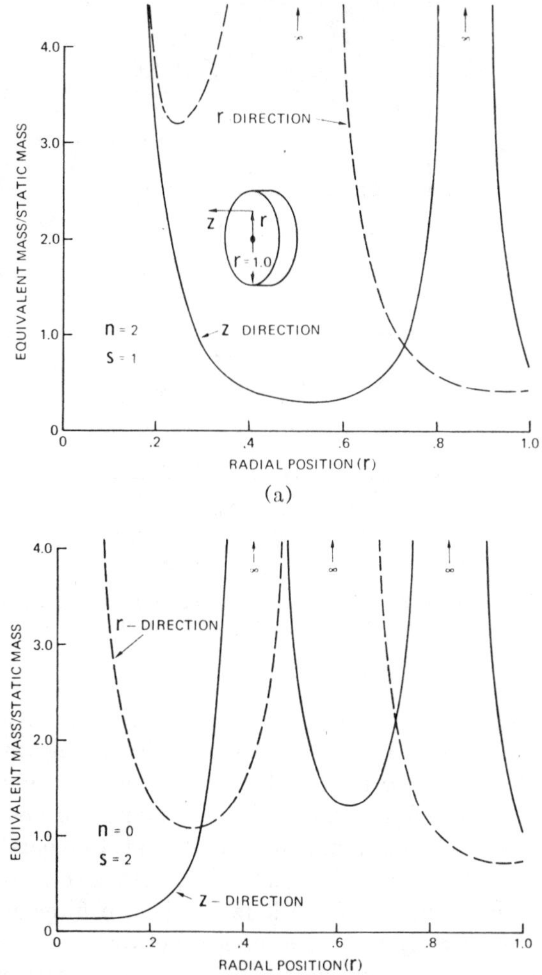

(a)

(b)

Fig. 20. Disk resonator equivalent mass in the radial r and axial z directions for the (a) 2-nodal diameters and 1-nodal circle, and (b) 2-nodal circle cases.

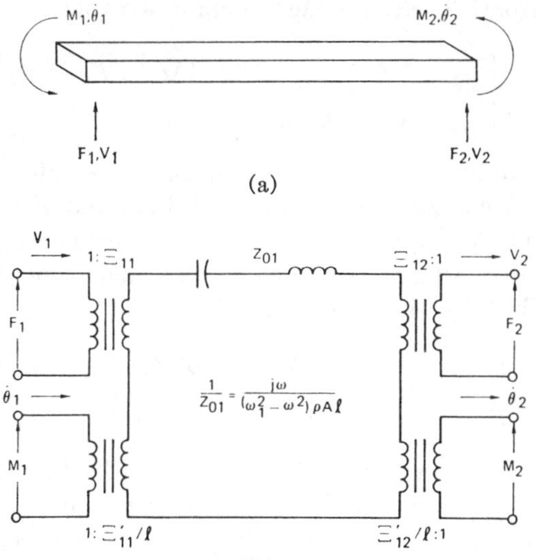

(a)

(b)

Fig. 21. Flexure mode bar. (a) Pictorial diagram. (b) Classical analogy (force-voltage) equivalent circuit representation.

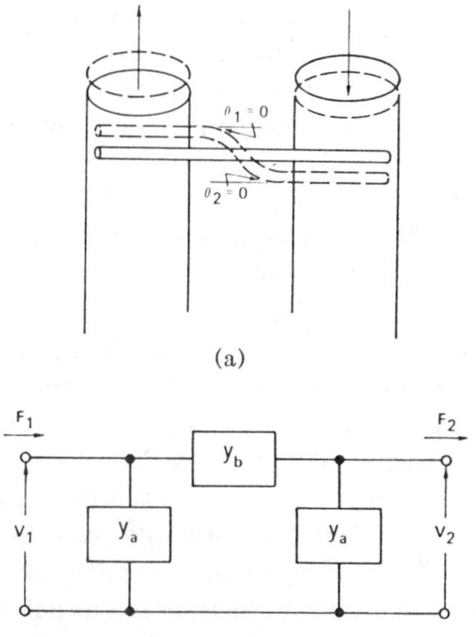

(a)

(b)

Fig. 22. (a) Longitudinal bar-resonator wire-coupling element (case of zero rotation). (b) π equivalent circuit.

tion, the two ports or points on the resonator are actually represented in the equivalent circuit by four ports, two for linear motion and two for rotation. As in the earlier disk case, a general equivalent circuit can be drawn without difficulty by parallel connecting networks of the form shown in Fig. 21(b) [49]. Konno's normalized function Ξ is similar to the ϕ function of (2) when the impressed bending moments are equal to zero.

The equivalent circuits of Fig. 21(b) and (3) are very useful in the case of a rod-wire filter where the rod resonator vibrates in an extensional mode [50]. Fig. 22(a) shows a coupling wire attached to the ends of two resonators. The coupling wire is driven in flexure, but there is no rotation of the wire at the points of contact with the resonator, i.e., $\theta_1 = \theta_2 = 0$. After some manipulation of (3), we can write

$$\begin{bmatrix} V_1 \\ F_1 \end{bmatrix} = \begin{bmatrix} H_6/H_7 & j(\omega l^3/\alpha^3 K)H_3/H_7 \\ (1/j)(\alpha^3 K/\omega l^3)(2H_1/H_7) & H_6/H_7 \end{bmatrix} \begin{bmatrix} V_2 \\ F_2 \end{bmatrix}. \quad (4)$$

The $ABCD$ matrix of (4) can be transformed to a y matrix, which, in turn, can be used to calculate y_a and y_b of the π-equivalent circuit of Fig. 22(b). We find that

$$\begin{aligned} Y_a &= j(K\alpha^3/\omega l^3)[(H_6 + H_7)/H_3] \\ Y_b &= -j(K\alpha^3/\omega l^3)(H_7/H_3). \end{aligned} \quad (5)$$

The coupling wire acts as a quarter-wavelength line when $H_6 = 0$. In this case, we see from (5) that $y_a = -y_b$

and from (4) after some manipulation that [50]

$$\begin{bmatrix} V_1 \\ F_1 \end{bmatrix} = \begin{bmatrix} 0 & j(\omega l^3/\sqrt{2}\,\alpha^3 K) \\ (1/j)(\sqrt{2}\,\alpha^3 K/\omega l^3) & 0 \end{bmatrix} \begin{bmatrix} V_2 \\ F_2 \end{bmatrix}. \quad (6)$$

An analysis of an extensional-mode resonator or coupling wire is somewhat less difficult than that of a flexural element. A rod or wire vibrating in an extensional mode acts as a simple transmission line and, thus, can be described by the *ABCD* matrix

$$\begin{bmatrix} V \\ F_1 \end{bmatrix} = \begin{bmatrix} \cos(\beta l) & jZ_0 \sin(\beta l) \\ j(\sin(\beta l)/Z_0) & \cos(\beta l) \end{bmatrix} \begin{bmatrix} V_2 \\ F_2 \end{bmatrix} \quad (7)$$

where $\beta = \omega\sqrt{\rho/E}$ and $Z_0 = (A\sqrt{\rho E})$. E is Young's modulus. Extensional-mode resonators and coupling elements have been described in considerable detail in [14] and [40].

Resonator Materials

Because of transducer bandwidth limitations, the presence of unwanted modes of vibration and competition with *LC* and ceramic filters, the ratio of bandwidth to center frequency in the case of mechanical filters is usually less than 10 percent but more commonly about 1 percent. This small fractional bandwidth requires the resonators to have a temperature coefficient of frequency of 1–10 ppm/°C, a corresponding low aging rate, and a Q value of at least 10 000.

The major contribution to the variation of frequency in a metallic alloy with temperature is the change in the stiffness or Young's modulus of the material. Iron-nickel alloys that contain either 27 or 44 percent Ni have a low-temperature coefficient of stiffness but are relatively unstable with regard to changes of the percentage of Ni. By adding chromium, the stability can be improved considerably, but the aging characteristics and Q values of these so-called Elinvar materials are not acceptable. The addition of titanium or beryllium to the Fe–Ni–Cr improves both aging and Q and, in addition, makes it possible to vary the temperature coefficient of resonant frequency material by cold work and heat treating [51].

In order to obtain a low temperature coefficient, the Fe–Ni–Cr and Be or Ti material is first solution annealed, then quenched, and then cold worked 15–50 percent. Next, the material is precipitation hardened (Ni is precipitated from a supersaturated solution by the Be or Ti) by heat treatment at 400°–675°C for at least two hours. The amount of cold work and temperature time determines the temperature coefficient of the material. Fig. 23(a) and (b) show temperature curves of Thermalast 5409 (Be) and Ni-Span C (Ti) after having been adjusted for best temperature coefficient. Both materials show frequency shifts of less than 50 Hz over a temperature range of 100°C at 500 kHz.

The addition of Be or Ti has the effect of reducing the aging rate to less than 1×10^{-7} ppm/week or approximately 25 Hz at 500 kHz over a period of 10 years. The Be or Ti also improves the Q of the resonators, which may vary from 10 000 to 25 000, depending on the amount of cold working and precipitation hardening. A value of 20 000 is typical for most applications and causes only a small amount of loss at the passband edges of a 1 percent fractional bandwidth filter. For instance, the response of a 3-kHz bandwidth filter at 200 kHz is practically that of a lossless network.

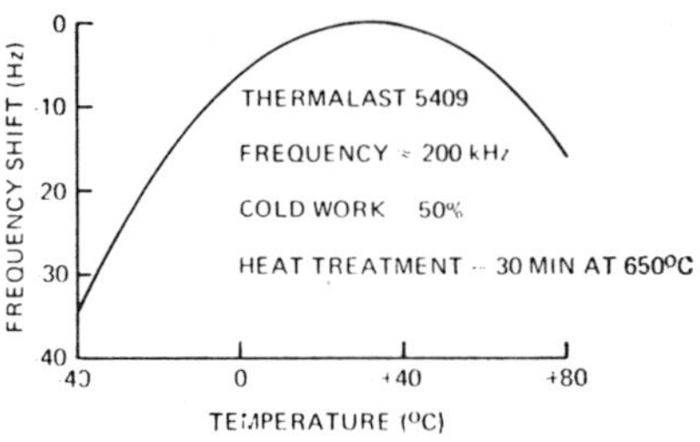

(a)

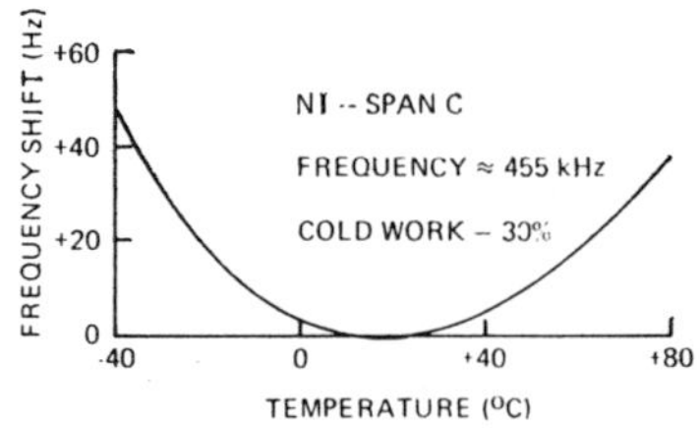

(b)

Fig. 23. Constant-modulus resonator characteristics. (a) Thermalast 5409 torsional rod resonator. (b) Ni-Span *C* flexure mode disk.

When adjusted for the best temperature coefficient, the resonator frequency shifts have little effect on the passband ripple. The passband-ripple variation will, for the most part, be determined by the characteristics of the transducer.

Transducers and Transducer Materials

Transducer Configurations

The most widely used transducers have been the simple magnetostrictive ferrite rod transducer (Fig. 5) and the Langevin ceramic–metal alloy transducer [Fig. 7(b)] for frequencies above 60 kHz and the composite or sandwich-type of ceramic–metal transducer at low frequencies (Fig. 8). Although these are the most popular, filters are being manufactured that use longitudinal mode ceramic rods, iron–nickel alloy wires, and various quartz crystal cuts.

An electromechanical transducer, regardless of the type, can be characterized by its resonant frequency, coupling coefficient, and static reactance. As an example, the magnetostrictive ferrite transducer shown in Fig. 24 can be defined by the mechanical resonant frequency f_1 (actually the mechanical resonance with the electrical terminals open circuited), the electromechanical coupling coefficient, which relates the electrical and mechanical parameters, and the inductance of the transducer coil L_1. If the transducer is directly attached to the end reso-

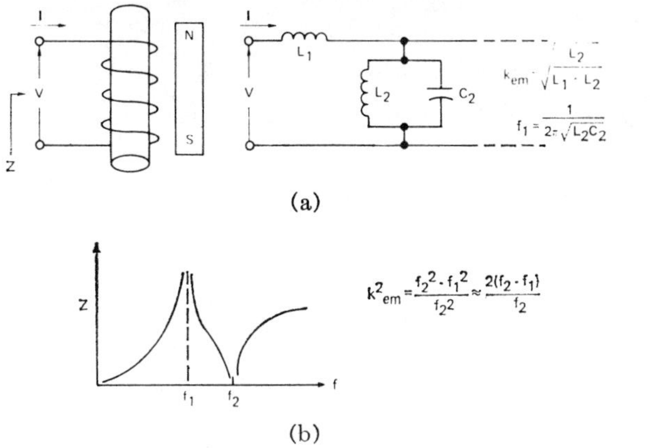

Fig. 24. Magnetostrictive transducer characteristics.

nator, it affects the resonant frequency of the combination in direct proportion to its relative equivalent mass. In the case of the electromechanical coupling coefficient, it is important that it be constant because it directly determines the transformation of the electrical terminating resistance into a mechanical termination that must be matched to the mechanical impedance of the filter. In most applications, the electrical inductance is resonated with a capacitor, which is used for temperature compensation in critical cases.

The filter designer has the task of finding the most stable transducer that will, in addition, properly match the electrical and mechanical networks. This involves choosing a material with a stable coupling coefficient first and then a configuration that reduces the transducer equivalent mass to a minimum so as to reduce the frequency shift with temperature. This can be done by decreasing the diameter of the ferrite rod or the thickness of the ceramic in a Langevin transducer. The size reduction is, in turn, limited by the sensitivity of the coil or static capacity, or in the case of wide-bandwidth filters by the fact that, in the limit, the electrical inductor–capacitor combinations actually become the end resonators.

Magnetostrictive Ferrite Transducers

The most widely used filter transducer material is a cobalt-substituted iron–nickel ferrite, which was specially developed for use in electromechanical filters [52]. Addition of 0.6 percent cobalt results in a highly stable coupling coefficient, whereas 1.0 percent cobalt results in a very small variation of frequency f_1 with temperature. Fig. 25 shows the variation of coupling coefficient and frequency f_2 with change in temperature for a practical design at 200 kHz [53]. In order to be able to reduce the effect of the transducer on the end resonator frequency, optimum values of permanent magnet bias and coil dimensions must be chosen so as to maximize the coupling coefficient. The coupling coefficient of 10 percent shown in Fig. 25, which is typical for a well-designed transducer, is also the limiting value of the filter fractional bandwidth. The coupling coefficient variation is approximately 5 percent, which is equivalent to a 10 percent change in the terminating resistance, which may or may not be excessive depending on the application. Variations of resonant frequency with temperature are on the order of 0.1 percent or approximately 500 Hz at 500 kHz over a temperature range of 100°C. Transducer coil Q varies from 10 to 40 or better, depending on whether a metal alloy or a ferrite shield is used.

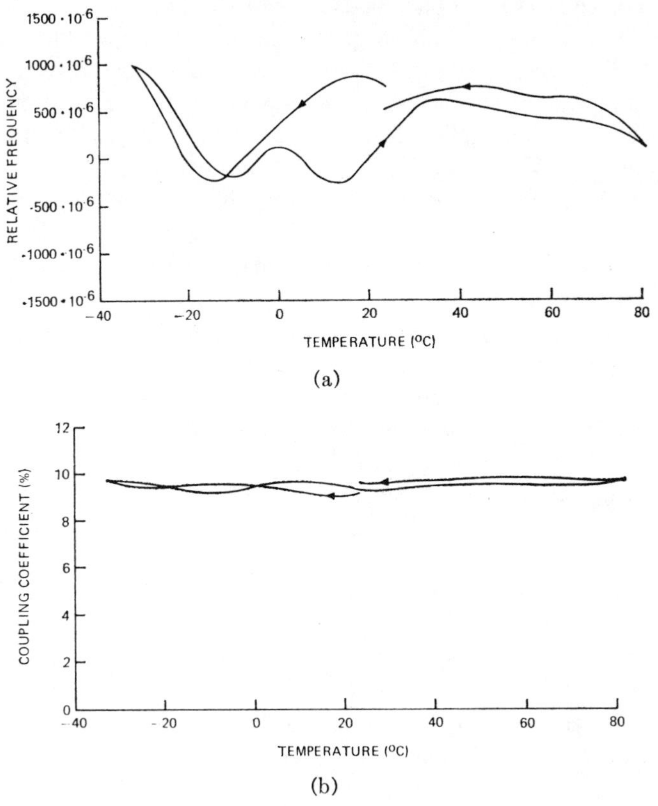

Fig. 25. Characteristics of Telefunken S3 ferrite transducers. (a) Frequency variation. (b) Coupling coefficient variation with change in temperature (longitudinal mode).

Some of the advantages of using magnetostrictive ferrite as a transducer material as opposed to a piezoelectric ceramic, for example, are ease of manufacturing a material with consistent properties, excellent stability of coupling and frequency with time as the result of using a stable permanent magnet bias, and simple capacitive tuning. In the case of high-performance channel filters, excellent matching between the filter and the external circuit can be achieved by the use of a series–parallel combination of tuning (and matching) capacitors. In the case of high-performance piezoelectric ceramic-transducer mechanical filters, transformers are used for both tuning and impedance matching, whereas in the case of inexpensive miniature filters, no tuning is necessary [54].

Piezoelectric Ceramic Transducers

Whereas in the U. S. and Germany, the most early transducer development was centered around magnetostrictive ferrite transducers, in Japan most initial work, in fact most current work, involves piezoelectric ceramic transducers. The barium–titanate material used in early designs suffered from large changes in coupling coefficient and frequency with temperature, which, in turn, resulted in large passband-ripple variations with temperature. The development of lead–zirconate–titanate (PZT) materials [55], which show vastly improved temperature characteristics, has made it possible for piezoelectric ceramic materials not only to compete with, but in a number of cases to replace, magnetostrictive ferrite transducers.

The principal advantage of the piezoelectric ceramic is its excellent electromechanical coupling coefficient. Whereas a bar-type ferrite transducer has a coupling coefficient of 10 percent, a side-plated bar of temperature-stable piezoelectric ceramic has a coupling coefficient between 15 and 25 percent. The higher coupling coefficient therefore makes it possible to use a minimum amount of ceramic material as compared to a highly temperature-stable metal disk alloy in the design of composite end resonator/transducer assemblies. In addition, the higher electrical Q of the piezoelectric material results in lower filter insertion loss and its greater inherent linearity results in lower intermodulation distortion. Fig. 26 shows the variation of frequency and coupling with temperature for the case of a PZT ceramic that has a coupling coefficient k_{13} of approximately 23 percent.

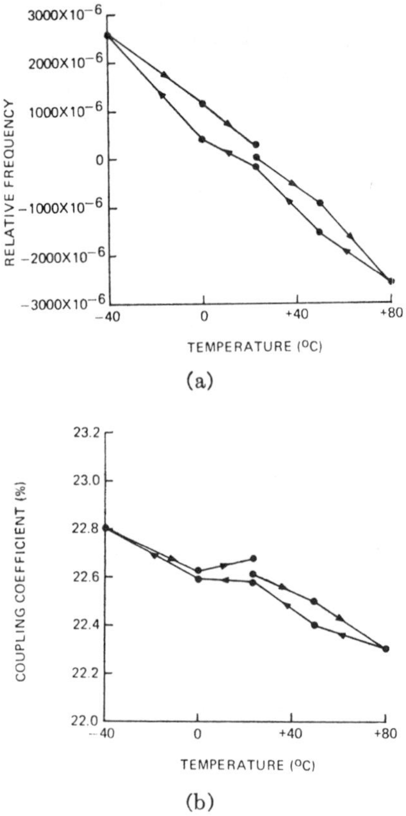

Fig. 26. Characteristics of PZT-type piezoelectric transducers (Murata). (a) Frequency variation. (b) Coupling coefficient variation with change of temperature.

The most widely used of the ceramic transducers is the Langevin type, which is in the form of a rod composed of a ceramic disk sandwiched between two metal alloy rods, Fig. 7(b). The high planar coupling coefficient of a thin disk, which is on the order of 25–40 percent (for stable materials), and the Langevin configuration make it possible to design resonator/transducer assemblies that have high mechanical Q and good temperature characteristics with only a small reduction in the coupling coefficient [54]. Development of this type of transducer for filter applications was started approximately 15 years ago by Tanaka [14] and Tagawa [56] in Japan. Because of the instability and low Q of most bonding materials such as epoxy, coupled with the fact that the bond has an appreciable thickness, the ceramic/metal alloy attachment problem is one of importance and has required a great deal of effort in its solution.

In the past few years, there has been a considerable amount of analytical work done in Germany on Langevin-type transducers [54], [57], and [58]. This work includes equations that describe the variation of coupling, frequency, Q, and temperature coefficient with changes in the relative thickness and position of the ceramic disk and has been applied to some new filter types at 455 kHz.

One of the most important characteristics of piezoelectric transducers is the capability of operating in various modes of vibration, as well as use in a large number of mechanical configurations.

A good example is shown in Fig. 8 where flexure modes are excited by composite ceramic/metal alloy transducers. In addition, the high electromechanical coupling coefficient of the ceramic makes it possible to design very stable wide-bandwidth (10–20 percent) low-frequency filters of this type.

A great deal of analytical work has been accomplished in Japan in the area of describing the electromechanical characteristics of composite bending transducer resonators. Konno's and Kusakabe's published work on this particular subject alone totals more than 200 pages and includes twisted bars, mass loaded bars, and bars that vibrate in nondegenerate modes, as well as simple bars driven with ceramic plates that only partially cover the upper and lower surfaces [59]. A very detailed analysis of a fully covered bar has been achieved by Okamoto *et al.* [60], and is summarized in [19].

Looking to the Future

We will look at the next few years based on the direction of our present technology. There will be breakthroughs of course such as the monolithic filter of some years ago, but these are difficult to predict, even if one is playing an active part in the development. Some of the more predictable areas are size reduction, an expanded use of low-frequency filters, an improvement of delay and ripple characteristics, the realization of multiresonator monolithic filters with finite attenuation poles, as well as various improvements in the discrete element mechanical filter as a result of the large amount of effort being expended on monolithic filters.

Size Reduction

In terms of mechanical filters, size reduction is still a relatively unexplored technology. In the case of disk-wire mechanical filters a size reduction from 5.0–1.3 cm^3 was made with relatively little effort and no breakthroughs in technology other than the use of a lower order mode of vibration. By making use of ceramic transducers and small diameter disks the volume can be reduced by 2:1 without a great deal of difficulty. At the present time, the longitudinal-mode bar, flexural-coupling rod design independently developed by Börner [54] and Konno [18], (see Fig. 27) is packaged in less than 1 cm^3. Although it may be somewhat more difficult (in comparison with the disk-wire filter) to reduce the size of this design because of the fixed length of the resonators, it is possible to do so by decreasing the diameter of the rods, but only to the point where flexural modes become a problem. Discrete element filters such as the disk or rod types actually have an advantage over monolithic structures (Fig. 28) in that relatively large plating surfaces are not needed to reduce both the filter impedance level and the effects of stray capacity. Disk and rod-type filters are low-impedance devices because of the high dielectric constant of PZT transducer materials, or in the case of magnetostrictive ferrites, the use of low-turn coils. In addition, only two transducers are needed, resonator spacing is commonly 0.025 cm regardless of filter bandwidth, and none of the volume is used for the purpose of reducing reflections from a boundary and very little is used for support.

It must be said that there are also some basic limitations as to how small we can build a discrete component mechanical filter. One, of course, is that of fabricating the parts and assembling them without excessive frequency shift and miscoupling. A second limitation, as we discussed earlier, is that of spurious modes due to having to fix one of the dimensions thus decreasing the frequency spacing of nearby bending modes [54]. Another size-reduction limitation relates to the nonlinearity of the transducer and resonator materials as a function of size. Yakuwa has made a very interesting comparison of ferrite core inductors, piezoelectric ceramics, nickel alloys, and quartz, showing the relationship between various loss factors and minimum size for a specific bandwidth filter [61]. A final consideration is the three-dimensional shape of most discrete element mechanical filters, which not only prevents the use of planar manufacturing processes but ultimately may be the most serious volume-reduction limitation.

Low Frequencies

The large amount of development work on low-frequency filters in Japan will most probably be felt in other parts of the world. Fig. 29 shows, for example, some of the wide variety of filters being manufactured by Fujitsu Ltd. Most applications of low-frequency mechanical filters involve selecting single tones, which in turn means that only one or two resonators are needed. Therefore, the selectivity is dependent solely on resonators that are coupled to the external mounting structure. This results in lower resonator Q and microphonic responses due to external vibrations.

The microphonics problem has been solved to some degree by the use of highly damped supports. This has made it possible to use fundamental modes of vibration,

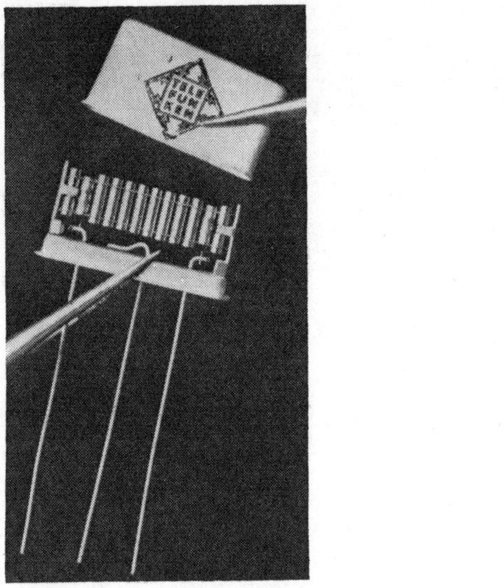

Fig. 27. Miniature 455-kHz rod-wire mechanical filter (1 cm^3).

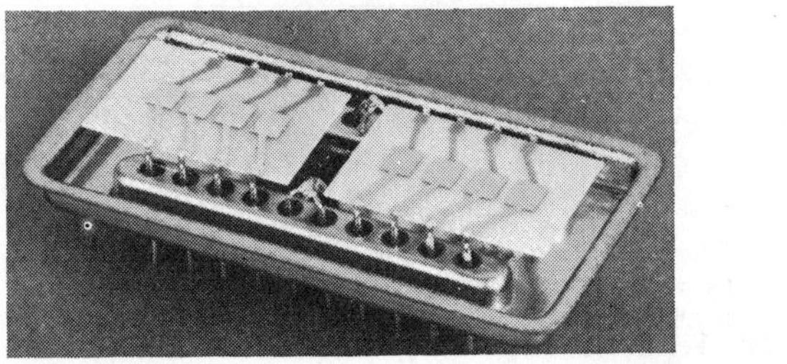

Fig. 28. Eight-pole monolithic quartz mechanical filter.

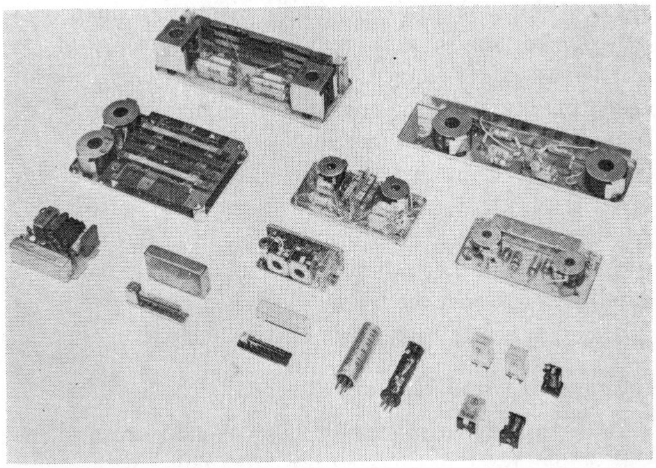

Fig. 29. Low-frequency mechanical filters (Fujitsu).

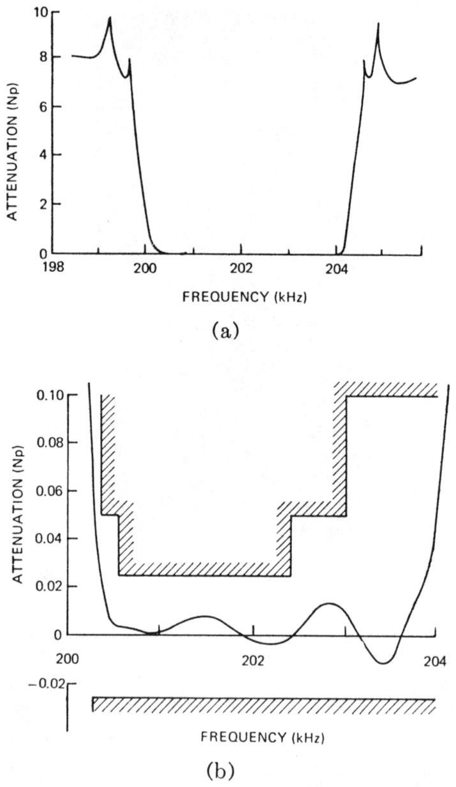

Fig. 30. Measured (a) stopband amplitude and (b) passband response of a high-performance rod-wire mechanical filter that employs double resonator bridging of the type shown in Fig. 14 (Telefunken).

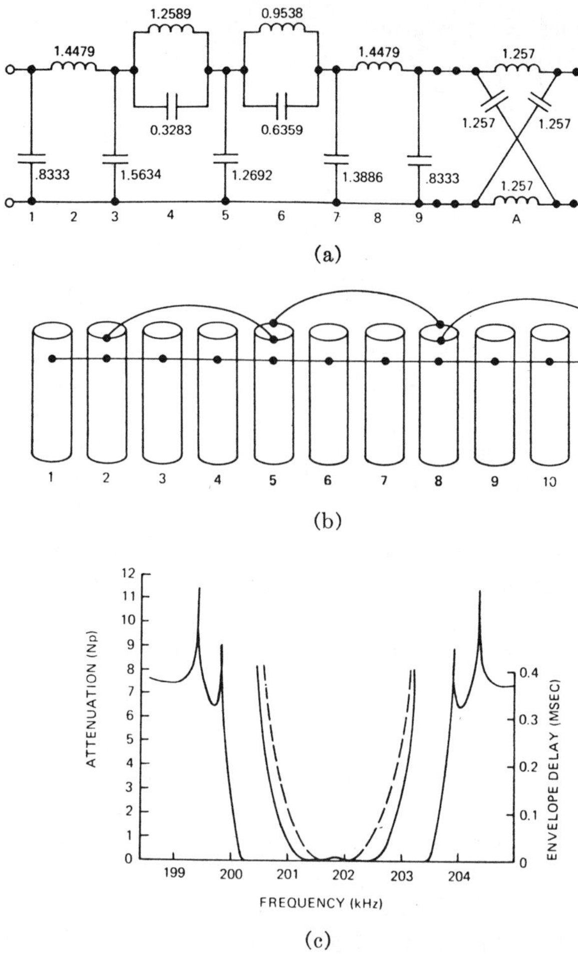

Fig. 31. Eleven resonator double bridging-wire filter with delay equalization. (a) Low-pass prototype. (b) Pictorial diagram of the torsional resonator longitudinal coupling-wire realization. (c) Calculated amplitude and delay response (dashed line corresponds to a nine-resonator design without delay correction).

which reduce both the complexity of the filter and its physical size. The problem of minimizing unwanted outputs due to impulse noise may be solved to some degree by the use of *H*-shaped resonators [62]. This type of filter consists of two identical vertical masses connected by a flexible web (the horizontal element of the *H*). A piezoelectric ceramic transducer causes the web to flex and the masses to rotate. Each mass is supported near its centroid, thus preventing the web from flexing when subjected to a shock impulse through the supports. A detailed analysis of the frequency response characteristics of this device has been made by Konno [63].

Improvement of Delay and Passband Ripple

It is possible to internally delay equalize a mechanical ladder filter by making use of bridging wires [25]. This has proved to be a very practical method [64], [65]. This design is based on the use of very exact quarter-wavelength long coupling wires. No frequency adjustments are needed after assembling the pretuned resonators, in the case of channel filters with pairs of attenuation poles (Fig. 30 and [66]) and in the case of channel filters with delay equalization (Fig. 31). Another practical method is to use mechanical resonators and coupling elements in an electrical lattice structure in much the same way as quartz resonators are used [67]. The work on lattice filters has resulted in some new computer-aided tuning techniques, which have also been reported by Yano [68]. In order to further improve the excellent response of voice channel filters, it will be advantageous in some cases to make use of on-line computers in the factory. The development of practical computer-assisted tuning methods may result from work presently being done on quartz monolithic mechanical filters.

Quartz Monolithic Mechanical Filters

In much the same way as the discrete element mechanical filter may benefit from monolithic mechanical filter technology, a number of concepts such as acoustic bridging will be applied to the monolithic filter. At the present time, attenuation poles are realized in two-resonator structures through capacitive bridging. The two-resonator sections are capacitively coupled to form a multiresonant filter having a somewhat equal-ripple response [69]. It is probable that exact designs with four or more resonators per plate and electrical bridging will soon become available.

It has already been shown that thickness modes can be trapped in much the same way as the shearing modes [70], which leads us to suspect that other new configurations will be developed in the next few years.

A large amount of technology has been devoted to the monolithic mechanical filter manufacturing process, for instance in areas such as the use of a laser to vary the coupling between electrode pairs and the fabrication of extremely flat and parallel crystal blanks. It is certain that this work will continue for a long period of time and should be applicable to mechanical filter technology in general.

Developments in Other Countries

Mechanical filter development outside of Japan, West Germany, and the U. S. has primarily been concentrated on distributed line filters of the type shown in Fig. 7(a). This has been the case in Great Britain [71], Poland [72], the USSR [73], and East Germany, where they have been manufactured for some years. Recent papers published in England have reflected an interest in transducer design [74] and low-frequency flexure mode resonators [75]. A study of flexure modes in mechanical filters is also a subject of interest in Czechoslovakia [76]. At this time, channel filters of the configuration of Fig. 5(b) are manufactured in East Germany (200 kHz) [77] and in Czechoslovakia (64–108 kHz) [78].

References

[1] R. Adler, "Compact electromechanical filters," *Electronics*, vol. 20, Apr. 1947, pp. 100–105.

[2] W. van B. Roberts and L. L. Burns, "Mechanical filters for radio frequencies," *RCA Rev.*, vol. 10, Sept. 1949, pp. 348–365.

[3] M. L. Doelz and J. C. Hathaway, "How to use mechanical I-F filters," *Electronics*, vol. 26, Mar. 1953, pp. 138–142.

[4] G. A. Campbell, "Physical theory of the electric wave-filter," *Bell Syst. Tech. J.*, vol. 1, Nov. 1922, pp. 1–32.

[5] R. V. L. Hartley, "Frequency selective transmission system," U.S. Patent 1 654 123, Dec. 1927.

[6] J. P. Maxfield and H. C. Harrison, "Methods of high quality recording and reproducing of music and speech based on telephone research," *Bell Syst. Tech. J.*, vol. 5, July 1926, pp. 493–523.

[7] W. P. Mason, "Wave transmission network," U.S. Patent 2 345 491, Mar. 1944.

[8] R. A. Sykes and W. D. Beaver, "High frequency monolithic filters with possible application to single frequency and single side band use," *Proc. Annu. Frequency Control Symposium*, Apr. 1966, pp. 288–308.

[9] Y. Nakazawa, "High frequency crystal electromechanical filter," *Proc. Annu. Frequency Control Symposium*, Apr. 1962, pp. 373–390.

[10] M. Börner and H. Schüssler, "Piezoelectric plate filter," West German Patent 1 274 675, Sept. 1964.

[11] M. L. Doelz, "Electromechanical filter," U.S. Patent 2 615 981, Oct. 1952.

[12] M. Börner, E. Kettel, and H. Ohnsorge, "Mechanische filter für die nachrichtentechnik," *Telefunken J.*, vol. 31, June 1958, pp. 105–114.

[13] M. Börner, E. Dürre, and H. Schüssler, "Mechanische Einseiten-bandfilter," *Telefunken J.*, vol. 36, May 1963, pp. 272–280.

[14] T. Tanaka and T. Inoguchi, "Studies on electronics materials and their applications," Division of Electronics Materials, Institute for Chemical Research, Kyoto Univ., Kyoto, Japan, 1959, pp. 22–25.

[15] M. Börner, "Biegeschwingungen in mechanischen filtern," *Telefunken J.*, vol. 31, June 1958, pp. 115–123; also, Sept. 1958, pp. 188–196.

[16] W. P. Mason and R. N. Thurston, "Compact electromechanical bandpass filter for frequencies below 20 kilocycles," *IRE Trans. Ultrason. Eng.*, vol. UE-7, June 1960, pp. 59–70.

[17] M. Konno, C. Kusakabe, and Y. Tomikawa, "Electromechanical filter composed of transversely vibrating resonators for low frequencies," *J. Acoust. Soc. Amer.*, Apr. 1967, pp. 953–961.

[18] M. Konno, "Theoretical considerations of mechanical filters" (in Japanese), *J. Inst. Elec. Commun. Eng. Jap.*, vol. 40, 1957, pp. 44–51.

[19] K. Yakuwa, "Characteristics of an electromechanical filter using piezolectric composite bending vibrators," *Electron. Commun. Jap.*, vol. 48, Nov. 1965, pp. 206–215.

[20] H. Albsmeier and W. Poschenrieder, "Electromechanical wave filter having resonant bars coupled to each other by torsion wires which also support bars," U.S. Patent 3 142 027, July 1964.

[21] K. Takahashi, "Low-frequency mechanical filters with U- or I-shaped couplers and resonators," *Electron. Commun. Jap.*, vol. 48, Nov. 1965, pp. 199–205.

[22] H. Yoda, "Quartz crystal mechanical filters," *Proc. Annu. Frequency Control Symposium*, May 1959.

[23] W. Poschenrieder and F. Schöfer, "Das elektromechanische quarzfilter-ein neues bauelement für die nachrichtentechnik," *Frequenz*, vol. 17, Mar. 1963, pp. 88–94.

[24] R. Adler, "Magnetostrictively driven mechanical wave filter," U.S. Patent 2 501 488, Mar. 1950.

[25] M. Börner, "Mechanische filter mit dämpfungspolen," *Arch. Elek. Übertragung*, vol. 17, Mar. 1963, pp. 103–107.

[26] R. A. Johnson, "A single-sideband disk-wire type mechanical filter," *IEEE Trans. Component Parts*, vol. CP-11, Dec. 1964, pp. 3–7.

[27] R. A. Johnson and R. J. Teske, "A mechanical filter having general stopband characteristics," *IEEE Trans. Sonics Ultrason.*, vol. SU-13, July 1966, pp. 41–48.

[28] B. Kohlhammer and H. Schüssler, "Berechnung allgeneiner mechanischer koppelfilter mit hilfe von äquivalenten schaltungen aus konzentrierten elektrischen schaltelementen", *Ber. AEG-Telefunken*, vol. 41, 1968, pp. 150–159.

[29] S. B. Cohn, "Direct-coupled-resonator filters," *Proc. IRE*, vol. 45, Feb. 1957, pp. 187–195.

[30] R. Saal and E. Ulbrich, "On the design of filters by synthesis," *IRE Trans. Circuit Theory*, vol. CT-5, Dec. 1958, pp. 284–327.

[31] R. A. Johnson, "Application of electro-acoustic models to the design of a miniaturized mechanical filter," *Rep. 6th Int. Congress on Acoustics*, vol. 4, Aug. 1968, pp. 93–96.

[32] I. Lucas, "Plattenförmige elektromechanische filter mit dämpfungspolen," *Arch. Elek. Übertragung*, vol. 17, 1963, pp. 249–255.

[33] J. W. S. Rayleigh, *The Theory of Sound*. New York: Dover, 1945, pp. 335–339.

[34] R. A. Johnson, "A twin tee multimode mechanical filter," *Proc. IEEE* (Corresp.), vol. 54, Dec. 1966, pp. 1961–1962.

[35] M. Konno, Y. Tomikawa, T. Takano, and H. Izumi, "Electromechanical filters using degeneration modes of a disk or ring," *Electron. Commun. Jap.*, vol. 52A, May 1969, pp. 19–28.

[36] K. Nakamura, H. Shimizu, and H. Watanabe, "Multi-channel electro-mechanical filters using orthogonal modes in a single mechanical system," *Rep. 6th Int. Congress on Acoustics*, vol. 4, Aug. 1968, pp. 109–112.

[37] M. Konno and Y. Tomikawa, "An electromechanical filter consisting of a flexural vibrator with double resonances," *Electron. Commun. Jap.*, vol. 50, Aug. 1967, pp. 64–73.

[38] F. Kündemund and K. Traub, "Mechanische filter mit Biegeschwingern," *Frequenz*, vol. 18, Sept. 1964, pp. 277–280.

[39] M. Konno, Y. Tomikawa, and H. Izumi, "Electro-mechanical filter consisting of a multiplex mode vibrator," *Rep. 6th Int. Congress on Acoustics*, vol. 4, Aug. 1968, pp. 117–120.

[40] R. A. Johnson, "Electrical circuit models of disk-wire mechanical filters," *IEEE Trans. Sonics Ultrason.*, vol. SU-15, Jan. 1968, pp. 41–50.

[41] H. Deresiewicz and R. D. Mindlin, "Axially symmetric flexural vibrations of a circular disk," *J. Appl. Mech.*, vol. 22, Mar. 1955, pp. 86–88.

[42] R. L. Sharma, "Dependence of frequency spectrum of a circular disk on Poisson's ratio," *J. Appl. Mech.*, vol. 24, Mar. 1957, pp. 53–54.

[43] M. Onoe and T. Yano, "Analysis of flexural vibrations of a circular disk," *IEEE Trans. Sonics Ultrason.*, vol. SU-15, July 1968, pp. 182–185.
[44] R. L. Sharma, "Equivalent circuit of a resonant, finite, isotropic, elastic circular disk," *J. Acoust. Soc. Amer.*, vol. 28, Nov. 1956, pp. 1153–1158.
[45] W. P. Mason, *Electromechanical Transducers and Wave Filters.* New York: Van Nostrand, 1948, pp. 291–297.
[46] E. Frymoyer, J. Klovstad, and R. A. Johnson, "Equivalent mass of a thick vibrating disk," unpublished.
[47] M. Onoe, "Contour vibrations of isotropic circular plates," *J. Acous. Soc. Amer.*, vol. 28, Nov. 1956, pp. 1158–1162.
[48] J. Näser, "Exakte berechnung der biegeresonanzen rechteckiger und zylindrischer stäbe," *Hochfreq. und Elektroakust.*, vol. 74, 1965, pp. 30–36.
[49] M. Konno and H. Nakamura, "Equivalent electrical network for the transversely vibrating uniform bar," *J. Acoust. Soc. Amer.*, vol. 38, Oct. 1965, pp. 614–622.
[50] M. Börner, "Mechanische filter mit biegekopplung," *Arch. Elek. Übertragung,* vol. 15, Dec. 1961, pp. 175–180.
[51] ——, "Magnetische werkstoffe in electromechanischen resonatoren und filtern," *IEEE Trans. Magnetics,* vol. MAG-2, Sept. 1966, pp. 613–620.
[52] C. M. van der Burgt, "Performance of ceramic ferrite resonators as transducers and filter elements," *J. Acoust. Soc. Amer.*, vol. 28, Nov. 1956, pp. 1020–1032.
[53] S. Schweitzerhof, "Über ferrite für magnetostriktive schwinger in filterkreisen," *Nachrichtentech Z.*, vol. 11, Apr. 1958, pp. 179–185.
[54] M. Börner and H. Schüssler, "Miniaturisierung mechanischer Filter," *Telefunken J.*, vol. 37, 1964, pp. 228–246.
[55] D. Berlincourt, B. Jaffe, H. Jaffe, and H. H. A. Krueger, "Transducer properties of lead titanate zirconate ceramics," *IRE Int. Convention Rec.*, vol. 7, pt. 6, pp. 227–240.
[56] Y. Tagawa and T. Hatano, *Design of Mechanical Filter and Crystal Filter Circuits* (in Japanese). Tokyo: Ohm, 1964, pp. 11–12.
[57] H. Schüssler, "Mechaniche Filter mit piezoelektrischen Wandlern," *Telefunken J.*, vol. 39, 1966, pp. 429–439.
[58] H. Albsmeier, "Über Antriebe für elektromechanische bandpässe mit piezoelektrischen Wandlern," *Frequenz,* vol. 19, Apr. 1965, pp. 125–133.
[59] C. Kusakabe, M. Konno, and Y. Tomikawa, "Resonant frequencies of transversely vibrating bar excited by electrostrictive transducer," *Electron. Commun. Jap.*, vol. 48, Nov. 1965, pp. 186–191.
[60] T. Okamoto, K. Yakuwa, and S. Okuta, "Low frequency electromechanical filter," *Fujitsu Sci. Tech. J.*, vol. 2, May 1966, pp. 53–86.
[61] K. Yakuwa and S. Okuta, "Miniaturization of the mechanical vibrator used in an electro-mechanical filter," *Rep. 6th Int. Congress on Acoustics,* vol. 4, Aug. 1968, pp. 97–100.
[62] H. Baker and J. R. Cressey, "H-shaped resonators signal upturn in tone telemetering," *Electronics,* Oct. 1967, pp. 99–106.
[63] M. Konno, K. Aoshima, and H. Nakamura, "H-shaped resonator and its application to electro-mechanical filters" (in Japanese), *Eng. Bull. Yamagata Univ.*, vol. 10, Mar. 1969, pp. 261–285.
[64] B. Kohlhammer, "Entwurf kanonischer koppelnetzwerke mit Anwendungen auf mechanische Filter," Ph.D. dissertation, Technische Hochschule, München, Germany, 1970.
[65] ——, "Berechnung, Bau und Untersuchung mechanischer polfilter," Diplom-Arbeit, Institut für HF-Technik und HF-Physik der Technischen Hochschule Karlsruhe, Munich, Germany, 1966.
[66] H. Schüssler, "Filter mit mechanischen resonatoren," *Bull. Schweizerischer Electrotechn. Verein* (Zurich), Mar. 1969, pp. 216–222.
[67] R. A. Johnson, "New single sideband mechanical filters," *1970 WESCON Tech. Papers,* Aug. 1970, pp. 1–10.
[68] K. Shibayama, "Electromechanical filters," *Electron. Commun. Jap.*, vol. 48, Nov. 1965, pp. 66–72.
[69] H. Yoda, Y. Nakazawa, and N. Kobori, "High frequency crystal monolithic (HCM) filters," *Proc. 23rd Annu. Symp. Frequency Control,* May 1969, pp. 76–92.
[70] S. Fujishima, S. Nosaka, and I. Ishiyama, "10 MC ceramic filters by trapped energy modes," *Proc. Acoust. Soc. Jap.*, Nov. 1966, p. 3.
[71] W. Struszynski, "A theoretical analysis of the torsional electron-mechanical filters," *Marconi Rev.*, 3rd quart., vol. 22, 1956, pp. 119–143.
[72] F. Kaminski, "The synthesis of electromechanical chain filters by means of equivalent circuits with distributed parameters" (in Polish), *Rozpr. Elektrotech.*, vol. 15, 1969, pp. 717–749.
[73] A. K. Losev, "Filters with multielement coupling," *Telecommun. Radio Eng.* (USSR), Jan. 1964, pp. 1–8.
[74] R. P. Walters, "Balanced action toroidal transducer for electromechanical filters," *Electron. Eng.*, Mar. 1969, pp. 238–245.
[75] Y. Kagawa and G. M. L. Gladwell, "Finite element analysis of flexure-type vibrators with electrostrictive transducers," *IEEE Trans. Sonics Ultrason.*, vol. SU-17, Jan. 1970, pp. 41–49.
[76] Z. Faktor, "Spurious vibrations in component parts of electromechanical filters and their investigation" (in Czech), *Slaboprodý Obzor,* vol. 30, 1969, pp. 444–450.
[77] L. Brier, "Der Entwurf von HF-Bandfiltern und mechanischen Filtern mit Dämpfungspolen nach dem Betriebsparameter verfahren," *Internationales Wissenschaftliches Kollegium,* Techniche Hochschule, Sept. 1965.
[78] J. Jungwirth, "Synteźa elektromechanických filtru podle teorie provoznich parametru" (in Polish), *Sborník Pracerýz-kumného Astavu Telemunikací Praha,* 1963, pp. 177–199.
[79] J. C. Hathaway and D. F. Babcock, "Survey of mechanical filters and their applications," *Proc. IRE,* vol. 45, Jan. 1957, pp. 5–16.
[80] M. Konno, "Electromechanical filters," *J. Acoust. Soc. Jap.* (in Japanese), vol. 14, 1958, pp. 241–262.
[81] ——, "Recent Electromechanical Filters," *J. Acoust. Soc. Jap.* (in Japanese), vol. 18, 1962, pp. 327–345.
[82] M. Börner, "Progress in electromechanical filters," *Radio Electro. Eng.*, vol. 29, Mar. 1965, pp. 173–184.
[83] M. Konno and Y. Tomikawa, "Electro-mechanical filters—Part 1, introduction" (in Japanese), *J. Inst. Electron. Commun. Eng. Jap.*, vol. 52, Mar. 1969, pp. 303–312.
[84] K. Yakuwa, "Electromechanical filters, Part 2, high frequency electromechanical filters" (in Japanese), *J. Inst. Electron. Commun. Eng. Jap.*, vol. 52, May 1969, pp. 568–577.
[85] T. Yuki and T. Yano, "Electro-mechanical filters—Part 13: Low frequency electro-mechanical filters" (in Japanese), *J. Inst. Electron. Commun. Eng. Jap.*, vol. 52, June 1969, pp. 727–732.
[86] G. S. Moschytz, "Inductorless filters: A survey—Part I: Electromechanical filters," *IEEE Spectrum,* vol. 7, Aug. 1970, pp. 30–36.

Mechanical Filters and Resonators

ROBERT A. JOHNSON, MEMBER, IEEE, AND ALFHART E. GUENTHER, SENIOR MEMBER, IEEE

I. INTRODUCTION

The mechanical filter is essentially a narrow bandwidth device with high resonator Q's and good temperature stability. Within the frequency range of application of mechanical filters (approximately a few hundred Hertz to 500 kHz), the 10 000–20 000 values of Q and 1 ppm/°C temperature characteristics of the resonators compare favorably to those of quartz. By a proper choice of center frequency, mechanical configuration, and process techniques, it is possible to manufacture mechanical filters with excellent passband and stopband characteristics on the same order as those of high-performance crystal-lattice filters. Filters are designed with 60dB–3dB bandwidth ratios as low as 1.3/1 and passband amplitude variations of ±0.1 dB.

Mechanical filters were first designed as IF filters for high performance receivers in the late 1940's [1]. Ten years later they were used as channel filters in telephone multiplex equipment [2]; this being about the time that work was started on low-frequency flexure-mode filters for signaling applications. Although the greatest demand for low-frequency mechanical filters is in the telecommunications field (including use in frequency-shift keying (FSK) modems), other applications such as sonar combsets, signaling for railroads and rapid transit systems, navigation equipment, and telemetry have developed.

To describe a mechanical filter is to describe a multitude of devices ranging from single-resonator tuning forks to 12-resonator disk-wire filters with attenuation-pole producing bridging wires [3]. Common to all mechanical filters are input and output electromechanical transducers, as well as acoustic (mechanical) coupling between the resonators when two or more resonators are used. So as to differentiate a mechanical filter from a monolithic filter, the coupling is assumed to be through a wire or "neck," rather than a nonelectroded region of a plate.

Transducers are most often piezoelectric ceramics or magnetostrictive ferrites, although quartz and iron-nickel alloy wires are also used on some filters. The transducer may be separate and act as an end mechanical resonator, or it may be directly attached to a constant modulus metal alloy to form a composite resonator. Resonators, within the filter structures, have various shapes such as bars, rods, plates, or disks and vibrate in extensional, torsional, flexural, and contour modes. The internal resonators are most often iron-nickel alloys with Be, Cr, Ti, or Mo additives used to improve the frequency versus temperature characteristics. Most often small diameter wires are used to couple adjacent resonators. Coupling wires are also used to bridge (or leapfrog) one or two resonators in order to realize attenuation poles both on and off the $j\omega$-axis so as to improve the stopband selectivity and the differential delay response. The coupling wire mode of vibration may also be torsional, extensional, or flexural, or some combination of these modes.

Fig. 1(a) shows a mechanical filter in block diagram form. The filter is terminated in a source and load resistance, while any shunt capacitance is considered to be part of the filter. The transducer may or may not be tuned, depending on the design concept used. The electromechanical conversion element is usually represented by a gyrator, in the piezoelectric case or an ideal transformer when a magnetostrictive transducer is used. Also included in the block diagram is the mechanical network composed of the resonators and coupling wires. Fig. 1(b) shows an electromechanical schematic diagram of a piezoelectric transducer filter with electrical circuit tuning and mechanical bridging.

R. A. Johnson is with the Collins Radio Group, Rockwell International, Newport Beach, Calif. 92663.
A. E. Guenther is with the Zentrallaboratorium für Nachrichtentechnik, Siemens AG, Munich, Germany.

II. ACOUSTIC AND EQUIVALENT CIRCUIT DESCRIPTIONS

Because of the remarkable analogy between mechanical and certain electrical structures, it is advantageous to develop electrical equivalent circuits. The mechanical filter designer then may utilize all of the well-proven and powerful tools of circuit theory. There are two problems to be overcome: first, the set-up of the elastomechanical differential equations and their integration satisfying all boundary conditions, and secondly, the representation of the obtained transmission ($ABCD$) matrix by lumped elements. Let us choose as an example of this process, the length-extensional transmission line which may be used either as a resonator or as a coupler.

Consider a differential length element dx of a uniform rod with cross section A, length l, Young's Modulus E, and density ρ. A stressing force $F(x,t)$ causes a displacement $u(x,t)$, the extension being $\partial u/\partial x$. From Hooke's Law and neglecting the Poisson contraction as a second-order effect, we obtain elastically

$$\frac{\partial F}{\partial x} = \frac{\partial}{\partial x}\left(AE\frac{\partial u}{\partial x}\right) = AE\frac{\partial^2 u}{\partial x^2}, \tag{1}$$

whereas the dynamic equation is given by Newton's law

$$\frac{\partial F}{\partial x} = \frac{\partial}{\partial x}(\rho A \cdot dx)\frac{\partial^2 u}{\partial t^2} = \rho A\frac{\partial^2 u}{\partial t^2}. \tag{2}$$

Equating (1) and (2) and assuming periodic motion

Reprinted from *IEEE Trans. Sonics Ultrason.*, vol. SU-21, pp. 244–256, Oct. 1974.

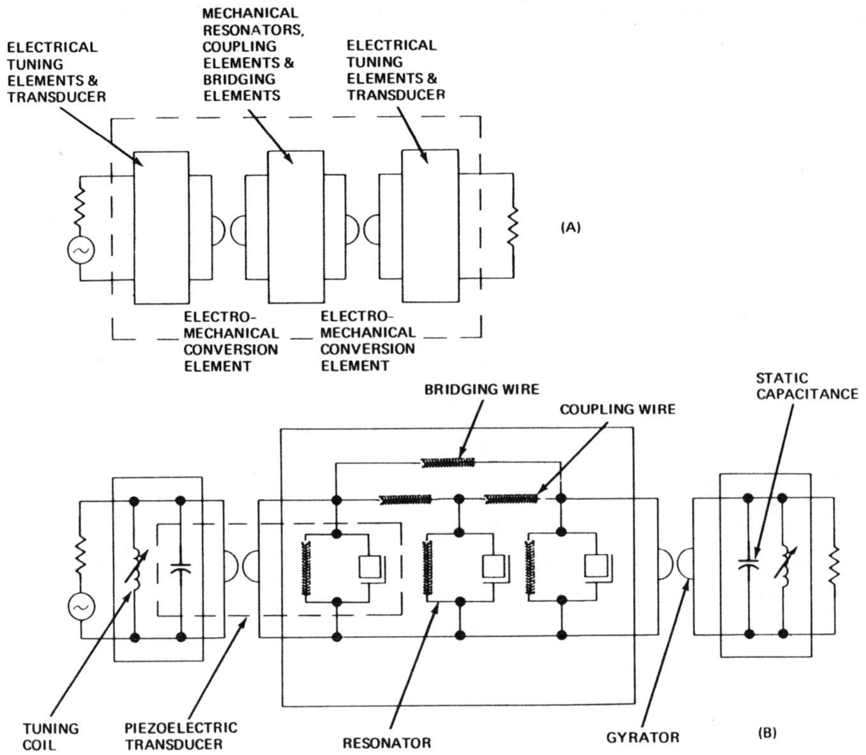

Fig. 1. Mechanical filter (a) block diagram; (b) electromechanical schematic diagram (piezoelectric transducer case).

$(u(x,t) = u(x) \exp (j\omega t))$ the integration yields

$$u(x) = P\cdot\cos\frac{\omega x}{c_0} + Q\cdot\sin\frac{\omega x}{c_0}, \tag{3}$$

where $c_0 = ((E/\rho)^{1/2})$, the phase (propagation) velocity of sound. The particle velocity is $v(x) = j\omega u(x)$, and the force $F(x) = AE[du(x)]/(dx)$. The arbitrary integration constants P and Q can be expressed by velocity and force at one end of the rod, v_0 and F_0, respectively.

Introducing the mechanical compliance $C_m = l/AE$ and the mass $M_m = \rho Al$ we obtain (after some algebra) the velocity and force at the other end in the matrix representation

$$\begin{bmatrix} v_1 \\ F_1 \end{bmatrix} = \begin{bmatrix} \cos\omega\,(M_mC_m)^{1/2} & j(C_m/M_m)^{1/2}\sin\omega(M_mC_m)^{1/2} \\ j(M_m/C_m)^{1/2}\sin\omega(M_mC_m)^{1/2} & \cos\omega(M_mC_m)^{1/2} \end{bmatrix} \begin{bmatrix} v_0 \\ F_0 \end{bmatrix} \tag{4}$$

which is completely analogous to the well-known transfer matrix of the electromagnetic transmission line.

Similar matrix expressions are obtained for all other kinds of vibration, such as torsion, shear, flexure, and coupled modes [4–16]. Instead of simple trigonometric functions, the matrix elements are usually more complicated transcendental functions, but the characteristic mobility (velocity to force ratio) is always the square root of compliance divided by mass (or moment of inertia in rotatory cases), and the determinant equals unity.

The eigenvalues characterize the state of forceless motion, hence they are given by the zeros of the C matrix element of the transmission matrix [17–23].[1] The series of natural frequencies of a length-extensional rod results from $\sin\omega(M_mC_m)^{1/2} = 0$, the resonances are $\omega_{0\nu} = \pi\nu/(C_mM_m)^{1/2}$ or $f_{0\nu} = (\nu c_0)/(2l)$, $\nu = 1,2,3\cdots$.

In the following, we derive electrical-circuitlike configurations and use electrical symbols, but we maintain the mechanical parameters. A resonator acts as a two-port when the couplers are attached at different points on the resonator, for instance, at both ends. The most commonly utilized resonator mode is the first symmetrical mode where the ends vibrate in opposite directions. Therefore, before evaluating the tee equivalent circuit branches, we extract from matrix (4) an ideal 1 : −1 transformer, as shown in Fig. 2. Expanding the trigonometric functions by partial fractions we obtain the branch mobilities

$$\frac{1}{X_1} = j(M_m/C_m)^{1/2}\tan\frac{\omega(M_mC_m)^{1/2}}{2} = \sum_{\nu=1}^{\infty}\frac{4\omega}{jC_m(\omega^2 - \omega_{0,2\nu-1}{}^2)} \tag{5a}$$

$$X_2 = j(C_m/M_m)^{1/2}\csc\omega(M_mC_m)^{1/2} = \frac{-1}{j\omega M_m} - \sum_{\nu=1}^{\infty}\frac{(-1)^{\nu}2\omega}{jM_m(\omega^2 - \omega_{0\nu}{}^2)}, \tag{5b}$$

the results of these expansions are shown in Fig. 3. Note that mass is represented by a capacitor and compliance by an inductor as is required when using the voltage-

[1] Some of the papers referred to obtain the eigenvalues by other means than matrix evaluation.

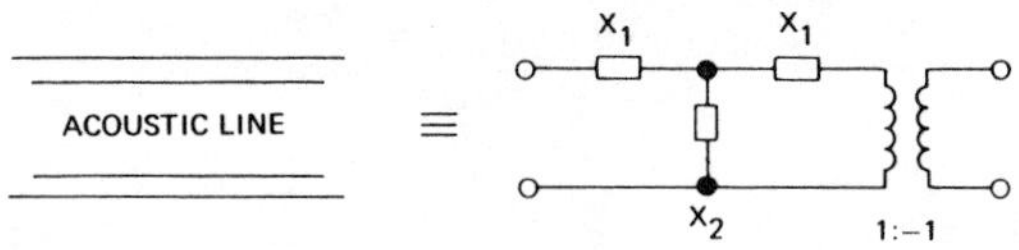

Fig. 2. Tee equivalent circuit of transmission line.

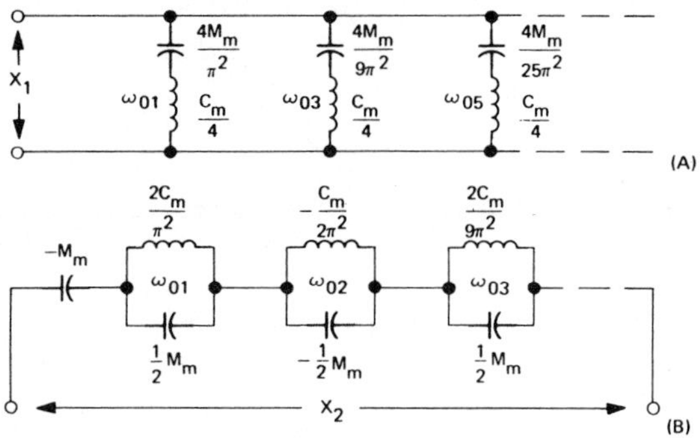

Fig. 3. Branches of transmission line tee equivalent circuit; (a) series branch, (b) shunt branch.

velocity mobility analog. In the one-port case the coupler arrives and leaves at the same point, so the mobility of the resonator is the open-end mobility of the line

$$X_0 = -j(C_m/M_m)^{1/2} \cot \omega(M_m C_m)^{1/2}$$

$$= \frac{1}{j\omega M_m} + \sum_{\nu=1}^{\infty} \frac{2\omega}{jM_m(\omega^2 - \omega_{0\nu}{}^2)}, \qquad (5c)$$

the equivalent circuit being the same as X_2 (Fig. 3), but with all elements positive; the effective mass is one half of the total mass. Narrowband equivalents [4–6, 8–10, 13, 15–18, 24] for the vicinity of ω_{01} are obtained by dropping all elements except those being resonant at ω_{01}. Suitable wideband (fractional bandwidth up to 30 percent) equivalent circuit configurations are shown in Fig. 4 [16]; for other representations and n-ports see [14, 24].

The coupler is always a two-port, but a pi equivalent circuit representation is preferable, as in Fig. 5. In the same fashion we obtain

$$X_1 = -j(C_m/M_m)^{1/2} \cot \frac{\omega(M_m C_m)^{1/2}}{2}$$

$$= \frac{2}{j\omega M_m} + \sum_{\nu=1}^{\infty} \frac{4\omega}{jM_m(\omega^2 - \omega_{0,2\nu}{}^2)}, \qquad (6a)$$

$$\frac{1}{X_2} = -j(M_m/C_m)^{1/2} \csc \omega(M_m C_m)^{1/2}$$

$$= \frac{1}{j\omega C_m} + \sum_{\nu=1}^{\infty} \frac{(-1)^\nu 2\omega}{jC_m(\omega^2 - \omega_{0\nu}{}^2)}. \qquad (6b)$$

The elements are shown in Fig. 6. Particularly interesting is the "short" coupler ($l < \lambda/8$) whose equivalent circuit sufficiently accurately is given by $1/X_1 = j\omega(M_m/2)$ and $X_2 = j\omega C_m$ in Fig. 7(a). The other important kind of coupler is the quarter-wave coupler, where $\omega_q = \omega_{01}/2$; at this frequency the coupler is an ideal impedance inverter. An often used zeroth order representation of this case

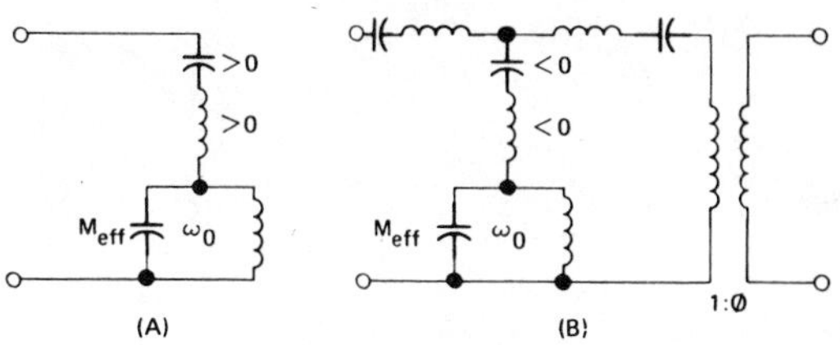

Fig. 4. Practical wideband equivalent circuits of (a) one-port and (b) two-port resonators.

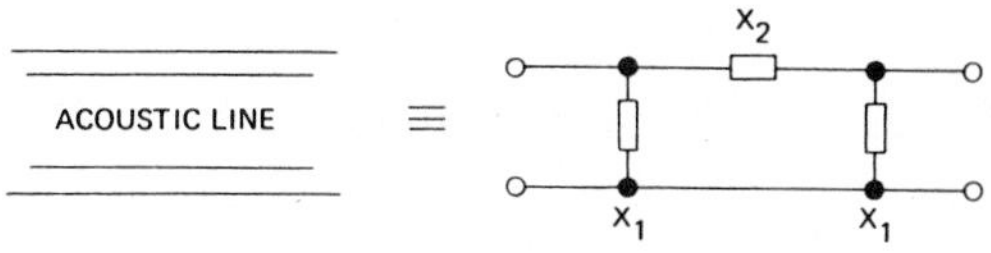

Fig. 5. Pi equivalent circuit of transmission line.

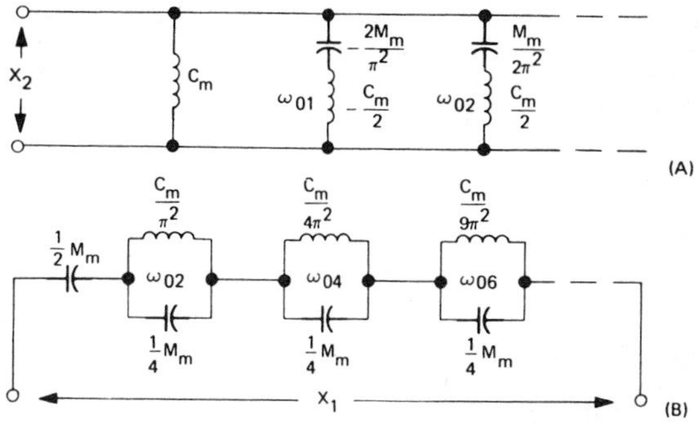

Fig. 6. Branches of transmission line pi equivalent circuit; (a) series branch, (b) shunt branch.

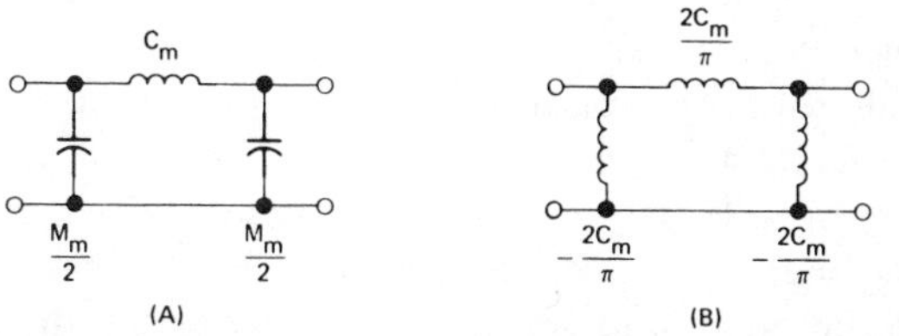

Fig. 7. Important coupler equivalent circuits; (a) short coupler, (b) narrowband equivalent circuit of quarter-wavelength coupler.

is a pi equivalent circuit with $X_2 = j2\omega C_m/\pi$, and $X_1 = -X_2$, as shown in Fig. 7(b).

In general, a resonator is designed for employing a distinct mode. Besides this mode, the resonator exhibits the full spectrum of overtones (Figs. 3 and 6), but moreover many other modes. The disk in Fig. 8 has its single-circle mode resonance (Fig. 8(a)) at 60.06 kHz and its three-diameter mode resonance (Fig. 8(b)) at 77.05 kHz. Since both modes are orthogonal and linear, i.e., the resultant vibration is a linear superposition of the two different vibrations without mutual influence, the narrowband equivalent circuit Fig. 8(c) describes the double resonance vibrator entirely. Therefore, we see a series connection of the single mode equivalent circuits, the transformer-turns ratios representing the amplitude ratios between the points of coupler access and a reference point [24], this ratio being negative for an odd number of nodes between these points.

Sometimes such second modes are welcome and utilized

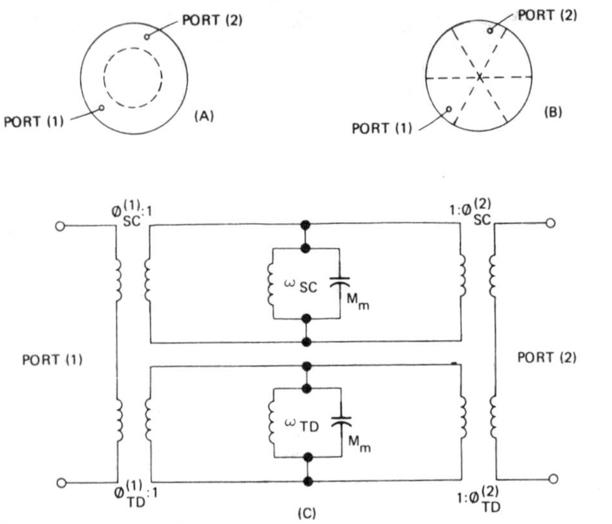

Fig. 8. Disk modes of vibration (nodal lines are dashed); (a) single-circle (SC) mode, (b) three-diameter (TD), (c) narrowband equivalent circuit for both modes.

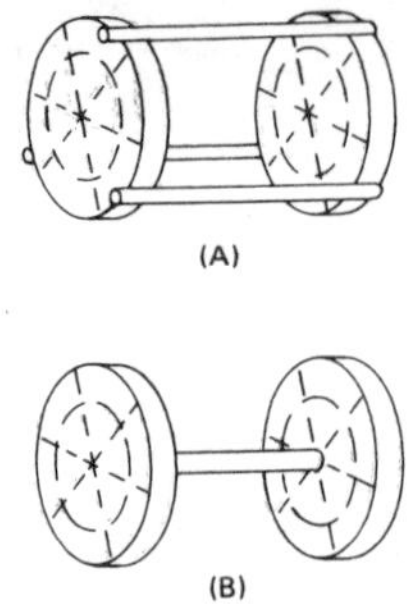

Fig. 9. Excitation and suppression of modes; (a) both modes are transferred, (b) 3-diameter mode is suppressed by coupler.

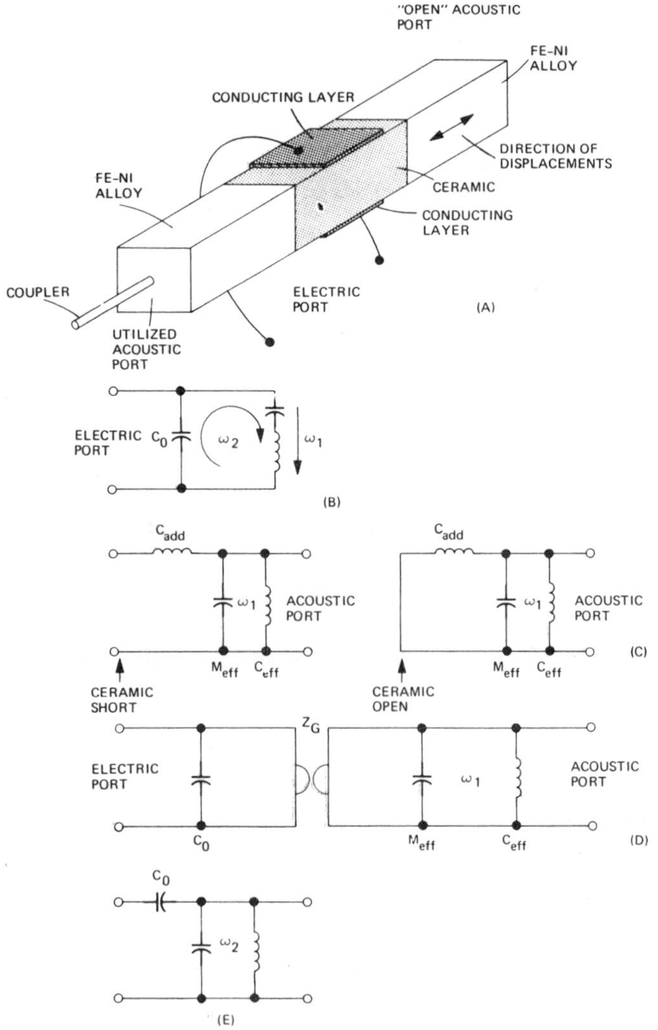

Fig. 10. Alloy/ceramic composite transducer resonator; (a) pictorial diagram, (b) impedance as viewed from electrical port, (c) impedance as viewed from acoustic port and dependent on electrical port, (d) narrowband equivalent circuit, (e) narrowband equivalent with gyrator removed.

for improvement of the filter performance [25–31], but most often they are spurious and must be suppressed. The coupler orientation (Fig. 9(a)) transfers both disk modes, whereas in Fig. 9(b) the coupler connects nodal points of the three-diameter mode which thus cannot be excited (in the equivalent circuit a nodal point means a turns ratio $\phi_{TD} = 0$). More critical than such additional resonator modes that can be predicted theoretically are vibrations of the whole filter structure. In spite of many theoretical [7, 9, 32] and experimental [33] investigations, a final solution of this problem has not yet been achieved.

As shown in the introduction, the transducing elements are resonators having electrical and mechanical ports. Such resonators may consist of magnetostrictive or piezoelectric material only, or they may be a composite of iron-nickel alloy and transducing material. Let us derive phenomenologically the equivalent circuit of a length-extensional composite resonator with crossed field (the field perpendicular to the mechanical displacement) ceramics as shown in Fig. 10(a). Only one of the two acoustic ports we assume to be utilized, the other is "open." The electrical port (the resonator being mechanically free) shows the response of the network of Fig. 10(b) which is an impedance zero at ω_1 and a pole at $\omega_2 > \omega_1 \cdot C_0$ is the static (i.e., the mechanical system is clamped) ceramic capacitance. At the mechanical (acoustic) port we observe the mobility behavior of a uniform bar, but with the different resonant frequencies ω_1 and ω_2, respectively, depending on the short or open-circuit condition of the ceramic. Since the mass cannot be altered by these port conditions, short circuiting the electrical port obviously increases the effective compliance C_{eff} thus lowering the frequency. We can model this effect by the additional compliance C_{add} connected in parallel to the resonator in the open-circuit case and being switched out of the circuit when short circuiting the ceramic, as shown in Fig. 10(c). This inverse behavior indicates the existence of a gyrator between both ports. Fig. 10(d) explains all measured effects. The gyrator impedance $Z_G{}^2 = C_{eff}/(C_0 k_{em}{}^2)$ relates the mechanical parameters to the corresponding electrical quantities, linked by the electromechanical coupling coefficient $k_{em}{}^2 \approx 2(\omega_2 - \omega_1)/\omega_1$. Approximating the gyrator by an impedance inverter and its lumped equivalent circuit we obtain the network of Fig. 10(e), which is the approximate three-element equivalent of Fig. 10(b).

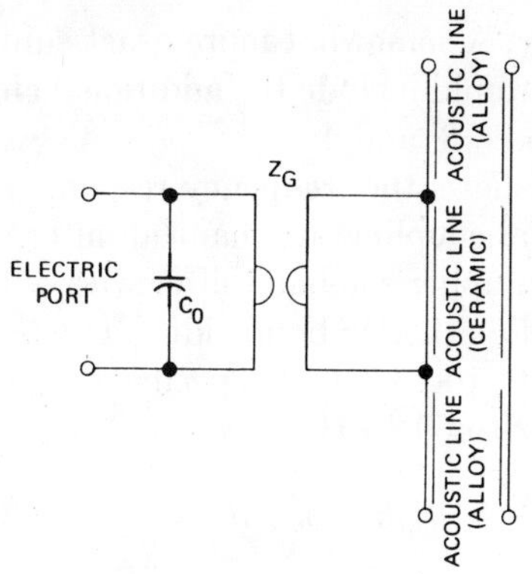

Fig. 11. Equivalent circuit of length-extensional composite alloy/ceramic resonator.

The result of the integration of the piezoelastic differential equations [4] is shown in Fig. 11, where the transmission lines are described by matrix (4). This applied technique yields Fig. 10(d) as a narrowband equivalent. Other cases of field orientation as well as the magnetostrictive effect are treated in [8, 34–44], the results showing acoustic transmission lines or their lumped element approximation, static capacitors or coils, and gyrators or ideal transformers between them.

III. CURRENT FILTER DESIGNS

In this section we will describe a number of often used mechanical filter configurations. In each case the transducers, resonators, and coupling elements can be represented as two-ports in much the same way as the extensional mode lines, disks, and transducers of the previous section. Some of the synthesis procedures used to obtain the filter realizations are described in Section IV.

Low-Frequency Flexure Modes

The lowest frequency mechanical filters are realized through the use of flexure-mode resonators, in the form of tuning forks. Tuning-fork mechanical filters are used in various narrow-band (1 percent or less) signaling, control and telemetry applications in the audio frequency range of 100 Hz to 3 kHz [45]. Coupling between resonators can be achieved acoustically [46] or electrically as shown in Fig. 12(a).

The two-resonator electrically-coupled filter operates in much the same way as a hybrid lattice filter where, in this case, the phase inversion is realized through reversing the polarity of one of the piezoelectric ceramic transducers. This reversal of polarity is represented by the phase inverter in Fig. 12(b). The resonator equivalent circuit is of the form shown in Fig. 10(e). In the case of acoustic coupling only two transducers are needed, the coupling being realized by a wire connecting adjacent prongs. The massive ends of the tuning forks are usually mounted in a silicone base. Multiple resonator filters are obtained by simply cascading more sections either electrically or acoustically.

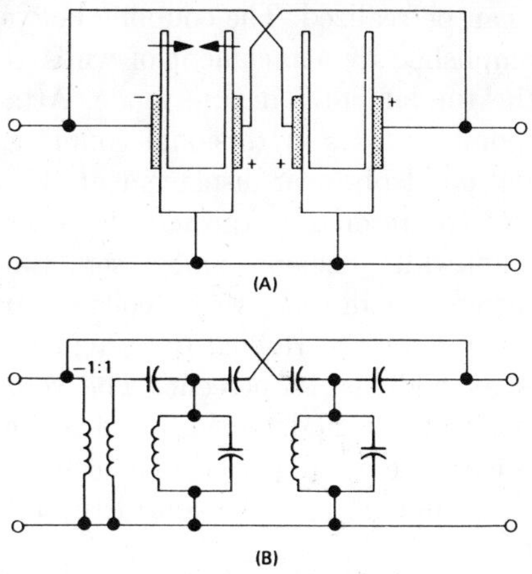

Fig. 12. Two-resonator tuning fork filter with electrical coupling; (a) pictorial diagram, (b) schematic diagram.

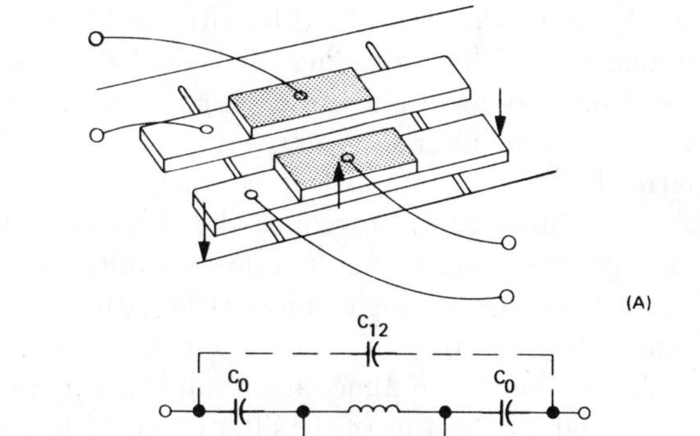

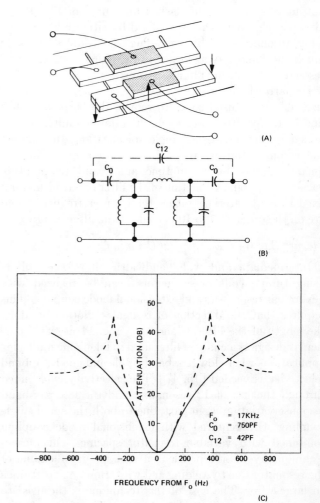

Fig. 13. Two-resonator flexure-mode filter with torsional coupling; (a) pictorial diagram, (b) schematic diagram including electrical bridging, (c) frequency response.

Flexure Bars With Torsional Coupling

In the 3 kHz to 50 kHz frequency region the prongs of the tuning forks are essentially straightened and shortened to form bar resonators as shown in Fig. 13(a) [14, 40, 45]. In this somewhat more simple configuration, the electromechanical coupling is increased and 2-percent wide filters can be realized. The coupling between resonators is accomplished by attachment of wires at the nodal points of the fundamental flexure mode. Attachment at these two points results in torsional coupling, the node being a nodal point of linear displacement. Wire coupling the centers of the resonators through flexure results in a maximum bandwidth response and is sometimes used at higher frequencies or in cases where coils are used on the input and output of the resonator to achieve fractional bandwidths as high as 10 percent. The resonators are wire supported at the nodal points and are either hard mounted to the base or supported in silicone rubber. The stiffness of the support must be considered in the design of the filter.

Fig. 13(b) shows an equivalent circuit of the 2-resonator filter. The effect of capacitive bridging from input to output is shown by the dashed line curve of Fig. 13(c). Greater selectivity can be realized by either adding more alloy resonators or by cascading the coupled pairs in much the same way as in polylithic designs (cascade of 2 resonator monolithic filter sections).

In narrowband designs (no tuning coil), a minimum size ceramic is used so as to reduce the frequency shift due to temperature and aging. The lower limit is determined by the passband ripple specification; the smaller the ceramic, the greater the passband ripple. The frequency shifts due to lifetime aging and temperature variations become a function of the filter bandwidth (for a fixed ripple, material) rather than center frequency and are on the order of 5 to 10 percent of the filter bandwidth.

Flexure Bar With Extensional Coupling

Intermediate and wide-bandwidth filters (i.e., filters using tuning coils or capacitors) can be realized with flexure bar resonators and extensional mode coupling. This requires that the direction of resonator motion be along the length of the filter as shown in Fig. 14(a). This particular design is a channel filter used in telephone multiplex applications where low passband ripple and good stopband selectivity is needed [15, 47]. The selectivity is obtained through the use of 12 resonators and the low passband response variation is, in part, due to both the small single coupling wire required with extensional mode coupling combined with variable resonator spacing. This design technique avoids the problem of variable mass loading of the resonator due to welding and wire trimming operations on large wires. The cost of this technique is the need to individually support each resonator. The filter of Fig. 14(a) is supported at the resonator nodal points, and the entire assembly is positioned in two solder boats. Also, a segment of material is removed from each resonator to remove the degeneracy of the flexural mode.

Piezoelectric ceramic transducers are used on the input and output of the filter and are inductively tuned as shown in the equivalent circuit of Fig. 14(b). The network is an ordinary ladder filter with no bridging. This very simple configuration lends itself to automated fabrication processes and eliminates the need to retune the resonators after assembly. A somewhat more exact equivalent circuit of Fig. 14(b) would include the additional circuit elements shown in Figs. 2, 3, and 4.

Fig. 14 (c) shows the frequency response of the flexure-bar/extensional-coupling mechanical filter. The filter is designed as an upper sideband filter with a 48 kHz carrier frequency and a 3.1 kHz bandwidth. The passband ripple specification is the CCITT specification of ± 0.11 dB from 48.6 kHz to 50.4 kHz.

Extensional-Mode Resonator Filters

One of the earliest mechanical filter designs employs both extensional (or torsional) resonators and coupling elements such as those described in the previous section [48]. These filters have been used as telephone channel filters at 48 kHz and in the 60 kHz to 108 kHz base-band frequency range [49] as well as for radio applications at 250 kHz and 455 kHz [50].

Fig. 15 shows a space reducing folded-line type of design which employs Langevin piezoelectric-ceramic alloy composite transducers [44, 49] and half-wavelength extensional-mode resonators. The coupling wires are less than one-eighth wavelength long, thus having the equivalent circuit representation of Fig. 7(a). The resonators are supported at the nodal points midway along their length by silicone rubber. Like the flexure-bar filter of the previous design, the transducers are tuned with coils, and the structure is a simple ladder; thus the filter has an equivalent circuit like that of Fig. 14(b) and a corresponding monotonic frequency response.

Torsional Rod With Extensional Mode Coupling

A filter type that has been used for radio applications for a number of years is the torsional-rod design shown in Fig. 16(a). A typical bandwidth is approximately 3 kHz; the carrier frequency is 200 kHz. The filter is driven with magnetostrictive ferrite transducers [35, 37], which are indirectly attached to the end torsional resonator. The inductance of the transducer coil is represented as a series inductor in the equivalent circuit of Fig. 16(b) in the same way as the static capacitance of the piezoelectric ceramic transducers were shown in Figs. 12 and 13. The main coupling wires are one-quarter acoustic wavelength long and are attached near the ends of the resonators, thus minimizing the wire diameter and mass loading. Wires are also welded at the nodal point, in the center of each resonator, and to studs connected to a base, thereby acting as a support for the filter structure.

Looking at Fig. 16(a), it should be noted that two of the wires are used to bridge across two resonators. These wires, which are shown as inductors in the equivalent circuit of Fig. 16(b), are used to realize the 4 attenuation poles in the curve of Fig. 16(c). The quarter-wavelength spacing between resonators results in 3/4 wavelength bridging wires. In the equivalent circuit, the bridging inductance should either have a negative value, or it should be in cascade with a $-1:1$ transformer (i.e., a

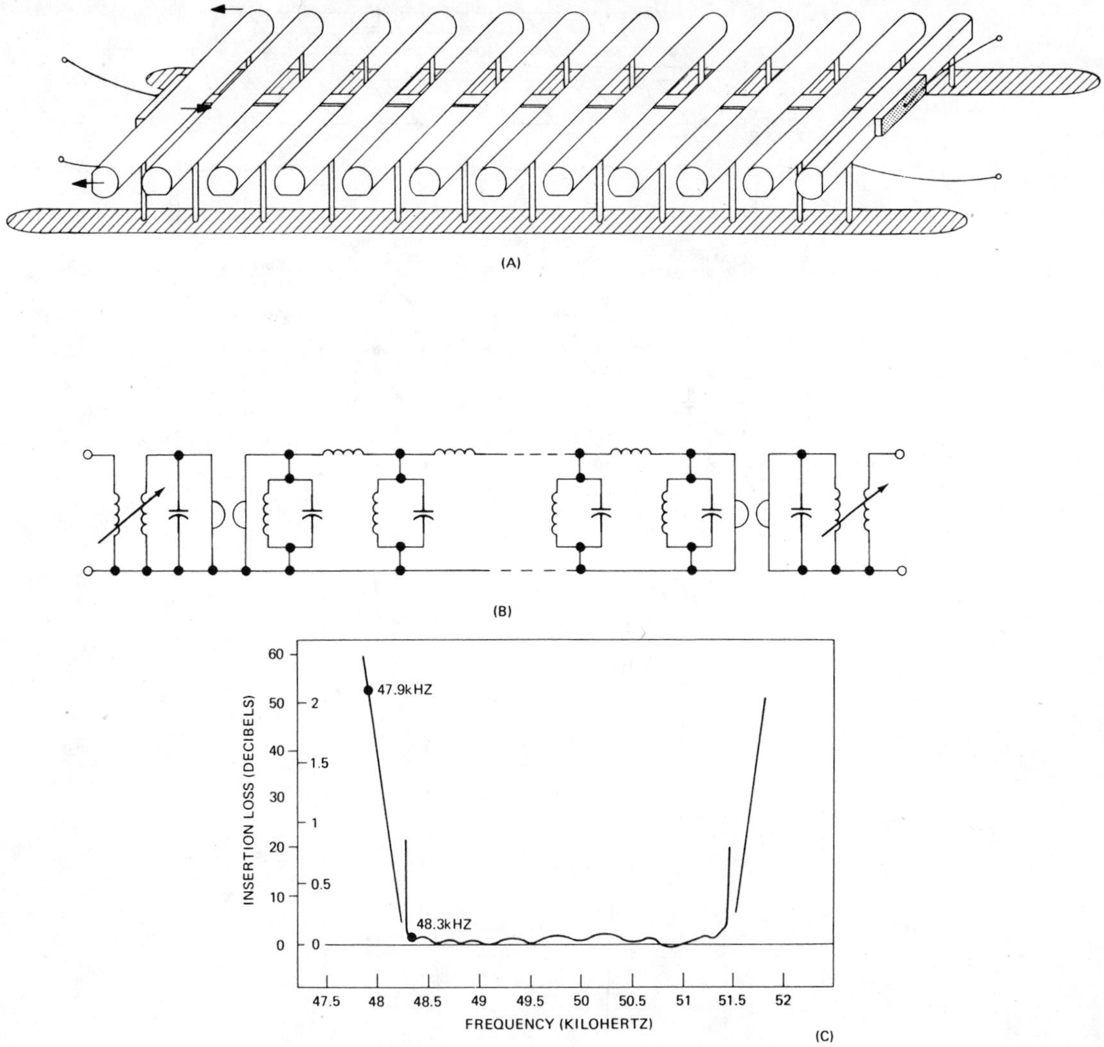

Fig. 14. 12-resonator flexure-mode filter with extensional mode coupling; (a) pictorial diagram, (b) schematic diagram, (c) frequency response of telephone channel filter.

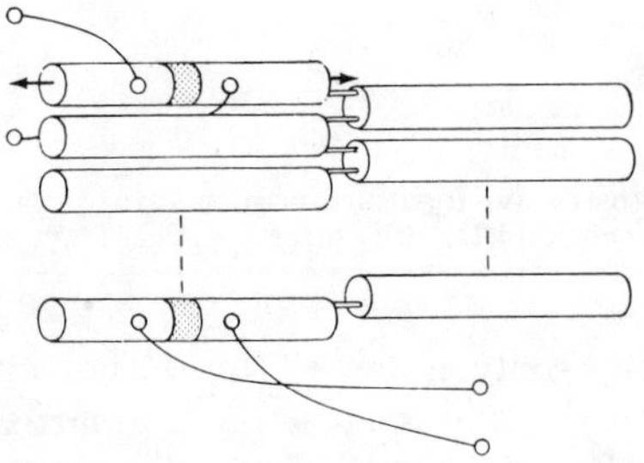

Fig. 15. Folded-line extensional mode filter with Langevin transducers.

phase inverter). The design of this type of filter is discussed later in the paper.

The torsional-rod filter type is not limited to the design described previously, but has been manufactured as a monotonic filter and has been designed with a multiplicity of wires, some spanning more than two resonators.

Flexural Disk-Wire Filter

One of the earliest mechanical filters is the disk-wire type [3], which is being manufactured at frequencies ranging from 60 kHz to 500 kHz and bandwidths from 500 Hz to 50 kHz. The wide center frequency range is the result of using both one- and two-nodal circle flexural-mode resonators [5, 24]. The filter shown in Fig. 17(a) could represent a filter at 100 kHz or 250 kHz, the disk resonators being approximately the same size (although the 100 kHz ferrite transducer is two and one half times as long as the 250 kHz). Like the torsional-bar filter, the magnetostrictive-ferrite transducer is indirectly attached with a small wire to the first disk resonator. Although the direct attachment is often used, the indirect method reduces unwanted spurious responses due to the ferrite-wire-disk combination acting like a wide bandwidth roofing filter.

The disk-wire mechanical filter is supported at each end by the coupling and bridging wires which are welded to a nonresonant metal ring. In this way, the coupling wires act as the total support for the filter, making the fabrication of the filter less costly, but at the expense of greater mass loading and variation of the resonator frequencies.

The filter of Fig. 17(a) makes use of two single-disk-

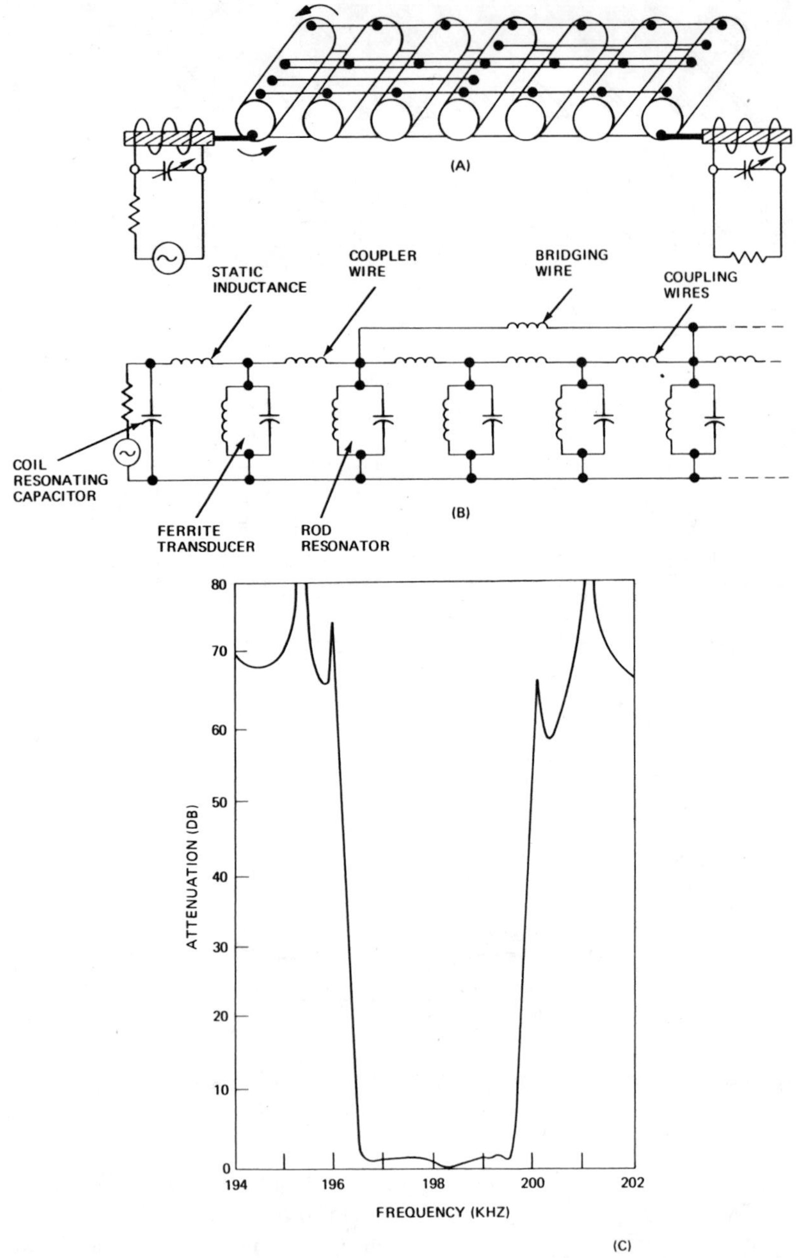

Fig. 16. Torsional mode filter with double resonator bridging and magnetostrictive ferrite transducers; (a) pictorial diagram, (b) schematic diagram, (c) frequency response of 200 kHz SSB filter.

bridging wires to realize the two upper-stopband attenuation poles of Fig. 17(b). The attenuation poles are used to obtain the best stopband response for the minimum absolute and differential delay. Fig. 17(b) shows the amplitude and delay responses of a new 6-resonator channel filter for telephone multiplex. The high-side audio response is further filtered with a low-pass filter in the audio band.

Other Filter Types

Mechanical filters have been designed using almost any conceivable resonator mode of vibration [45]. Examples are radial-mode spheres, concentric-shear disks, face-shear and extensional-mode plates, zig-zag shaped flexural-mode tuning forks, and flexure-mode H-shaped mass-webs, to name only a few. Transducers have included magnetstrictive alloy plates and quarter-wavelength wires as well as a variety of ceramic and quartz-crystal rods, disks, bars, plates, and cylinders used with alloy resonators in composite and wire-coupled configurations.

IV. DESIGN PROCEDURES

Fig. 18 recapitulates the narrow-band equivalent circircuits of resonators, couplers, and transducers. Since couplers are always located between resonators, obviously we can join the coupler and resonator masses, whereas the quarter-wave coupler acts as an inverter. Removing the transducer gyrators, the static shunt capacitors pass over to series arm inductors, and external shunt coils then become capacitors in series with these inductors.

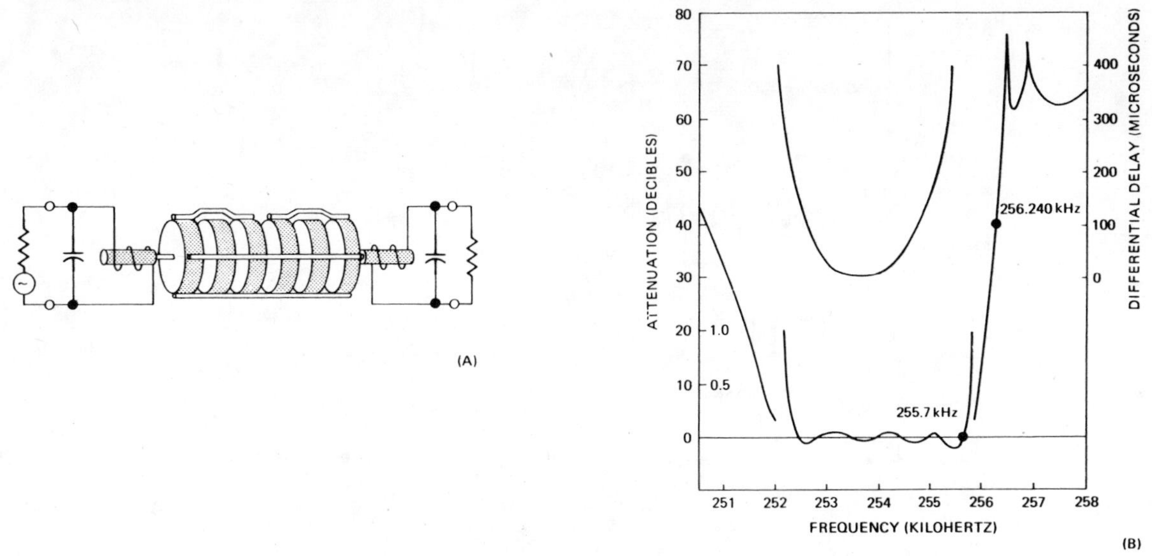

Fig. 17. Disk-wire mechanical filter with single-disk bridging; (a) pictorial diagram, (b) frequency response of telephone channel filter.

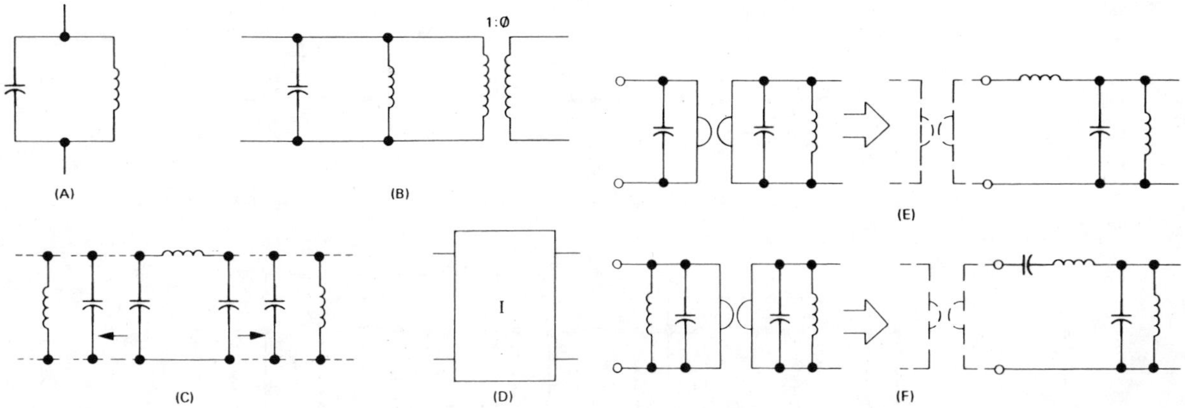

Fig. 18. Narrowband equivalent circuits; (a) resonator as one-port, (b) resonator as two-port, (c) short coupler (coupler and resonator masses joined), (d) coupler as ideal inverter, (e) transducer resonator before and after gyrator removal, (f) like (e) but with external shunt coil.

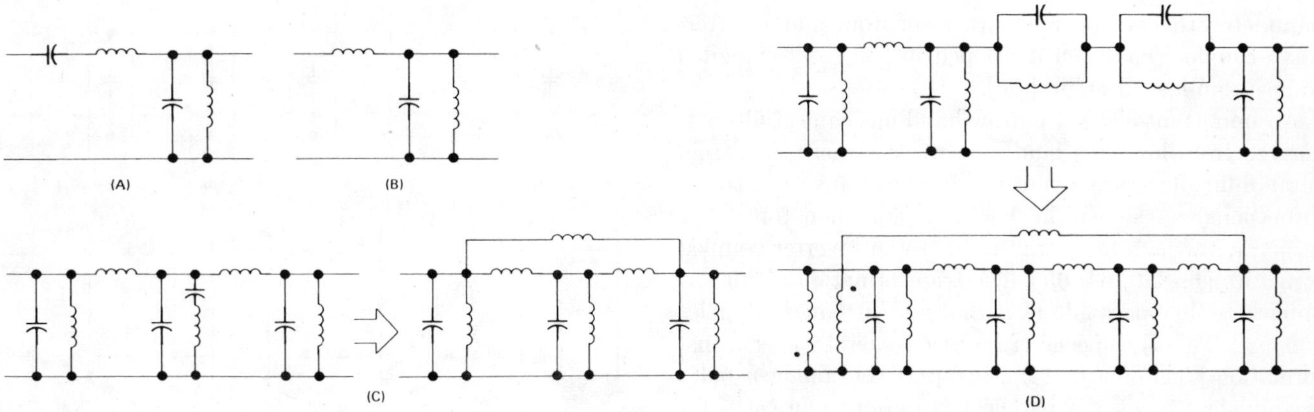

Fig. 19. Image-parameter half sections, combined sections, and transformations (Cauer classifications) (a) $2b^*b$, (b) $1a^*b$, (c) $1ba^* + 1a^*b_i + 1b_ia^* + 1a^*b$, (d) $1ba^* + 1a^*b + 2bd^* + 2d^*b$.

Therefore, the building blocks of mechanical filters are grounded parallel resonance circuits, series arm inductors, ideal inverters, and ideal transformers.

Much of the design work involves equivalent network transformations. Some popular image-parameter sections and their mechanically realizable equivalents [16, 51–53] are shown in Fig. 19. More difficulties are inherent in insertion loss theory designs, where immediately realizable circuits seldom result. Fig. 20 shows some helpful transformations as well as an application [54, 55]. An image

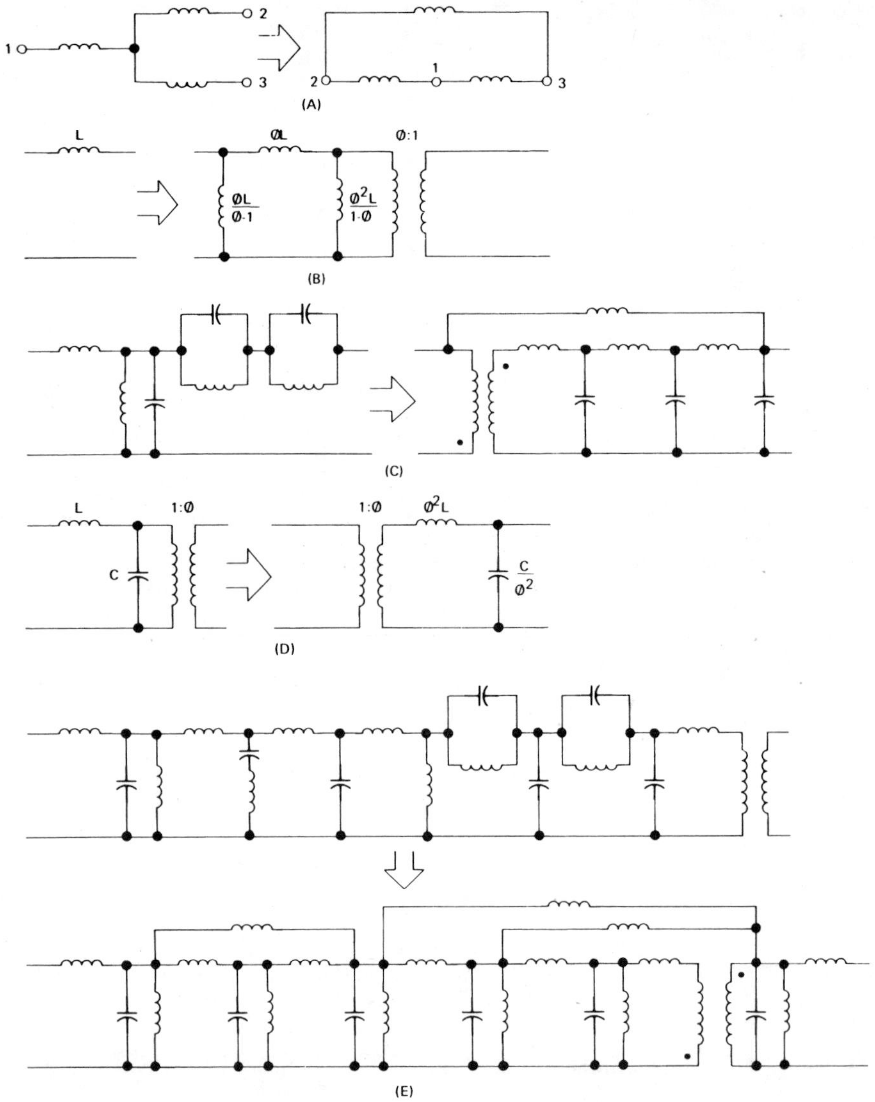

Fig. 20. Equivalent network transformations; (a) tee to delta, (b) Norton transformation, (c) Kuenemund transformation, (d) immittance transformation by ideal transformers, (e) application of result of insertion-loss theory design (intermediate steps not shown).

parameter theory for two-port resonators and quarter-wave couplers has been developed in [56] and extended to insertion loss filters in [57].

An important class of narrow-band mechanical filters is derived from low-pass ladder networks, where the structural difficulties are overcome by ideal inverter transformations, as shown in Fig. 21. The monotonic loss response low-pass filter transforms to an inverter coupled structure [6, 8, 9, 64, 65]. The transformation of Fig. 22 applies to low-pass filters with finite attenuation poles [55, 58]. This is a special case of the general Cauer transformation [59] of Fig. 23: a two-port containing n poles is equivalent to a series-parallel connection of an $(n-1)$-pole two-port and an ideal transformer. Successive applications of this transformation results in an attenuation pole free ladder circuit with n bridging branches [60, 61]. Also, each low-pass pole becomes one bandpass pole in each stopband.

Direct treatment of a four resonator bridged topology is given in [62, 63]. Finally, in using the narrowband equivalent of the inverter we obtain from the inverter

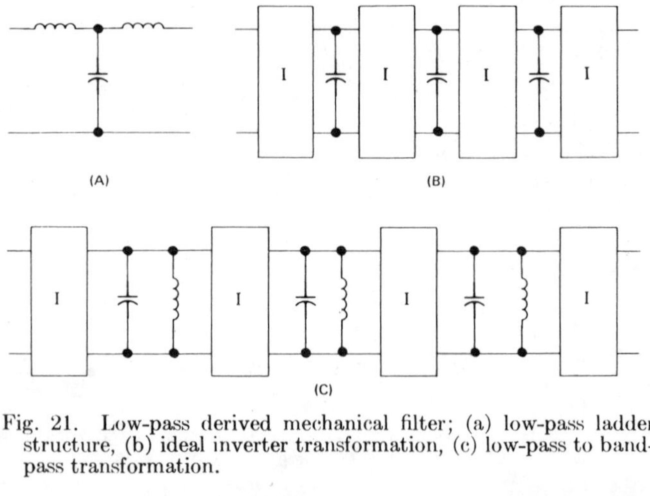

Fig. 21. Low-pass derived mechanical filter; (a) low-pass ladder structure, (b) ideal inverter transformation, (c) low-pass to band-pass transformation.

structures (Fig. 24) short coupler filter design data [10, 44].

In wideband filter design, because of the mutual dependence of the equivalent circuit elements, conventional methods fail, and modern optimization techniques replace

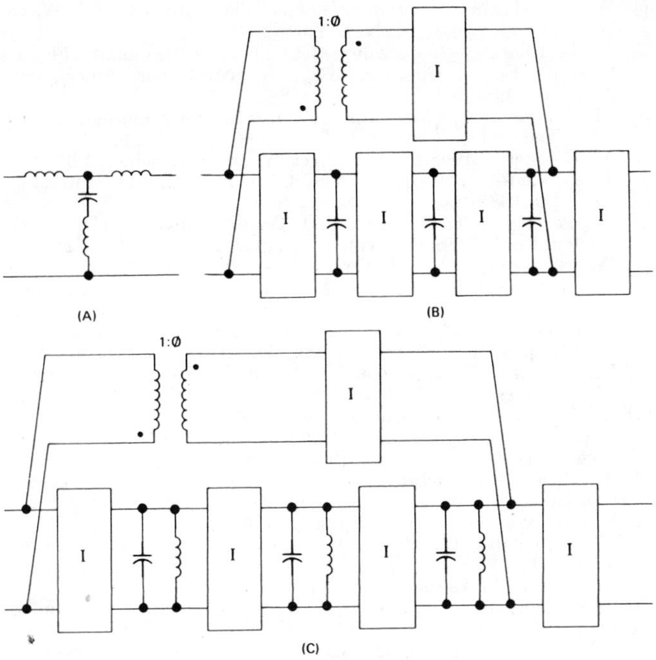

Fig. 22. Low-pass derived mechanical filter with finite poles; (a) low-pass ladder structure, (b) ideal inverter transformation, (c) low-pass to bandpass transformation.

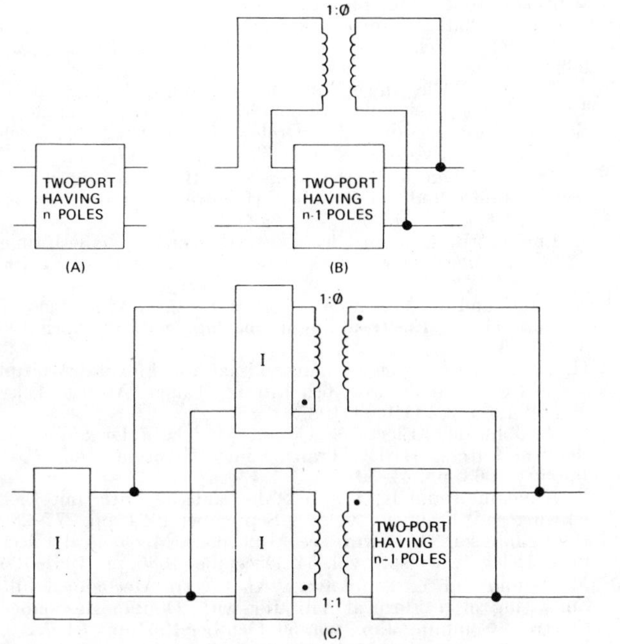

Fig. 23. Cauer transformation; (a) initial circuit, (b) Cauer transformation, (c) inverter transformation.

them. The requisite starting values, the choice of which is the crucial problem of all optimization, are easily obtained from image parameters [66].

For satisfying group delay specifications without cascaded equalizers, two different concepts have been developed (Fig. 25). In Fig. 25(a) is shown a hybrid transformer circuit, with mechanical reactances in the lattice arms [67], which is able to realize attenuation poles at complex frequencies. These complex poles are also produced by bridging couplers (Fig. 25(b)) where the couplers

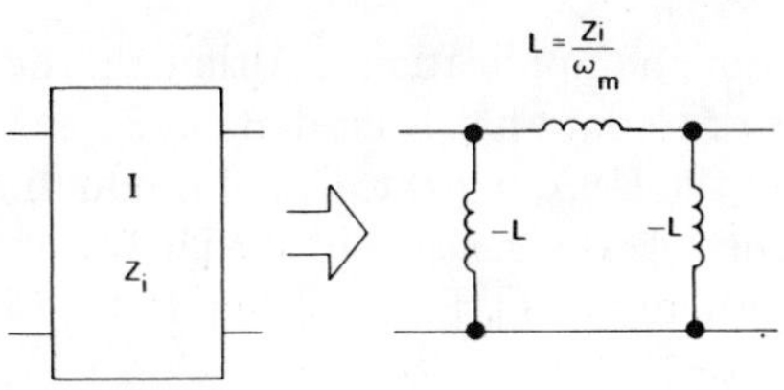

Fig. 24. Narrowband equivalent circuit of ideal inverter.

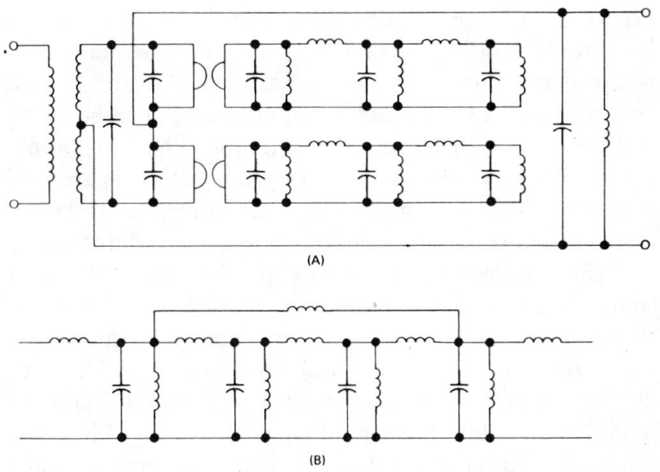

Fig. 25. Group delay equalization networks; (a) mechanical lattice filter, (b) in-phase bridged section.

are in phase [16, 58, 60, 62]. The design data may be obtained from classical procedures as well as from optimization.

Of the numerous technological requirements to be considered in the design of mechanical filters, we will mention only two examples. First, the ease of fabrication suggests the use of resonators with equal cross sections throughout the filter. Because of the resonance condition $f_0 = c/2l$, valid for extensional and torsional mode vibrators, equal cross sections are attained when $M_m f_0$ = constant for all resonators; bending mode resonators [66] require $M_m^2 f_0$ = constant. Narrowband filter designs meet those conditions by aid of the Norton transformation.

Sometimes, e.g. with the flexure-bar type channel filter, the frequency where the impedance is to become a maximum is prescribed. In moving two reflection coefficient zeros to complex frequencies, it is possible to influence the location of the maximum and thus meet this unusual requirement. Because of the complicated relation between impedance and loss characteristics, this procedure is not straightforward, but iterative [68].

V. THE FUTURE

Applications

Future applications will continue to be heavy in the area of communications (radio, telephone, and telegraph) but will also reach further into navigation equipment (Omega, for instance) sonar systems (fishing fleets), transportation system controls (as in Japan), and test equipment. The widening development, production, and use of mechanical filters throughout the world is resulting in an expansion of applications. Although the greatest amount of work to date has been done in Japan, Western Germany, and the USA, a great deal of production and/or development has also been accomplished in Czechoslovakia [69], East Germany [63], France [70], Poland [71], and the USSR [72].

Automation

Coming generations of currently fabricated filters are expected to be considerably smaller in size, while the absolute fabrication tolerances will be much the same. This will require tuning of already assembled filters. Most probably this adjustment, by sandblasting or laser shots, will be done with an on-line computer. This may apply to present filter types with extremely tight passband response specifications. Possible adjustment criteria are the short circuit or open-end resonances of the filters [73].

Todays adjustment routines applied to couplers suffer from the lack of one degree of freedom; the coupling strength can only be reduced by removing material. A computer controlled local heat treatment applicable to distinct coupler materials achieves increased Youngs' modulus and thus greater coupling strength [74]. These same automation and on-line computer controls may make delay compensated mechanical-lattice filters economically feasible [67].

Material Progress

The mechanical filter center frequency and minimum bandwidth frequency application range is primarily limited by the temperature, aging and Q characteristics of the transducer and resonator materials. In general, mechanical filters have a 0.1 percent minimum fractional bandwidth, but, in the specific case of SSB voice-band-width filters, the practical limitation is presently around 0.5 percent, thus, a center frequency of 500 kHz. Work is being done on new resonator alloys of niobium-titanium and niobium-zirconium [75, 76] that have very flat temperature characteristics but may be difficult to weld. The use of improved alloys and lithium tantalate transducers [77] may make it possible to extend the mechanical filter frequency range into the MHz region.

Improvements are also being made in the stability of piezoelectric ceramic materials by the addition of manganese, antimony, bismuth, and sodium to the binary solid solution of lead, zirconimum, and titanium (PZT) [78]. These new materials maximize the ratio of $k_{em}^2/(\Delta f_r/(f_r \cdot \Delta T))$ where Δf_r is the change in the resonant frequency due to a temperature or elapsed time change ΔT. k_{em}^2 is the electromechanical coupling coefficient of the material. By maximizing k_{em}^2, a minimum size ceramic can be used in a composite resonator.

REFERENCES

[1] R. Adler, "Compact electromechanical filters," *Electronics*, vol. 20, Apr. 1947, pp. 100–105.
[2] J. C. Hathaway and D. F. Babcock, "Survey of mechanical filters and their applications," *Proc. IRE*, vol. 45, Jan. 1957, pp. 5-16.
[3] R. Johnson, M. Börner and M. Konno, "Mechanical Filters—A Review of Progress," IEEE Trans. Sonics Ultrason., vol. SU-18, July 1971, pp. 155–170.
[4] W. P. Mason, "Electromechanical Transducers and Wave Filters," van Nostrand, New York 1942.
[5] R. L. Sharma, "Equivalent Circuit of a Resonant, Finite, Isotropic Elastic Circular Disk," J. Acoust. Soc. Amer., vol. 28, November 1956, pp. 1153–1158.
[6] M. Börner et al., "Mechanische Filter für die Nachrichtentechnik," Telefunken-Z., vol. 31, June 1958, pp. 105–114.
[7] M. Börner, "Biegeschwingungen in mechanischen Filtern," Telefunken-Z., vol. 31, June 1958, pp. 115–123 and September 1958, pp. 188–196.
[8] W. Poschenrieder, "Physik und Technik elektromechanischer Wellenfilter," Frequenz, vol. 12, August 1958, pp. 317–325.
[9] W. Struszinski, "A Theoretical Analysis of the Torsional Electro-Mechanical Filters," Marconi Rev., vol. 22, Fall 1959, pp. 119–143.
[10] E. Trzeba, "Elektromechanische Vierpole als Kopplungsfilter," Hochfrequenzt. u. Elektroak., vol. 69, no. 3, 1960, pp. 108–117.
[11] M. Börner, "Mechanische Filter mit Biegekopplung," Arch. Elek. Übertr., vol. 15, April 1961, pp. 175–180.
[12] M. Börner, "Transformation der Kopplung in mechanischen Filtern mit beidseitig frei schwingenden Biegeresonatoren," Arch. Elek. Übertr., vol. 16, July 1962, pp. 355–358.
[13] M. Börner, "Mechanische Filter mit biegegekoppelten Biegeresonatoren und kopplungsfreie Biegefilter," Arch. Elek. Übertr., vol. 16, September 1962, pp. 459–464.
[14] M. Konno and H. Nakamura, "Equivalent Electrical Networks for the Transversely Vibrating Uniform Bar," J. Acoust. Soc. Amer., vol. 38, October 1965, pp. 614–622.
[15] F. Künemund, "Channel Filters with Longitudinally Coupled Flexural Mode Resonators," Siemens Forsch.-u. Entwickl.-Ber., vol. 1, no. 4, 1972, pp. 325–328.
[16] A. E. Günther, "High-Quality Wide-Band Mechanical Filters—Theory and Design," IEEE Trans. Sonics Ultrason., vol. SU-20, October 1973, pp. 294–301.
[17] J. Näser, "Kurze Darlegung der Theorie elektromechanischer Filter mit Plattenresonatoren," Hochfrequenzt. u. Elektroak., vol. 71, October 1962, pp. 123–132.
[18] I. Lucas, "Plattenförmige elektromechanische Filter mit Dämpfungspolen," Arch. Elek. Übertr., vol. 17, May 1963, pp. 230–236.
[19] R. Straube, "Die Anwendung der Störungsrechnung auf freie Biegeschwingungen von Stäben bei Berücksichtigung der Schubverformung und der Drehträgheit," Wiss. Z. Techn. Univ. Dresden, vol. 12, May 1963, pp. 1173–1176.
[20] J. Näser, "Exakte Berechnung der Biegeresonanzen rechteckiger und zylindrischer Stäbe," Hochfrequenzt. u. Elektroak., vol. 74, 1965, pp. 30–36.
[21] I. Lucas, "Berechnung der Eigenfrequenzen quaderförmiger Biegeschwinger," Arch. Elek. Übertr., vol. 20, January 1966, pp. 64–70.
[22] M. Onoe and T. Yano, "Analysis of Flexural Vibrations of a Circular Disk," Electron. Commun. Jap., vol. 51, April 1968, pp. 33–36.
[23] H. Jumonji, "Analysis of Longitudinal and Flexural Multiple-Mode Resonators," Proc. 6th Intern'l. Congr. Acoust. Tokyo, August 1968, pp. G105–G108.
[24] R. A. Johnson, "Electrical Circuit Models of Disk-Wire Mechanical Filters," IEEE Trans. Sonics Ultrason., vol. SU-15, January 1968, pp. 41–50.
[25] F. Künemund and K. Traub, "Mechanische Filter mit Biegeschwingern," Frequenz, vol. 18, September 1964, pp. 277–280.
[26] R. A. Johnson, "A Twin Tee Multimode Mechanical Filter," Proc. IEEE (Corresp.), vol. 54, December 1966, pp. 1961–1962.
[27] M. Konno and Y. Tomikawa, "An Electro-Mechanical Filter Consisting of a Flexural Vibrator with Double Resonance," Electron. Commun. Jap., vol. 50, October 1967, pp. 64–73.
[28] K. Nakamura *et al.*, "Multi-Channel Electromechanical Filters Using Orthogonal Modes in Single Mechanical System," Proc. 6th Int. Cong. Acoust. Tokyo, August 1968, pp. G109–G112.
[29] Y. Tomikawa and M. Konno, "Electromechanical Filter Consisting of a Triple-Resonance Vibrator of Longitudinal and Bending Resonance Modes," Electron. Commun. Jap., vol. 52, April 1969, pp. 58–60.
[30] M. Konno *et al.*, "Electromechanical Filters Using Degeneration Modes of a Disk or Ring," Electron. Commun. Jap., vol. 52, July 1969, pp. 19–28.
[31] Y. Tomikawa *et al.*, "Electromechanical Filters Consisting of Longitudinal- and Bending-Resonance Mode Vibrators," Electron. Commun. Jap., vol. 52, November 1969, pp. 1–8.
[32] M. Börner and H. Schüssler, "Miniaturisierung mechanischer Filter," Telefunken-Z., vol. 37, Fall 1964, pp. 228–246.
[33] Y. Tsuzuki *et al.*, "The Holographic Investigation of Spurious Modes of Mechanical Filters," Electron. Commun. Jap., vol. 54, May 1971, pp. 31–38.
[34] A. Lenk, "Die Vierpolersatzschaltbilder der elektromechanischen Wandler, Teil 2," Acustica, vol. 6, 1956, pp. 303–316.

[35] C. M. van der Burgt, "Performance of Ceramic Ferrite Resonators as Transducers and Filter Elements," J. Acoust. Soc. Amer., vol. 28, November 1956, pp. 1020–1032.
[36] W. P. Mason and R. N. Thurston, "A Compact Electromechanical Band-Pass Filter for Frequencies Below 20 Kilocycles," IRE Trans. Ultrason. Engng., vol. UE-7, June 1960, pp. 59–70.
[37] C. Kurth, "Magnetostriktive Wandler als selektive Vierpole," Frequenz, vol. 14 August 1960, pp. 272–288.
[38] H. Albsmeier, "Über Antriebe für elektromechanische Bandpässe mit piezoelektrischen Wandlern," Frequenz, vol. 19, April 1965, pp. 125–133.
[39] C. Kusakabe *et al.*, "Resonant Frequencies of Transversely Vibrating Bar Excited by Electrostrictive Transducer," Electron. Commun. Jap., vol. 48, November 1965, pp. 186–191.
[40] K. Yakuwa, "Characteristics of an Electromechanical Filter Using Piezoelectric Composite Bending Vibrators," Electron. Commun. Jap., vol. 48, November 1965, pp. 206–215.
[41] H. Jumonji and M. Onoe, "Analysis of Piezoelectric Multiple-Mode Resonators Vibrating in Longitudinal and Flexural Modes," Electron. Commun. Jap., vol. 51, May 1968, pp. 35–42.
[42] H. Schüssler, "Darstellung elektromechanischer keramischer Wandler als Dickenscherschwinger mit piezoelektrischer und piezomagnetischer Anregung," Arch. Elek. Übertr., vol. 22, August 1968, pp. 399–406.
[43] M. Onoe, "General Equivalent Circuit of Piezoelectric Transducers Vibrating in Thickness Modes," Electron. Commun. Jap., July 1972, pp. 39–45.
[44] R. A. Johnson, "Mechanical Bandpass Filters," in "Modern Filter Theory and Design" (Ed. G. C. Temes and S. K. Mitra), John Wiley & Sons, New York 1973.
[45] M. Konno, "Electromechanical Resonators and Applications," Corona Publ., Tokyo 1974.
[46] K. Takahashi, "Low-Frequency Mechanical Filters With U- or I Shaped Couplers and Resonators," Electron. Commun. Jap., vol. 48, Nov. 1965, pp. 199–205.
[47] H. Kopp, "A Mechanical Filter Channel Bank," IEEE Trans. on Commun., vol. COM-20, Feb. 1972, pp. 64–67.
[48] W. van Roberts and L. L. Burns, "Mechanical Filters for Radio Frequencies," RCA Rev., vol. 10, Sept. 1949, pp. 348–365.
[49] K. Shibayama, "Electromechanical Filters," Electron. Commun. Jap., vol. 48, Nov. 1965, pp. 66–72.
[50] H. Schüssler, "Mechanische Filter mit piezoelektrischen Wandlern," Telefunken-Zeitung, vol. 39, no. 3, 1966, pp. 429–439.
[51] H. Albsmeier, "Mechanische Bandpässe mit mechanisch erzeugten, frei wählbaren Dämpfungspolen," Frequenz, vol. 17, December 1963, pp. 442–448.
[52] R. A. Johnson, "A Single-Sideband, Disk-Wire Type Mechanical Filter," IEEE Trans. Comp. Parts, vol CP-11, December 1964, pp. 3–7.
[53] T. Okamoto *et al.*, "Low Frequency Electromechanical Filter," Fujitsu Sci. Techn. J., vol. 17, May 1966, pp. 53–86.
[54] R. A. Johnson and R. J. Teske, "A Mechanical Filter Having General Stopband Characteristics," IEEE Trans. Sonics Ultrason., vol. SU-13, July 1966, pp. 41–48.
[55] F. Künemund, "Dimensionierung überbrückter Bandpässe mit Dämpfungspolen," Frequenz, vol. 24, July 1970, pp. 190–192.
[56] C. Kurth, "Eine Wellenparametertheorie für mechanische Vierpole in Kompressions—oder Torsionsschwingungen," Nachrichtent., vol. 9, November 1959, pp. 490–503.
[57] C. Kurth, "Anwendung der Betriebsparametertheorie bei der Berechnung von Siebschaltungen, die sich aus in Kette geschalteten Leitungsstücken zusammensetzen," Frequenz, vol. 16, December 1962, pp. 482–495.
[58] B. Kohlhammer and H. Schüssler, "Berechnung allgemeiner mechanischer Koppelfilter mit Hilfe von äquivalenten Schaltungen aus konzentrierten elektrischen Schaltelementen," Wiss. Ber. AEG-Telefunken, vol. 41, no. 3, 1968, pp. 150–159.
[59] W. Cauer, "Theorie der linearen Wechselstromschaltungen," Akad. Verl.-Ges. Becker & Erler KG., Leipzig 1941.
[60] K. Wittmann *et al.*, "Dimensionierung reflexionsfaktor- und laufzeitgeebneter versteilerter Filter mit Überbrückungen," Frequenz, vol. 24, October 1970, pp. 307–312.
[61] B. Kohlhammer, "Ein neuartiges Entwurfsverfahren zur Synthese von mechanischen Filtern und Gyrator-Filtern," Wiss. Ber AEG-Telefunken, vol. 43, no 3/4, 1970, pp. 170–177.
[62] M. Börner, "Mechanische Filter mit Dämpfungspolen," Arch. Elek. Übertr., vol. 17, March 1963, pp. 103–107.
[63] L. Brier, "Der Entwurf von HF-Bandfiltern und mechanischen Filtern mit Dämpfungspolen nach dem Betriebsparameterverfahren," X. Intern. Wiss. Kolloqu. TH Ilmenau (Digest of Papers), Summer 1965, pp. 73–83.
[64] H. D. Piper, "Mechanische Filter," Telefunken Z., vol 32, December 1959, pp. 279–283.
[65] M. Börner, "Berechnung Mechanischer Filter," Elektron. Rundsch., vol. 15, January 1961, pp. 11–14.
[66] A. Günther, "Bemerkungen zum Entwurf breitbandiger mechanischer Filter," Nachrichtent. Z., vol. 25, August 1972, pp. 345–351.
[67] R. A. Johnson, "New Single Sideband Mechanical Filters," WESCON Techn. Papers, vol. 14, August 1970, sec. 10/1, pp. 1–10.
[68] A. E. Günther, "Electromechanical Filters—Satisfying Additional Demands," Proc. 1973 IEEE Int. Symp. Circuit Theory, April 1973. pp. 142–145.
[69] V. Sobotka and J. Trnka, "Die Benutzung von Digitalrechnern beim Entwurf elektromechanischer Filter für die Grosserienfertigung," Frequenz, vol. 26, June 1972, pp. 177–182.
[70] R. Bosc and P. Loyez, "Design of an Electromechanical Filter at 128 kHz in a Two Step Modulation System," Proc. 1974 IEEE Intern. Symp. on Circuits and Systems, April 1974, pp. 111–114.
[71] F. Kamiński, "Synthesis of an Electromechanical Filter With Simple Couplers," Archiwum Elektrotechniki, vol. 22, no. 3, 1973, pp. 525–537 (in Polish).
[72] S. Kogan and A. Stepanov, "Electromechanical Filters for Long Distance Communications Systems," Vopr. Radioelektron., TPS, 1967, No. 7.
[73] H. Albsmeier and A. Günther, "Verfahren zum Abgleich eines mechanischen Filters," German Published Patent Appl. 2048125, September 30, 1970.
[74] A. Günther, "Verfahren zum Abgleich mechanischer Filter," German Published Patent Appl. 2047899, September 29, 1970.
[75] H. Albert and I. Pfeiffer, "Anomalien der Temperaturabhängigkeit des Elastizitätsmoduls von Niob-Zirkonium-Legierungen und reinem Niob," Z. Metallkde, vol. 58, no. 5, 1967, pp. 311–316.
[76] H. Albert and I. Pfeiffer, "Uber die Temperaturabhangigkeit des Elastizstatsmoduls von Niob-Titan-Legierungen," Z. Metallkde, vol. 63, no. 3, 1972, pp. 126–131.
[77] J. Hannon, P. Lloyd and R. Smith, "Lithium Tantalate and Lithium Niobate Piezoelectric Resonators in the Medium Frequency Range With Low Ratios of Capacitance and Low Temperature Coefficients of Frequency," IEEE Trans. Sonics Ultrason., vol. SU-17, Oct. 1970, pp. 239–246.
[78] T. Akashi, "Piezoelectrics for Superceding Ferrites in Telecommunications," IEEE Trans. Mag., vol. MAG-8, Sept. 1972, pp. 705–707.

Section I-B
Filter Synthesis

Although there is a broad cross section of papers in this section, we omitted a number of others because they were not published in English. Of particular value are the papers of Kohlhammer which deal with wire bridging [1], [2], but also of interest are older papers by Trzeba [3] and Kurth [4] which relate to monotonic-stopband (polynomial) filters.

The first paper in this section, "The Design of Electro-Mechanical Filters for Telecommunications," contains a good summary of a monotonic-stopband mechanical filter design procedure. In addition, the Jungwirt paper deals with the problem of trying to maintain the greatest uniformity and a minimum number of parts in order to reduce manufacturing costs. This problem is sometimes solved through the use of network transformations (such as the Norton transformation) or alternatively by optimization. In this paper, however, it was done through an iterative solution of the network equations.

The next four papers deal with the problem of realizing attenuation poles at finite frequencies through the use of bridging wires. Other methods of obtaining attenuation poles through the use of additional shunt or series branch elements in ladder configurations are possible, but they have not proved to be practical in a manufacturing environment. The paper "The Design of Mechanical Filters with Bridged Resonators" is a survey paper that acquaints the reader with a variety of electrical and mechanical topologies and methods of realizing bridged networks. Guenther's paper, which follows, provides an explanation of the use of bridging wires to improve the differential delay response of a mechanical filter. In addition, he discusses the effect of the electrical terminating tuned circuits on the reflection-coefficient zeros of the transfer function. This relationship is used to design a filter with good temperature stability and prescribed driving-point impedance characteristics. The paper "A Torsional-Mode Pole-Type Mechanical Channel Filter" is excellent from the viewpoint of providing the reader with a detailed description of the synthesis and realization of a specific bridging-wire type filter. Although the authors have limited themselves to a single filter, the coupling, tuning, bridging, and fabrication concepts discussed in the paper can be generalized to other mechanical filter designs.

In the final paper of this section, Cucchi and Molo discuss a computational method for obtaining bridging realizations of any general form. In addition to the synthesis procedure, the paper describes a means of obtaining finite attenuation poles at low frequencies by mass loading the bridging wires. Each mass contributes phase shift to the greater than 180° needed between the input and output of the wire for realizing attenuation poles above and below the filter passband.

References

[1] B. Kohlhammer and H. Schüssler, "Berechnung allgemeiner mechanischer Koppelfilter mit Hilfe von äquivalenten Schaltungen aus konzentrierten elektrischen Schaltelementen," *Wiss. Ber. AEG-Telefunken*, vol. 41, no. 3, pp. 150–159, 1968.

[2] B. Kohlhammer, "Ein neuartiges Entwurfsverfahren zur Synthese von mechanischen Filtern und von Gyrator-Filtern," *Wiss. Ber. AEG-Telefunken*, vol. 43, no. 3, pp. 170–177, 1970.

[3] E. Trzeba, "Elektromechanische Vierpole als Kopplungsfilter," *Hochfrequenzt. u. Elektroak*, vol. 69, no. 3, pp. 108–117, 1960.

[4] C. Kurth "Eine Wellenparametertheorie für mechanische Vierpole in Kompressions-oder Torsionsschwingungen," *Nachrichtentech. Z.*, vol. 9, pp. 490–503, Nov. 1959.

THE DESIGN OF ELECTRO-MECHANICAL FILTERS FOR TELECOMMUNICATIONS

Jaroslav JUNGWIRT, *TESLA — Strašnice, Prague*

The article presents a method for the design of electromechanical filters based on an approximation of lossless filter transmission properties by LC models of quasi-polynomial type. Elements of the models are determined by employing any one of the standard techniques for the design of LC filters (e.g. by frequency transformation of the lowpass). An iterative approach to the solution of the equations set has been used to avoid the necessity of dealing with higher order algebraic equations.

Electro-mechanical filters (EMF) find their use mainly in carrier telephony where parameters of the filters largely determine the performance and costs of the systems. Methods of designing these filters have received considerable attention in recent years, as evidenced by the great number of papers published on this subject, e.g. [2], [4], [5], [8], [10]. In these papers it is assumed that the lengths of the coupling elements are multiples of a quarter wavelength for a selected mode of vibrations (or the elements behave as if their lengths were multiples of $\lambda/4$), and that the filter is of the Tchebysheff or Butterworth type. This results in a considerable degree of freedom when choosing the basic parameters of the filter.

It is more difficult to design a filter with coupling elements shorter than $\lambda/4$, when constraints are imposed on some of its parameters. For example, we may wish to design a filter having a maximum possible degree of uniformity and exhibiting satisfactory attenuation characteristics. Some of its parameters may be interrelated according to certain arithmetic or logic expressions. In this case, the transfer function of the filter cannot be approximated fulfilling simultaneously all the additional requirements, even if the structure of the circuit equivalent of the filter is known.

In certain individual cases the design can be accomplished numerically on a digital computer using an appropriate optimization procedure. However, such an approach is not suitable when a whole set of filters is to be designed with a high degree of physical dimension unification.

In the TESLA Telecommunications Research Institute a unified method for the design of EMF has been developed which enables through proper modification of the electric equivalent circuit of the filter to approximate its transmission properties in a narrow frequency band by a transformable *LC* model even when the filter contains coupling sections of various lengths.

In what follows, an equivalent circuit for the EMF will be discussed, into which the so called positive inverter notion will be introduced and considered in conversion of the equivalent circuit into an *LC* model. Then a modified equivalent circuit will be derived and, by means of an example, the iterative synthesis of an EMF will be demonstrated.

ELECTRICAL EQUIVALENT OF ELECTRO-MECHANICAL FILTERS

Let us assume that the mechanical elements of the EMF can be represented as four-terminal or two-terminal networks (if one end of the element is clamped or vibrates freely). To be able to design an electrical equivalent of an EMF, let us state some fundamental relations. We shall make use of the well-known analogies of the first kind, that is, $F \sim U$, $v \sim I$, where F, U, v and I denote force, voltage, velocity and current respectively. A section of lossless transmission line is normally represented by a π or T network, the elements of which can be determined by solving the relevant wave equations (see e.g. [4]).

We shall restrict our discussion to uniformly tapered mechanical elements, this being sufficient for an illustration of the method. Then, such an element can be represented by the networks shown in *Figs 1 (a)* to *1 (d)*.

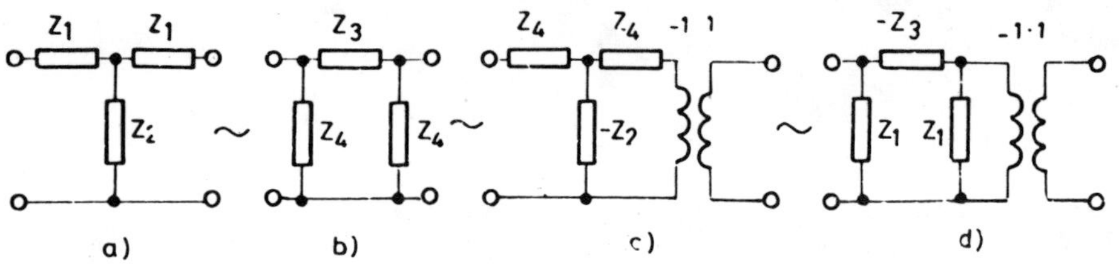

Fig. 1. *Electric two-port equivalents of a section of uniform transmission line*

The impedances are specified by the following formulas:

$$Z_1 = jZ_0 \tan\frac{\Theta}{2}, \quad (1a)$$

$$Z_2 = -j\frac{Z_0}{\sin\Theta}, \quad (1b)$$

$$Z_3 = jZ_0 \sin\Theta, \quad (1c)$$

$$Z_4 = -jZ_0 \cot\frac{2}{\Theta}, \quad (1d)$$

where Z_0 is the wave impedance and $\Theta = \omega l/v_0$, where l is the length of the equivalent transmission line, v_0 is the velocity of propagation of the particular vibration mode.

In lumped-parameter representation of the above impedances, that is, in approximation of the trigonometric functions by the rational ones, we usually require that the lumped equivalent represents the line with sufficient accuracy only around the center frequency ω_0 of the passband. Then, it is sufficient to take only one or two *L* or *C* elements. The conditions from which the element values can be calculated are usually chosen so that

$$Z_e(\omega_0) = Z_m(\omega_0), \quad (2)$$

$$\left.\frac{\partial Z_e(\omega)}{\partial\omega}\right|_{\omega=\omega_0} = \left.\frac{\partial Z_m(\omega)}{\partial\omega}\right|_{\omega=\omega_0} \quad (3)$$

where $Z_e(\omega)$ is the impedance of the *LC* equivalent and $Z_m(\omega)$ represents the mechanical impedance of the network in *Fig. 1*. In one-element approximation, condition (2) is used, while both conditions (2) and (3) are used when impedances (1) are represented by two elements. From the equivalent circuits of *Fig. 1* we choose those for which the mechanical impedances can be represented by realizable electrical impedances. For example, in the region where $l < \lambda_0/4$ circuit *1(a)* is suitable, where $\lambda_0/4 < l < \lambda_0/2$ circuit *1(c)* should be used, for $l \doteq \lambda_0/2$ circuit *1(c)* or *1(d)* would be appropriate, etc. The electro-mechanical transducers can be represented in a similar way.

Only two types of electro-mechanical transducers are practically feasible — the magnetostrictive and piezoelectric ones. Electrical equivalents for the two types (longitudinal vibrations) can be found in [8], [9].

For practical purposes, equivalent circuits for the transducers can be further simplified so that they become identical for both types, as shown in *Fig. 2*. The two-port denoted as EMC is characterized by the following equation for the magnetostrictive transducer:

$$\begin{bmatrix} U \\ I \end{bmatrix} = \begin{bmatrix} 0 & \alpha_m \\ \frac{1}{\alpha_m} & 0 \end{bmatrix} \cdot \begin{bmatrix} F \\ v \end{bmatrix}, \quad (4)$$

and for the piezoelectric transducer by

$$\begin{bmatrix} U \\ I \end{bmatrix} = \begin{bmatrix} \alpha_P & 0 \\ 0 & \frac{1}{\alpha_P} \end{bmatrix} \cdot \begin{bmatrix} F \\ v \end{bmatrix}, \quad (5)$$

where α_M, α_P are constants specifying the process of energy conversion.

A voltage generator with an internal resistance R_{eg} is seen at the output of the two-port EMC as a source of mechanical force *F* with an internal resistance R_{mg}. Therefore, in subsequent discussion we shall regard the terminals *1, 1'* on the mechanical side of the transducer as input (or output) terminals of the filter terminated by constant mechanical resistances.

The equivalent circuit of the transducer is virtually a half section of a band-pass filter which may form a relatively wide band-width network or it may be incorporated into the overall equivalent circuit of the filter.

To achieve the relative bandwidth of the filters $B > 1$ to 2 %, it is usually necessary to add external *LC* circuits which can be transformed by the constants α_M or α_P to the mechanical side *1, 1'* of the transducer as mechanical impedances.

There is a wide variety of EMF using various transducers, resonators, couplers and modes of vibrations. Since it is impossible to deal with all of them here, we choose a filter with a certain structure, e.g. consisting of torsional vibrators *(Fig. 3)* which finds its application mainly as a channel filter in carrier telephony systems with direct modulation. The resonators work as mechanical two-terminal networks which can be represented according to equations (1), (2) and (3) by series resonant circuits.

Let us represent the couplers by their electrical equivalents according to *Fig. 1(a)* — so far without introducing *LC* elements. The external electrical circuits are transformed to mechanical terminals *1, 1'* in accordance with equation (5), where they form parallel resonant circuits together with mechanical capacitances C_0 (see *Fig. 2*).

The electrical equivalent circuit of the couplers contains impedances which can be expressed for the k-th section as

$$Z_{cs_k} = jZ_{o_k} \tan \frac{\Theta_k}{2}, \quad (6)$$

$$Z_{cp_k} = \frac{-jZ_{o_k}}{\sin \Theta_k}. \quad (7)$$

The resulting equivalent circuit of the entire filter is presented in *Fig. 4* and can be used e.g. for computer evaluation of the filter attenuation. At this stage it is not necessary to express explicitly the electrical equivalents of the resonators using equations (2) and (3); a symbolic representation of the resonators will be sufficient.

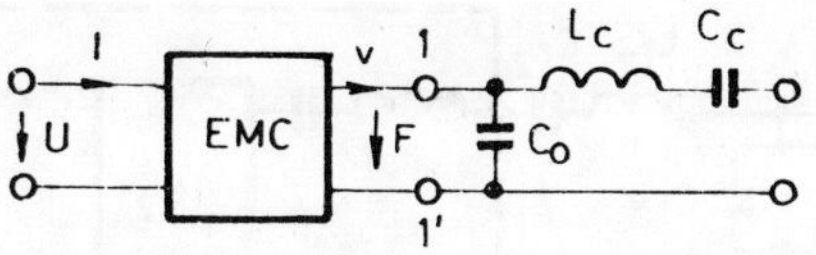

Fig. 2. *A simplified equivalent circuit for an electro-mechanical transducer with longitudinal vibrations: L_c, C_c — equivalent elements of mechanical resonant circuit*

THE USE OF THE IMPEDANCE INVERTER IN THE SYNTHESIS OF ELECTROMECHANICAL FILTERS

In theoretical investigation of an EMF it is useful to introduce a passive two-port featuring properties of the positive impedance inverter. It can be described by a set of chain parameters. The corresponding matrix equation is

$$\begin{bmatrix} U_1 \\ I_1 \end{bmatrix} = \begin{bmatrix} 0 & jZ_{inv} \\ j/Z_{inv} & 0 \end{bmatrix} \cdot \begin{bmatrix} U_2 \\ I_2 \end{bmatrix}, \quad (8)$$

where Z_{inv} is a real constant impedance.

When dealing with mechanical networks, F and v or the twisting moment M and angular velocity ψ instead of U and I should be used, respectively. Z_{inv} will have the dimension of the pertinent mechanical resistance.

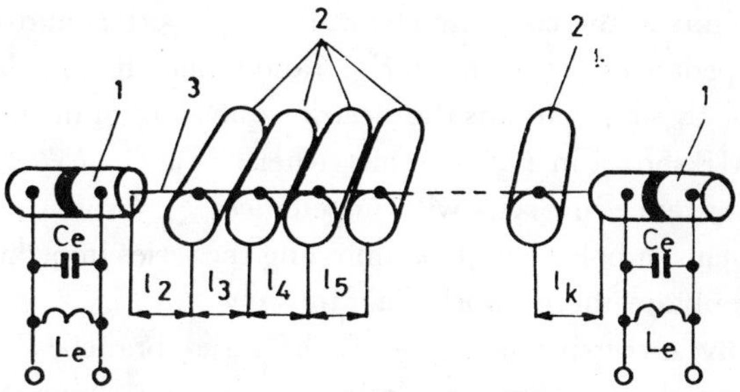

Fig. 3. *Physical construction of the filter used for illustration of the design method: 1 — transducers with longitudinal vibrations, driven by a piezo-ceramic wafer; 2 — $\lambda_0/2$ torsional resonators; 3 — coupling elements with longitudinal vibrations; l_2 to l_k are the lengths of the couplers, L_e, C_e are external electrical circuit elements*

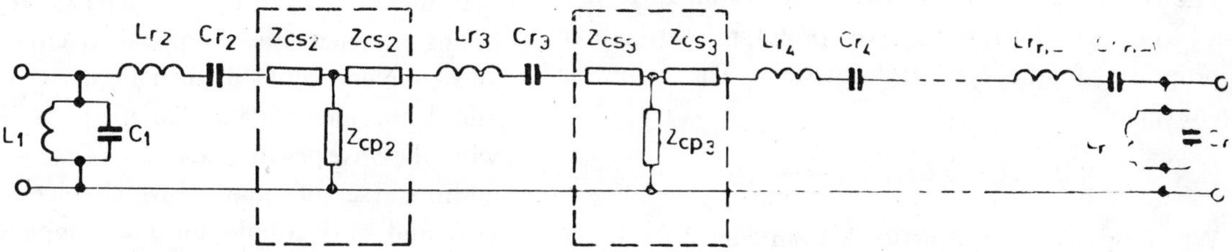

Fig. 4. *Electric equivalent of the filter shown in Fig. 3: L_1, C_1, L_n, C_n — electric elements transformed to the mechanical side; Lr_i, Cr_i — elements representing the resonators and transducers*

For an inversion of impedances the familiar relation holds

$$Z_{in} = \frac{Z_{inv}^2}{Z}. \quad (9)$$

The equivalent circuit in *Fig. 1(a)* can be modified by splitting the series impedances as shown in *Fig. 5*. Then the following equalities hold:

$$Z_{c_k} = -jZ_{o_k} \cot \frac{\Theta_k}{2}, \quad (10)$$

$$Z_{cp_k} = \frac{-jZ_{o_k}}{\sin \Theta}, \quad (11)$$

where Z_{o_k} is the wave impedance of a section of the transmission line.

If we choose the length of all couplers e.g. $l = \lambda_0/4$ for the center frequency of the filter $\omega = \omega_0$, the impedance

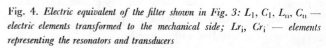

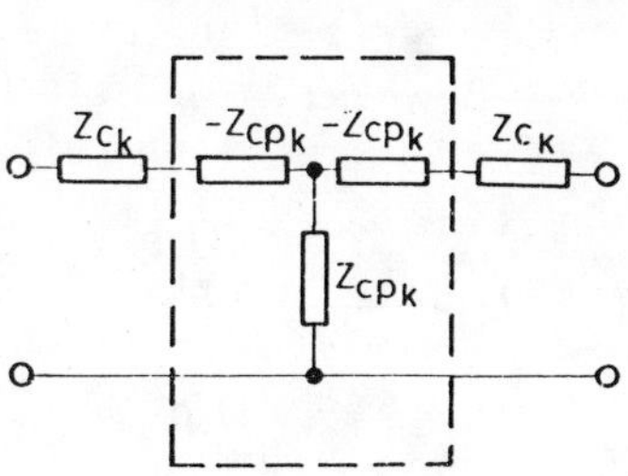

Fig. 5. *Representation of a coupling section by series impedances and an inverter*

Z_{cp_k} will be almost constant in the vicinity of ω_0, namely

$$Z_{cp_k} \doteq -jZ_{o_k}. \quad (12)$$

The two-port of *Fig. 5* consisting of impedances Z_{cp_k} and $-Z_{cp_k}$ can then be regarded as an impedance inverter with $Z_{inv} = Z_{o_k}$. By breaking up the equivalent

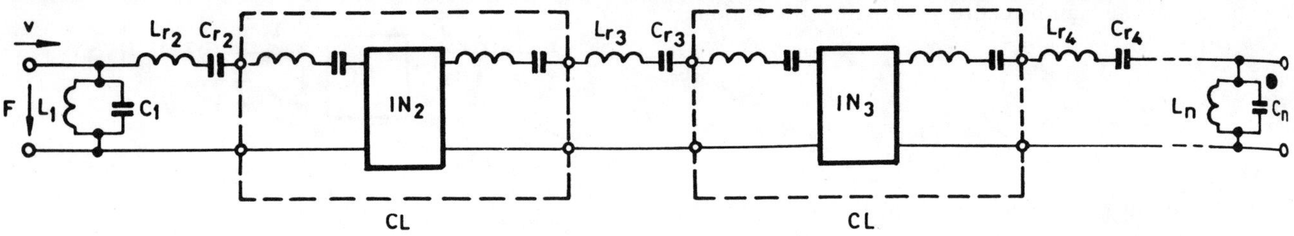

Fig. 6. *Electric equivalent of a filter having its coupling elements represented by series impedances and inverters: CL — coupling sections, IN_i — inverters*

circuits of the coupling elements into inverters and series impedances according to *Fig. 5* and replacing impedances Zcs_k by series circuits the electric equivalent of the overall EMF shown in *Fig. 6* results, where $IN_2, \ldots IN_k$ denote impedance inverters with impedances Z_{inv} which will be given symbols ρ_2 to ρ_k. Combining the series impedances, we obtain the network diagram shown in *Fig. 7.*

By successive inversion of the series branches of the network according to relation (9), we can eliminate the inverters and obtain the common ladder network [6], [10]. For an even number of inverters we get a symmetrical band-pass filter shown in *Fig. 8(a)*, for an odd number of inverters an antisymmetrical band-pass filter with an *IN* inverter in cascade results — *Fig. 8(b)*— and, in both cases, dual equivalents can be obtained.

The two filters are in fact LC models of an EMF and can be obtained from the lowpass prototypes through an appropriate frequency transformation. The following equality holds:

$$Ls_i \,.\, Cs_i = Lp_i \,.\, Cp_i = \frac{1}{\omega_0^2}\,. \qquad (13)$$

When converting the network from *Fig. 7* to an *LC* model, we get relations between respective elements of the two configurations; for example, in the case of the symmetrical bandpass filter, we have

$$L_1 = Lp_1, \qquad (14a)$$
$$L_2 = Ls_1.$$

For odd $i \leq n - 2$, we get

$$L_i = \frac{1}{\omega_0 Lp_i} \cdot \frac{\prod_{j=1}^{\frac{i-1}{2}} \rho_{2j}^2}{\prod_{j=1}^{\frac{i-3}{2}} \rho_{2j+1}^2}, \qquad (14b)$$

while for even $i \leq n - 3$, we obtain

$$L_i = Ls_i \,.\, \prod_{j=1}^{j=\frac{i-2}{2}} \frac{\rho_{2j+1}^2}{\rho_{2j}^2}, \qquad (14c)$$

and similarly for the final elements of the network, we have

$$L_{n-1} = \frac{1}{N^2} \,.\, Ls_{(n-1)}, \qquad (14d)$$
$$L_n = \frac{1}{N^2} \,.\, Lp_n.$$

The transformation ratio of the transformer is given by

$$N^2 = \prod_{j=1}^{j=\frac{i-3}{2}} \frac{\rho_{2j}^2}{\rho_{2j+1}^2}. \qquad (14e)$$

Inductances L_2 to L_{n-1}, which have resulted from combining the elements of equivalent circuits for resonators and couplers, could in turn be derived from the mechanical parameters of the components — as it was done with the inverter impedances. This would give rise to additional sets of equations which, together with equations (13) and (14), would form a complete set needed for calculating the filter parameters. Since the set of equations contains more unknown variables than the number of equations, values of some of them can be chosen. However, the variables are subject to certain constraints, as will be shown later.

In conclusion, an EMF with quarter-wave couplers can be represented by an *LC* model of a transformable bandpass filter. Such a model is quite convenient to deal with, the most difficult task being the proper selection of some parameters and solution of the set of equations.

The design of the filter in *Fig. 4* is, however, much more involved when the lengths of the coupler $l \neq \lambda/4$.

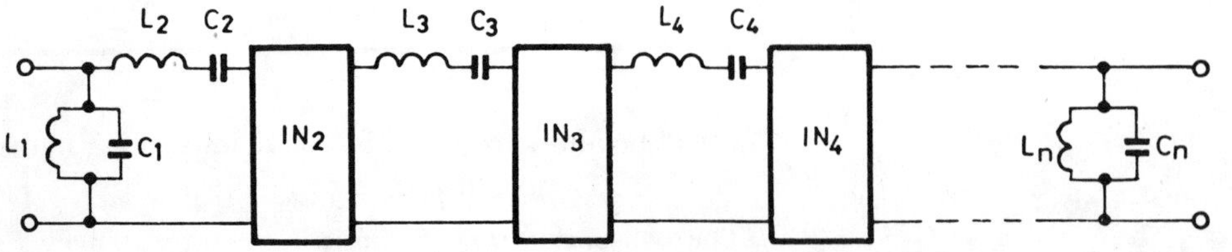

Fig. 7. *The network from Fig. 6 with its series impedances combined*

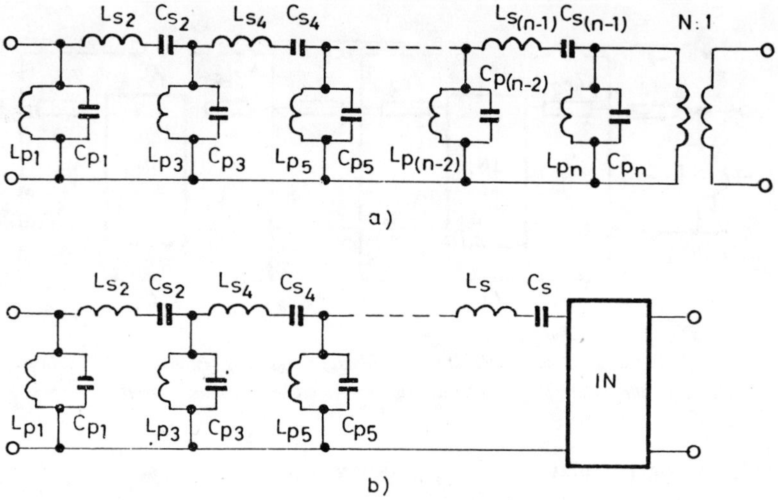

Fig. 8. *A frequency-transformable model of EMF: a — symmetrical band-pass filter; b — antisymmetrical band-pass filter*

For example, in case when $l < \lambda/8$ and the passband is relatively narrow, using relations (1), (2) and (3), or *Fig. 1(a)* we can — as a first approximation — replace the series branches of the equivalent four-terminal networks by inductances and the shunt arms by capacitances. Then, combining impedances in the series arms, we obtain an electric equivalent of the filter shown in *Fig. 9* which is a lumped-constant approximation to the EMF in a narrow frequency band.

The composite loss of a filter is given by

$$b = \frac{1}{2}\ln(1 + |\varphi(p)|^2). \qquad (15)$$

The function $\varphi(p)$ of the complex variable $p = \sigma + j\omega$ can be expressed for this particular case according to (3) as

$$\varphi = H\frac{\prod_{k=1}^{k=n}(p^2 - p_k^2)}{p^3}, \qquad (16)$$

where n is the order of the filter, p_k are the roots of $\varphi(p)$ lying on the imaginary axis, H is a real constant.

The frequency transformation of this bandpass filter is not always of the immittance type. Therefore, when approximating the transfer function or $\varphi(p)$, each case must be treated individually. Moreover, if the quantities involved are correlated (e.g. when equal wave impedances of all resonators, or lengths and/or impedances of all couplers are required), such an approximation is not generally feasible. In order to make the design of such a bandpass filter generally possible, as it is in the case of $\lambda/4$ couplers, we have to start off from the frequency transformable bandpass filter shown in *Fig. 8* and find its approximate equivalent according to *Fig. 7*.

At the cost of introducing some error, the equivalent for the coupler can — for a relatively narrow frequency band — be broken down into series impedances and inverters according to *Fig. 7* even when the lengths of the couplers are not equal to $\lambda/4$ at the center frequency ω_0. Then the inverter impedance will be frequency dependent over the passband, but in order to facilitate the design of the filter we shall assume it constant and equal to the value of Z_{cp} for $\omega = \omega_0$. Denoting the constant impedance of the inverter ρ_k, we have

$$\rho_k = \frac{Z_{0k}}{\sin\frac{\omega_0 l_k}{v_0}}, \qquad (17)$$

where l_k is the length of the k-th coupler.

A MODIFIEC EQUIVALENT CIRCUIT

In accordance with the foregoing discussion we can alter the configuration of the network in *Fig. 4* by breaking it down into inverters and series impedances. These impedances can in turn be combined with impedances of the resonators to yield the impedances Z_{s_i} in the network of *Fig. 10* which will be called a modified equivalent network of the EMF. Approximation of the branches Z_s by series resonant circuits according to (2) and (3) yields two sets of equations,

$$\left.\frac{\delta Z_{s_i}}{\delta\omega}\right|_{\omega=\omega_0} = j2L_i, \qquad (18)$$

and

$$Z_{s_i}(\omega_0) = 0. \qquad (19)$$

From this we get

$$C_i = \frac{1}{\omega_0^2 L_i}. \qquad (20)$$

Relations (14), (17), (18) and (19) form a complete

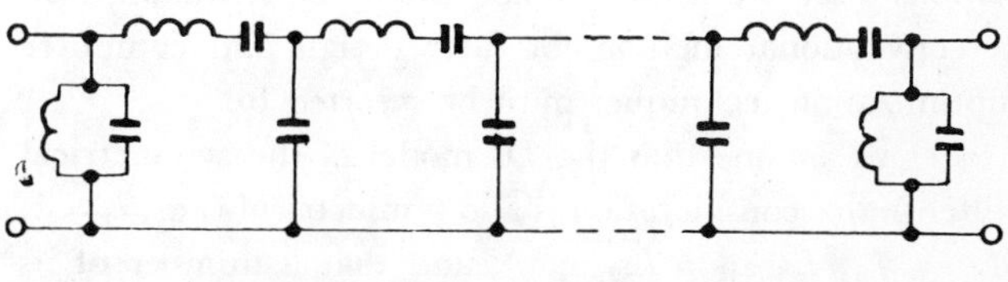

Fig. 9. *An equivalent circuit of an EMF with $l < \lambda_0/4$ couplers*

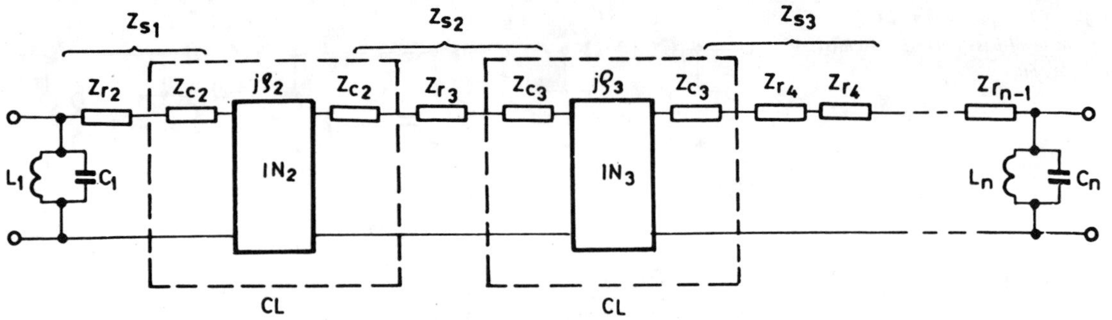

Fig. 10. *A modified equivalent circuit of an EMF: Zr_i — impedances representing resonators and transducers; Zc_i — series impedances resulting from decomposition of the networks representing the couplers; ρ_i — inverter impedances; CL — coupling elements*

set of equations for calculation of the unknown parameters, that is, the length l_k and wave impedances Zo_k of the coupling elements, resonant frequencies (which are implicitly included in expressions for Zr_i) and wave impedances of the resonators, circuit elements L_1, C_1, L_n, C_n (from which by electro-mechanical conversion electric parameters of the external electrical elements can be obtained), and two sets of auxiliary variables L_2 to L_{n-1} and ρ_k.

Now we have less equations than unknowns so that a number of variables can be chosen, the variables being usually subject to some constraints. For example, it may be required that some of the resonators or coupling elements have the same wave impedance or length, or the filter exhibit lengthwise symmetry. Because of relations that exist among the equations, the redundant variables cannot be chosen arbitrarily, but a possible set of constraints must be selected.

The set of equations can then be basically utilized in the synthesis of the EMF in two ways:

(a) Choose proper constraints to make the set of equations complete, find solutions for the set of equations.

(b) Using mathematical programming find the most suitable constraints for the unspecified variables.

It follows from the analysis of the set of equations that the choice of some constraints will have an effect on the electrical model of the filter. For example, if we require that a part of the filter be a cascade of uniform sections (i.e. we want equal wave impedances and resonant frequencies of the resonators, and the same lengths and wave impedances of the couplers), then the corresponding LC model must also consist of uniform networks connected in a chain. The converse does not hold, since a cascade of uniform networks can be realized by a cascade of nonuniform sections. The LC model fulfilling the conditions discussed above cannot usually be synthesized by a conventional method for filter design and computer optimization techniques must be resorted to.

Let us assume that the LC model of the symmetrical filter under consideration is also symmetrical (i.e. $N = 1$, $L_n = L_1$, $Ls_{n-1} = Ls_2$, etc.), and that a number of its elements recur, e.g. $Ls_4 = Ls_6 = \ldots = Ls_{n-3}$, $Lp_5 = Lp_7 = \ldots Lp_{n-4}$. The second assumption means that a part of the LC network is a cascade of uniform sections.

By substitution of the above conditions into the complete set of equations, the number of equations will be reduced. We shall be left with the following sets of equations:

$$\left.\begin{aligned} L_1 &= Lp_1, \\ L_2 &= Ls_2, \\ L_3 &= \frac{\rho_2^2}{\omega_0^2 Lp_3}, \\ L_4 &= \frac{\rho_3^2}{\rho_2^2} \cdot Ls_4 = \frac{\rho_2^2 \rho_4^2}{\rho_3^2 \omega_0^2 Lp_5}; \end{aligned}\right\} \quad (21)$$

$$\left.\begin{aligned} L_2 &= \frac{1}{2}\frac{\partial}{\partial\omega}(Zc_2 + Zr_2)\Big|_{\omega=\omega_0}, \\ L_3 &= \frac{1}{2}\frac{\partial}{\partial\omega}(Zc_2 + Zc_3 + Zr_3)\Big|_{\omega=\omega_0}, \\ L_4 &= \frac{1}{2}\frac{\partial}{\partial\omega}(2Zc_4 + Zr_4)\Big|_{\omega=\omega_0}, \end{aligned}\right\} \quad (22)$$

where Zr_i are impedances of the respective resonators, Zr_2 represents the mechanical side of the converter:

$$\left.\begin{aligned} \rho_2 &= \frac{Zo_2}{\sin\frac{\omega_0 l_2}{v_0}}, \\ \rho_3 &= \frac{Zo_3}{\sin\frac{\omega_0 l_3}{v_0}}, \\ \rho_4 &= \frac{Zo_4}{\sin\frac{\omega_0 l_4}{v_0}}; \end{aligned}\right\} \quad (23)$$

$$\left.\begin{aligned} Zc_2 + Zr_2|_{\omega=\omega_0} &= 0, \\ Zc_2 + Zc_3 + Zr_3|_{\omega=\omega_0} &= 0, \\ 2Zc_4 + Zr_4|_{\omega=\omega_0} &= 0, \\ L_1 C_1 &= \frac{1}{\omega_0^2}. \end{aligned}\right\} \quad (24)$$

Impedances Zr_2, Zr_3 and Zr_4 include the unknown wave impedances of the resonators Zor_2, Zor_3, Zor_4 and the resonant frequencies ωr_2, ωr_3, ωr_4. Since the number of equations is still larger than the number of unknowns, another five constraints can be specified. For example, we can choose l_2, l_3, l_4 and the ratio Zor_2/Zor_4 together with the condition that $Zor_3 = Zor_4$, or require that $Zo_2 = Zo_3 = Zo_4$, $Zor_3 = Zor_4$ and specify the length of the elements ($Zor_3 = Zor_4$) and the ratio Zor_2/Zor_4.

Equations (21), (22), (23) and (24) can be solved by successive elimination of variables which raises the degree of the remaining equations. This is impractical and an alternative approach to the problem relying on iterative procedure has been sought. One way which proved successful in practice is to form several sets of equations, by separation of variables. The dependent variables in one group of equations should constitute a set of independent variables in another group of equations. This enables the calculated values of the independent variables in one group of equations to be used for more accurate evaluation of independent variables in the other group, and vice versa. Let us apply the method to solution of equations (21) to (24).

From (21) we express the variables ρ_i as functions of L_i,

$$\begin{aligned} \rho_2 &= f_2(L_2, L_3 L_4), \\ \rho_3 &= f_3(L_2, L_3, L_4), \\ \rho_4 &= f_4(L_2, L_3, L_4). \end{aligned} \tag{25}$$

Processing equations (22) and substituting the results from (23) and (24) into (22), we obtain another group of quadratic equations, by whose solution we arrive at the following:

$$\begin{aligned} L_{22} &= f_{22}(\rho_2, \rho_3, \rho_4), \\ L_{33} &= f_{33}(\rho_2, \rho_3, \rho_4), \\ L_{44} &= f_{44}(\rho_2, \rho_3, \rho_4). \end{aligned} \tag{26}$$

The computation starts up with supplying initial values for L_2, L_3 and L_4 into equations (25). The resulting values for the variables ρ_i are then inserted into (26) to obtain more accurate values for the variables L_i which are in turn used in (25), etc. Mechanical parameters of the filter are then calculated from the auxiliary variables L_i, ρ_i.

In practice, when the initial values were specified with an accuracy of about 5 %, calculation of the unknown variables to 6 or 7 significant digits required only 2 to 3 iterations.

CONCLUSION

The method of EMF design presented in the paper makes it possible, requiring only modest computational effort, to design the non-transformable narrow-band filters in a way similar to the one used in designing filters whose attenuation characteristics feature geometrical or arithmetical symmetry. It enables a unified treatment of EMFs utilizing various modes of vibrations or unlike sections of nonuniform transmission lines. The resultant set of equations, which serves to find the parameters of the filter, lends itself to sensitivity and tolerance analysis of EMF and provides better understanding of relations among the physical parameters and elements of the filter.

The iterative method proved to be the most efficient approach to the solution of the set of equations. The method described in the paper has been used in research and development of electro-mechanical channel filters for TESLA KPK 12 equipment [7]. All the torsional resonators in the bank of 12 filters covering the 60 to 108 kHz frequency band are of uniform diameter and the coupling sections are of the same length. This high degree of uniformity of the filter elements results in reduced production costs and the possibility of compiling a catalogue of the filters with the aid of a computer.

In certain cases of relatively broad-band filters or when requirements on the attenuation characteristic in the passband are severe, some correction procedures must be used. The description of these procedures is, however, outside the scope of this paper.

REFERENCES

[1] Albsmeier, H.: Ein Vergleich der Realisierungsmöglichkeiten EM-Kanalfilter im Frequenzbereich 12 kHz bis 10 MHz. Frequenz, vol. 25, 1971, No. 3, pp. 74—79

[2] Börner, M.: Mechanische Filter für die Trägerfrequenztechnik. Nachrichtentechnik Fachberichte, vol. 19, 1960, pp. 34—37

[3] Cauer, W.: Theorie der linearen Wechselstromschaltungen. 1954

[4] Jungwirt, J.: Náhradní schémata EMF s rovnoměrně uloženými válečkovými rezonátory. (Equivalent circuits for EMF with uniformly spaced roller resonators. In Czech.) Sborník prací TESLA VÚT, Prague, 1962

[5] Jungwirt, J.: Syntéza EMF podle teorie provozních parametrů. (Synthesis of EMF based on the theory of operating parameters. In Czech.) Sborník prací TESLA VÚT, Prague, 1963

[6] Jungwirt, J.: Návrh EMF s relativně krátkými vazebními členy. (The design of EMF with relatively short coupling sections. In Czech.) Research Report, TESLA VÚT, Prague, 1967

[7] The Electromechanical Filters of TESLA Strašnice. TESLA Electronics, vol. 5, 1972, No. 2, pp. 57—59

[8] Mason, W. P.: Electromechanical Transducers and Wave Filters. 1958, 2-nd ed., D. Van Nostrand Co.

[9] Schüssler, H.: Mechanische Filter mit piezoelektrischen Wandlern. Telefunken-Zeitung, vol. 39, 1966, No. 314, pp. 429—439

[10] Struczynski, W.: A Theoretical Analysis of the EMF. Marconi Review, 1959, II. Quarter, pp. 119—143

THE DESIGN OF MECHANICAL FILTERS WITH BRIDGED RESONATORS

Robert A. Johnson
Collins Radio Group-Rockwell International
Newport Beach, California

Summary

An overview of the state-of-the-art of realizing attenuation poles through the use of wire bridging is presented. Included are acoustic methods of realizing phase inverters, synthesis techniques and examples of presently designed filters. Possibilities of future work are outlined.

Introduction

Primary interest in bridging-wire realizations has centered around the need to reduce both absolute and differential delay while maintaining a specified stopband attenuation. The first designs were introduced about 10 years ago and involved simple bridging across one and two resonators for the purpose of realizing $j\omega$-axis attenuation-poles.[1] At this time, a proposal was made that the bridging also be used to realize non-minimum-phase sections for delay equalization; but these filters have not been realized in manufacturing. Although other methods of realizing finite-frequency attenuation poles have been used, none have been as effective as the bridging technique. The other methods have included the use of multiple resonator modes in ladder networks and lattice structures.

One of the drawbacks of bridging-wire filters is their greater complexity; another is their greater sensitivity to variations in resonator frequencies, adjacent resonator coupling and in some cases welding tolerances. Further work needs to be done in analyzing the sensitivities of various configurations, possibly using methods like those of Schüssler[2] and Sobotka[3].

Bridging Topologies

The earliest mechanical filters to use bridging techniques were mechanically coupled quartz plates which used input-to-output capacitor coupling to obtain attenuation poles above and below the filter passband.[1] These were followed by three-resonator and four-resonator bridging filters. An example of a four-resonator filter is shown in Fig. 1.

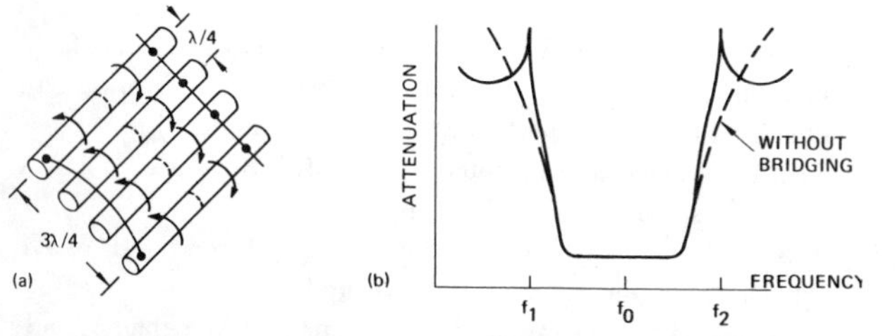

Fig. 1 - Torsional mode mechanical filter; (a) pictorial diagram showing nodal points (dashed lines) and direction of motion at the lower passband edge and (b) amplitude response.

In this design, the torsional-mode resonators are directly coupled by a one-quarter wavelength (extensional mode of vibration) wire and bridged by a three-quarter wavelength wire. At the attenuation-pole frequencies f_1 and f_2, the vibration amplitude contribution due to the bridging wire is equal to that of the adjacent resonator coupling (but opposite in phase) and cancellation takes place. In this case the number of resonators bridged is two ($n = 2$) and the bridging wire acts as a phase inverter (-1:1 transformer) as well as a compliance (spring). Table 1 lists the results of changing either phase or the number of bridged resonators. Table 1 applies to cases where the reactances of the bridging and coupling elements have the same sign and when there is no phase inversion under the bridging. In addition to the simple bridging techniques shown in Table 1, there are other methods available such as the use of multiple-wire bridging, twin-tees and mass loading; all of which may additionally use electrical input-to-output bridging.

n	PHASE	ATTENUATION POLE LOCATION
EVEN	−1:1	jω AXIS – ABOVE & BELOW PASSBAND
EVEN	+1:1	COMPLEX – RIGHT & LEFT HALF-PLANES
ODD	−1:1	jω AXIS – BELOW FILTER PASSBAND
ODD	+1:1	jω AXIS – ABOVE FILTER PASSBAND

Table 1 - Attenuation-pole locations for different n and phase conditions.

Fig. 2 shows some of the bridging methods described in the mechanical filter literature. The dashed lines in Fig. 2 represent nodal points and the heavy dots represent wire-to-resonator welds; the bracketed numbers are reference papers. The terms in-phase or out-of-phase regions correspond to the direction of motion of regions on the resonators at the lower passband edge frequency (where adjacent resonators vibrate in-phase). The coupling in all cases is predominately through the extensional mode of vibration of the bridging and coupling wires and unless noted, the length of the wire is less than one-half wavelength. The diagrams are probably self-explanatory, with a few exceptions such as the twin-tee with double resonator bridging in the disk flexure case (the center resonator has two nearby modes of vibration thus acting like two single-mode disks). If you have difficulty picturing the resonator mode in the different diagrams, remember that the wires are all acting as extensional (longitudinal) mode elements; thus defining the direction of motion of the resonators. From Fig. 2 you will note the various means of obtaining phase inversion; namely through connecting points that are moving in opposite directions or through using long wires where the opposite ends of the wires are out-of-phase. The mass loaded case involves reducing the wire length by

Reprinted from *Proc. 1975 IEEE Int. Symp. on Circuits and Syst.*, Apr. 21–23, 1975, pp. 313–316.

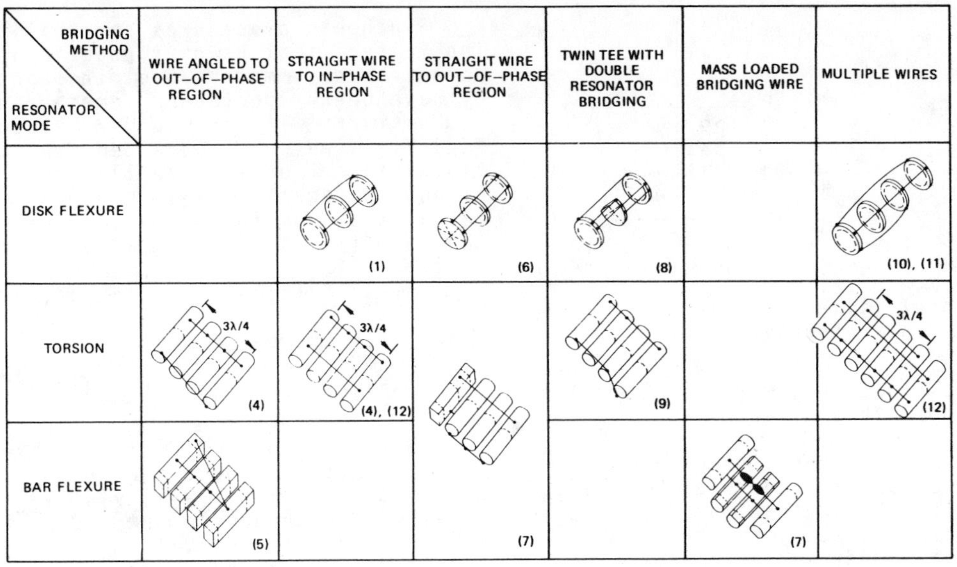
BRIDGING METHOD
RESONATOR MODE
WIRE ANGLED TO OUT–OF–PHASE REGION
STRAIGHT WIRE TO IN–PHASE REGION
STRAIGHT WIRE TO OUT–OF–PHASE REGION
TWIN TEE WITH DOUBLE RESONATOR BRIDGING
MASS LOADED BRIDGING WIRE
MULTIPLE WIRES
DISK FLEXURE
TORSION
BAR FLEXURE
(1)
(6)
(8)
(10), (11)
3λ/4
(4)
(4), (12)
(9)
(12)
(5)
(7)
(7)

Fig. 2 - Mechanical filter resonator-bridging realizations.

stringing masses on the bridging wire thus obtaining the required phase inversion.

Mechanical Realizations & Equivalent Circuits

Fig's 3 and 4 illustrate, in greater detail, means of obtaining phase inversion. Fig's 3(a) and 3(b) show the attachment of two wires at out-of-phase sections of flexure-mode and torsion-mode resonators. Making use of the mobility analogy, which preserves the topology of the electrical and mechanical systems[13], we obtain Fig. 3(c) where the mechanical resonator is represented by a parallel tuned circuit and the coupling wires by inductors. The placement of the phase inverter in cascade with one of the other inductors is arbitrary. Fig. 4 shows lumped-element equivalent circuits of short lines (less than $\lambda/8$), quarter-wavelength and three quarter-wavelength mechanical lines (wires propagating acoustic extensional or torsional waves). Generally, short or quarter-wavelength wires are used for adjacent resonator coupling, whereas the means of obtaining phase inversion is dependent on the mechanical configuration and filter center frequency. At frequencies below 200kHz, the use of greater than half-wavelength wires for phase inversion is usually not practical because of the long acoustic wavelengths involved so connections to out-of-phase portions of a resonator must be used.

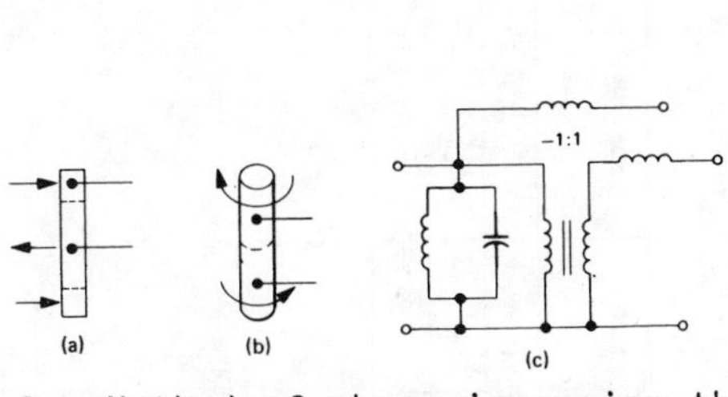

Fig. 3 - Method of phase inversion through attachment of coupling wire at out-of-phase sections of a (a) flexure-mode resonator; (b) torsion-mode resonator; (c) electrical equivalent circuit.

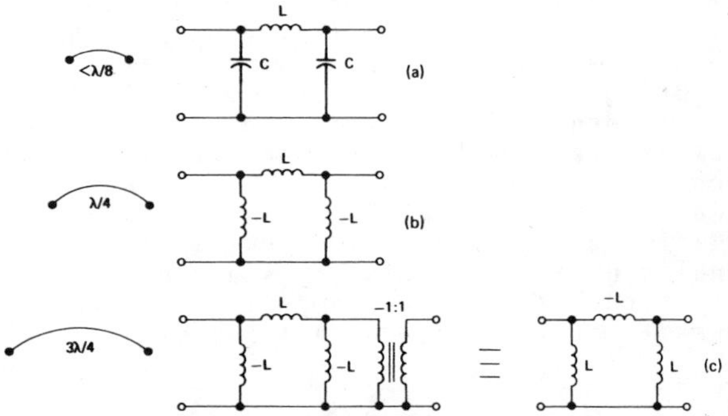

Fig. 4 - Electrical equivalent circuits of a (a) short acoustic line, (b) one-quarter wavelength line and (c) three-quarter wavelength line.

A good example of the use of both double and single-resonator bridging is the telephone channel filter described by Yakuwa[7] and shown in Fig. 5. The filter is driven in flexure by a composite piezoelectric ceramic attached to a nickel-alloy bar. The internal rod-resonators vibrate in a torsional mode. The center wire is attached to nodal points and is used as a support. The bridging and coupling wires attached to the end resonators vibrate out-of-phase but are attached to points on the torsional resonators that are in-phase (note that one is attached to the top of a resonator and the other is connected to the bottom of the other half of the resonator). The two bridging sections realize a pair of upper and lower attenuation-poles and a lower-stopband pole as shown in Fig. 5.

Synthesis of Bridged Sections

Considering the multitude of mechanical filter bridging configurations (Fig. 2), we can correctly assume that there are many synthesis techniques involved in realizing these networks. Table 2 is designed to show a wide variety of these methods. The references

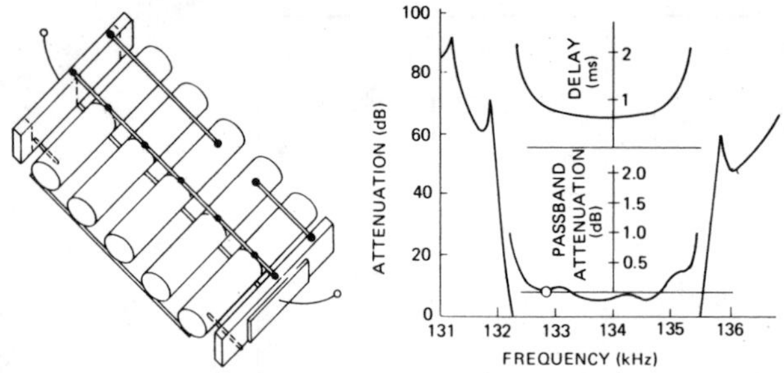

Fig. 5 - Channel filter with two-resonator and one-resonator bridging.

shown in brackets lead to a more complete list of papers dealing with similar methods.

The most common synthesis techniques involve the realization of ladder networks which are transformed to a bridging topology, the realization of bridged sections (from polynomials) within a ladder network and the realization of multiple bridging-wire sections based on the Cauer transformation shown in Fig. 6. In the Cauer case, impedance inverters (λ/4 and 3λ/4) are added in cascade to the left-hand-side of the network. The λ/4 inverter is moved into the network both as a bridging and coupling element. The process is then repeated until the attenuation poles are realized. This procedure for realizing bridging elements is outlined in detail in Refs. 15 and 17. Bell has developed a realization of a similar form, but with additional coupling elements, by starting from a lattice realization and comparing lattice-arm values to bisected ladder arm terms[18].

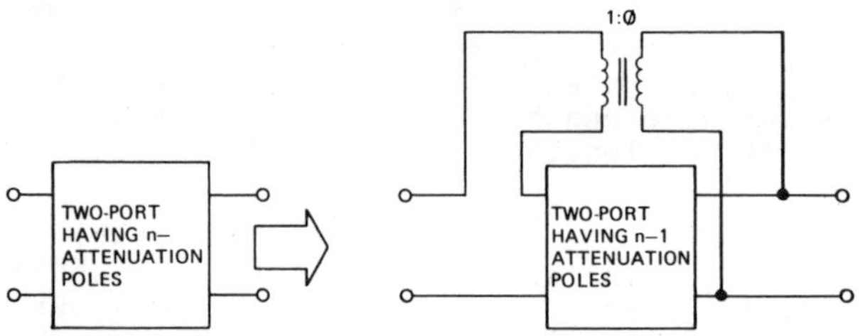

Fig. 6 - Cauer transformation.

Synthesis procedures for realizing bridging filters that have attenuation poles that do not have symmetry about the center frequency have been reported by Temes[10] and Günther[20]. Temes starts with the ladder network shown in Fig. 7(a), which is obtained from a conventional insertion-loss realization. The short-circuit parameters (y_{11}, y_{12}, y_{22}) are then calculated from the element values and have the form (i,j=1,2),

$$y_{ij}=(-1)^{i+j}(a_{ij}S^4+b_{ij}S^2+c_{ij})/(dS^5+eS^3+f)$$

where, for example,

$$a_{11}=L_4(L_2+L_3)+L_5(L_2+L_3+L_4).$$

Next a bridging inductor L_a is removed from the network and a new set of short circuit matrix elements y_{11}, y_{12}, y_{22} are found. Finally, the three element shunt circuit is converted to its equivalent and then a tee-to-pi transformation is made on the series inductor and coupling elements L_1' and L_2' to obtain a network with bridging inductor L_b.

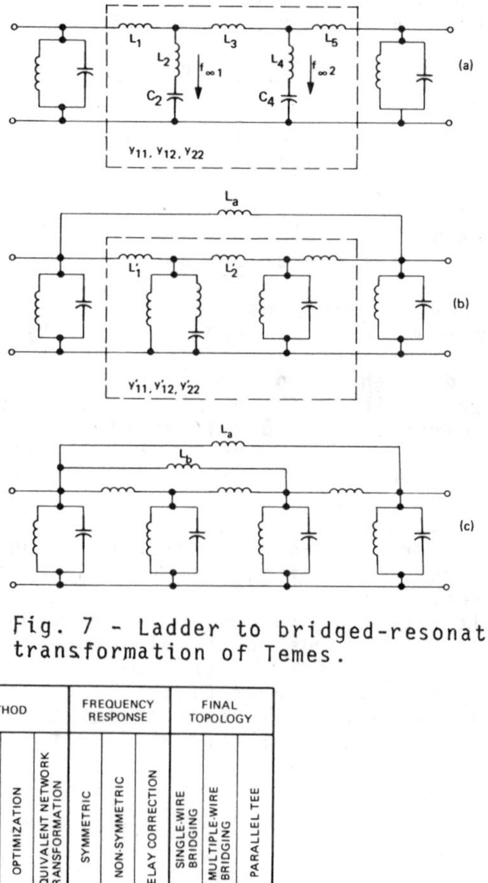

Fig. 7 - Ladder to bridged-resonator transformation of Temes.

SYNTHESIS / PAPER	FREQ. PLANE: LOWPASS	FREQ. PLANE: BANDPASS	STARTING POINT: POLYNOMIAL	STARTING POINT (NETWORK): LADDER	STARTING POINT (NETWORK): LATTICE	STARTING POINT (NETWORK): BRIDGED TOPOLOGY	METHOD: IMAGE PARAMETER	METHOD: INSERTION LOSS	METHOD: OPTIMIZATION	METHOD: EQUIVALENT NETWORK TRANSFORMATION	FREQUENCY RESPONSE: SYMMETRIC	FREQUENCY RESPONSE: NON-SYMMETRIC	FREQUENCY RESPONSE: DELAY CORRECTION	FINAL TOPOLOGY: SINGLE-WIRE BRIDGING	FINAL TOPOLOGY: MULTIPLE-WIRE BRIDGING	FINAL TOPOLOGY: PARALLEL TEE
BÖRNER (4)	X					X	X			X	X		X	X		
KOHLHAMMER & SCHÜSSLER (14)	X			X						X	X		X	X		
KOHLHAMMER (15)	X		X					X			X		X		X	
KÜNEMUND (16)	X	X		X						X	X			X		
WITTMANN ET. AL. (17)	X		X					X			X		X		X	
TEMES (10)		X	X					X				X	X		X	
GÜNTHER (11) (20)		X				X			X			X	X		X	
BELL (18)	X				X			X				X			X	
YANO (9)		X		X						X		X				X
LEE (19)		X		X						X	X			X		

Table 2 - Summary of bridging techniques.

The synthesis of Yanos' filter[9] is described by Nakauchi in these proceedings. Lees' work[19] is an effort to realize attenuation poles in monolithic crystal filters by using bridging capacitors.

Future Work

There is still a great deal of work that can be done both on the university and industry levels in the area of resonator bridging. There seems to be no limit on the number of mechanical configurations that can be used to realize a particular topology and I am sure there are many topologies that have not yet been investigated in terms of being realized mechanically. This brings us to the question of which configurations are best.

As in the case of active filters, one criteria for evaluating mechanical filter networks is the sensitivity of the frequency response to variations of network elements. In addition, there is the question of which configurations result in the most stable element values both in terms of temperature and aging as well as production processes. A method of analyzing mechanical filters, in terms of the effect of production variables on the element values of the network model and element value variations on the frequency response, has been outlined[3]; but few results have been shown in the literature. Some work showing a comparison of a multiple bridging wire design to a monotonic no-bridging design has been discussed[2], but little work has been done to try to establish more general results. There are numerous other ways of realizing attenuation poles such as through use of ladder-arm poles and zeros as well as lattice structures. These methods should also be compared to the monotonic or finite attenuation pole-filters.

Acknowledgements

I want to thank Gabor Temes for sharing his unpublished work, Kazuo Yakuwa for providing me with some key material and Alfhart Günther for his valuable suggestions.

References

1. R. A. Johnson, M. Börner and M. Konno, "Mechanical filters-a review of progress," IEEE Trans. Sonics Ultrason., vol. SU-18, July 1971, pp. 155-170
2. H. Schüssler, "Filters for channel bank filtering with mechanical resonators, quartz crystals and gyrators," Proc. 1974 Int. Symp. Circuits and Syst. (San Francisco), Apr. 1974, pp. 106-110.
3. V. Sobotka and J. Trnka, "The use of digital computers in modelling of transfer characteristics of serially produced electromechanical filters," Proc. 1971 SSCT, Tale, Low Tatra Mountains, Czechoslovakia, pp. 14-1 to 14-20.
4. M. Börner, "Mechaniche Filter mit Dämpfungspolen," Arch. Elek. Übertragung, vol. 17, Mar. 1963, pp. 103-107.
5. H. Albsmeier, "Ein Vergleich der Realisierungsmöglichkeiten elektromechanischer Kanalfilter im Frequenzbereich 12kHz bis 10MHz," Frequenz, vol. 25, Mar. 1971, pp. 74-79.
6. R. A. Johnson, "A mechanical filter having general stopband characteristics," IEEE Trans. Sonics Ultrason., vol. SU-13, July 1966, pp. 41-48.
7. K. Yakuwa, "High performance mechanical filters are coming to be footlighted," Nikkei Electronics, Dec. 2, 1974, pp. 76-127.
8. R. A. Johnson, "A twin tee multimode mechanical filter," Proc. IEEE, vol. 54, no. 12, Dec. 1966, pp. 1961-1962.
9. T. Yano, T. Futami and S. Kanazawa, "New torsional mode electromechanical channel filter," 1974 IEEE European Conf. on Circuit Theory and Design (London), July 1974.
10. G. C. Temes, "Asymmetrical loss-pole mechanical filter," U.S. Patent 3,725,828, Apr. 3, 1973.
11. A. Günther, "Electro-mechanical filters-satisfying additional demands," Proc. 1973 IEEE Int. Symp. Circuit Theory, Apr. 1973, pp. 142-145.
12. R. A. Johnson and A. E. Günther, "Filters and resonators, Part III. Mechanical filters and resonators," IEEE Trans. Sonics Ultrason., vol. SU-21, Oct. 1974, pp. 244-256.
13. D. F. Sheahan and R. A. Johnson, "Crystal and mechanical filters," IEEE Trans Circuits and Syst., vol. CAS-22, Feb. 1975, pp. 69-89.
14. B. Kohlhammer and H. Schüssler, "Berechnung allgemeiner mechanischer Koppelfilter mit Hilfe von äquivalenten Schaltungen aus konzentrierten elektrischen Schaltelementen," Wiss. Ber. AEG-Telefunken, vol. 41, no. 3, 1968, pp. 150-159.
15. B. Kohlhammer, "Ein neuartiges Entwurfsverfahren zur Synthese von mechanischen Filtern und von Gyrator-Filtern," Wiss. Ber. AEG-Telefunken, vol. 43, no. 3, 1970, pp. 170-177.
16. F. Künemund, "Dimensionierung überbrückter Bandpässe mit Dämpfungspolen," Frequenz, vol. 24, no. 6, 1970, pp. 190-192.
17. K. Wittmann, G. Pfitzenmaier and F. Künemund, "Dimensionierung reflexionsfaktor- und laufzeitgeebneter versteilerter Filter mit Überbrückungen," Frequenz, vol. 24, no. 10, 1970, pp. 307-312.
18. H. C. Bell, "Canonical lowpass prototype network for symmetric coupled-resonator bandpass filters," Electronics Letters, vol. 10, no. 13, June 1974, pp. 265-266.
19. M. S. Lee, "Equivalent network for bridged crystal filters," Electronic Letters, vol. 10, no. 24, Nov. 1974, pp. 507-508.
20. A. E. Günther, "High-quality wide-band mechanical filters theory and design," IEEE Trans. Sonics Ultrason., vol. SU-20, Oct. 1973, pp. 294-301.

ELECTRO-MECHANICAL FILTERS - SATISFYING ADDITIONAL DEMANDS

Alfhart E. Günther
Zentrallaboratorium für Nachrichtentechnik
Siemens AG, D-8000 München 70, Germany

Abstract

Physical and manufacturing considerations limit electro-mechanical circuit structure selection and the well-proven LC-filter technology synthesis procedures are inadequate for wide-band mechanical filter design. Group delay requirements can be achieved with complex attenuation poles while complex reflection zeros provide temperature compensation and input impedance manipulation.

1. Introduction

While electro-mechanical filters (MF) possess striking advantages over filters constructed from lumped passive and active elements they have one serious drawback: the filter designer has only limited freedom in choosing circuit structures. An additional restriction is placed upon the designer because of manufacturing considerations, since filter elements with constant and homogeneous cross sections are preferred.

From the point of view of design there is a considerable difference between filters with small (<0.01) and wide (>0.05) fractional bandwidths. In the range from 0.01 to 0.05 the design method is dependent on the specified tolerances.

The analogs of the mechanical elements, given by the electro-mechanical mobility-analogy, are presented in Fig.1 [1]. The ideal transformer can appear only in the two port equivalent of a resonator and the transformer ratio is the ratio of vibration amplitudes between the energy input point (1) and output point (2) on the vibrator. The sign of the transformer ratio depends only on the number of nodal points enclosed by (1) and (2): an even number of nodes yields a positive, an odd number a negative, transformer ratio.

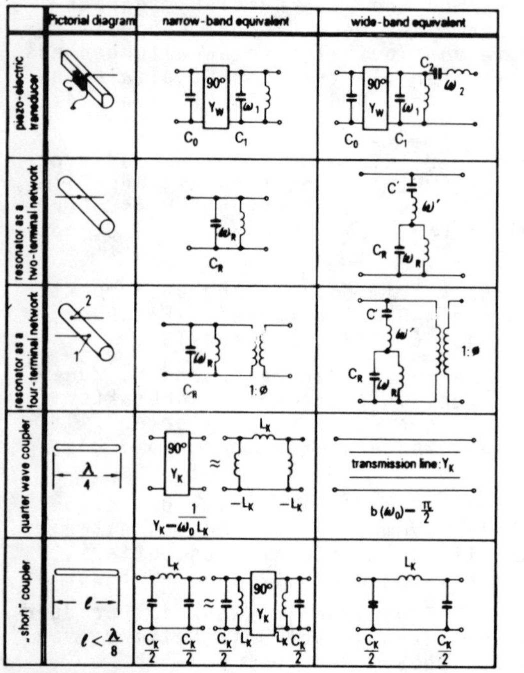

Fig. 1

No difficulties exist in the theoretical treatment of small bandwidth MFs. The design problems can be directly equated to those of either ideally coupled, inductor coupled, or some microwave filters [2].

By bypassing or bridging couplers [3,4] it is possible to produce finite attenuation poles as well as circuit sections containing an allpass factor. Thus for small bandwidths powerful methods for approximating a prescribed attenuation and phase response are at the disposal of filter designers.

Although for larger bandwidths the same physical principles exist, the exact procedures of synthesis fail. The two attenuation poles in the resonator analog have a constant frequency ratio to the vibrator resonance, dependent only on geometric dimensions. For inclusion of these poles into the design procedure of the characteristic function knowledge of the vibrator resonances would be required. In such cases the circuit design is performed by iterative optimization [5].

Both the exact and iterative design methods often result in element values, which in turn, place unusual demands on the manufacturing technology. For other problems, e.g. where requirements on the impedance performance are specified, these procedures give no answer.

2. Group delay equalization

The flattening of the group delay of filters with lumped elements is normally performed by cascading the network with a delay equalizer, where, because the equalizer impedance is frequency independent, the design of filter and equalizer may be performed independently. In [6] a mechanical arrangement, alleged to be the equivalent of such a cascade of filtering and equalizing networks, is presented. This equivalence is a case of degeneration and mechanically realizable only with a transducer coupling coefficient k=1 (the available electrostrictive materials only achieve k<0.7). Fig. 2 shows the mechanical equivalent of a delay equalizing X-section, assuming quarter wave couplers. Because energy input into a coupler is only possible by a resonator or a transducer, such a configuration is only realizable in the interior of a filter

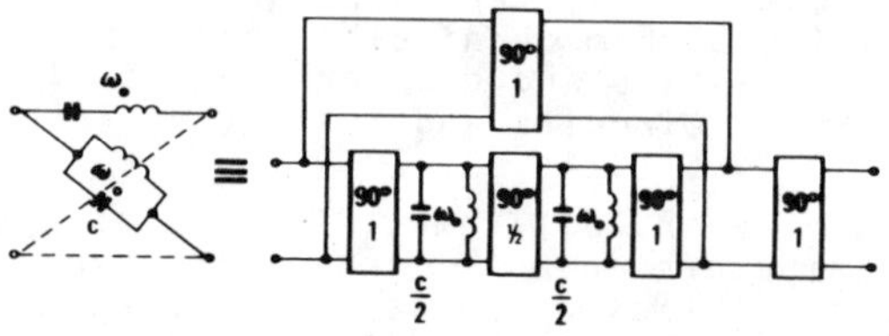

Fig. 2: Coupling filter equivalent to an X-section allpass (Z=1Ω)

Reprinted from *Int. Symp. on Circuit Theory 1973*, Apr. 9–11, 1973, pp. 142–145.

network, and hence it cannot be designed independently from the filter. The minimum network for mechanical realization with an allpass factor in the transfer function as well as its bandpass equivalent are shown in Fig. 3.

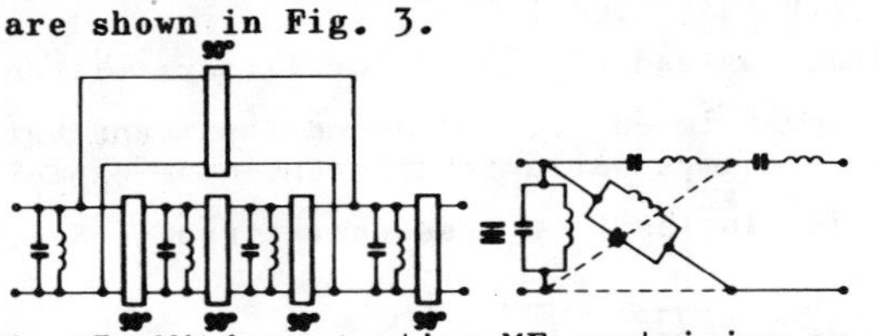

Fig. 3: Minimum section MF containing an allpass and its bandpass filter equivalent

Competing with the delay equalization by allpass factors are the proposals of Bennett [7] and Watanabe and co-workers [8] to provide attenuation poles at complex frequencies, eliminating the requirements for allpass factors in the filter. Watanabe pointed out the higher economy of such circuits and

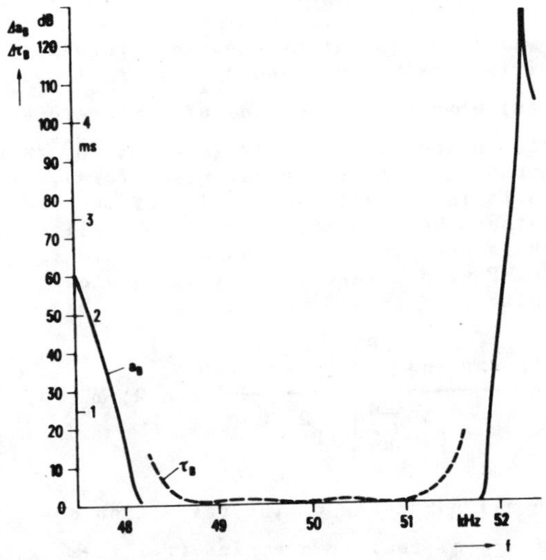

Fig. 4: Attenuation and group delay of filters with delay equalization

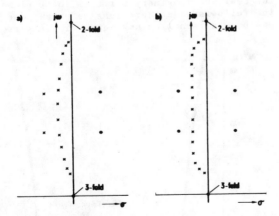

Fig. 5: Pole-zero-diagrams of filters with group delay equalization
(a) Allpass factor in the transfer function
(b) Attenuation poles at complex frequencies

indeed this method yields advantages. Fig. 4 presents the attenuation and group delay of a mechanically realizable bandpass of 13 resonant circuits with two cascaded X-sections (a total of 17 resonators). We obtain the same response with 16 resonant circuits by providing two attenuation quadrupoles. The pole-zero-diagrams of both networks are given in Fig. 5.

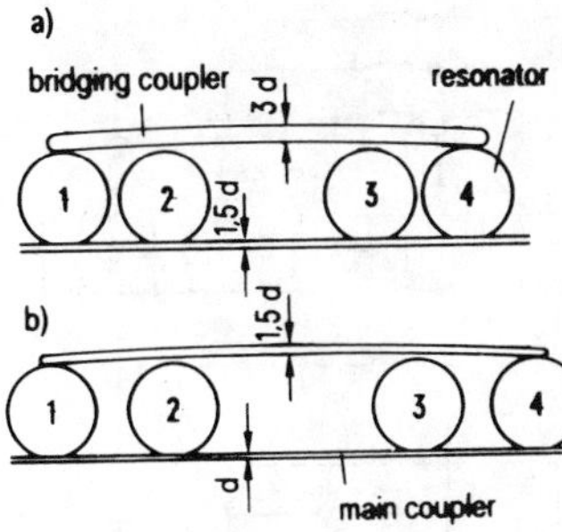

Fig. 6: Approximate dimensions of the mechanical realizations of delay equalizing sections
(a) Filter section containing allpass
(b) Filter section producing quadrupole of attenuation

Disregarding the saving of one resonator, the Bennett-Watanabe method yields better dimensions for the mechanical construction with short couplers (Fig. 6). The very thick bridging wire required for realization of an allpass response requires great manufacturing accuracy and is a source of spurious resonances.

Further applications of this method in the microwave filter field and synthesis procedures also applicable to MFs are presented in [9,10].

Another very interesting way of mechanically realizing simultaneous attenuation and group delay requirements was given by Johnson [11] by including an allpass factor in the filter transfer function. Johnson then developed a lattice filter with mechanical reactances in the lattice arms.

For wide-band filters these points of view are also valid, but three complications exist:

1. The quasi-symmetry, which exists with respect to the center frequency in a narrow-band filter derived from a lowpass prototype vanishes and the group delay assumes a distinct slope [12],

2. the change of curvature above the resonance in the wide-band analog of the resonator (Fig. 1) additionally influences this group delay tendency,

3. as a consequence of point (1), to guarantee the necessary number of degrees of freedom for a structure like Fig. 6 the bridging 1-4 must be accompanied by at least a further one, 1-3 or 2-4.

As already mentioned in the introduction no exact synthesis procedures for solving these wide-band problems exist to date and only optimization techniques are available. Particularly advantageous are such procedures which represent the network functions as implicit functions of the network elements [13,14].

Whereas LC filter technology tries to find coil saving designs (hence the development of parametric filter theory), MF technology aims to attain the simplest possible coupler construction. While LC filters require one coil or capacitor for each additional attenuation pole, MFs need only one additional coupling wire. We can take advantage of this fact for better delay equalization. Fig. 7a shows a filter section with attenuation poles at complex frequencies produced by the bridging wires 1-6 and 2-5. Instead of the wanted delay response, curve (a) in Fig. 8, the inclined curve (b) re-

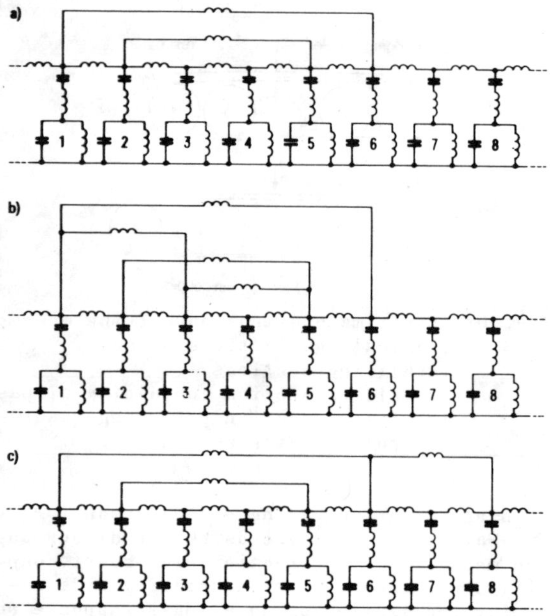

Fig. 7: Part of an MF with bridging couplers
(a) Four attenuation poles at complex frequencies by the couplings 1-6 and 2-5, two degrees of freedom
(b) Four poles as (a), but four degrees of freedom by the additional couplers 1-3 and 3-5
(c) Four poles at complex frequencies and one finite pole by the additional coupling 6-8

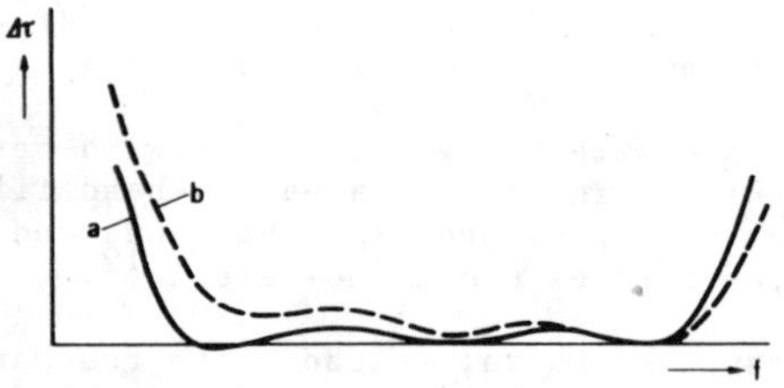

Fig. 8: Group delay vs frequency of a wideband MF
(a) desired response
(b) distorted group delay

sults. The slope can be removed by further bridgings as 1-3 and 3-5 in Fig. 7b, but results in crossing couplers. The same effect on the group delay originates from a bridging 6-8, Fig. 7c, producing a finite pole in the upper stopband; with suitable choice of the pole frequency, after Chebyshev flattening of the passband attenuation, the delay response becomes horizontal.

3. Temperature compensation

Often MFs are terminated with electric end circuits. The reasons for this practise may be reduction of the necessary electro-mechanical coupling coefficient ($k^2 = \frac{1}{1+\omega_1^2 C_o C_1/Y_w^2} \sim B_o^2$, B_o denoting the fractional bandwidth, instead of $k^2 \sim B_o$ for filters without coils), lower impedance, balanced input and better temperature compensation. The transducer capacity C_o is included in the L_z C_z resonant circuit (Fig. 9).

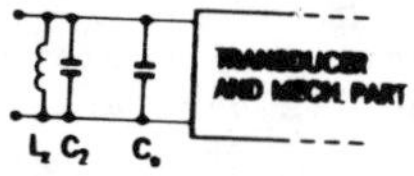

Fig. 9: MF with electric end circuit (L_z coil, C_z capacitor, C_o transducer capacity)

While the temperature coefficients of the mechanical elements are about 10^{-6}/K, the ceramic capacitance has $\alpha_{Co} \approx 2.5 \cdot 10^{-3}$/K.

To maintain a constant resonant frequency over a band of temperature α_{Lz} and α_{Cz} are suitably chosen. As, however, the spread of temperature coefficients in the end circuits is about 10^{-4}/K (much greater than the α of the steel resonators), it is desirable to increase the bandwidth of the end circuits so that their contribution to the filter selectivity is decreased. The characteristic function of an MF with n resonators and two electric end circuits is

$$K(p) = \frac{(p^4 + rp^2 + q) \prod_{\lambda=1}^{n} (p^2 + \omega_{o\lambda}^2)}{k \cdot p^3 \prod_{\mu=1}^{m} (p^2 + \omega_{\infty\mu}^2)}; \quad m \leqq n-2; \; \omega_{o\lambda}^2 > 0.$$

For $r^2 > 4q$ all n+2 zeros of the reflection coefficient are at real frequencies ($r = \omega_{n+1}^2 + \omega_{n+2}^2$, $q = \omega_{n+1}^2 \omega_{n+2}^2$), for $r^2 < 4q$ there are two zeros at complex frequencies ($r = 2(\beta^2 - \alpha^2)$, $q = (\alpha^2 + \beta^2)^2$, α and β are the real and imaginary parts in the p-plane). For the following considerations we presuppose that the passband attenuation has equal ripple perform-

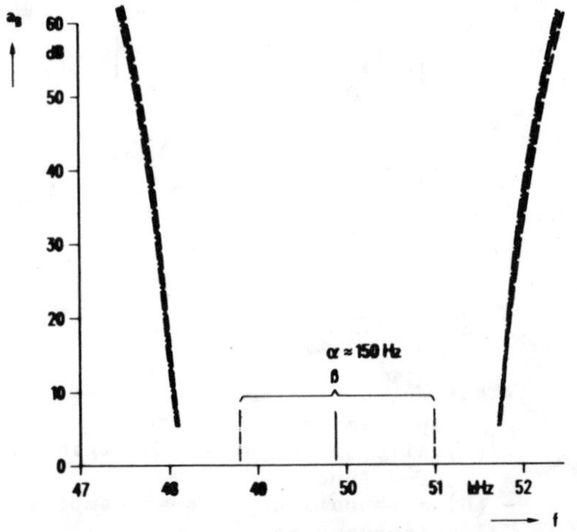

Fig. 10: Attenuation vs frequency of an MF (12 resonators) with two reflection zeros at complex frequencies ($p_o = \pm\alpha \pm \beta$)

ance whether there are complex zeros or not. We recognize that the greater α the greater the bandwidths of the end circuits, while the bandwidth of the other circuits is only little changed. The steepness of the cutoff edges mainly depends on the magnitude of the imaginary part β: the steepness of both edges is reduced, but mainly of the edge to which the distance $|\omega_g - \beta|$ is smaller (Fig. 10). A similar approach, applied to quartz-bandpass lattice sections, is given in [15].

4. Manipulation of the input impedance

Filter circuits with the impedance becoming zero at frequencies zero and infinity have above and below the passband a remarkable impedance maximum. It is well-known that the impedance maxima move away from the passband as the attenuation ripples decrease. Sometimes a secondary requirement is specified, that is, the frequency at which the impedance is to be a maximum. Attempting to achieve this maximum by varying the ripples may be uneconomic because of the general loss in stopband steepness. It is more favourable to work with reflection coefficient zeros at complex frequencies: In the upper stopband the higher β ($\omega_{-g} < \beta < \omega_g$; ω_{-g}, ω_g denoting the band edges) the higher the frequency of the impedance maximum; the reduction in stopband attenuation is thus minimized (Fig. 11). An analogous consideration applies in the lower stopband.

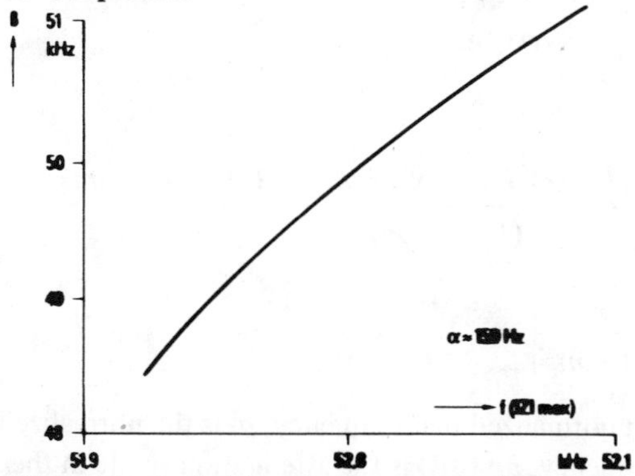

Fig. 11: Relation between complex reflection zero and input impedance maximum

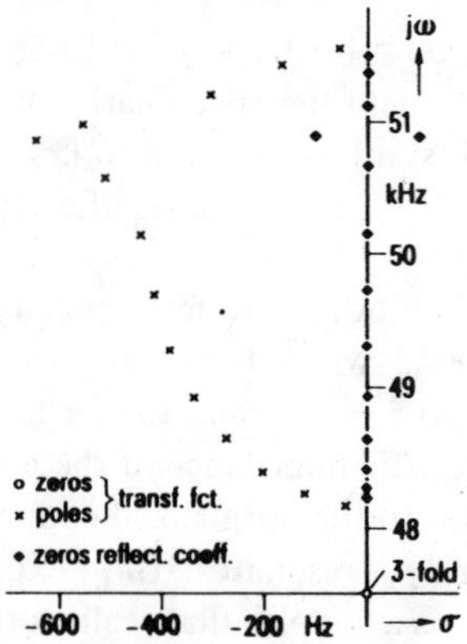

Fig. 12: Pole-zero-diagram of an MF with complex reflection zeros

We emphasize that by suitable choice of such a complex zero of the reflection coefficient the tasks of impedance manipulation and temperature compensation can be solved simultaneously. The pole-zero-diagram of such a filter with a Chebyshev passband characteristic is shown in Fig. 12.

Conclusion

The above presented methods for delay equalization, temperature compensation, and impedance manipulation are rather pragmatic more than systematic. But the increasing importance of MFs in carrier frequency technology is a challenge to the analyst to develope a homogeneous and systematic theory of the mechanical wide-band filter.

References

1 A.Günther, "High-quality wide-band filters of ultrasonic resonators -- technique and design," abstract in IEEE Trans. Sonics Ultrason., vol. SU-19, no. 3, July 1972, p. 406; complete text to be published.

2 P.Richards, "Resistor-transmission-line circuits," Proc. IRE, vol. 36, Feb. 1948, pp. 217 - 220.

3 E.Trzeba, "Lineare Bandfilter-Schaltungen der Hochfrequenztechnik, Teil III," Nachrichtentechnik, vol. 12 (1962), no. 8, pp. 297 - 302.

4 M.Börner, "Mechanische Filter mit Dämpfungspolen," AEÜ, vol. 17 (1963), no. 3, pp. 103 - 107.

5 A.Günther, "Bemerkungen zum Entwurf breitbandiger mechanischer Filter," NTZ, vol. 25, no. 8, pp. 345 - 351.

6 B.Kohlhammer and H.Schüßler, "Berechnung allgemeiner mechanischer Koppelfilter mit Hilfe von äquivalenten Schaltungen aus konzentrierten elektrischen Schaltelementen," Wiss. Ber. AEG Telefunken, vol. 41 (1968), no. 3, pp. 150 - 159.

7 B.J.Bennett, "Synthesis of electric filters with arbitrary phase characteristics," IRE Conv. Rec. 1953, part 5, pp. 19 - 26.

8 M.Hibino, Y.Ishizaki, and H.Watanabe, "Design of Chebyshev filters with flat group-delay characteristics," IEEE Trans. Circuit Theory, vol. CT-15, no. 4, Dec. 1968, pp. 316 - 325.

9 K.Wittmann, G.Pfitzenmaier, and F.Künemund, "Dimensionierung reflexionsfaktor- und laufzeitgeebneter versteilerter Filter mit Überbrückungen," Frequenz, vol. 24 (1970), no. 10, pp. 307 - 312.

10 G.Pfitzenmaier, "Realisierung reflexionsfaktor- und laufzeitgeebneter versteilerter Mikrowellenbandpässe mit Überbrückungen," Frequenz, vol. 26 (1972), no. 1, pp. 19 - 26.

11 R.A.Johnson, "New single-sideband mechanical filters," 1970 WESCON Techn. Papers 10-1, Aug. 1970, pp. 1 - 10.

12 H.Blinchikoff, "Derivative of group delay at band center," IEEE Trans. Circuit Theory (Correspondence), vol. CT-17, Nov. 1970, pp. 636 - 637.

13 Y.Ishizaki and H.Watanabe, "Iterative Chebyshev approximation technique for network design," J. Inst. Electr. Eng. Japan, vol. 50 (1967), no. 4, pp. 672 - 676.

14 A.Günther, "Synthese mechanisch realisierbarer Schaltungen," 1967; intern. Siemens report.

15 H.Betzl, "Ein Beitrag zur Berechnung von eingliedrigen Quarz-Brückenbandpässen mittlerer Bandbreite nach der Betriebsparametertheorie," Frequenz, vol. 19 (1965), no. 5, pp. 206 - 209.

A Torsional-Mode Pole-Type Mechanical Channel Filter

KEN-ICHI SAWAMOTO, SHIGEO KONDO, NOBORU WATANABE, KAZUO TSUKAMOTO, MINORU KIYOMOTO, AND OSAMU IBARAKI

Abstract—A torsional-mode pole-type mechanical filter is studied, with the aim of further miniaturization and economy of channel translating equipment. New torsional-mode transducers have been developed. These are built of piezoelectric ceramics disks with a high coupling constant k_{15} and of bars of a constant elastic modulus alloy. As the capacitance ratio of the transducer is about 9–11, the filter does not need inductors to compensate for the clamped capacitance of the transducers.

The distances between the resonators are adequately determined, therefore, the couplers are constructed of only few types of single straight wire lines. Finite attenuation poles are produced by two oblique bridging wires.

In the passband region, the attenuation ripples of the filter fulfill the desired specification (CCITT 1/20) over the temperature range of 5°C to 45°C.

Delay distortion and minimum group delay time are nearly equal to that of the *LC* channel filter being used in Japan.

The filter contains 7 torsional-mode resonators and 2 torsional-mode transducers. The filter volume is about 5.3 cm^3 and its weight is 13 g. These values are about 1/10 those of the *LC* channel filter.

I. INTRODUCTION

HITHERTO, the channel filter in Japan was constructed from an inductor L and a capacitor C. Research on ferrite cores and capacitors, of which the filters are constructed, has reached a great height of perfection. Therefore, it seems to be impossible to play an active part in further improving its cost/performance ratio. On the other hand, mechanical filters are constructed from resonators, transducers and couplers, made from Ni-Fe alloys (constant elastic modulus alloys; for example Ni-Span C). The quality factor of a mechanical resonator is about 100 times higher than that of an *LC* resonator, the resonator frequency characteristics are stabilized by adequate heat treatment and furthermore, the alloy cost is cheap. Thus small-sized low-cost, and excellent quality filters are expected.

Recently, channel filters with bending mode [1], [2] and torsional mode [3], [4] vibrations were reported. A torsional-mode nonfinite attenuation pole-type mechanical filter was also studied by the authors [5], aiming at further miniaturization and economy. This filter did not need inductors to compensate for the clamped capacitance of the transducers. Attenuation ripples in the passband region fulfilled the desired specification (1/20 CCITT) over the temperature range of 5°C to 45°C. However, because of large number of resonators needed to meet the stopband requirement the minimum group delay time in the passband region was about 1.2 ms, which was about 2 times longer than that of the *LC* channel filter, or the pole-type mechanical filter. This paper concerns the pole-type mechanical filter design process and the results obtained.

II. POLE-TYPE MECHANICAL FILTER DESIGN PROCESS

Pole-type mechanical filters were already proposed by M. Börner [6], B. Kohlhammer [7], R. A. Johnson [8], and T. Yano [4]. Finite attenuation poles of these filters were the result of additional bridging wires acting as couplers. First a 9th order low-pass filter is dealt with; then it is transformed to a mechanical filter according to B. Kohlhammer, *et al.* [7].

The transmission function $H_9(x)$, for a ninth-order Chebyshev type low-pass filter, is given as

$$H_9^2(x) = 1 + (10^{0.1\,bp} - 1)\,Y_9^2(x) \tag{1}$$

$$Y_9(x) = \operatorname{Im}\frac{(\sqrt{1-x^2}+jx)^5(\sqrt{1-x^2}+jm_1x)^2(\sqrt{1-x^2}+jm_2x)^2}{(1-x^2/\rho_1^2)(1-x^2/\rho_2^2)} \tag{2}$$

and

$$\rho_i^2 = 1/(1-m_i^2) \tag{3}$$

where x is a normalized real frequency, ρ_i is the normalized ith pole frequency, bp (dB) is the attenuation ripple in the passband, and Im in (2) represents the imaginary part of (2) [9].

If $bp = 0.01$ dB, $\rho_1 = 1.30$, and $\rho_2 = 1.65$, then $H_9(x)$ satisfies the desired channel filter specification in Japan. A low-pass filter circuit is synthesized from $H_9(x)$. The resultant circuit is shown in Fig. 1. The values of each element are shown in Table I.

An equivalent circuit of a 7-element pole-type mechanical filter is given in Fig. 2, where the parallel *LC* resonators numbered from 2 to 8 correspond to mechanical resonators, where the characteristic impedance of these resonators is Z, and $k_{i,j}$ corresponds to the normalized coupling coefficient between the ith and jth resonator. $Gk_{i,j}$ expressed by an imaginary gyrator, is a coupler that realizes the coupling coef-

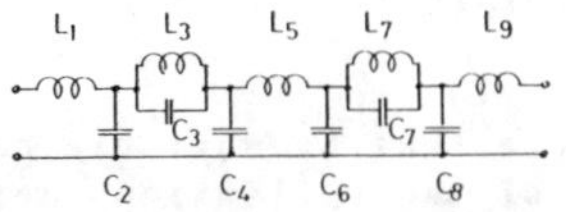

Fig. 1. A pole-type 9th-order *LC* low-pass Chebyshev filter. The pole frequency ρ_1 is 1.30 and ρ_2 is 1.65.

Manuscript received June 20, 1975; revised September 29, 1975.
The authors are with Nippon Telegraph and Telephone Public Corp., Electrical Communication Laboratory, Tokyo 180, Japan.

Reprinted from *IEEE Trans. Sonics Ultrason.*, vol. SU-23, pp. 148–153, May 1976.

TABLE I
POLE-TYPE 9TH-ORDER LOW-PASS FILTER ELEMENT VALUES AS SHOWN IN FIG. 1

L_1	0.82308	C_2	1.01872
L_3	0.96998	C_3	0.61003
L_5	1.96594	C_4	1.39853
L_7	1.35177	C_6	1.54868
L_9	0.82308	C_7	0.27173
		C_8	1.21230

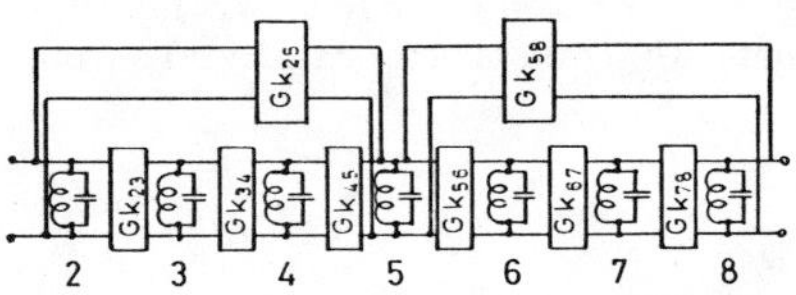

Fig. 2. An equivalent 7-element pole-type mechanical filter circuit.

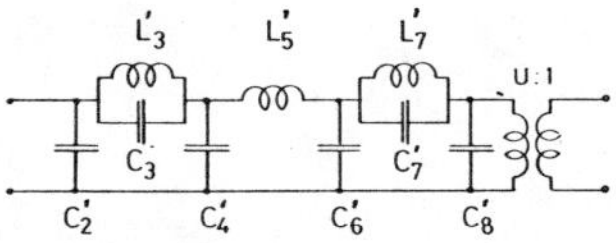

Fig. 3. A low-pass filter transformed from the mechanical filter according to B. Kohlhammer *et al.*

ficient $k_{i,j}$. $Gk_{2,5}$ or $Gk_{5,8}$ is an overcoupler constructed by a bridging wire.

According to B. Kohlhammer [7], the circuit shown in Fig. 2 is transformed to the circuit shown in Fig. 3. It corresponds to the circuit shown in Fig.1, if L_1 and L_9 are deleted and $U = 1$. The characteristic impedance of each mechanical resonator Z is given as

$$1/Z = C_2 + (C_3 \cdot C_4)/(C_3 + C_4). \tag{4}$$

The normalized coupling coefficient $k_{i,j}$ and turns ratio U of the transformer are expressed as shown in Table II, where the expression of from $k_{2,3}$ to $k_{7,8}$ containing the overcoupling coefficient are given from [7]. The derivation of $k_{1,2}$ and $k_{8,9}$ are given as follows. In a ladder type low-pass filter with a cutoff frequency of 1.0, the coupling coef-

TABLE II
NINTH-ORDER MECHANICAL FILTER NORMALIZED COUPLING COEFFICIENT $k_{i,j}$ AND TRANSFORMER TURN RATIO U

i,j	$k_{i,j}$
1,2	$1/\sqrt{L_1[C_2 + C_3 \cdot C_4/(C_3 + C_4)]}$
2,3	$1/[(1 + C_3/C_4)\sqrt{L_3[C_2 + C_3 \cdot C_4/(C_3 + C_4)]}\,]$
3,4	$1/\sqrt{L_3(C_3 + C_4)}$
4,5	$1/\sqrt{L_5(C_3 + C_4)}$
5,6	$1/\sqrt{L_5(C_6 + C_7)}$
6,7	$1/\sqrt{L_7(C_6 + C_7)}$
7,8	$1/[(1 + C_7/C_6)\sqrt{L_7[C_8 + C_6 \cdot C_7/(C_6 + C_7)]}\,]$
8,9	$1/\sqrt{L_9[C_8 + C_6 \cdot C_7/(C_6 + C_7)]}$
2,5	$-1/[(1 + C_4/C_3)\sqrt{L_5[C_2 + C_3 \cdot C_4/(C_3 + C_4)]}\,]$
5,8	$-1/[(1 + C_6/C_7)\sqrt{L_5[C_8 + C_6 \cdot C_7/(C_6 + C_7)]}\,]$
U	$-\sqrt{[C_2 + C_3 \cdot C_4/(C_3 + C_4)]}/\sqrt{[C_8 + C_6 \cdot C_7/(C_6 + C_7)]}$

ficient $k_{i,i+1}$ is generally given as $1/\sqrt{L_i \cdot C_{i+1}}$. Considering the expressions $k_{i,j}$ shown in Table II, C_2' and C_8' in the circuit shown in Fig. 3 may be expressed as

$$C_2' = C_2 + (C_3 \cdot C_4)/(C_3 + C_4)$$
$$C_8' = C_8 + (C_6 \cdot C_7)/(C_6 + C_7) \tag{5}$$

in case of this pole-type mechanical filter. Therefore, $k_{1,2}$ and $k_{8,9}$ are given as

$$k_{1,2} = 1/\sqrt{L_1 \cdot C_2'}$$
$$= 1/\sqrt{L_1 \cdot [C_2 + (C_3 \cdot C_4)/(C_3 + C_4)]}$$
$$k_{8,9} = 1/\sqrt{L_9 \cdot C_8'}$$
$$= 1/\sqrt{L_9 \cdot [C_8 + (C_6 \cdot C_7)/(C_6 + C_7)]} \tag{6}$$

in the case of $U = \pm 1$. From the values of L_i, C_j in Table I, U becomes -1. An actual coupling coefficient $K_{i,j}$ is calculated from $k_{i,j}$ as

$$K_{i,j} = k_{i,j} \cdot (\Delta f/f) \tag{7}$$

where Δf is the passband width, for 0.01-dB ripple, equaling 3.22 kHz, and f is the center frequency of the passband. Setting the carrier frequency to 112 kHz, f becomes 113.9 kHz. The normalized Q value q_1 for the input terminal is equal to L_1. For the output terminal $q_9 = L_9$. $k_{i,j}$, $K_{i,j}$, U, and q_i are shown in Table III.

III. A TORSIONAL-MODE TRANSDUCER

From a miniaturization standpoint, it is desirable to use a torsional-mode transducer similar to the resonator. Fig. 4 shows the manufacturing process of the developed torsional-mode transducer. Two half-circle disks, with electrodes on each surface, are formed by ultrasonic punching from a piezoelectric ceramic plate having a large coupling factor k_{15} of more than 60 percent. The poling direction is parallel to the diameter, as shown by the arrow in Fig. 4(a). The two disks are placed in a circle, where the poling directions of the disks are reciprocally inverse, as shown in Fig. 4(b). Then the disks are bonded by epoxy adhesive between two equal length rods of Ni-Fe alloy. The diameter of the rods is the same as that of the disks, as shown in Fig. 4(c).

Thin short wires are spot-welded vertically at each center of the end circle of the rods, as shown in Fig. 4(c). These wires act not only as supports but also as input or output terminals. The thickness of the ceramics is 1.6 mm, the

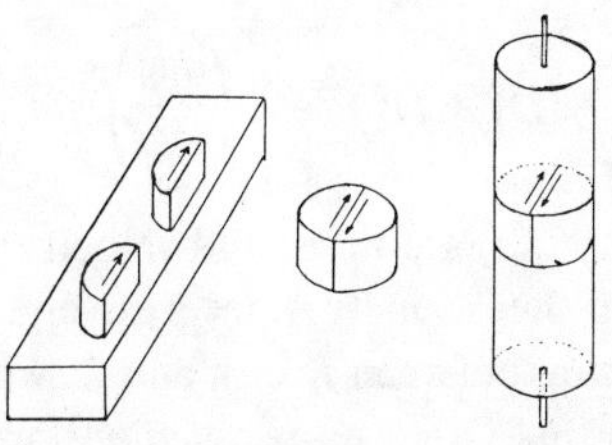

(a) (b) (c)

Fig. 4. Torsional-mode transducer manufacturing process. (a) Piezoelectric ceramics. (b) Two half-circle ceramics disks. (c) A torsional-mode transducer.

clamped capacitance is about 47 pF at 1 kHz, and the capacitance ratio γ is 9-11 at an average resonant frequency of 112 kHz. The resonant frequency deviation from 5°C to 45°C is about 150 Hz at 112 kHz. Because of the small value of γ, an inductorless mechanical filter can be constructed.

IV. Coupling Between Two Resonators

A torsional-mode resonator is a rod made from the Ni-Fe alloy. The length of each resonator is 13 mm and the diameter is $2D_1 = 3$ mm or $2D_2 = 3.05$ mm. After suitable heat treatment in a vacuum, the average Q value of these resonators becomes about 3×10^4 and the resonant frequency temperature dependence $(\delta f/f)/°C$ is 1-2 ppm/°C. The coupler is of the same alloy as the resonators. Its diameter is $2d$.

Typical coupling states of two resonators are shown in Fig. 5(a) and Fig. 5(b). Fig. 5(a) corresponds to a coupling state of adjacent resonators and Fig. 5(b) corresponds to an overcoupling (bridging) state, where the diameters of the two outside resonators $2D_2$ are slightly larger than the inside resonators $2D_1$. Therefore, an overcoupler does not contact the inside wires. A coupler is spot-welded at the same position l_a on the two resonators and supports are welded at the midpoints of both sides of the resonators. The distance between resonators is L. Resonant frequencies of both resonators are equal to f_r before coupling.

In the coupling state, the two observed resonant frequencies f_1 and f_2 (corresponding to the zeros of the input impedance), center frequency f_c, coupling coefficient K, and frequency shift η, are defined as follows:

$$f_c = (f_1 + f_2)/2$$
$$K = (f_2 - f_1)/f_r \cong (f_2 - f_1)/f_c$$
$$\eta = (f_c - f_r)/f_r. \tag{8}$$

The K and η for the coupling state of adjacent resonators are expressed as (9) and (10) [5]

$$K = \frac{2d^2 V_l \rho_l}{\pi D_1^2 D_2^2 V_t \rho_t} \cdot \sqrt{(D_1^2 + D_2^2)^2 \cos^2\left(\frac{\omega L}{V_l}\right) + 4D_1^2 D_2^2 \sin^2\left(\frac{\omega L}{V_l}\right)} \cdot \left(\frac{\cos^2\left(\frac{\omega l_a}{V_t}\right)}{\sin\left(\frac{\omega L}{V_l}\right)}\right) \tag{9}$$

$$2\eta = \frac{2d^2 V_l \rho_l}{\pi D_1^2 D_2^2 V_t \rho_t} \cdot (D_1^2 + D_2^2) \cos^2\left(\frac{\omega l_a}{V_t}\right) \cot\left(\frac{\omega L}{V_l}\right) \tag{10}$$

where V_l, V_t, and ρ_l, ρ_t are the sound velocity and density of the coupler and the resonators, respectively. Fig. 6 shows a typical relationship between K or η and L, where $l_a = 2.94$ mm, $2d = 0.278$ mm. Other numerical constants are also shown in Fig. 6. When the radii of two resonators are equal to D_1, observed values are shown as the symbol "Δ." In the case of different radii, values are shown as the symbol

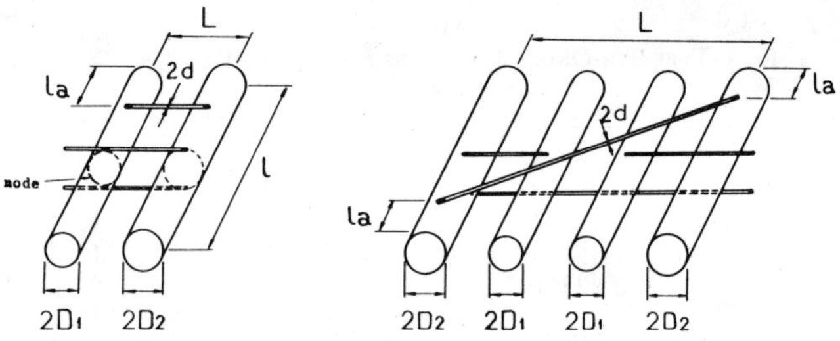

Fig. 5. Coupling state between two resonators. (a) Adjacent resonators coupling state. (b) Overcoupling state by an oblique coupler.

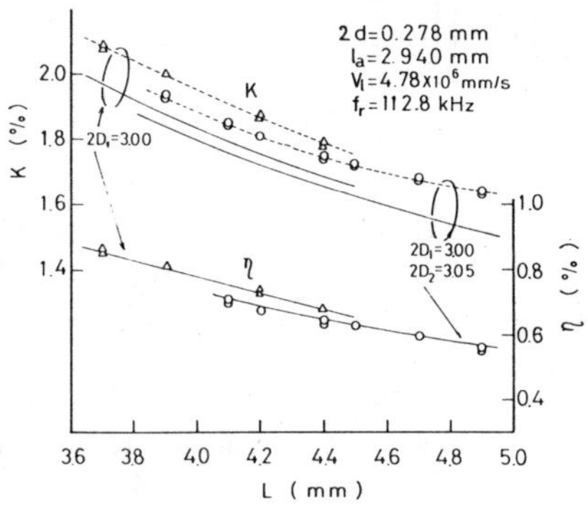

Fig. 6. Relation between coupling coefficient K or frequency shift η and distance L between resonators.

"o." Solid lines are calculated from (9) or (10). With respect to K, the solid lines are approximately in parallel with the observed ones. A possible reason for the coupling error was that the coupler length was too short, compared with length L, by the length of the welded areas. Based a microscopic observation, the parts of the welded area on the coupler were the same order as the shortened length. The required coupling coefficient $K_{i,j}$ from Table III is, therefore, fulfilled when an adequate distance L is set. For example, when $K_{3,4} = 2.025$ percent, then $L = 3.83$ mm. In respect to η, solid lines fit observed ones.

In case of an overcoupling, the distance L must be replaced by the length of the overcoupler. As the overcoupler is connected on the oblique, the coupling coefficient sign is changed. Experimentally obtained results were similar to those of adjacent resonators.

TABLE III
Values of $k_{i,j}$, $K_{i,j}$, q_i and U

i,j	$k_{i,j}$	$K_{i,j}$ (%)
1,2	0.91744	2.5936
2,3	0.58844	1.6635
3,4	0.71644	2.0254
4,5	0.50324	1.4227
5,6	0.52861	1.4944
6,7	0.63748	1.8022
7,8	0.60903	1.7218
8,9	0.91743	2.5936
2,5	−0.18029	−0.5097
5,8	−0.08861	−0.2505

$q_1 = q_9 = 0.82308$, $U = -1.00000$

The resonant frequency f_b of the bending mode standing wave on the coupler must not be near that of the resonator. If f_b is near f_r, the measured coupling coefficient deviates from the experimentally obtained curves shown in Fig. 6. f_b is given as

$$f_b = \frac{d}{4\pi}\left(\frac{\alpha_m}{L}\right)^2 V_l \tag{11}$$

where α_m is a constant value depending on the order of overtone (m). Therefore, the length l_a and L are selected on the basis of such considerations.

V. Tuning the Individual Resonator Resonant Frequency

For simplication, a coupling state of three resonators is considered. If the coupler length L is equal to $\lambda/4$, the coupler acts as an imaginary gyrator, where λ is a wavelength of a longitudinal mode propagating through the coupler. In this case, an equivalent mechanical filter circuit is shown in Fig. 7(a), where three resonators are expressed as series resonant circuits having the same resonant frequency ω_c. ω_c values

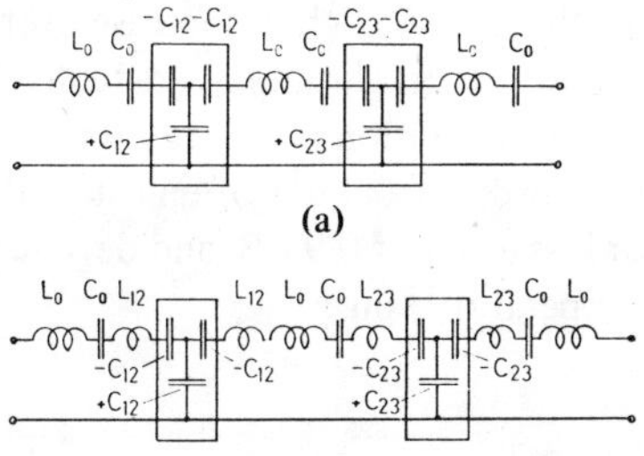

Fig. 7. Equivalent circuits for coupler state at 3 resonators. (a) Coupler length is equal to $\lambda/4$. This is the imaginary gyrator coupling case. (b) Coupler length is shorter than $\lambda/4$.

equaling the center frequency and $K_{1,2}$ or $K_{2,3}$, are given as follows

$$\omega_c = 1/\sqrt{L_0 \cdot C_0}, \quad K_{1,2} = C_0/C_{1,2}, \quad K_{2,3} = C_0/C_{2,3}. \tag{12}$$

In the case of a shorter coupler than $\lambda/4$, the shorter L is, the larger $K_{i,j}$ and $\eta_{i,j}$ become. In a mechanical filter, the center frequency must be set at a constant value ω_c. In the case when resonators number 1 and number 2 (each with the same resonant frequency, ω_1) are coupled with each other, $K_{1,2}$ and $\eta_{1,2}$ are obtained, where $K_{1,2}$ coincides with the theoretically obtained value. The resultant frequency that resonator number 1 is to be tuned to is given as follows

$$\omega_1 = \omega_c/(1 + \eta_{1,2}). \tag{13}$$

This could also be the tuning frequency of resonator number 2 except for the influence of the coupling between 2 and 3. In coupling resonator number 2 with number 3, having the same resonant frequency, ω_3, the effect of the coupling between 2 and 3 on resonator number 2 is expressed as

$$\omega_3 = \omega_c/(1 + \eta_{2,3}). \tag{14}$$

As the ω_1 or ω_3 is lower than ω_c by the amount of $\eta_{i,j}$, this can be simulated in three series resonant circuits by adding inductors $L_{i,j}$ on both sides of the imaginary gyrators, as shown in Fig. 7(b). From this equivalent circuit, the tuning frequency for resonator number 2 is given as

$$\omega_2 = 1/\sqrt{[C_0 \cdot (L_0 + L_{1,2} + L_{2,3})]}$$
$$= \omega_c/(1 + \eta_{1,2} + \eta_{2,3}) \tag{15}$$

because of $\eta_{i,j} \ll 1$. In a coupling state of more than three resonators, a similar treatment is carried out; thus the resonant frequency of the nth number resonator is set at ω_n, where

$$\omega_n = \omega_c/[1 + \eta_{n-1,n} + \eta_{n,n+1} + \delta_{pn}\eta_{n,n\pm 3}] \tag{16}$$

where $\eta_{n-1,n}$ is the frequency shift in the case of the coupling of the $(n-1)$th and nth resonator, δ_{pn} is the Kronecker delta function, and $p = 2,5$ or 8 corresponds to the overcoupling. Because the length of the overcoupler is less than $\lambda/2$, the sign of $\eta_{n,n\pm 3}$ becomes negative.

A diagram of the pole-type mechanical filter is shown in Fig. 8. The diameter of resonators number 2,5, or 8 is slightly

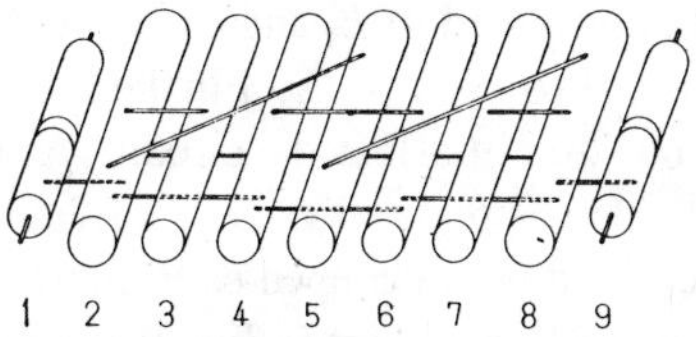

Fig. 8. Pole-type mechanical filter diagram.

larger than that of the others. Oblique bridging wires are used as the overcouplers. The couplers between adjacent resonators (2 to 4, 4 to 6, and 6 to 8) are shown on the underside of the resonators, and are constructed of single straight lines only.

The transducer length is about 10.5 mm, shorter than that of the resonators, which equals 13 mm. In order to assist in miniaturization, the symmetrical filter construction with respect to the support wires on the resonator, is effective, when the welded position of the coupler on the transducer is different from l_a on the adjacent resonator. In such a symmetrical case, the tuning frequency given by (16) is not applicable. In the symmetrical structure, the two pole frequencies of the input impedance for the coupled pair are higher than the frequencies of the asymmetrical coupling state, where welding positions l_a are identical with each other. If the resonant frequency of the transducer is lowered from the set value obtained from (16) by about 150 Hz and that of the adjacent resonator is raised from that value by about 125 Hz, in a symmetrical coupling state, not only the zero but also the pole frequencies coincide with the zeros and poles for the asymmetrical coupling state. Revised values of about 150 Hz and 125 Hz are obtained experimentally from minimizing the passband ripples of the filter attenuation.

VI. Transducer's Clamped Capacitance Compensation

The capacitance ratio γ of the torsional mode transducer is about 9–11, as mentioned before. The inductor on both the input and output terminals is not needed to compensate

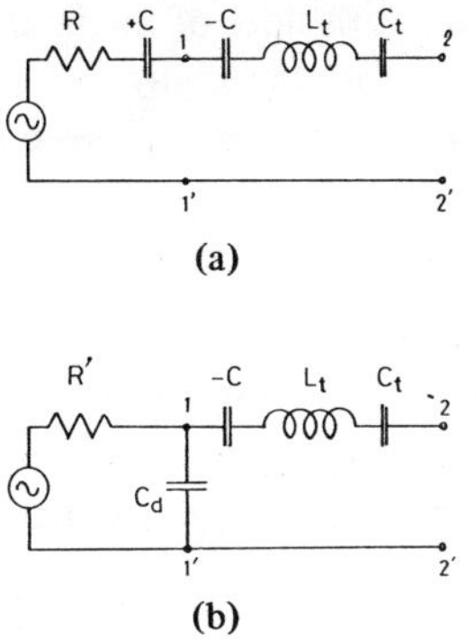

Fig. 9. Transducer's clamped capacitance compensation. (a) Normal terminal region equivalent circuit. (b) From the left side of the terminal, 1, 1′ is equivalently replaced with the transducer clamped capacitance C_d.

for the clamped capacitance C_d of the transducer, but it is necessary to change the tuning frequency of the transducer as discussed in the preceding section. Fig. 9(a) shows an equivalent circuit of the terminal region, where R is the terminating resistance and the transducer is expressed as a series resonant circuit composed of L_t and C_t, with its resonant frequency ω_t equal to $1/\sqrt{L_t \cdot C_t}$. Furthermore, $+C$ and $-C$ are added, but do not affect the circuit. Assuming that the circuit is divided at points 1, 1′; the left-hand side is equivalently replaced with clamped capacitance C_d and terminal resistance R', as shown in Fig. 9(b). Therefore, the following relations are obtained

$$R' = R[1 + 1/(\beta^2 x^2)]$$
$$\gamma = C_d/C_t = x/(1 + \beta^2 x^2) \tag{17}$$

where $x = C/C_t$ and $\beta = \omega_t \cdot C_t \cdot R$.

A normalized input or output terminal Q value (q_1 or q_9) is equal to L_1 or L_9, as mentioned before, and is related to the realized Q value of the terminal, as follows

$$Q = q_1 \cdot (f/\Delta f) = \omega_t \cdot L_t/R = 1/(R \cdot \omega_t \cdot C_t) = 1/\beta. \tag{18}$$

From (18), β is a constant value for a fixed bandwidth and passband ripple. x is then obtained as follows

$$x = \frac{1}{(2\beta)}\left[\frac{1}{(\beta\gamma)} + \sqrt{\left(\frac{1}{(\beta\gamma)^2} - 4\right)}\right]. \tag{19}$$

From the equivalent circuit, as shown in Fig. 9(b), the tuning frequency of the transducer changes from ω_t to ω_t', equaling that of the series circuit L_t, C_t, and $-C$. Then, the new tuning frequency ω_t' is obtained as follows

$$\omega_t' = \sqrt{\frac{1}{L_t}\cdot\left(\frac{1}{C_t} - \frac{1}{C}\right)} = \omega_t \cdot \sqrt{(1 - 1/x)}. \tag{20}$$

On the other hand, the terminal resistance R' is approximately given from (18) and (17), as follows,

$$R' = \frac{\Delta f \cdot \gamma}{2\pi f^2 q_1 C_d}[1 + 1/(\beta^2 \cdot x^2)] \tag{21}$$

where $f = 3.22$ KHz, $f = 113.9$ KHz.

VII. Pole-Type Mechanical Filter Characteristics

The filter volume is about 5.3 cm^3, 1/10 that of the *LC* channel filter now used in the channel translating equipment in Japan, and it weighs about 13 g. Fig. 10 shows the external appearance of the filter. Terminating impedances are about 13-14 kΩ. These values coincide fairly well with the value calculated from (21).

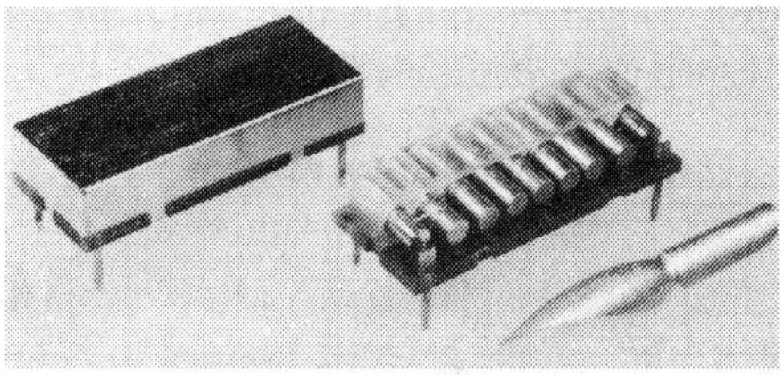

Fig. 10. Torsional-mode pole-type mechanical channel filter.

Typical attenuation frequency responses are shown in Fig. 11. In the passband region, the ripples are within CCITT 1/20 requirements over the temperature range of from 5°C to 45°C. However, in case of sudden cooling from room temperature to 5°C, the ripples slightly increase over that specification during about the first hour and gradually recover to the original values. This condition is caused by the piezoelectric ceramics of the transducers, and is often called thermal shock effect. The flat loss is about 0.9 dB, and deviates about 0.15 dB over that temperature range.

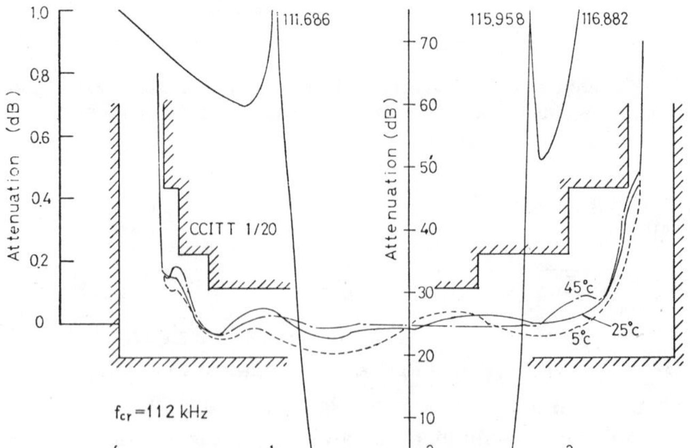

Fig. 11. Pole-type mechanical filter attenuation frequency response.

Two attenuation poles are observed in the upper stopband, however only one pole is obtained in the lower stopband. Although the reason is not clear, it may be caused by impedance lowering at the lowest pole frequency due to a spurious vibration mode. The attenuation rise on both skirt regions fulfilled the desired specification in Japan. Fig. 12 shows the attenuation characteristics of the filter over the wide frequency range of from 50 kHz to 450 kHz. The spurious responses observed at 66 kHz and 154 kHz correspond to the funda-

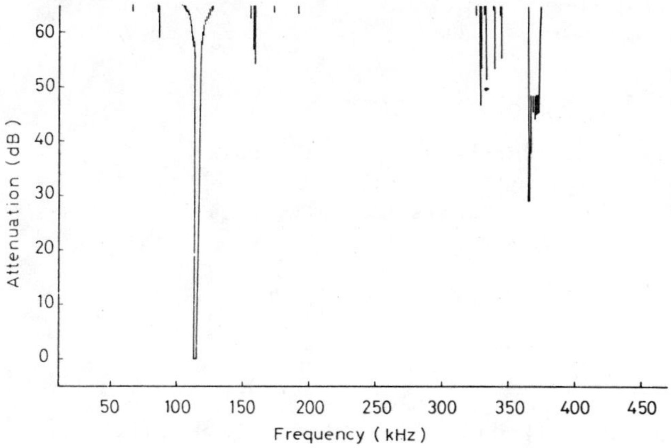

Fig. 12. Filter attenuation characteristics over a wide frequency range.

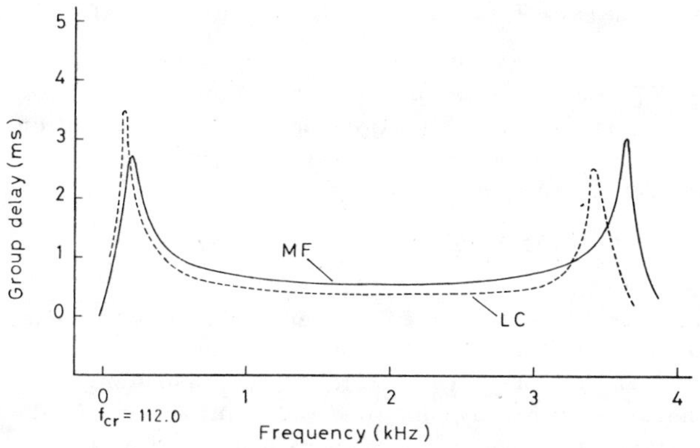

Fig. 13. Mechanical filter and *LC* channel filter group delay distortions.

mental and second-harmonic bending mode of vibration of the nearest neighboring resonators to the transducers. The coupler from the input transducer is spot-welded at the fundamental bending mode nodal point on the number 2 resonator. Another coupler from the output transducer is welded at the second-mode nodal point on the number 8 resonator. Using this method of coupling the spurious responses are suppressed about 20 dB. Fig. 13 shows group delay distortion of both the mechanical and *LC* filters. Delay distortion of the mechanical filter is slightly larger than that of the *LC* filter.

VIII. Conclusion

A torsional-mode pole-type mechanical channel filter was studied. Finite attenuation poles are produced by two oblique bridging wires. Torsional-mode transducers are developed. The capacitance ratio of these transducers is about 9-11; therefore, the filter does not need inductors to compensate for the transducer's clamped capacitances. The filter volume is about 5.3 cm^3 and it weighs 13 g. Attenuation ripples in the passband fulfill the desired specification (CCITT 1/20). Delay distortion and minimum group delay time are nearly equal to that of the *LC* channel filter.

Acknowledgment

The authors wish to express their appreciation to Dr. K. Takahara, Director of Yokosuka Electric Communication Laboratory, NTT., Dr. S. Hamada, Director of Electric Equipment Development Division, Dr. A. Kawamata, Deputy Director of Electric Equipment Development Division, Dr. R. Kaneoya, Deputy Director of Technology Division, and Dr. K. Kawashima, Director of Circuit Component Section, NTT., for their continual encouragement.

References

[1] H. Albsmeier, "Ein vergleich der realisierungsmöglichkeiten elektromechanischer Kanalfilter im Frequenzbereich 12 kHz bis 10 MHz," *Frequenz*, 25, p. 74, (1971).
[2] K. Yakuwa, S. Okuda, and M. Yanagi, "Development of new channel bandpass filters," *Proc. ISCAS/74*, p. 100, (1974).
[3] H. Schüßler, "Filters for channel bank filtering with mechanical resonators," *Proc. ISCAS/74*, p. 106, (1974).
[4] T. Yano, T. Futami, and S. Kanazawa, "New torsional mode electromechanical channel filter," presented at 1974 IEEE European Conf. on Circuit Theory and Design. London, July, 1974.
[5] K. Sawamoto, E. Sasaki, S. Kondo, T. Ashida, and K. Shinozaki, "A torsional mode mechanical filter," *Trans. IECE. of Japan*, 57-A, p. 575, (1974).
[6] M. Börner, "Mechanischer filter mit Dämpfungs polen," *A.E.U.*, 17, p. 103, (1963).
[7] B. Kohlhammer and H. Schüßler, "Berechnung allgemeiner mechanischer Koppel filter mit—," *Wiss. Ber. AEG. Telefunken*, 41, p. 150, (1968).
[8] R. A. Johnson and W. A. Winget, "FDM equipment using mechanical filters," *Proc. ISCAS/74*, p. 127, (1974).
[9] H. Watanabe, "Approximation theory for filter networks," *IRE Trans. on Circuit Theory*, p. 341, (1961).

BRIDGING ELEMENTS IN MECHANICAL FILTERS: DESIGN PROCEDURE AND AN EXAMPLE OF NEGATIVE BRIDGING ELEMENT REALIZATION

S. Cucchi, F. Molo
FDM Division Laboratories
Telettra S.p.A. Milan, Italy

Abstract

A computational method to obtain bridged resonator filters is given: the peaks of attenuation are at real frequencies without any constraint on their position. A realization example of negative bridging element by a mass loaded wire is also presented.

INTRODUCTION

Bridged configurations are of great interest for mechanical filters to realize attenuation peaks at finite frequencies. With a specified attenuation requirement on a filter, the insertion of attenuation peaks allows a reduction in number of resonators, thus giving a sensitivity decrease and a flatter group delay in the passband, what is of particular interest in the application of mechanical filters to FDM channel bank. The paper of A. J. Johnson [1] gives a general view of the problem and contains also a large number of references.

The procedure to be explained starts from a ladder network, as obtained by the largely used synthesis programs, applying equivalent transformations to it. The filters here considered have a minimum number of capacitors (the mobility analogy is supposed to be used for the electrical equivalent circuit of a mechanical filter) and 3 + 2n attenuation peaks at infinite frequency and 1 at zero frequency. The procedure can anyway be extended to more general types of filters.

COMPUTATION OF BRIDGED STRUCTURES

The ladder network of fig. 1.a) can have attenuation peaks at frequencies below the passband, at the expense of some negative series inductances. The position of the parallel inductance is unimportant. As first step the nodes, where only inductances are connected, can be eliminated by a generalization of the π -star transformation: the remaining nodes, where capacitances are connected, are linked together in every possible way (fig. 1.b). To the Y matrix of the new network the bilinear transformation:

$$A^T Y A \qquad (1)$$

is applied, where A is a diagonal matrix of real numbers, $x_i = \sqrt{C_o/C_i}$, that makes all the capacitances equal to a value C_o. The new network has an unchanged topology but contains inductances in parallel to the capacitances.

A further transformation of type (1) is applied, but now the A matrix must satisfy the following conditions:

1) the capacitances must remain in the same position and of the same value;
2) the input-output behavior of the network must remain unchanged;
3) the number of bridging elements must be reduced to a minimum.

Splitting the Y matrix in the inductive and capacitive parts Y_L and Y_c ($Y_c = p\,C_o[1]$), relation (1) becomes

$$A^T Y_L A + A^T Y_C A = A^T Y_L A + A^T A Y_C$$

Condition 1) above is fulfilled if $A^T A = [1]$ (A is therefore a real orthogonal matrix); considering also condition 2), the consequence is that A must have zeros on the first and on the last rows and columns except for a 1 on the main diagonal [2].

The maximum number of elements to be zeroed in the Y_L matrix by a bilinear transformation with orthogonal A matrix (satisfying the before mentioned conditions) is twice (due to the symmetry of Y_L) the number:

$$\frac{(n-3)(n-2)}{2}$$

which is the difference between the number of elements in A ($[n-2]^2$) and the orthogonality conditions on A ($(n-1)(n-2)/2$). As a consequence, in a network with n resonators and a maximum number of attenuation peaks there are in general at least 2n - 3 coupling elements. As finding a matrix A satisfying condition 3) is difficult, a step by step procedure, as used in reducing a symmetrical matrix to a diagonal or tridiagonal form [3], is applied to Y_L. This implies the repeated application of the Jacobi transformation in which the orthogonal matrix J is a unity matrix, except for the four elements on the crossing of rows i, j with the columns of same indexes, given by:

$$b_{ii} = b_{jj} = \cos\varphi \qquad \left(-\frac{\pi}{2} < \varphi < \frac{\pi}{2}\right)$$
$$b_{ij} = -b_{ji} = \sin\varphi$$

To fulfill condition 2) must be: $i, j \neq 1, n$

With a single Jacobi transformation the elements a_{ri}, a_{rj} of a matrix, ($r \neq i, j$) are transformed into the following:

$$a'_{ri} = a_{ri}\cos\varphi + a_{rj}\sin\varphi$$
$$a'_{rj} = a_{ri}\sin\varphi + a_{rj}\cos\varphi \qquad (2)$$

One of these elements a'_{rk} can be put to zero by a suitable choice of the angle φ. The step by step procedure can be quite general since an orthogonal matrix A can be decomposed into the product of Jacobi matrices. Indeed a general matrix A can be reduced to a lower triangular matrix T by the following transformation:

$$A \prod_i \prod_j J_{ij} = T \qquad \begin{pmatrix} i = 1 \ldots n-1 \\ j = i+1 \ldots n \end{pmatrix} \qquad (3)$$

where Jij is a Jacobi matrix applied to columns i, j to make zero the element a_{ij} of matrix A (relations (2) are still valid). The matrices Jij must be

Reprinted from *Proc. 1976 IEEE Int. Symp. on Circuits and Systems*, Apr. 27-29, 1976, pp. 746-749.

applied in the order given by the variability of indexes and the order of the product symbols π. Multiplying the two members of (3) by their transposed the relation $TT^T = [1]$ is found, from which it is easy to see that follows: $T = [\pm 1]$, and so (neglecting the -1 elements that correspond to 1: -1 transformers, which can be introduced in the final structure):

$$A^T = \underset{i}{\pi} \; \underset{j}{\pi} \; J_{ij}$$

Some bridging element reduction procedure are next given. Choosing the couples (2, 3), (2, 4) ... (2, n-1) as indexes i, j in the Jacobi transformations, the elements from 3 to n-1 of the first row of the Y_L matrix are made zero. Also the elements symmetric to the main diagonal are reduced to zero. A similar step is applied to other rows moving from outside to inside of the matrix and also eventually up and down to the middle point of the matrix. In every row the possible zeros are one less the zeros of the preceding one and on the same columns. The procedure can also start making zero the elements of the last row. Fig. 2 explains the application of the procedure to a six resonator filter, giving two of the many possible configurations. On the right of the matrices the order of application of the process to the rows is given. The possible configurations for the maximum number of attenuation peaks (n-2), contain all the adjacent resonators coupling elements, and the number of additional bridging elements is n-2 except for particular positions of the attenuation peaks. As a general rule there is a bridging element from input to output and the subsequent internal bridging element has a common node with the preceding one and bridges the maximum number of resonators: same rule for the more internal elements.

In a second type of procedure, when the Jacobi transformations are applied in the same order as above, the element a_{ij} can be reduced to zero (what implies a solution of a second degree equation in tg φ, which has always real roots for a symmetrical matrix). If this procedure is iteratively applied (see [3]) the final result is a diagonal matrix with non zero elements also in the first and last columns and rows. This corres ponds to a "parallel T" network as described in [4]. The application of this procedure to a single row leads to the separation of a single T structure from the filter. The procedure is not iterative for those sections of a filter implementing only a couple of attenuation peaks.

A third type of procedure can be obtained considering that, after the removal of the more external bridging element, the remaining network has an unchanged degree but one less attenuation peak. This must allow splitting the filter in two separate bridged structures below the input-output bridge. Each one of the two subnetworks can realize an odd number of attenuation peaks only if an odd number of these peaks is on the real or imaginary axis. To split the Y matrix the following iterative procedure has proved to be efficient:

1) The elements from 3 to n-1 of the first row are made zero using the first described procedure.

2) Jacobi transformations of indexes 3, r (r=s+1, ...n-1) are applied, making every time the ratio of the elements (2, r), (3, r) equal to the ratio of the elements (2, n), (3, n) of the last column. It is possible to realize the proportionality only for the transformations having real solutions for the second degree equation in tg φ, which solves the problem. A series of similar transformations is applied increasing every time the indexes 2, 3 by one, till indexes s-1, s are reached. In the second degree equations in tg φ the positive or negative sign of the descriminant can be chosen, without changing it in the subsequent iterations, and the transformations having no real solution in tgφ are tried again in the following step (see 3) below).

3) As every transformation alters one or more of the already constructed proportionalities, the procedure in point 2) is repeated till the ratios of elements are invariant to subsequent iterations. If all the proportionalities have been realized, the network can be split around the node s.

4) To split the network around a node s, Jacobi transformations are applied to the indexes (2, 3) (3, 4) till (s -1, s), making every time zero the elements on the row of lower index, on the right of column s. This is possible because the zero condition for an element (angle φ) is valid also for the other elements on the same row.

After step 4) a zero rectangular matrix (except for the input-output element) is created above and on the right of the (s, s) element. The two subnetworks can be processed with any one of the above procedures.

The above procedures can also be applied in a mixed manner to the parts of the matrix of a network, still to be reduced in the number of bridging elements. In the final network the capacitances can be again changed in values by a diagonal matrix, and negative adjacent couplings made positive without affecting the topology. When the number of attenuation peaks is not maximum, these can be implemented in a single or separate sections. The above transformations allow to obtain a great variety of configurations, in particular all those given in the literature on mechanical filters, and have also interest for a larger class of filters (LC, active RC, etc.).

AN EXAMPLE OF NEGATIVE BRIDGING ELEMENT IMPLEMENTATION

Mechanical filters with flexural or torsional resonators, a longitudinally vibrating coupling wire, and bridging elements, have been widely studied for application to FDM channel bank filtering. As this type of filters is generally realized at relatively low frequencies (less than 150 KHz) there are difficulties in realizing negative bridging elements, which are required by attenuation peaks in the lower stop-band. Many solutions are known to solve this problem [1]. In this section a structure is presented, which has been shortly outlined before in a more general context [5].

The structure, depicted in fig. 3.a) and conceived for application to a flexural resonator filter with longitudinal coupling wire, is built by a longitudinally vibrating wire loaded with two masses. The structure connected to the resonators in points of maximum vibration amplitude, is not critical against small longitudinal and transversal translations with regard to the filter structure, has a good realization repeatibility and is not excited with flexural spurious oscillations. The structure is supposed to be symmetric and the electrical equivalent circuit follows, as usual, the mobility analogy. As all the mechanical elements are supposed to have dimensions much smaller than a half-wave lenght, they can be represented as lumped elements as in fig. 3.a).

With symbols and definitions of fig. 3.a) the network admittances are given by:

$$Y_{11} - Y_{12} = \frac{pL_1}{p^2 + \omega_1^2}$$

$$\frac{1}{Y_{12}} = pL_o \frac{(p^2 + \omega_1^2)(p^2 + \omega_2^2)}{\omega_1^2 \omega_2^2}$$

The term $1/pY_{12}$ can be considered as a frequency dependent inductance L_{eq} (fig. 3.b), and the negative inductance $-L_k$ to be realized can be identified, owing to the small relative bandwidth of mechanical filters, with the minimum of L_{eq}, to be located at passband mid frequency ω_o of the filter. A possible set of design equations is the following:

$$L_o/L_k = \frac{1 - 2x}{x^2} \qquad (x = L_1/L_o < 0.5)$$

$$C_1/C_k = \frac{x(1 - x)}{(1 - 2x)^2} \qquad (C_k = 1/\omega_o^2 L_k)$$

$$(\omega_o/\omega_1)^2 = \frac{1 - x}{1 - 2x}$$

The behavior of these functions is given in fig. 4. Inductances and capacitances are simply related to the diameter of the wire and to the weight of the masses. As too small diameters of the wire or too big masses are to be avoided, the optimal range of x is about from 0.3 to 0.4.

The structure has been used in the laboratory realization of a 48÷52 KHz mechanical channel bank filter, having ten flexural resonators, a longitudinal coupling wire and two attenuation peaks (attenuation curve of the filter in fig. 5). The two peaks have been implemented using the mechanical structure and its equivalent electrical configuration of fig. 6. For a particular relative frequency position of the attenuation peaks, the internal bridging element, which as seen before is in general present, becomes zero. The wire is 15 mm long with a diameter of 0.17 mm. On this, two cupper masses of 23 mg weight are mounted at a relative distance of 3 mm ($x = 0.4$). After the assembling, the small cupper-cylinders are squeezed in the direction of the wire to make them adhere to the wire itself.

From the realization point of view it is useful to know the sensitivities of the bridging element against variations of parameters having a simple or direct connection with mechanical parameters, such as: weight of the masses (C_1), diameter of the wire (L_o), the distance of the masses (x) and a rigid longitudinal translation T of the bridging element against the structure of the filter. The above sensitivities are:

$$S_{C_1}^{L_k} = S_T^{L_k} = 0$$

$$S_{L_o}^{L_k} = 1$$

$$S_x^{L_k} = \frac{2(1 - x)}{1 - 2x}$$

The structure has a good behavior except for the sensitivity $S_x^{L_k}$, which can reach large values, and coincides with the functions $(\omega_o/\omega_1)^2$ of fig. 4, apart from the factor 2.

REFERENCES

[1] R.A. JOHNSON: "The design of mechanical filters with bridged resonators", Proc. ISCAS 1975

[2] W. CAUER: "Vierpole", Elektrische Nachrichtentechnik, Vol. 6, July 1962, pp. 272÷282

[3] E. DURAND: "Solutions numériques des équations algébriques", Tome II, Masson et C.; Paris, 1961

[4] Y. NAKAUCHI: "A synthesis of parallel ladder circuits using equivalent transformation techniques" Proc. ISCAS 1975

[5] F. MOLO: "FDM Channel bank filtering: LC or one of its alternatives?", Proc. ISCAS 1974 (oral presentation).

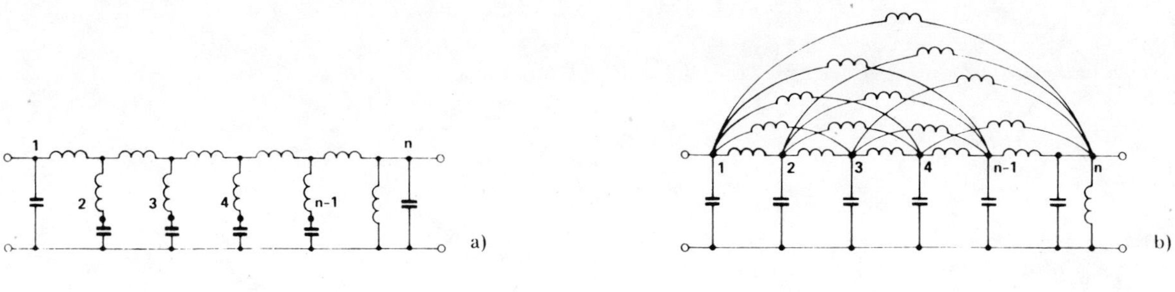

Fig. 1 — a) Starting ladder network
b) Network equivalent to a) after elimination of inductive nodes.

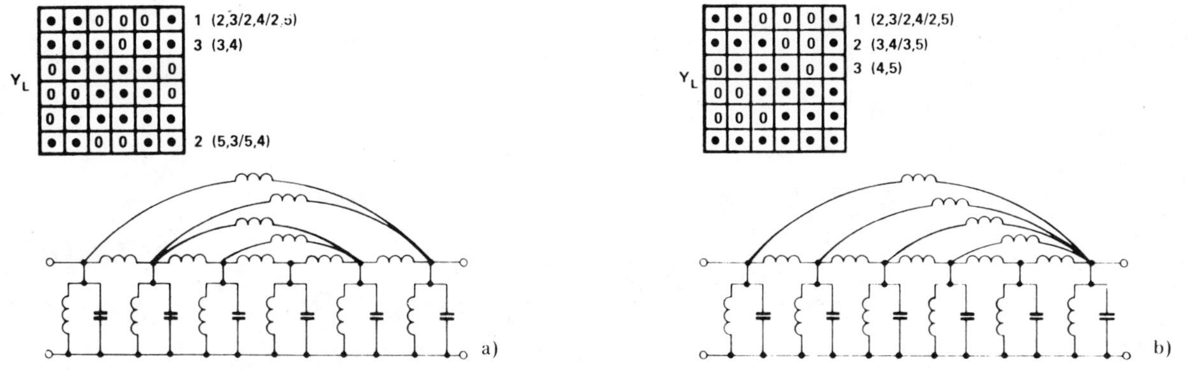

Fig. 2 — Two possible final structures after redundant bridging elements elimination, with position of zeros in the final matrix.

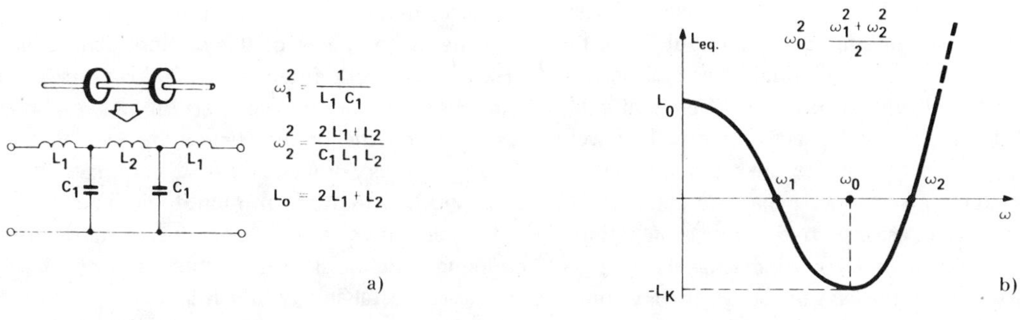

Fig. 3 — a) Negative coupling element
b) Frequency behavior of $1/p\,Y_{12}$ of the network a).

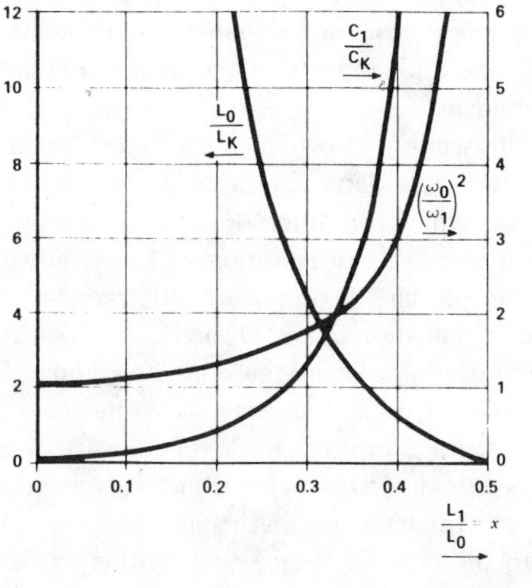

Fig. 4 — Plot of design equations for the structure of fig.3a) as functions of the parameter x.

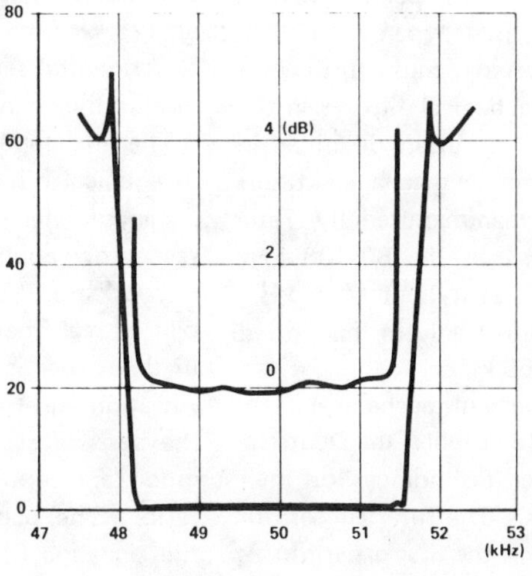

Fig. 5 — Attenuation curve of a 48/52 kHz FDM channel bank mechanical filter.

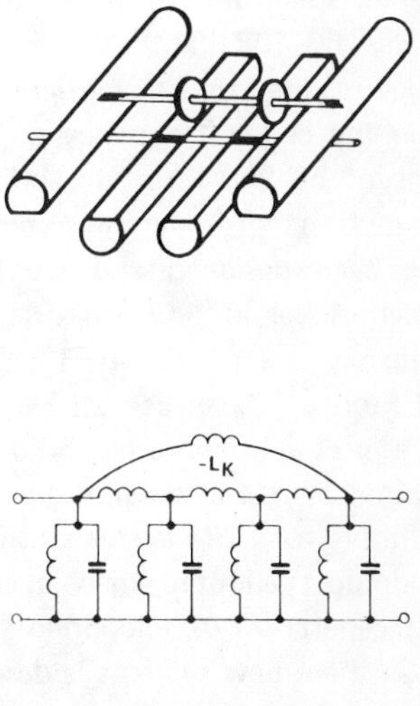

Fig. 6 — Mechanical and equivalent electrical realization of two attenuation peaks.

Section I-C
Channel Bank Filter Design

Although voice-channel mechanical filters have been used in FDM telephone systems since 1955, it has only been since the early 1970's that mechanical channel filters have found worldwide acceptance. Before this time, most channel filters manufactured by telecommunication companies outside of the United States were designed around inductors and capacitors. The *L-C* filters had the advantage of lower differential delay distortion for the same stopband selectivity, but they were large and costly to manufacture due to the many hand operations involved. A trend toward more channels per bay and reduced production costs resulted in a number of national telephone systems deciding in favor of mechanical channel and signaling filters and, in the case of West Germany, pilot-tone filters as well.

The filters described in the four papers of this section all use a premodulation scheme where only one filter type is used. A second modulation step is used to translate the filtered spectrum to its desired frequency band which is usually 60–108 kHz. Building one filter type rather than twelve results in a substantial reduction in manufacturing cost when the center frequency of the mechanical filter is in the range of 50–250 kHz. In addition to the filters described in this section, torsional and extensional mode mechanical filters have been designed and are being manufactured by Tesla (Czechoslovakia) and Kokusai (Japan) in the 60–108 kHz frequency range, by RFT (East Germany) at 200 kHz, and Fujitsu (Japan at 128 kHz. Collins Radio in the United States uses a disk-wire technology at 256 kHz.

A major breakthrough in the acceptance of mechanical channel filters for FDM systems came in 1972 when the Deutsche Bundespost adopted mechanical filter technology for their next generation of telephone systems. The filter chosen for use in these new systems is described in the first paper of this section. Albsmeier *et al.* discuss the premodulation scheme and the basic design concepts involving the mechanical configuration. They also discuss its equivalent circuit, stability, and manufacturability. In addition, they describe the two-pole and three-pole mechanical filters used for signaling.

In the second paper of this section, Schuessler compares 50 kHz flexural-mode filters with 200 kHz torsional-mode filters, and he also makes a comparison between the monotonic response and the finite attenuation pole type of filters. His Figure 6 is very interesting as it compares the passband and stopband sensitivities of monotonic and finite attenuation pole-type filters to changes in resonator frequencies and couplings. Schuessler also compares mechanical filter designs to quartz crystal and gyrator filters.

The Bosc and Loyez paper describes the work done in France on an extensional-mode resonator design that uses a zigzag geometry with the coupling wire axis set at 18° from the axis of the resonator. Although this is a short paper, the authors touch on the subjects of the premodulation scheme at 128 kHz, the mechanical structure, material specifications, experimental results, and a description of the signaling filters used in the channel modem.

The final paper in this section, "New Torsional Mode Electromechanical Filter," discusses a design proposed for use in the Japanese telephone system. The filter uses torsional-mode resonators coupled in twin-tee configurations. The twin-tee networks are used to obtain a frequency dissymmetrical pair of real-frequency attenuation poles. Of particular interest in this paper is the discussion of how the resonator mode of vibration and the 120 kHz carrier frequency were selected on the basis of size, loss-factor, linearity (as a function of size, mode, and input levels), and the resonator length to diameter ratio. The paper also discusses the mechanical coupling of the signaling filter to the channel filter which results in the elimination of a transducer.

Some Special Design Considerations for a Mechanical Filter Channel Bank

HANS ALBSMEIER, ALFHART E. GÜNTHER, AND WILHELM VOLEJNIK

***Abstract*—The technical concept realized by the channel bank is optimum with respect to a variety of requirements. Considerations of size and fabrication technology recommend a frequency of 50 kHz for the mechanical filter. The general concept of the modulator suggests a filter design with tuned conventional transformers. Since subsequent adjustment of the assembled mechanical part of the filter is undesirable, the provision of finite attenuation poles has been abandoned at the expense of adding two extra resonantors. The design imposes only modest demands as to the reproducibility of the mechanical couplings. By tuning the transformers it is possible to correct minor production tolerances.**

A special design of the channel and associated signal filter results in a very low temperature dependence and permits to connect both filters directly in parallel.

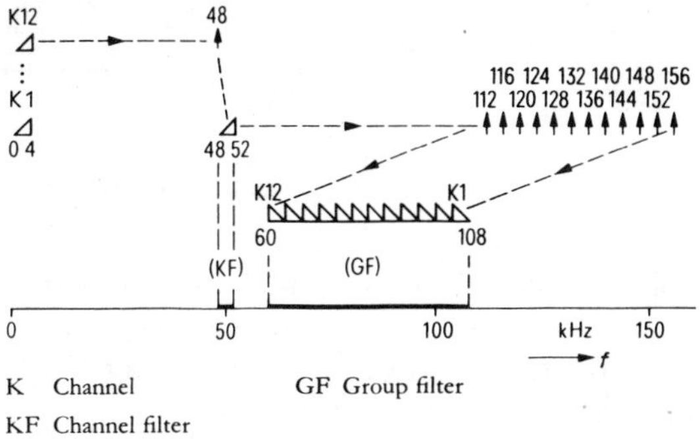

Fig. 1. Modulation scheme for formation of channel base group.

I. PREMODULATION CONCEPT

THE CHANNEL MODEM provides the first step of modulation from the voice-frequency to carrier spectrum as well as demodulation. It is an important objective criterion in judging the state of the art of communications systems. The principal features and specifications of the channel modem with electromechanical filters are reviewed in [1], [2]. This paper emphasizes some of the developmental and technological problems and their solution.

As elucidated [1], [2], 12 VF telephone channels are multiplexed to the basegroup by premodulation. The system recommendations specified by CCITT exclude premodulation between 60 and 108 kHz because this frequency range is reserved for the basegroup [3]. It was found that particularly easy conditions are attained when the operating frequencies of the channel filters are chosen below 60 kHz and when the channel filters are designed with piezoelectrically driven bending resonators of Thermelast®, connected by short length-extensional couplers. Thermelast® is a special nickel–iron alloy with certain additions.

The concept discussed enables us to avoid any kind of internal or external loss compensation and yields a number of further important advantages over other system concepts.

As is well known, the CCITT recommendations prescribe a channel filter bandwidth of 3.1 kHz independent of the applied carrier frequency. The chosen system concept leads to fractional bandwidths essentially larger than 2 percent, so reducing the demands as to the Q values of the resonator material, as well as making the bandedge distortions uncritical. At the same time the influence of the temperature coefficients of the resonators is reduced to a negligible quantity, even though the temperature sweep is 55 K.

Since flexural vibration is the lowest possible mode, a resonator driven in this mode has no spurious resonances below this resonance. The geometry of bending resonators for lower frequencies is more practical.

Since the length-extensional mode of a coupling wire has the highest possible impedance, it is optimum; the impedances of flexural and torsional modes are much lower. Thus a spurious flexural or torsional mode can often be eliminated by ohmic losses. It should here be emphasized that the application of short ($<\lambda/8$) length-extensional couplers yields the significant benefit of two degrees of freedom (length and diameter), so improving the fabrication conditions with respect to the non-continuous spectrum of available wire diameters. The final choice was a premodulation carrier frequency of 48 kHz. The channels are translated from the premodulation stage to the range of the basegroup with the aid of one of twelve carriers between 112 and 156 kHz (Fig. 1).

II. CIRCUIT DESIGN

Fig. 2 shows the block diagram of the channel modem. Owing to the premodulation, there are two modulation-stages in each direction of transmission.

The main objective in developing the modulators for the channel modem was to find the most economical solution to a given set of technical specifications. The optimum solution is the one which minimizes the overall cost, of which the cost of the modulators is only one factor.

Manuscript received October 1, 1973; revised February 20, 1974.
The authors are with the Zentrallaboratorium für Nachrichtentechnik, Siemens Aktiengesellschaft, Germany.
® Registered trademark of Vacuumschmelze AG, Hanau, Germany.

Reprinted from *IEEE Trans. Circuits and Syst.*, vol. CAS-21, pp. 511–516, July 1974.

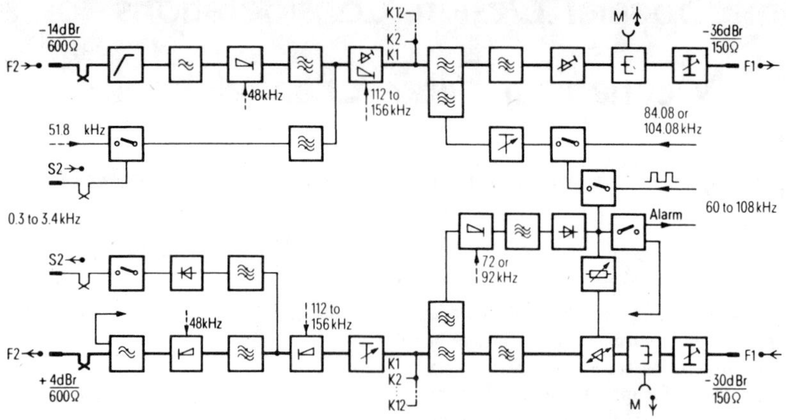

Fig. 2. Block diagram of channel modem (K = channel).

This solution was not easy to find because many disturbing side effects had to be eliminated by suitable circuit design. Crosstalk, for instance, caused many difficulties and its elimination was time-consuming.

Circuit design could not, of course, be oversimplified at the cost of more complicated alignment in the factory. It was found possible to design the modulators such that no alignment was necessary, so avoiding the otherwise relatively high cost of alignment.

Only two complementary transistor types (BCW 60 and BCW 61 p-n-p, n-p-n) are used throughout the channel modem except for the two switching transistors of the electronic switch of the outband signal receiver. The use of only two types of transistors means "quantity use" and, consequently, reduced cost.

The resistors are fabricated using thick-film technology. The low cost per thick-film resistor suggests the use of circuits in which resistors outnumber other components. However, the requirements placed on the resistor accuracy should not be too high: where possible, 2 percent should not be exceeded according to manufacturing cost.

The technical aspects which governed the choice of circuits and their characteristics were the typical modulator requirements such as linearity, noise, stability, balance, etc.

Loss minimization was also essential. With the high packing density of the channel modem, the excess temperature inside the equipment is a prime consideration in life calculations.

The channel modem contains four different modulators.

A. Transmitting-End Premodulator

This modulator translates the audio input signal to the premodulation band 48-52 kHz. Moderate requirements have to be met with regard to the harmonic distortion of this modulator. On account of this and the dissipation requirements, a passive balanced modulator was selected. Two complementary transistors are the switching elements.

B. Transmitting-End Channel Modulator

The modulator which translates the input signal from the premodulation band to the basegroup, is an active single push-pull modulator. The high differential impedance of the collectors makes it possible to combine the twelve channels of a basegroup such that the twelve channel modulator outputs are simply connected in parallel. They have a common output impedance which is at the same time the termination for the succeeding group filter.

The carrier terminal of the modulator can be connected to either of the two phases of the balanced carrier voltage. The objective is a random distribution of the terminals between phase and antiphase to avoid the periodical voltage peaks which might result from the addition of a series of outband signaling voltages.

C. Receiving-End Channel Demodulator

The frequency band of a basegroup (60-108 kHz) is applied to the input of this modulator. This band is translated with one of the carrier frequencies between 112 and 156 kHz so that the channel to be received is available in the band of 48-52 kHz.

The broad input frequency band of this modulator demands low harmonic distortion. The conversion gain should be as high as possible to avoid the need for a channel amplifier. For this reason, an active double push-pull modulator was selected.

For level adjustment, the gain of each channel at the receiving end should be variable by ± 3 dB. The modulator front panel must embody an in-service gain control. A capacitive voltage divider with a variable capacitor for changing the division ratio offers a reliable long-term solution. In order to keep the division ratio independent of frequency, the voltage source must have a relatively low impedance and the voltage tap a relatively high impedance with respect to the divider reactance.

Such a divider circuit serves as the channel demodulator input. Since the reactance of the divider capacitances is very high as compared with the characteristic impedance of the group filter, the twelve inputs of the channel modulators of a group can be connected in parallel without giving rise to intolerable interaction. The voltage is picked off the divider via a transistor stage with a high input impedance (300-kΩ bootstrap). The balanced signal voltage is transferred from the transistor stage output to the modulator via a resistor network, which isolates the carrier feeder line from the signal path.

D. *Demodulation from the Premodulation Range*

The primary objective concerning the final circuit configuration of the second modulator at the receiving end was to select gain and output level such that there would be no need for a succeeding channel amplifier. With a relative output level of +4 dBm, it became necessary to have a conversion gain of 25 dB and a maximum output level of the wanted signal of about +10 dBm. This modulator consumes a large portion of the total dc power of the channel modem. The self-heating caused by this high power consumption calls for high modulator efficiency. For all these reasons an active double push-pull modulator was selected.

In order to reduce the dc power consumption by a factor of about 2, the source impedance at the modulator output is formed by combined negative current/voltage feedback rather than a resistor. The leakage inductance of the output transformer is built out to form a low-pass filter to suppress the unwanted modulation products; the filter's capacitors present a short circuit to these sidebands so that the collector dc voltage can be fully driven by the wanted sideband.

III. ELECTROMECHANICAL CHANNEL FILTER

A. *Preliminary Considerations*

Possible realizations of the filter concept described in Section I are the disk-wire filter [4] and the rod-wire filter [5]. The latter configuration permits very rugged construction: a rod driven in its first bending mode has two nodal points of motion which are used as points for attaching supporting wires. Since the couplers are therefore not required to assist filter stability, they can be designed for optimum coupling. The axes of inertia of a uniform rod depend on random inhomogenities. Distinct directions can be obtained, and the two eigenfrequencies can be separated sufficiently by slightly flattening a part of the perimeter of the rod.

The main design objective was to find a configuration which can be easily manufactured. Thus all steel resonators and all couplers were designed with equal cross sections and differing only in length. Furthermore, no final adjustment after welding should be necessary. Since the modulator requires a balanced low-impedance input, the end circuits of the filter are composed of tapped coils and capacitors, including the static capacitance of the transducers.

The transducer resonators utilize the transverse piezoelectric effect. A small ceramic plate soldered to the plane side of a beveled steel rod forces the resonator to vibrate in the flexural mode. This represents the simplest possible configuration.

Since the group delay specifications are less stringent than the attenuation specifications, the filter designer has the choice between a filter with monotonically increasing stopband attenuation or a filter with finite attenuation poles [6] provided by mechanical and/or electrical bridging. Although the filter with finite attenuation poles (reported in [7]) is fascinating from the engineering aspect, this elegant solution was abandoned because of the fabrication reasons already noted.

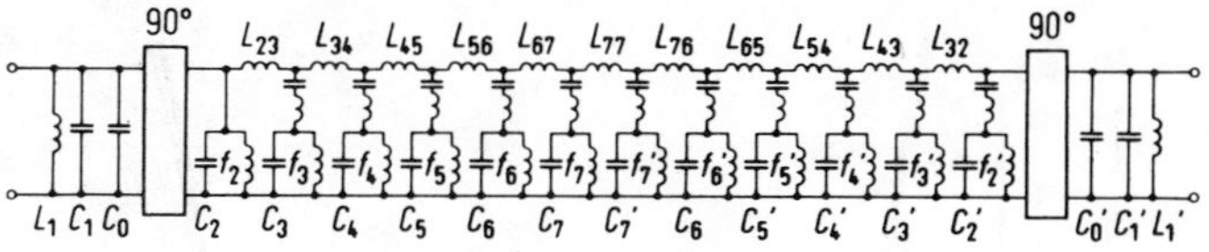

Fig. 3. Equivalent circuit of channel filter.

B. *Fundamental Design of the Equivalent Circuit*

The four-element equivalent circuit of a bending rod with tangential coupler access is derived in [8]. One of its features is the pole-zero ratios depending only on the geometry of the resonator, the squared ratios being the construction constants of the resonator type ($a_1 = \omega_{-\infty}^2/\omega_0^2$, $a_2 = \omega_{+\infty}^2/\omega_0^2$). It has proven practical to regard the short coupler as a massless spring and to include its mass into the masses of the adjacent resonators. Thus the resonator is considered to be not a mere rod but a rod loaded with two half-coupler masses; its resonant frequency is lowered by this mass loading.

Since the input and output specifications of the filter are the same, a symmetric filter structure will logically best meet all performance and fabrication requirements. A filter of the 14th degree satisfies all specifications with a good margin. Thus we arrive at the equivalent circuit of Fig. 3 in which (1) and (1′) are the electrical end circuits, C_0 and C_0' the static ceramic capacitances, (2) and (2′) the transducer resonators, (3)···(3′) the ten steel resonators, and $L_{23}\cdots L_{32}$ the couplers; the ideal inverters in the transducer equivalent circuits are necessary because of the chosen electromechanical mobility analogy [7]. The requirement for equal cross sections for all steel resonators is satisfied by the relation

$$C_\nu^2 \cdot \omega_\nu = \text{constant} \tag{1}$$

for all circuits (3)···(3′), i.e., the designer must meet this additional demand. The design procedure, being an iterative optimization technique described in [7], satisfies (1) by an approximate Norton transformation. The design data are the bandedge frequencies (3-dB points) of 48 210 and 51 485 Hz, a reflection coefficient of 10 percent (i.e., a ripple $\Delta a = 0.045$ dB), and the construction constants of the resonators $a_1 = 0.66$ and $a_2 = 3.16$.

C. *Improvement of Temperature Stability*

While the temperature coefficients of the mechanical elements are <10 ppm/K, that of the ceramic capacitance is $\alpha_{C0} \approx 2500$ ppm/K. To maintain a constant resonant frequency between 10 and 65°C, α_{L1} and α_{C1} have to be suitably chosen. However, as the spread of temperature coefficients in the end circuits is about 10 ppm/K (much

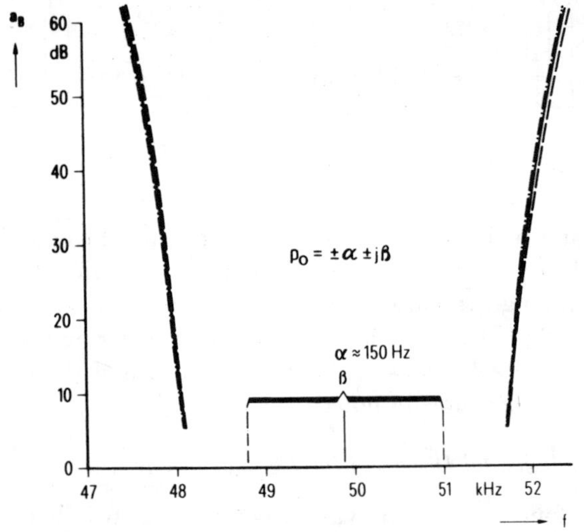

Fig. 4. Attenuation of mechanical filter with two reflection zeros at complex frequencies.

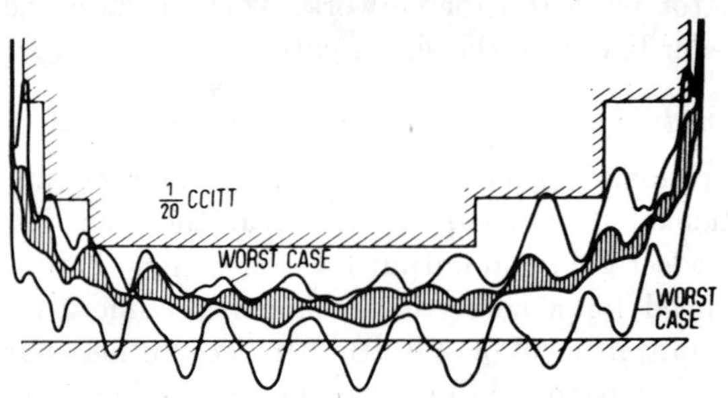

Fig. 5. Temperature performance (insertion loss versus frequency). Instabilities: $\alpha_{L1} = 150 \pm 60$ ppm/K, $\alpha_{C1} = -1500 \pm 250$ ppm/K, $\alpha_{C0} = 2500 \pm 250$ ppm/K; $T = \pm 30$ K.

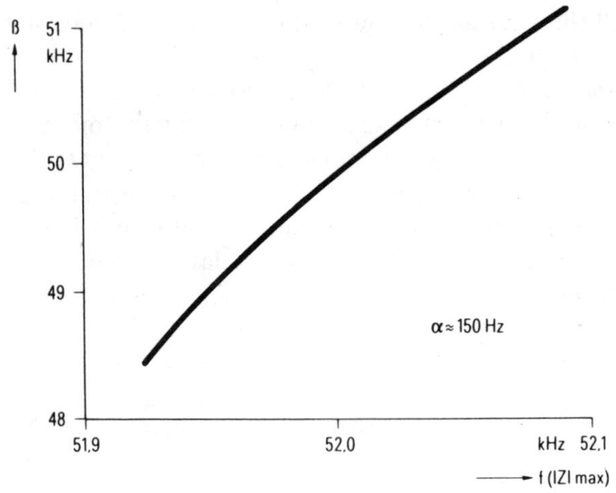

Fig. 6. Impedance maximum with complex reflection zeros.

greater than the α of the steel resonators), it is desirable to increase the bandwidth of the end circuits so as to reduce their contribution to the filter selectivity and hence to the filter tolerances. The characteristic function of our filter [9] is

$$K(p) = \frac{(p^4 + rp^2 + q) \prod_{\lambda=1}^{12} (p^2 + \omega_{0\lambda}^2)}{k \cdot p^3}; \qquad \omega_{0\lambda}^2 > 0 \quad (2)$$

(p denotes the complex frequency $\sigma + j\omega$), the additional poles and zeros originating from the four-element equivalent circuit of the steel resonators being neglected here.

For $r^2 > 4q$ all 14 zeros of the reflection coeficient are at real frequencies, for $r^2 < 4q$ there are two zeros at complex frequencies:

$$r = 2(\beta^2 - \alpha^2)$$

$$q = (\alpha^2 + \beta^2)^2$$

where α and β are the real and imaginary parts in the p plane. In the following considerations, we assume the passband attenuation to have equiripple performance irrespective of whether there are complex zeros (actually the ripple will deviate somewhat from this shape because of the 48.8-kHz reference frequency of the CCITT specifications, see later). It is found that the larger the α, the larger the bandwidths of the end circuits, whereas the bandwidth of the other circuits remains much the same. The steepness of the skirt depends mainly on the magnitude of the imaginary part β: the steepness of both skirt sides is reduced, but the one nearer to β is reduced more (Fig. 4).

Increasing the end circuit bandwidth by 30 percent reduces the temperature sensitivity of the filter sufficiently (Fig. 5): the hatched areas show the filter performance now expected over a 60-deg temperature sweep, and the outer boundary lines denote the worst case of temperature coefficient combinations ($\alpha_{L1} = 150 \pm 60$ ppm/K, $\alpha_{C1} = -1500 \pm 250$ ppm/K, $\alpha_{C0} = 2500 \pm 250$ ppm/K).

D. Adjustment of Input Impedance

The wiring of channel and signal filters, explained in Section II, requires the channel filter impedance maximum to appear at the signal filter center frequency of 51.84 kHz. The location of this impedance maximum is not an objective but a result of the conventional design procedure. With the design data given in Section IIIB, i.e., with all reflection zeros at finite frequencies, we find this maximum at ≈51.6 kHz. It is well known, however, that the impedance maxima move farther from the passband as the attenuation ripples decrease. Attempting to place the maximum where it is required by varying the ripples may prove uneconomic because of the general loss in skirt steepness. It is more favorable to work with reflection coefficient zeros at complex frequencies: in the upper stopband the higher the β ($\omega_{-g} < \beta < \omega_g$; ω_{-g}, ω_g denoting the bandedges), the higher the frequency of the impedance maximum; the reduction in stopband attenuation is thus minimized (the example in Fig. 6 is associated with a 5-percent reflection coefficient design, whereas the final filter has a reflection coefficient of 10 percent). An analogous consideration would apply in the lower stopband, but there are no impedance specifications.

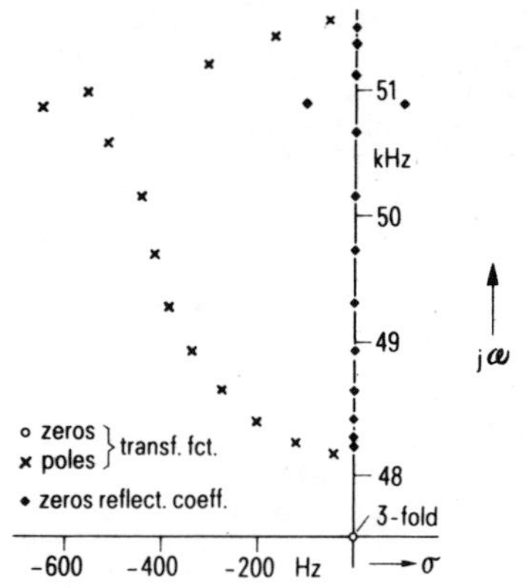

Fig. 7. Pole-zero pattern of mechanical channel filter.

We emphasize that by choosing a suitable complex zero of the reflection coefficient the problems of impedance control and temperature compensation will be solved simultaneously. The pole-zero diagram of the transfer function of such a filter with an equiripple passband characteristic is shown in Fig. 7; it is distinctly different from the familiar elliptical pole distribution of a normal Chebyshev filter.

E. Fabrication

The fabrication concept provides for automatic tuning of the resonators (length 16 mm, diameter 3.5 mm) to within 2 Hz. The pretuned resonators are automatically welded to a single wire (diameter 0.28 mm) whose sections between the resonators are the individual couplers. The only final adjustment is the tuning of the end coils.

While the attenuation minimum of a filter is the usual reference in the tolerance specifications, in the CCITT case it is the attenuation at 800 Hz off carrier, i.e., here at 48.8 kHz. The tolerance stability of the filter is improved if the two ripples closest to 48.8 kHz are about 0.03 dB instead of the usual 0.045 dB. Assuming reasonable fabrication tolerances, we obtain the probable deviations by analysis (Fig. 8). The calculations are performed statistically by the Monte Carlo method (1000 games) and analytically by element sensitivities, the results of both methods being in good agreement: 99.7 percent of the fabricated filters are to be expected to fall within 1/30 and 100 percent within 1/15 of the CCITT scheme. This excellent result could be slightly improved by tuning the end coils, but this degree of freedom can also be used to place the impedance maximum at the specified frequency. Fig. 9 shows the passband attenuation of an arbitrarily selected filter at the temperatures of 10, 25, and 65°C.

IV. ELECTROMECHANICAL SIGNAL FILTERS

A. Receiver Filter

The attenuation specifications have to be met with three resonators (two of them transducer resonators) without electrical end circuits. Again the resonators vibrate in the flexural mode, but the couplers, being fastened at the nodal points of the resonators, apply the torsional mode. This construction seems to be optimum since length-extensional couplers would be either very long or extremely thin. Nevertheless much investigation was necessary in order to find a design without spurious modes. Since the skirt specifications are tighter for the upper skirt side, an attenuation pole is provided there by electrical bridging. This bridging does not require an additional capacitor but is achieved by suitably designing the transducer resonators, utilizing the mutual capacitances of the ceramic slabs and of the connecting wires.

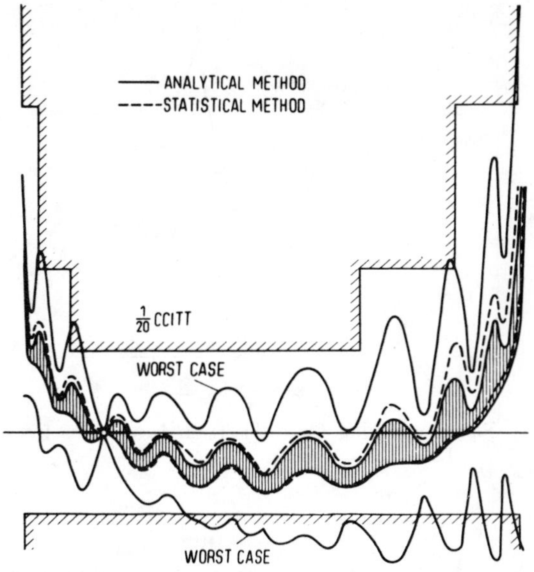

Fig. 8. Fabrication tolerances (insertion loss versus frequency).

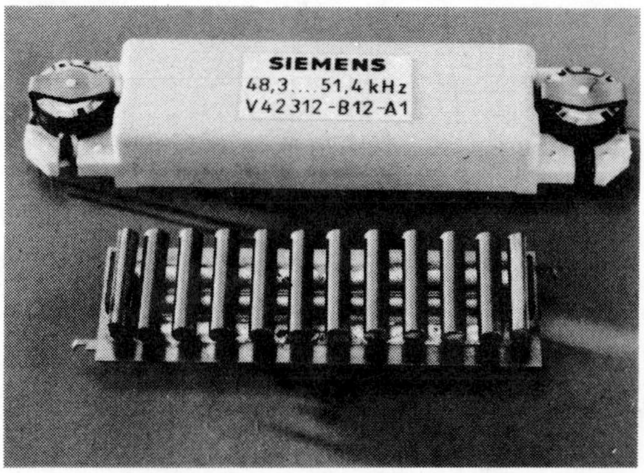

(a)

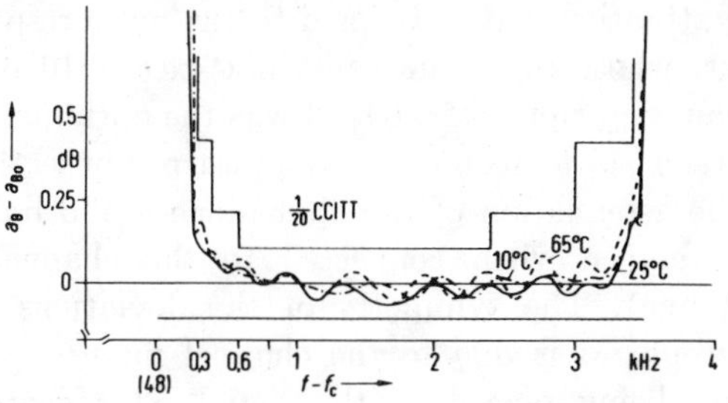

(b)

Fig. 9. Mechanical channel filter 50 kHz, temperature response of insertion loss in passband (measured data).

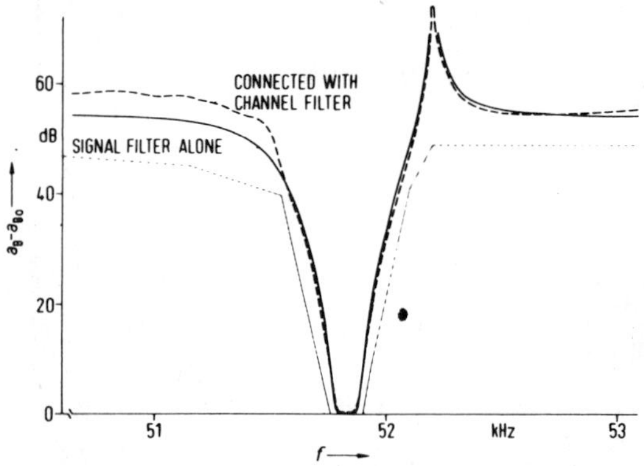

Fig. 10. Signal receiving filter (insertion loss versus frequency).

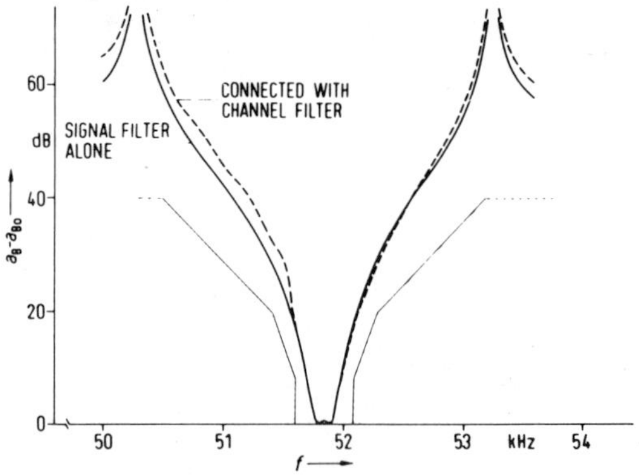

Fig. 12. Signal transmitting filter (insertion loss versus frequency).

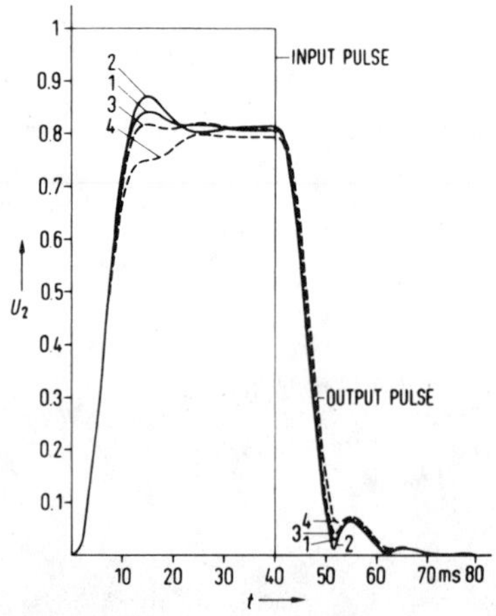

Fig. 11. Pulse response of signal receiving filter signal. Carrier frequencies: 51815 Hz(1), 51825 Hz(2), 51850 Hz(3), 51860 Hz(4).

Some attention had to be paid to the pulse response of the filter, whose overshoot must not exceed 10 percent. An attenuation ripple of 0.04 dB was the optimum tradeoff between skirt steepness and pulse response. Fig. 10 shows the attenuation of the filter alone (solid line) and its performance when connected to the channel filter (dashed line). The symmetry of the deviations in the nearer stopband is due to the channel filter impedance maximum being placed at the signal filter center frequency. Fig. 11 shows the pulse response for four different carriers. It should be added that any deviation in the impedance maximum location would increase the overshoot.

B. Transmitter Filter

The same considerations apply to the transmitter filter with two instead of three resonators owing to the less stringent specifications. As a result of the even number of electrically bypassed resonators there are two finite attenuation poles. Fig. 12 shows the stopband performance of the filter analogous to Fig. 10. The pulse response is less critical and is therefore not shown.

REFERENCES

[1] H. Kopp, "A mechanical filter channel bank," *IEEE Trans. Commun.* (Corresp.), vol. COM-20, pp. 64–67, Feb. 1972.

[2] K. Ey, F. Hornung, and W. Volejnik, "Channel modem features electromechanical filters," *Siemens Rev.*, vol. 39, pp. 239–298, July 1972.

[3] H. Albsmeier, "Ein Vergleich der Realisierungsmöglichkeiten elektromechanischer Kanalfilter im Frequenzbereich 12 kHz bis 10 MHz," *Frequenz*, vol. 25, pp. 74–79, Mar. 1971.

[4] R. A. Johnson, "New single sideband mechanical filters," WESCON 1970 Tech. Papers, vol. 14, sec. 10-1, pp. 1–10.

[5] K. Traub, "Mechanische statt elektrischer Schwingkreise in Trägerfrequenzfernsprechsystemen," CONSTRONIC 1972, Digest of Papers, pp. 139–152, Apr. 1972.

[6] R. A. Johnson, "Mechanical filters," in *Proc. 1973 IEEE Int. Symp. Circuit Theory*, pp. 402–405, Apr. 1973.

[7] A. E. Guenther, "High-quality wide-band mechanical filters theory and design," *IEEE Trans. Sonics Ultrason.*, vol. SU-20, pp. 294–301, Oct. 1973.

[8] F. Künemund, "Channel filters with longitudinally coupled flexural mode resonators," *Siemens Forsch.- u. Entw.-Ber.*, vol. 1, no. 4, pp. 325–328, 1972.

[9] A. E. Günther, "Electromechanical filters—satisfying additional demands," in *Proc. 1973 IEEE Int. Symp. Circuit Theory*, pp. 142–145. Apr. 1973.

FILTERS FOR CHANNEL BANK FILTERING WITH MECHANICAL RESONATORS, QUARTZ CRYSTALS AND GYRATORS

Hans Schüßler
AEG - TELEFUNKEN Research Institute
U l m (Donau) Germany

Abstract

Filters with mechanical resonators are discussed as a new component in carrier frequency transmission systems. Main points of technical progress were the optimisation of transducers and of production tolerances. New filter concepts with grounded gyrators are promising.

1. INTRODUCTION

The trials of new component and circuit concepts for analog telecommunication techniques have had an influence on the transmission circuitry. First of all work on mechanical filters was successful. The latest results of work on filter and vibration theory and the progress in material development and production techniques allowed the introduction of mechanical filters as channel filters. In the near future the question will arise whether networks with semiconductor components, e.g. monolithic gyrators, could successfully replace filters with mechanical resonators.

2. SYSTEM AND FILTER CONCEPTS

More than ten years ago a number of patent applications with different types of mechanical filters as channel filters were published (1), (2), (3), (4). A summary of our ideas together with the year of the patent application is shown in Fig. 1. Filters compatible in size with present equipment have been recently developped (5), (6), (7), (8). Why has such a long period of time elapsed? First the problems of economic production and the guarantee of the realibility must be solved. Furthermore, a change in the modulation system was necessary. CCITT Specifications are given for the primary group. The proposed and used modulation diagram for the racks has been influenced by the adapted dialling systems and equipments. In Western Germany out-band signalling is needed for the EMD office. A premodulation system with a carrier frequency of 200 kHz was proposed (2). The channel unit shown in Fig. 2 consists of two channel filters and one additional signalling filter each for the signalling transmitter and receiver. Some years later a premodulation frequency of 48 kHz was suggested (9). The two concepts differ only as regards frequency for the filters and for 14 carriers.

3. FILTER DESIGN

A survey of the specs for the channel and signalling filter is given in Fig. 3. The out-band signalling is the reason for a steeper transition characteristic at the carrier end and at the upper band edge than is usually used in the USA. For two different filter approximations the transmission curves, with the normalized quality factor q as a parameter, are given. Polynominal filters of the degree 12 have been materialised as mechanical filters for 48 kHz carrier frequency, and a monolithic crystal filter for 8 MHz. Filters of the degree 10 with two finite pairs of poles have been built for 200 kHz. It could be seen that the transmission curve with q = 80, e.g. monolithic crystal filter for 8 MHz and a Q factor of 200 000, could not keep the defined specifications, and we have, because there was no prospect of increasing this value by the requisite factor 4, concentrated our work on mechanical filters with steel resonators at lower frequencies. At carrier frequencies of 48 or 200 kHz normalized q factors much larger than 250 are reached without difficulty.

3.2. MECHANICAL FILTERS

A photo of materialised mechanical filters together with a monolithic crystal filter, which has been developed as an IF filter for 10.7 MHz in FM equipment, is given in Fig. 4 (10). At the lower frequency of 48 kHz we use bending resonators with a longitudinally vibrating coupling wire for the channel

Reprinted from *Proc. 1974 IEEE Int. Symp. on Circuits and Syst.*, Apr. 22-25, 1974, pp. 106-110.

filter and bending resonators and couplers for the two different signalling filters (8). At 200 kHz the filter is designed with torsional resonators and longitudinal vibrating wires (7). A detailed comparison of the system and filter concepts did not show a significant advantage for either of them. Therefore, we joined at the request of the German Post Administration, with the proposals of other companies for a 48 kHz-system. Measured filter characteristics for both types of filter are summerized with the temperature dependance of the characteristics in Fig. 5.

4. TOLERANCE PROBLEMS

4.1 SENSITIVITIES

The refinement of the calculation methods of mechanical filters with the main points in vibrator and filter modelling was the aim of a great deal of scientific work (11), (12), (13). At this juncture we will give some results of sensitivity investigations for the two examined mechanical filters T 12 10 and P 10 05. The calculated maximum sensitivities in the passband and the sensitivities at carrier frequency for all the coupling coefficients and resonators frequencies are summerized in Fig. 6 (14). In the passband the polynomial filter T 12 10 shows, in accordance with the higher degree and the larger reflexion coefficient, higher sensitivities than the polefilter P 10 05. The bridging couplings K_{25} and K_{26} are comparably less sensitive than the coupling between adjacent resonators. In the stopband there is another tendency. The polynomial filter has balanced sensitivity values which are much lower than the values of the polefilter. Especially high sensitivity values are associated with the frequencies of the bridged resonators and the coupling coefficients between these resonators as well as for that bridging coupling which is responsible for the pair of finite poles nearest the band edge. A summary of calculated sensitivities in the form of median values is shown in Table 1. The values confirm the results of Fig. 6. We note for the polefilter sensitivity values lower by factor 2 in the passband, whereas values higher by factor 2 result in the transition band.

4.2 MONTE CARLO CALCULATION

Monte Carlo calculations have been performed to study the tolerance problems before starting the production. With a simulation program the attenuation characteristic was calculated with statistically added tolerances of all the tuning frequencies, coupling and quality factors and the matching impedances. To optain a simple calculation we used a Gaussian distribution for all parameter values (15). In Table 2 some results of the investigations on the two mentioned earlier filter concepts are listed. The boundaries for 5 % to 95 % of all filters are given for the passband and at carrier frequency. With comparable spreading of frequencies and coupling factors there are only small differences in the boundaries. The high normalized quality factor q = 600 for the 48 kHz filter guarantees a higher margin of safety in the passband. The results of the Monte Carlo studies help to fix the admissible variations in the different production steps and the necessary tolerances for the temperature coefficients and aging values of the resonators, transducers and matching circuits.

4.3 COUPLING WIRES

An additional point of view to the tolerance investigations is the length sensitivity of the coupling factor and resonator detuning as a function of the absolute length of the coupling wires. The transition from the quarter-wave-length coupler of the 200 kHz filter to a $\lambda/30$ length of the coupler at 48 kHz increases the sensitivities by more than factor 4 so that the disadvantage of higher relative accuracy in frequency tuning at 200 kHz is compensated by the lower sensitivity of the coupler length.

5. TRANSDUCERS

Results in material research were the basis of the introduction of mechanical filters. An indication should be given about the outcome of ceramic and transducer development which was startet 15 years ago and which has now produced useable transducers for mechanical filters. The important points were small variations of transducer data as well as the temperature and time stability. Some possibilities of adjusting the data of a composite transducer may be taken from Fig. 7. A symmetrical arrangement of the exciting ceramic disc with a disc thickness of 0,75 of the total length of the transducer, shows the maximum electromechanical coupling factor which exceeds the coupling factor K_{33} of the ceramic material. Transducers for mechanical filters need a high quality factor, high stability values and low input impedance. Therefore, the ceramic data are modified by the steel parts of the transducers to produce higher quality factors. Three materialised transducers and the measured coupling and quality factor are shown in Fig. 8. A comparison of the flexural transducers shows the advantage of the longitudinally excited transducers, which allows the highest cutting factor for a given electromechanical coupling factor of the material. For coil-free matching of the 48 kHz filters the measured coupling factors are more than factor 2 too low. Further results of the stability investigations given in Table 3 show good values and the fact that the connection technique for combining ceramic and steel pieces is now also perfect.

6. GYRATOR FILTER

The direct replacement of coils by a monolithic gyrator and a capacitor was not successful because of the difficulties in designing floating gyrators and of the power and noise problems in such filters. New proposals for gyrator filter networks as shown in Fig. 9 work exclusively with grounded gyrators (16), (17). The three circuits are equivalent as regards transmission characteristic. The well-known disadvantage of high sensitivity, which is inherent in active filters, are not observed for these new networks. In Table 4 a comparison of the sensitivities of a conventional filter with two filters of the new concept do not show any significant difference in sensitivity for the capacitors, and only slighty higher values for the gyration resistors. If there should be a chance to surpass this new adopted technology of mechanical filters in carrier frequency equipment, a lot of progress will be needed in the design of low power and low cost grounded gyrators and in production of thin film circuitry for RC networks (18).

7. ACKNOWLEDGEMENTS

For making available the results of theoretical invertigations I am obliged to Messrs. B. Birn and B. Kohlhammer and for results of measurements on transducers and filters I am greatly indebted to Messrs. J. Deckert, P. Güls and W. Hirsch.

8. REFERENCES

(1) C.E.Lane: Crystal Channel Filters for the Cable Carrier System
Bell Syst. Techn. J. 17 (1938) p. 125

(2) M. Börner: Mechanische Filter für die Trägerfrequenztechnik
NTZ 19 (1960) S. 34

(3) R.A.Johnson: A single-sideband disk-wire type mechanical filter
IEEE Trans. CP-11 (1964) pp. 3-7

(4) F. Künemund, K. Traub: Mechanische Filter mit Biegeschwingern
Frequenz 18 (1964) S. 277-280

(5) R.A.Sykes and W.D.Beaver: High frequency monolithic filters with possible application to single side band use,
Proc. Annu. Frequency Control Symp. (1966) pp. 288-308

(6) H. Albsmeier: Ein Vergleich der Realisierungsmöglichkeiten elektromechanischer Kanalfilter im Frequenzbereich 12 kHz bis 10 MHz
Frequenz 25 (1971), S. 74-79

(7) H. Schüßler: Consideration about Channel Filters for a new Carrier Frequency System with Mechanical Filters
Proc. Annu. Frequency Control Symp. (1971) p. 262

(8) Techn. Mitt. AEG-TELEFUNKEN (1974) Beiheft Trägerfrequenztechnik

(9) W. Haas: Möglichkeiten des Ersatzes von Induktivitäten und LC-Schwingkreisen in zukünftigem System der Multiplextechnik
Frequenz 19 (1965) S. 297-307

(10) G.R.Kohlbacher: The Design of Compact Monolithic Crystal Filters for Portable Telecommunications Equipment
Proc. Annu. Frequency Control Symp. (1972) p. 187

(11) F. Künemund: Channel Filters with longitudinally coupled flexural resonators
Siemens Forsch.- u.Entw.-Ber. 1 (1972) S. 325-328

(12) H. Kampfhenkel: Iterative Synthese symmetrischer elektromechanischer Filter
NTZ 26 (1973) S. 401-407

(13) Alfhart E. Günther: Electro-Mechanical Filters Satisfying Additional Demands
Proc. IEEE Intern. Symp. on Circuit Theory (1973) p. 142

(14) B. Kohlhammer: Vergleich verschiedener elektromechanischer Kanalfilter mit Hilfe von Empfindlichkeitskriterien
Frequenz 26 (1972) S. 169-176

(15) V. Sobotka und J. Trnka: Die Benutzung von Digitalrechnern beim Entwurf elektromechanischer Filter für die Großserienfertigung
Frequenz 26 (1972) S. 177-188

(16) B. Kohlhammer: The Design of a New Type of Gyrator Filters
Proc. Electronic Components Conference (1971) pp. 348-357

(17) H. Glöckler: Über Realisierungsmöglichkeiten beim Entwurf von Gyrator-C-Filtern
AEÜ (1974) S. 15-24

(18) D.Blom, J.O.Voormann: Noise and Dissipation of Electronic Gyrators
Philips Res. Rep. 26 (1971) p. 103

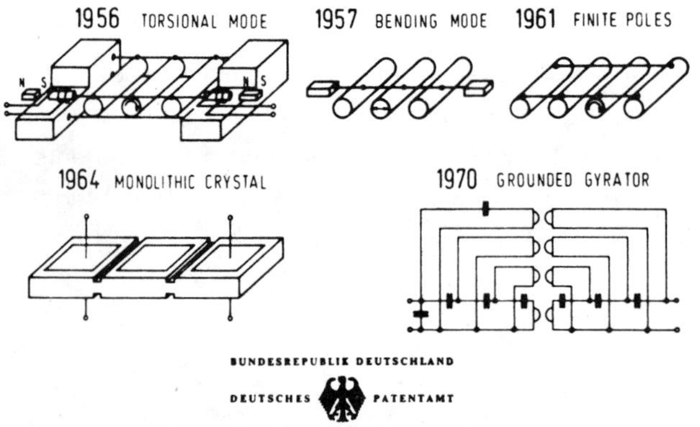

Fig. 1: Patent Applications about Different Kinds of Filter Concepts with Mechanical Resonators, Quartz Crystals and Gyrators.

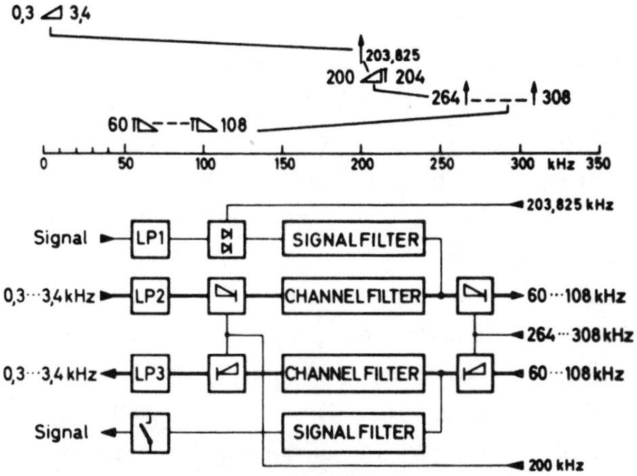

Fig. 2: Modulation System and Channel Unit for a Premodulation System with Mechanical Filters Using Torsional Resonators.

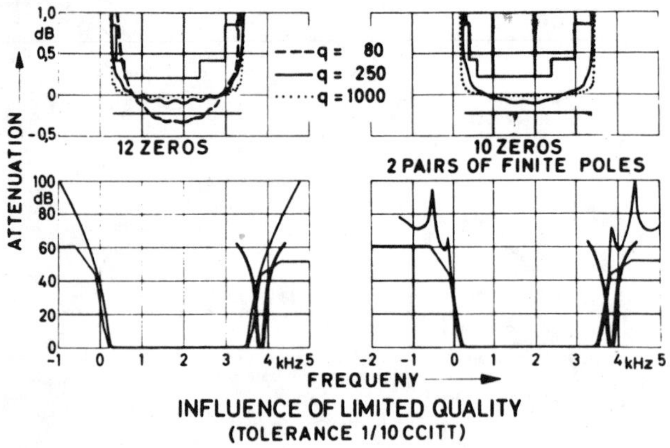

Fig. 3: Influence of Limited Quality Factor of the Resonators on the Linear Distortions. Normalized Quality Factor q as a Parameter.

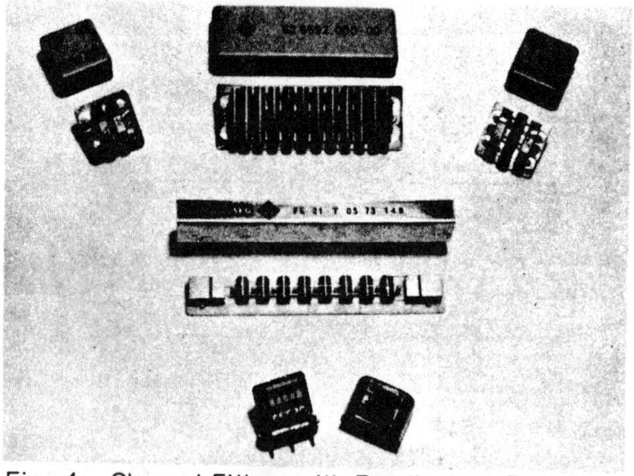

Fig. 4: Channel Filters with Bending and Torsional Resonators and a Monolithic Quartz Crystal Filter (10.7 MHz) for FM Equipment.

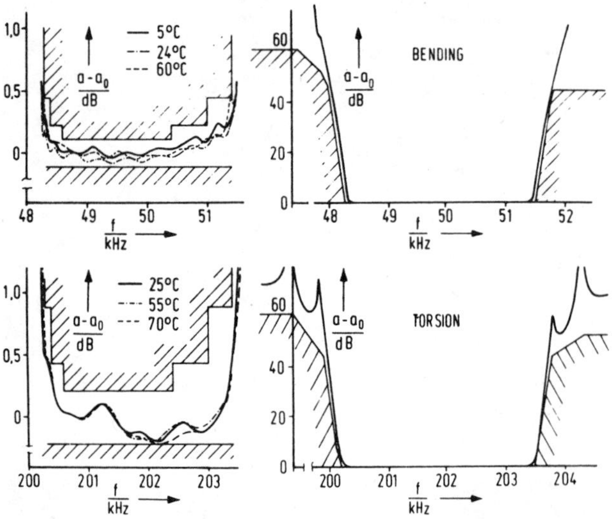

Fig. 5: Measured Transmission Characteristic of Channel Filters with Bending Resonators (10 Resonators) and with Torsional Resonators (8 Resonators and 2 Bridging Wires).

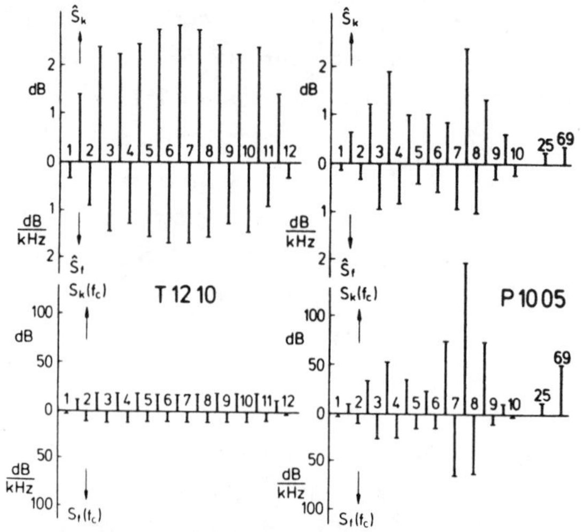

Fig. 6: Maximum Values of Sensitivity in the Passband (0.3 to 3.4 kHz) and Sensitivity at Carrier Frequency f_C as a Function of Detuning of Resonators and of Deviations of the Coupling Factors (Filters T 12 10 and P 10 05).

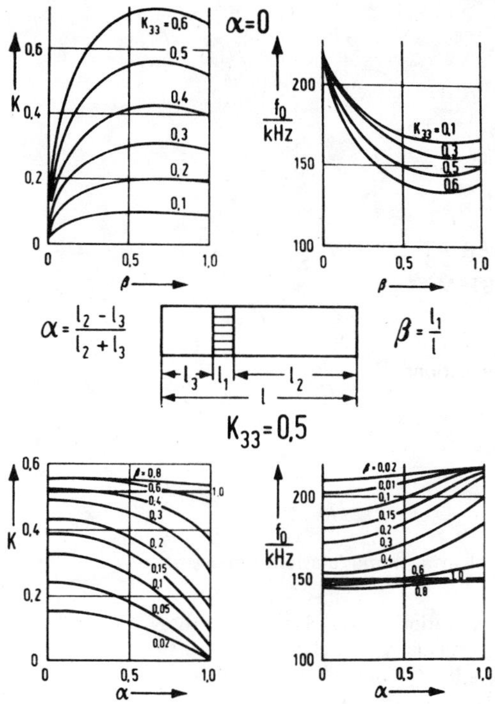

Fig. 7: Design of Composite Transducers with Longitudinal Excitation with the Thickness and the Location of the Ceramic Disc as Parameters.

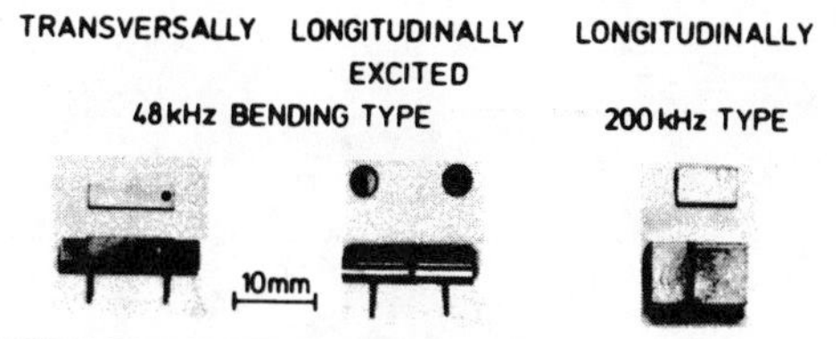

48 kHz bending type (transversally)	48 kHz bending type (longitudinally excited)	200 kHz type (longitudinally)
CERAMIC DISC		
K_{eff} = 23.7 %	K_{eff} = 25 % (K_p = 42 %)	K_{eff} = 18 %
Q = 715	Q = 435	Q = 540
TRANSDUCER		
K_{eff} = 11 %	K_{eff} = 15 %	K_{eff} = 24 %
Q = 1800	Q = 1000	Q = 1580

Fig. 8: Composite Transducers for the Excitation of Bending and Longitudinal Vibrations.

TABLE 1: MEDIAN VALUES OF THE MAXIMUM SENSITIVITIES IN THE PASSBAND AND OF THE SENSITIVITIES AT CARRIER FREQUENCY f_c.

Filter characteristic	P 10 05 (10 Zeros + 2 Finite Poles)	T 12 10 (12 Zeros)
Passband (0,3-3,4 kHz)		
$\overline{S}_K$	0.88 dB	1.84 dB
$\overline{S}_f$	0.47 dB/kHz	0.96 dB/kHz
Transition band - $f = f_c$		
$\overline{S}_K$	40.8 dB	18.2 dB
$\overline{S}_f$	18 dB/kHz	9.4 dB

TABLE 2: MONTE CARLO CALCULATIONS FOR TWO TYPES OF CHANNEL FILTERS WITH COMPARABLE TOLERANCE VALUES. ATTENUATION VALUES FOR 5% TO 95% OF ALL FILTERS.

Filter characteristic		P 10 05 200 kHz, q=320	T 12 10 48 kHz, q=525
Passband:			
0.3 kHz	< 0.44 dB	0.29... 0.35	0.25... 0.33
1.5 kHz	>-0.11 dB	-0.08...-0.36	-0.05...-0.02
Transitionband:			
0 kHz	> 30 dB	33.6... 35.6	33.4... 34.5
Tolerance Values:		$f_{1,10}$: 16 Hz	$f_{1,12}$: 15 Hz
		$f_{2...9}$: 5 Hz	$f_{2...11}$: 8 Hz
		$K_{12,910}$: 0.75%	$K_{12,1112}$: 0.75%
		$K_{23...89}$: 0.25%	$K_{23,1011}$: 0,5 %
			$K_{34...910}$: 0.25%
		$K_{25,69}$: 0.5 %	

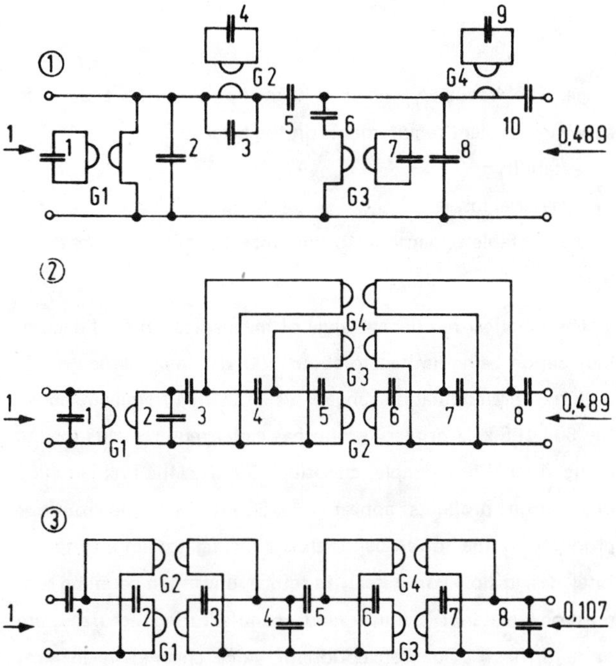

Fig. 9: Different Configurations of Gyrator Filters

TABLE 3 : TEMPERATURE AND TIME STABILITY OF COMPOSITE TRANSDUCERS AND THE DRIVING CERAMIC DISCS

	Wished Stability of Transducers		Realized Values for Transducers	Realized Values for Ceramic Discs
$TC(f) \cdot 10^6\ K$	≤	2	- 2	- 20
$TC(k) \cdot 10^6\ K$	≤	200	- 200	- 500
$TC(C) \cdot 10^6\ K$	≤	3000	+ 2600	+ 2600
$\frac{\Delta f}{f} \cdot 10^3 \cdot$ Decade	≤	0.2	+ 0.15	+ 2.0
$\frac{\Delta K}{K} \cdot 10^3 \cdot$ Decade	≤	3	- 2.5	- 10
$\frac{\Delta C}{C} \cdot 10^3 \cdot$ Decade	≤	10	- 10	- 10

TABLE 4: MEDIAN VALUES OF SENSITIVITIES IN THE PASSBAND (20 - 24 kHz) FOR DIFFERENT TYPES OF GYRATOR BANDPASS FILTERS P 08 20.

Elements	Floating Gyrators	Grounded gyrators: Double Bridging	Grounded gyrators: Single Bridging
	dB	dB	dB
C_1	2.5	2.5	2.5
C_2	2.3	2.6	3.4
C_3	2.7	2.2	1.5
C_4	2.3	1.8	4.5
C_5	0.8	2.9	5.2
C_6	4.1	2.7	2.3
C_7	4.8	2.9	4.8
C_8	0.7	2.4	2.5
C_9	0.5		
C_{10}	2.3		
R_{G1}	5.0	5.1	5.6
R_{G2}	6.7	5.3	8.3
R_{G3}	9.5	18.8	11.8
R_{G4}	5.0	17.8	11.5

DESIGN OF AN ELECTROMECHANICAL FILTER AT 128 kHz
IN A TWO STEP MODULATION SYSTEM

R. Bosc and P. Loyez
Centre National d'Etudes des Télécommunications
Issy-les-Moulineaux (France)

Abstract

To meet the overall channel specifications of the voice channel in a premodulation system, an electromechanical filter has been developped.

It is composed of 13 resonators, wire-coupled, in extensional vibration mode. Transducers use magnetostrictive ferrite rods. The advantages of such a design and, particularly, the choice of the first modulation frequency are discussed. A brief description of a signalling filter is added with informations about experimental results complying with the 1/10 CCITT requirements.

INTRODUCTION

French FDM systems are presently based on LC filter technology [1] which has been widely used in European countries. However, due to the economical importance of the channel bank in the long distance communications, there has been a continuous trend toward lower cost and size reduction.

Filters, being the major technical and economical part of the channel bank, have received the greatest research and development effort [2].

Many filter technologies have been studied in that scope, LC of course, quartz monolithic [3], active, mechanical...[4,5]

Among others, mechanical filters can meet, with a single device, almost the overall channel specifications. They seem also economically attractive since automatization can adapt to mass production, also needed in the fast growing french communications market.

LC filters of the preceding pregroup system cannot be replaced by mechanical filters without modifying the system organization [6].

The choice of the first modulation frequency is strongly dependent on filter and material considerations [7], frequency translation being no more a critical problem, due to the availability of new active modulators operating in a wide frequency range. The frequency choice is then the result of a trade - off between stringent requirements on the filter :

– stability

– material losses

– acceptable dimension, to minimize the size of the channel unit.

The practical frequency range of mechanical filters for communication being limited to about 200 kHz, among the various preferred premodulation range a system with a frequency above the 60 - 108 kHz primary group has easier requirements for the group filter. For example, choosing 128 kHz, the first jamming modulation products appear at 128 kHz with no spurious products in the lower band, thus allowing to simplify group filter design (low-pass filter). In return, unwanted coupling suppression, material stability and manufacturing accuracy are more difficult. Our conception of voice channel mechanical filter, with extensional resonators leading to a small number of critical parameters in a close correlated experimental/theoretical model (fig. 1), is also satisfying with respect to machining tolerances between 108 and 150 kHz.

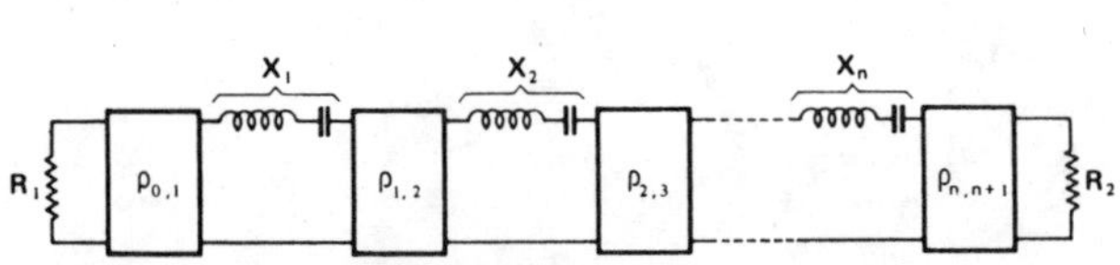

Fig. 1. Approximative equivalent circuit of the channel filter

Reprinted from *Proc. 1974 IEEE Int. Symp. on Circuits and Syst.*, Apr. 22–25, 1974, pp. 111–114.

The choice of 128 kHz for the first modulation frequency was finally a compromise between technical limitations, operation convenience (128kHz is available in new terminals) and size reduction achievement with a folded geometry.

Intrinsic material instability for the signalling filter could be deemphasized with a 2 hybrid metal-ceramic resonator filter at 32 kHz meeting the signalling channel specifications. This solution demands an additional inexpensive modulator which offers compatibility for the two out-of-band signalling frequencies (0 and 3825 Hz).

Figure 2 shows the block-diagram of a channel unit including alternate signalling filters.

Filter design

The transmission tolerances, stated by CCITT recommendations, can be satisfied by polynomial filters with 13 resonators and Q factor exceeding 10 000. Temperature and time stability must be achieved between 10° C and 50° C and for a life time of 20-30 years. It implies that the drift of the center frequency of voice and signalling filters should not be greater than 10 Hz in the two above-mentioned service conditions.

Our design, with λ/2 rod resonators and λ/4 coupling wires, acting as impedance inverters, has been investigated during 8 years.

On the equivalent network (fig. 1), valid in a well-known narrow band approximation, it can be seen that the Chebyshev behaviour is implemented through different couplings between identical resonators [8, 9]. In our case, the coupling wires diameters vary in the .7 - .9 mm range, with a .2 dB inband ripple synthesis.

The rod resonators are approximately 20 mm long with a 4.5 mm diameter. Diameter tolerances are 1 μm for the couplers and 10 μm for the resonators in order to keep these parameters influence well below the required coupling coefficient accuracy which should be roughly around 1 p. cent.

Coupling wires are built in the resonators, in a zig-zag geometry, designed with a 18° angle to reduce the overall length and to minimize spurious vibration modes of the filter assembly.

Rejection level greater than 60 dB has been experimentally observed over the specified modulation range, actually from 0 to 500 kHz.

Optimum results assume that :

— resonator tuning is made within ± 3 Hz
— Q factor is greater than 10 000
— soldering process is carefully controlled.

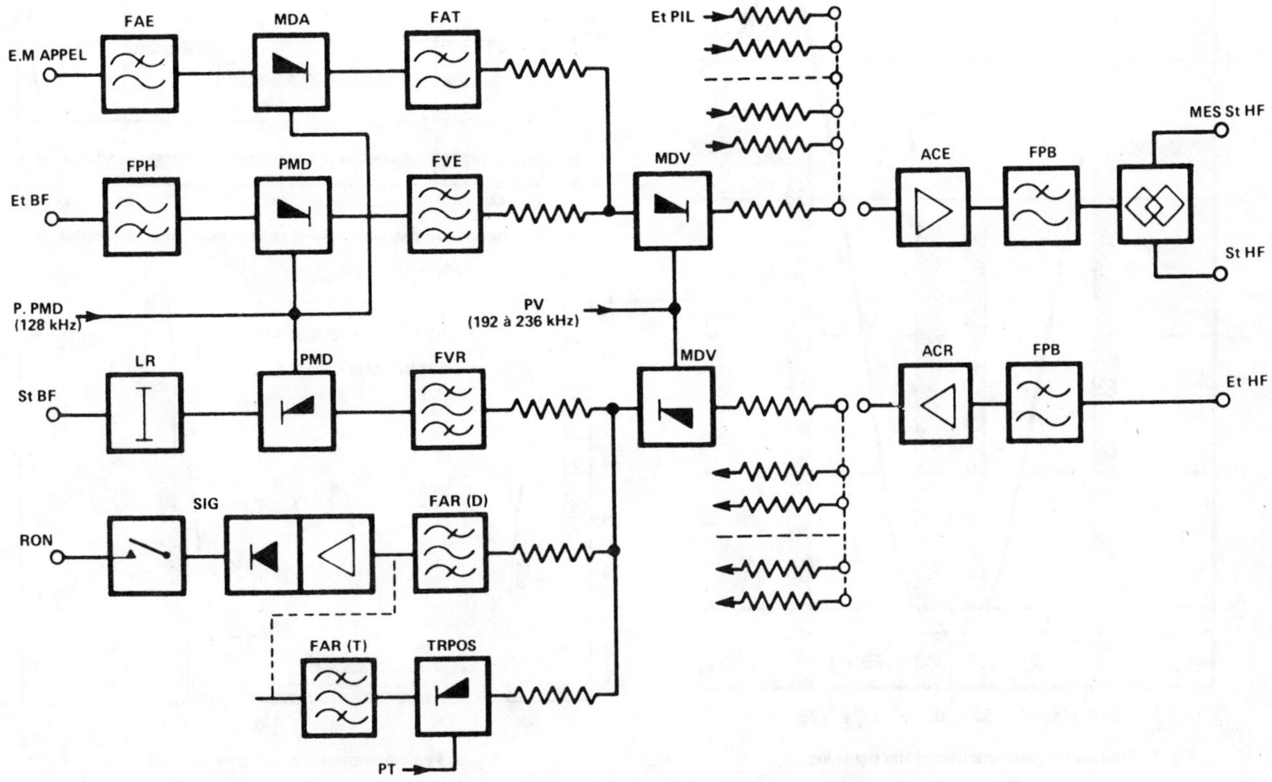

Fig. 2. Basic circuit of the channel translating unit with premodulation at 128 kHz

The extensional transducer is a λ/2 ferrite rod tuned with an accuracy lower than for resonators (± 20 Hz). The electromechanical coupling coefficient (approx. 18 %) is sufficient to keep the filter response weakly dependent on transducer characteristics.

To get low matching impedance to external circuits, the transducer coil is tuned with a series-capacitor.

Size reduction on the channel unit can also be achieved by superposing the Em / Rec filters, in a 110 cm^3 box. The mechanical structures are supported by 4 clips including ground connection.

Signalling filter

Integral specification for signalling channel is satisfied by a two pole band-pass filter with flexural resonators and torsional coupling wires.

Metal-ceramic hybrid resonators are designed to accommodate the moderate stability and Q factor of piezoceramic to need for high effective electromechanical coupling, i.e. reasonable impedance level.

The selectivity is increased with an external coupling capacitor giving two attenuation poles near the pass-band.

Material specifications

Temperature stability and high Q of mechanical resonators are the result of great improvements concerning special Ni Fe alloys such as Elinvar, Durinval..., which offer, after optimum annealing treatment, stability around 1 p.p.m between 10° C and 50° C and Q factor higher than 13 000.

The overall ripple in the voice channel filter depends on the transducers stability. Good temperature behaviour is achieved with a special ferrite rod whose frequency drift is less than 10 p.p.m between 10° C and 50° C.

The composite transducers in the signalling filter use piezoelectric ceramics with an electromechanical coupling coefficient approximatively 20 %, and temperature stability typically 100 p.p.m.

Experimental results

Figures 3 to 5 show experimental filter responses. Deviations

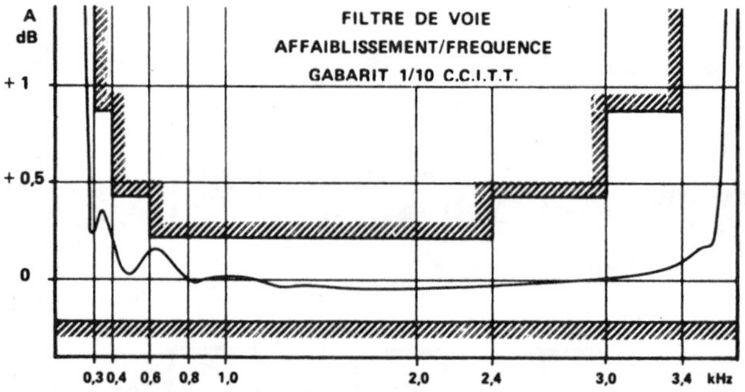

Fig. 3. Attenuation characteristic of the channel filter (Example)

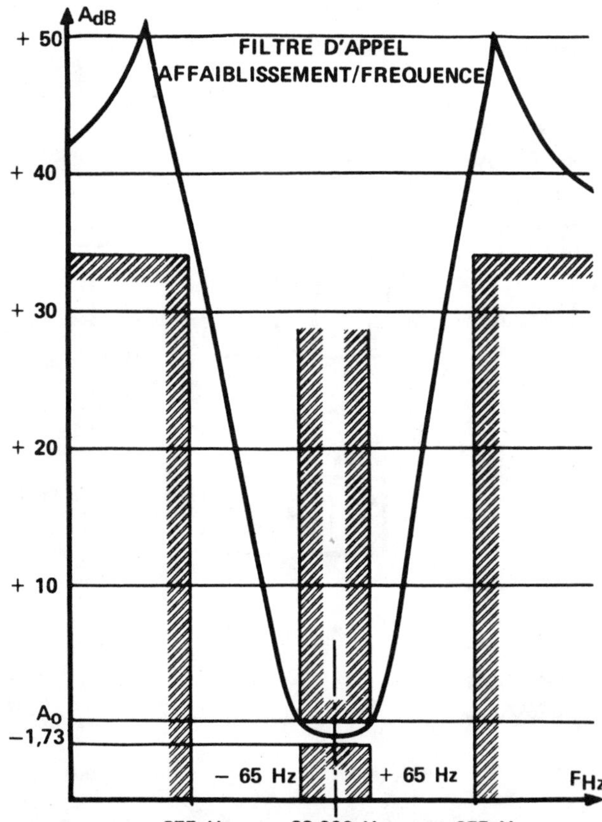

Fig. 4. Attenuation characteristic of the signalling filter

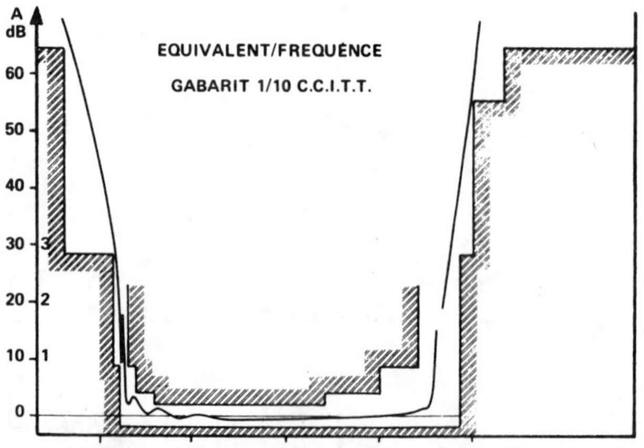

Fig. 5. Typical over-all channel characteristic

during temperature and aging tests are less than .2 dB in pass-band, 1 dB in attenuation band. The variation versus frequency of the equivalent of a transmitting receiving equipment is kept within 1/10 of limits recommended by the CCITT. A typical overall channel characteristic is shown on figure 5.

CONCLUSION

Though the project has not yet been directed towards ultimate miniaturization, significant unit size reduction has been obtained, since 300 Em/Rec channels could take place in a frame, instead of 144 with the present technology.

The essential goal was to design a filter ensuring a good compromise between model accuracy and element sensitivity which should lead to favourable technical and economical manufacturing conditions. This has been tested in a small prototype production which is the first approach to the industrial step.

REFERENCES

[1] Câbles et Transmission 25e A - N° 4 (oct. 71) - Equipements de transmission Matériel Sotelec 1970.

[2] Carl F. Kurth, Analog and Digital Filtering in Multiplex Communication Systems - IEEE Trans. Circuit Theory. CT 20 N° 4 (July 1973).

[3] P. Lloyd, Monolithic crystal filters for FDM, Proc. 25 th Frequency Control Symposium (Atlantic City - 26-28 April 1971).

[4] H. Albsmeier - Ein Vergleich der Realisierungmöglichkeiten elektromechanischer Kanalfilter im Frequenzbereich 12 kHz bis 10 MHz - Frequenz 25, n° 3, 1971.

[5] R.A. Johnson, M. Borner and M. Konno. - Mechanical filters, a review of progress. - IEEE Trans. on Sonics and Ultrasonics, vol. SU 18, n° 3, (July 1971).

[6] R. Bosc, F. Collombat, P. Loyez. - Application de nouvelles technologies de filtrage à un système 12 voies. Câbles et Transmission N° 1, Janvier 1973.

[7] R. Bertin, G. Duval et S. Talmasky. - Considérations sur la formation du groupe primaire de base. - Câbles et Transmission, 23e A., n° 3, Juillet 1969.

[8] P. Amstutz. - Filtres à bande étroite. - Câbles et Transmission, 21e A., n° 2, Avril 1967.

[9] M. Dishal. - Two New Equations for the Design of Filters. - Electrical Communication - Déc. 1973.

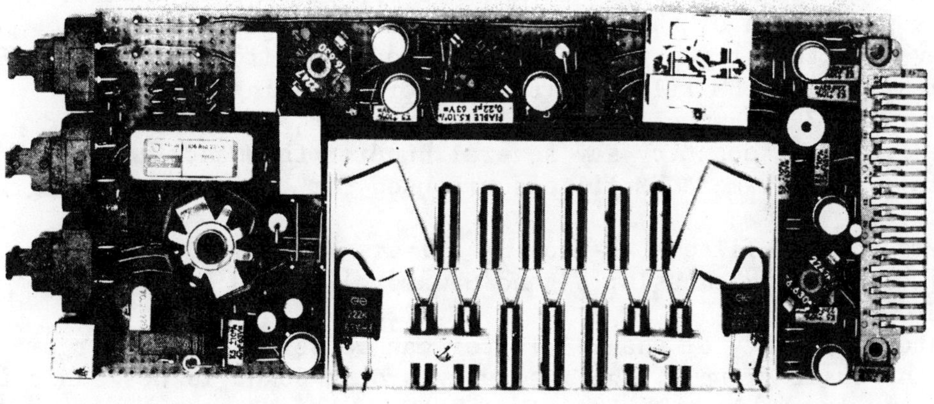

Fig. 6. Photograph of a channel unit prototype with electromechanical filters (voice channel filters are fitted in the same case)

NEW TORSIONAL MODE ELECTROMECHANICAL CHANNEL FILTER

T. Yano, T. Futami and S. Kanazawa

Introduction:

One of the most important applications of the filter design theory is found in the design of channel filter, since it is widely used in carrier telephone transmission systems. Various efforts are continuously made for developing new channel filters, aiming simultaneous achievment of small size and reduced manufacturing cost. At present, there are four types of realization: inductor-capacitor(LC) filter, monolithic crystal filter, electromechanical(EM) filter and active RC filter.

Success of development of Super Neferrite cores made it possible to reduce the size of LC filters to about a half in comparison with ones now in production[1].

Regarding EM filters, an important problem is the choice of vibration mode of the resonators and channel filter frequency. Recently, Sawamoto et al.[2] proposed a channel EM filter in torsional mode, having a center frequency a little higher than the basic group B.

In this paper, comparison of the modes is made both theoretically and experimentally, leading to the conclusion that the Sawamoto's solution is much advantageous compared with other proposed EM filters. Methods for realizing EM filters with attenuation poles and for realizing EM wave separating filters are also described. The Monte Carlo analysis technique was extensively used throughout the work. Finally, data on trial prototype filters are given.

Vibration Modes of Resonators:

There can be three different modes, ie. bending(B), torsional(T) and longitudinal(L), in realizing channel EM filters.

In Nippon Electric Co., various EM filters including above three modes are in production. T-mode EM filters were first developed aiming a premodulation and direct modulation channel filter about 10 years ago based on Tanaka's work[3]. For use in channel translating equipment, however, mass production of these EM filters was difficult from the technical level at that time, therefore they found application in the SSB radio equipment. B-mode EM filters are used for signaling current extraction for the 3x4 channel translating equipment, and have already been produced in large quantity say several hundred thousands in the passed several years. L-mode EM filters are used for carrier supplying and pilot extraction.

Regarding channel filters, B-mode EM filters were developed by Albsmeier[4], T-mode by Schüssler[5] and Sawamoto[2], and L-mode by Tagawa and Glowinski[6].

The center frequency of channel filter can be set either lower or higher than the basic group B. The frequency around 50kHz is desirable in the former case. In the latter case, though there are no essential restrictions of the frequency from modulation scheme, it should be separately discussed on two cases, slightly higher than the basic group B, for instance 120kHz, and far higher, for instance 200kHz.

In order to decide which vibration mode is most suitable for our purpose, the following factors must be taken into account.

(i)Let resonator diameter be d and the length be ℓ, then the resonance

The authors are with Transmission Division, Nippon Electric Co., Ltd. Kawasaki, Japan

Reprinted with permission from *1974 European Conf. on Circuit Theory and Design*, IEE Conf. Publ. no. 116, July 23-26, 1974, pp. 121-126.

frequency f_r of three vibration modes are

$$
\begin{aligned}
f_r &= K_b d/\ell^2 && \text{for B-mode resonators} \\
& K_t/\ell && \text{for T-mode resonators} \qquad (1) \\
& K_\ell/\ell && \text{for L-mode resonators}
\end{aligned}
$$

where K_b, K_t and K_ℓ are frequency constants, which take the value of 3800, 1450 and 2450 (kHz·mm), respectively.

At first, let us compare T-mode resonators and L-mode resonators. From (1), length ℓ of L-mode resonators is about 1.7 times longer than that of T-mode resonators, if the same frequency is assumed, while, Q's and temperature stabilities of both vibration mode resonators are about the same. Therefore, T-mode resonators are superior to L-mode resonators.

The comparison hereafter will be restricted to T-mode and B-mode resonators.

If we assume a 50kHz T-mode resonator, the length becomes 30mm, which is too long. T-mode resonators should be applied to frequencies above 100kHz.

(ii) Let resonator Q be Qr, material Q be Qm, and the inverse of acoustical loss be Qa, then the following relation holds.

$$1/Qr = 1/Qm + 1/Qa \qquad (2)$$

1/Qa can be estimated from Q values both in the air and in vacuum. Though it depends extremely on environments, representative data show $1/Qa = 5.6\times10^{-5}$ in B-mode, and $1/Qa = 7.1\times10^{-7}$ in T-mode. The difference is due to the fact that B-mode is a kind of outer vibration, while T-mode is inner one.

The influence of resonator loss on the frequency responce of channel filters is proportional to loss factor $2\pi f_r/Qr$ (δ). Let us compare the Q's of different frequencies in δ. Upper bound of δ for non-predistorted channel filters is 31 (rad/sec). If practical value of Qm=33,000 is assumed, we have from (2)

$$
\begin{aligned}
\delta &= 27.1 && \text{in 50kHz B-mode} \\
& 65.0 && \text{in 120kHz B-mode} \\
& 23.4 && \text{in 120kHz T-mode} \qquad (3) \\
& 39.0 && \text{in 200kHz T-mode}
\end{aligned}
$$

50kHz B-mode and 120kHz T-mode resonators have sufficient Qr, while, 120kHz B-mode and 200kHz T-mode is insufficient. 120kHz T-mode is the most advantageous of all.

Then, let us compare 50kHz B-mode resonators and 120kHz T-mode resonators.

(iii) One of the factors against miniaturization of filters is dependence of resonator characteristics on vibration amplitude. Measurements on B-mode and T-mode resonators were made for various source voltages using magnetostrictive bridge. Then vibration amplitudes of resonators were calculated from the source voltages, using equivalence of consumption powers in the electrical equivalent circuit and in mechanical vibration. The results are

	resonator dimensions (mm)	resonance frequencies (kHz)	amplitude range (micron r.m.s.)	change of frequency (Hz)	change of δ (rad/sec)
B-mode	$d=3.5, \ell=16.9$	47.51	0.0015∼0.047	-0.2	+4
	$d=2.5, \ell=14.2$	51.14	0.0019∼0.060	-0.4	+6
T-mode	$d=3.5, \ell=13.4$	108.5	0.0012∼0.038	-0.3	+2
	$d=2.7, \ell=12.8$	113.4	0.0015∼0.048	-0.4	+2

where the amplitude range corresponds to -30∼0dBm of the filter input level. Change of δ in B-mode is about two or three times larger than that in T-mode.

(iv) The ratio of length to the diameter of a resonator is desired to be close to unity by the following reasons. First, the severest dimensional precision of resonators is required for the smaller of either length or diameter. Secondly, the voluminal efficiency of resonator compared with the filter volume increases as the ratio approaches to unity. The ratios of diameter to length are

$$\frac{d}{\ell}=\left(\frac{4}{\pi}\right)^{\frac{1}{5}}K_b^{-\frac{3}{5}}\cdot f_\gamma^{\frac{2}{5}}\cdot V^{\frac{1}{5}} \quad \text{for B-mode}, \quad \frac{d}{\ell}=\left(\frac{4}{\pi}\right)^{\frac{1}{2}}K^{-\frac{3}{2}}\cdot f_\gamma^{\frac{3}{2}}\cdot V^{\frac{1}{2}} \quad \text{for T-mode} \tag{4}$$

where V and f_γ denote the volume and resonance frequencies of the resonator, respectively. Let f_γ=50kHz for B-mode and f_γ=120kHz for T-mode, then

$$d/\ell = 7.8\times10^{-2}\times V^{\frac{1}{5}} \quad \text{for B-mode}, \quad d/\ell = 2.7\times10^{-2}\times V^{\frac{1}{2}} \quad \text{for T-mode} \tag{5}$$

This fact shows that B-mode is advantageous in $d<1.9$, while T-mode is advantageous in $d>1.9$. The former, however, seems difficult to apply under the present and near future level of precise production technology.

From the above discussion, we have the conclusion that T-mode EM filter with center frequency around 120kHz is the most advantageous of all. Furthermore, the T-mode is highly advantageous in realizing pole type filter and EM wave separating filter[7][8].

Pole Type Filter:

The introduction of attenuation poles at the finite frequencies reduces the number of resonators and group delay distortion in the pass-band.

Wellknown methods using additional couplers between nonadjacent resonators in a cascade resonator-coupler structure[9][5] have the restrictions on the numbers or the location of the attenuation poles. Another method proposed by Johnson[10] makes use of multimode resonators which are equivalent to symmetrical twin tee circuits. Unfortunately it cannot be applied to our purpose, because it is limited to symmetrical structure.

The method proposed here uses general parallel-tee circuits, in which an arbitrary number of resonators are connected in parallel. By an appropriate application of this general structure, the sensitivities at the pass-band and stop-band can be balanced.

[Lemma] A nessesary and sufficient condition that a two-port N can be realized by an EM filter with resonator-coupler structure is that

1.N can be decomposed into cascade sections of two types denoted type-A and type-B

2.Type-A and type-B must appear alternately and the both ends must be type-A

3.Z matrix of each of type-A sections is

$$\begin{bmatrix} z_{11} & z_{12} \\ z_{12} & z_{22} \end{bmatrix} = \frac{zp}{p^2+q^2}\begin{bmatrix} 1 & 1 \\ 1 & 1 \end{bmatrix} \tag{6}$$

which is obviously representing parallel resonance circuit

4.Y matrix of each of type-B section is

$$\begin{bmatrix} y_{11} & y_{12} \\ y_{12} & y_{22} \end{bmatrix} = \frac{1}{p}\begin{bmatrix} \alpha_0 & \beta_0 \\ \beta_0 & \gamma_0 \end{bmatrix} + p\begin{bmatrix} \alpha_\infty & \beta_\infty \\ \beta_\infty & \gamma_\infty \end{bmatrix} + \sum_{i=1}^{n}\frac{p}{p^2+q_i^2}\begin{bmatrix} \alpha_i & \beta_i \\ \beta_i & \gamma_i \end{bmatrix} \tag{7}$$

where $\alpha_0, \gamma_0, \alpha_\infty, \gamma_\infty, \alpha_i, \gamma_i$: real positive numbers, $\beta_0, \beta_\infty, \beta_i$: real numbers, $\beta_i \neq 0$, and $\alpha_i\gamma_i - \beta_i^2=0$ (compact)

The decomposition is illustrated in fig.1.

The simplest form of type-B section is composed of one inductor. In that case the two-port becomes nonpole filter. General structure of type-B section is shown in fig.2, which can be translated into parallel tee circuit by formulas shown in table 1. The parallel tee circuit can be realized as a combination of resonators and couplers as shown in fig.3.

The conditions of the lemma are satisfied for almost all practical two ports, then the design problem is to find a suitable decomposition.

Another problem which occurs in the realization of mechanical structure lies in realizing 1:-1 ideal transformer. In T-mode EM filters, 1:-1 transformer can be easily realized as shown in fig.4.

EM Wave Separation:

In the channel translating equipment, a channel filter and a signaling filter must be connected in parallel at one end. It is more advantageous to connect them mechanically than electrically, since only three transducers are needed instead of four, and total size can be reduced.

The following condition must be satisfied for two filters connected in parallel to constitute a separation network.

1. The absolute value of input impedance of each filter must be large enough at the pass-band of the other filter.

2. Real part of the filters must be approximately constant at their pass-bands and they must be equal.

By carefully adjusting pass-band ripple design parameters of the channel filter, the absolute value of its input impedance can be made large enough at the pass-band frequency of the signaling filter.

Signaling filter is designed to have a series impedance at the input end, in order to have high input impedance at the pass-band of the channel filter.

The adoption of the T-mode resonator is advantageous also in realizing channel-signaling EM wave separating filter, because the same manner of coupling can be used throughout the wave separating filter even in the case of connecting relatively wide band filter, e.g. channel filter, and very narrow band filter, e.g. signaling filter.

Results on Trial Production:

A number of trial T-mode EM filters were built and tested. The carrier frequency of 112kHz was selected as an example. Fig.5(a) shows a non-pole type channel EM filter, which have eleven T-mode resonators and two T-mode transducers. The filter requires impedance matching transformers at each of the input and output ends. Fig.6 illustrates representative frequency responce of the non-pole type filter. Result of trial production of this non-pole type filters in small quantity shows that they meet 1/20 of the CCITT recomended value with good margin.

Some pole type channel filters were also designed with the objectivs of approximating their group delay distortion characteristics to the current LC filter. The channel filter and a signaling filter are connected mechanically to constitute an EM wave separating filter. Fig.5(b) shows the wave separating filter. The channel filter part was constructed with nine resonators and two transducers, while the signaling filter part with three resonators and one transducer. The total volume reduction of 15% was attained by introducing the EM wave separation. The channel filter part contains two parallel-tee circuit sections, each producing an attenuation poles, or a total of two poles. The attenuation shapes of the channel filter part and the signaling filter part are shown in fig.7 and fig.8, respectively. The pass-band response of fig.9 has been proven to meet 1/20 of the CCITT recomended value throughout the temperature range 0-60°C. The group delay distortion characteristic of fig.10 shows a good agreement with designed value.

Conclusion:

Channel EM filters of various modes and frequencies were compared both theoretically and experimentally. The EM filter with the torsional mode resonators having its center frequency around 120kHz seems to be the best solution. At the same time, design methods for pole type EM filter and EM wave separating filter were discussed. Validity of the design methods was verified by trial models.

Acknowledgment:

The authors wish to thank Mr.Kaneoya and his co-workers of Electrical Communication Laboratories, Nippon Telephone and Telegraph Public Corporation for their useful discussions.

They are also indebted much to those who collaborated in their special fields.

References:

[1] I.Ishihara et al. "Miniaturized LC channel filter" presented at the IECE Conf. of Japan. Tokyo; No 109, March 1973(in Japanese)
[2] K.Sawamoto et al. "Electromechanical filter using torsional mode resonators" presented at the IECE Conf. of Japan. Tokyo; No 67, March 1973(in Japanese)
[3] T.Tanaka,"Torsional electromechanical filter",Japan Patent S.34-8453, Sep.1956.
[4] H.Albsmeier,"Ein Vergreich··· 12kHz bis 10MHz",Frequenz,vol.43, pp.170-177,Mar. 1973.
[5] H.Schüssler ,"Consideration about Channel Filters ··· ",Proc.of the 25th Ann. Sym. on Freq. Con.,p262-270, Apr.1971.
[6] A.Growinski,"Microacoustique et Telecommunications",Londe Electrique, vol.52,fasc.2,pp.54-67,fev.1972.
[7] E.Kura and T.Yuuki,"Mechanical filter",Japan Patent S.39-894,Feb.1964.
[8] M.Onoe and T.Yano,"Electromechanical wave separating filters",Proc. 20th Electronic Component Conference, May 1970
[9] M.Börner ,"Mechanische Filter mit Dämpfungspolen",A.E.Ü.,vol.17, pp.103-107,Mar.1972.
[10]R.A.Johnson,"A twin tee multimode mechanical filter",Proc.IEEE.,vol.54, pp.1961-1962,Dec.1966.

	Formulae	Original circuits	Transformed circuits
(a)	$L_o^{(1)} = \frac{L_o}{\lvert\phi_o\rvert(\lvert\phi_o\rvert-1)}$ $L_o' = \frac{L_o}{\lvert\phi_o\rvert}$ $L_o^{(2)} = \frac{L_o}{1-\lvert\phi_o\rvert}$	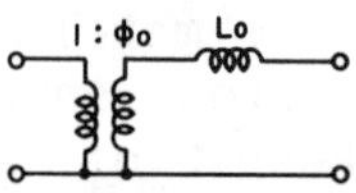	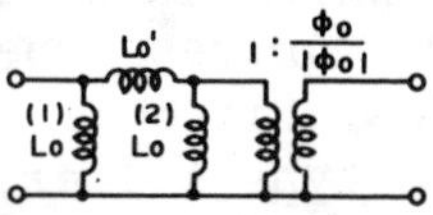
(b)	$L_i^{(1)} = \frac{1}{\omega_0}\sqrt{\frac{L_i}{C_i'}}$ $L_i^{(2)} = \frac{1}{\lvert\phi_i\rvert} L_i^{(2)}$ $L_i' = \left\{\frac{1}{(L_i^{(2)}\omega_0)^2 C_i} - \frac{1}{L_i^{(1)}} - \frac{1}{L_i^{(2)}}\right\}^{-1}$	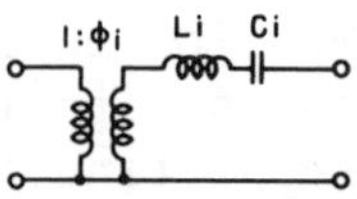	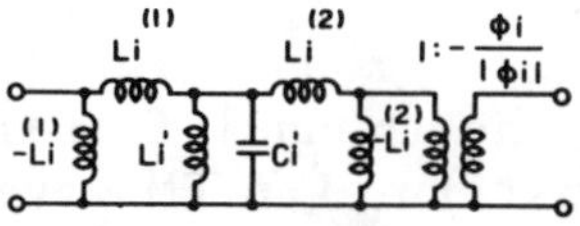
(c)	$L_\infty^{(1)} = -\frac{1}{\lvert\phi_\infty\rvert(\lvert\phi_\infty\rvert+1)\omega_0^2 C_\infty}$ $L_\infty' = \frac{1}{\lvert\phi_\infty\rvert\omega_0^2 C_\infty}$ $L_\infty^{(2)} = -\frac{1}{(\lvert\phi_\infty\rvert+1)\omega_0^2 C_\infty}$	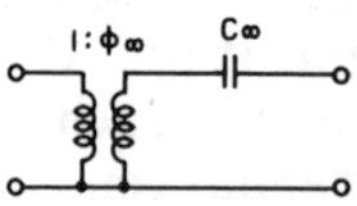	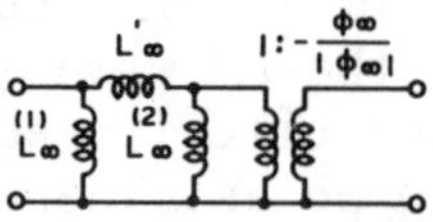

Table.1

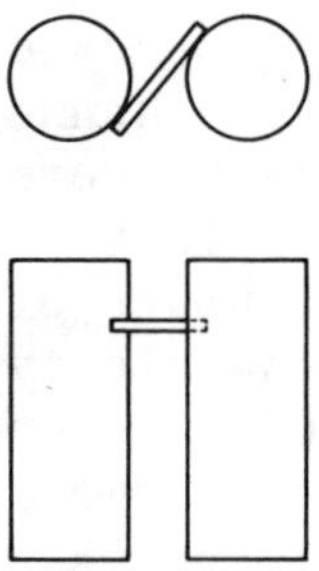

Fig.4 Realization of 1:-1 ideal transformer

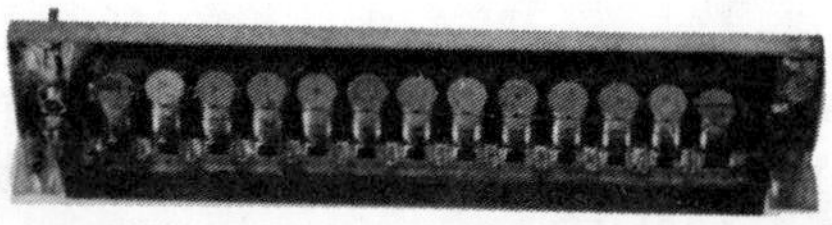

Fig.5(a) Non-pole type EM filter

Fig.5(b) Wave separating filter having two poles in its channel filter part

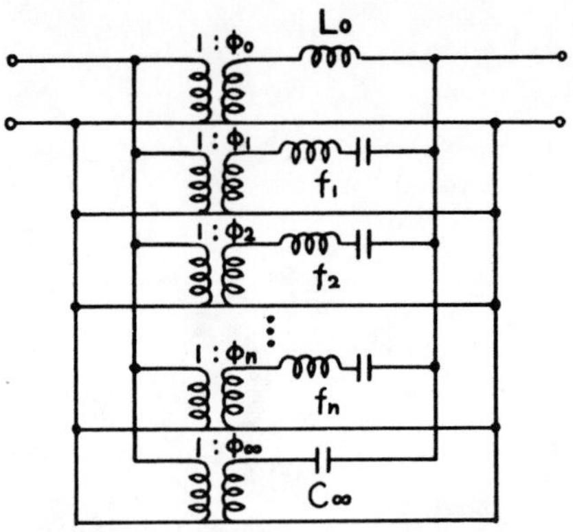

Fig.2 General description of type-B circuit

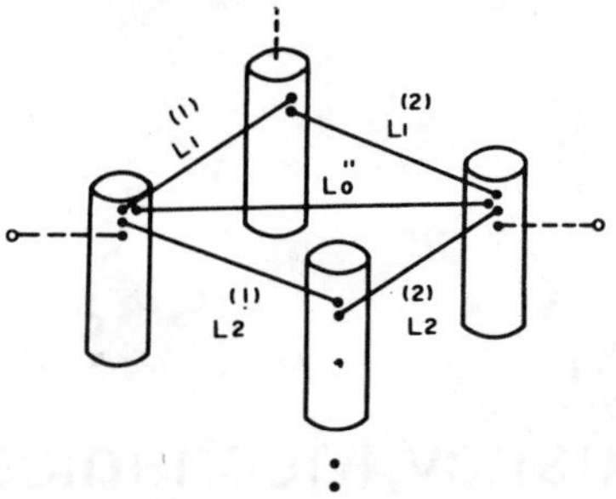

Fig.3 Realization of parallel tee circuit by resonators and couplers

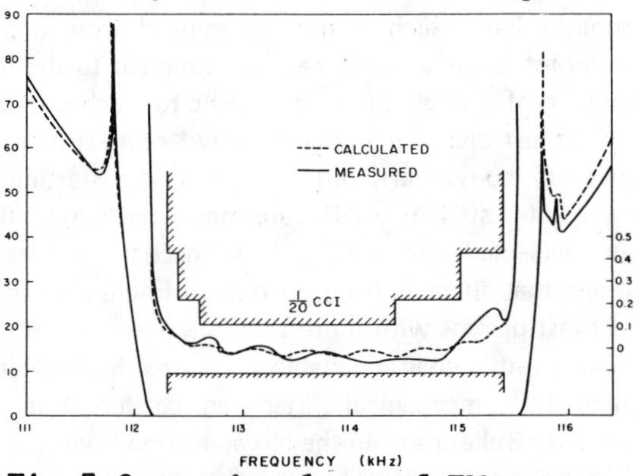

Fig.7 Operating loss of EM wave separating filter shown in fig.5(b) (channel filter part)

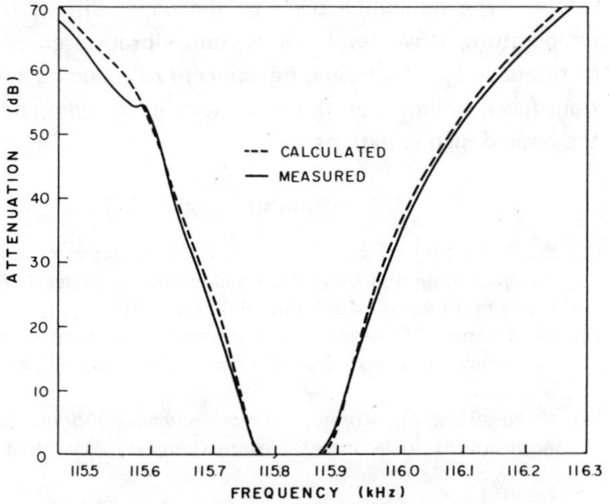

Fig.8 Operating loss of EM wave separating filter shown in fig.5(b) (signaling filter part)

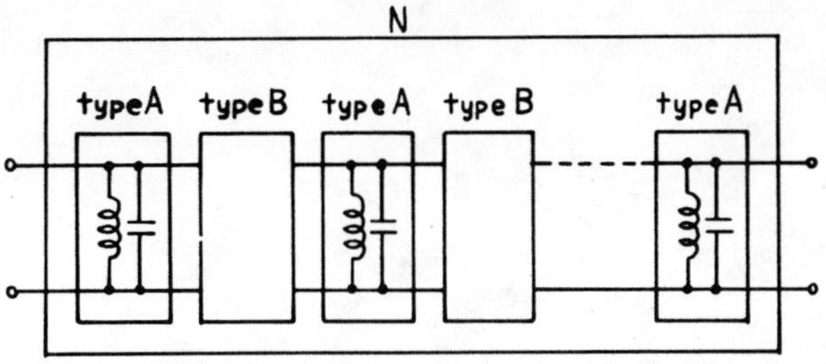

Fig.1 Two port N which can be realized by an EM filter

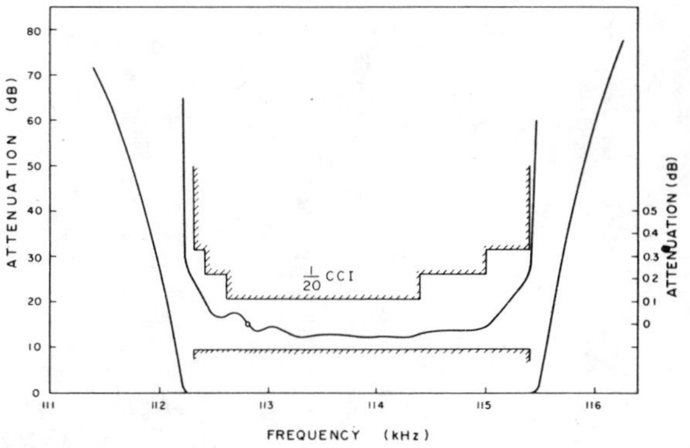

Fig.6 Operating loss of non-pole type EM filter shown in fig.5(a)

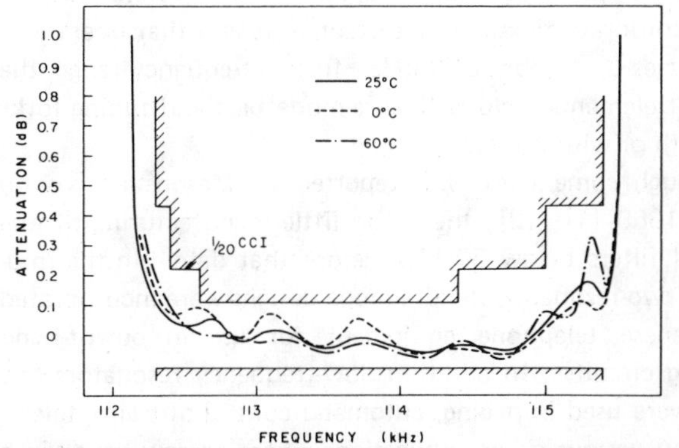

Fig.9 Pass-band responce of EM wave separating filter shown in fig.5(b) (channel filter part)

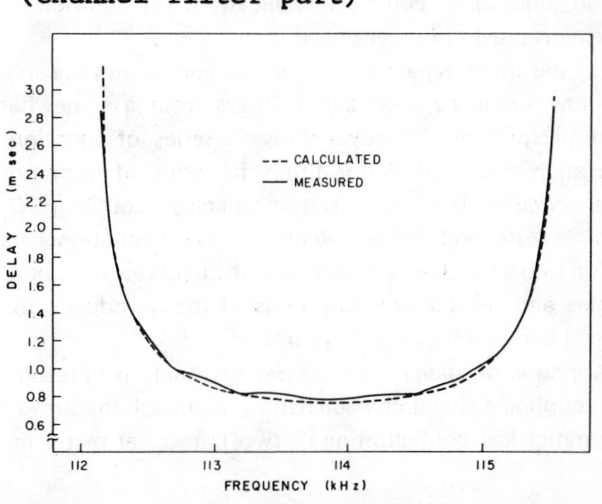

Fig.10 Group delay time of EM wave separating filter shown in fig.5(b) (channel filter part)

Section I-D
Low-Frequency Mechanical Filters

The term "low-frequency mechanical filters" refers to filters that incorporate flexure-mode resonators and that operate at frequencies up to about 50 kHz. In this frequency range, the resonant elements include flexure-mode bars and tuning forks used with or without mass loading.

Although some work was reported by Mason and Konno before 1960 [1], [2], there was little manufacturing of mechanical filters below 50 kHz before that date. In the mid-1960's, two-resonator, flexure-mode designs were incorporated in Japanese telephone equipment for use in out-of-band signaling circuits. In addition, low-frequency resonators and filters were used in paging, automatic control of trains, telemetry, and various other applications where narrow bandwidth signals needed to be selected. In the early 1970's, these filters were introduced in the United States for use in FSK, Omega navigation, and sonar comb-set applications, in addition to those systems previously mentioned.

Each of the three papers in this section approaches the subject of low-frequency mechanical filters from a somewhat different perspective. Yakuwa shows a series of equations used to analyze two-resonator filters in terms of resonator electrical circuit elements, electromechanical coupling, Q, and temperature and time stability. These equations are useful for understanding the relationships between the above parameters and the material constants of the transducer, the alloy metal bar, and the coupling wires.

The Konno *et al.* paper is a more detailed analysis of resonator and coupling element characteristics. Although the authors use a symmetrical configuration of two transducer plates per resonator bar, which is not economical from a production standpoint, their analysis can be extended to the more practical case of a single piece of ceramic for each resonator. This paper is just one of many written by Konno and his Yamagata University co-workers, but it can act as a starting point for people interested in bar-flexure mode mechanical filters. We have neglected the tuning fork mechanical filter for the reasons that there is little written in English on this subject, and most designs with more than one resonator use electrical coupling rather than mechanical. A wealth of material on the tuning fork mechanical filter can be found in Yamagata University Bulletins or in the Nagai–Konno book [3].

The final paper describes the characteristics of production-type, low-frequency mechanical filters. The paper begins with the same basic design equations used for monolithic crystal filters. The remaining parts of the paper discuss the aging, temperature, drive level, shock, and vibration characteristics of filters designed by using the concept of reducing the ceramic transducer volume to the minimum value compatible with the basic design equations.

References

[1] W. P. Mason and R. N. Thurston, "Compact electromechanical bandpass filter for frequencies below 20 kilocycles," *IRE Trans. Ultrason. Eng.*, vol. UE-7, pp. 59–70, June 1960.

[2] M. Konno, "Theoretical considerations of mechanical filters" (in Japanese), *J. Inst. Elec. Commun. Eng. Jap.*, vol. 40, pp. 44–51, 1957.

[3] K. Nagai and M. Konno, *Electro-Mechanical Vibrators and Their Applications* (in Japanese). Corona Company, Mar. 1974, ch. 9.

CHARACTERISTICS OF AN ELECTROMECHANICAL FILTER USING PIEZOELECTRIC COMPOSITE BENDING VIBRATORS

Kazuo Yakuwa, Member

Fujitsu Limited, Kawasaki

SUMMARY

The electromechanical filter constructed with bending metal vibrators coupled torsionally at their nodal points is suitable for use in the frequency range from 300 cycles to several tens of kilocycles. The electromechanical elements for such a filter can consist of a composite piezoelectric transducer and vibrator bonded together, and such a construction is analyzed. The equivalent circuit constants, the quality Q, and the temperature and time stability are compared to experimental results. Based on this study and the objective of using such filters in various communication equipments, the characteristics of a practical design, such as the center frequency, fractional pass bandwidth, input impedance, etc., are examined. It is shown that with the construction described, an electromechanical filter with a fractional pass banwidth up to about 20% can be designed for practical use in the frequency range from 300 c/s to 30 kc/s. That the input impedance characteristic can be designed arbitrarily is also shown.

1. Introduction

Important characteristics of the electromechanical filter are its performance as a bandpass filter of superior stability and a cutoff property based on its low loss (high Q). Its comparatively simple construction is an additional asset. Although it has been put into practical use in the frequency range from 50 kc to 500 kc, the necessity for an electromechanical filter which operates below 50 kc and on down to the audio-frequency range, accompanies the development of complex, economical, and miniaturized communication equipments. In particular, in the range from audio frequencies up to about 10 kc, due to the needed reactances and the high loss of a narrow-band LC filter, it is difficult to manufacture a conventional miniature filter with good cutoff characteristics. The practical use of a low-frequency electromechanical filter is one solution to this problem.

Although an electromechanical filter using torsionally coupled bending vibrators, as shown in Fig. 1, which is suitable for operation at a frequency below 50 kc has been proposed [1], a piezoelectric transducer combined with it yields the advantages of simple construction and easy fabrication. Hence, such a low-frequency mechanical filter has been put into practical use with applications in various telecommunication and control systems [2].

In the present paper, the characteristics of this electromechanical filter using a composite transducer and vibrator are discussed. Because of manufacturing simplicity and economy, an analysis of the case where the piezoelectric transducer entirely covers one surface of the vibrator is presented.

2. Transducer

The piezoelectric composite transducer for such a mechanical filter is shown in Fig. 2. This transducer is composed of a metal vibrator with a rectangular cross section bonded to a piezoelectric ceramic plate which has a thin electrode attached to its upper side. Although several published papers [3—5] are concerned with some characteristics of a

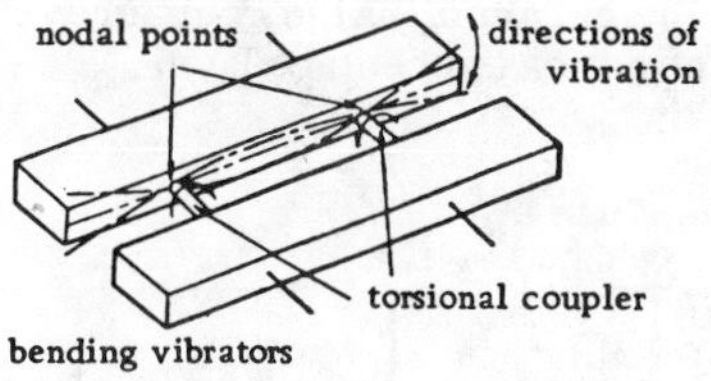

Fig. 1. An electromechanical filter using torsionally coupled bending vibrators

Reprinted with permission from *Electron. Commun. Japan*, vol. 48, pp. 206-215, Nov. 1965. Reprinted with the consent of Scripta Publishing Co., 1511 K St., N.W., Washington, DC 20005.

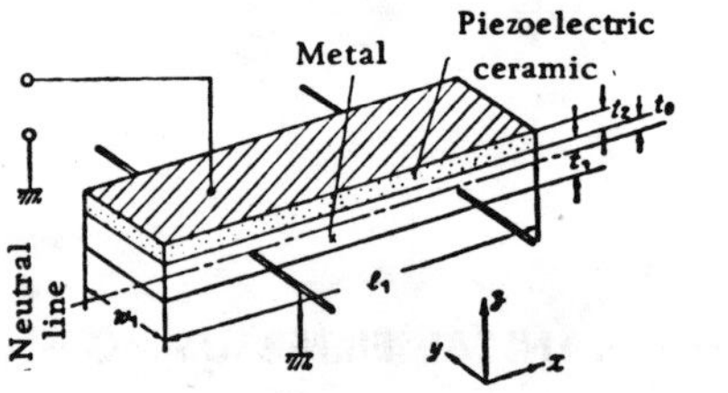

Fig. 2. Construction of transducer

composite transducer with piezoelectric ceramic plates completely or partially bonded to both surfaces of the metal vibrator, a study treating the composite transducer with piezoelectric ceramic plate bonded only to one of the main surfaces of the metal vibrator has not been found. As basic information for the design of such a mechanical filter and to judge its critical characteristics, electrical and mechanical equivalent circuit constants will be found first.

2.1 Equivalent Circuit Constants

The equivalent circuit of a piezoelectric transducer near the resonant frequency can be expressed as in Fig. 3(a). Here C_d is the damped capacitance of the piezoelectric ceramic plate; m, s, and r are the equivalent mechanical constants of the transducer, respectively representing mass, stiffness and mechanical resistance; and A_n is the electromechanical coupling coefficient. Transforming this equivalent circuit into another equivalent electrical circuit, Fig. 3(b) is obtained. The relationship of m, s and r to L, C and R is as follows:

$$\left.\begin{aligned} L&=\frac{m}{A^2}\\ C&=\frac{A^2}{s}\\ R&=\frac{r}{A^2}\end{aligned}\right\} \tag{1}$$

Although the following is based on the analysis of Mason and Thurston [4], it differs in that it gives consideration to the change in position of the neutral surface from the center of the vibrator, which takes place due to the difference between the Young's moduli of the metal and the ceramic. Also, our analysis omits the effect of the contact layer.

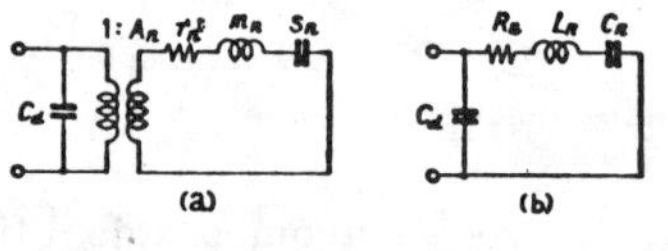

Fig. 3. Equivalent circuits of transducer

Notation for the dimensions of the parts is shown in Fig. 2. Symbols for the material constants are as follows:

Y_M: Young's modulus of metallic material

Y_D: Young's modulus of piezoelectric ceramic material

ρ_M: Density of metallic material

ρ_D: Density of piezoelectric ceramic material

$\epsilon : \epsilon_{33}{}^T(1-k_{31}{}^2)$, longitudinally damped permitivity

$B : k_{31}{}^2/d_{31}(1-k_{31}{}^2)$

$k_{31}{}^2 : d_{31}{}^2/\epsilon_{33}{}^T S_{11}{}^E=\epsilon B^2/Y_D$, longitudinal static coupling coefficient

d_{31}: Piezoelectric constant of piezoelectric ceramic

The short-circuit resonant angular frequency ω_n, the open-circuit resonant angular frequency ω_{n0}, and the equivalent mechanical constants m_n, s_n and r_n at the vibrator terminals can be expressed as follows:

$$\omega_n{}^2=\frac{\alpha_n{}^4 t_1{}^2 Y_M}{12 l_1{}^4 \rho_M}\cdot\frac{1}{2\left(1+\frac{Y_D}{Y_M}u\right)^2\left(1+\frac{\rho_D}{\rho_M}u\right)}$$
$$\left[\left(1-\frac{Y_D}{Y_M}u^2\right)^2\left(1-\frac{Y_D}{Y_M}\right)+\left(1+2\frac{Y_D}{Y_M}u+\frac{Y_D}{Y_M}u^2\right)^2\right.$$
$$\left.+\left(1+2u+\frac{Y_D}{Y_M}u^2\right)^2\frac{Y_D}{Y_M}-\frac{6\epsilon B^2}{Y_M}\left(1+\frac{Y_D}{Y_M}u\right)(1+u^2)u\right] \tag{2}$$

$$\omega_{n0}{}^2=\frac{\alpha_n{}^4 t_1{}^2 Y_M}{12 l_1{}^4 \rho_M}\cdot\frac{\left[\left(1-\frac{Y_D}{Y_M}u^2\right)^2\left(1-\frac{Y_D}{Y_M}\right)+\left(1+2\frac{Y_D}{Y_M}u+\frac{Y_D}{Y_M}u^2\right)^2+\frac{Y_D}{Y_M}\left(1+2u+\frac{Y_D}{Y_M}u^2\right)^2\right]}{2\left(1+\frac{Y_D}{Y_M}u\right)^2\left(1+\frac{\rho_D}{\rho_M}u\right)} \tag{3}$$

$$m_n=\frac{M}{4}=\frac{(\rho_M t_1+\rho_D t_2)wl_1}{4} \tag{4}$$

$$s_n=\omega_n{}^2\cdot m_n \tag{5}$$

$$r_n=\frac{\omega_n\cdot m_n}{Q_m} \tag{6}$$

Here, $u = t_2/t_1$, α_n is the root of the equation $\tan\frac{\alpha}{2}+\tanh\frac{\alpha}{2}=0$, Q_m is the mechanical Q of the transducer and subscript n represents the node number.

The admittance Y looking into the electric

terminals is

$$\dot{Y}=j\omega C_d\left\{1+\frac{4\epsilon wB^2}{\alpha^2 K}t_2(t_2+2t_0)^2\sum_1^{\infty}\frac{\beta_n}{\left(1-\frac{\omega_n^2}{\omega^2}\right)}\right\} \tag{7}$$

where

$$\beta_n=\tan\frac{\alpha_n}{2}\tanh\frac{\alpha_n}{2}=-\tan^2\frac{\alpha_n}{2} \tag{8}$$

$$K\equiv\frac{w}{3}[Y_M\{t_0^3+(t_1-t_0)^3\}+Y_D\{(t_2-t_0)^3-t_0^3\}]$$
$$-\frac{w\epsilon B^2}{4}t_2(t_2+2t_0)^2 \tag{9}$$

From the above, the equivalent constants of the composite transducer can be found to be as follows:

$$C_d=\frac{\epsilon wl_1}{t_2} \tag{10}$$

$$L_n=-\frac{1}{4\epsilon^2B^2\beta_n\alpha_n^2}\cdot\frac{(\rho_M t_1+\rho_D t_2)l_1^3}{(t_2+2t_0)^2w^2} \tag{11}$$

$$C_n=\frac{1}{\omega_n^2 L_n}=-\frac{4\epsilon^2B^2\beta_n}{\alpha_n^2}\cdot\frac{l_1(t_2+2t_0)^2w^2}{K} \tag{12}$$

$$R_n=\frac{\omega_n L_n}{Q_{mn}} \tag{13}$$

From Eqs. (1), (4), and (11) the electromechanical coupling coefficient of the transducer becomes

$$A_n=\alpha_n\left|\tan\frac{\alpha_n}{2}\right|\cdot\frac{\epsilon B(t_2+2t_0)w^{3/2}}{l_1} \tag{14}$$

The values of α_n, $\tan\frac{\alpha_n}{2}$, and β_n are shown in Table 1.

Table 1

Values of α_n, β_n, and $\tan\frac{\alpha_n}{2}$

n	1	2	3
α_n	4.7300	10.9956	17.2783
$\tan\frac{\alpha_n}{2}$	−0.9825	-0.9999_6	-0.9999_9
β_n	−0.9653	-0.9999_3	-0.9999_0

2.2 Capacitance Ratio

The capacitance ratio r is defined as

$$r=C_d/C_1 \tag{15}$$

This quantity has a relation to the resonant frequency f_1 and the antiresonant frequency f_{1a}, which are found from the impedance characteristics looking into the electrical terminals of the transducer, as follows:

$$\frac{f_{1a}-f_1}{f_1}=B_T\fallingdotseq\frac{1}{2r} \tag{16}$$

Ordinarily, the capacitance ratio of the piezoelectric vibrator is measured from the above relation. If C_d is known, C_1 and L_1 can be determined from the following equations. Since the fundamental vibration mode is the main action of the low-frequency mechanical filter, we will limit the treatment to this in the following:

$$C_1=\frac{C_d}{r}\fallingdotseq 2C_dB_T \tag{17}$$

$$L_1=\frac{r}{\omega_1^2C_d}\fallingdotseq\frac{1}{2B_TC_d\omega_1^2} \tag{18}$$

The quantity B_T given by Eq. (16) is tentatively called the fractional pass band of the transducer. It can be very easily found from first-order measurement results. Related to this value, the fractional pass band of the mechanical filter can be determined; hence, it is one of the parameters which show the quality of the transducer.

Now, from Eqs. (10), (12), and (15), the capacitance ratio of the transducer in Fig. 2 becomes

$$\frac{1}{r}=\frac{C_1}{C_d}=-\frac{16\beta_1}{\alpha_1^2}\cdot\frac{1}{\dfrac{\frac{4}{3}\frac{Y_D}{\epsilon B^2}\left[\{(t_2+t_0)^3-t_0^3\}+\frac{Y_D}{Y_M}\{t_0^3+(t_1-t_0)^3\}\right]}{t_2(t_2+2t_0)^2}-1} \tag{19}$$

Substituting $\frac{\epsilon B^2}{Y_D}=k_{31}^2$, $\frac{t_2}{t_1}=u$, $t_0=t_1\left(1-\frac{Y_D}{Y_M}u^2\right)\Big/2\left(1+\frac{Y_D}{Y_M}u\right)$ into the equation, we have

$$\frac{1}{r}=-\frac{16\beta_1}{\alpha_1^2}\cdot\frac{1}{\dfrac{\frac{1}{6k_{31}^2}\left[\left(\frac{Y_M}{Y_D}-1\right)\left(1-\frac{Y_D}{Y_M}u^2\right)^2+\left(1+2u+\frac{Y_D}{Y_M}u^2\right)^2+\frac{Y_M}{Y_D}\left(1+2\frac{Y_D}{Y_M}u+\frac{Y_D}{Y_M}u^2\right)^2\right]}{u(1+u)^2\left(1+\frac{Y_D}{Y_M}u\right)}-1} \tag{20}$$

As to the combination of metallic and piezoelectric materials, when an Fe-Ni-Mo alloy and lead zirconium titanate ceramic are used respectively, the physical constants are as in Table 2. Relations of B_T, u and Y_D/Y_M are shown in Fig. 4 for $B_T(\fallingdotseq 1/2r)$ of the

Table 2

Physical Constants of the Composite Transducer Materials

Material	ρ (g/cm^3)	E (dyne/cm^2)	k_{31}^2
Fe-Ni-Mo alloy	8.15	16.9×10^{11}	
Lead zirconium titanate ceramic	7.5	7.25×10^{11}	0.1

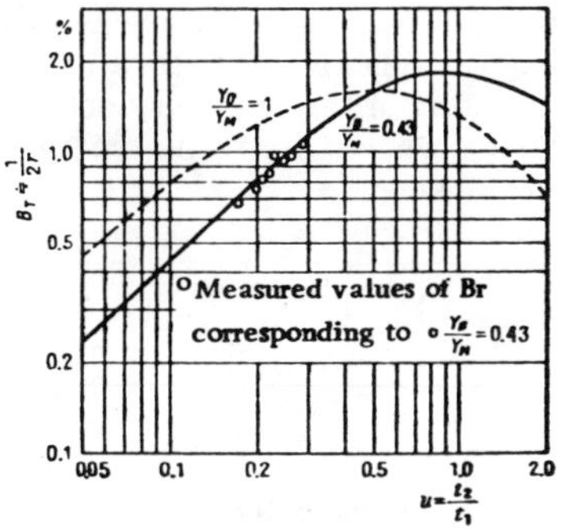

Fig. 4. Relations of B_T, u and $\frac{Y_D}{Y_M}$

fundamental vibration, corresponding to thickness ratio u.

For $u \doteqdot 0.85$ ($t_2 = 0.85\ t_1$) and the combination of Fe-Ni-Mo and lead zirconium titanate, $B_{T\,max} = 1.84\%$ or $r_{min} = 27.5$ can be obtained, since the variation of B_T with respect to u is very slight. Hence, if the thickness ratio of the materials is chosen in this vicinity, a wide pass band can be obtained.

From the measured values of B_T of the experimental vibrator, as shown in Fig. 4, it can be seen that they are in good agreement.

2.3 Values of Q

The metallic material used in the transducer has a Q of the order of 10^4, but the mechanical Q of the piezoelectric ceramic material is only 10^2 to 10^3. The mechanical Q of the composite device composed of both materials is greatly reduced from that of the single metallic material. The equations for evaluating the equivalent mechanical Q of the composite transducer from the material constants and dimensions of the elements will now be derived.

The energies of elastic dissipation of the motion of the vibrator with internal friction can be expressed as follows:

$$K=\frac{1}{2}\rho A\int_0^l\left(\frac{\partial\xi}{\partial t}\right)^2 dx \tag{21}$$

$$U=\frac{1}{2}YAk^2\int_0^l\left(\frac{\partial^2\xi}{\partial x^2}\right)dx \tag{22}$$

$$P=\frac{1}{2}\eta Ak^2\int_0^l\left(\frac{\partial^3\xi}{\partial t\partial x^2}\right)dx \tag{23}$$

Where ρ is the density, Y is Young's modulus, η is the direct viscosity coefficient, k is the radius of gyration of the cross section, ξ is the bending of the bar in the x direction, and A is the cross sectional area of the bar. Supposing ξ has a factor of $e^{j\omega t}$, then P becomes

$$P=\frac{1}{2}\eta A\omega^2k^2\int_0^l\left(\frac{\partial^2\xi}{\partial t^2}\right)^2 dx=\frac{\eta\omega^2}{Y}U \tag{24}$$

From the definition of Q_m, we have

$$\frac{1}{Q_m}=\frac{1}{\omega}\cdot\frac{P_{max}}{U_{max}}=\frac{\eta\omega}{Y} \tag{25}$$

From the above relations, if we denote each quantity related to metallic and piezoelectric materials by suffixes M and D respectively, then Q_c of the composite transducer can be found from the following equation:

$$\frac{1}{Q_c}=\frac{1}{Q_M}\cdot\frac{1+\frac{Q_M}{Q_D}\cdot\frac{(U_D)_{max}}{(U_M)_{max}}}{1+\frac{(U_D)_{max}}{(U_M)_{max}}} \tag{26}$$

By using the corresponding flexural rigidity (= YAk^2),

$$K_M=\frac{Y_M w}{3}\{t_0^3+(t_1-t_0)^3\} \tag{27}$$

$$K_D=\frac{Y_D w}{3}\{(t_2+t_0)^3-t_0^3\} \tag{28}$$

the ratio of elastic energies for each part can be given simply by the following equation:

$$\frac{U_D}{U_M}=\frac{K_D}{K_M}=\frac{Y_D}{Y_M}u\cdot\left\{\frac{\left(\frac{Y_D}{Y_M}\right)^2u^4+2\frac{Y_D}{Y_M}u^3+4u^2+6u+3}{3\left(\frac{Y_D}{Y_M}\right)^2u^4+6\left(\frac{Y_D}{Y_M}\right)^2u^3+4\left(\frac{Y_D}{Y_M}\right)^2u^2+2\frac{Y_D}{Y_M}u+1}\right\} \tag{29}$$

From Eqs. (25) and (28), the relation between Q_c/Q_M and u are shown in Fig. 5 for Q_D = 1000 and Q_M = 5000, and also for Q_D = 100 and Q_M = 5000 with the Fe-Ni-Mo alloy and lead zirconium titanate. It is clear that Q_c decreases sharply as the thickness of the piezoelectric ceramic plate increases. For

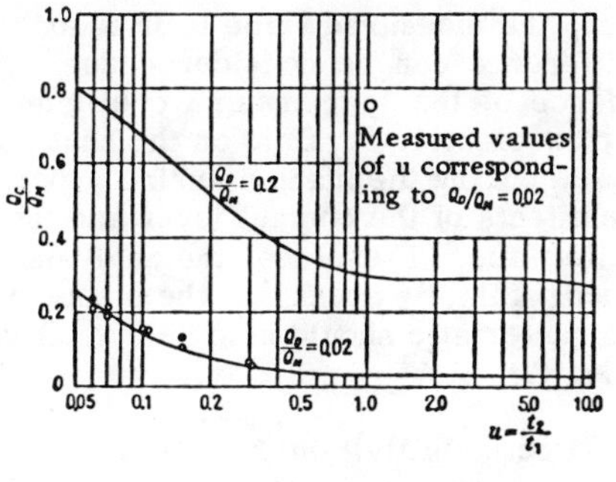

Fig. 5. Relation between Q_C/Q_M and u

the purpose of comparison, the measured values are also shown in the figure.

2.4 Stability

(a) Temperature stability. Although the resonant frequency of the composite transducer is determined from the dimensions and material constants of the composite material by Eq. (3), Young's modulus or the length or both, will vary with temperature, and these are the main causes of temperature variation of the resonant frequency. In order to find the temperature stability of the resonant frequency, Eq. (3) is differentiated with respect to temperature:

$$2\frac{1}{\omega_n}\cdot\frac{d\omega_n}{dT}=2\frac{1}{\omega_{nM}}\cdot\frac{d\omega_{nM}}{dT}+A\left(\frac{1}{Y_D}\cdot\frac{dY_D}{dT}-\frac{1}{Y_M}\cdot\frac{dY_M}{dT}\right)+B\left(\frac{1}{t_2}\cdot\frac{dt_2}{dT}-\frac{1}{t_1}\cdot\frac{dt_1}{dT}\right) \qquad (30)$$

Again,

$$2\frac{1}{\omega_{nM}}\cdot\frac{d\omega_{nM}}{dT}=\frac{1}{Y_M}\cdot\frac{dY_M}{dT}-\frac{1}{t_1}\cdot\frac{dt_1}{dT} \qquad (31)$$

where ω_{nM} represents the resonant angular frequency of the metal vibrator, and

$$A\equiv\frac{\left[-\frac{Y_D}{Y_M}\left\{1-\frac{Y_D}{Y_M}u^2\right\}^2\left\{1+3u^2-4\frac{Y_D}{Y_M}u^2\right\}+3\left\{1+2\frac{Y_D}{Y_M}u+\frac{Y_D}{Y_M}u^2\right\}^2\frac{Y_D}{Y_M}u(2+u)+\frac{Y_D}{Y_M}\left\{1+2u+4\frac{Y_D}{Y_M}u^2\right\}\left\{1+2u+\frac{Y_D}{Y_M}u^2\right\}^2\right]}{\left[\left\{1-\frac{Y_D}{Y_M}u^2\right\}^2\left(1-\frac{Y_D}{Y_M}\right)+\left\{1+2\frac{Y_D}{Y_M}u+\frac{Y_D}{Y_M}u^2\right\}^3+\frac{Y_D}{Y_M}\left\{1+2u+\frac{Y_D}{Y_M}u^2\right\}^3\right]}-\frac{3}{1+\frac{Y_M}{Y_D}\cdot\frac{1}{u}} \qquad (32)$$

$$B=\frac{\left[-6\left(1-\frac{Y_D}{Y_M}\right)\frac{Y_D}{Y_M}u^2\left\{1-\frac{Y_D}{Y_M}u^2\right\}^2+6\left\{1+2\frac{Y_D}{Y_M}u+\frac{Y_D}{Y_M}u^2\right\}^2\frac{Y_D}{Y_M}u(1+u)+6\frac{Y_D}{Y_M}u\left\{1+\frac{Y_D}{Y_M}u\right\}\left\{1+2u+\frac{Y_D}{Y_M}u^2\right\}^2\right]}{\left[\left\{1-\frac{Y_D}{Y_M}u^2\right\}^3\left(1-\frac{Y_D}{Y_M}\right)+\left\{1+2\frac{Y_D}{Y_M}u+\frac{Y_D}{Y_M}u^2\right\}^3+\frac{Y_D}{Y_M}\left\{1+2u+\frac{Y_D}{Y_M}u^2\right\}^3\right]}-\frac{3}{1+\frac{Y_M}{Y_D}\cdot\frac{1}{u}}+\frac{2}{1+\frac{\rho_M}{\rho_D}\cdot\frac{1}{u}} \qquad (33)$$

Furthermore, $\frac{1}{Y_D}\cdot\frac{dY_D}{dT}$, $\frac{1}{Y_M}\cdot\frac{dY_M}{dT}$, $\frac{1}{t_2}\frac{dt_2}{dT}$, $\frac{1}{t_1}\cdot\frac{dt_1}{dT}$ each represents the linear expansion coefficient and the temperature coefficient of Young's modulus of the piezoelectric material and the metal respectively. These are designated β_D, β_M, α_D, α_M, respectively. Values of these coefficients are shown as in Table 3. Since, for the ceramic the linear expansion coefficient is less than the temperature coefficient of Young's modulus by a ratio of the order of 10^{-1}, an approximate evaluation of the total stability is obtained by taking into account only the temperature coefficient of Young's modulus. By neglecting α_D and α_M from equation (30), the following relation can be obtained:

$$\frac{1}{f_n}\cdot\frac{df_n}{dT}\doteqdot\frac{1}{2}(1-A)\beta_M+\frac{A}{2}\beta_D \qquad (34)$$

The value of A of Eq. (32) for the same combination of Fe-Ni-Mo and lead zirconium titanate was found to be as shown in Fig. 6.

Table 3

Stability Constants of the Materials

	Temp. coef. of Young's Modulus	Coef. of linear (longitudinal) expansion	Variations in aging
Fe-Ni-Mo alloy	10×10^{-6}/°C	8.8×10^{-6}/°C	$<2.3\times10^{-4}$/decade
Zirconium titanate ceramic	-200×10^{-6}/°C	7.5×10^{-6}/°C	$<6\times10^{-4}$/decade

When $u\to0$, $A=0$ (for the metal vibrator); for the piezoelectric ceramic vibrator, when $u\to\infty$, $A=1$. Corresponding to each case the following results can be obtained:

$$\left(\frac{1}{f_n}\cdot\frac{df_n}{dT}\right)_M\doteqdot\frac{1}{2}\beta_M,\ \left(\frac{1}{f_n}\cdot\frac{df_n}{dT}\right)_D\doteqdot\frac{1}{2}\beta_D \qquad (35)$$

If the signs of β_M and β_D are different, there exists a value of u which will make the

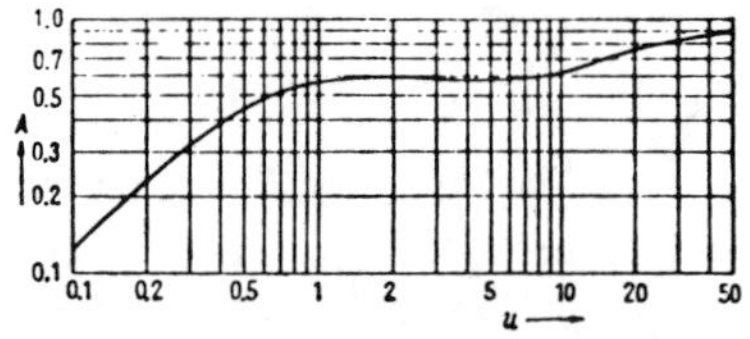

Fig. 6. Relation between A and u

frequency-temperature coefficient of the composite transducer zero. Such a condition is given by the following equation.

$$A \fallingdotseq \frac{1}{1-\frac{\beta_D}{\beta_M}} \tag{36}$$

The relationships between u and the frequency-temperature coefficient ϵ were calculated for four combinations of β_D, and β_M. These are shown in Fig. 7. From these curves, the condition for zero temperature coefficient is found to be in the vicinity of $u = 0.15$, if $\beta_D = -50 \times 10^{-6}$ °/C and $\beta_M = +10 \times 10^{-6}$ °/C. However, if $\beta_D = -100 \times 10^{-6}$ °/C, the condition of zero temperature coefficient does not exist for practical dimensions. But, the frequency-temperature coefficient for the typical piezoelectric vibrator in the range below $u \fallingdotseq 0.5$ can be made better than -25×10^{-6} °/C. In an LC filter (generally a ferrite inductor and a styroflex capacitor are used), the temperature coefficient of the total resonant frequency is better than roughly $\pm 50 \times 10^{-6}$ °/C. However, the temperature stability of the composite transducer is better than that of the LC filter.

The experimental transducer at 5 kc with $u \fallingdotseq 0.265$, $\beta_M \fallingdotseq 18 \times 10^{-6}$ °/C, and $\beta_D \fallingdotseq 300 \times 10^{-6}$ °/C, has a frequency-temperature coefficient of 38×10^{-6} °/C according to Eq. (34);

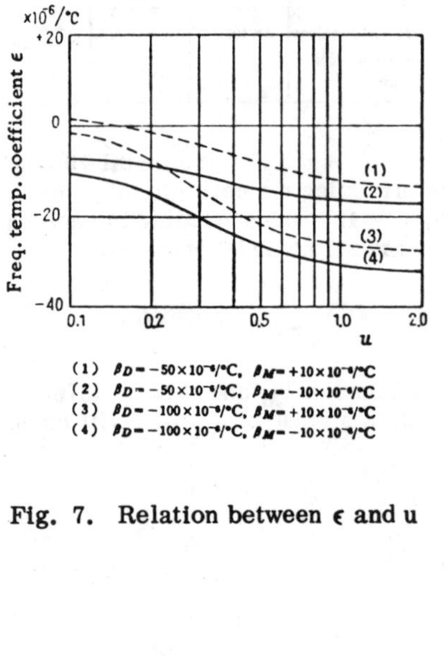

Fig. 7. Relation between ϵ and u

however, the measured value is 18×10^{-6}° /C. The difference can be considered due to the dependence of the temperature coefficient of the piezoelectric ceramic on the heat treatment used during the manufacturing processes, and the effects of the contact layer and the supporting line. Therefore, the analysis was continued. Strictly speaking, the piezoelectric characteristic should also be considered from Eq. (2).

(b) Time stability. In the previous results, if the temperature variable T is replaced by the time, a thorough study of the aging stability of the composite transducer is possible. In general, for a vibrator made from one material, the resonant frequency will change as follows as the dimensions of the vibrator and Young's modulus of the material vary:

$$\frac{\Delta f}{f_0} = \frac{\Delta t}{t_0} + \frac{1}{2}\cdot\frac{\Delta Y}{Y_0} - \left(2\frac{\Delta l}{l_0} + \frac{1}{2}\cdot\frac{\Delta \rho}{\rho_0}\right) \tag{37}$$

where suffix 0 denotes the reference value. In this equation, since there is no time variation in density and since also the time variation in dimensions (such as oxidation of the metal surface) is neglected, the resonant frequency has a time variation mainly due to the time variation of Young's modulus (strictly speaking the effect of the support should also be considered). In other words,

$$\frac{\Delta f}{f_0} \fallingdotseq \frac{1}{2}\cdot\frac{\Delta Y}{Y_0} \tag{38}$$

On the one hand, the time variation of the resonant frequency of the mechanical vibrator increases linearly with respect to the logarithm of time. It can be expressed in the following approximate relation, as is known from experience:

$$\frac{\Delta f}{f_0} \fallingdotseq \alpha' \log_{10} T \tag{39}$$

where α' is the time variation in a decade (ten units of time). From Eqs. (38) and (39),

$$\frac{\Delta Y}{Y_0} = 2\alpha' \log_{10} T \tag{40}$$

is obtained. For the case of the composite transducer of this paper corresponding to Eq. (30), we obtain from Eq. (38) the following:

$$\frac{\Delta f_n}{f_n} = \frac{1}{2}(1-A)\frac{\Delta Y_M}{Y_M} + \frac{A}{2}\cdot\frac{\Delta Y_D}{Y_D} \tag{41}$$

By substituting the relationship of Eq. (40), we obtain

$$\frac{\Delta f_n}{f_n} = \{(1-A)\alpha_M' + A\,\alpha_D'\}\log_{10} T \tag{42}$$

The quantities α'_M and α'_D in Eq. (42) are the time variation rates in a decade for the metal and piezoelectric ceramic, respectively. Accordingly, the rate of time variation in a decade for the composite transducer is as follows:

$$\alpha_c' = (1-A)\alpha_M' + A\,\alpha_D' \tag{43}$$

Using $\alpha'_M \doteqdot 2.3 \times 10^{-5}$ and $\alpha'_D \doteqdot 60 \times 10^{-5}$ for the Fe-Ni-Mo alloy and the lead zirconium titanate ceramic, α'_c is obtained as shown in Fig. 8. Since both quantities must have a positive sign, there will be no mutual compensation as is the case for the frequency-temperature coefficient. In general, because $\alpha_D' > \alpha_M'$, $0 < A < 1$, it follows that $\alpha_D' > \alpha_c' > \alpha_M'$ and the aging stability of the composite transducer is improved compared to the stability of the piezoelectric ceramic itself. In particular, within the range $u < 0.5$, we obtain $\alpha'_c < 3 \times 10^{-4}$/decade. The measured results are shown in Fig. 8 for comparison.

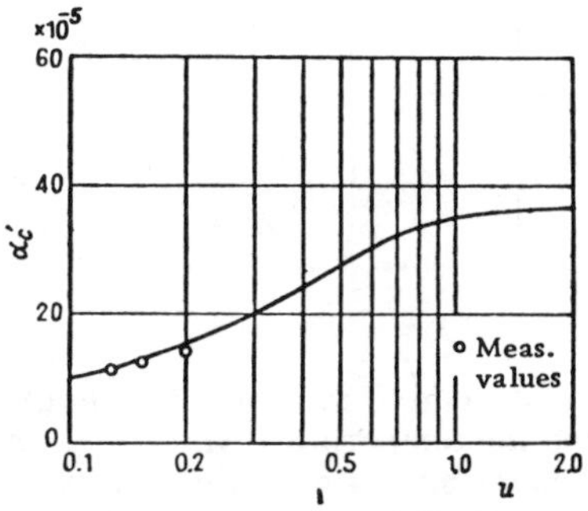

Fig. 8. Relation between α'_c and u

In the above investigation, although the effect of the intermediate layer was not considered, if the contact agent is a resin that hardens with heat treatment, it may be possible that hardening occurs gradually, and its effect on frequency cannot be simply neglected. Hence, this effect might well be investigated again.

3. Characteristics of an Electromechanical Filter using Piezoelectric Composite Bending Vibrators

Because the input impedance becomes higher as the frequency decreases in the piezoelectric transducer, the low-frequency mechanical filter using such a transducer involves a severe problem in matching the transducer to the electrical circuit. Especially for application in the low-frequency range, a comparatively wide-band filter is required, so that a frequency above several kc/s is presently considered appropriate for this type of device.

Here, using the results of the previous section for reference, we will examine the frequency range and the pass band which can be realized by the mechanical filter of Fig. 1, using the vibrator with a piezoelectric ceramic transducer bonded to one of the main surfaces.

3.1 Center Frequency

The relations between transducer dimensions and resonant frequency which are calculated from Eq. (3) are shown in Fig. 9. The transducer shown in Fig. 2 can be realized for such properties, generally within the frequency range from 300 c/s to 30 kc/s for which the bending vibrator is practical. Accordingly, the design of an electromechanical filter constructed as shown in Fig. 1 is possible.

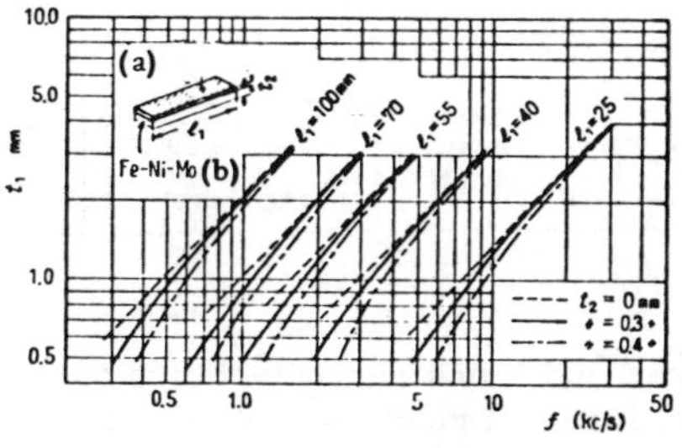

Fig. 9. Relation between resonant frequency and dimensions of the composite transducer

(a) Lead zirconium titanate ceramic
(b) Alloy

3.2 Fractional Pass Band

In general, the widening of the pass band of a mechanical filter is limited by the characteristic B_T of the transducer. This is inversely proportional to the minimum capacitance ratio, which in turn depends on the method of construction and the material constants of the transducer, a constraint which is added in determining the values of circuit elements.

Three basic forms of transducer configurations, including the equivalent electrical circuit, can be considered as shown in Table 4. By resonating the electrical circuit, the pass band of the transducer can be widened. (The circuits in the table are based on the assumption

Table 4

Various Transducer Circuit Configurations and Their Pass Bandwidths Input Impedances

		Circuit configuration	Specific pass bandwidth	Max. spec. band B_{Fmax}	Input image impedance
1	3-element circuit, π end	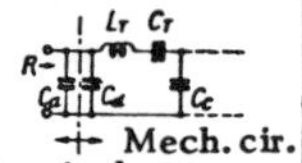 Mech. cir. Electrical circuit Ref. [7]	$B_F = \frac{\Delta f}{f_0} \approx \frac{C_T}{(C_a+C_d)}$ $B_{Fmax} \approx \frac{C_T}{C_d} \approx 2B_T$	3.7%	
2	Constant-K circuit, π end	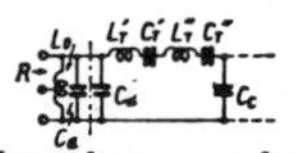 Impedance can be transformed by use of L_0 [7]	$B_F = \frac{\Delta f}{f_0} \approx \sqrt{\frac{2C_T}{C_a+C_d}}$ $B_{Fmax} = \sqrt{2\frac{C_T}{C_d}} \approx 2\sqrt{B_T}$	27	
3–1	3-element circuit, T end (1)	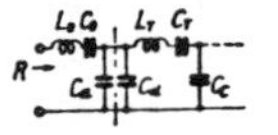	$B_F = \frac{\Delta f}{f_0} \approx \frac{2C_T}{C_a+C_d}$ $B_{Fmax} \approx \frac{2C_T}{C_d} \approx 4B_T$	7.4	
3–2	3-element circuit, T end (2)	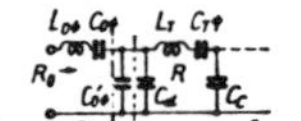 Impedance can be transformed by Norton transformation $\phi^2 = \frac{R}{R_0}$	$B_F = \frac{\Delta f}{f_0} \approx 2\phi\left\{1+\frac{1}{4}\left(1-\frac{1}{\phi}\right) \times \left(\frac{f_{+1}^2}{f_{-1}^2}\right)\right\}\frac{C_{T\phi}}{C_{a\phi}'+C_d}$ $B_{Fmax} \approx 4\phi B_T$	7.4ϕ	
3–3	3-element circuit, T end (3)	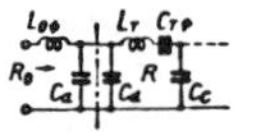 (As above)	$C_{a\phi} \to \infty$ in case of 3-2 by Norton transformation $B_{Fmax} \approx 2\sqrt{B_T}$	27	

that the input and output characteristics are identical and that all mechanical elements are transformed into electrical equivalents; also, element losses are neglected. Each element value of the equivalent circuits can be determined by the same method used for the ordinary filter. The condition for obtaining the maximum fractional pass band from a given transducer, which is regarded as a mechanical filter, is that the additional capacitance C_a connected in parallel with the damped capacity C_d of the transducer becomes zero. Based on this condition, the relationships between the fractional pass band ($B_T \cong 1/2\ r$) and the maximum value $B_{F\,max}$ are as shown in column 3 of the table.

In the case of the composite transducer of Fig. 2, employing Fe-Ni-Mo alloy and lead zirconium titanate ceramic, the fractional pass band of the mechanical filter shown in the fourth column of Table 4 can be realized for various configurations. This results from the maximum value of $B_T \fallingdotseq 1.84\%$ obtained by setting $u \fallingdotseq 0.85$ from Fig. 4.

However, in practice, although the additional capacitance C_a is zero, the value of C_d of the ceramic changes greatly with temperature (about 3×10^{-3} /°C), and characteristic variation arises due to mismatching which accompanies the temperature variation. Accordingly, under such a condition we have to add a large stable capacitance C_a to capacitance C_d in order to improve the temperature stability, which seems to reduce B_T.

Although a fractional pass band of 20% was realized experimentally, it could be improved further with better material characteristics. In addition, the capacitance ratio will decrease to some extent due to partial bonding; this has some effect on widening the pass band.

Furthermore, the level of input and output impedance can be adjusted arbitrarily by utilizing an electrical circuit connected to the transducer. Hence, the frequency characteristics can be designed for a constant-K circuit with either a T or a π termination.

4. Experimental Example

The design results of this type of filter using two vibrators (each with a transducer) are as follows:

Center frequency	1270 c/s
Pass band	160 c/s

Input and output impedance 600 Ω

4.1 Vibrator Constants

The piezoelectric ceramic plate must of course be bonded to the vibrator to furnish a composite vibrator. From the results in Sect. 2, the dimensions of the transducer are as shown in Fig. 10. A metal plate of 1.09 x 4 x 70 mm was used. However, since a long enough piezoelectric ceramic was not available to be affixed to the whole surface, a piezoelectric ceramic sheet of 0.4 x 4 x 56 mm was used, and an approximate design made. The measured values of the constants are shown in Table 5.

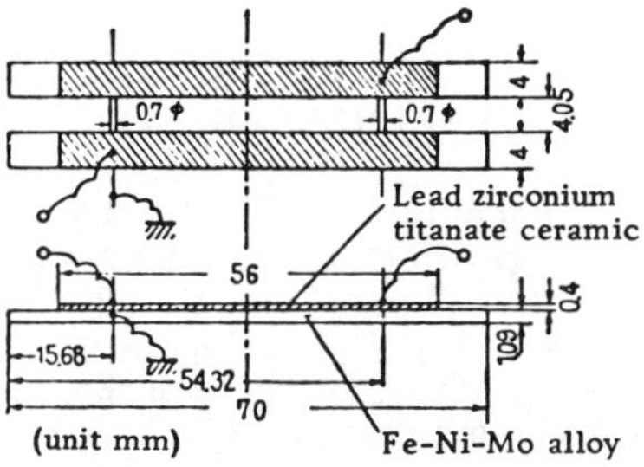

Fig. 10. Dimensions of a electromechanical filter using composite transducers; φ = diameter

4.2 Circuit Design

Using the mechanical equivalent constants in Table 5, the dimensions of the coupler can be easily designed. As for the electrical circuit, taking the equivalent inductance of the transducer as the reference, the elements are found according to the image-parameter method. The operations, including Norton's transformation, give the performance shown in Fig. 11. The experimental example was designed for the maximum pass band obtainable for an impedance of 600 Ω. The number of circuit elements was reduced by eliminating C in the LC series resonant arms at both the input and output terminals. Of course, if the impedance level is raised higher than 600 Ω, the pass band can be widened.

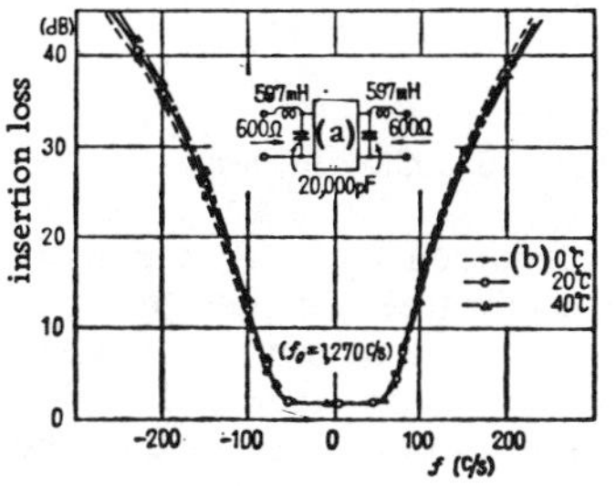

Fig. 11. Insertion loss characteristics and electrical circuit of electromechanical filter
(a) Mechanical circuit (b) Temperature

4.3 Characteristics

The insertion loss and its temperature characteristics of this mechanical filter are shown in Fig. 11. The loss in the pass band is small, and good general characteristics are obtained. If we examine the variation of the center frequency of the 3-dB pass band, it is found to be about $+6 \times 10^{-5}$ /°C. Hence, the electrical circuit has a good effect, and the pass band is extremely stable. An example of input and output impedance characteristics is shown in Fig. 12.

Table 5

Mechanical and Electrical Constants of the Tested Composite Transducer

	m (gr)	s (dyne/cm)	r (dyne/kine)	s_e (dyne/cm)*	Dimen. of coupler (mm)	
1	0.782	4.59×10^7	2.26×10^3	4.62×10^8 (per coupler)**	0.7φ × 4.05	
2	0.782	4.59×10^7	2.09×10^3			
	C_d (pF)	L_e (H)	C_e (pF)	f_1 (c/s)	f_{1a} (c/s)	Q
1	6150	74.7	237	1220.8	1243.4	265
2	6310	75.0	236	1212.2	1234.3	289

*Stiffness of torsional coupler.
**Two couplers were used in the mechanical filter.

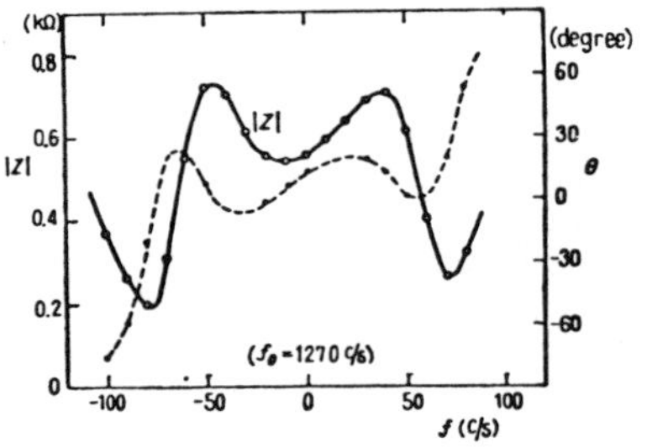

Fig. 12. Input and output impedance characteristics

5. Conclusions

The electromechanical filter constructed with bending vibrators coupled torsionally at their nodal points is suitable as a wave filter in the frequency range from 300 cycles to several tens of kilocycles. As the electromechanical transducer for such filter, the piezoelectric composite transducer composed of the metal vibrator and a piezoelectric ceramic plate bonded to one of the main surfaces of the metal vibrator is analyzed, and the equivalent circuit constants, the Q, and the time and temperature stabilities are shown in comparison with experimental results. It is shown that in this construction, the electromechanical filter with a fractional pass band up to about 20% can be designed for practical use in the frequency range from 300 cps to 30 kc/s.

The stability of the new type is better than for an LC tuned circuit; its Q is also higher; and a better filter characteristic can be realized.

Furthermore, the input and output impedance can be easily matched to the electrical circuit; the impedance-frequency characteristics can be transformed arbitrarily so as to become either high or low; hence, it can be easily applied to communication equipment. Examples are low-frequency carrier-wave filters, carrier band-pass filters, or channel filters for various telecommunication equipment. It is most useful in applications where the characteristics of narrow band and high selectivity are required.

Furthermore, if the electrical-to-mechanical conversion efficiency and the temperature stability of the piezoelectric ceramic, or in particular the temperature stability of the transfer coefficient were raised, it could be expected that the scope of the mechanical filter would be extended.

Finally, the author would like to express his gratitude to Mr. Ikuta, director of the division of transmission of Fujitsu, Ltd.; Mr. Yamazaki, deputy director of the division of transmission techniques; Mr. Okamoto, chief of the section of transmission components; the network research group; and those who cooperated in the experiments, for their very helpful advice. Also thanks are expressed to Prof. Konno of Yamagata University.

Submitted May 31, 1965

REFERENCES

1. Kusakabe and Konno: Electromechanical filter composed of transversely vibrating bars connected by torsionally vibrating bars, Jour. I.E.C.E., Japan, 47, 12, p. 1845 (Dec. 1964); available in English in E.C.J., same date, p. 15.
2. Yakuwa, Okuta and Tanaka: Low frequency mechanical filter, Fujitsu, 15, p. 386 (1964).
3. Nukiyama and Suzuki: Vibration modes of M-type piezoelectric vibrators and their electroacoustical constants, Jour. I.E.C.E., Japan, 17, 231, p. 367 (June 1942).
4. W.P. Mason and R.N. Thurston: A compact electromechanical band-pass filter for frequencies below 20 kc, IRE Trans., UE-7, 2, p. 59 (1960).
5. Takahashi: Analysis of electroacoustic metal vibrators used as filter and oscillating elements, Papers of Technical Group on Circuit Theory, I.E.C.E., Japan (May 1964).
6. Yakuwa: On the maximum pass band of the low-frequency mechanical filter, Natl. Conv. Record of I.E.C.E., Japan, No. 93 (1964).
7. Konno: The electromechanical filter, Kyoritsu Book Co., (May 1959).

Electromechanical Filter Composed of Transversely Vibrating Resonators for Low Frequencies

MASASHI KONNO

Department of Electrical Engineering, Faculty of Engineering, Yamagata University, Yonezawa, Japan

CHIHARU KUSAKABE

Department of Technology, Faculty of Education, Yamagata University, Yamagata, Japan

YOSHIRÔ TOMIKAWA

Department of Electrical Engineering, Faculty of Engineering, Yamagata University, Yonezawa, Japan

An electromechanical bandpass filter that is suitable for low frequencies is described. The mechanical filter consists of two transducer–resonators and two couplers, of which the transducer–resonators vibrate transversely and are coupled at their nodal points with the torsionally vibrating couplers. Each transducer–resonator consists of two pieces of electrostrictive material with a basis reed sandwiched between them. Also, in this paper, the effects of impedance of electrostrictive material on the basis reed are analyzed. In order to analyze the mechanical vibrating system, mechanical network theory is adopted.

INTRODUCTION

An electromechanical filter exhibits excellent properties, such as high Q, compactness, durability, and low insertion loss.

This paper deals with the method of design and the experimental results of the mechanical filter, consisting of the transversely vibrating resonators coupled one to another by the torsionally vibrating slender rods.

For this type of mechanical filter, Mason and Thurston[1] have reported on the filter, utilizing the second mode of resonators. Parallel with their development, the authors[2,3] have been executing the development of the filter, utilizing the first mode of resonators.

The electromechanical transducers used here are all electrostrictive ones of the higher-harmonic suppression type. The impedance of the electrostrictive materials bonded to drive the basis reed can not be neglected, in general, as against the impedance of the basis reed. The paper also deals with the analysis of such a thin transducer–resonator as mentioned above.

I. STRUCTURE OF THE MECHANICAL FILTER

The structure of the electrostrictive mechanical filter is shown in Fig. 1. The transducer–resonators vibrate in the first mode and are coupled at the nodal points with the torsionally vibrating couplers. The vibration energy of one resonator is transmitted to the other resonator

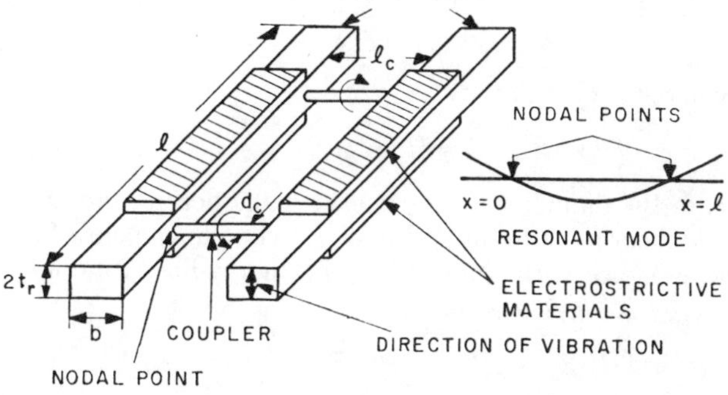

FIG. 1. Structure of a mechanical filter.

[1] W. P. Mason and R. N. Thurston, "A Compact Electromechanical Band-Pass Filter for Frequencies below 20 Kilocycles," IRE Trans. Ultrasonics Eng. 7, No. 2, 59–70 (June 1960).

[2] M. Konno, "Theoretical of Mechanical Filters," J. IECE Japan 40, No. 1, 44–51 (Jan. 1957).

[3] C. Kusakabe and M. Konno, "Electromechanical Filter Composed of Transversely Vibrating Bars Connected by Torsionally Vibrating Bars," J. IECE Japan 47, No. 12, 1845–1854 (Dec. 1964).

Reprinted with permission from *J. Acoust. Soc. Amer.*, vol. 41, pt. 2, pp. 953-961, Apr. 1967.

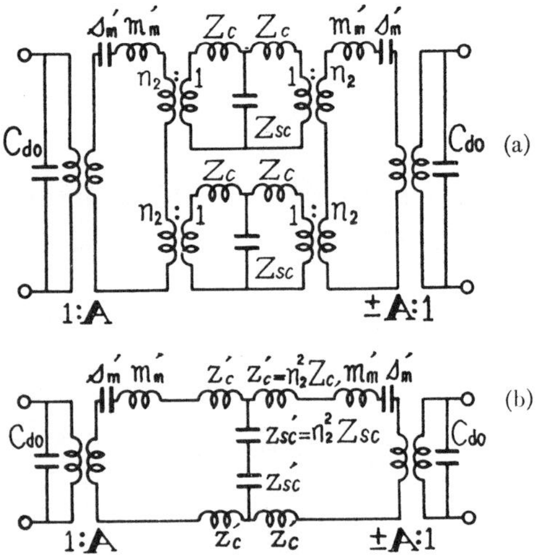

FIG. 2. (a) Equivalent circuit of mechanical filter. (b) Modified equivalent circuit.

through the torsional impedance of the coupler. Electrostrictive materials are bonded on both sides of the basis reed, and are polarized to excite the first mode of the transverse vibration.

Figure 2(a) shows the equivalent circuit of the mechanical filter, utilizing the mth mode of the resonator and Fig. 2(b) shows the modified equivalent circuit. In Fig. 2, C_{d0} is the damping capacitance, **A** is the force factor, and the sign of the ideal transformer ratio ($\pm$**A**:1) is determined by the polarity of polarization of the electrostrictive materials. Z_c, Z_{sc}, m_m', s_m', and n_2 are discussed in Sec. II.

II. EQUIVALENT-CIRCUIT CONSTANTS

A. Coupler

The torsional vibration of the coupler is analogous to an electrical transmission line. When the length of the coupler is less than $\frac{1}{8}$ of the wavelength λ at the operating frequency, the impedances Z_c and Z_{sc} in Fig. 2 are given as follows:

$$Z_c = Z_{0c}\tanh(\gamma_c l_c/2) \doteqdot j\omega(J_c\rho_c l_c/2),$$
$$Z_{sc} = Z_{0c}\,\mathrm{cosech}(\gamma_c l_c) \doteqdot (1/j\omega)(J_c G_c/l_c), \tag{1}$$

or

$$z_c' = n_2^2 Z_c \equiv j\omega m_c,$$
$$z_{sc}' = n_2^2 Z_{sc} \equiv s_{sc}/j\omega, \quad s_{sc} = n_2^2 S_{sc}, \tag{2}$$

where

$$Z_{0c} = J_c(\rho_c G_c)^{\frac{1}{2}}, \quad J_c = (\pi/32)d_c^4, \quad \gamma_c = j\omega(\rho_c/G_c)^{\frac{1}{2}}, \tag{3}$$

d_c is the diameter, l_c the length, ρ_c the density, and G_c the shear modulus of the coupler. n_2 is defined below for a resonator with length l: the mth normalized function $\Xi_{m(X)}$, $X = x/l$ (cgs units):

$$n_2 = (1/l)(\Xi_{m(X_c)}'/\Xi_{m(X_d)}), \quad \Xi_{m(X)}' = \partial\Xi_{m(X)}/\partial X, \tag{4}$$

where the origin of the x axis is the point at the lefthand end of the resonator, X_d $(=x_d/l)$ is the driving point, and X_c is the point at which the coupler is connected.

TABLE I. $\Xi_{1(X)}$ and $\Xi_{1(X)}'$ in the both ends free of the basis reed.

X	$\Xi_{1(X)}$	$\Xi_{1(X)}'$	X	$\Xi_{1(X)}$	$\Xi_{1(X)}'$
0.00	2.000	−9.294	0.30	−0.544	−6.342
0.05	1.535	−9.275	0.35	−0.833	−4.971
0.10	1.074	−9.147	0.40	−1.039	−3.467
0.15	0.624	−8.828	0.45	−1.172	−1.753
0.20	0.195	−8.264	0.50	−1.216	0.000
0.25	−0.200	−7.439			

When the coupler is a slender rod, $2m_c \ll m_m'$ and $l_c < \lambda/8$, the effect of the coupler may be regarded as stiffness only. Consequently, the static torsional stiffness $S_{sc}(=G_cJ_c/l_c)$ may be employed. The resultant equivalent circuit is given as in Fig. 3, where

$$s_{sc}' = 2n_2^2 S_{sc} = 2\frac{\pi}{32}\cdot\frac{d_c^4}{l_c}\cdot\frac{G_c}{l^2}\cdot\frac{\Xi_m'^2(X_c)}{\Xi_m^2(X_d)}. \tag{5}$$

B. Resonator

In Fig. 2, the equivalent mass m_m' and the equivalent stiffness s_m' of the transducer–resonator are given as follows:

$$m_m' = M_r'/\Xi_m^2(X_d),$$
$$s_m' = \omega_m'^2 m_m', \tag{6}$$

where the subscript m denotes the mode of vibration; M_r' is the total mass of the transducer–resonator, and ω_m' is the mth apparent resonant angular frequency.

If the electrostrictive material is much thinner than the basis reed and its impedance is neglected, the resultant mechanical impedance of the transducer–resonator may be approximately equal to the impedance of the basis reed only. Then $\omega_m' \doteqdot \omega_m$, $m_m' \doteqdot m_m$, $s_m' \doteqdot s_m$, and

$$\omega_m = \frac{\alpha_m^2}{l^2}R\left(\frac{E_r}{\rho_r}\right)^{\frac{1}{2}}, \quad s_m = \omega_m^2 m_m, \tag{7}$$
$$\alpha_{m=1} = 4.730, \qquad \alpha_{m=3} = 10.9956, \cdots,$$

where m_m is the equivalent mass of the basis reed, and α_m is the eigenvalue of the basis reed, E_r, ρ_r, and R are the Young's modulus, the density, and the radius of gyration, respectively.

The values of $\Xi_{1(X)}$ and $\Xi_{1(X)}'$ in the both ends free of the basis reed are given in Table I, and, in the event that the impedance of electrostrictive material is negligible, the point X_c connecting the coupler is 0.224.

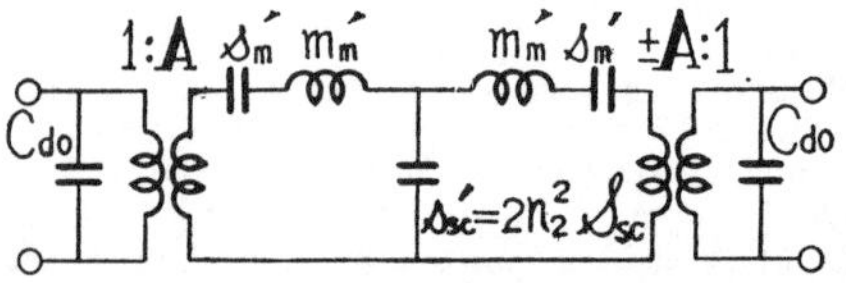

FIG. 3. Equivalent circuit of mechanical filter where the length of coupler is smaller than $\frac{1}{8}$ the wavelength and $2m_c \ll m_m'$.

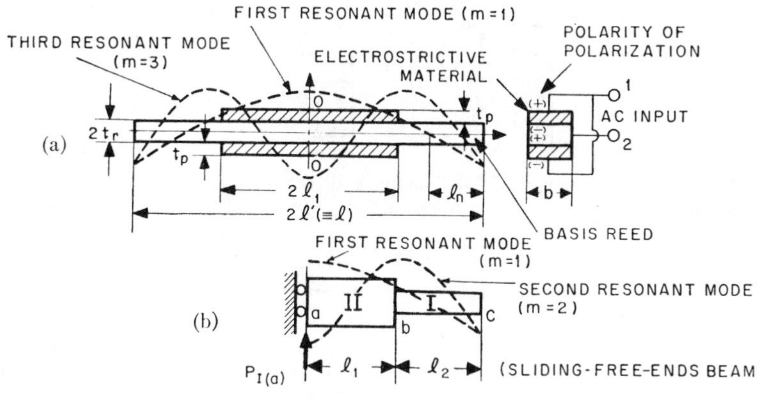

FIG. 4. (a) Electrostrictive transducer–resonator. (b) "Sliding-free-ends beam" vibrating transversely.

C. Electrostrictive Transducer–Resonator

When the impedance of the electrostrictive material is not negligible, the resonant frequency and the vibration mode of the transducer–resonator differ from those of the basis reed only, and then the condition of the harmonic suppression and the position of the nodal points change.

1. Apparent Resonant Frequency

Figure 4(a) shows the structure of the transducer–resonator. The electiostrictive materials are polarized as shown in Fig. 4(a), and the upper and lower electrodes are connected with each other electrically. Accordingly, the even modes of the transverse vibration are excited, but the odd modes are not.

As the central point OO in Fig. 4(a) is regarded as the "sliding end" of the mechanical beam, the analysis of the vibration ($m=1$, 3) of Fig. 4(a) is to be accomplished as the analysis of the vibration ($m=1$, 2) in the "sliding-free ends beam" in Fig. 4(b).

In Fig. 4(b), Part I indicates the portion at which the electrostrictive materials are bonded and Part II indicates the portion of the basis reed only.

The equivalent mechanical network[4] of Fig. 4(b) can be shown in Fig. 5, where P is the impressed force, M the impressed bending moment, $v(\equiv j\omega\eta)$ the velocity of displacement, and $\dot{\theta}(\equiv j\omega\theta)$ the velocity of angular displacement.

Now, if $P=P\epsilon^{j\omega t}$ and $M=M\epsilon^{j\omega t}$, the following equations[5] are given:

At Part I:

$$\begin{Bmatrix} v_{\mathrm{I}(a)} \\ v_{\mathrm{I}(b)} \\ \dot{\theta}_{\mathrm{I}(b)} \end{Bmatrix} = \begin{Bmatrix} y_{aa'} & y_{ab'} & \bar{\mathbf{Y}}_{ab'} \\ y_{ba'} & y_{bb'} & \bar{\mathbf{Y}}_{bb'} \\ \mathbf{Y}_{ba'} & \mathbf{Y}_{bb'} & Y_{bb'} \end{Bmatrix} \begin{Bmatrix} P_{\mathrm{I}(a)} \\ P_{\mathrm{I}(b)} \\ M_{\mathrm{I}(b)} \end{Bmatrix} \tag{8}$$

At Part II:

$$\begin{Bmatrix} v_{\mathrm{II}(b)}=v_{\mathrm{I}(b)} \\ \dot{\theta}_{\mathrm{II}(b)}=\dot{\theta}_{\mathrm{I}(b)} \end{Bmatrix} = \begin{Bmatrix} y_{aa} & \bar{\mathbf{Y}}_{aa} \\ \mathbf{Y}_{aa} & Y_{aa} \end{Bmatrix} \begin{Bmatrix} P_{\mathrm{II}(b)}=-P_{\mathrm{I}(b)} \\ M_{\mathrm{II}(b)}=-M_{\mathrm{I}(b)} \end{Bmatrix}. \tag{9}$$

The driving-point impedance $z_i(=P_{\mathrm{I}(a)}/v_{\mathrm{I}(a)})$ is calculated with Eqs. 8 and 9, and, when $z_i=0$ is set, the frequency equation is

$$(y_{aa}+y_{bb'})(Y_{aa}+Y_{bb'})-(\bar{\mathbf{Y}}_{aa}+\bar{\mathbf{Y}}_{bb'})(\mathbf{Y}_{aa}+\mathbf{Y}_{bb'})=0, \tag{10}$$

or

$$\begin{aligned}(\cosh\alpha_1\cdot\sin\alpha_1+\sinh\alpha_1\cdot\cos\alpha_1)(1+\cosh k\alpha_1\cdot\cos k\alpha_1)+2\gamma^{\frac{3}{4}}\beta^{\frac{1}{4}}\sinh\alpha_1\cdot\sin\alpha_1(\sinh k\alpha_1\cdot\cos k\alpha_1-\cosh k\alpha_1\cdot\sin k\alpha_1)\\ +2\gamma^{\frac{1}{4}}\beta^{\frac{3}{4}}\cosh\alpha_1\cdot\cos\alpha_1(\sinh k\alpha_1\cdot\cos k\alpha_1+\cosh k\alpha_1\cdot\sin k\alpha_1)+\gamma\beta(\cosh\alpha_1\cdot\sin\alpha_1+\sinh\alpha_1\cdot\cos\alpha_1)\\ \times(\cosh k\alpha_1\cdot\cos k\alpha_1-1)-2\gamma^{\frac{1}{2}}\beta^{\frac{1}{2}}\sinh k\alpha_1\cdot\sin k\alpha_1(\cosh\alpha_1\cdot\sin\alpha_1-\sinh\alpha_1\cdot\cos\alpha_1)=0,\end{aligned} \tag{11}$$

where

$$\begin{aligned}&\alpha_1\equiv(\omega'^2\rho_1A_1/K_1)^{\frac{1}{4}}\cdot l_1,\quad \alpha_2=k\alpha_1,\quad k=\beta^{\frac{1}{4}}\cdot\gamma^{-\frac{1}{4}}\cdot\lambda,\quad \lambda=l_2/l_1,\\ &\beta=\rho_2A_2/\rho_1A_1,\quad \gamma=K_2/K_1,\quad \rho_1A_1=2b(t_p\rho_p+t_r\rho_r),\quad \rho_2A_2=2bt_r\rho_r,\\ &K_1=\tfrac{2}{3}b\{E_pt_p(t_p^2+3t_pt_r+3t_r^2)+E_rt_r^3\},\quad K_2=\tfrac{2}{3}bE_rt_r^3,\end{aligned} \tag{12}$$

the subscripts 1, 2, p, and r indicate the Part I, the Part II, the electrostrictive material, and the basis-reed metal, respectively. A is the sectional area and the symbols l, t, and b are indicated in Fig. 4. The calculation values and the experimental results are shown in Fig. 6 for the first apparent resonant frequency and in Fig. 7 for the third one.

The electrostrictive material used here is a ceramic containing lead zirconate and lead titanate ($\rho_p=7.5$

[4] M. Konno and H. Nakamura, "Equivalent Electrical Network for the Transversely Vibrating Uniform Bar," J. Acoust. Soc. Am. 38, No. 4, 614–622 (Oct. 1965).

[5] C. Kusakabe, M. Konno, and Y. Tomikawa, "Resonant Frequencies of Transversely Vibrating Bar Excited by Electrostrictive Transducer," J. IECE Japan 48, No. 11, 1938–1943 (Nov. 1965).

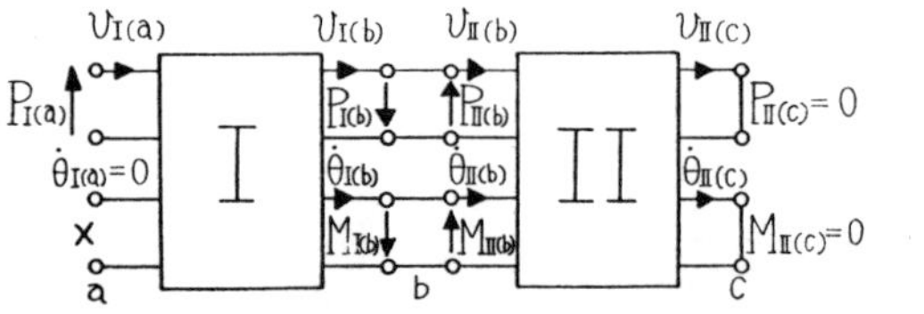

FIG. 5. Equivalent network of "sliding-free-ends beam" shown in Fig. 4(b).

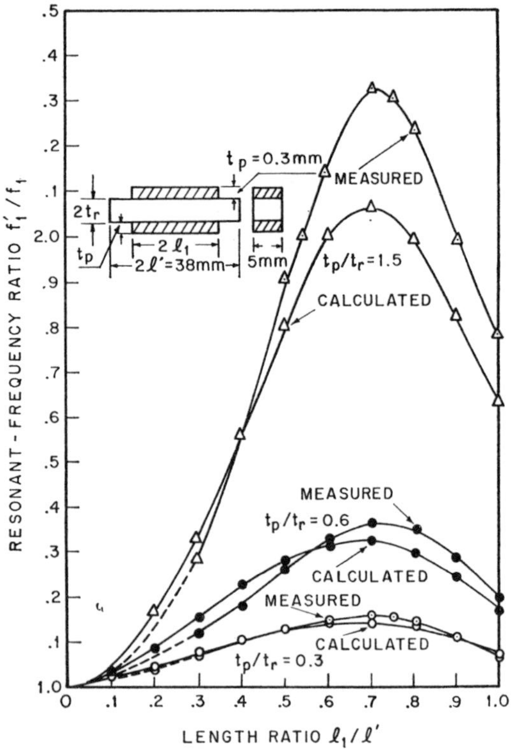

FIG. 6. Frequency ratios as a function of the length ratio ($f_1=1.476$ kcps).

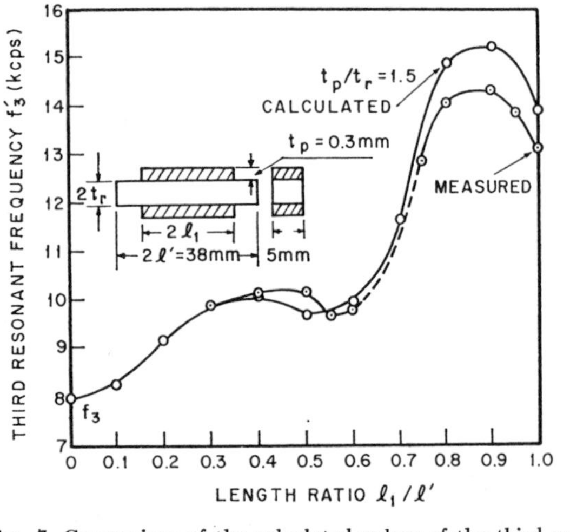

FIG. 7. Comparison of the calculated values of the third resonant frequency with the measured ones.

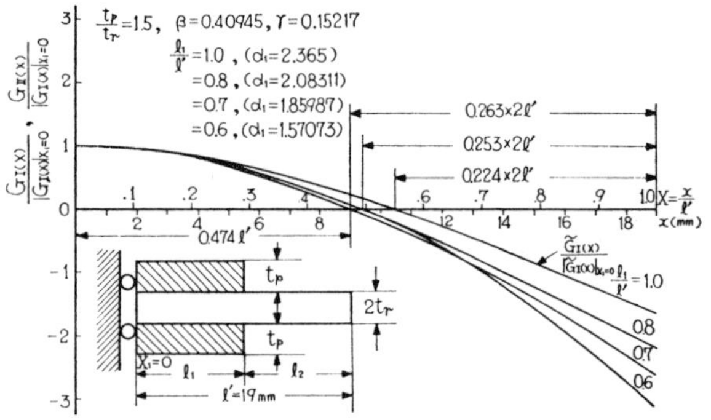

FIG. 8. Calculated values of vibration mode in the first resonant frequency as a function of the length ratio.

g/cm³, $E_p=0.8\times10^{12}$ dyn/cm²); the material of the basis reed is steel alloy ($\rho_r=7.8$ g/cm³, $E_r=2.1\times10^{12}$ dyn/cm²).

In conclusion, then, it appears that—

- When the length $2l'$ and the length ratio l_1/l' are kept constant, the frequency deviation from the first natural frequency f_1 of the basis reed increases correspondingly with the thickness ratio t_p/t_r.
- When the length $2l'$ and the thickness ratio t_p/t_r are kept constant, respectively, the first apparent resonant frequency f_1' has the maximum value at the length ratio $l_1/l' \doteqdot 0.7$.

2. *Vibration Mode of the Electrostrictive Transducer–Resonator*

From the equivalent network shown in Fig. 5, the velocity $v_{I(X)}$ and $v_{II(X)}$ at any points of the transducer–resonator in Fig. 4(b) are given as follows[6]:

At Part I:

$$v_{I(X)}=y_{Xa'}P_{I(a)}+y_{Xb'}P_{I(b)}+\bar{Y}_{Xb'}M_{I(b)}. \tag{13}$$

At Part II:

$$v_{II(X)}=y_{Xb}P_{II(b)}+\bar{Y}_{Xb}M_{II(b)}. \tag{14}$$

Accordingly, from Eqs. 8, 9, and 13, $v_{I(X)}$ becomes

$$v_{I(X)}=\left\{y_{Xa'}+y_{Xb'}\frac{\{\mathbf{Y}_{ba'}(\bar{\mathbf{Y}}_{aa}+\bar{\mathbf{Y}}_{bb'})-y_{ba'}(Y_{aa}+Y_{bb'})\}}{\Delta}+\bar{\mathbf{Y}}_{Xb'}\frac{\{y_{ba'}(\mathbf{Y}_{aa}+\mathbf{Y}_{bb'})-\mathbf{Y}_{ba'}(y_{aa}+y_{bb'})\}}{\Delta}\right\}P_{I(a)}, \tag{15}$$

$$\Delta=(y_{aa}+y_{bb'})(Y_{aa}+Y_{bb'})-(\bar{\mathbf{Y}}_{aa}+\bar{\mathbf{Y}}_{bb'})(\mathbf{Y}_{aa}+\mathbf{Y}_{bb'}).$$

If we set $\eta_{I(X)}=v_{I(X)}/j\omega$, substitute each element value of short-circuit admittance matrices, and put the resonant condition $\Delta=0$ into the numerator of Eq. 15,

[6] See Appendices B and C.

the resonant mode in Part I is expressed by the numerator $G_{\mathrm{I}(X)}$ of $\eta_{\mathrm{I}(X)}$; that is,

$$G_{\mathrm{I}(x)}=\mathbf{B}\cos\alpha_1X_1+\mathbf{C}\cosh\alpha_1X_1, \tag{16}$$

where

$$\begin{aligned}\mathbf{B}=&(C_1s_1+S_1c_1)(1-C_2c_2)\gamma^{\frac{1}{4}}\beta^{\frac{3}{4}}+(C_1s_1-S_1c_1-2C_1S_1)S_2s_2\gamma^{-\frac{1}{4}}\beta^{\frac{1}{4}}\\&-C_1(c_1+C_1)(S_2c_2+C_2s_2)\gamma^{-\frac{1}{2}}\beta^{\frac{1}{2}}-S_1(s_1-S_1)(S_2c_2-C_2s_2),\\ \mathbf{C}=&(C_1s_1+S_1c_1)(1-C_2c_2)\gamma^{\frac{1}{4}}\beta^{\frac{3}{4}}+(C_1s_1-S_1c_1+2c_1s_1)S_2s_2\gamma^{-\frac{1}{4}}\beta^{\frac{1}{4}}\\&-c_1(c_1+C_1)(S_2c_2+C_2s_2)\gamma^{-\frac{1}{2}}\beta^{\frac{1}{2}}+s_1(s_1-S_1)(S_2c_2-C_2s_2),\end{aligned} \tag{17}$$

$c_1=\cos\alpha_1$, $c_2=\cos\alpha_2$, $s_1=\sin\alpha_1$, $s_2=\sin\alpha_2$, $C_1=\cosh\alpha_1$, $C_2=\cosh\alpha_2$, $S_1=\sinh\alpha_1$, $S_2=\sinh\alpha_2$, $X_1=x/l_1$.

The subscripts 1 and 2 denote Part I and Part II in Fig. 4(b).

Similarly, from Eqs. 8, 9, and 14, the resonant mode in Part II is given as

$$G_{\mathrm{II}(X)}=\frac{(S_1c_1+C_1s_1)}{2\gamma^{\frac{1}{2}}\beta^{\frac{1}{2}}}(\mathbf{B}'\cos k\alpha_1X_2+\mathbf{C}'\cosh k\alpha_1X_2+\mathbf{D}'\sin k\alpha_1X_2+\mathbf{E}'\sinh k\alpha_1X_2), \tag{18}$$

where $X_2=x/l_2$,

$$\begin{aligned}\mathbf{B}'=&(s_1+S_1)(S_2c_2-C_2s_2)\gamma^{\frac{1}{2}}\beta^{\frac{1}{2}}-(c_1+C_1)(S_2s_2-C_2c_2-1)\gamma^{-\frac{1}{4}}\beta^{\frac{1}{4}}\\&+(c_1-C_1)(S_2s_2-C_2c_2+1)\gamma^{\frac{1}{4}}\beta^{\frac{3}{4}}-(s_1-S_1)(S_2c_2-C_2s_2),\\ \mathbf{C}'=&(s_1+S_1)(S_2c_2-C_2s_2)\gamma^{\frac{1}{2}}\beta^{\frac{1}{2}}+(c_1+C_1)(S_2s_2+C_2c_2+1)\gamma^{-\frac{1}{4}}\beta^{\frac{1}{4}}\\&+(c_1-C_1)(S_2s_2+C_2c_2-1)\gamma^{\frac{1}{4}}\beta^{\frac{3}{4}}+(s_1-S_1)(S_2c_2-C_2s_2),\\ \mathbf{D}'=&(s_1+S_1)(S_2s_2+C_2c_2-1)\gamma^{\frac{1}{2}}\beta^{\frac{1}{2}}+(c_1+C_1)(S_2c_2+C_2s_2)\gamma^{-\frac{1}{4}}\beta^{\frac{1}{4}}\\&-(c_1-C_1)(S_2c_2+C_2s_2)\gamma^{\frac{1}{4}}\beta^{\frac{3}{4}}-(s_1-S_1)(S_2s_2+C_2c_2+1),\\ \mathbf{E}'=&(s_1+S_1)(S_2s_2-C_2c_2+1)\gamma^{\frac{1}{2}}\beta^{\frac{1}{2}}-(c_1+C_1)(S_2c_2+C_2s_2)\gamma^{-\frac{1}{4}}\beta^{\frac{1}{4}}\\&-(c_1-C_1)(S_2c_2+C_2s_2)\gamma^{\frac{1}{4}}\beta^{\frac{3}{4}}+(s_1-S_1)(S_2s_2-C_2c_2-1),\end{aligned} \tag{19}$$

In particular, if $l_2=0$ (e.g., the case in which the electrostrictive materials are bonded over on the whole surface of the basis reed), Eq. 16 is shown as follows:

$$\tilde{G}_{\mathrm{I}(X)}=(S_1s_1-C_1c_1-1)\cos\alpha_1X_1-(S_1s_1+C_1c_1+1)\cosh\alpha_1X_1. \tag{20}$$

The calculation values of Eqs. 16, 18, and 20 for the first mode are shown in Fig. 8; the measured values compared with the theoretical value of the nodal points are given in Table II.

It is obvious that the nodal point in the first mode moves inwardmost at $l_1/l'\doteqdot 0.7$, at which the apparent resonant frequency has the maximum value. Figure 9 shows the calculated values for the second mode, with $t_p/t_r=1.5$, 0.6, and 0.3 in Fig. 4(b).

TABLE II. Values of ratio $l_n/2l'$ for $t_p/t_r=1.5$, 0.6, and 0.3. l_n is the length from the right end of the transducer-resonator with length $2l'$ to the nodal point.

l_1/l'	$t_p/t_r=1.5$ meas	$t_p/t_r=1.5$ calc	$t_p/t_r=0.6$ meas	$t_p/t_r=0.6$ calc	$t_p/t_r=0.3$ meas	$t_p/t_r=0.3$ calc
1.0	0.224	0.224	0.224	0.224	0.224	0.224
0.9	0.239		0.233		0.231	
0.8	0.250	0.253	0.242	0.242	0.235	0.234
0.7	0.263	0.263	0.246	0.245	0.241	0.241
0.6	0.258	0.253	0.243	0.244	0.236	0.238
0.5	0.250		0.240	0.242	0.235	0.237
0.4	0.240		0.237		0.234	

3. Force Factor and the Third-Harmonic Suppression

The displacement current generated in the electrodes of Fig. 4(a) becomes

$$I_d=\frac{j\omega 2b(t_p+2t_r)e}{l_1}\left|\frac{\partial G_{\mathrm{I}(X)}}{\partial X_1}\right|_0^1=\frac{j\omega 2b(t_p+2t_r)e\alpha_1(-\mathbf{B}\sin\alpha_1+\mathbf{C}\sinh\alpha_1)}{l_1}, \tag{21}$$

where e is the piezoelectric constant and α_1 is the root of Eq. 11.

Meanwhile, the velocity v_s at the central point OO of Fig. 4(a) is given as

$$v_s=j\omega|G_{\mathrm{I}(X)}|_{X_1=0}=j\omega(\mathbf{B}+\mathbf{C}). \tag{22}$$

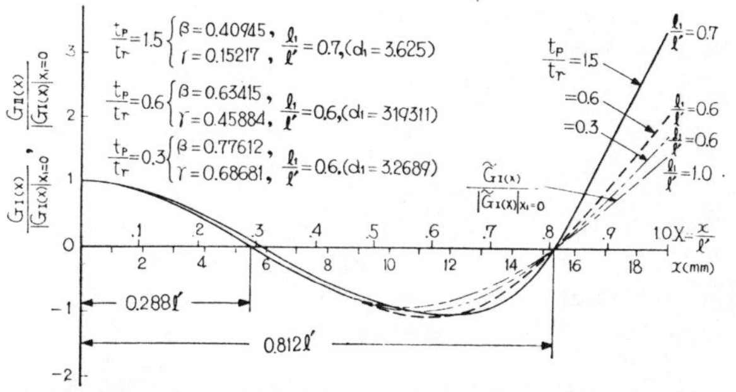

FIG. 9. Calculated values of vibration mode in the second resonant frequency as a function of the thickness ratio.

Then, the force factor **A** at the central position of the transducer–resonator is

$$\mathbf{A}=\frac{I_{\rm d}}{v_s}=\frac{2b(t_p+2t_r)e\alpha_1(-\mathbf{B}\sin\alpha_1+\mathbf{C}\sinh\alpha_1)}{l_1(\mathbf{B}+\mathbf{C})}. \quad (23)$$

Similarly, if $l_2=0$,

$$\tilde{\mathbf{A}}=\frac{b(t_p+2t_r)e\alpha_1\{(S_1s_1-C_1c_1-1)\sin\alpha_1+(S_1s_1+C_1c_1+1)\sinh\alpha_1\}}{l_1(1+C_1c_1)}. \quad (24)$$

Figure 10 shows the force-factor ratio $\mathbf{A}_3/\mathbf{A}_{30}$ in the third mode, where $\mathbf{A}_3$ is the force factor for $l'>l_1$ and $\mathbf{A}_{30}$ is for $l'=l_1$.

It is obvious that the third mode is to be eliminated in the case of the length ratio $l_1/l'\doteqdot 0.67$ for the thickness ratio $t_p/t_r=1.5$, $l_1/l'\doteqdot 0.6$ for $t_p/t_r=0.6$, and $l_1/l'\doteqdot 0.58$ for $t_p/t_r=0.3$, respectively. [And, if the electrostrictive material is much thinner than the basis reed, $l_1/l'=0.55$ for all t_p/t_r (Ref. 7).]

III. DESIGN OF MECHANICAL FILTER

The equivalent circuit of the electrostrictive mechanical filter is shown in Fig. 2 for nonadditional inductance and in Fig. 11 for the additional inductance $L_{\rm d}$ and externally added capacitance $C_{\rm d0'}$ at both ends, where $L_m'=m_m'/\mathbf{A}^2$, $1/C_m'=s_m'/\mathbf{A}^2$, $1/C_{\rm c}=s_{\rm sc}'/\mathbf{A}^2$. Accordingly, the design formulas for the bandpass filter of Fig. 11 are given as follows:

$$\omega_{\rm c2}^2=\left(1+2\frac{C_m'}{C_{\rm c}}\right)\Big/L_{\rm d}C_{\rm d},$$

$$\omega_{\rm c1}^2=1\Big/L_{\rm d}C_{\rm d}\left(1+\frac{2C_m'}{C_{\rm c}}\right), \quad (25)$$

$$\omega_0^2=(\omega_{\rm c1}\cdot\omega_{\rm c2})=1/(L_{\rm d}C_{\rm d}),$$

$$C_{\rm c}/C_m'=\frac{-\{\Delta(\omega_{\rm c1}+\omega_{\rm c2})-2\omega_0^2\}+\{\Delta^2(\omega_{\rm c1}+\omega_{\rm c2})^2+4\omega_0^4\}^{\frac{1}{2}}}{\Delta(\omega_{\rm c1}+\omega_{\rm c2})}, \quad (26)$$

$$\Delta=\omega_{\rm c2}-\omega_{\rm c1},$$

$$L_{\rm d}=2L_m'C_m'\Big/C_{\rm c}\left(2+\frac{C_{\rm c}}{C_m'}\right)(H),$$

$$C_{\rm d}=(C_{\rm d0}+C_{\rm d0}')=\frac{C_{\rm c}}{2}\left(1+\frac{C_{\rm c}}{C_m'}\right)(F), \quad (27)$$

$$\omega_m'^2=1/(L_m'C_m')=\omega_0^2\Big/\left(1+\frac{C_m'}{C_m'+C_{\rm c}}\right), \quad (28)$$

where $\omega_{\rm c1}$ and $\omega_{\rm c2}$ are the lower and upper cutoff angular frequencies, respectively, and ω_m' (i.e., ω' of Eq. 12) is the resonant angular frequency of the transducer–resonator.

The fractional bandwidth B and the terminating electrical resistance $R_{\rm E}$ are given as follows:

$$B=\Delta/\omega_0\doteqdot 4\frac{C_m'}{C_{\rm c}}[\omega_0/(\omega_{\rm c1}+\omega_{\rm c2})]. \quad (29)$$

$$R_{\rm E}=\left|\frac{\omega_0}{C_{\rm d}}[1/(\omega_0^2-\omega_{\rm c1}^2)(\omega_0^2-\omega_{\rm c2}^2)]^{\frac{1}{2}}\right|.\ (\Omega) \quad (30)$$

From Eqs: 28, 26, 12, and 5, the dimensions of the transducer–resonator and the coupler can be decided. On the other hand, C_m and L_m in Fig. 12 for the multiple sections are given as follows, respectively:

$$C_m=C_m'\left(1+\frac{C_m'}{C_{\rm c}}\right),\quad L_m=L_m',$$

$$\omega_m^2=1/(L_mC_m)=\omega_0^2\Big/\left(1+\frac{2C_m'}{C_{\rm c}}\right). \quad (31)$$

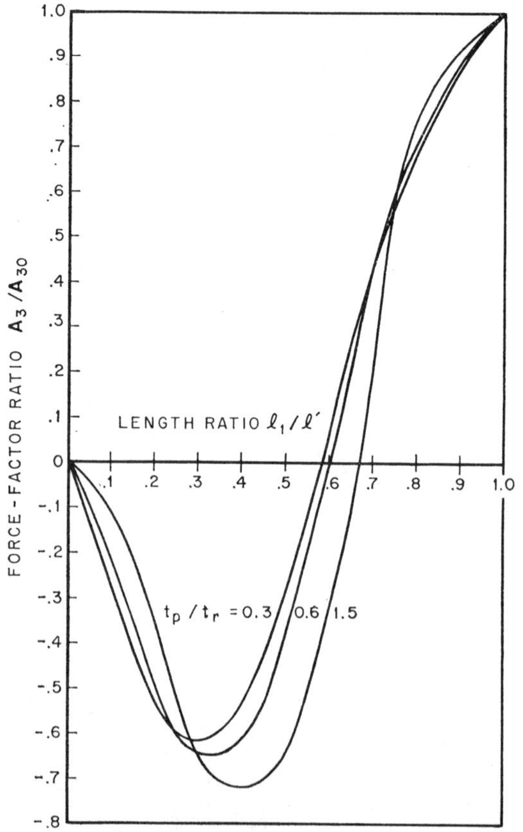

FIG. 10. Force-factor ratios in the third resonant frequency as a function of the length ratio.

[7] H. Nukiyama and T. Suzuki, "On the Mode of Vibration and Electroacoustic Constants of the Piezo Electric Vibrator (Magnoscope Type)," J. IECE Japan **231**, 367–381 (June 1942).

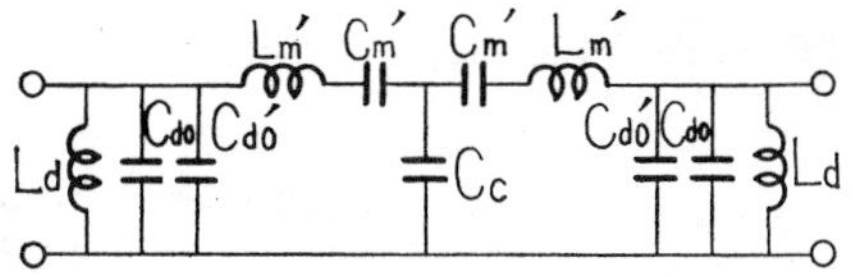

FIG. 11. Equivalent circuit of an electrostrictive mechanical filter.

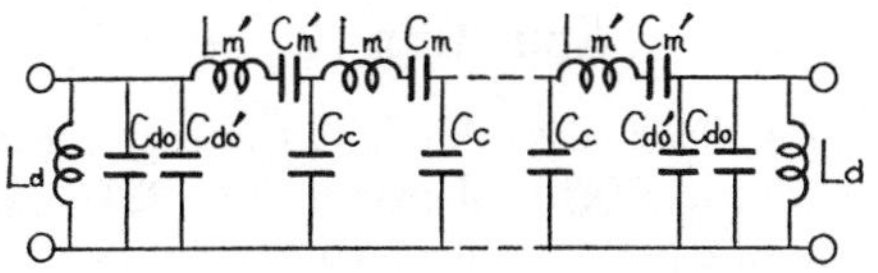

FIG. 12. Equivalent circuit of an electrostrictive mechanical filter (multiple-cascade connection).

IV. EXPERIMENTAL RESULTS ON THE MECHANICAL FILTER

Curve (a) of Fig. 13 is one of the experimental results where the specifications for the design are given as $f_0=6.3$ kcps, $f_{c1}=6.2$ kcps, $f_{c2}=6.4$ kcps, $\Delta_f=f_{c2}-f_{c1}=200$ cps, and where, moreover, the third mode is suppressed.

It is important in the trial production that the characteristics of two transducer–resonators coincide. In our experiment, the force factor **A** and the apparent resonant frequency f_1' were $\mathbf{A}=4.0028\times10^5$ (esu) and $f_1'=6.24$ kcps for the transducer at the input side, and $\mathbf{A}=4.1476\times10^5$ (esu) and $f_1'=6.26$ kcps for the output side, respectively.

The calculated values by employing the value of **A** at the input side were $L_d=7.4$ mH, $C_d=86\,330$ pF, and $R_E=9.23$ kΩ. The experimental results were, however, given as $L_d=9$ mH, $C_d=70\,540$ pF, and $R_E=15$ kΩ, respectively.

By adding a condenser C_a between Terminals 1 and 2 as shown in Fig. 13, two attenuation peaks can be obtained. Curve (b) of Fig. 13 is one of the experimental results. Curve (c) of Fig. 13 shows the experimental result by the simple method of design, in which case the L_d and C_d differ much from the above theoretical values, respectively, but only the condition of center angular frequency $\omega_0^2=1/(L_dC_d)$ is kept. Compared with Curve (a) of Fig. 13, the characteristic curve of (c) is inferior to that of (a), but the calculation for the design is simpler.

V. CONCLUSIONS

The design procedure and the experimental results of the electrostrictive mechanical filter of the torsional coupler type have been illustrated.

The apparent resonant frequency and the vibration mode of the transducer–resonator—of which the impedance of the electrostrictive material are not negligible in comparison with the one of the basis reed—are described, and then the method of the higher harmonic suppression is made clear.

The analysis of the mechanical vibrating system was accomplished with the mechanical network theory, which is very advantageous to the analysis of the composite system combined with the unit elements.

(a) $L_d=9$ mH. $C'_{do}=67\,800$ pF. $R_E=15$ kΩ.

(b) $C_a=180$ pF. $L_d=9$ mH. $C'_{do}=67\,800$ pF. $R_E=15$ kΩ.

(c) $L_d=30$ mH. $C'_{do}=18\,500$ pF. $R_E=10$ kΩ.

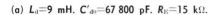

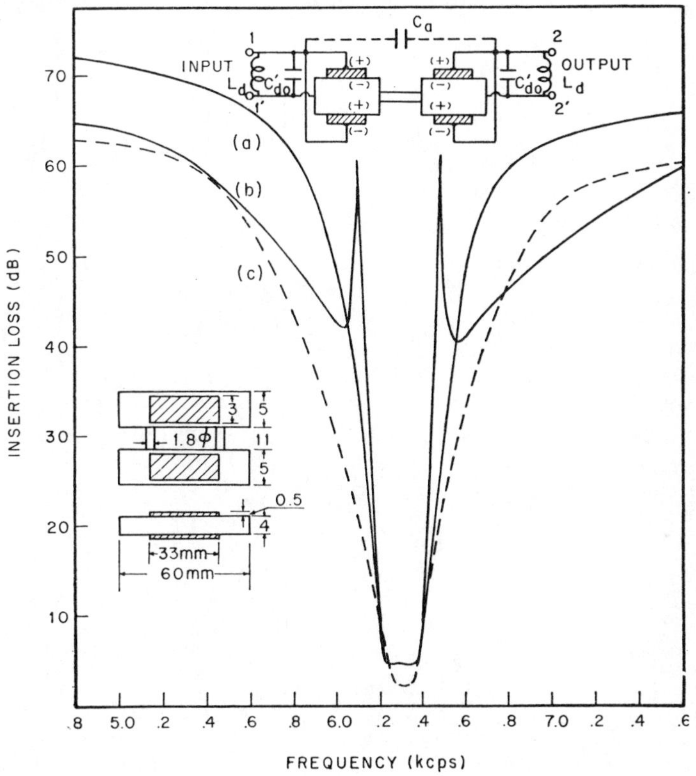

FIG. 13. Insertion-loss curves of an electrostrictive mechanical filter.

Appendix A. Velocity v_X in the Transversely Vibrating Beam

The equivalent network of the transversely vibrating beam in Fig. A-1(a) is shown in Fig. A-1(b), and the fundamental equation[A1] is given as follows:

$$\begin{bmatrix} \begin{Bmatrix} -P_X \\ -M_X \end{Bmatrix} \\ \\ \begin{Bmatrix} v_X \\ \dot{\theta}_X \end{Bmatrix} \end{bmatrix} = [\Gamma_{Xa}] \begin{bmatrix} \begin{Bmatrix} P_a \\ M_a \end{Bmatrix} \\ \\ \begin{Bmatrix} v_a \\ \dot{\theta}_a \end{Bmatrix} \end{bmatrix}, \tag{A1}$$

where

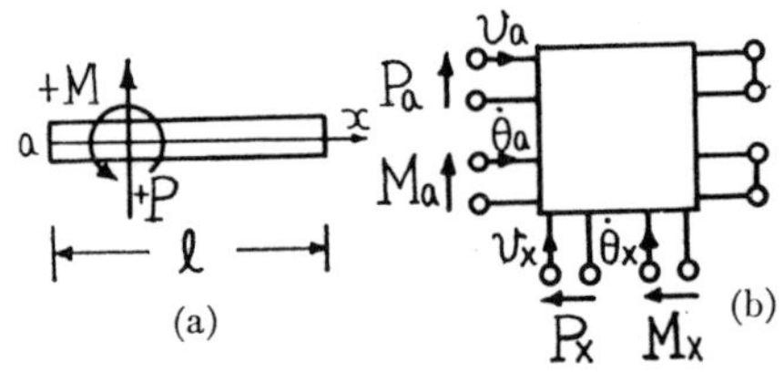

FIG. A-1. (a) Transversely vibrating beam. (b) Equivalent network.

$$[\Gamma_{Xa}] = \tfrac{1}{2} \begin{bmatrix} (c_X+C_X) & (s_X-S_X)\underline{\alpha}' & (s_X+S_X)\underline{\alpha}^3 & -(c_X-C_X)\underline{\alpha}^2 \\ -(s_X+S_X)/\underline{\alpha}' & (c_X+C_X) & (c_X-C_X)\underline{\alpha}^2 & (s_X-S_X)\underline{\alpha} \\ -(s_X-S_X)/\underline{\alpha}^3 & (c_X-C_X)/\underline{\alpha}^2 & (c_X+C_X) & (s_X+S_X)/\underline{\alpha}' \\ -(c_X-C_X)/\underline{\alpha}^2 & -(s_X+S_X)/\underline{\alpha} & -(s_X-S_X)\underline{\alpha}' & (c_X+C_X) \end{bmatrix}, \tag{A2}$$

where

$$c_X \equiv \cos\alpha X, \quad s_X \equiv \sin\alpha X, \quad C_X \equiv \cosh\alpha X, \quad S_X \equiv \sinh\alpha X, \quad X = x/l,$$

$$\underline{\alpha} = \frac{K}{j\omega}\frac{\alpha}{l}, \quad \underline{\alpha}^2 = \frac{K}{j\omega}\frac{\alpha^2}{l^2}, \quad \underline{\alpha}^3 = \frac{K}{j\omega}\frac{\alpha^3}{l^3}, \quad \underline{\alpha}' = \frac{\alpha}{l}, \quad \underline{\alpha}^4 = \frac{\rho A}{K}\omega^2 l^4, \tag{A3}$$

$K(=EI)$ is the flexural rigidity, and I is the second moment of area of the cross section about the neutral axis. Then,

$$v_X = \frac{1}{2}\left\{-(s_X-S_X)\frac{P_a}{\underline{\alpha}^3} + (c_X-C_X)\frac{M_a}{\underline{\alpha}^2} + (c_X+C_X)v_a + (s_X+S_X)\frac{\dot{\theta}_a}{\underline{\alpha}'}\right\}. \tag{A4}$$

[A1] Ref. 4, p. 620

Appendix B. v_X in the "Sliding–Free-Ends Beam"

The equivalent network of Fig. B-1(a) is shown in Fig. B-1(b), and the following equations are given:

$$\begin{bmatrix} v_a \\ v_b \\ \dot{\theta}_b \end{bmatrix} = \begin{bmatrix} y_{aa'} & y_{ab'} & \bar{\mathbf{Y}}_{ab'} \\ y_{ba'} & y_{bb'} & \bar{\mathbf{Y}}_{bb'} \\ \mathbf{Y}_{ba'} & \mathbf{Y}_{bb'} & Y_{bb'} \end{bmatrix} \begin{bmatrix} P_a \\ P_b \\ M_b \end{bmatrix}, \tag{B1}$$

$$M_a = -\frac{1}{Y_{aa}}(\mathbf{Y}_{aa}P_a + \mathbf{Y}_{ab}P_b + Y_{ab}M_b),$$

where

$$y_{aa'} = \frac{-(1+Cc)j\omega l^3}{(Sc+Cs)K\alpha^3}, \qquad y_{bb'} = \frac{(-2Cc)j\omega l^3}{(Sc+Cs)K\alpha^3},$$

$$y_{ab'} = y_{ba'} = \frac{-(c+C)j\omega l^3}{(Sc+Cs)K\alpha^3}, \qquad \bar{\mathbf{Y}}_{bb'} = \mathbf{Y}_{bb'} = \frac{(Cs-Sc)j\omega l^2}{(Sc+Cs)K\alpha^2},$$

$$\bar{\mathbf{Y}}_{ab'} = \mathbf{Y}_{ba'} = \frac{(s-S)j\omega l^2}{(Sc+Cs)K\alpha^2}, \qquad Y_{bb'} = \frac{(2Ss)j\omega l}{(Sc+Cs)K\alpha}, \tag{B2}$$

$$\mathbf{Y}_{aa} = \frac{(Ss)j\omega l^2}{(1-Cc)K\alpha^2}, \qquad \mathbf{Y}_{ab} = \frac{(c-C)j\omega l^2}{(1-Cc)K\alpha^2},$$

$$Y_{ab} = \frac{-(s+S)j\omega l}{(1-Cc)K\alpha}, \qquad Y_{aa} = \frac{-(Sc+Cs)j\omega l}{(1-Cc)K\alpha},$$

$$c \equiv \cos\alpha, \quad s \equiv \sin\alpha, \quad C \equiv \cosh\alpha, \quad S \equiv \sinh\alpha.$$

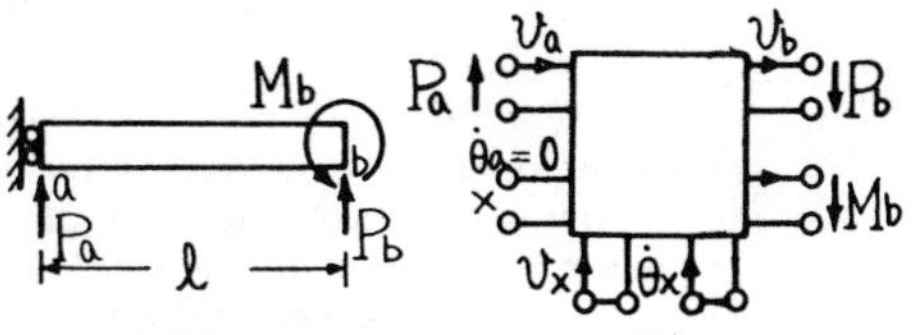

FIG. B-1. (a) "Sliding-free-ends beam" vibrating transversely. (b) Equivalent network.

Putting Eqs. B1 and B2 into Eq. A4,

$$v_X = y_{Xa}{}'P_a + y_{Xb}{}'P_b + \overline{\mathbf{Y}}_{Xb}{}'M_b, \tag{B3}$$

where

$$\begin{aligned}
y_{Xa'} &= \frac{j\omega l^3}{2(Sc+Cs)K\alpha^3}\{-(Sc+Cs)(s_X - S_X) + (Ss - Cc - 1)c_X - (Ss + Cc + 1)C_X\},\\
y_{Xb'} &= \frac{-j\omega l^3}{(Sc+Cs)K\alpha^3}(C\cdot c_X + c\cdot C_X),\\
\overline{\mathbf{Y}}_{Xb'} &= \frac{j\omega l^2}{(Sc+Cs)K\alpha^2}(s\cdot C_X - S\cdot c_X).
\end{aligned} \tag{B4}$$

Appendix C. v_X in the "Both-Ends-Free Beam"

The equivalent network of Fig. C-1(a) is shown in Fig. C-1(b), and the following equation[C1] is given:

$$\begin{bmatrix} v_b \\ \dot{\theta}_b \\ v_c \\ \dot{\theta}_c \end{bmatrix} = \begin{bmatrix} y_{bb} & \overline{\mathbf{Y}}_{bb} & y_{bc} & \overline{\mathbf{Y}}_{bc} \\ \mathbf{Y}_{bb} & Y_{bb} & \mathbf{Y}_{bc} & Y_{bc} \\ y_{cb} & \overline{\mathbf{Y}}_{cb} & y_{cc} & \overline{\mathbf{Y}}_{cc} \\ \mathbf{Y}_{cb} & Y_{cb} & \mathbf{Y}_{cc} & Y_{cc} \end{bmatrix} \begin{bmatrix} P_b \\ M_b \\ P_c \\ M_c \end{bmatrix}. \tag{C1}$$

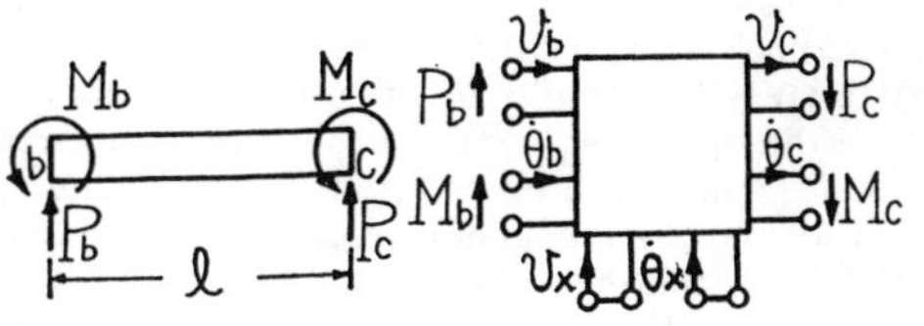

FIG. C-1. (a) "Both-ends-free beam" vibrating transversely. (b) Equivalent network.

Putting Eq. C1 into Eq. A4,

$$v_X = y_{Xb}P_b + \overline{\mathbf{Y}}_{Xb}M_b + y_{Xc}P_c + \overline{\mathbf{Y}}_{Xc}M_c, \tag{C2}$$

where

$$\begin{aligned}
y_{Xb} &= \frac{j\omega l^3}{2(1-Cc)K\alpha^3}\{(Sc - Cs)(c_X + C_X) + (Ss + Cc - 1)s_X + (Ss - Cc + 1)S_X\},\\
\overline{\mathbf{Y}}_{Xb} &= \frac{j\omega l^2}{2(1-Cc)K\alpha^2}\{(Ss - Cc + 1)c_X + (Ss + Cc - 1)C_X - (Sc + Cs)(s_X + S_X)\},\\
y_{Xc} &= \frac{-j\omega l^3}{2(1-Cc)K\alpha^3}\{(s - S)(c_X + C_X) - (c - C)(s_X + S_X)\},\\
\overline{\mathbf{Y}}_{Xc} &= \frac{-j\omega l^2}{2(1-Cc)K\alpha^2}\{(c - C)(c_X + C_X) + (s + S)(s_X + S_X)\}.
\end{aligned} \tag{C3}$$

[C1] Ref. 4, p. 616.

CHARACTERISTICS OF LOW FREQUENCY MECHANICAL FILTERS

D. P. Havens and P. Ysais
Collins Radio Group
Division of Rockwell International
4311 Jamboree Road
Newport Beach, California 92663

ABSTRACT. The frequency characteristics and stability of low frequency mechanical filters using ceramic transducers bonded to a nickel-iron alloy bar have been investigated. The filters are designed from a 3.5 to a 70 kHz center frequency with bandwidths of 0.2 to 1.5 percent of that frequency. The filters were subjected to shock, vibration, and accelerated aging and then tested for changes in their frequency response. Their temperature and drive level dependence was also measured.

Introduction

In recent years low frequency flexure mode mechanical filters have developed from an experimental concept[1,2,3] into a production item.[4,5,6,7] They provide a durable low cost device with a narrow-band response in the 3.5 to 70 kHz frequency region.[7,8] The increase in both usage and interest in this filter has motivated us to provide more information on its characteristics. The filters that will be described are those produced by Rockwell International; but the infomation is, for the most part, applicable to other mechanical filters of similar construction.

The device consists of two metal alloy bars bonded to piezoelectric ceramic transducers and coupled mechanically with wires which also act as the supporting structure. See Fig. 1. The bars are made out of a constant modulus nickel-iron alloy whose temperature coefficient is adjusted by heat treatment to help compensate for the high positive temperature coefficient of the ceramic. The ceramic is a binary solid solution of $PbTiO_3$ - $PbZrO_3$ with the additive $Pb(Co_{1/3}\text{-}Nb_{2/3})O_3$ (U.S. Patent No. 3544469). The composite resonators operate in the flexure mode with wires coupling the bars torsionally. Fig. 2 shows the equivalent circuit of the filter.[9,10]

The fact that the wires act both as the supporting structure as well as the coupling mechanism creates a limit on how narrow a bandwidth can be achieved. That limit is approximately two-tenths of one percent. The complete limits of the operating range of the filter, in terms of bandwidth and frequency, are shown in Fig. 3. The upper and lower limits of the frequency range are determined by the size of the resonators. Below 3.5 kHz they become too large to work with and above 70 kHz they become too small. These size limitations are set more by present tooling and packaging than by any other consideration. The bar lengths presently being used are between 10 mm and 40 mm. The upper limit on the bandwidth is determined by the electromechanical coupling coefficient of the resonators. It is possible to increase this limit if inductors are used to tune out some or all of the static capacitance of the ceramic; however, this adds considerably to the size and cost of the filter.

These limits for the bandwidth-frequency range of low frequency mechanical filters are only approximate. When deciding if a design is possible, other factors must be taken into consideration. For example, if the shock and vibration requirements are not severe, it is possible to build narrower filters. However, the narrower the bandwidth, the greater is the problem of aging and temperature stability. What follows is a report on what changes would be expected in a typical low frequency mechanical filter when subjected to such environmental conditions as shock, vibration, temperature extremes and aging.

Basic Design Considerations

Because the stability of the nickel-iron alloy bar is used to compensate for the relative instability of the ceramic transducer, it is best to design the filter with the least amount of ceramic. However, there are limits to how much the ceramic size can be reduced.

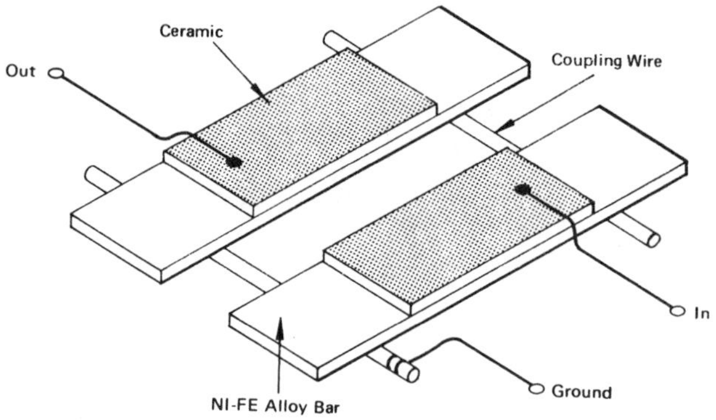

Figure 1 - Low Frequency Mechanical Filter

The basic narrowband design equation is:

$$\Delta f = B_m/q_1$$

where

Δf = Composite resonator pole-zero spacing
B_m = Maximum filter bandwidth
q_1 = Normalized end-section Q(see Ref. 11)

In order to minimize the amount of ceramic needed for a specific alloy bar size, i.e., minimize the ceramic-to-metal ratio, set B_m equal to the 3 dB bandwidth. The normalized Q is a function of the specified passband ripple and the number of resonators. For example, for a two resonator filter with 1.0 dB passband ripple, q_1 would equal 1.82. Therefore, for a 50 Hz bandwidth filter with 1.0 dB ripple the pole-zero spacing is 27.5 Hz. The filter can be designed with a greater pole-zero spacing, but at the expense of a larger (longer or thicker) ceramic transducer which, in turn, results in a less stable filter.

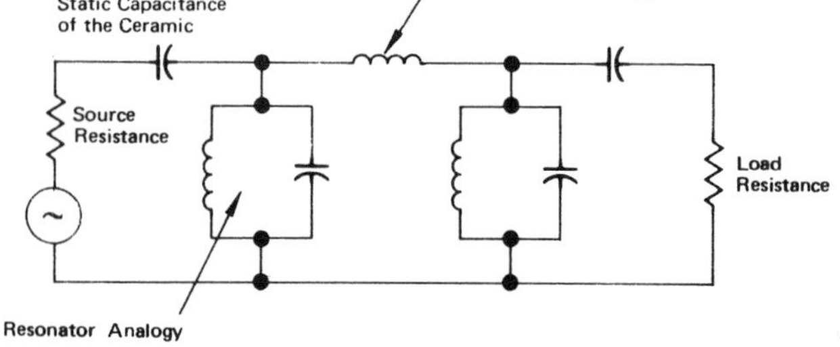

Figure 2 - Equivalent Circuit of a Low Frequency Mechanical Filter

"1974 Ultrasonics Symposium Proceedings,
IEEE Cat. #74 CHO 896-ISU"

Reprinted from *1974 Ultrasonics Symp. Proc.*, Nov. 11-14, 1974, pp. 599-602.

Aging

One source of information on the reliability of a device is from examining the actual usage. Fifty-five percent of the filters that were returned, came back because the 3 dB points moved and fourteen percent because the ripple increased. Both of these problems are due to aging.

The low frequency mechanical filter has four components and processes which are likely to contribute to aging problems: the nickel-iron alloy bar, the ceramic transducer, the solder bond between the bar and transducer and the welds that connect the coupling wires to the bars. When heat treated properly, the nickel-iron alloy bar is very stable. Because its mass is much greater than any of the other components, the stability of the bar helps to compensate for the instability of the ceramic and solder bond.

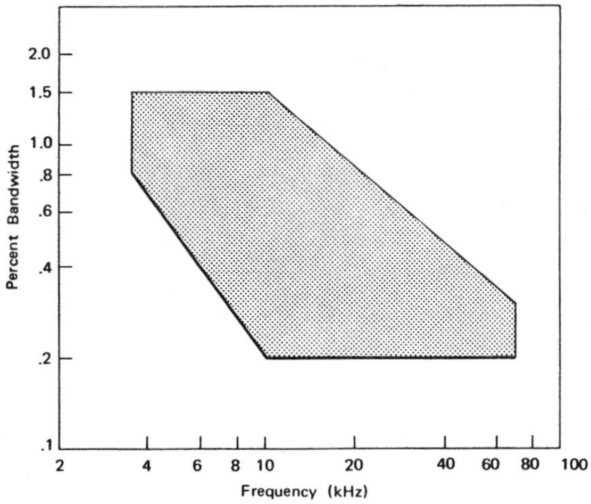

Figure 3 - Practical Design Limits of Low Frequency Mechanical Filters

Because the aging characteristics of low frequency mechanical filters are dependent on the ratio of the ceramic transducer mass to the nickel-iron alloy bar mass, the rate of aging for various types of filters is not only dependent on the center frequency, but perhaps, to a greater extent, on the bandwidth of the filter. As shown earlier, the narrower the bandwidth, the smaller the ceramic transducer need be. This means that the narrower bandwidth filters will age, as a percentage of their center frequency, less than wider bandwidth filters at the same frequency.

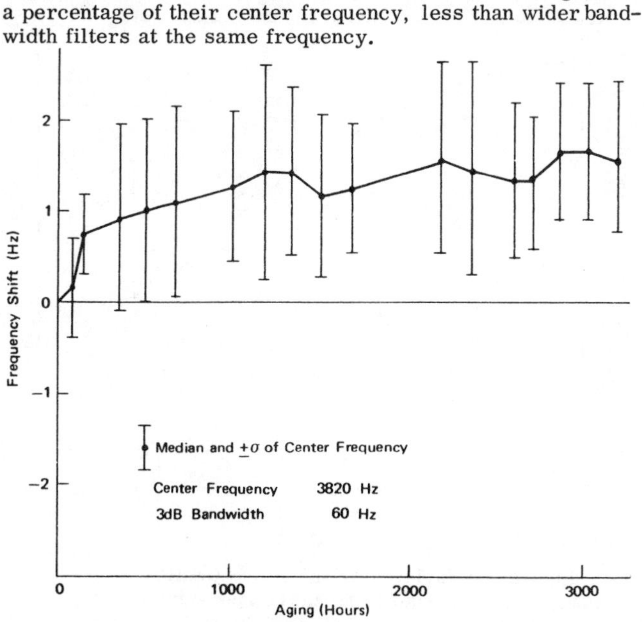

Figure 4 - Aging of Center Frequency

In order to obtain a quantitative idea of the relationship between aging and the ceramic length, twelve resonators were built. Six of the resonators used an 11.4 mm (.450 inch) long ceramic and six used a 3.8 mm (.150 inch) ceramic. The nickel-iron alloy bars were all approximately the same dimension. Because of the difference in ceramic length, the resonators with the longer ceramic had a slightly higher frequency, 10.3 kHz, than the 9.5 kHz resonators with the smaller ceramic. The resonators were aged for over 4000 hours at a temperature of 85° C. The elevated temperature was used to try and accelerate aging. The results showed that the average absolute shift in frequency for the six resonators with 11.4 mm ceramic was 2.85 Hz or 280 ppm. The shift for the resonator with the short ceramic was 1.36 Hz or 140 ppm. The maximum 3 dB bandwidth obtainable, assuming a 0.3 dB ripple, for a filter using the resonators having 11.4 mm ceramic would be 200 Hz; for the resonators with 3.8 mm ceramic, it would be 45 Hz.

Along with the aging of the twelve resonators, twelve filters were also subjected to controlled aging. The average shift in the center frequency is shown in Fig. 4. From the graph it can be seen that the filters aged an average of about +1.5 Hz (390 ppm). The 3 dB bandwidth changed by 0.5 Hz. The length of ceramic used on the filters was 11.4 mm. The reason the filters changed percentage-wise more than the resonators can be related to the fact that the filters were at a lower frequency and required a much thinner alloy bar; i.e., the ceramic to alloy bar mass ratio was greater.

By looking at the aging curves it is seen that most of the aging occurs within the first 300 hours. These filters (as well as the resonators) had been subjected to a standard pre-aging processes. They were temperature cycled from -55° C to 95° C for 23 hours; the temperature being switched every two hours. The fact that most of the aging does occur within the first 300 hours indicated that more pre-aging might improve stability. However, there are questions about the effects that rapid temperature changes have on ceramic materials. Our temperature cycling procedure, while relieving stresses in the solder bond between the ceramic and nickel-iron bar, may be inducing time dependent changes in the ceramic transducers.

Temperature Effects

There are two types of effect seen when low frequency mechanical filters are subjected to temperature changes. The first type is the normal non-time dependent changes in frequency response characteristics due to the temperature coefficients of the components in the filters. This type of temperature dependence occurs when the rate of change in temperature is slow. The second type of effect is that which is dependent on the rate of change in temperature.

The temperature characteristics of the nickel-iron alloy bar is shown for various heat treatments in Fig. 5. By using a heat treatment that results in a negative temperature coefficient, it is possible to compensate for the large positive temperature coefficient of the ceramic. Different heat treatments are used depending on the amount of ceramic used. A long ceramic would need a bar heat treated near 900° F, where, a short ceramic would need a bar heat treated at 1050° F. The temperature behavior of the ceramic is shown in Fig. 6 and that of the filter in Fig. 7.

It will be noted that the ceramic does not retrace the same path when the temperature is lowered as it did when the temperature was raised. This effect is primarily due to varying the temperature too rapidly. Rapid temperature variation causes changes in the structure of the ceramic transducers which give rise to a softening effect and ultimately a lower frequency. More information on this effect and the use of additives to the ceramic to help correct it, can be found elsewhere and will not be covered in detail here.[12, 13]

The lowering of the resonant frequency with rapid temperature variation is not permanent. The filters tend to drift back to their original values with time. The rate of temperature change needed to produce any appreciable effect must be greater than 5°C per minute. When the filters are packaged the resonators are thermally isolated enough that the effect is rarely seen even when the external temperature is changed suddenly.

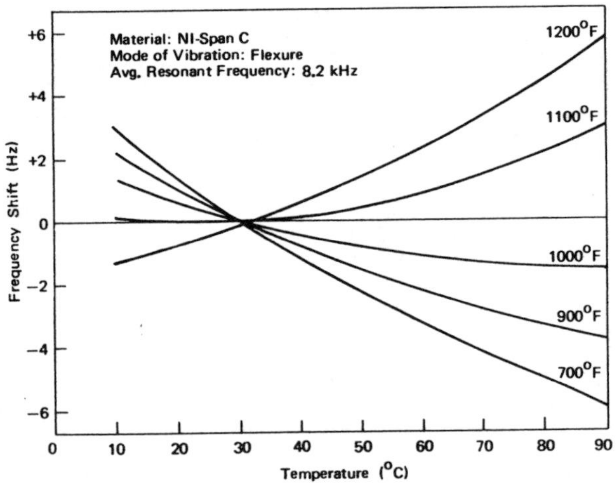

Figure 5 - Temperature Variation of Nickel-Iron Alloy Bars for Various Heat Treatments

Drive Level

Low frequency mechanical filters remain very linear with drive level up to 0.5 v, at which point they rapidly become non-linear. See Fig. 8. The shift in the center frequency at a ten volt level is enough to move most filters out of specification. Also, along with the non-linear effects, excessive drive level causes time dependent changes in the ceramic material, as did the rapid temperature variation. These changes cause the center frequency of the filter to be lower after being driven at a high voltage level. For a 32 v drive level the shift was 200 ppm. Once the drive level is reduced the filter response begins to return to its original frequency. After three days the center frequency had returned to one-half of its original value.

Shock and Vibration

In many applications, electrical components are subjected to severe shock and vibration. Low frequency mechanical filters have proven to be very resilient.

In order to test the shock and vibration reliability of the low frequency mechanical filters, Military Specification 202, methods 201 and 213C were used. Under these methods the filters were subjected to a six millisecond 100 G shock and two hours of vibration at 10 G's for each axis. The frequency of vibration was swept from 10 Hz to 55 Hz. The results varied greatly from one filter type to another. In general, the narrower the bandwidth or the lower in center frequency, the more the filter response would change. If a filter type requires the ability to withstand a severe environment, there are design options that can be considered. By increasing the spacing between resonators, it is possible to maintain the same bandwidth by increasing the diameter of the coupling wires. This increases the strength of the filter.

Most filter types survived the shock and vibration tests with only minimal changes in their response characteristics. Even the more delicate filters showed only small variations. As an example of the changes that occurred in one of the structurally weaker types, the center frequency decreased 0.8 Hz and the bandwidth narrowed by 1.0 Hz. This filter was designed for use in the central office of a telephone system. It was later redesigned for greater structural strength. A stronger design showed changes of less than a few tenths of a Hertz when subjected to the same shock and vibration levels. The point of destruction was reached for a vibration level between 12 and 18 G's, or a shock of 1200 G's for 0.5 millisecond.

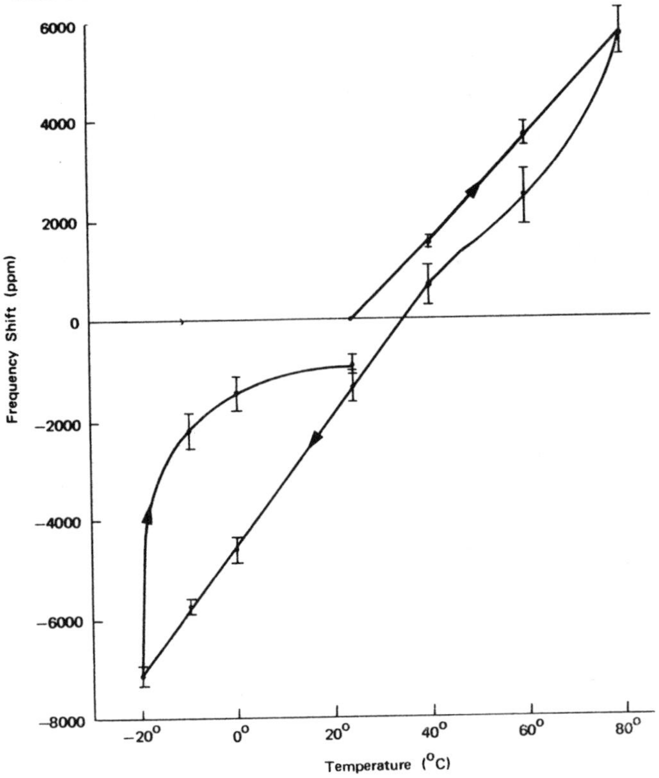

Figure 6 - Variation of Piezoelectric Ceramic Frequency with Temperature

Often it is not only important that a filter survive a specific vibration level, but it is also important that it retains the proper response characteristics during the vibration. While the attenuation response showed little change, there was some variation of phase and some microphonic effects during vibration. These are important characteristics in applications like Omega navigation systems.

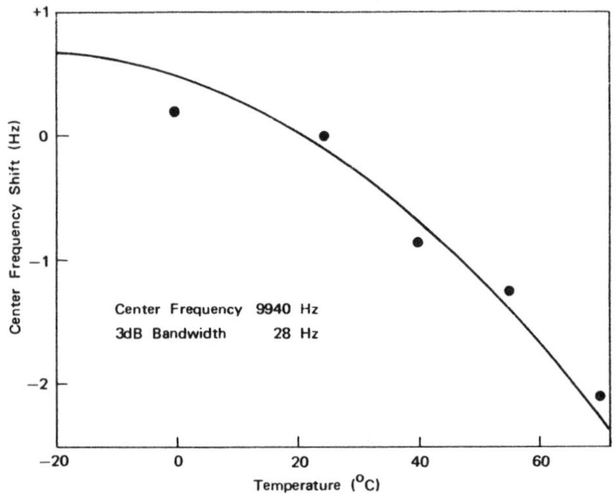

Figure 7 - Variation of the Filter Center Frequency with Temperature

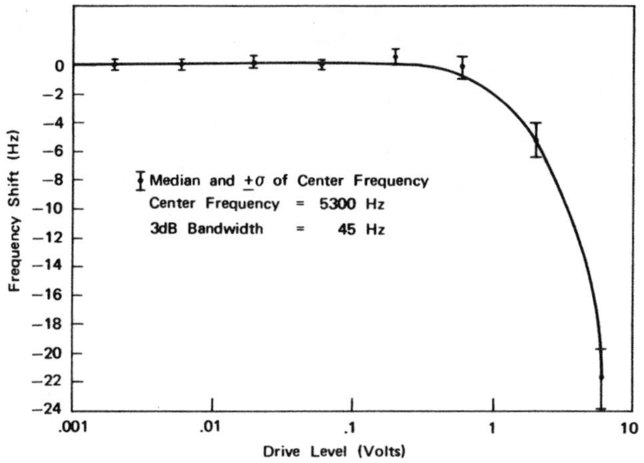

Figure 8 - Variation in Center Frequency with Drive Level

The variation of phase of an Omega filter as a function of vibration frequency is shown in Fig. 9. It will be noticed that there is a peak deviation point around 400 Hz. The reason for the peak is that the entire filter structure has a resonance at this point. This structural resonance also has an effect on the microphonic noise level. At a 2 G vibration level, the noise in the passband is below one millivolt for vibration frequencies below 200 Hz. From 200 Hz the noise level gradually increases to about 10 millivolts at 400 Hz and then declines back to 3 millivolts. In applications where these levels are intolerable, special structural design has been incorporated into the equipment in order to dampen the vibrations around 400 Hz.

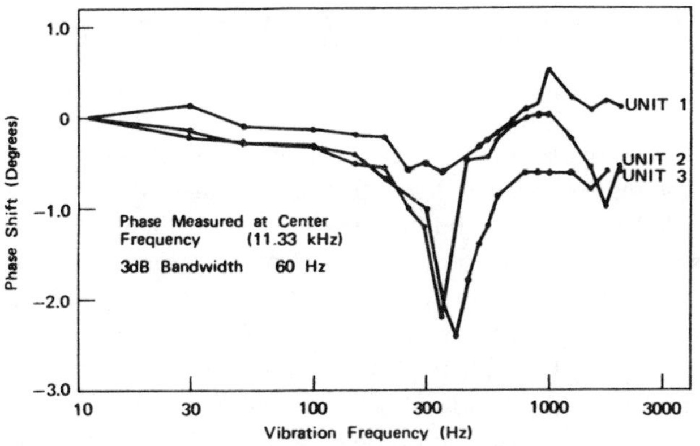

Figure 9 - Variation in Phase with External Vibration (10 G Level)

Conclusion

By bonding ceramic transducers onto nickel-iron alloy bars, stable and reliable filter characteristics have been achieved. The high stability of the alloy bars helps compensate for instabilities in the ceramic; but the ceramic, at the present time, remains the limiting factor. As newer and more stable ceramics are developed, low frequency mechanical filters will also improve.

Acknowledgement

The authors are especially indebted to Mr. R. A. Johnson for valuable discussions, and his assistance with the manuscript.

References

1. M. Börner, "Biegeschwingung in Mechanischen Filten," Telefunken J., Vol. 31, June 1958, pp. 115-123; also Sept. 1958, pp. 188-196.
2. M. Konno, "Theoretical Considerations of Mechanical Filters," J. Inst. Elec. Commun. Eng. Jap., Vol. 40, 1957, pp. 44-51.
3. M. Konno, C. Kusakabe, and Y. Tomikawa, "Electromechanical Filter Composed of Transversely Vibrating Resonators at Low Frequencies, " The Journal of the Acoustical Society of America, Vol. 41, No. 4 (Part 2), April 1967, pp. 953-961.
4. H. Albsmeier, A. E. Günther, and W. Volejnik, "Some Special Design Considerations for a Mechanical Filter Channel Bank," IEEE Trans. Circuits and Systems, CAS 21, July 1974.
5. R. A. Johnson and W. A. Winget, "FDM Equipment Using Mechanical Filters," Inter. Symposium on Circuits and Systems, April 1974, pp. 127-131.
6. K. Yakuwa, S. Okuda, Y. Kasai, and Y. Katsuba, "Reliability of Electromechanical Filters," FUJITSU, Vol. 24, No. 1, 1973, pp. 172-179.
7. T. Okamoto, K. Yakuwa, and S. Okuta, "Low Frequency Electromechanical Filters," Fujitsu Sci. Tech. J., Vol. 2, May 1966, pp. 53-86.
8. R. A. Johnson and W. D. Peterson, "Build Compact Narrowband Circuits with Low Frequency Mechanical Filters," Electronic Design 3, Feb. 1973, pp. 60-64.
9. M. Konno and H. Nakamura, "Equivalent Electrical Network for the Transversely Vibrating Uniform Bar," J. Acoust. Soc. Amer., Vol. 38, Oct, 1965, pp. 614-622.
10. Y. Kagawa and G. M. L. Gladwell, "Finite Element Analysis of Flexure Type Vibrators with Electrostrictive Transducers," IEEE Trans. Sonics Ultrason., Vol. SU-17, Jan. 1970, pp. 41-49.
11. "Reference Data for Radio Engineers, Fifth Edition," edited by H. P. Westman, Howard W. Sams & Co., Inc., Indianapolis, Sec. 8, pp. 4-15.
12. M. Takahashi, "Space Charge Effect in Lead Zirconate Titanate Ceramics Caused by the Addition of Impurities," Japanese Journal of Applied Physics, Vol. 9, No. 10, Oct. 1970, pp. 1236-1246.
13. M. Takahashi, F. Yamauchi, and S. Takahashi, "Stabilization of Resonance Frequencies in Piezoelectric Ceramic Resonators Against Sudden Temperature Change," Proceedings of the 28th Annual Symposium on Frequency Control, May 1974, pp. 109-116.

Section I-E
Use of Multiple-Mode Resonators

The use of a single mechanical filter element to realize two or more resonators is of interest from the standpoints of size, cost, and reliability, in much the same way as the planar device is important in the field of microelectronics and modern crystal filters. Although a number of papers have been written on this subject, as applied to mechanical filters, there has been little or no production of these devices. The reason for this has undoubtedly been the cost of fabrication as compared to conventional wire coupling. A filter which may possibly be included in this category, but which has been manufactured, is the three-prong tuning fork filter which has two natural resonances and requires no wire coupling. It is hoped that the reading of this section will stretch the imagination of the reader and result in technology fallout in some field of filter design.

The first paper demonstrates a method of obtaining four natural resonances and two attenuation poles with a three-disk design. Basic to both of these papers is the concept of causing a degenerate mode (two or more natural modes at the same frequency) to split by perturbing the shape of the element so as to obtain two natural resonances at specified frequencies. The splitting is accomplished by removing a segment from the middle disk. In addition, a bridging wire is used to obtain attenuation poles above and below the filter passband. The resultant twin-tee network is similar to that described by Yano in an earlier section.

The Konno–Tomikawa paper makes use of a flexural-mode chamfered bar to obtain two coupled natural resonances. The paper shows equations, equivalent circuits, and curves for use in the design of this type of filter.

A Twin Tee Multimode Mechanical Filter

Recent papers dealing with crystals and crystal filters have shown how it is possible to make use of more than one resonance associated with each crystal plate [1], [2]. In this letter, we will describe a mechanical filter of the disk-wire type [3], [4] which also uses more than one resonator natural frequency as well as a technique called "mode decoupling." A double resonance is produced by the removal of a segment of a disk resonator, the mode decoupling by special coupling wire orientation. The design is based on lattice-to-twin tee transformations. The lattice prototype is capable of realizing nonminimum phase characteristics; hence the resultant twin tee can be designed to have delay compensating behavior. The two filters that have been designed and constructed operate at the lower end of the 64–108 kHz voice channel spectrum but the theory developed is applicable with slight modifications at 455 kHz and 500 kHz as well.

Basically, this filter makes use of the fact that either the addition of mass or the removal of mass from the edge of a disk type resonator produces an additional natural mode of vibration for each existing diameter mode [4], [5]. The disk resonators are coupled in such a way that we can associate one diameter mode natural frequency with one wire (or more wires if they are placed in the same relative position with respect to a nodal point) and the additional diameter mode natural resonance with the other wire(s). In addition, a phase inverter is realized when the coupling wires (associated with a certain mode) entering and leaving a disk are in adjacent sectors, as shown in Fig. 1.

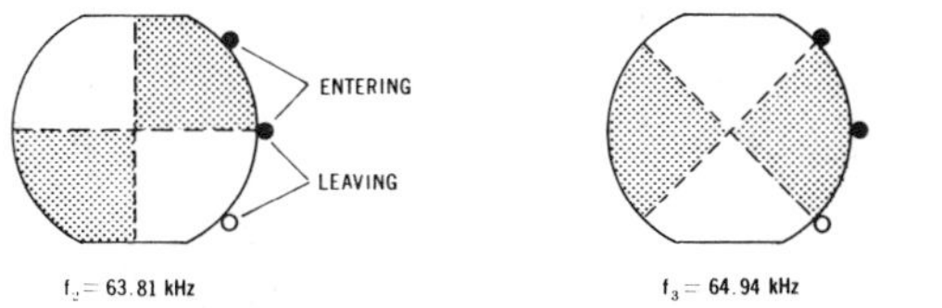

Fig. 1. Two diameter mode segmented disk resonators.

Figure 2 shows a pictorial representation and schematic diagram of a four "pole" filter employing the principles described above. The end resonantors include extensional mode ferrite transducers. The electrical analogy of the mechanical system is the force-current or mobility type as opposed to the force-voltage or impedance analogy [6].

The schematic diagram of Fig. 2(b) has been reduced in complexity by eliminating all modes of vibration well removed in frequency from the filter passband as well as f_3 and f_2 modes (see Fig. 1) associated with the f_2 and f_3 shunt arms, respectively. This mode decoupling associated with the f_2 and f_3 frequencies is a result of coupling wires being placed at nodal points as shown in Fig. 1.

Element values can be obtained for the electrical analogy by transforming a single section lattice network of the type shown in Fig. 3(a) to that of 3(b) and 3(c) which can be converted in turn to 2(b). Single section lattice element values were found by making use of a general stopband, equal ripple passband, filter-design program [7]. Image parameter design techniques could also have been used [8].

The shunt capacitors in the arms of the lattice network of Fig. 3(a) can be made equal by setting the number of half poles of attenuation at infinity equal to 3. The shunt capacitors as well as an arbitrary amount of shunt inductance are removed from the lattice, to form Fig. 3(b). The lattice of Fig. 3(b) is then split into two lattices, the arms of one of these being interchanged as the result of adding a phase inverter. The lattice structures of Fig. 3(c) are then converted into ladders to form the twin tee of Fig. 2(b). The final step in the design is to apply Norton transformations to the center resonators so as to obtain realizations having prescribed disk resonator mass ratios.

Figure 4 shows the frequency response characteristics of a mechanical filter design that was based on the above transformations. A number of wire size adjustments were needed to obtain this response, possibly because of the sensitivity of the filter response to coupling wire orienta-

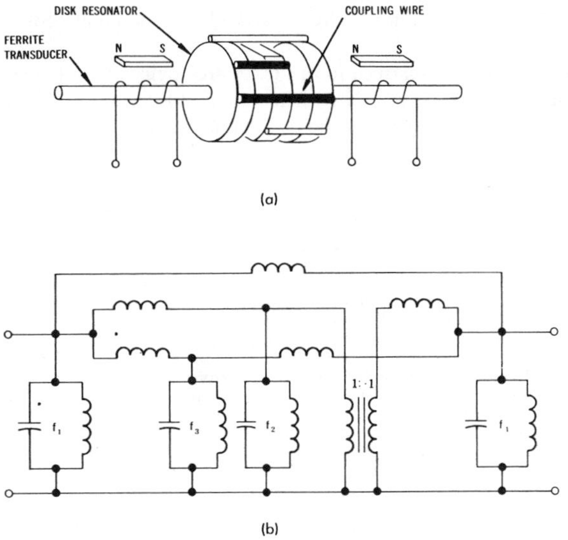

Fig. 2. Four pole, multimode mechanical filter (a) pictorial representation and (b) electrical analogy.

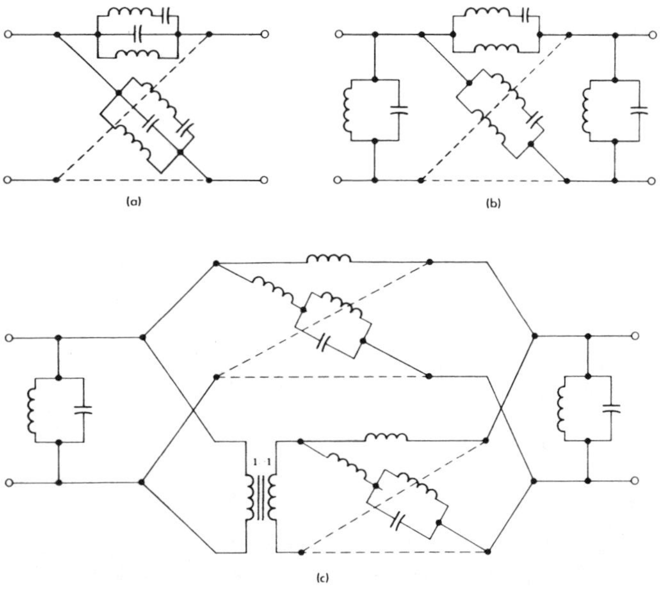

Fig. 3. (a) Four pole lattice prototype. (b) and (c) Intermediate steps in obtaining the network of Fig. 2(b).

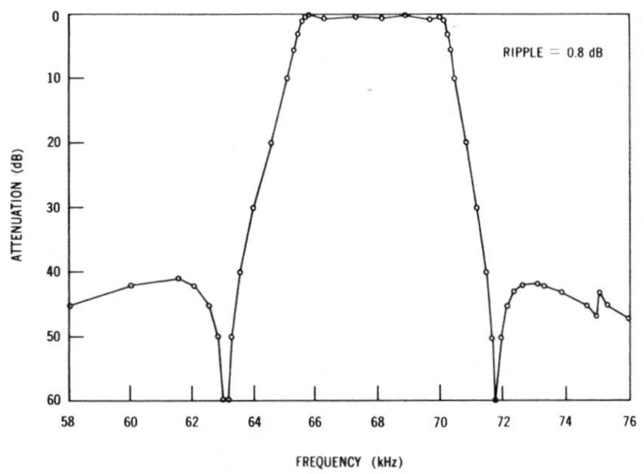

Fig. 4. Actual frequency response characteristics of a twin tee multimode mechanical filter.

Manuscript received September 6, 1966.

Reprinted from *Proc. IEEE*, vol. 54, pp. 1961–1962, Dec. 1966.

tion. At the present time we have not built any nonminimum phase filters.

The techniques described in this letter are applicable to higher frequencies, for instance 500 kHz, where a one-circle, two-diameter mode could be used in the place of the two-diameter mode.

ROBERT A. JOHNSON
Collins Radio Co.
Newport Beach, Calif.

REFERENCES

[1] R. A. Sykes and W. D. Beaver, "High frequency monolithic crystal filters with possible application to single frequency and single side band use," *1966 Proc. 20th Annual Frequency Control Symposium*, to be published.
[2] M. Onoe, H. Jumonji, and N. Kobori, "High frequency crystal filters employing multiple mode resonators vibrating in trapped energy modes," *1966 Proc. 20th Annual Frequency Control Symposium*, to be published.
[3] R. A. Johnson, "A single-sideband, disk-wire type mechanical filter," *IEEE Trans. on Component Parts*, vol. CP-11, pp 3–7, December 1964.
[4] R. A. Johnson and R. J. Teske, "A mechanical filter having general stopband characteristics," *IEEE Trans. on Sonics and Ultrasonics*, vol. SU-13, July 1966.
[5] J. W. S Rayleigh, *The Theory of Sound*, vol. 1. New York: Dover, 1945, pp. 334–339.
[6] M. Konno and H. Nakamura, "Equivalent electrical network for the transversely vibrating uniform bar," *J. Acoust. Soc. Am.*, vol. 37, pp. 614–622, December 1965.
[7] J. A. C. Bingham, "A new method of solving the accuracy problem in filter design," *IEEE Trans. on Circuit Theory*, vol. CT-11, pp. 327–341, September 1964.
[8] D. Indjoudjian and P. Andrieux, *Les Filtres a Cristaux Piézoélectriques*. Paris: Gauthier-Villars, 1953, pp. 85–93.

AN ELECTROMECHANICAL FILTER CONSISTING OF A FLEXURAL VIBRATOR WITH DOUBLE RESONANCES

Masashi Konno and Yoshiro Tomikawa, Members

Faculty of Engineering
Yamagata University, Yonezawa

SUMMARY

The construction, design and experimental results of an electromechanical filter which is effective for center frequencies below 50 kHz are reported. The filter utilizes two resonances occurring in a flexural vibrator having rectangular cross section and free ends. Since just one such vibrator gives a bandpass filter characteristic, it is expected to be of practical advantage in the creation of small filters. The two directions of vibration excited by electrostrictive driver attached to the surface of the vibrator are first confirmed and then an equivalent-circuit representation of the vibrator is attempted. On the basis of these results, the design and limitations of an electromechanical filter are discussed. A method of constructing the filter in polarized form and a method of coping with low-frequency problems are reported. The possibility of increasing the passband and some of the problems it involves are also reported.

1. Introduction

The electromechanical filter (abbreviated M. F.), with its high Q, superior attenuation characteristic and small size, is widely used at present. The recent multiplexing of various communication networks calls for M. F.'s which are effective for frequencies below 50 kHz. As an example of M. F. for relatively low frequency bands (below 50 kHz), this paper reports an M. F. which utilizes two resonances occurring in a free-ended, transverse-vibrating flexural vibrator. Since a bandpass filter characteristic can be obtained from just one vibrator, such a vibrator has a big advantage in constructing a small filter. We shall call a flexural vibrator in which two resonances occur "double-resonance vibrator." The flexural vibrators with elliptical cross section and square cross section reported by Kimura [1] and the flexural vibrator with orthohexagonal cross section reported by Abe and Tanaka [2] have contributed to the study of such flexural vibrators but have not yet been put into practical forms. Although the report by F. Künemund et al. [3] of Germany on the chamfering of a vibrator with square cross section is interesting, there are many points which still are not clear in the equivalent-circuit representation and the design of their vibrator. Thus there are problems yet to be solved before it is put to actual use. We have conducted experiments with such vibrators and obtained a relatively good bandpass filter characteristic with just one double-resonance vibrator. The configuration of the double-resonance vibrator which is driven with an electrostrictive material bound to its surface will be reported first, followed by the direction of vibration, the resonant frequencies and so forth. Then the vibrator will be represented by an equivalent circuit. The force coefficient will be experimentally checked, followed by a discussion of the design of M. F. using the double-resonance vibrator and its construction in polar form. It will be shown that a stepped configuration makes it possible to develop an M. F. for low frequencies. Problems involved in increasing the passband will be also discussed.

2. Double-Resonance Vibrator

The double-resonance vibrator dealt with in this paper is a transverse-vibrating type having rectangular cross section as shown in Fig. 1 (a). If it is driven by means of an electrostrictive material bound to its surface, two vibrations occur as shown by the arrows. With this kind of vibrator the preferred

Reprinted with permission from *Electron. Commun. Japan*, vol. 50, pp. 64–73, Aug. 1967. Reprinted with the consent of Scripta Publishing Co., 1511 K St., N.W., Washington, DC 20005.

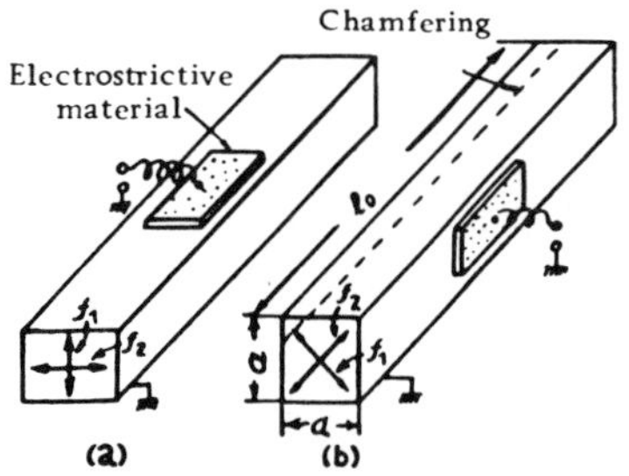

Fig. 1. Flexural vibrators with double resonance

directions of vibration are presumably determined by the shape of the cross section, internal nonuniformity of the vibrator and so forth. If, therefore, we cause the geometrical moment of inertia taken in a certain direction to change greatly by chamfering one of the corners of a thin rod having square cross section [shown by the dashed line in Fig. 1 (b)] as described by Künemund et al., we obtain double resonances ($f_1 < f_2$). Although Künemund et al. reported that the direction of vibration is the same for both Figs. 1 (a) and (b), we have observed different directions of vibration, the driving force having been applied in different ways.

2.1 Methods of Driving and Directions of Vibration

Figure 2 shows the directions of vibration of the vibrator ends at various frequencies observed under a microscope for different vibrators.

The directions of vibration for vibrators having square cross section [Fig. 2(b)] were

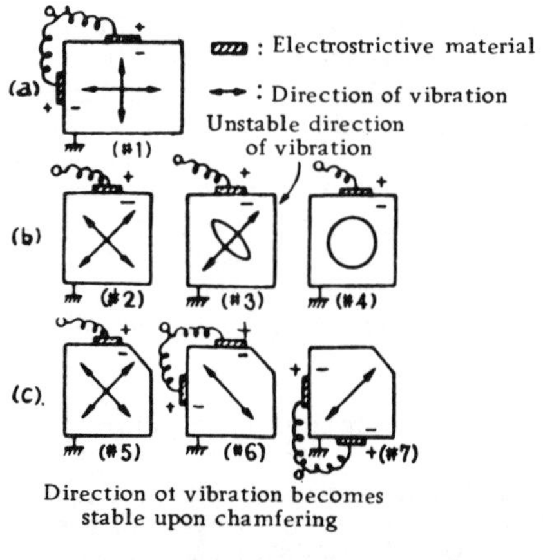

Fig. 2. Directions of vibration of the same vibrators shown in Fig. 1

unstable in general with the driving achieved as shown. The directions of vibration for chamfered vibrators [Fig. 2(c)] were stable and the vibration frequency could be controlled easily. Even for chamfered vibrators, however, vibrations become unstable if the two resonant frequencies are too close to each other, and so in practice unstable modes should be avoided. For this reason, vibrators grouped in Fig. 2 (c) were chosen for study.

2.2 Amount of Chamfering and Resonant Frequencies

The required $\Delta f (= f_2 - f_1)$ may be obtained by chamfering the double-resonance vibrators by such methods as shown in Fig. 3 (a) — (c). Figures 4 — 6 show the relationships of the amount of chamfering δ (= h/a) to the resonant frequency ratios f_1/f_0 and f_2/f_0 and to $\Delta f/f_2$ (rate of change of resonant frequency). These relationships will serve as a guide for design, which will be referred to later. From the point of view of frequency control, the methods of chamfering shown in Fig. 3 (a) and (b) are desirable. The method of chamfering shown in Fig. 3 (c) is advantageous for small

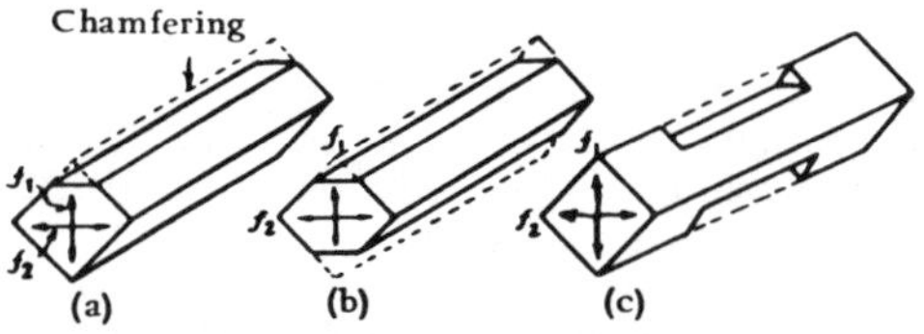

Fig. 3. Some methods of chamfering the vibrator in Fig. 1 (b)

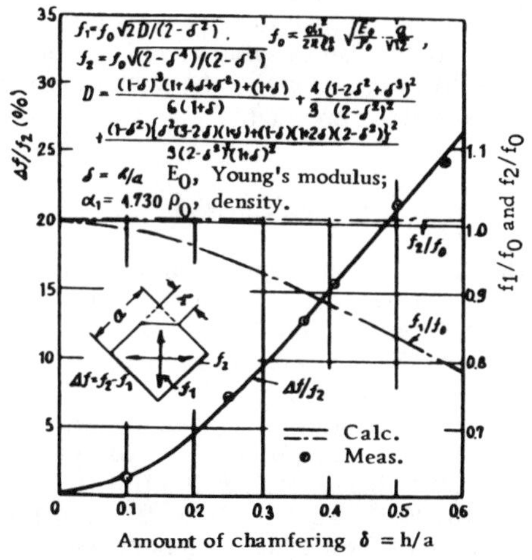

Fig. 4. f_1/f_0, f_2/f_0 and $\Delta f/f_2$ of the vibrator in Fig. 3 (a) as functions of $\delta = h/a$

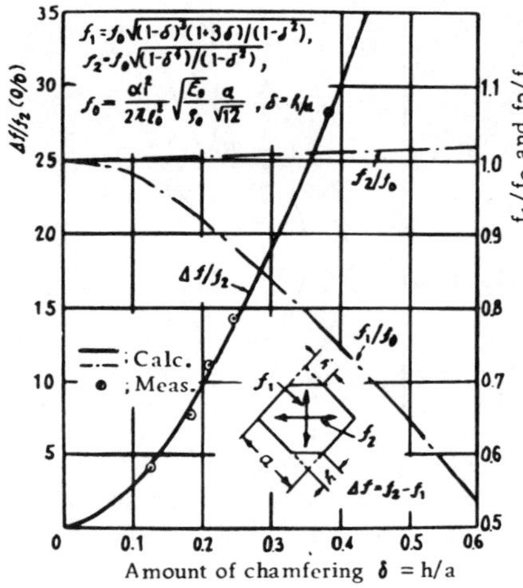

Fig. 5. f_1/f_0, f_2/f_0 and $\Delta f/f_2$ of the vibrator in Fig. 3 (b) as functions of $\delta = h/a$

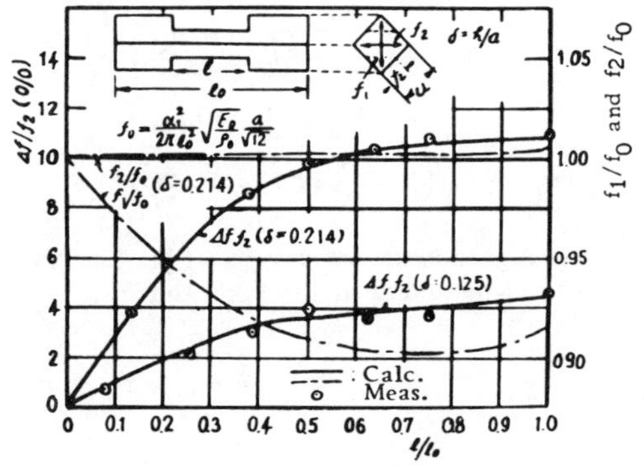

Fig. 6. f_1/f_0, f_2/f_0 and $\Delta f/f_2$ of the vibrator in Fig. 3 (c) as functions of l/l_0

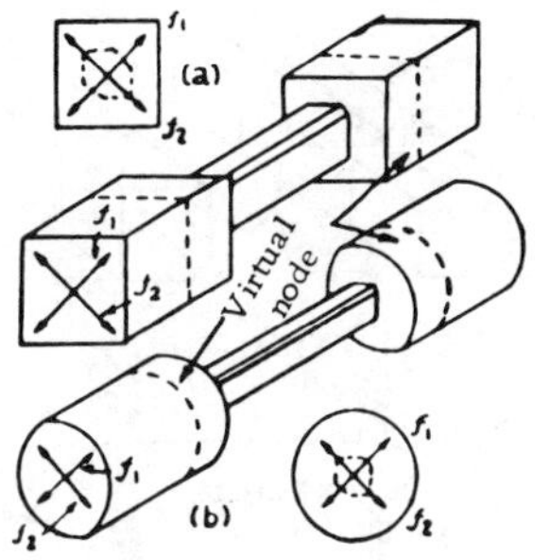

Fig. 7. Some double-resonance vibrators suited for use at lower frequencies

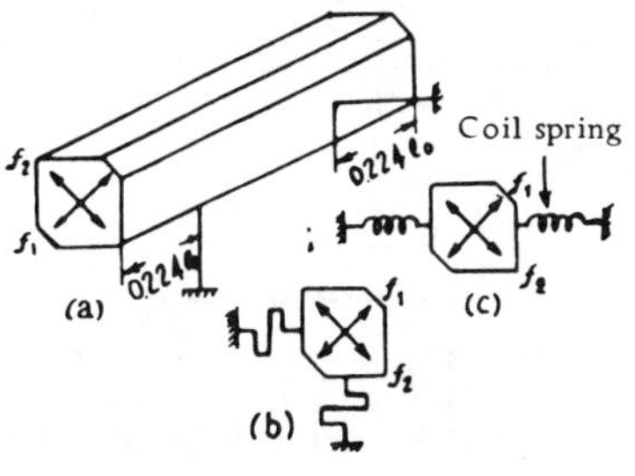

Fig. 8. Some methods of supporting the double-resonance vibrator (in the first mode)

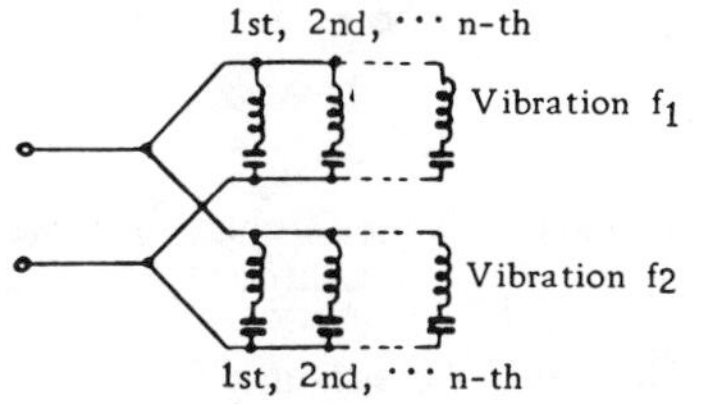

Fig. 9. The mechanical equivalent circuit of the double-resonance vibrator

filters and for low frequencies and may lead to the variations shown in Fig. 7. The vibrators should be supported as gently as possible at their virtual nodes as shown in Figs. 8 (a) — (c). These methods of support were adopted for the present study.

3. Equivalent Circuit Representation

In representing a double-resonance vibrator by an equivalent circuit, coupling between the two vibrations (f_1 and f_2) generally becomes a problem. With the double-resonance vibrators used for the present study there also were cases where some coupling was noted which was due to asymmetry of the cross-sectional shape, nonuniformity of the material, etc., but the degree of coupling was very small. That is to say, the vibrations with resonant frequencies f_1 and f_2 may be considered to be independent. The driving-point mechanical impedance is represented by the equivalent circuit of Fig. 9. The double-resonance vibrators with electrostrictive materials attached [Fig. 10 (a)] are represented by the equivalent circuit of Fig. 10 (b), the sign of the force coefficient (A) being determined by the polarity and the position of the electrostrictive material attached (see Fig. 2 and Fig. 18).

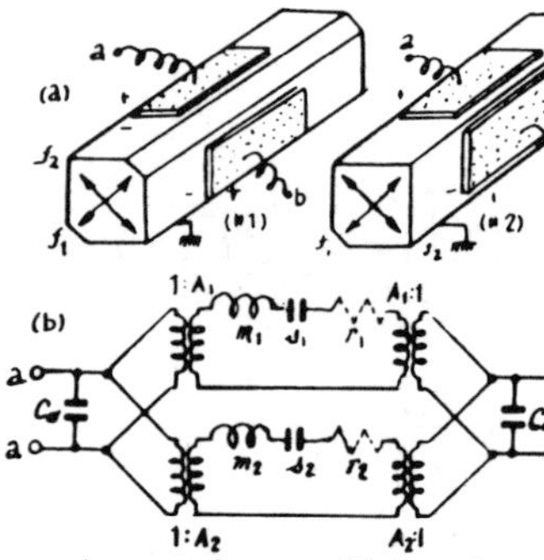

Λ_1, Λ_2: force coefficients for vibrations with frequencies f_1 and f_2

Fig. 10. The electrical equivalent circuit of the double-resonance vibrator driven electrostrictively (A is the force factor)

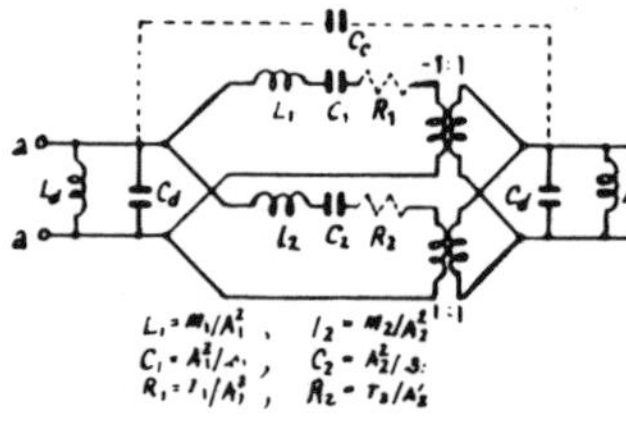

Fig. 11. Equivalent circuit of electromechanical filter consisting of double-resonance vibrator having parallel inductances at both end terminals

4. M.F. Using Double-Resonance Vibrator

4.1 Method of Synthesis

Figure 10 shows a single double-resonance vibrator forming a bandpass filter circuit. In this section we shall discuss the parallel insertion of inductances L_d as shown in Fig. 11 in order to increase the bandwidth (insertion of bridge capacitance C_c will be discussed in Sect. 5 and series insertion of inductances L_d in Appendix A1).

4.2 Cutoff Angular Frequencies and Fractional Bandwidth

The short-circuit impedance Z_s and the open-circuit impedance Z_f of the circuit of Fig. 11 ($C_c = 0$) are given by

$$\left.\begin{aligned} Z_s &= j\omega L_d(1-\omega^2/\omega_{s0}^2)/(1-\omega^2/\omega_{s\infty1}^2) \\ &\quad \cdot(1-\omega^2/\omega_{s\infty2}^2), \\ Z_f &= j\omega L_d(1-\omega^2/\omega_{f0}^2)/(1-\omega^2/\omega_{f\infty1}^2) \\ &\quad \cdot(1-\omega^2/\omega_{f\infty2}^2), \end{aligned}\right\} \tag{1}$$

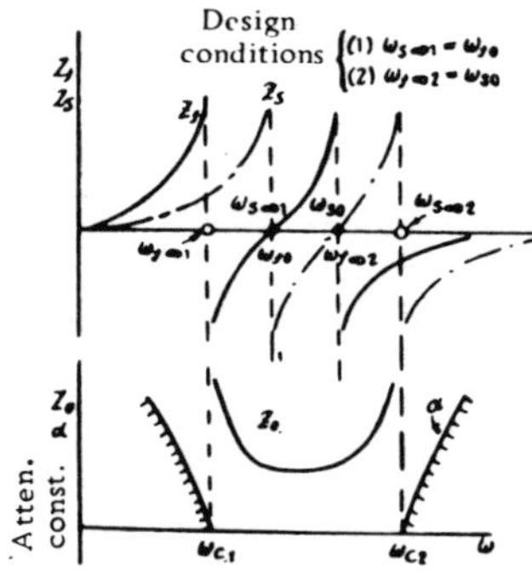

Fig. 12. Characteristics of the electromechanical filter shown in Fig. 11 ($C_c = 0$)

where

$$\begin{aligned} &\omega_{s0}^2 \equiv \omega_2^2 = 1/L_2C_2, \quad \omega_{f0}^2 \equiv \omega_1^2 = 1/L_1C_1, \\ &\omega_d^2 = 1/L_dC_d \end{aligned} \tag{2}$$

$$\left.\begin{aligned} \omega_{s\infty1}^2 \\ \omega_{s\infty2}^2 \end{aligned}\right\} = \omega_2^2\left\{\left(\frac{1}{2}+\frac{\omega_d^2}{2\omega_2^2}+C_2L_d\,\omega_d^2\right) \mp\sqrt{\left(\frac{1}{2}+\frac{\omega_d^2}{2\omega_2^2}+C_2L_d\,\omega_d^2\right)^2-\frac{\omega_d^4}{\omega_2^4}}\right\} \tag{3}$$

$$\left.\begin{aligned} \omega_{f\infty1}^2 \\ \omega_{f\infty2}^2 \end{aligned}\right\} = \omega_1^2\left\{\left(\frac{1}{2}+\frac{\omega_d^2}{2\omega_1^2}+C_1L_d\,\omega_d^2\right) \mp\sqrt{\left(\frac{1}{2}+\frac{\omega_d^2}{2\omega_1^2}+C_1L_d\,\omega_d^2\right)^2-\frac{\omega_d^4}{\omega_1^4}}\right\} \tag{4}$$

The design conditions are, as Fig. 12 shows, $\omega_{s\infty1} = \omega_{f0}$ and $\omega_{s\infty2} = \omega_{s0}$, that is,

$$\left.\begin{aligned} L_d &= (\omega_2/\omega_1-\omega_1/\omega_2)^2/2\left(\frac{1}{L_1}+\frac{1}{L_2}\right) \\ C_d &= 2(C_1+C_2)/(\omega_2/\omega_1-\omega_1/\omega_2)^2 \end{aligned}\right\} \tag{5}$$

The cutoff angular frequencies ω_{c1}, ω_{c2} ($\omega_{c1} < \omega_{c2}$) are given by

$$\omega_{c1} = \omega_{f\infty1} = (\omega_1/\omega_2)\omega_d, \quad \omega_{c2} = (\omega_2/\omega_1)\omega_d \tag{6}$$

and the fractional bandwidth B by

$$B = \frac{\omega_{c2}-\omega_{c1}}{\omega_0} = (\omega_2/\omega_1-\omega_1/\omega_2) = \sqrt{2(C_1/C_d+C_2/C_d)} \tag{7}$$

where $\omega_0(=\sqrt{\omega_{c1}\cdot\omega_{c2}})$ is the center angular frequency. The characteristic impedance Z_0 at ω_0 is given by

$$Z_0 = \sqrt{Z_s\cdot Z_f}_{\,\omega=\omega_0} = \sqrt{\frac{L_d}{C_d}}\Big/(\omega_2/\omega_1-\omega_1/\omega_2) \tag{8}$$

4.3 Force Coefficient

(a) Design Formula. Figure 13 shows how the electrostrictive materials are arranged. We assume that the vibration mode is not altered by attachment of the electrostrictive materials. Since exact calculation of the force coefficient for a vibrator with

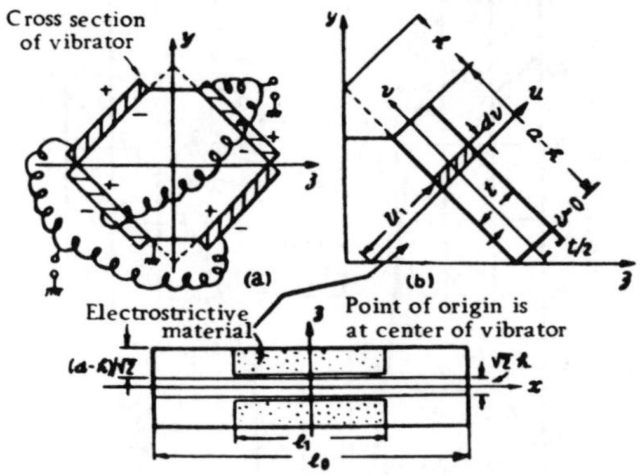

Fig. 13. Arrangement of piezoelectric materials on the double-resonance vibrator

electrostrictive material is difficult, the coefficient was approximately calculated in a manner similar to that for a common acoustic vibrator with bimorph type of electrostrictive material [4]. The following formulas giving values in rough agreement with the measured values were derived.

$$\left.\begin{aligned}&\text{Force coefficient for } f_1 \text{ vibration at vibrator end}\\&A_1 = \frac{e'}{\sqrt{2}}\frac{\alpha_1}{l_0}a^2\left(1-\frac{h}{a}\right)^2 u(X_1)\\&\text{Force coefficient for } f_2 \text{ vibration at vibrator end}\\&A_2 = \frac{e'}{\sqrt{2}}\frac{\alpha_2}{l_0}a^2\left(1-\frac{h^2}{a^2}\right)u(X_1)\\&u(X_1) = \left\{\frac{\sinh(\alpha_1 X_1/2)}{\cosh(\alpha_1/2)} - \frac{\sin(\alpha_1 X_1/2)}{\cos(\alpha_1/2)}\right\},\\&X_1 = l_1/l_0\end{aligned}\right\} \quad (9)^*$$

where e' is the piezoelectric constant. This e', which can differ from the piezoelectric constant e for a bimorph electrostrictive material, was experimentally corrected to obtain $e'/e \fallingdotseq 1.1$ [5].

(b) Electrode Arrangement for $A_1 = A_2$. As the bandwidth becomes relatively wide, there occur cases where the damping characteristic becomes asymmetrical for $A_1 \neq A_2$, as will be described later. In order to improve this, the electrostrictive materials may be bound to the vibrator surface in a symmetrical arrangement so that $A_1 = A_2$, as shown in Fig. 14 (and Fig. 17). The force coefficient A_e for this case is given by

$$A_e = \frac{e'}{\sqrt{2}}\frac{\alpha_1}{l_0}a^2\left(1-2\frac{h}{a}\right)u(X_1) \quad (10)^*$$

---

*See Appendix A2 for derivation of these formulas and an experimental check.

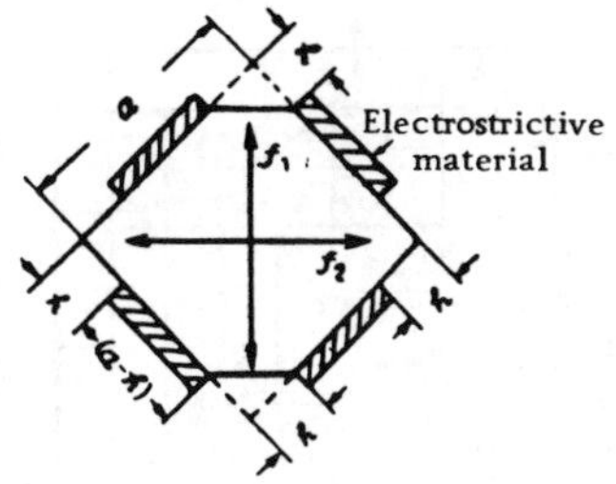

Fig. 14. Arrangement of piezoelectric materials for $A_1 = A_2$

4.4 Amount of Chamfering and Fractional Bandwidth

Using the force coefficient given in the preceding section, we obtain the capacitance ratio $r_i = C_{d0}/C_i$ (i = 1, 2) for vibrator 1 of Fig. 10 (a), for example, with

$$\left.\begin{aligned}&\text{total mass of vibrator}\\&M_0 = (a^2-h^2)\rho_0 l_0 + 4t\rho_p(a-h)l_1\\&\text{angular resonant frequencies}\\&\omega_1^2 \fallingdotseq \alpha_1^4 E_0 I_1/\rho_0(a^2-h^2)l_0^4,\\&I_1 = \frac{a^4}{12}(1-\delta)^3(1+3\delta)\\&\omega_2^2 \fallingdotseq \alpha_1^4 E_0 I_2/\rho_0(a^2-h^2)l_0^4,\\&I_2 = \frac{a^4}{12}(1-\delta^4)\end{aligned}\right\} \quad (11)$$

as

$$r_i^{-1} = 4A_i^2/\omega_i^2 M_0 C_{d0}, \quad C_{d0} \fallingdotseq 2\epsilon(a-h)l_1/4\pi t \quad (12)$$

where ϵ and ρ_p are the dielectric constant and the density, respectively, of the electrostrictive material while C_{d0}, E_0 and ρ_0 are the damping capacitance, Young's modulus and the density, respectively, of the vibrator itself. From Eqs. (7) and (12) [with $C_d = C_{d0}$], the fractional bandwidth B_r determined by the capacitance ratio of the vibrator itself is given for $A_1 \neq A_2$ by

$$\left.\begin{aligned}&B_r = \sqrt{\frac{48\pi}{E_0\epsilon}}\cdot\frac{e'}{\alpha_1}\cdot\frac{u(X_1)}{\sqrt{X_1}}\sqrt{\frac{q}{k}}\sqrt{\frac{1}{(1+3\delta)}+\frac{(1+\delta)}{(1+\delta^2)}}\\&k = \left\{1+\frac{\rho_p l_1 2q}{\rho_0 l_0(1+\delta)}\right\}, \quad q = 2t/a\end{aligned}\right\} \quad (13)$$

and for $A_1 = A_2$ by

$$\left.\begin{aligned}&B_r = \sqrt{\frac{48\pi}{E_0\epsilon}}\cdot\frac{e'}{\alpha_1}\cdot\frac{u(X_1)}{\sqrt{X_1}}\cdot\sqrt{\frac{q}{k'}}\cdot\sqrt{\frac{(1-2\delta)}{(1-\delta)}}\cdot\\&\quad\cdot\sqrt{\frac{1}{(1+3\delta)(1-\delta)^2}+\frac{1}{(1+\delta)(1+\delta^2)}}\\&k' = \left\{1+\frac{\rho_p l_1 2q(1-2\delta)}{\rho_0 l_0(1-\delta^2)}\right\}\end{aligned}\right\} \quad (14)$$

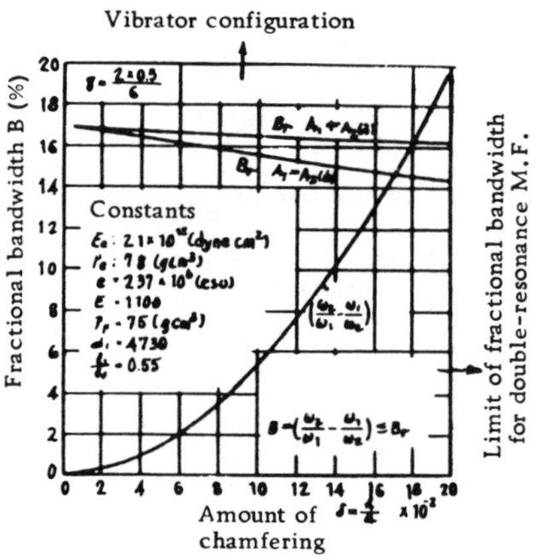

Fig. 15. One example of the limit of the fractional bandwidth as a function of $\delta = h/a$

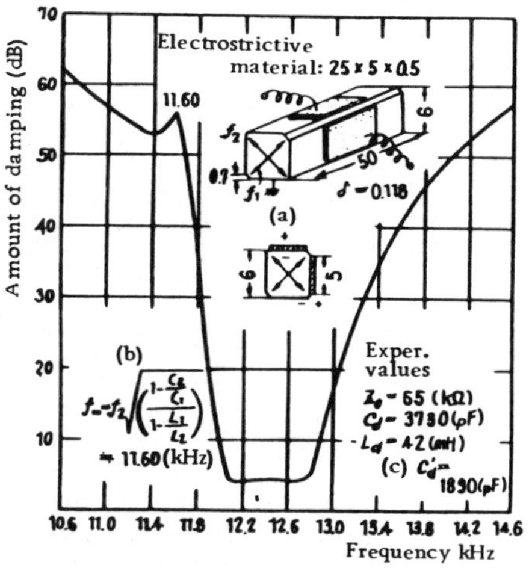

Fig. 16. Insertion loss characteristic of filter employing double-resonance vibrator ($A_1 \neq A_2$) and having parallel inductances at both end terminals

a) Vibrator: soft iron; b) Computed value of damping pole; c) Added capacitance.

The bandwidth B for which it is possible to synthesize an M. F. is given by

$$B = (\omega_2/\omega_1 - \omega_1/\omega_2) \fallingdotseq (\sqrt{I_2/I_1} - \sqrt{I_1/I_2})$$
$$= \left(\sqrt{\frac{(1+\delta)(1+\delta^2)}{(1-\delta)^2(1+3\delta)}} - \sqrt{\frac{(1-\delta)^2(1+3\delta)}{(1+\delta)(1+\delta^2)}}\right) \leq B_r \qquad (15)$$

In case of $B < B_r$ an externally added capacitance (C_d') becomes necessary. Figure 15 shows an example of the relationship of the amount of chamfering to the fractional bandwidth and also shows the design limit.

4.5 Examples of Design and Trial Construction

Given the center angular frequency ω_0 and the fractional bandwidth B, we know the necessary amount of chamfering from Eqs. (6) and (7) and from Figs. 4 and 5 (or Fig. 15) and can thereby design the double-resonance vibrator. Figures 16 and 17 show some results of measurement taken on the vibrator of Fig. 10 (a). Parts (a) and (b) of Table 1 show the measured and calculated values of the cutoff angular frequencies, etc., for the two vibrators shown in Figs. 16 and 17. As is clear from the measurement results, a damping pole occurs owing to $A_1 \neq A_2$ (that is, $L_1 \neq L_2$) for the test sample of Fig. 16 and the damping characteristic becomes asymmetrical, whereas the characteristic curve (a) of Fig. 17 is less asymmetrical than the former owing to $A_1 \fallingdotseq A_2$ and shows a desirable result. The characteristic curve (b) in Fig. 17 will be referred to later.

5. Synthesis in Polar Form with Capacitance C_c Connecting Input and Output Terminals

5.1 Design Conditions for Cutoff Angular Frequencies

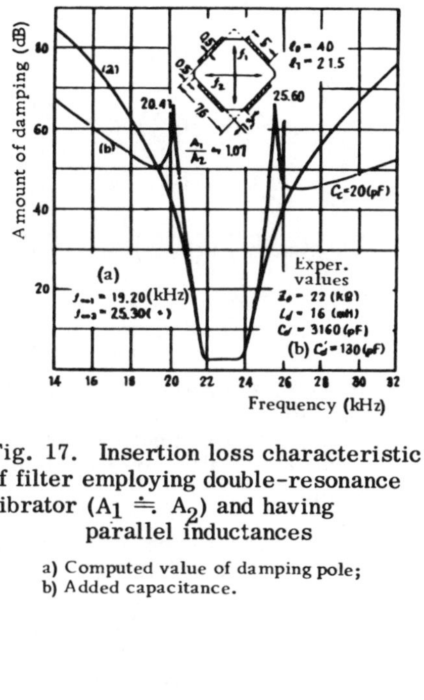

Fig. 17. Insertion loss characteristic of filter employing double-resonance vibrator ($A_1 \fallingdotseq A_2$) and having parallel inductances

a) Computed value of damping pole; b) Added capacitance.

From the equivalent circuit of Fig. 11, with C_c inserted to connect the input and the output terminals, the design conditions become

$$\left.\begin{aligned} C_d &= [2(C_1+C_2)/(\omega_2/\omega_1 - \omega_1/\omega_2)^2] \\ &\quad + [2C_c/\{(\omega_2/\omega_1)^2 - 1\}] \\ L_d &= (\omega_2/\omega_1 - \omega_1/\omega_2)^2/2\left\{\frac{1}{L_1} + \frac{1}{L_2}\right. \\ &\quad \left. + C_c(\omega_2{}^2 - \omega_1{}^2)\right\} \end{aligned}\right\} \qquad (16)$$

Table 1 (a). Measured values of constants for double-resonance vibrator

	f_1(kHz)	f_2(kHz)	L_1(H)	L_2(H)	C_1(pF)	C_2(pF)
Test sample of Fig. 16	12.113	12.568	48.75	26.03	3.54	6.18
Test sample of Fig. 17	22.218	23.561	4.28	4.93	12.00	9.25
	Q_1	Q_2	R_1(kΩ)	R_2(kΩ)	C_{d0}(pF)	
Test sample of Fig. 16	527	423	7.04	4.85	1840	
Test sample of Fig. 17	880	860	0.84	1.00	9030	

Table 1 (b). Values of constants computed from Table 1 (a)

	f_{c1}(kHz)	f_{c2}(kHz)	L_d(mH)	C_d(pF)	B(%)	Z_0(kΩ)
Test sample of Fig. 16	12.09	12.82	44.2	3732	7.2	49.2
Test sample of Fig. 17	21.74	24.19	15.8	3082	11.7	19.3

and the cutoff angular frequencies or the characteristic impedance become

$$\left.\begin{aligned} &\omega_{c1}=(\omega_1/\omega_2)\omega_d,\ \omega_{c2}=(\omega_2/\omega_1)\omega_d\Big/\sqrt{\left(1+\frac{2C_c}{C_d}\right)} \\ &Z_0=\sqrt{\frac{L_d}{C_d}}\Big/\left(\frac{\omega_2}{\omega_1}-\frac{\omega_1}{\omega_2}\sqrt{1+\frac{2C_c}{C_d}}\right) \end{aligned}\right\} \qquad (17)$$

When the capacítance C_c to be inserted is small, the design conditions, the cutoff angular frequencies and the characteristic impedance are not very different from those for the case where C_c is not considered [see Eq. (5)].

5.2 Damping Poles and Polarity of Electrostrictive Material

The damping pole angular frequencies ω_∞ are given, with the short-circuit and the open-circuit impedances of the mid-section circuit of Fig. 11 equated to each other, by

$$\left.\begin{matrix}\omega_{\infty1}^2\\ \omega_{\infty2}^2\end{matrix}\right\} =\frac{\omega_1^2}{2}\Bigg(\left\{\left(\frac{\omega_2}{\omega_1}\right)^2\left(1+\frac{C_2}{C_c}\right)+\left(1-\frac{C_1}{C_c}\right)\right\} \mp\omega_2^2\sqrt{\left\{\frac{1}{\omega_1^2}-\frac{1}{\omega_2^2}+\frac{1}{C_c}\left(\frac{\sqrt{C_2}}{\omega_1}-\frac{\sqrt{C_1}}{\omega_2}\right)^2\right\}\left\{\frac{1}{\omega_1^2}-\frac{1}{\omega_2^2}+\frac{1}{C_c}\left(\frac{\sqrt{C_2}}{\omega_1}+\frac{\sqrt{C_1}}{\omega_2}\right)^2\right\}}\Bigg) \qquad (18)$$

Under the conditions

$$\left\{\frac{1}{\omega_1^2}+\frac{1}{\omega_2^2}+\frac{1}{C_c}\left(\frac{C_2}{\omega_1^2}-\frac{C_1}{\omega_2^2}\right)\right\}>0,$$

$$\left\{1+\frac{C_2-C_1}{C_c}\right\}>0 \qquad (19)$$

two damping poles always appear. When an ideal transformer (1 : -1) is inserted only in the f_2 ($f_1 < f_2$) branch of the equivalent circuit of Fig. 11, the subscripts 1 and 2 in Eqs. (18) and (19) should be interchanged. The damping poles for such a case should be studied further. The test sample of Fig. 19 in the next section is an example. Varying C_c from 10 to 100 pF did not cause damping poles to appear. From the foregoing discussion we know that in order for the damping poles to occur easily the polarity of the electrostrictive material should be such (for example, as arranged in Fig. 18) that only the ideal transformer in the lower resonant-frequency branch has phase inversion ($-A_1 : A_1$). Curve (b) of Fig. 17 is a result of measurement for the case where $A_1 \fallingdotseq A_2$.

6. M. F. Using Stepped Double-Resonance Vibrator

The vibrator configurations of Fig. 7 are effective for making small filters and for low

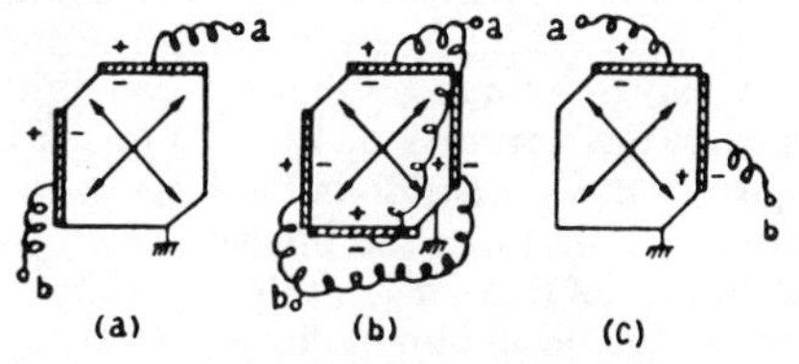

Fig. 18. Arrangement of piezoelectric materials for the filter shown in Fig. 11

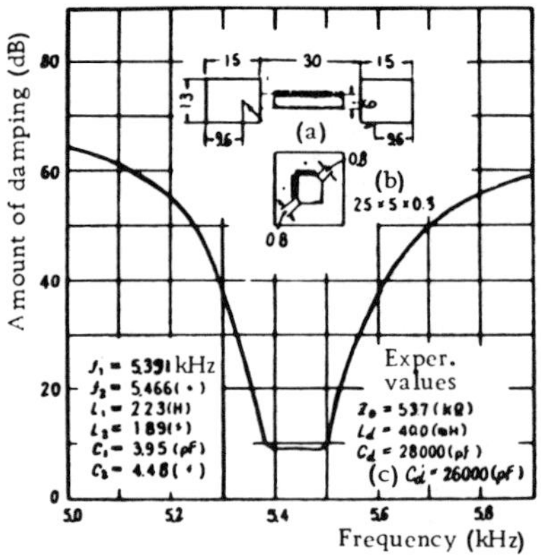

Fig. 19. Insertion loss characteristic of filter employing the vibrator shown in Fig. 7 (a)

a) Points of support (where nodes occur); b) Electrostrictive material; c) Added capacitance.

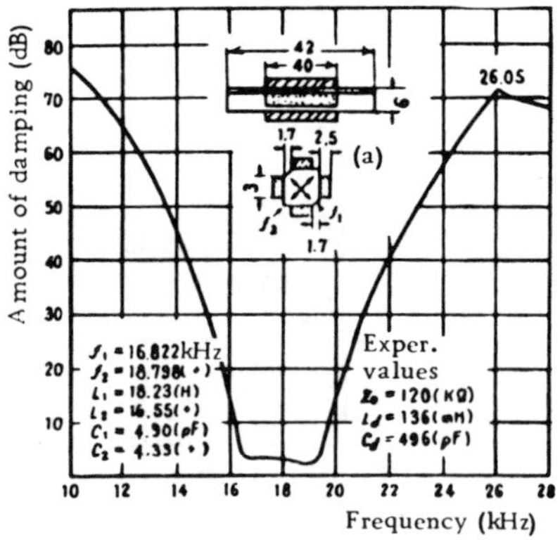

Fig. 20. Insertion loss characteristic of wideband filter employing double-resonance vibrator

a) Electrostrictive material VPZ-7 made by Tohoku Metals Co.

frequencies. Figure 19 shows the result of measurement taken on one such vibrator, so designed that its nodes occur in the thick parts on the ends and these points (where the nodes occur) were given soft support. If, however, the middle section is made so thin as to make the vibrator good for low center frequencies, the force coefficient becomes small and the insertion loss due to L_d becomes very large, which is a defect of this configuration.

7. Wideband M. F. Design Example and Its Problems

Up to the previous section we studied cases where the thickness of the electrostrictive materials is sufficiently small compared to that of the vibrator itself. To increase the bandwidth the first thing to do is to use, from Eqs. (13) and (14), electrostrictive material having a large dielectric constant e, but another simple method is to make the material thick and the capacitance ratio small. Figure 20 shows the result of measurement under such conditions. It is seen that an increase of about 20% in the bandwidth would be possible even with the electrostrictive materials now commercially available. However, for fractional bandwidths of more than 20% the design formulas of Sect. 4.2 are not valid. The main reason for this seems to be that the constants for the equivalent circuits of Fig. 10 (b), etc., then require values other than those at each resonance. Further, as the electrostrictive material becomes thicker its thickness can no longer be neglected in comparison with that of the vibrator itself and hence the resonant frequencies, positions of the nodes or the vibration modes should be reconsidered. Also, as the vibrator becomes thicker (a becomes larger) and shorter, the resonant frequency constant α_1 becomes smaller than 4.730 and of course requires correction.

8. Conclusion

The double-resonance vibrator was described and an example of the relatively good bandpass filter characteristic obtained with just one such vibrator was presented. Thickening of the electrostrictive material attached to the vibrator and thereby improvement of the capacitance ratio showed that a considerable increase in bandwidth can be expected even with electrostrictive materials now commercially available. Since this kind of M. F. can be realized with only one vibrator, it is expected to find practical use. Although the present study was limited to the utilization of the first mode of two transverse vibrations, the third mode (or the n-th mode in general) of these vibrations may also be used. Further, the n-th mode of the transverse vibration and the first mode of the longitudinal vibration may be used together [6] for designing even better M. F.'s. The authors wish to report on this at a later date.

Acknowledgements. In closing the authors wish to express their deep appreciation to Professor Onoue of University of Tokyo for his advice and discussion concerning the coupling of transverse-vibrating rods, to Messrs. Okamoto and Yagua of Fujitsu Ltd. for their discussions and to Mr. Sotoda of Yamagata University for his cooperation in the experiments. The authors' thanks are also due to the members of Technical Group on Barium Titanate.

Submitted November 24, 1966

REFERENCES

1. Kimura: Study of double-resonance vibrator filter, Dissertation for Ph.D., Tohoku University (Feb. 1957).
2. Abe, Tanaka et al.: Polyphase flexural vibrator using barium titanate ceramics, Annual Report of Technical Group on Barium Titanate, IEE, Japan, II-10-40, p. 19 (1955).
3. F. Künemund und K. Traub: Mechanishe Filter mit Biegeshwinger, Frequenz, Bd 12, Nr 19 (1964).
4. Nukiyama and Suzuki: Vibration modes of M-type piezoelectric vibrators and their electroacoustic constants, Jour. I.E.C.E., Japan, 17, 231, p. 367 (June 1942).
5. Konno et al.: Mechanical filters using double-resonance vibrators, Conv. Record of the Acoustical Society of Japan, 1-1-4 (Nov. 1966).
6. Konno et al.: Electromechanical filter consisting of electrostrictive composite vibrators, Memoirs of Yamagata University Faculty of Engineering, 8, 1, p. 107 (Jan. 1964).

APPENDICES

A1. Cutoff Angular Frequencies and Fractional Bandwidth of M.F. with L_d Inserted in Series

From the equivalent circuit of Fig. A1, the design conditions for this case ($C_c = 0$) are given by

$$\left.\begin{aligned} C_d &\doteq 2(C_1+C_2)\Big/\left(\frac{\omega_2}{\omega_1}-\frac{\omega_1}{\omega_2}\right)^2 \\ L_d &\doteq \left(\frac{1}{L_1}+\frac{1}{L_2}\right)\left(\frac{\omega_2}{\omega_1}-\frac{\omega_1}{\omega_2}\right)^2\Big/2\left(\frac{1}{L_2}\frac{\omega_2}{\omega_1}+\frac{1}{L_1}\frac{\omega_1}{\omega_2}\right)^2 \end{aligned}\right\} \quad \text{(A1)}$$

The cutoff angular frequencies ω_{c1}, ω_{c2} and the fractional bandwidth B are given by

$$\left.\begin{aligned} \omega_{c1} &= \left(\frac{\omega_1}{\omega_2}\right)\omega_d\Big/\sqrt{1+\frac{2C_2}{C_d}} \doteq \left(\frac{\omega_1}{\omega_2}\right)\omega_d, \ \left(\frac{2C_2}{C_d}\ll 1\right) \\ \omega_{c2} &= \left(\frac{\omega_2}{\omega_1}\right)\omega_d\Big/\sqrt{1+\frac{2C_1}{C_d}} \doteq \left(\frac{\omega_2}{\omega_1}\right)\omega_d, \ \left(\frac{2C_1}{C_d}\ll 1\right) \end{aligned}\right\} \quad \text{(A2)}$$

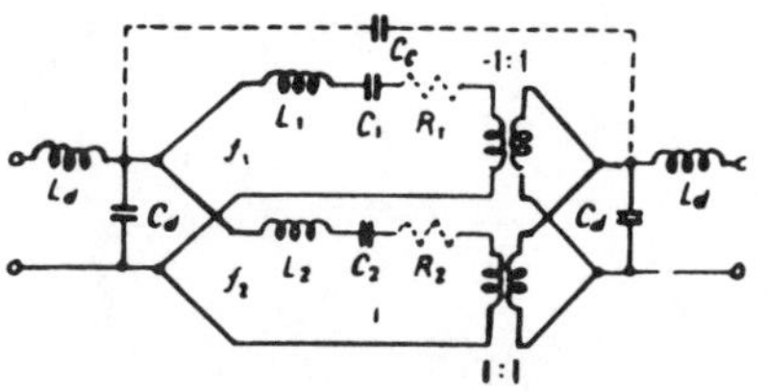

Fig. A1. Equivalent circuit of filter employing double-resonance vibrator and having series inductances at both end terminals

and

$$\begin{aligned} B &= \frac{\omega_{c2}-\omega_{c1}}{\omega_0} = (\omega_{c2}-\omega_{c1})/\sqrt{\omega_{c1}\cdot\omega_{c2}} \\ &= \left(\frac{\omega_2}{\omega_1}\sqrt[4]{\frac{1+\frac{2C_2}{C_d}}{1+\frac{2C_1}{C_d}}}-\frac{\omega_1}{\omega_2}\sqrt[4]{\frac{1+\frac{2C_1}{C_d}}{1+\frac{2C_2}{C_d}}}\right) \doteq \left(\frac{\omega_2}{\omega_1}-\frac{\omega_1}{\omega_2}\right) \\ &\leq \sqrt{2\left(\frac{C_1}{C_d}+\frac{C_2}{C_d}\right)} \end{aligned} \quad \text{(A3)}$$

These do not differ very much from those for the circuit with L_d inserted in parallel. However, the characteristic impedance Z_0 at ω_0 is given by

$$\begin{aligned} Z_0 &= \sqrt{\frac{L_d}{C_d}}\left[\left\{\left(\frac{\omega_2}{\omega_1}\right)\Big/\sqrt{1+\frac{2C_1}{C_d}}\right\} - \left\{\left(\frac{\omega_1}{\omega_2}\right)\Big/\sqrt{1+\frac{2C_2}{C_d}}\right\}\right] \\ &\doteq \sqrt{\frac{L_d}{C_d}}\left(\frac{\omega_2}{\omega_1}-\frac{\omega_1}{\omega_2}\right) \end{aligned} \quad \text{(A4)}$$

and is considerably smaller than that for the circuit with L_d inserted in parallel [(see Eq. (8)]. Figure A2 shows a measurement result for a test sample which is the same as the one in Fig. 17 and the characteristic curve (b) is an example of synthesis in polar form.

A2. Derivation of Formula for Calculating Force Coefficient and Its Experimental Check

The mean value of distortions $\overline{S}$ necessary for obtaining the force coefficient can be calculated, for example, for the u-axis in Fig. 13

Dimensions of sample: same as those of sample of Fig. 17

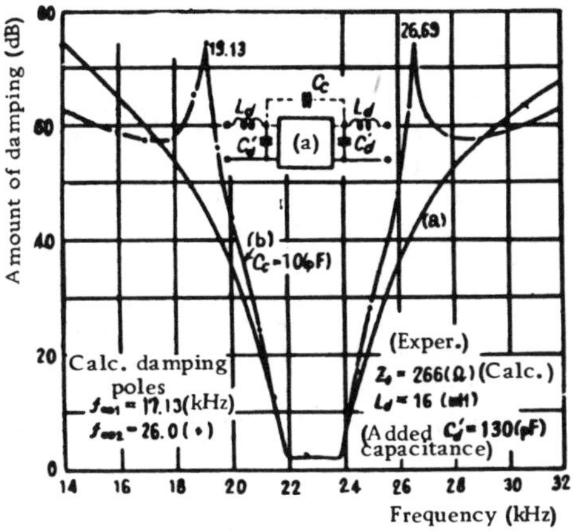

Fig. A2. Insertion loss characteristic of the filter shown in Fig. A1

a) Double-resonance vibrator.

for f_1 vibration, and the displacement current I_d can be computed for the v-axis, which gives $\bar{S}$. Thus, we have

$$\bar{S}=\frac{1}{t}\int_{u_1}^{u_1+t}\frac{u}{\sqrt{2}}\left(\frac{\partial^2\zeta}{\partial x^2}\right)du=\left(\frac{t}{2\sqrt{2}}+\frac{a}{2\sqrt{2}}-z\right)\frac{\partial^2\zeta}{\partial x^2} \qquad \text{(A5)}$$

and

$$I_d=j\omega 2\int_{-l_1/2}^{l_1/2}dx\int_0^{(a-h)}(e'\bar{S})dv$$
$$=j\omega\frac{e'}{\sqrt{2}}\frac{a^2\alpha_1}{l_0}\left(1-\frac{h}{a}\right)^2 2P\left\{\sinh\left(\frac{\alpha_1 X_1}{2}\right)\cos\left(\frac{\alpha_1}{2}\right)-\sin\left(\frac{\alpha_1 X_1}{2}\right)\cosh\left(\frac{\alpha_1}{2}\right)\right\} \qquad \text{(A6)}$$

The velocity of the vibrator ends is given by

$$v_f=j\omega[\zeta(X)]_{X=1/2}=j\omega 2P\cosh\left(\frac{\alpha_1}{2}\right)\cos\left(\frac{\alpha_1}{2}\right) \qquad \text{(A7)}$$

where displacement ζ is given by

Table A-1. Comparison of measured and calculated values

		No. 1	No. 2	No. 3	No. 4	No. 5	No. 6
Test sample dimensions	$a-h$	3	3	3	5	3	3
	a	5	5	4.8	6	5	5
	t	1	2	2	2	2.5	3
Meas. A_2 (e.s.u.)		7.61 $\times 10^6$	7.41 $\times 10^6$	6.98 $\times 10^6$	11.72 $\times 10^5$	7.25 $\times 10^6$	7.61 $\times 10^6$
Calc. A_2 (e.s.u.)		7.20 $\times 10^6$	7.20 $\times 10^6$	6.79 $\times 10^6$	12.00 $\times 10^6$	7.20 $\times 10^6$	7.20 $\times 10^6$

Temperature at measurement, 14 — 16°C; electrostrictive material, VPZ-7 made by Tohoku Metals Co.; $e = 2.37 \times 10^6$ (e.s.u); cement, Super-cemedine; $\alpha_1 = 4.730$.

$$\zeta=P\left\{\cosh(\alpha_1 X)\cos\left(\frac{\alpha_1}{2}\right)+\cos(\alpha_1 X)\cosh\left(\frac{\alpha_1}{2}\right)\right\} \qquad \text{(A8)}$$

$$X=\frac{x}{l_0},\quad X_1=\frac{l_1}{l_0},\quad \alpha_1=4.730$$

From these the force coefficient A_1 at the vibrator end is given by

$$A_1=\frac{I_d}{v_f}=\frac{e'}{\sqrt{2}}\frac{a^2\alpha_1}{l_0}\left(1-\frac{h}{a}\right)^2 u(X_1) \qquad \text{(A9)}$$

Coefficients A_2 and A_e [(Eq. (10)] can be derived in similar ways. For the bimorph electrostrictive material the force coefficient was computed from the admittance circle by means of Eq. (4) [attachment of electrostrictive material over entire surface]. For the double-resonance vibrator, Eq. (9) for A_2 with $X_1 = 1$ yielded $e'/e = 1.1$. Next, by means of these formulas the force coefficients were calculated for different kinds of vibrators with electrostrictive material (attached over their whole surface).

Table A1 compares these calculated values with the actually measured values. Although the measured values in the table show some dispersion, they do not differ by more than 6% from the calculated values. Therefore, Eqs. (9) and (10) are considered satisfactory, for practical purposes, for computing the force coefficient.

Section I-F
Models and Circuit Element Descriptions

The mechanical filter engineer has the choice of designing and analyzing his filter using differential equations that describe the entire system or alternatively approximating the network with discrete coupled elements. The first method is possible with simple, low-degree systems, but it is too complex in the case of most practical filters. The second method, although not exact, lends itself well to visualization of the system. It also allows us to construct the electrical equivalent circuit of the system and to then analyze that circuit using general purpose computer analysis programs. The first two papers in this section discuss the formation of lumped element electrical equivalent circuit models of distributed parameter disks and bars. The Yakuwa-Okuda paper focuses on the linearity limits of resonators as a function of size, whereas the final paper describes a method for analyzing costs as a function of component tolerances.

The Johnson paper is somewhat tutorial in nature, and it describes equivalent circuits of mechanical resonators coupled with wires. The resonator equivalent circuits are quite general, and they are based on lumped "equivalent" masses and compliances at points on the resonators where coupling wires are attached. The coupling wires are described with both distributed and lumped equivalent circuits. Practical application of the theory is also included in the paper.

The Kuenemund paper shows the derivation of a four-element resonator equivalent circuit model for flexure-mode bar resonators. The admittance versus frequency curves for the resonator, as described by solutions from the differential equations of the bar, are compared with those obtained from the lumped-element equivalent circuit.

For a specified input power level and degree of nonlinearity, there is a minimum filter volume. Yakuwa and Okuda show that this volume is a function of filter bandwidth, vibration mode, and measured nonlinearity constants of the material. A flowchart for the calculation of the minimum volume is included.

The final paper in this section is primarily concerned with defining the tolerance versus cost problem in the production of mechanical filters. Emphasis is on statistical methods of testing the data used in the analysis, as well as the analysis itself. Details of the relationships between the properties of the filter elements and the electrical equivalent circuit values used in the statistical analysis are discussed in [1] below.

REFERENCES

[1] V. Sobotka and J. Trnka, "The use of digital computers in modelling of transfer characteristics of serially producted electromechanical filters," in *Proc. SSCT 71*, Tale, Czechoslovakia, 1971.

Electrical Circuit Models of Disk-Wire Mechanical Filters

ROBERT A. JOHNSON, MEMBER, IEEE

Abstract—A new lumped parameter electrical circuit model of a disk-wire mechanical filter is described. The disk resonator model is based on a knowledge of its natural frequencies and its equivalent mass at various points; the coupling-wire model is based on transmission line concepts. In order to maintain a topological correspondence between the mechanical and electrical circuits, the mobility analogy has been used throughout.

The new model has provided a means for understanding the effect of coupling-wire orientation on the selectivity of simple disk-wire filters. It is shown that variations in wire position affect the equivalent mass of "diameter" modes, causing changes in the finite attenuation poles located near the filter passband. Also discussed are problems associated with using "disk-pair" data in the design process. A solution based on an analysis of the improved filter model is discussed.

Introduction

Electrical circuit models can be very helpful in the design and manufacture of mechanical filters. In some cases, the model can be used as part of the filter design process or, in other cases, simply as an analysis tool. Models can be used in conjunction with digital computer network analysis programs or as a visual help in solving problems and creating new configurations. The models discussed in this paper are oriented towards disk-wire-type mechanical filters but are applicable to many other configurations as well.

The most popular electrical circuit models of mechanical systems have been those based on the "classical" force-voltage analogy.[1]–[4] This type of model which is characterized by series LC resonators is adequate in terms of its relationship with the differential equations of the system and is very helpful when used to describe mechanical filters employing certain types of piezoelectric transducers. A somewhat troublesome problem with this type of model is that it does not relate well to the concept of force being a "through" variable and velocity or displacement an "across" variable.[5]–[7]

The force-through velocity-across concept is very helpful in constructing mechanical schematic diagrams[8],[9] that can be directly transformed into electrical models having the same topology. This transformation is based on the "mobility analogy" where the through variables (force and current) are analogous quantities.[5] If one wishes to retain the force-voltage analogy, he may 1) find the dual of the electrical circuit based on the mobility analogy or 2) write the appropriate mechanical differential equations, convert these to electrical equations,

Manuscript received August 21, 1967.
The author is with Collins Radio Company, Newport Beach, Calif.

Reprinted from *IEEE Trans. Sonics Ultrason.*, vol. SU-15, pp. 41-50, Jan. 1968.

and then form a schematic diagram. In either case, the topological equivalence of the electrical and mechanical systems is lost when the force-voltage analogy is used. Therefore, in this paper, networks based on the force-current mobility analogy will be used exclusively.

Although simple electrical circuits based on the mobility analogy have been used to describe mechanical filters,[10]–[12] there has been no attempt to discuss the mobility model problem in general terms. The first section of this paper will deal with resonator equivalent circuits primarily from a mechanical viewpoint. Subsequent sections on coupling elements, measurements, and applications will be oriented towards the electrical equivalent circuit model. The topological correspondence of the mechanical and electrical networks makes the problem of conversion from one to the other almost trivial.

Resonator Equivalent Circuits

Although we are specifically concerned with the disk-wire type of mechanical filter, most of the material in this section can be applied to any distributed element resonator. The only restrictions we place on the resonator is that it be linear and bilateral so that superposition and reciprocity hold and that it be lossless. Since each natural mode of a linear resonator can be excited independently,[13],[14] we can write [see (25) in the Appendix noting that the letter j, when not used as a mode-identifying subscript, is the operators $\sqrt{-1}$]

$$v_i = (\omega/j) \sum_{j=1}^{\infty} [M_{ij}(\omega^2 - \omega_j^2)]^{-1} f_i, \tag{1}$$

where v_i is the velocity at point i on the resonator, M_{ij} is the "equivalent" mass of mode j corresponding to the frequency ω_j, and f_i is the applied force at i. Dividing v_i by f_i, we recognize the remaining expression as a partial fraction expansion of a reactance function with each term corresponding to a tuned circuit in a Foster-type network.[15] When force is considered the "through" variable, we obtain the mechanical driving-point immittance model shown in Fig. 1. K_{ij} is the equivalent stiffness of the mode j at the point i. One point that might be a source of confusion is the fact that the masses M_{ij} are not connected to a common reference. This is a point worthy of further discussion.

Pictorial and schematic diagrams of a simple system of springs and masses are shown in Fig. 2(a). The velocities v_1 and v_2 correspond to displacements x_1 and x_2 measured with reference to a common ground plane (inertial reference). The fact that both masses have a single inertial reference leads to the "grounding" of the reference terminal of each mass in the schematic diagram. Let us now look at Fig. 2(b). It is not difficult to show that the driving-point immittance characteristics of Fig. 2(b) are identical to those of Fig. 2(a). The parallel tuned circuits resonate at the frequencies,

$$\omega_1 = (K/M)^{1/2} \text{ and } \omega_2 = [(K + 2k)/M]^{1/2} \tag{2}$$

which correspond to the natural frequencies of the network. We are now brought back to the question of why the two masses $2M$ are not connected to the same common reference. Rather than throw the question out by saying that equivalent circuits for driving-point impedances may not have to follow the same rules as complete multiterminal circuits such as Fig. 2(a), let us attempt to establish a simple rule or set of rules to cover this case.

Since we can consider the oscillation of a linear circuit as a superposition of noninteracting modes of vibration,[13] we might consider trying to separate the coordinates associated with each mode of vibration. This problem is treated in numerous texts that deal with oscillations of linear systems.[16] The solution is to apply a linear transformation to the original coordinates x_1 and x_2 that results in a set of normal coordinates

$$\begin{bmatrix} y_1 \\ y_2 \end{bmatrix} = \begin{bmatrix} 1 & 1 \\ 1 & -1 \end{bmatrix} \begin{bmatrix} x_1 \\ x_2 \end{bmatrix} \tag{3}$$

which are uncoupled in the sense that when the system vibrates at frequency ω_1, only y_1 is finite while y_2 is zero. At frequency ω_2, y_2 is finite and y_1 is zero. Therefore in Fig. 2(b), we show each mass tied to a different reference point. These points then act as references for a new system of velocities $(v_1 + v_2)/2$ and $(v_1 - v_2)/2$ associated with the new normal coordinates y_1 and y_2. Generalizing this result, we can say that it is possible to find an independent reference point for each mass M_{ij} in the circuit of Fig. 1 in the sense that this point or node acts as a reference for a new coordinate that is only associated with the mode j. Physically, we might consider each reference as a set of nodal lines associated with a particular mode of vibration such as those shown in Fig. 3(a).

In Fig. 3(a) we are considering not only two modes of vibration but two points of measurement as well. At the resonant frequency of the first mode, the two-diameter mode, there is a fixed relationship between the velocities v_{11} and v_{21}, according to (18) and (20) in the Appendix. Since the kinetic energy of the vibrating disk does not change when we change measurement points, we would then expect that the equivalent masses of the disk, at points 1 and 2, would also have a fixed relationship [see (26) in the Appendix]. In fact, the equivalent masses are inversely proportional to the square of their respective velocities. We would expect that these constraints on the equivalent masses and velocities would lead to the use of ideal coupling elements (levers or transformers) in the mechanical equivalent circuit model.[17] This is indeed the case as is shown in Fig. 3(b). The velocity ratio ϕ_{ij} is considered a lever arm length ratio in mechanical networks and a transformer turns ratio in electrical networks. Therefore, defining ϕ_{ij} as the ratio of primary velocity to secondary velocity, we can write a set of equations to describe the two-port of Fig. 3(b);

$$\begin{bmatrix} v_1 \\ v_2 \end{bmatrix} = \begin{bmatrix} \phi_{11}^2 z_1 & \phi_{11}\phi_{21} z_1 \\ \phi_{11}\phi_{21} z_1 & \phi_{21}^2 z_1 \end{bmatrix} \begin{bmatrix} f_1 \\ f_2 \end{bmatrix}, \tag{4}$$

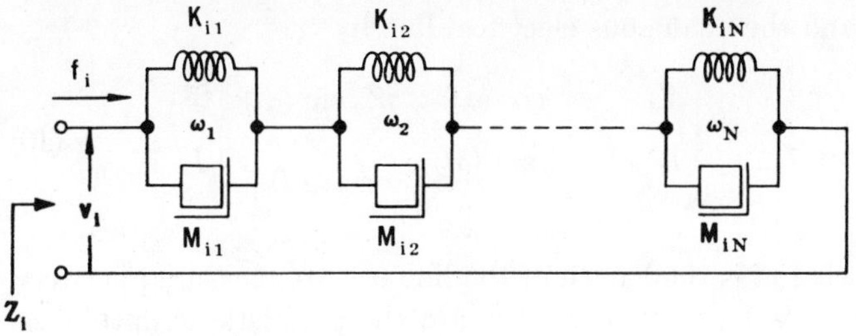

Fig. 1. Mechanical driving-point equivalent circuit of a disk resonator.

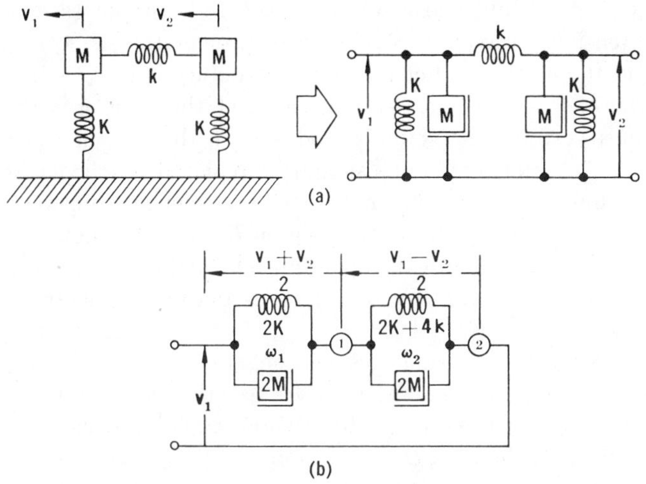

Fig. 2. Spring-coupled resonator pair; (a) pictorial diagram and schematic diagram, and (b) driving-point immittance schematic diagram.

where z_1 is the mobility (ratio of velocity across to force through) $(\omega/j)M_1^{-1}(\omega^2 - \omega_1^2)^{-1}$ of the parallel tuned circuit. Relating the input and output quantities of an *ABCD* transmission type of matrix, we obtain the equations,

$$\begin{bmatrix} v_1 \\ f_1 \end{bmatrix} = \begin{bmatrix} \phi_{11}/\phi_{21} & 0 \\ 1/(\phi_{11}\phi_{21}z_1) & \phi_{21}/\phi_{11} \end{bmatrix} \begin{bmatrix} v_2 \\ -f_2 \end{bmatrix}. \quad (5)$$

Since it has been necessary to make use of ideal transformers in the multiport problem, possibly they can also be used in driving-point immittance models. Considering a single point (or port) $i = 1$ and two modes $j = 1, 2$ of the disk of Fig. 3(a), it seems logical that instead of representing the network in the manner shown in Fig. 1, the circuit of Fig. 3(c) can be used in its place. It should be said that the distributed parameter resonator problem we are considering is very similar to a lumped spring-mass problem involving more than one inertial reference; for instance, a mass that has both linear and angular motion. Alternate ways of representing networks such as Fig. 3(c) in multireference problems have included the use of mutual mass,[8] levers,[2] or constraint equation "oblongs."[9] The velocity v_1 is related to the force f_1 by

$$v_1 = (\phi_{11}^2 z_1 + \phi_{12}^2 z_2) f_1. \quad (6)$$

Making use of the results of (4) and (6), we can express the velocities of an *M*-port, *N*-mode resonator as a func-

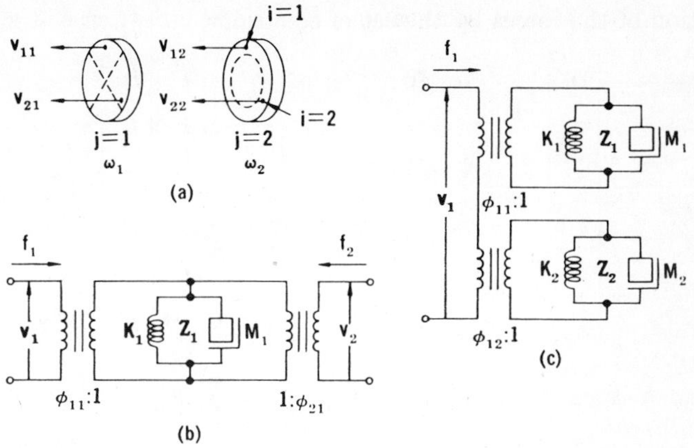

Fig. 3. Two-port, double-mode resonator; (a) two-diameter and one-circle modes of vibration, (b) the disk as a two-port near frequency ω_1, and (c) the driving-point immittance schematic at point (port) $i = 1$.

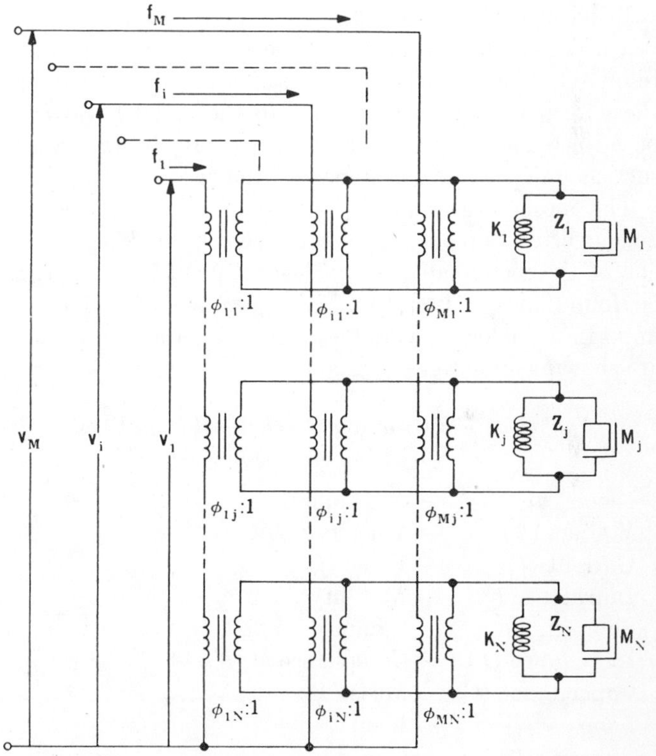

Fig. 4. Generalized schematic diagram of an *M*-port, *N*-mode resonator.

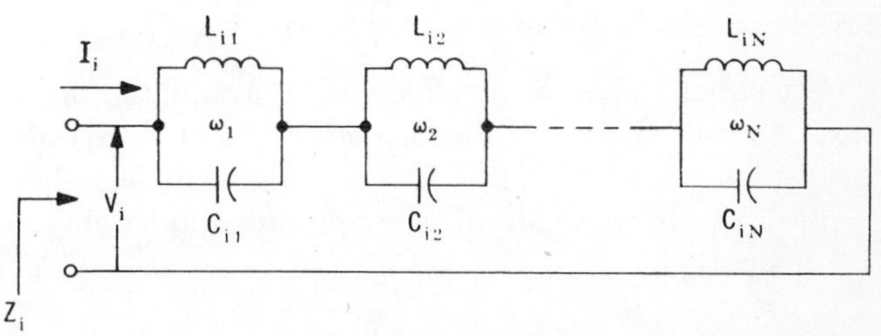

Fig. 5. Electrical driving-point equivalent circuit of the disk resonator of Fig. 1.

tion of the forces by the set of equations,

$$\begin{bmatrix} v_1 \\ v_2 \\ \vdots \\ v_i \\ \vdots \\ v_M \end{bmatrix} = \begin{bmatrix} z_{11} & z_{12} & \cdots & z_{1i} & \cdots & z_{1M} \\ z_{21} & z_{22} & \cdots & z_{2i} & \cdots & z_{2M} \\ \vdots & \vdots & & \vdots & & \vdots \\ z_{i1} & z_{i2} & \cdots & z_{ii} & \cdots & z_{iM} \\ \vdots & \vdots & & \vdots & & \vdots \\ z_{M1} & z_{M2} & \cdots & z_{Mi} & \cdots & z_{MM} \end{bmatrix} \begin{bmatrix} f_1 \\ f_2 \\ \vdots \\ f_i \\ \vdots \\ f_M \end{bmatrix} \quad (7)$$

$$z_{11} = \phi_{11}^2 z_1 \cdots + \phi_{1j}^2 z_j \cdots + \phi_{1N}^2 z_N = \sum_{j=1}^{N} \phi_{1j}^2 z_j$$

$$z_{1i} = \phi_{11}\phi_{i1} z_j \cdots + \phi_{1j}\phi_{ij} z_j \cdots + \phi_{1N}\phi_{iN} z_N = \sum_{j=1}^{N} \phi_{1j}\phi_{ij} z_j,$$

etc. A generalized resonator model having M-ports and N natural resonances is shown in Fig. 4.

In terms of disk-wire mechanical filters, each port may be considered the point of attachment of a wire to the disk resonator. The velocity ratios (which are dependent on the wire placement with respect to the nodal lines) can be adjusted so as to make the M_j's any desired value such as unity or the static mass of the disk, etc.

The mechanical resonator model of Fig. 4 can easily be converted to its electrical equivalent by making use of one of the many tables of analogous quantities that can be found in the literature.[5],[7],[8] Some of the more important analogous variables, network parameters, etc., are shown as follows.

Analogous Electrical and Mechanical Quantities (Mobility Analogy)

Electrical	→ Mechanical
Voltage (V)	→ Velocity (v)
Current (I)	→ Force (f)
Impedance (Z)	→ Mobility (z)
Short circuit	→ Clamped point
Inductance (L)	→ Compliance ($1/K$)
Capacitance (C)	→ Mass (M)
Loop	→ Loop

Making use of this list of analogous quantities, the mechanical driving-point immittance model of Fig. 1 can be converted to its electrically analogous form shown in Fig. 5.

Coupling-Wire Descriptions

The characteristics of small diameter coupling wires (rods) are very similar to the characteristics of electrical transmission lines.[1],[18],[19] The coupling wire can be described by the mechanical equations (assuming there is no dissipation)

$$\begin{bmatrix} v_1 \\ f_1 \end{bmatrix} = \begin{bmatrix} \cos(\alpha l) & jz_0 \sin(\alpha l) \\ j\dfrac{\sin(\alpha l)}{z_0} & \cos(\alpha l) \end{bmatrix} \begin{bmatrix} v_2 \\ f_2 \end{bmatrix} \quad (9)$$

and the analogous electrical line by

$$\begin{bmatrix} V_1 \\ I_1 \end{bmatrix} = \begin{bmatrix} \cos(\alpha l) & jZ_0 \sin(\alpha l) \\ j\dfrac{\sin(\alpha l)}{Z_0} & \cos(\alpha l) \end{bmatrix} \begin{bmatrix} V_2 \\ I_2 \end{bmatrix}, \quad (10)$$

where l is the length of the line or wire measured between points 1 and 2, Z_0 and z_0 are characteristic immittances, and α is the propagation constant (angular frequency divided by the velocity of propagation of the wave).

The electrical analogy can be used to describe both extensional and shear waves in the rod. In the case of extensional waves, the velocity of propagation is $\sqrt{E/\rho}$ and the mechanical characteristic mobility is $(A\sqrt{\rho E})^{-1}$, where A is the cross-sectional area of the wire, ρ is the density, and E is Young's modulus. In the case of shear waves, Young's modulus should be replaced by the shear modulus. Keeping in mind the relationships between analogous electrical and mechanical systems (i.e., the topologies are identical, capacitance is analogous to mass, inductance to compliance, etc.), let us analyze the electrical transmission line analogy.

A very useful transmission line model is the pi network shown in Fig. 6(a). The y's are the short-circuited admittance parameters of the two-port and can easily be found from the $ABCD$ description of (10). Since

$$\begin{aligned} y_{11} &= D/B = (1/jZ_0)\cot(\alpha l) \\ y_{12} &= -1/B = j/[Z_0 \sin(\alpha l), \end{aligned} \quad (11)$$

we can use a half-angle formula to obtain the shunt arm admittance,

$$y_{11} + y_{12} = (j/Z_0)\tan(\alpha l/2). \quad (12)$$

When the coupling wire is less than one-eight wavelength, the shunt arm can be replaced by a capacitance $C_{ST}/2$ which is analogous to one-half of the static mass of the wire. The series arm can be replaced, as shown in Fig. 6(b), by an inductance L which is analogous to the compliance l/AE extensional mode case.

A very interesting case is when the length of the coupling wire is a quarter wavelength. It is not difficult to show that the equivalent circuit of Fig. 6(c) approximates a quarter-wavelength line by equating matrix elements in (10) and the $ABCD$ matrix of Fig. 6(c). Thus, when l is a quarter wavelength, $\alpha l = \pi/2$ and

$$\begin{bmatrix} 0 & jZ_0 \\ j/Z_0 & 0 \end{bmatrix} = \begin{bmatrix} 0 & j\omega L \\ j/\omega L & 0 \end{bmatrix}. \quad (13)$$

Therefore, $L = Z_0/\omega$ which is analogous to $2/\pi$ times the compliance l/AE of the wire.

A generalized equivalent circuit model could be constructed by replacing $-y_{12}$ and $y_{11} + y_{12}$ of Fig. 6(a) by lumped circuit equivalents. A somewhat different approach is to treat the transmission line like a two-port, multiple-mode resonator. The open-circuit driving-point

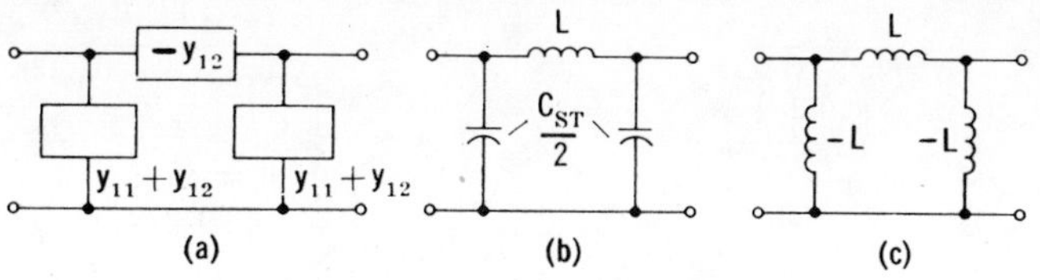

Fig. 6. Pi-type coupling-wire descriptions; (a) generalized two-port admittance description, (b) "short line" equivalent circuit, and (c) quarter-wavelength equivalent circuit.

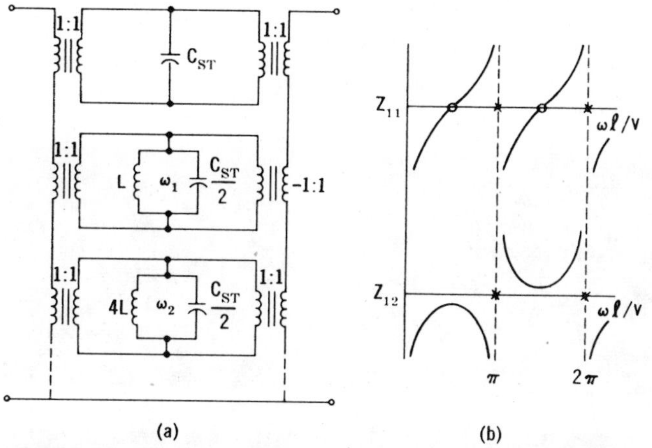

Fig. 7. (a) Generalized coupling-wire equivalent circuit for a single-wave propagation mode, and (b) open-circuit driving-point and transfer impedance characteristics.

impedance is simply

$$z_{11} = A/C = -jZ_0 \cot (\alpha l). \tag{14}$$

At low frequencies we see from (14) or Fig. 6(b) that the line looks like a shunt capacitor C_{ST} which is analogous to the static mass of the wire. Additional parallel tuned circuits such as those shown in Fig. 4 can be found by inverting (14) and equating the slope of $(1/z_{11})$, at integral numbers of half wavelengths, to the slope of the admittance of a parallel tuned circuit at the same frequency.[1] Thus, from (15) (v is the velocity of propagation)

$$d[(j/Z_0) \tan (\omega l/v)]/d\omega = d[(\omega_1^2 - \omega^2)/(j\omega/C)]/d\omega, \tag{15}$$

we can obtain the result that $C = C_{ST}/2$. The capacitor C is resonant with the inductance $(n)^2L$ where n is the number of half wavelengths associated with each resonance and L is analogous to $2/\pi^2$ times the compliance of the wire.

Having found expressions for the element values of the lumped driving-point immittance model and knowing that the velocities at opposite ends of a resonant rod are equal in magnitude and phase when n is even (and opposite in phase when n is odd), we can construct the model of Fig. 7(a). The complexity of this model can be reduced somewhat by removing the transformers on the left-hand side of Fig. 7(a) and series connecting the static capacity and tuned circuits.

Corresponding to the model are sketches of the open-circuited driving-point impedance z_{11} and transfer impedance z_{12} shown in Fig. 7(b). It is important to note that there are no transfer impedance zeros. This fact seems to be in conflict with the test data of Fig. 8 which shows

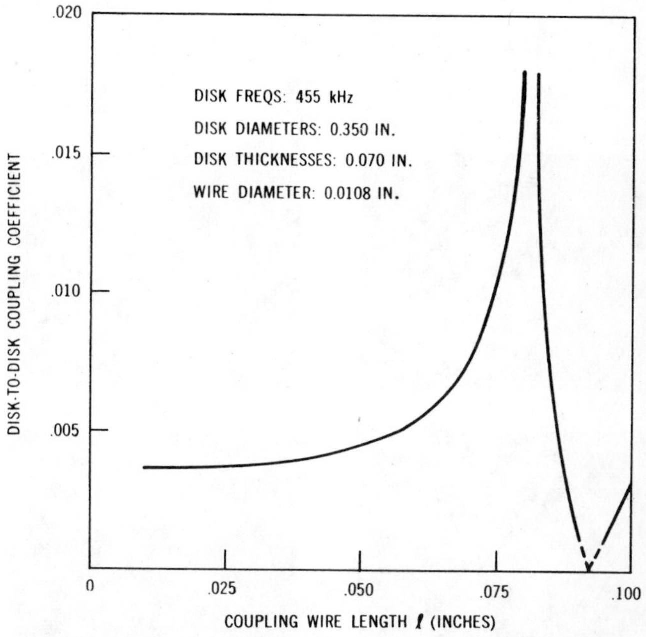

Fig. 8. The effect of multiple modes of propagation on resonator coupling (three symmetrically spaced nickel coupling wires).

the variation of coupling between two disk resonators as a function of coupling-wire length. If it is assumed that there is only one type of wave being propagated, then there is nothing in Fig. 7(b) that would explain the extremely small value of coupling at 0.092 inches. If two or more modes are being propagated (which seems likely because of the relatively large thickness of the disk), we can represent the wire as a parallel combination of the equivalent circuits of Fig. 7(a).[6] For instance, there would be one network for the shear mode, one for the extensional mode, etc. Adding the y parameters of (11) corresponding to each mode, we will obtain a full matrix description of the coupling wire. Since the shear-mode propagation constant is not equal to that of the extensional mode, zeros of y_{12} (for the entire wire) and z_{12} will result.

Disk and Coupling-Wire Measurements

Although it is sometimes helpful to know that more than one type of wave is being propagated through the coupling wires, this fact is not of much use in finding purely analytical expressions for calculating coupling-wire diameters and lengths needed to satisfy specific design requirements. The primary reason for this difficulty is that in attaching a coupling wire to a disk resonator, the most consistent results have been obtained by welding; this process creates gross distortions in the shape of

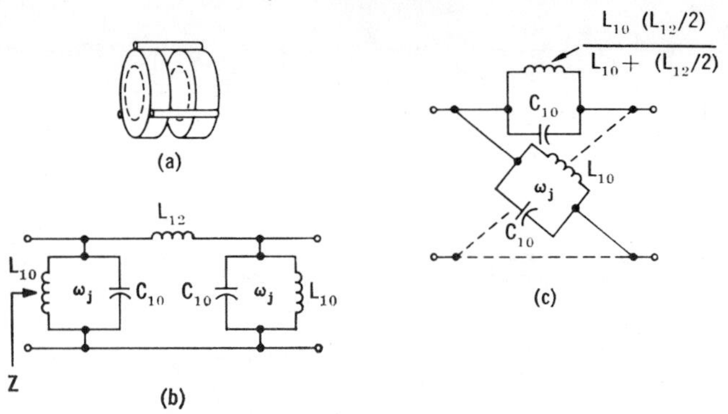

Fig. 9. (a) Wire-coupled disk pair, (b) single-mode single-wire model, and (c) lattice equivalent of (b).

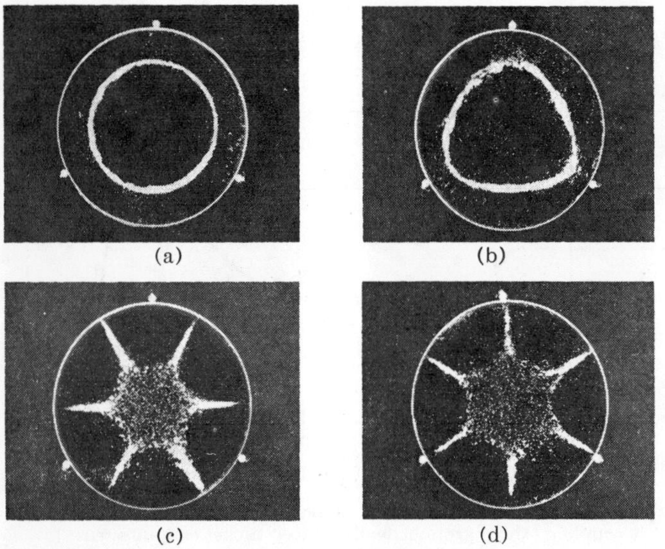

Fig. 10. Photographs of lycopodium powder nodal patterns of a two identical disk set. The coupling-wire orientation is symmetrical. The modes are (a) the single-circle mode, $F_1 = 60.06$ kHz; (b) a combination single-circle and low-frequency three-diameter mode, $F_2 = 63.38$ kHz; (c) the low-frequency three-diameter mode, $F_3 = 77.05$ kHz; and (d) the high-frequency three-diameter mode, $F_4 = 78.05$ kHz.

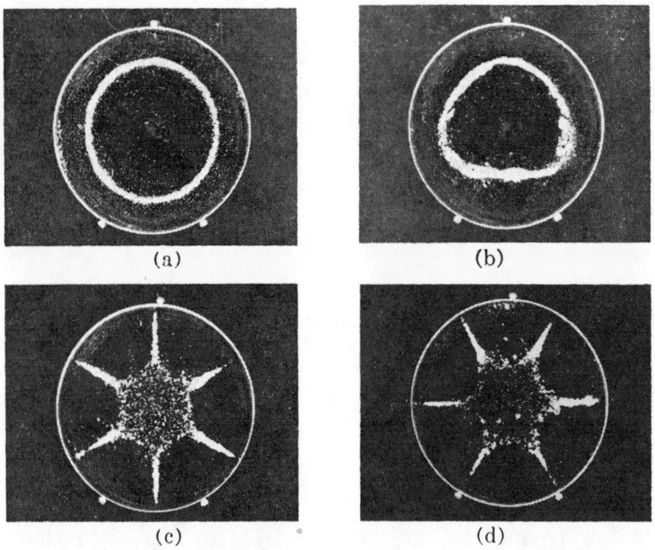

Fig. 11. Coupled disk-pair nodal patterns corresponding to an unsymmetrical coupling-wire orientation. The modes are (a) the single-circle diameter mode, $F_1 = 59.95$ kHz; (b) a combination of two-diameter, single-circle, and three-diameter modes, $F_2 = 68.03$ kHz; (c) the low-frequency three-diameter mode, $F_3 = 77.27$ kHz; and (d) the high-frequency three-diameter mode, $F_4 = 77.59$ kHz.

the wire. These distortions make any exact mathematical analysis practically impossible. Because of this problem (along with the fact that very little work has been done in the area of diameter mode vibrations of thick disks), a great deal of effort has gone into finding empirical techniques for measuring disk and wire characteristics.

The most basic measurement is that of finding the natural frequencies of a disk resonator. This is accomplished by making use of the magnetostrictive property of the Ni–Span *C* iron–nickel disk alloy. The disk is placed in the magnetic field of a coil which is driven at a particular frequency or it may act as the resonant element of an oscillator circuit. Placement of the disk in such a way as to confine the greatest amount of flux to the upper or lower surfaces of the disk makes it possible to excite all bending modes. The electromechanical coupling is small so the measured resonances are very close to the actual natural resonant frequencies of the disk. In addition to the single disk meaurements, the same test equipment can be used to measure the characteristics of wire-coupled pairs of disks.

Measurements of wire-coupled disk pairs, such as those shown in Fig. 9(a), are extremely helpful in determining relative values of coupling-wire stiffness and disk equivalent mass. If we only consider a narrow frequency range about one of the natural resonances ω_j and if the coupling wires are short and symmetrically spaced with respect to the nodal lines (the orientation is arbitrary in the case where there are only nodal circles), a general disk-pair equivalent circuit based on Figs. 4 and 6(a) can be reduced to that of Fig. 9(b).

The model of Fig. 9(b) has been used for a number of years by mechanical filter designers.[11],[20],[21] The two natural frequencies of the disk pair can easily be found by applying Bartletts' bisection theorem to the ladder of Fig. 9(b) to obtain the lattice of Fig. 9(c). The natural frequencies, which are the open-circuit driving-point impedance poles, are just the resonant frequencies of the parallel tuned circuits. It is not difficult to show that the bandwidth of a constant-k type of image parameter design that contains sections like that of Fig. 9(b)[22] will be inversely proportional to C_{10} and equal to twice the difference of the natural frequencies of the wire-coupled disk-pair model. Therefore, it should be possible for the designer to estimate coupling-wire lengths and diameters needed to satisfy certain filter specifications by building and measuring simple disk pairs. This technique works very well for narrow bandwidth filters and filters that have special coupling-wire orientations that minimize the effect of the higher-frequency mode adjacent to the desired one or two circle mode.

Fig. 10(a) shows the nodal pattern of one of the disks of a disk pair that has been excited at the lower natural resonance of the pair. Because the two disks were tuned to the same single-circle mode resonant frequency, we see from Fig. 9 that this frequency corresponds to the lower resonance. Therefore, as we would expect, the nodal pattern is circular.

The second natural resonance occurs at a frequency between the single-circle mode of Fig. 10(a) and the three-diameter modes of vibration shown in Figs. 10(c) and 10(d).[11] At this second natural mode, the amplitude distribution is a linear combination of the single-circle and three-diameter amplitudes. The relative amplitude of each contribution is fixed and independent of the means of excitation (because a natural resonance has been excited). The nodal pattern corresponding to the second mode is shown in Fig. 10(b). It is apparent that the three-diameter mode of Fig. 10(c) has caused the circular pattern to be distorted considerably.

Fig. 11 shows a set of disk-pair nodal patterns corresponding to a different coupling-wire orientation. The symmetrical orientation of Fig. 10 places the wires midway between the three-diameter nodal lines resulting in the maximum possible excitation of this mode [see (18) and (19) in the Appendix]. As the two lower wires are brought closer together, the excitation of the lower three-diameter mode as shown in Fig. 11(c) is decreased and the velocity of the upper mode contribution Fig. 11(d) is increased. The resultant second natural resonance of the disk pair is shown in Fig. 11(b). Note that not only has the amplitude distribution varied considerably but the frequency as well.

The filter designer is now presented with the problem of estimating the coupling-wire dimensions needed to obtain a specified filter bandwidth through the use of data that are very much influenced by the adjacent three-diameter mode. This problem is usually side-stepped by constructing four disk "quads" having a maximum natural frequency difference equal to 0.85 percent of the filter bandwidth. Thus, a very close estimate can be obtained, but at the expense of a more complex coupled-disk set.

The coupled-disk problem is closely related to a problem that puzzled mechanical filter design engineers for many years: that of accounting for the variation in filter selectivity with coupling-wire orientation.

Amplitude Response Variation with Changes in Coupling-Wire Orientation

The frequency response of a seven-disk voice channel filter is shown in Fig. 12. This particular filter which employs three symmetrically oriented short coupling wires is able to achieve a high degree of selectivity on the carrier side of the response without the use of "bridging wires."[11],[21],[23] The asymmetrical response cannot be realized with the simple model of Fig. 9(b) so use must be made of a more general equivalent circuit based on Fig. 4 which takes into account modes of vibration adjacent to the desired single-circle mode.

The two modes of interest in this particular case are the single-circle and the three-diameter modes of Fig. 13. Because of the symmetrical coupling-wire orientation, other nearby modes are not excited to as great an extent as the three-diameter mode shown. As the positions of the coupling wires on the disk vary, the other modes of vibration must be considered. Because of the symmetry of the wires with respect to the nodal lines, the three coupling wires [each of which can be represented by the short line equivalent circuit of Fig. 6(b)] can be reduced to a single wire. If the equivalent masses of the wires (shunt capacitors) are ignored for the sake of simplifying the analysis, a disk and its attached wires can be represented by the equivalent circuit of Fig. 14.

Since the actual filter is composed of a cascade of seven disks coupled by the three wires, infinite attenuation points will be produced at frequencies where the impedance of the shunt arms of the network are equal to zero. Referring to Fig. 14, this frequency can be easily shown to be

$$\omega_\infty = [(\mathbf{M}\omega_2^2 + \omega_1^2)/(\mathbf{M} + 1)]^{1/2}, \tag{16}$$

where $M = C_2/C_1$. Although the noise level of the test equipment prevents us from seeing the infinite attenuation points of the curve of Fig. 12, they can be seen when the number of sections is reduced.

As the two lower coupling wires shown in Fig. 13 are either brought closer together or are separated, the complexity of our model increases considerably. First, we can no longer assume that the only modes that need to be considered are those shown in Fig. 13. In fact, there are two modes in the ω_2 frequency region that will be excited.[18]

Secondly, because of the variation of the displacement around the circumference of the disk, each wire will have a different effect on the circuit. Therefore, the three wires can no longer be combined into one as was done in Fig. 14. Because of the additional modes and the coupling-wire complexity, it is necessary to make use of the general

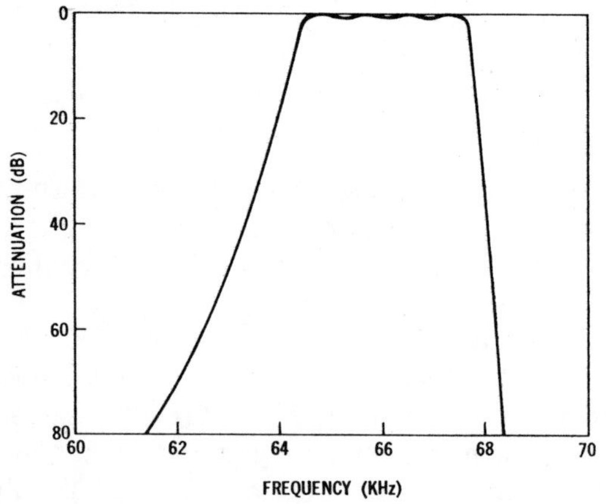

Fig. 12. Frequency response of a conventional seven-disk voice channel filter. The steep carrier side response is a result of symmetrical coupling-wire orientation.

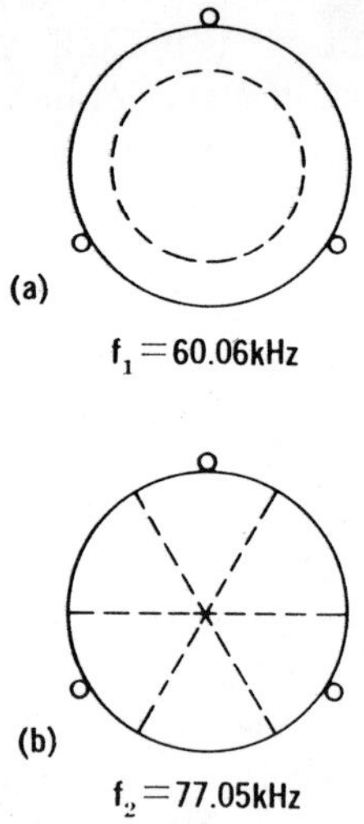

Fig. 13. Important disk modes of vibration (seven-disk voice channel filter); (a) single-circle mode, and (b) low-frequency three-diameter mode.

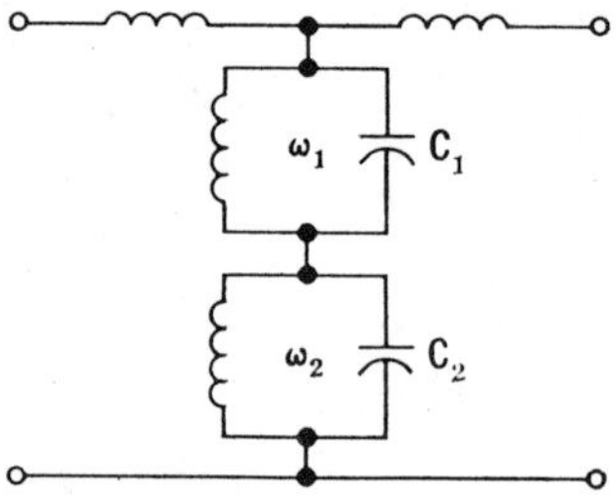

Fig. 14. Double-mode disk, symmetrical coupling-wire equivalent circuit.

resonator equivalent circuit model of Fig. 4 and the short wire circuit of Fig. 6(b). This means that the velocity ratios ϕ_{ij} must be found.

Because of the small amount of work that has been done in the area of calculating the equivalent mass of a thick disk vibrating in a diameter mode, a few basic laboratory measurements must be made in order to be able to calculate the velocity ratios ϕ_{ij}. These measurements consist of coupling a pair of disks with a single wire or its equivalent and measuring the frequency differences Δf_j between the resultant coupled natural frequency pairs, i.e., the two natural frequencies of the disk pairs associated with each mode j of the disks individually. Since it has been shown that equivalent mass is proportional to velocity squared and is inversely proportional to Δf_j, we can easily relate the maximum velocities $v_{j\,\max}$ associated with each mode.

In addition to finding the relative maximum velocities of each mode, it is necessary to find, for each wire, the ratio of the velocity at the point where the coupling wire is attached to the disk to the maximum velocity point on the circumference of the disk. These ratios can easily be found from (18) and (19) in the Appendix if the value of α which relates to the angular position of the diameter nodal lines is known. The orientation of the nodal lines is determined by the position of the coupling-wire masses because of the effect of wire mass on frequency and the fact that the orientation of the diameter lines will be such that the natural frequencies of the disk will take on the maximum and minimum possible values.[11],[18]

The problem of determining nodal-diameter orientation can be solved by assuming a position of the nodal lines and calculating the disk frequency shift due to the mass loading effect of the coupling wires. The orientation is then changed and a new frequency is calculated. In a few trials the nodal-line positions corresponding to the maximum and minimum natural frequencies can be found. Analytical expressions that relate frequency, nodal-diameter orientation, and coupling-wire orientation can be found by making reference to Fig. 15.

Fig. 15(a) shows one-half of the equivalent circuit of a three-wire two-disk network which is excited in the frequency region near ω_j. The effect of each wire mass, represented by the capacitors C, on the natural frequencies of the disks can be determined by moving each capacitor "through" its respective transformer. The resultant equivalent circuit is shown in Fig. 15(b). a_j is simply the sum of the ϕ_{ij}^2's where

$$\phi_{ij} = v_{ij}/v_{j\,\max} = \cos\,[n(\theta_i - \alpha_j)], \qquad (17)$$

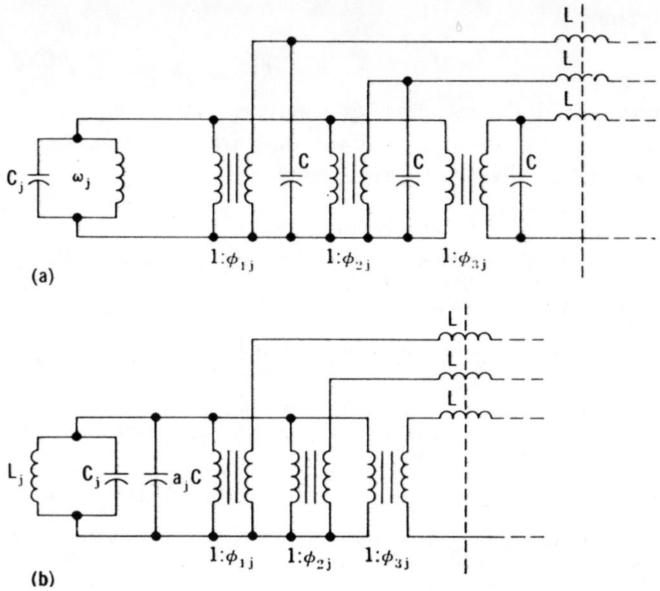

Fig. 15. Equivalent circuit (only half is shown) of a three-wire, disk pair near the frequency ω_j; (a) with the equivalent wire mass connected to the wire, and (b) with the masses brought through the transformers and combined.

where n is the number of nodal diameters, θ_{ij} is the angle between the point i and an arbitrary reference, and α_j is the angle between the reference and a maximum velocity point corresponding to an assumed nodal pattern of mode j. The maximum velocity point bisects the arc between the nodal diameters, as shown in Fig. 16.

The maximum frequency mode corresponds to a nodal pattern that results in a minimum value of a_jC, which is apparent when a_jC is combined with C_j in Fig. 15(b). The C_j's are related by the Δf_j measurements discussed previously. In the case of three symmetrically spaced coupling wires (spaced 120 degrees apart) and a two-diameter mode of vibration, the nodal-diameter orientation is indeterminate. This is an unusual case, the sum of the ϕ_{ij}^2's being equal to 1.5 for all values of θ and both two-diameter mode natural frequencies being coincident.

The same basic concepts that have been applied to the case of the one-circle mode voice channel filter can be applied to two-circle mode filters that operate in the 250 to 600 kHz frequency region. Because the adjacent modes are the two-diameter, one-circle and the six-diameter modes, symmetrical coupling-wire spacing again results in a lower sideband response.

Conclusions

A basic understanding of the effect of coupling-wire orientation on filter selectivity has been achieved by making use of a new disk resonator equivalent circuit model. The new disk and coupling-wire models described are lumped element approximations to the actual distributed systems and are based on the mobility analogy. Use of the mobility analogy allows the circuit designer to interchange mechanical and electrical circuit concepts freely

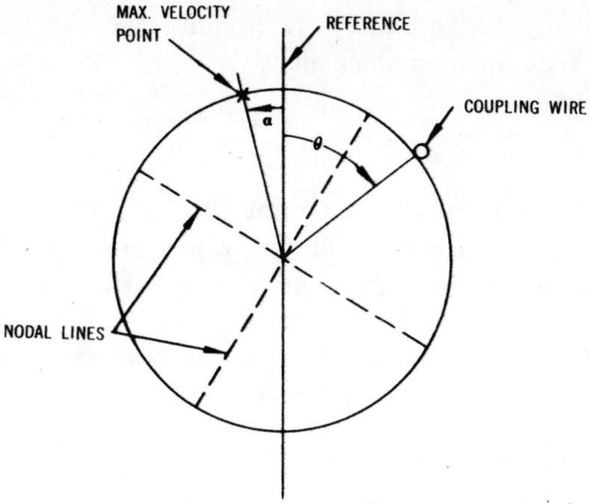

Fig. 16. Disk resonator described by (17).

because of the topological identity of the two systems.

The new model has been helpful in the design of mechanical filters that employ diameter modes for phase inversion,[11] as a means of reducing the number of disks needed to realize a specified number of natural frequencies,[23] and in the design of the wire orientation dependent single-sideband filters described in this paper. The model has also been useful in the design of narrowband filters employing "nonresonant" decoupling disks and in understanding and controlling spurious responses.

The new model, though very complex, should be helpful in determining resonator coupling and frequency sensitivity to changes in coupling-wire length and orientation. This work, which can be done on a digital computer, will be useful in both filter design and manufacture.

Appendix

The following equations are written in such a way as to give the reader a better understanding of how both velocity and equivalent mass vary across the surface of a disk. These equations have essentially been derived by Rayleigh[18] and Skudrzyk,[14] but in a somewhat different form. We will only consider the case where the disk resonators are linear, lossless, and time invariant.

A thin disk resonator vibrating freely at the frequency of one of its natural modes nm has a normalized velocity distribution,

$$v(n, m, \theta, r) = B(n, m, \theta, r) = E(n, m, \theta)F(n, m, r), \quad (18)$$

where n and m are the number of nodal diameters and nodal circles, respectively; θ and r are the polar coordinates of a point on the surface of the disk, and

$$E(n, m, \theta) = \cos (n\theta + \alpha) \quad (19)$$

$$F(n, m, r) = J_n(kr) + \lambda J_n(jkr), \quad (20)$$

where α, k, and λ are functions of the natural mode nm. In addition, J denotes a Bessel function and j is $\sqrt{-1}$. Note that we have dropped any time dependency.

Since any forcing function, in our case the force f, excites each mode independently,[13] we can write the expression,

$$v_{ij}(l, \omega) = [A_j(l, f, \omega)B_{ij}]f, \tag{21}$$

where i corresponds to the point (θ, r) subscript j is an integer representing the mode mn, while l corresponds to a set of dimensions and the density of the disk and ω is the angular frequency of the force f. $A_j(l, f, \omega)$ which is a function of the force distribution (not the amplitude of the force, since we assume linearity) can be written,

$$A_j(l, f, \omega) = -j\omega K_j(l, f)/[\omega^2 - \omega_j^2]. \tag{22}$$

Since each of the natural modes j are excited independently, we can write an expression for the velocity at a point i as

$$v_i(l, \omega) = \left[\sum_{j=1}^{\infty} A_j(l, f, \omega)B_{ij}\right]f. \tag{23}$$

The nature of the disk-wire mechanical filter is that the forcing functions act as point sources even though the disks and wires themselves are distributed parameter-type elements. Therefore, let us look at the velocity at point i due to a force applied at the same point. Remembering that $\mathbf{K}$ is a function of the physical properties of the disk and that the force distribution is that of an applied force at i, from (22) and (23), we can write

$$v_i = \left[(\omega/j) \sum_{j=1}^{\infty} B_{ij}\mathbf{K}_{ij}/(\omega^2 - \omega_j^2)\right]f_i \tag{24}$$

or an equivalent expression corresponding to a lumped spring-mass system,

$$v_i = (\omega/j) \sum_{j=1}^{\infty} [M_{ij}(\omega^2 - \omega_j^2)]^{-1}f_i, \tag{25}$$

where M_{ij} is the jth mode equivalent mass of the disk at the point i. The equivalent mass can be found by calculating the kinetic energy T of the disk corresponding to a particular mode and then dividing this quantity by $v_{ij}^2/2$. The total kinetic energy of the disk is just the velocity (18) squared times one-half the mass per unit area integrated across the surface of the disk. The equivalent mass concept is justified by the fact that an observer at the driving-point i cannot tell the difference between a disk and a simple spring-mass system when the system is vibrating near the natural frequency ω_j.

Since

$$M_{ij} = T/(v_{ij}^2/2) \tag{26}$$

normalized curves showing the variation of equivalent mass across the surface of the disk are merely curves of reciprocal velocity (18) squared.

References

[1] W. P. Mason, *Electromechanical Transducers and Wave Filters.* New York: Van Nostrand, 1948.

[2] H. Olsen, *Dynamical Analogies.* New York: Van Nostrand, 1943.

[3] R. L. Sharma, "Equivalent circuit of a resonant finite, isotropic elastic circular disk," *J. Acoust. Soc. Am.*, vol. 28, pp. 1153–1158, November 1956.

[4] M. Konno and H. Nakamura, "Equivalent electrical network for the transversely vibrating uniform bar," *J. Acoust. Soc. Am.*, vol. 38, pp. 614–622, October 1965.

[5] F. A. Firestone, "A new analogy between mechanical and electrical systems," *J. Acoust. Soc. Am.*, vol. 4, pp. 249–267, January 1933.

[6] H. M. Trent, "Isomorphisms between oriented linear graphs and lumped physical systems," *J. Acoust. Soc. Am.*, vol. 27, pp. 500–527, May 1955.

[7] G. F. Paskusz and B. Bussell, *Linear Circuit Analysis.* Englewood Cliffs, N. J.: Prentice-Hall, 1963.

[8] M. F. Gardner and J. L. Barnes, *Transients in Linear Systems.* New York: Wiley, 1942.

[9] H. M. Trent, "On the construction of schematic diagrams for mechanical systems," *J. Acoust. Soc. Am.*, vol. 30, pp. 795–800, August 1958.

[10] M. Börner, "Mechanische Filter Mit Dämpfungspolen," *Archiv Elektr. Ubertrag*, vol. 17, pp. 103–107, March 1963.

[11] R. A. Johnson and R. J. Teske, "A mechanical filter having general stopband characteristics," *IEEE Trans. Sonics and Ultrasonics*, vol. SU-13, pp. 41–48, July 1966.

[12] F. Künemund and K. Traub, "Mechanische Filter mit Biegeschwingern," *Frequenz*, vol. 18, pp. 277–280, September 1964.

[13] C. A. Desoer, "Modes in linear circuits," *IRE Trans. Circuit Theory*, vol. CT-7, pp. 211–223, September 1960.

[14] E. J. Skudrzyk, "Vibrations of a system with a finite or an infinite number of resonances," *J. Acoust. Soc. Am.*, vol. 30, pp. 1140–1152, December 1958.

[15] R. M. Foster, "A reactance theorem," *Bell Sys. Tech. J.*, vol. 3, pp. 259–267, April 1924.

[16] L. A. Pipes, *Applied Mathematics for Engineers and Physicists.* New York: McGraw-Hill, 1958.

[17] H. M. Trent, "On the conceptual necessity and use of perfect couplers in schematic diagrams," *J. Acoust. Soc. Am.*, vol. 31, pp. 326–332, March 1959.

[18] J. W. S. Rayleigh, *The Theory of Sound.* New York: Dover, 1945.

[19] W. P. Mason, *Physical Acoustics and the Properties of Solids.* New York: Van Nostrand, 1958.

[20] R. A. Johnson, "Mechanical filters for FM mobile applications," *IRE Trans. Vehicular Communications*, vol. VC-10, pp. 32–37, April 1961.

[21] R. A. Johnson, "A single-sideband, disk-wire type mechanical filter," *IEEE Trans. Component Parts*, vol. CP-11, pp. 3–7, December 1964.

[22] *Reference Data for Radio Engineers.* New York: Internat'l Telephone and Telegraph Corp., 1956.

[23] R. A. Johnson, "A twin tee multimode mechanical filter," *Proc. IEEE (Correspondence)*, vol. 54, pp. 1961–1962, December 1966.

Channel Filters with Longitudinally Coupled Flexural Mode Resonators

F. Künemund

0. Introduction

The great technical and economic significance of single-sideband filters for voice channel selection in carrier transmission systems has already been emphasized in [1 to 6] and some of the aspects substantiated at length. It has also been shown that mechanical filters are well suited to this specific application. Extensive development projects of our organization have resulted in the mechanical channel filter for a carrier frequency of 48 kHz treated in [1 to 4]. This filter consists largely of 12 flexural mode resonators interconnected in the middle by a single longitudinally vibrating coupling wire (Figure 1). Resonators and

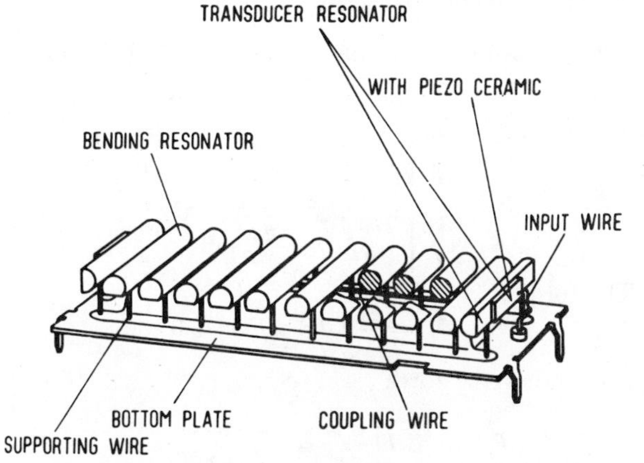

Figure 1. Arrangement of flexural mode resonators within the mechanical channel filter

couplers are made of drawn round stock of the ferronickel alloy THERMELAST 5409; compound resonators made of Thermelast and an electrostrictive ceramic material are used as electromechanical transducers.

1. Basic Development Work

Any new product such as our mechanical channel filter is necessarily the result of cooperation between various departments of our organization. Thus the ferronickel alloy was developed at the Vacuumschmelze GmbH in Hanau and the electrostrictive ceramic at our research laboratories in Munich. In consequence the materials experts were able to modify the ferronickel alloy and the ceramic according to the desired application and the filter designers were able to take full account of the modifiability of the materials.

The channel filter has not only to meet relatively stringent electrical requirements but must at the same time also have a certain geometry and be inexpensive to manufacture [1 to 4]. To avoid having to overcome too many difficulties right at the beginning, we started by designing mechanical filters having only a few resonators. 15 different types of carrier, signal and pilot filters resulted. In order that the work should not be confined to pure research, the filters were designed specifically for use in various carrier transmission systems.

The electromechanical transducers of all these filters were made of electrostrictive ceramic material and in some of them compound flexural mode resonators designed on the same principle as the transducer resonators of the channel filter were already used [7].

In developing our channel filters we soon changed after initial experiments with plate-shaped longitudinally vibrating resonators to flexural mode resonators in which we first sought to make use of two orthogonal modes. The principal characteristics of flexural mode resonators which are longitudinally coupled in the middle are, however, already given in [8], viz. selectable filter geometry and the avoidance of unwanted modes.

Subsequently we also discontinued the development of filters with steepened attenuation characteristics and channel filters for frequencies above 108 kHz and between 12 and 24 kHz. Finally, the chosen realization of the filter with coils in the terminating electrical circuits proved to be more favorable for the design of the channel converter than the mechanical filter without coils.

These optimization measures, which also included the choice of mid-band frequency and filter geometry, were carried out in close cooperation with the system developers and designing engineers, as well as the respective departments for planning, calculation, coordination, and, in particular, production.

2. Filter Synthesis

The reason why no mechanical filters using flexural mode resonators with bandwidths as wide as in our channel filter have so far been heard of is possibly due in part to the fact that such filters are more difficult to design than narrowband filters with longitudinal

Manuscript received on September 17, 1971

Dipl.-Ing. Friedrich Künemund, München
Siemens AG, Zentrallaboratorium für Nachrichtentechnik

or torsional resonators. In order to represent the behavior of flexural mode resonators in mathematical terms it is known to be necessary to use not only trigonometric but also hyperbolic functions. The rather unwieldy equation which results, e.g. for the ratio of force to velocity in the middle of a flexural mode resonator where the coupling wire acts tangentially can no longer be replaced in the case of filters of wider relative bandwidth—about 7% in the present instance—by the reactance function of a simple resonant circuit with two reactances. The difference in frequency response between these two functions disallows the use of simple reactance functions in designing a filter in which extreme demands have to be met in the passband and the stopband.

In order to overcome this problem in synthesizing a filter with rational functions, an equivalent circuit with four reactances was chosen for each flexural mode resonator. The frequency response of the flexural resonators can in this way be matched to within considerably less than one part in a thousand not only within the passband but also for the frequencies in the adjoining stopband for which attenuation specifications have to be met. Only two of the four reactances can however be chosen freely, the two others being associated with the two selectable reactances in correspondence with the frequency response of the flexural mode resonators.

Thus it is not possible to realize a conventional filter synthesis on the basis of insertion loss theory. A description of how such a synthesis can be accomplished using an iterative program is therefore given in [9], in which the improvements are obtained by analysis. The equivalent circuit of the compound resonators—one of the essential requirements for the synthesis described in [9]—is given in [10] and [11]. The calculated frequency response of the flexural mode resonators will be compared first with the measured values and then with the frequency response of the equivalent circuit composed of four reactances mentioned above.

3. Frequency Response of Flexural Mode Resonator and Equivalent Circuit

For calculating the behavior of the flexural mode resonator, the differential equation [12] published by Poisson in 1829 for a nonslender flexural resonator will be used as a point of departure

$$\frac{\partial^4 w}{\partial x^4} - \frac{\varrho}{E}\frac{\partial^4 w}{\partial x^2 \partial t^2} + \frac{\varrho q}{E J}\frac{\partial^2 w}{\partial t^2} = 0$$

or

$$\frac{d^4 w}{d x^4} + \omega^2 \frac{\varrho}{E}\frac{d^2 w}{d x^2} - \frac{\varrho q}{E J}\omega^2 w = 0,$$

where the second term represents the influence of rotational inertia.

The coordinate system is chosen corresponding to Figure 2 such that the x axis is parallel to the longitudinal axis of the flexural mode resonator, whereas

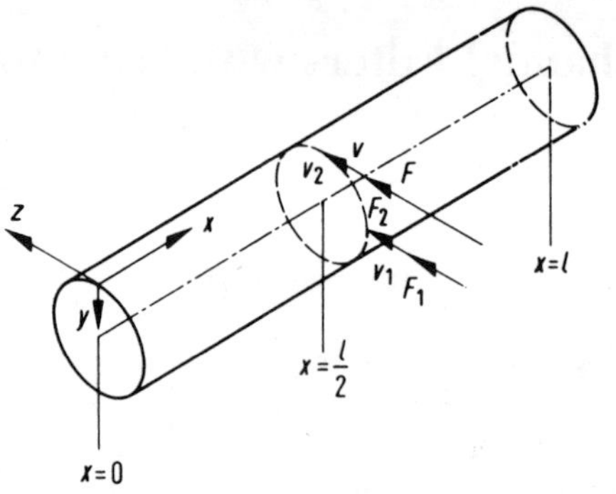

Figure 2. Flexural mode resonator in the Cartesian coordinate system

the deflection w of the resonator has to be in the direction of the z axis. The frequency f yields $\omega = 2\pi f$; ϱ and E denote the density and modulus of elasticity of the resonator and q and J the cross section and equatorial moment of inertia.

The conventional setup

$$w = \alpha \cosh k_1 x + \beta \sinh k_1 x + \gamma \cos k_2 x + \delta \sin k_2 x$$

yields

$$k_1 = \sqrt{\sqrt{\left(\frac{\omega^2 \varrho}{2E}\right)^2 + \frac{\varrho q}{E J}\omega^2} - \frac{\omega^2 \varrho}{2E}}\,;$$

$$k_2 = \sqrt{\sqrt{\left(\frac{\omega^2 \varrho}{2E}\right)^2 + \frac{\varrho q}{E J}\omega^2} + \frac{\omega^2 \varrho}{2E}}\,.$$

The torque

$$M_1 = -E J \partial^2 w / \partial x^2 = 0$$

and transverse force

$$F_1 = -E J \left[\frac{\partial^3 w}{\partial x^3} - \frac{\varrho}{E}\frac{\partial^3 w}{\partial x \partial t^2}\right] = 0$$

at the point $x = 0$ yield $\beta k_2 = \delta k_1$ and $\alpha k_1^2 = \gamma k_2^2$. Given a resonator length $x = l$, the excitation at the middle of the resonator is $\mathrm{d}w/\mathrm{d}x = 0$ at the point $x = l/2$.

This and the abridgments

$$\sinh k_1 l/2 = S; \qquad \cosh k_1 l/2 = C;$$

$$\sin k_2 l/2 = s; \qquad \cos k_2 l/2 = c$$

yield

$$\delta = \alpha \frac{k_1^2 s - k_1 k_2 S}{k_2^2 c + k_1^2 C}\,.$$

The ratio of force to velocity at the point $x = l/2$ for a resonator length $x = 2l/2$ and $v_1 = \mathrm{j}\omega w$ and the abridgments

$$\tanh k_1 l/2 = T; \qquad \tan k_2 l/2 = t$$

leads to

$$2\frac{F_1}{v_1} = \mathrm{j}\frac{E J}{\omega}\frac{k_1^2 + k_2^2}{k_1 k_2}\frac{k_1^3 t + k_2^3 T}{\frac{k_1^4 + k_2^4}{2 k_1^2 k_2^2} + \frac{1}{C c} - \frac{k_2^2 - k_1^2}{2 k_1 k_2} T t}$$

(the corresponding equation for the slender flexural mode resonator is obtained with $k_1 = k_2$).

Using the force-current analogy, let this be termed the admittance Y_1 for which the two vectors for force

$\boldsymbol{F}_1$ and velocity $\boldsymbol{v}_1$ are in the same plane as the elastic line of the resonator.

In order to approach the arrangement in the channel filter it is further necessary to consider the tangential excitation. Starting from the transmission line equations for a torsioned rod with G as the shear modulus and J_p as the polar moment of inertia of the rod cross section,

$$\frac{\partial \frac{\partial \varphi}{\partial t}}{\partial x} = \frac{1}{G J_\mathrm{p}} \frac{\partial M_2}{\partial t}; \qquad \frac{\partial M_2}{\partial x} = J_\mathrm{p} \varrho \frac{\partial \frac{\partial \varphi}{\partial t}}{\partial t},$$

a simple transformation with R as the radius of the circular rod cross section yields

$$\frac{\partial \boldsymbol{v}_2}{\partial x} = \frac{R^2}{G J_\mathrm{p}} \frac{\partial \boldsymbol{F}_2}{\partial t}; \qquad \frac{\partial \boldsymbol{F}_2}{\partial x} = \frac{J_\mathrm{p}}{R^2} \varrho \frac{\partial \boldsymbol{v}_2}{\partial t}.$$

This, together with the ratio of the vectors $\boldsymbol{F}_2$ and $\boldsymbol{v}_2$ that act tangentially in the middle of the resonator, yields the admittance

$$Y_2 = \mathrm{j}\pi R^2 \sqrt{G\varrho} \tan \sqrt{\frac{\varrho}{G}} \cdot l/2\omega.$$

The series connection of the two admittances Y_1 and Y_2 yields the unknown admittance

$$Y = \frac{Y_1 Y_2}{Y_1 + Y_2}.$$

Hence, after several transformations with $v_\mathrm{L} = \sqrt{E/\varrho}$ and $v_\mathrm{T} = \sqrt{G/\varrho}$ as the propagation velocity for longitudinal waves and torsional waves respectively, and with $m = \pi R^2 l \varrho$ as the mass of the resonator and $\eta = \omega/\omega_0$ the normalized frequency:

$$Y = \mathrm{j} m \omega_0 \frac{\Omega_0}{l\left(1 + \frac{\eta}{Cc} + \Omega_1 + \Omega_2\right)}$$

with

$$\Omega_0 = \frac{1}{2}\left(\frac{v_\mathrm{L} R}{\omega_0}\right)^2 \sqrt{1 + \left(\frac{R\omega_0}{4 v_\mathrm{L}}\eta\right)^2} \, (k_1^3 t + k_2^3 T),$$

$$\Omega_1 = \frac{R\omega_0}{4 v_\mathrm{L}} \eta^2 \left(\frac{R\omega_0}{2 v_\mathrm{L}} - T t\right),$$

$$\Omega_2 = \frac{\omega_0 \Omega_0}{v_\mathrm{T} \tan\left(\frac{\omega_0}{v_\mathrm{T}} \frac{l}{2} \eta\right)}.$$

The other frequency-dependent terms were already given as the abridgments k_1, k_2, t, T, c and C.

In the numerical evaluation of the admittance Y, the values $v_\mathrm{L} = 4.595 \cdot 10^6$ mm/s, $v_\mathrm{T} = 2.8325 \cdot 10^6$ mm/s measured for the resonator material Thermelast 5409 were used along with the dimensions $l = 16.73$ mm and $R = 1.75$ mm. The lowest finite frequency at which $(k_1^3 t + k_2^3 T) = 0$ and thence $Y = 0$ calculates under the given assumptions at $f = f_0 = 48$ kHz, which was chosen as a reference frequency.

The admittance normalized to $m\omega_0$ is represented in Figure 3 as a solid curve plotted over the normalized frequency η. Corresponding to the normalization results: $Y = 0$ at the point $\eta = 1$. The two poles marked in Figure 3 lie at $\eta_{\infty 1} = 0.777$ and $\eta_{\infty 2} = 2.055$. The calculated pole frequencies for the slender flexural

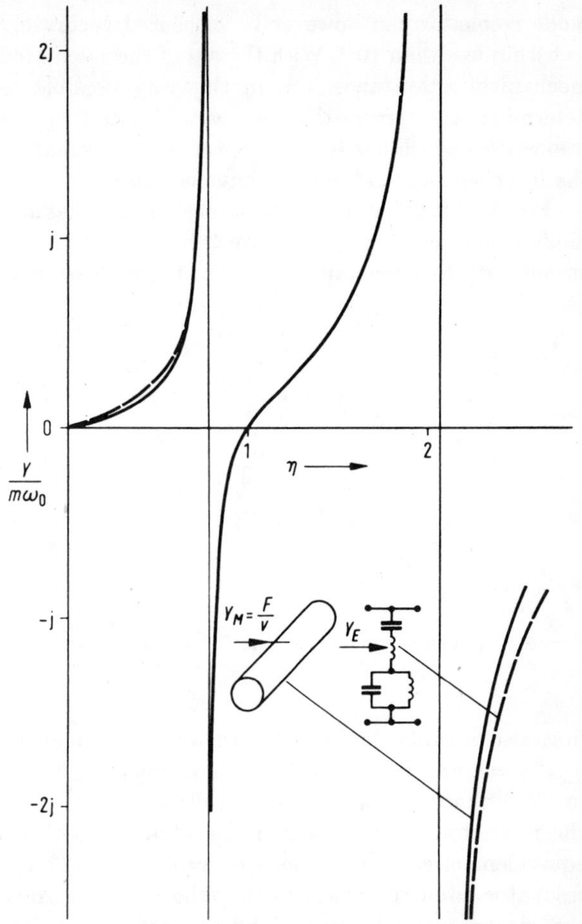

Figure 3. Admittance of flexural mode resonator and its equivalent circuit

resonator when excited in the same plane as the elastic line are given for $\eta_{\infty 1} = 0.63$ and $\eta_{\infty 2} = 3.94$ to allow comparison. The corresponding frequencies for the tangentially excited slender flexural resonator, where the rotational inertia is accordingly neglected, lie at $\eta_{\infty 1} = 0.796$ and $\eta_{\infty 2} = 2.2$. This means that the influence of the tangential excitation exceeds by far that of the rotational inertia.

The values for $\eta_{\infty 1}$ and $\eta_{\infty 2}$ measured on a flexural mode resonator in the channel filter agree in measuring accuracy with the previously given values although, among other things, the resonator holder and the flat section along one generatrix have been neglected. The measuring error for $\eta_{\infty 1}$ was estimated to be 0.5% and that for the more remote pole $\eta_{\infty 2}$ to be 5%.

A simple calculation indicates that an error of, say, 5% for $\eta_{\infty 2}$ will reduce the agreement between the measurement and the calculation and cause an error of the admittance of about 1%, e.g. at the point $\eta = 1.1$. The resonator mass and the two other characteristic frequencies $\eta = \eta_{\infty 1}$ and $\eta = 1$ were here assumed to agree. Accuracy of this order is naturally not sufficient for designing a filter. At the discrete selectable frequencies in the main region of interest in the vicinity of $\eta = 1$, where agreement with the equivalent circuit has to be enforced, the admittance of the flexural mode resonator can however be measured accurately to within less than 10^{-5}. With the aid of the calculated mechanical admittance it is in this way possible to determine the error that appears between these frequencies which has to be accepted when calculating the filter with the aid of the equivalent circuit.

For phantoming the admittance of the flexural mode resonator shown in Figure 3 by the equivalent circuit with four reactances given in Figure 4 we may

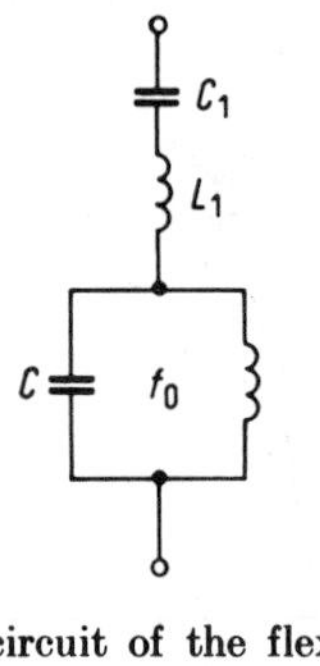

Figure 4. Equivalent circuit of the flexural mode resonator

first also assume the three characteristic frequencies $\eta_{\infty 1}$, $\eta = 1$, and $\eta_{\infty 2}$ for the reactance function. As the fourth unknown quantity, let the first derivatives of the resonator admittance and the admittance of the equivalent circuit be assumed to agree at $\eta = 1$. The resonator admittance will in the following be termed the mechanical admittance and denoted Y_M and the admittance of the equivalent circuit termed the electrical admittance and denoted Y_E.

The first derivative of the mechanical admittance at the point $\eta = 1$ comes to:

$$\left.\frac{\mathrm{d}Y_M}{\mathrm{d}\eta}\right|_{\eta=1} = \mathrm{j}\, m\omega_0 \cdot \frac{\dfrac{R^2\sqrt{1+(R\omega_0/4v_L)^2}}{4\,l\,(k_2^2-\omega_0^2/2\,v_L^2)}\cdot(\Omega_3+\Omega_4)}{1+1/Cc+\Omega_1}$$

with

$$\Omega_3 = \left(\frac{4}{R^2} - k_1^2\right)\left(3\,k_1 t + \frac{l k_2^3}{2\,k_1}\,(1-T^2)\right)$$

and

$$\Omega_4 = \left(\frac{4}{R^2} + k_2^2\right)\left(3\,k_2 T + \frac{l k_1^3}{2\,k_2}\,(1+t^2)\right).$$

$\eta = 1$ must likewise be inserted in the equation given for Ω_1. A numerical evaluation yields:

$$\left.\frac{\mathrm{d}Y_M}{\mathrm{d}\eta}\right|_{\eta=1} = \mathrm{j}\cdot 1.547\,\omega_0 m.$$

For the equivalent circuit in Figure 4

$$Y_E = \mathrm{j}\omega_0\, C\,(\eta_{\infty 2}^2-1)\,(1-\eta_{\infty 1}^2)\cdot\frac{\eta\,(1-\eta^2)}{(\eta^2-\eta_{\infty 1}^2)(\eta^2-\eta_{\infty 2}^2)}$$

and

$$\left.\frac{\mathrm{d}Y_E}{\mathrm{d}\eta}\right|_{\eta=1} = \mathrm{j}\cdot 2\omega_0 C, \text{ thence } C \triangleq 0.7735\text{ m},$$

and

$$\frac{Y_E}{\omega_0 m} = \mathrm{j}\cdot 0.98815\cdot\frac{\eta\,(1-\eta^2)}{(\eta^2-\eta_{\infty 1}^2)(\eta^2-\eta_{\infty 2}^2)}\,.$$

The admittance Y_E, which is likewise referred to $\omega_0 m$, is represented in Figure 3 as a dashed line. Differences between the mechanical and the electrical admittance appear in the chosen scale solely in the frequency regions below $\eta_{\infty 1}$ and above $\eta_{\infty 2}$. At the boundaries of the frequency range of particular interest at $\eta = 0.95$ and $\eta = 1.1$ the error of the electrical representation is about 1%. However, if the pole frequencies of Y_E are chosen slightly different from the pole frequencies of Y_M, an error to within less than 10^{-4} can be achieved throughout the stated

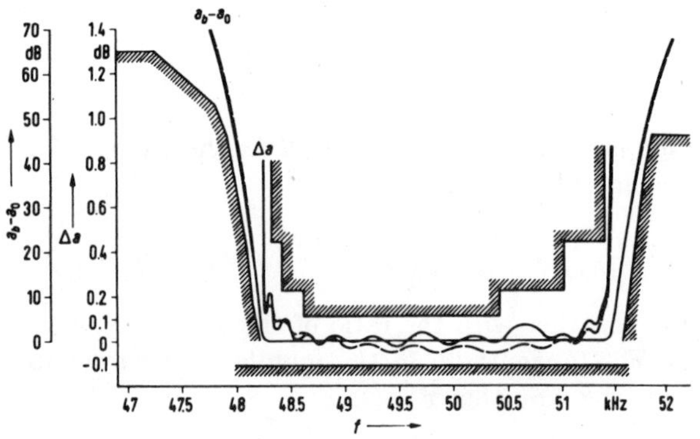

Figure 5. Measured and calculated insertion loss of a mechanical channel filter

frequency range. For the insertion loss requirements which here have to be met, errors of such low order are negligible.

In Figure 5 the measured insertion loss of a mechanical filter represented by a solid line is compared with the calculated response represented by a dashed line. It will be noted that the filter meets the requirements with a good safety margin. The differences between the two curves are due mainly to tolerances in the mechanical structure.

References

1. Kopp, H.; Mehr, A.: Channel Modem with Electromechanical Filters. Siemens Rev. 38 (1971), pp. 297—300
2. Albsmeier, H.: Ein Vergleich der Realisierungsmöglichkeiten elektromechanischer Kanalfilter im Frequenzbereich 12 kHz bis 10 MHz. Frequenz 25 (1971), pp. 74—79
3. Künemund, F.: Materials for High-grade Electromechanical Frequency Filters. Conference 19. 8. 1971, Perth, Australia. Materials for the Electrical and Electronic Industries
4. Kopp, H.: Entwicklungstendenzen in der Trägerfrequenztechnik. VDE-Fachber. 26 (1970), pp. 145—149
5. Reiner, H.: Technologien kommender Gerätegenerationen der Nachrichtentechnik. Elektrotechn. Z. A 91 (1970), pp. 697—704
6. Johnson, R. A.; Börner, M.; Konno, M.: Mechanical Filters — A Review of Progress. IEEE Trans. Sonics Ultrason., Vol. SU-18, No. 3, July 1971
7. Poschenrieder, W.; Albsmeier, H.: French Patent No. 1,281,734, application of 26. 2. 60
8. Künemund, F.; Traub, K.: Mechanische Filter mit Biegeschwingern. Frequenz 18 (1964), pp. 277—280
9. Günther, A.: Bemerkungen zum Entwurf breitbandiger mechanischer Filter. Nachr.-techn. Z. 25 (1972), pp. 345—351
10. Albsmeier, H.: Über Antriebe für elektromechanische Bandpässe mit piezoelektrischen Wandlern. Frequenz 19 (1965), pp. 125—133
11. Mason, P.: Physical Acoustics, I Part A. New York: Academic Press 1964, p. 238
12. Poisson, S. D.: Sur le mouvement des corps élastiques. Paris Mém. 8 (1829), p. 463

DESIGN OF MECHANICAL FILTERS USING RESONATORS WITH MINIMIZED VOLUME

K. Yakuwa and S. Okuda
FUJITSU LIMITED, Kawasaki, Japan

ABSTRACT

Resonant frequency and loss of a mechanical vibrator changes in relation to the driving power. Such change is determined by the volume, vibration mode, and material of the vibrator. Determination method of the minimum volume of vibrators for a mechanical filter is discussed. The volume of a mechanical filter using the vibrators proved to be far smaller than a L-C filter.

1. INTRODUCTION

The electromechanical filter which uses a mechanical vibrator is characterized by compactness as well as high performance and economy.

It is proposed here that the minimum volume of a mechanical vibrator can be determined from the nonlinear characteristics of the material of the vibrator attributed to elastic strains, considering that, as a general property of a mechanical vibrator, the resonant frequency and mechanical Q of the vibrator decrease as the driving power of the vibrator is increased.

2. STRAIN ENERGY AND ENERGY LOSS OF THE VIBRATOR

The equivalent circuit of a mechanical filter is shown in Fig. 1.
The relation between the input power P and the strain energy U stored in the mechanical vibrator at its center frequency can be expressed in the following when loss is neglected.

$$P = \frac{1}{2}(\omega_{+1} - \omega_{-1})U \qquad (1)$$

where ω_{+1} and ω_{-1} are the cutoff angular frequencies of the filter.
The strain energy U of the mechanical vibrator is given by the following equation.

Longitudinal mode and bending mode:

$$U = \frac{1}{2}E\iiint S^2 dV \equiv \bar{u}V$$

Torsional mode: (2)

$$U = \frac{1}{2}G\iiint S^2 dV \equiv \bar{u}V$$

where E is the modulus of elasticity, G modulus of rigidity, S the strain and V the volume of the vibrator.
$\bar{u}$ means the average strain energy density of the vibrator, and is equivalent to u/V.
Meanwhile, the energy loss D per cycle of the vibrator is expressed in the following way.

$$D = \pi\eta\omega\iiint S^2 dV, \qquad (3)$$

Reprinted from *Proc. 1976 IEEE Int. Symp. Circuits and Syst.*, Apr. 27–29, 1976, pp. 790–793.

where η is a friction coefficient.

The mechanical Q or $(\tan\delta)^{-1}$ of the vibrator is related to U and D as given below.

$$\frac{1}{Q} = \frac{D}{U} = \tan\delta \qquad (4)$$

3. NONLINEARITY OF THE VIBRATOR

The resonant frequancy and mechanical Q of mechanical vibrators decrease, in general, as its input power increases. This is due to the change in the modulus of elasticity or rigidity. The following relations hold.

$$\left.\begin{aligned} E &= Eo(1 + \beta_1 S^{\gamma 1}), \\ G &= Go(1 + \beta_1 S^{\gamma 1}), \end{aligned}\right\} \qquad (5)$$

$$\eta = \eta o(1 + \beta_2 S^{\gamma 2}), \qquad (6)$$

where Eo is modulus of elasticity at S=0, Go modulus of rigidity at S=0, ηo a friction coefficient at S=0, β_1, β_2, γ_1 and γ_2 are the nonlinearity constants of the vibrating material.

The strain energy U stored in the vibrator is obtained by equations (2) and (5), and expressed as $Uo + \Delta U$. As a result, if the vibrating mode is unchanged the rate of change ε_1 of the resonance frequency of the vibrator is expressed as follows.

$$\varepsilon_1 = \frac{\Delta\omega}{\omega} = \frac{\Delta U}{2Uo} = \frac{\beta 1}{2} \frac{\iiint S^{(2+\gamma 1)} dV}{\iiint S^2 dV}$$

$$\equiv \psi 1 \bar{u}^{\frac{\gamma 1}{2}} \qquad (7)$$

where ψ_1 is a coefficient determined from the nonlinearity constants β_1 and γ_1 of the material of the vibrator and the vibration mode. As the equation (7) indicates, the frequency stability of the vibrator is improved as ψ_1 becomes small. The values of ψ_1 of a few vibrators are shown in Tab. 1.

Similarly the nonlinearity of $\tan\delta$ is expressed with the following equation.

$$\varepsilon_2 = \Delta\tan\delta / \tan\delta o = \psi_2 \bar{u}^{\frac{\gamma 2}{2}} \qquad (8)$$

Nonlinear characteristics were measured for the vibrators shown in Tab. 2. The relation between the average strain energy density $\bar{u}$ and the rate of change of frequency, ε_1, or of $\tan\delta$, ε_2, is shown in Fig. 2. Resulting values of the nonlinearity coefficients ψ_1 and ψ_2 of each vibrator material and the nonlinearity constants β_1, γ_1 and γ_2 are given in Tab. 3.

In Fig. 2 and Tab. 3 the results of measurements of nonlinear characteristics of a pot-shaped ferrite core (of 14 mmø x 8 mm) are given for reference.

4. DETERMINATION OF THE MINIMUM VOLUME OF THE MECHANICAL VIBRATOR

If such system requirements for a mechanical filter as the center frequency, passband width, allowable changes of transmission characteristics and input level are given, the minimum volume of the mechanical vibrator is given by equation (9).

Namely $\bar{u}_{max}$ can be known by equations (7) and (8) once allowances against ε_1 and ε_2 are given; $\bar{u}_{max}$ thus obtained and equations (1) and (2) induce the following equation (9) for the minimum volume.

$$V_{min} = \frac{2P}{\bar{u}_{max}(\omega_{+1} - \omega_{-1})} \quad (9)$$

The flow chart to determine the minimum volume of a mechanical filter is shown in Fig. 3.

An example is shown as follows:

(1) Frequency: 48 kHz

(2) Bandwidth: 3.25 kHz

(3) Maximum input level: -10 dBm

(4) Vibrating mode: Bending

(5) Material: Fe-Ni-Cr-Ti alloy

(6) $\varepsilon_1 \leqq 2 \times 10^{-4}$

(7) $\varepsilon_2 \leqq 20\%$

Then the calculated minimum volume is 2.45×10^{-6}cc from condition (6); likewise 2.99×10^{-2}cc from (7). The larger figure of the two determines the minimum volume. The dimensions 0.98 x 3.05 x 10.0 mm would be a reasonable solution.

5. CONCLUSION

The nonlinear relationship between the strain energy and the resonant frequency or loss in a mechanical vibrator used for a mechanical filter has been clarified in relation to the vibration mode, volume, and nonlinearity constants of the material. From the results, a method to determine the minimum volume of the vibrator used in a mechanical filter has been established so that the change of transmission characteristics can be kept within a allowable range.

REFERENCE

[1] K. Yakuwa et al "Miniaturization of the mechanical vibrator used in an electromechanical filter". Reports of 6th ICA, G-3-5, August 1968.

[2] K. Yakuwa et al "On the relation between the volume of mechanical vibrators used in an electromechanical filter and the allowable input level". Technical group Meeting of Ultrasonics of I.E.C.E of Japan US67-10 (1967-7)

Mode	Ψ_1	$\Psi_1(\gamma_1 = 1)$
L 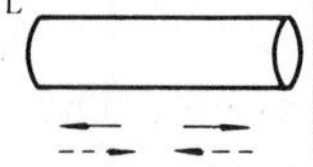	$\beta_1 \left(\frac{4}{E_0}\right)^{\frac{\gamma_1}{2}} \frac{1}{\ell} \int_0^{\ell} (\sin \frac{\pi}{\ell} x)^{2+\gamma_1} dx$	$\frac{8\beta_1}{3\pi\sqrt{E_0}}$
T 	$\frac{4\beta_1}{4+\gamma_1} \left(\frac{8}{G_0}\right)^{\frac{\gamma_1}{2}} \frac{1}{\ell} \int_0^{\ell} (\sin \frac{\pi}{\ell} x)^{2+\gamma_1} dx$	$\frac{32\sqrt{2}\beta_1}{15\pi\sqrt{G_0}}$
B_1 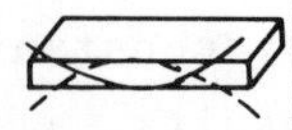	$\frac{3\beta_1}{2(3+\gamma_1)} \left(\frac{6}{E_0}\right)^{\frac{\gamma_1}{2}} \frac{1}{\ell} \int_0^{\ell} \left\{A\right\}^{2+\gamma_1} dx$	$1.2208 \cdot \frac{\beta_1}{\sqrt{E_0}}$
B_2 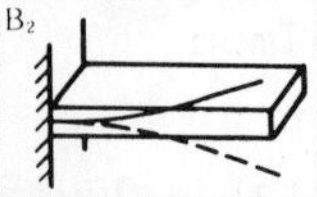	$\frac{3\beta_1}{2(3+\gamma_1)} \left(\frac{6}{E_0}\right)^{\frac{\gamma_1}{2}} \frac{1}{\ell} \int_0^{\ell} \left\{B\right\}^{2+\gamma_1} dx$	$1.3575 \cdot \frac{\beta_1}{\sqrt{E_0}}$

$$\left\{A\right\} = \cos\alpha_1 X + \cos h\alpha_1 X + \frac{\cos\alpha_1 - \cos h\alpha_1}{\sin\alpha_1 - \sin h\alpha_1}(\sin\alpha_1 X - \sin\alpha_1 X)$$

$$\left\{B\right\} = \cos\alpha_1 X - \cos h\alpha_1 X + \frac{\cos\alpha_1 + \cos h\alpha_1}{\sin\alpha_1 + \sin h\alpha_1}(\sin\alpha_1 X + \sin\alpha_1 X)$$

$$X = \frac{x}{\ell}$$

Table 1 Non-linearity coefficient of resonant frequency, ψ_1

Material		Size $t \times w \cdot \ell$ (mm)	Density ρ (gr/cm^3)	Elas. Const. E_0 (dyne cm^2)	Res. Freq. f_0 (KHz)	Mechanical Q
PZT	(1)	1.05 · 5.02 · 25.0	7.48	$5.75 \cdot 10^{11}$	55.275	90
	(2)	1.00 · 6.20 · 35.0	7.45	$8.20 \cdot 10^{11}$	47.704	1,240
Fe Ni Cr Ti Alloy	(1)	1.14 · 9.70 · 100	8.04	$1.95 \cdot 10^{12}$	0.57456	5,120
	(2)	1.96 · 9.70 · 100	8.04	$1.95 \cdot 10^{12}$	0.98250	12,000
	(3)	3.0[illegible] · 9.70 · 100	8.04	$1.95 \cdot 10^{12}$	1.55093	21,200
Quartz (X cut)		1.68 · 3.20 · 19.5	2.55	$8.16 \cdot 10^{11}$	145.000	111,000

Table 2 Samples of vibrators

Material		Ψ_1	β_1	γ_1	Ψ_2	γ_2
PZT	(1)	-3.4×10^{-4}	-3.03×10^{2}	1	2.4×10^{-2}	1
	(2)	6.2×10^{-5}	-6.61×10	1	5.7×10^{-3}	1
Fe Ni Cr Ti Alloy		1.0×10^{-6}	-1.14	1	1.2×10^{-2}	1
Quartz (X cut)		6.8×10^{-8}	-7.2×10^{-2}	1	5.0×10^{-3}	1
Ferrite core ($A_L = 250$)		9.5×10^{-4}	——	1	——	—

Table 3 Measurement results of non-linearity coefficients, ψ_1 and ψ_2, non-linearity constants, β_1; γ_1 and γ_2

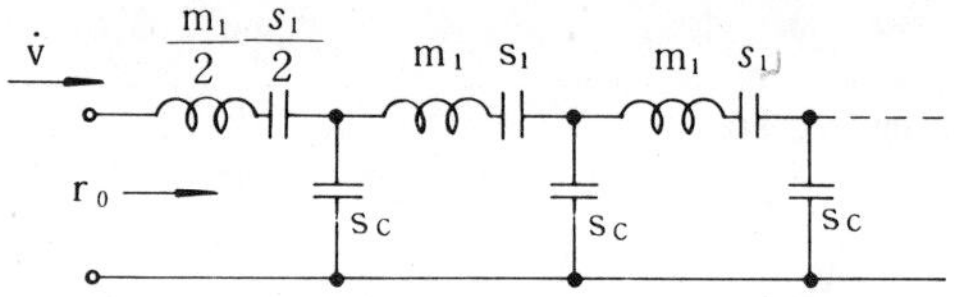

Fig. 1 The equivalent circuit of a mechanical filter

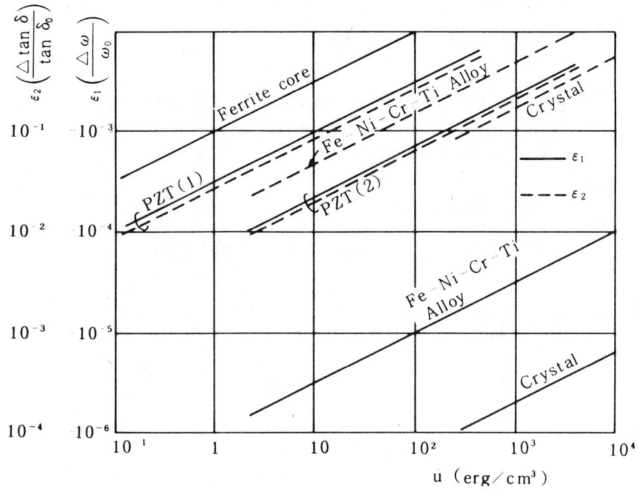

Fig. 2 Non-linearity characteristics of vibrator materials

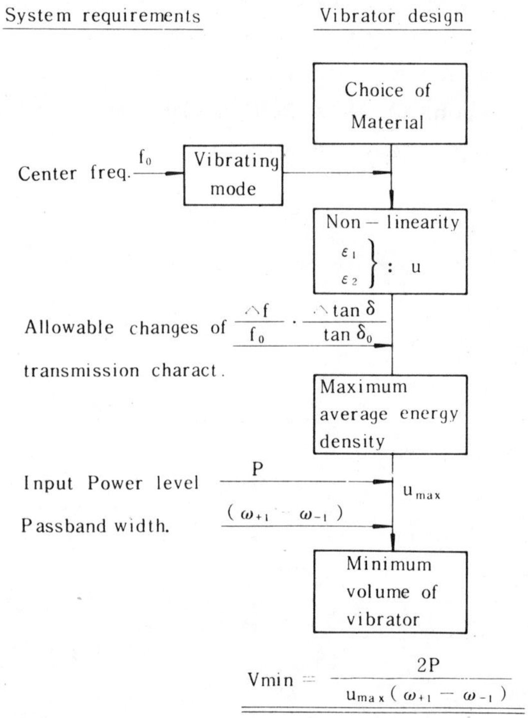

Fig. 3 Design flow of miniaturized vibrators

COMPUTER MODELLING OF MF TRANSFER CHARACTERISTICS

V.SOBOTKA
Czech Technical University
Prague, Czechoslovakia

J.TRNKA
Tesla Strašnice
Prague, Czechoslovakia

Abstract

This paper describes some methods for computer-aided design of mechanical filters (MF), namely synthesis of MF, statistical evaluation of input parameters, tolerance analysis, economical use of materials, cost optimization etc.
These metods are to be applied to MF with torsional resonators but most of them are more universal.

1. Synthesis of MF

The start for following discussions is the example of synthesis of MF based on the approximation of its transfer function by LC equivalent. The frequency location of MF often sets the coupling length between resonators to less than $\lambda/4$. From the technological standpoint there is an effort to reach the maximum of identical elements (resonators and couplings), so it is impossible to apply the effective loss parameters method as in ref. [1,2]. As usual, the lowpass prototype is taken as an estimate and computer-optimized according to ref.[6,7]. After frequency transformation the quasipolynomial LC equivalent is obtained. In ref.[3] the basic system of nonlinear equations is derived for :

a) fixed or pre-computed variables (central frequency, bandwidth, value of lowpass prototype elements, terminating resistances, coupling lengths)
b) independent (optional) variables (impedances of resonators and transducers and their couplings, propagation velocities, transducer ratio)
c) dependent (computed) variables (resonant frequencies, transducer coupling length, impedances of resonator couplings, electrical inductance)

The computation is performed iteratively. The items ad c) are printed in the catalogue form, the items ad a) and b) representing parameters. All resonators have the same impedance, the frequencies of most resonators are also equal.

2. Statistical testing of input parameters

Denote **k** values of arbitrary response of MF by

$$Y_j = \xi(X_1, X_2, \ldots, X_r;\ f_j)\quad j = 1,2,\ldots,k \qquad (1)$$

the symbols $X_h (h = 1,2,\ldots,r)$ representing all parameters necessary for the computation of characteristic (1). From analytical point of wiew X_h represent input variables while $f_j (j=1,2,\ldots k)$ are the frequency points in which the evaluation of function (1) is processed. So that we may test material properties (i.e.all input variables X_h) and obtain the tolerance analysis of MF including production philosophy the following subroutines have been written as parts of a program system for design of MF :

a) Statistical analysis of precision of measuring methods.
b) Statistical tests of measured data with automatic error detection.
c) Determination of density functions of input parameters.
d) Determination of correlations among input parameters.

3. Tolerance analysis of MF characteristics

Tolerance regions of MF characteristics are simulated according to the principal scheme in Fig.1. In the block A the random choice of input parameters

$$X_h \ (h = 1,2,\ldots,r)$$

is made with arbitrary distributions. Also correlations

Reprinted from *Proc. 1975 IEEE Int. Symp. on Circuits and Syst.*, Apr. 21–23, 1975 are acceptable. In the block B the values

$$Y_j \ (j = 1,2,\dots,k)$$

of the characteristic (1) are computed and put into storage C. This procedure is n-times repeated. The series of quantities Y_{ij} $(i = 1,2,\dots,n)$ is evaluated statistically in the block D for each j. The computer prints namely the plot of means $\bar{Y}_j$, lower Y_{jM+} and upper Y_{jM-} limit values, the production reject considered to be α (Fig.3).

The performance of the block A is in the Fig.2. In the block uniformly distributed random numbers $G_h \in \langle 0,1\rangle$ are generated. In the 2nd block the G_h are transformed into random numbers $H_h \in \langle 0,1\rangle$ with desired distributions. In the 3rd block the H_h are transformed into physical dimensions of input variables X_h.

Special interests in the block A :

a) The exact simulation of required density functions of input parameters is in accordance with the sets of measured data converted into histograms. The normalized histograms are then approximated by the best of the series of standard distributions stored in a computer. For the block 2 in the Fig.2 it is represented by the type of distribution "TD".

b) The same process may be applied to various correlations among the input parameters caused by functional relations or certain type of manufacturing process. For the block 2 in Fig.2 they are represented by the code "C".

In the block D it proved to be helpful to use the nonparametric statistical methods for determination of upper Y_{jM+} and lower Y_{jM-} $(j = 1,2,\dots,k)$ limit values of the characteristic (1). If the requiered reject α is determined with the given probability $P \geq \gamma$ it is necessary (according to [4]) to simulate the MF n-times

$$n = 1 + \frac{1-\alpha}{2\alpha}\,\chi^2_{1-\gamma}(4) \qquad (2)$$

so that for the upper Y_{jM+} and lower Y_{jM-} values of this simulated set might hold

$$Y_{jM+} \leq Y_{jmax} \qquad y_{jM-} \geq Y_{jmin} \qquad (3)$$

$\chi^2_{1-\gamma}(4)$ is the critical value of the χ^2 distribution with four degrees of freedom.

This model of tolerance analysis of MF fulfils all requirements for tolerance analysis in industrial use : precision, flexibility, simple data handling etc. For details cf [5].

4. Economical use of materials for MF

If the tolerance regions of analysed response of MF computed for a required production reject do not satisfy the prescribed tolerance diagram :

a) a different material is used having the smaller dispersion of physical parameters and/or a different technology is employed

b) existing material and technology are saved but the tolerances of appropriately chosen physical parameters are made tighter in correspondence with sensitivies $\partial Y/\partial X_h$ and with production costs.

After tolerance tightening the high original dispersion of physical parameters may cause a higt reject of MF components. The computer turns this reject to profit by dividing the rest of material into classes of the width equal to tolerance of corresponding physical parameters. A new design can be made for each class if it has an economically sufficient number of samples.

5. Optimization of production costs

Notice the density function $\emptyset\,(Y_j)$ of the analyzed characteristic (1) in the frequency points F_j, $(j = 1,2,\dots,k)$ in the Fig.3. The limit values Y_{jM-} and Y_{jM+} are fixed by the prescribed tolerance diagram. Having extended the tolerances of input network parameter X_h, the dispersion of the function $\emptyset\,(Y_j)$ arises. (Function a → b → c in Fig.3)

To make the explanation more simple suppose all components have equal percentual tolerances τ. Then both the cost per one MF and density function $\emptyset\,(Y_j)$ depend on τ. The cost of production set of extent <u>n</u> MF equals to $P_S = n.P_F$. The cost of mechanical filters not satisfying the tolerance diagram at frequency F_j (production reject) is

$$P_\alpha = n\,P_F\,\alpha = n\,P_F\left[1 - \int_{Y_{jM-}}^{Y_{jM+}} \emptyset\,(Y_j)\,dY_j\right] \qquad (4)$$

This cost makes the cost of the satisfying MF increase to

$$P_{MF} = \frac{P_F}{1-\alpha}$$

In Fig.4 P_{MF} has a distinct minimum P_{Fmin} for certain optimal tolernace τ_0.

In general, the components of actual MF have different tolerances. Then the value P_{Fmin} is found with a computer in all critical points of analyzed characteristic.

6. Example

As an example of a convenient use of above mentioned methods we present the channel bank filtr MF 123 TESLA. The use of piezoceramic transducers and torsional resonators enabled us to reach the volume of 18 cm^3. The tolerance diagram and tolerance regions of the effective loss calculated for production reject $\alpha = 0{,}05$ are presented in the Fig.5.

7. Conclusion

Considering the small extent of Proceedings only the main problems of computer-aided industrial design of MF have been presented. It is necessary to point out that the realization of discussed methods will ever be influenced by construction of MF and by technological efficiency and business interests of the producer.

References

1 M.Börner : Mechanische Filter für die Träger-frequenztechnik. Nachrichtentechnik Fachbereite, Bd 19 (1960), S.34 bis 37

2 J.Jungwirt : Synthesis of EMF with the Use of Effective Loss Parameters. Proceedings TESLA VÚT, 1960, Prague (In Czech)

3 J.Jungwirt : The Design of EMF´s for Telecommunications. TESLA Electronics, Vol.7, No.1, March 1974, pp 3 ÷ 9

4 J.Machek : Nonparametric Limits and Intervals. Research Report, TESLA Strašnice, 1970, Prague (In Czech)

5 V.Sobotka, J.Trnka : The Use of Digital Computers in Modelling of Transfer Characteristics of Serially Producted Electromechanical Filters. Proceedings of the SSCT 71, 1971, Tále, Czechoslovakia

6 V.Sobotka, J.Trnka : Die Benutzung von Digitalrechern beim Entwurf elektromechanischer Filter für die Großserienfertigung. Frequenz, 1972, No.6. S. 177 - 182.

7 K.Vagenknecht : Optimalisation of the Low Pass Prototype. Research Report, TESLA Strašnice, 1970, Prague (In Czech).

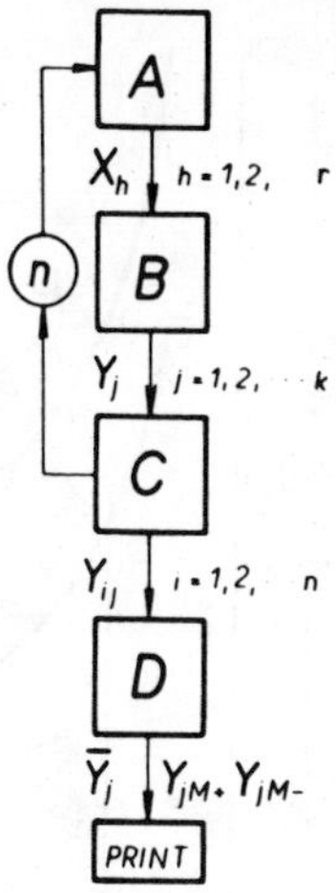

Fig.1 Principal block scheme of tolreance analysis model

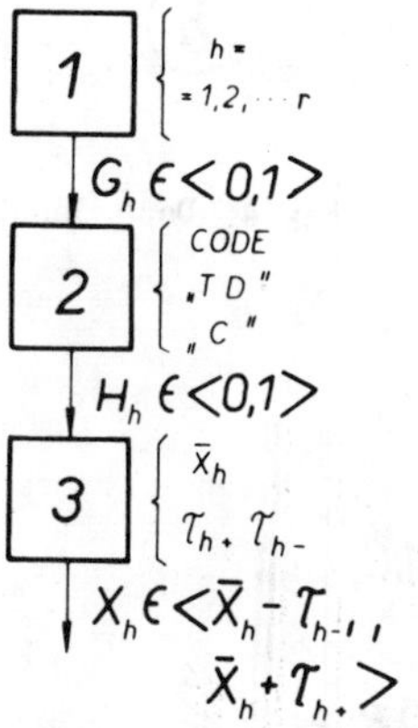

Fig.2 Modelling of input parameters of MF

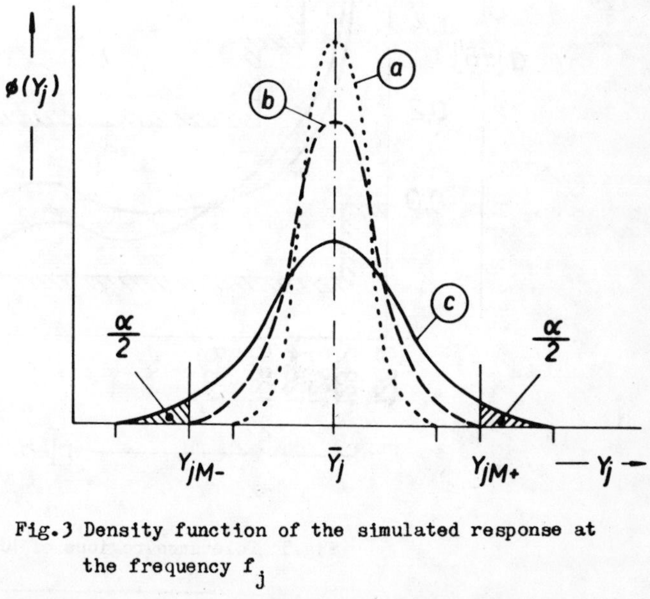

Fig.3 Density function of the simulated response at the frequency f_j

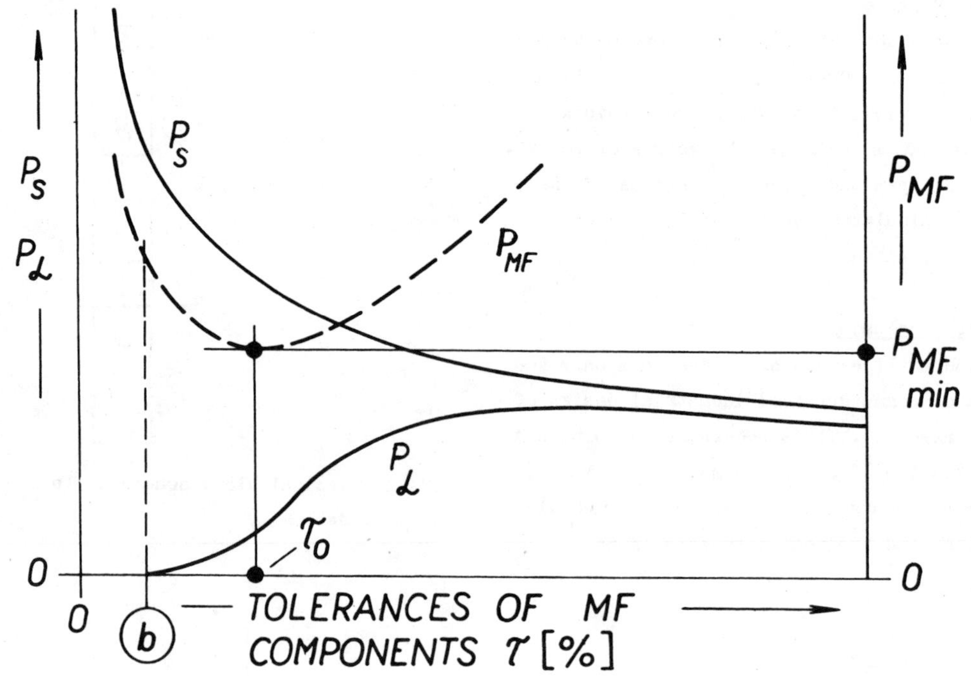

Fig.4 Dependence of the cost of MF on component tolerances

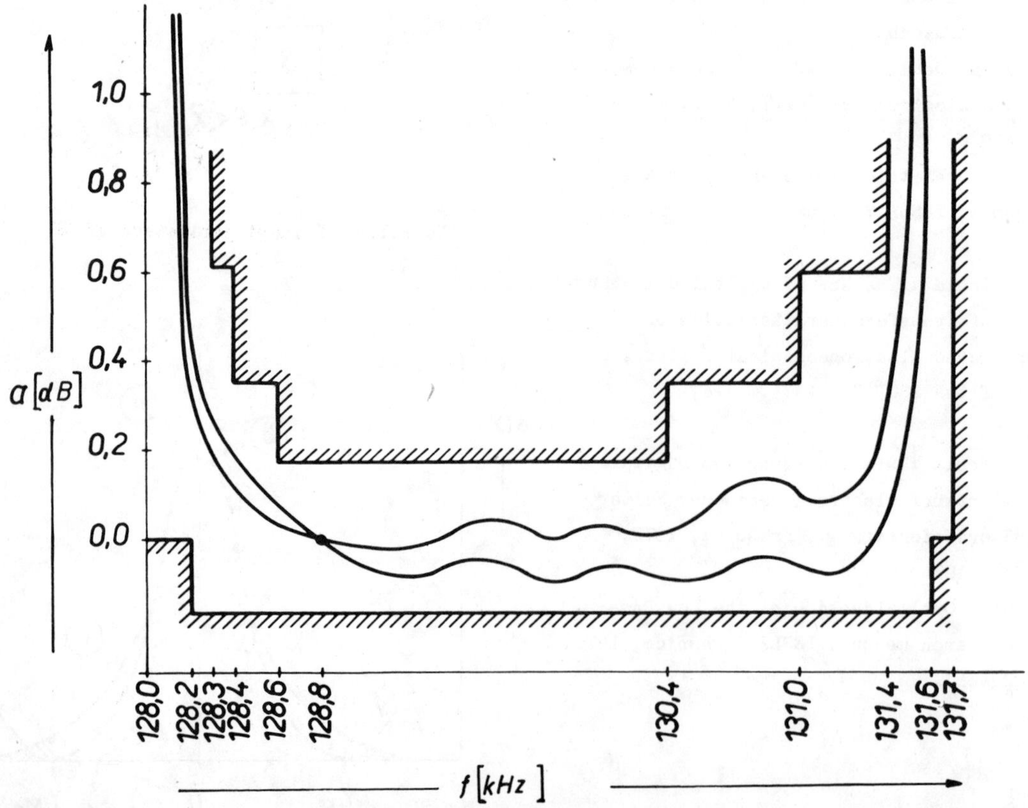

Fig.5 Tolerance regions of MF 128 kHz Tesla Czechoslovakia

Part II
Crystal Filters

Section II-A
Historical Overview

Crystal filters have been used in electronic equipment for over 40 years, so, therefore, they are quite old in terms of the age of the electronic industry itself. For many years the types of filter designs made with crystals and the techniques required to produce them did not change very much. The crystals were generally used as discrete resonators to form a filter in conjunction with a number of other components. In the early 1960's, however, there occurred a number of developments that were to have a profound effect on crystal filter technology. These developments made possible a range of high-frequency, high-order crystal filters that theoretically would not require the use of any other components. It is this class of post-1960 filters that we call modern crystal filters, and they are the subject of this collection of papers.

Modern crystal filters utilize two important principles that are called energy trapping and acoustic coupling. These terms apply specifically to resonators consisting of a piezoelectric material such as quartz that is sandwiched between two metal electrodes. It is customary to use the so-called AT-cut of quartz for the piezoelectric material, and when this is done, the resonator vibrates in a thickness-shear mode and its resonant frequency is stable with temperature variations. The resonant frequency is, however, sensitive to the mass of metal contained in the electrodes and fine tuning of the resonant frequency is accomplished by adjusting this mass. Energy trapping means that the mechanical vibrating energy of a resonator can be confined to the area immediately underneath the electrodes themselves. This is done by having a sufficient mass in the electrodes. Application of this relatively simple concept permitted crystal engineers to concentrate on optimizing the desired resonator characteristics and not to have to spend a great deal of time solving problems caused by edge effects and mounting techniques.

The technique of acoustical coupling is accomplished by placing two resonators side by side on the same piece of quartz and simply controlling the degree of energy trapping in each resonator. Energy is, therefore, allowed to couple from one resonator to the other, and the degree of coupling obtained is determined by the amount of the trapping and the separation between the resonators.

Application of these principles can result in filters produced from a number of different piezoelectric materials but, with the possible exception of PZT ceramic, quartz is the only material used in large quantities. Use of quartz does impose restrictions on the frequency range and the bandwidth that can be used. The frequency of operation is confined to frequencies greater than 5 MHz because that is the most practical range for AT-cut quartz resonators. The principles can still be applied at lower frequencies, but since the size and cost of the crystal increase with decreasing frequency, there is not much incentive to use crystals at these lower frequencies. One could say it was fortunate that a new technology for high-frequency crystals opened up at approximately the same time that inexpensive high-frequency transistors were opening up applications where there was a need for just this kind of filter. In practice, however, technologies tend to develop side by side, with one borrowing techniques from the other. This has been the case between the semiconductor industry and the crystal industry. In the areas of cutting, lapping, mask making, and vapor deposition, similar techniques are used in both areas, and in some cases even the same process equipment is used in the two industries.

The papers in this collection have been gathered together in various subgroups that are considered to be relevant to the subject of crystal filters. In a task like this, it is always difficult to make a choice between different papers, but a choice does have to be made in order to keep the book down to a reasonable size. Selections have been made on the basis of papers that will be useful to designers and users of crystal filters; consequently, many significant papers that pertain to the mathematics of resonator operation and design have been omitted. The filter designer and user can, however, obtain good insight into the mathematics of resonator theory from the book by Tiersten [1]. In order to aid in the understanding of this book, it is recommended that one also consult an additional text by Nye [2], who gives a good understanding of terminology which is not normally part of the filter designer's lexicon.

The first grouping of papers are of historical interest, and they show how the main ideas that are commonly accepted today were actually arrived at. As is true with most scientific discoveries, the ideas were arrived at by a slow, somewhat stumbling process where many people made contributions. It is with the benefit of hindsight that we can see the progression of ideas that contributed to building up our present knowledge of the subject.

In 1962, Nakazawa reported on the first practical devices that used acoustically coupled resonator pairs. Nakazawa recognized that these devices contained two modes of vibration, so he cascaded them to build relatively complex filters at 10.7 MHz. The theories of energy trapping and acoustical coupling were not well understood at that time, so consequently there is some confusion in the paper that was caused by his results showing a relationship between the degree of interresonator coupling and the crystal blank diameter. From subsequent knowledge of the subject, we now know that the relationship shown can be explained by the fact that at that time it was customary to vary the electrode diameter and the blank diameter together while preserving the same ratio between them. Variation of coupling is then fully consistent with a variation in the electrode diameter.

In 1963, Shockley, Curran, and Koneval reported their results on a program to develop a filter technology based on mass loaded piezoelectric ceramic resonators. The program was aimed at producing a number of resonators on the same piece of ceramic material and to mass load and separate them to the extent that there would be little mechanical coupling between them. In the course of this work, they offered a theory of energy trapping to account for the observed phenomena and they utilized an analogy with the cutoff effect in waveguides. It was subsequently pointed out, however, that this useful analogy had been first used ten years earlier by Mortley [3].

It remained for Onoe in Japan to piece together the energy trapping theory with the experimental observations of Nakazawa. This was done in a series of papers in Japan during 1964 and 1965 [4]-[7] and later in the English language version that is reprinted here. This unified theory, formed by joining together the theories of energy trapping and of acoustic coupling, is the basis of the ideas used to design modern crystal filters. Subsequent developments followed rapidly, and the first modern crystal filters were offered for sale in the United States in 1965.

The next major development was the recognition by Sykes and his co-workers of the importance of these discoveries to the telephone industry. The telephone industry is a large user of complex filters, and the greatest usage is confined to the channel filter required to form a single sideband in frequency division multiplex transmission systems. Various technologies have been used to build these filters, and they vary from discrete crystals to inductors and capacitors to mechanical filters and even active filters. In theory, at least, the principle of acoustical coupling offered the possibility of placing a large number of resonators in tandem on a single piece of quartz and using automated equipment to produce low-cost channel filters by this means.

The papers by Beaver published at the time of this effort to produce a telephone channel filter attracted a great deal of interest, and some of the empirical results that he obtained were in turn used as a starting point by many other people who subsequently commenced working in the field. We see, therefore, that by 1967 the main ideas had been formulated and work had already started on applying these ideas to the most promising applications. Subsequent papers highlight some of the problems that arose in these applications and the techniques that were used to combat them.

References

[1] H. F. Tiersten, *Linear Piezoelectric Plate Vibrations*. New York: Plenum, 1969.

[2] J. F. Nye, *Physical Properties of Crystals*. Oxford, England: Oxford Univ. Press, 1957.

[3] W. S. Mortley, "Frequency modulated quartz oscillators for broadcasting equipment," *Proc. IEE* (London), vol. 1048, pp. 239–253, Dec. 1956.

[4] M. Onoe, Paper XIII-70-437, Papers of Barium Titanate Appl. Res. Comm., IECE, Japan, pp. 113–114, July 1964.

[5] M. Onoe and N. Kobori, "The theory of coupling between laterally spaced energy-trapped resonators," Paper XIII-70-438, *ibid.*, pp. 115–116, July 1964.

[6] ——, Paper XIII-71-450, *ibid.*, pp. 195–196, July 1964.

[7] M. Onoe and H. Jumonji, "Analysis of piezoelectric resonators working in trapped-energy modes," in *Proc. Ultrason. Res. Comm.*, IECE, Japan, Feb. 1, 1965.

HIGH FREQUENCY CRYSTAL ELECTROMECHANICAL FILTERS

Y. Nakazawa

Toyo Communication Equipment Co., Ltd., Japan

ABSTRACT

At the 13th Annual Symposium(1) we reported the development and performance characteristics of 455kc quartz crystal electromechanical filters. Other reports(2) have discussed high frequency electromechanical filters which make use of a single crystal converted into a two-transducer unit by means of divided plating, the coupling being controlled to some extent by cutting away or otherwise changing portions of the unplated section.

This paper is presented on a high frequency crystal electromechanical filter in which AT-Cut quartz crystals of thickness shear modes are used and two vibration modes are effectively utilized.

In general, thickness shear crystals have numerous resonance frequencies, and the modes, except for the fundamental, have been unavailable as unwanted modes. However, by utilizing a selected unwanted mode, electromechanical filters can be constructed by present methods.

This report describes, in detail, the principle of construction and also the experimental results of its application, the methods of suppressing the unwanted vibration, and the relation between the diameter of AT-Cut crystal blanks and pass band-width from 10kc to 50kc with a center frequency of 10.7 Mc.

By proper combination of some electrical elements with these electromechanical filters, it is possible to construct a polarized band pass filter, and if these filters are connected in cascade, crystal electromechanical filter of good shape factor can be produced easily.

INTRODUCTION

In general, AT-cut plates have numerous possible frequency vibrations. Of them, the principal ones given by the group of thickness shear modes are largely independent of the shape and dimensions of the plate, while frequency vibrations given by the so-called "higher flexural mode group" are seriously affected by the shape and dimensions of the plate, being rapidly increased with decreasing dimensions. (3) (4) (5)

Fig. 1 shows typical examples of these vibration behaviors as measured with a spectral response equipment. The crystal vibrator used in the measurement is an ordinary AT-cut circular plate having the frequency of 10.7 Mc. The diameter of the plate is 9mm, and that of electrode is 5mm.

The fundamental thickness-shear mode gives a major thickness shear vibration which is utilized for crystal oscillators, crystal filters, etc. The high inharmonic overtone modes and shear flexural modes result in unwanted response in general.

Reprinted with permission from *Proc. 16th Ann. Symp. on Frequency Control*, Apr. 25-27, 1962, pp. 373-390, sponsored by the U.S. Army Electronics Command, Ft. Monmouth, NJ.

The high frequency electromechanical filter described in this paper makes use of two particular vibrations of fundamental mode and second inharmonic mode among these thickness shear vibration groups, for the purpose of constructing a band pass filter.

ANALYSIS

A simplified schematic of the high frequency mechanical filter is shown in Fig. 2. A divided plate, applied on either side of a circular or rectangular crystal, forms an electrode. The dividing direction and shape of the electrode are determined taking into account selected mode of vibration and equivalent inductance (or equivalent capacitance) and also with consideration for the elimination of unwanted response.

In the divided circular crystal of this figure, terminals 1 and 2 denote input and terminals 3 and 4 denote output, respectively.

When the measurement is made at the terminals 1 and 2, or 3 and 4 with spectral response equipment, the fundamental mode and second inharmonic mode in the X-direction, (these modes corresponding to the mode designations (1,1,0) and (2,1,0) of the rectangular crystal), develop very strongly as shown in Fig. 3, where transmission direction lies in the X-direction.

When transmission direction lies in the Z-direction, these strong resonances are fundamental mode and second inharmonic mode in Z-direction (these modes corresponding to the mode designation (1,1,1) of the rectangular crystal).(3)(4)(5)(6)(7)

The vibration of fundamental mode is resonance of two-terminals formed by connecting terminals 1 and 3,2 and 4, together, respectively, and that of the second inharmonic mode is resonance of two-terminals formed by connecting terminals 1 and 4, 2 and 3, together, respectively.

Now, consider the fundamental mode of the same divided-electrode crystal with reference to Fig. 4 (a) where the transmission direction lies in X-direction and terminals, 1 and 2, and 3 and 4 denote input and output, respectively. The crystal vibrator may be replaced by the equivalent electrical circuit of Fig. 4 (b) which represents a lattice type circuit comprising capacitances C_0, C_{13}, C_{14}, C_{23} and C_{24} and equivalent electrical constants L_1 and C_1.(8)(9) The inductance L_1 is the equivalent electrical inductance of the crystal being excited in the fundamental mode while the terminals 1 and 3, and 2 and 4 are respectively connected together to form a two-terminal circuit.

As for the second inharmonic mode of this crystal, the equivalent electrical circuit derived from the measurement at terminals, 1 and 2 or 3 and 4, in the same manner as described above, becomes as shown in Fig. 4 (c). Here again, it is a lattice circuit comprising capacitances C_0, C_{13}, C_{14} and C_{24} and equivalent electrical inductance and capacitance, L_1' and C_1'. The inductance L_1' is the equivalent electrical inductance of the crystal being excited in the second inharmonic mode while the terminals 1 and 4, and 2 and 3 are respectively connected together to form a two-terminal circuit. The values of calculated L_1 and L_1' are approximately the same, and therefore, this crystal makes the construction of symmetrical band pass filter feasible.

The two frequency vibrations described above are developed essentially in one same plate of crystal. Thus, by the integration of input terminals 1 and 2 and output terminals 3 and 4, a lattice type filter as shown in Fig. 4 (d) is constructed. If impedance of series arm comprising $2L_1'$, $C_1'/2$ and $C_0/2+C_{13}$ and impedance of lattice arm comprising $2L_1$, $C_1/2$ and $C_0/2+C_{14}$ are denoted by Z_A and Z_B respectively, the image transfer constant $\tanh\frac{\theta}{2}=\sqrt{\frac{Z_B}{Z_A}}$ and the image impedance $Z_0=\sqrt{Z_A\cdot Z_B}$ are obtained. as is well known.

A point of $Z_AZ_B = 1$ is the ideal pass point, and inside the pass band, Z_A and Z_B should have at least dissimilar signs. A point of infinite attenuation is given by $Z_A = Z_B$ and inside the attenuation region, Z_A and Z_B should be of the same sign. By the application of the reactance theorem, Z_A and Z_B are then expressed as $Z_A=j\omega H\frac{\omega^2-\omega_0{}^2}{\omega^2(\omega^2-\omega_1{}^2)}$ and $Z_B=j\omega H\frac{\omega^2-\omega_{-1}{}^2}{\omega^2(\omega^2-\omega_0{}^2)}$, respectively, where H is a constant, ω_1 is the upper cut-off angular frequency and ω_{-1} is the lower cut-off angular frequency, that is, $\omega_{-1}<\omega_0<\omega_1$. The pass band width is given by $\Delta=\omega_1-\omega_{-1}$.

With reference to Fig. 4, ω_{-1} corresponds to the resonant angular frequency of L_1C_1, ω_0 to the anti-resonant angular frequency of Z_B and ω_1 to the anti-resonant angular frequency of Z_A.

As mentioned above, a band pass filter can be constructed using a single divided-electrode plate of crystal, and the filter thus constructed is a constant K type one, because the attenuation pole is so far in the attenuation region owing to $L_1 \fallingdotseq L_1'$ and $C_{13} \fallingdotseq C_{14}$. If, however, a capacitance C_p is inserted between terminals, 1 and 3, as shown in Fig. 5 (a), it is easy to obtain a polarized filter. The equivalent electrical circuit for this case is shown in Fig. 5(b), where C_p is inserted in parallel with the series arm Z_A and attenuation pole is produced at such frequency at which the condition $Z_A = Z_B$ is satisfied.

EXPERIMENTAL RESULTS

Frequency difference between fundamental mode and second inharmonic mode in the X-direction

Fig. 6, shows the frequency difference of two modes of divided-electrode crystals of varying dimensions, as measured with a spectral response equipment. The electrode is divided into two in X-direction, and excitation is applied through terminals, 1 and 2. The difference in resonance frequencies of two vibrations thus developed is plotted in Fig. 7, which indicates that the frequency difference increases with decreasing crystal diameter. From this plot, an experimental formula

$$\Delta f=3k\left(\frac{y_0}{d}\right)^h f_0$$

is obtained. In this formula, Δf is the frequency difference between fundamental mode and second inharmonic mode in the X-direction, f_0 is the fundamental frequency, d is the diameter of crystal, y_0 is its thickness, and k and h are the constants which are introduced to compensate for the fact that the crystal vibrator is too thin to exhibit perfect (1,1,0) and (2,1,0) modes corresponding to that of the rectangular crystal and that it is circular. Experimentally, these values are found as $k \fallingdotseq 57$ and $h \fallingdotseq 2.7$, respectively.

High frequency electromechanical filter

A photo in Fig. 7 shows an electromechanical filter comprising a circular AT-cut divided-electrode crystal of 10.7 Mc center frequency, having a slit for the suppression of unwanted response. The electrodes of a single crystal have three terminals. The electrode on one side is commonly grounded, and the one on the other side is divided equally into two parts, one of which is the input terminal and the other is the output terminal. Fig. 8 shows typical measured attenuation curves of these simplest electromechanical filters having various pass band widths. Attenuation characteristics curves of electromechanical filters terminated by designed resistances at the input and output are shown in Fig. 8. In case the termination resistances are greatly capacitive, proper L,C tank circuit must be connected across the termination resistances to improve the ripple of pass band region. Fig. 9 shows the results obtained on the filters with two and three sections, each of 30 kc band width, in cascade connection. From this figure it is seen that the effective attenuation increases with increasing number of sections.

Polarized attenuation band pass filter

Fig. 10 shows a polarized attenuation band pass filter constructed by inserting an electrostatic capacity between terminals, 1 and 3, of a divided-electrode crystal. As clearly seen in this figure, the smaller the value of the electrostatic capacity, C_p, the farther off the cut-off frequency is the frequency of attenuation pole and the larger is the effective attenuation Conversely, the larger C_p, the nearer the cut-off frequency is the frequency of attenuation pole and the smaller is the effective attenuation.

Narrow band pass filter

As already seen in Fig. 6, a narrow band pass filter is feasible with increased diameter of the crystal.

Also, if the fundamental mode and second inharmonic mode in Z-direction effectively utilized, narrow band pass filter can be produced.

However, if the dimension in Z-direction is extremely reduced, suppression of unwanted response is achieved and at the same time, narrow band pass filter is realized. Experiment using a divided-electrode crystal having dimensions $x_0 = 8$ mm, $y_0 = 0.157$ mm and $z_0 = 1.795$ mm indicates that the diference in two resonant frequencies of this crystal is small enough to permit construction of a filter having the band width of about 10 kc as shown in Fig. 11.

Spurious response

Since in this filter only two particular vibrations of thickness shear mode group are utilized, it is natural that the other vibrations should appear as spurious response. To prevent these spurious vibrations from being electrically excited, special considerations are given to the area and shape of electrodes. Further, as shown in Fig. 8, the crystal plate is given a slit to suppress inharmonic vibration in Z-direction. Application of mechanical loss at supporting point of clip mount is also effective for the suppression of unwanted response of the crystal. By these means, spurious response can be suppressed to within reasonable limits.

Termination resistance

For the crystal filter having the center frequency of 10.7 Mc, termination resistance corresponding to the characteristic impedance described before is 1 to 5 k ohms approximately, although it varies somewhat the pass band width and the electrical inductances, L_1 and L_1'. Also, if the termination resistances are greatly capacitive, it must be connected across the termination resistances by proper LC tuning tanks.

Temperature characteristic

This electromechanical filter has essentially the same temperature characteristic as AT-cut crystal, because it is based on the major resonance frequency of AT-cut crystal and its inharmonic overtone. The temperature characteristic curve is a cubic curve of very excellent nature. With practically no use of other electrical parts, it is possible to realize a filter which withstands use at such high temperature as nearly 150°C.

Insertion loss

As seen in the figures so far given, insertion loss of 30 kc filter is about 0.5 to 2 db for one section. For two sections and three sections, it is about 2 to 3 db and 3 to 5 db, respectively, but the loss is ascribed mostly to the mismatching between sections.

Size and weight of filter

A photo in Fig. 12 shows the appearance of a prototype electromechanical filter; an HC-6/U holder is also shown for comparison. The dimensions are 20 mm long, 12.5 mm wide and 18 mm high, and the weight is about 7 grams.

CONCLUSION

It was presented that a band pass filter could be constructed by the use of two thickness shear vibrations of one divided-electrode crystal. Although the description was made for the crystal filter having the center frequency of 10.7 Mc, used generally as intermediate frequency for FM communication, the theory is applicable for any frequency vibration so far as the thickness shear modes are concerned.

The crystal filter according to this theory is very small, light and inexpensive compared with the conventional crystal filter.

Being essentially an AT-cut crystal, the filter is excellent in temperature characteristic and in stability. In particular, substantial freedom from any electrical parts has made the filter capable of withstanding elevated temperature, and so, it will find extensive application in future.

ACKNOWLEDGEMENT

The author wishes to express his appreciation to Dr. H. Fukuyo, Tokyo Institute of Technology, Mr. S. Fukunaga and Mr. H. Sato, and Dr. H. Yoda, Toyo Communication Equipment Co., for their contributions to this study.

CAPTIONS OF FIGURES

REFERENCES

1. H. Yoda, "Quartz Crystal Mechanical Filter," 13th Annual Symposium, 1959.

2. C.R. Mingins, A.D. Frost, R.W. Perry, L.A. Howard, D.W. Macheod and G.A. Larson, "An Investigation of the Characterisitics of Electromechanical Filter" Supplemental Report, Contract No. DA 36-039-SC-5402, 1954.

3. I. Koga, H. Fukuyo and J.E. Rhodes, "Modes of Vibration the Probe Method" 13th Annual Symposium, 1959.

4. R.D. Mindlin and H. Dersiewicz, "Thickness-Shear and Flexural Vibrations of a Circular Disk", Jour. Appl. Phys., Vol. 25, No. 10, Oct., 1954.

5. H. Ekstein, "High Frequency Vibration of Thin Crystal Plates," Phys., vol. 68, No. 182, Jul., 1945.

6. H. Fukuyo, "Investigation of the spectra of circular AT-Cut Quartz Crystal Plates," Project No. A-402-1 ~ 13.

7. H. Fukuyo, "Quartz Crystal Studies and Measurement," Contract No. DA-36-039-SC-78910.

8. W.P. Mason, "Electromechanical Transducers and Wave Filters," D. Van Nostrand, 1952.

9. W.P. Mason and R.A. Sykes, "Electric Wave Filters Employing Quartz Crystal with Normal and Divided Electrodes," BSTJ, vol. 19, 1940.

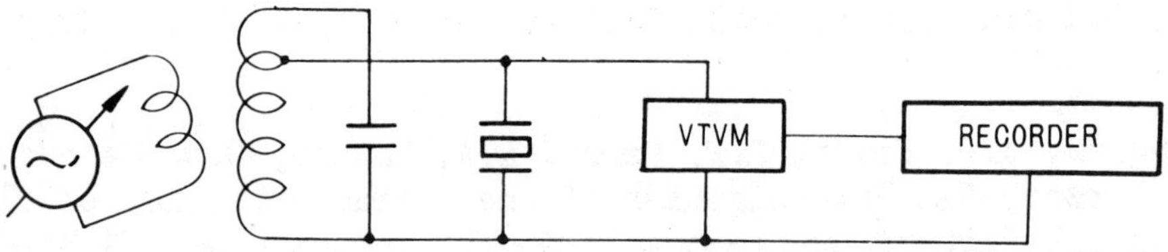

SCHEMATIC OF SPECTRAL RESPONSE EQUIPMENT

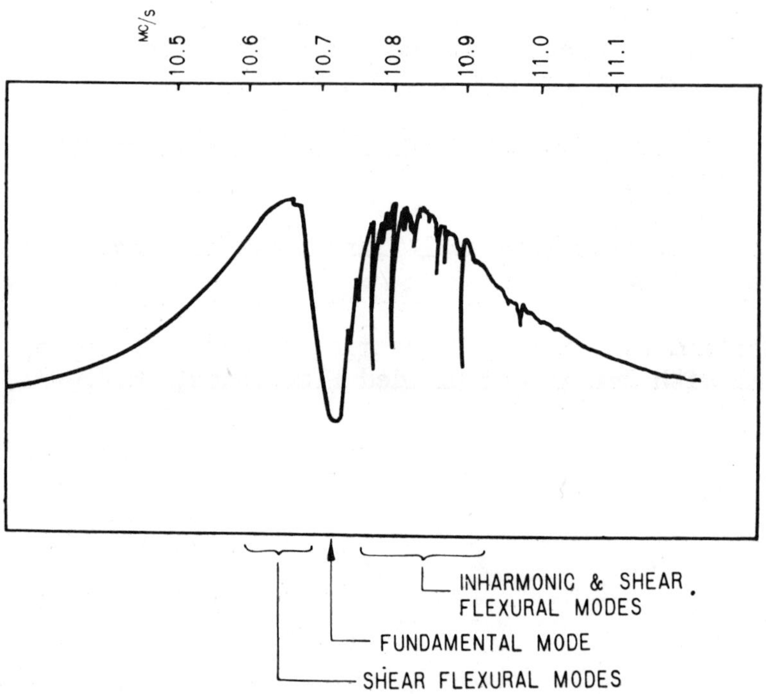

Fig. 1 Frequency Spectral response of normal 10.7 Mc/s AT-cut quartz crystal resonator.

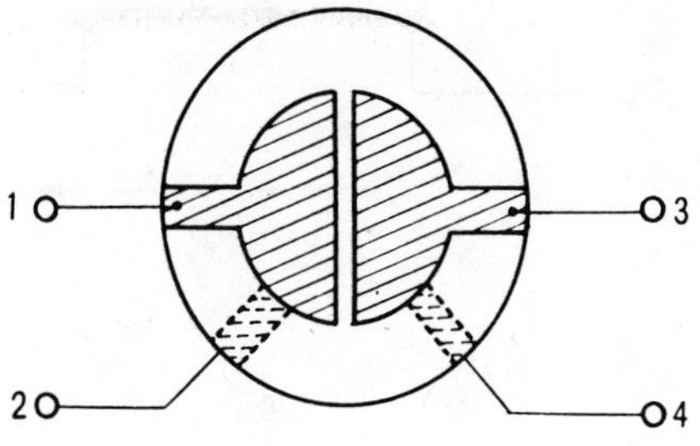

Fig. 2 Divided electrode crystal

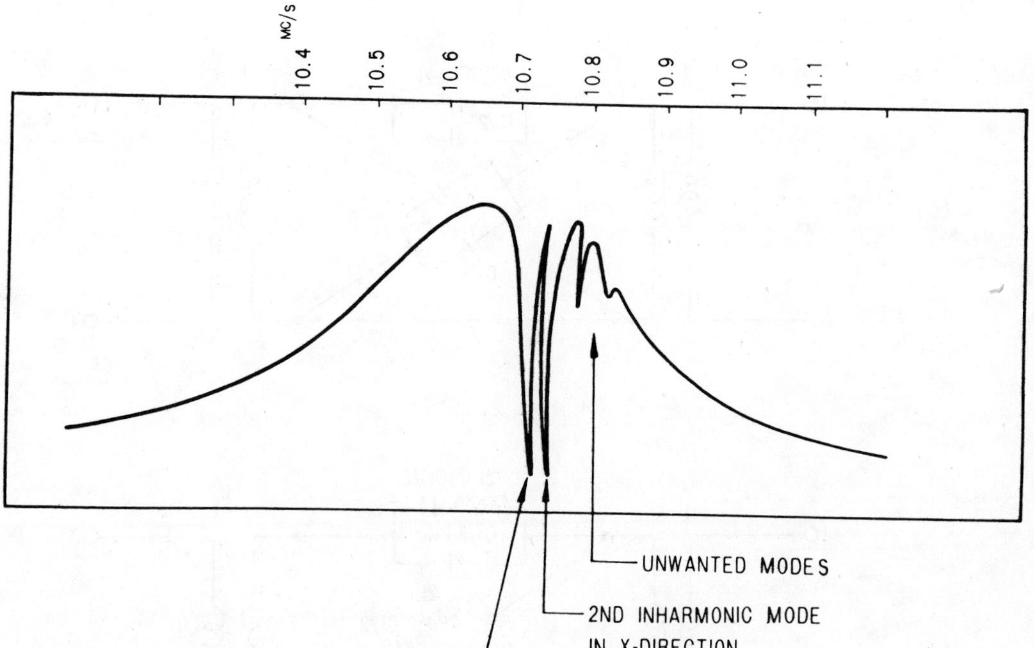

Fig. 3 Typical frequency spectral response of divided electrode crystal, as measured between input terminals, 1 and 2.

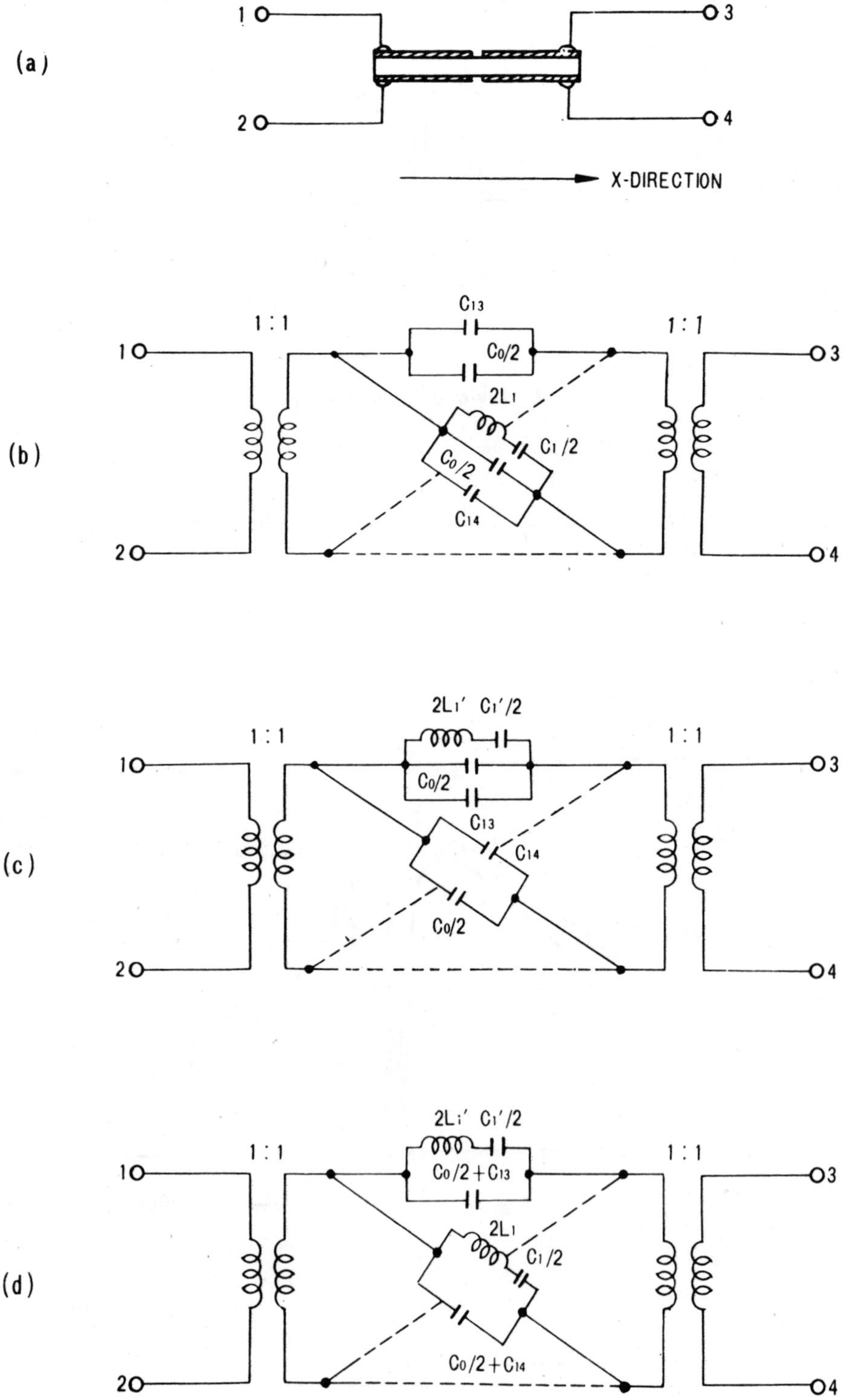

Fig. 4 (a) Divided electrode crystal (b) Electrical equivalent circuit for fundamental mode (c) Electrical equivalent circuit for second inharmonic mode (d) Electrical equivalent circuit containing fundamental mode and second inharmonic mode.

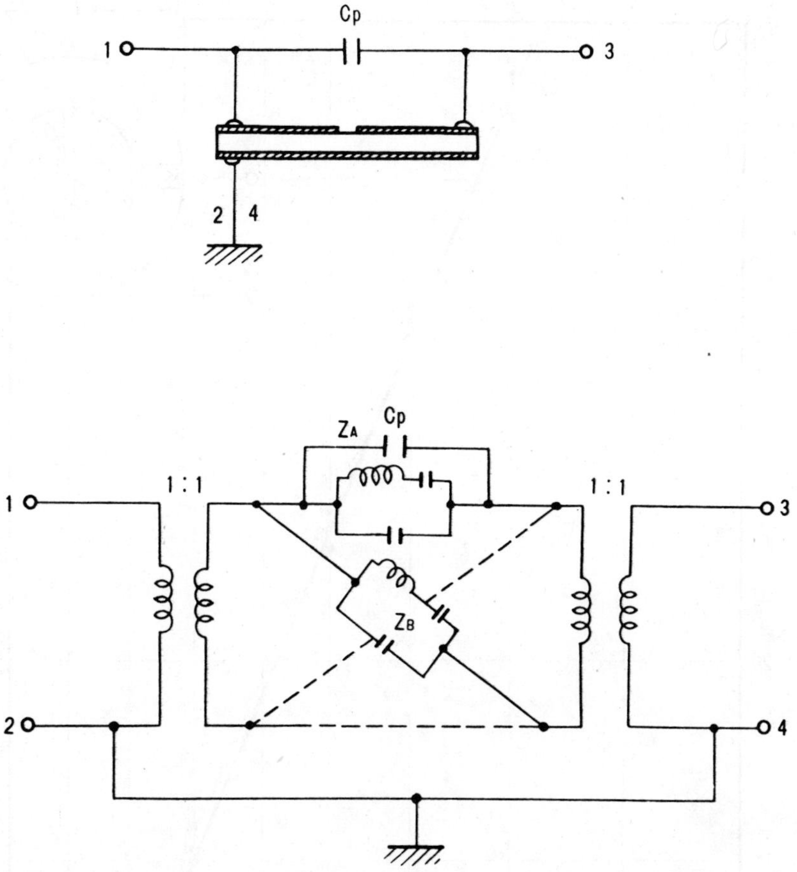

Fig. 5 Electrical equivalent circuit of polarized filter constructed by insertion of electrical static capacity.

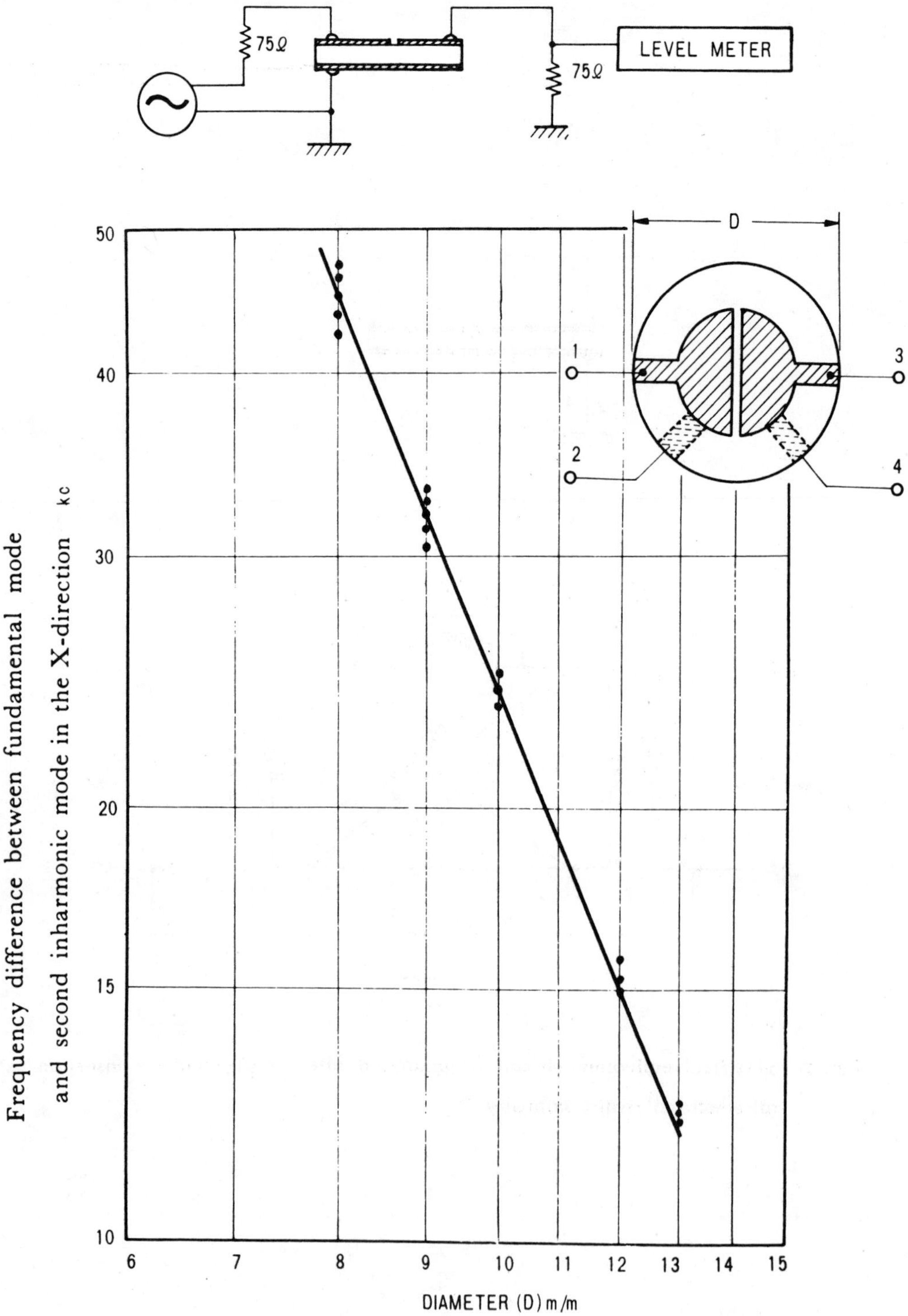

Fig. 6 Bandwidth V.S diameter of circular crystal.

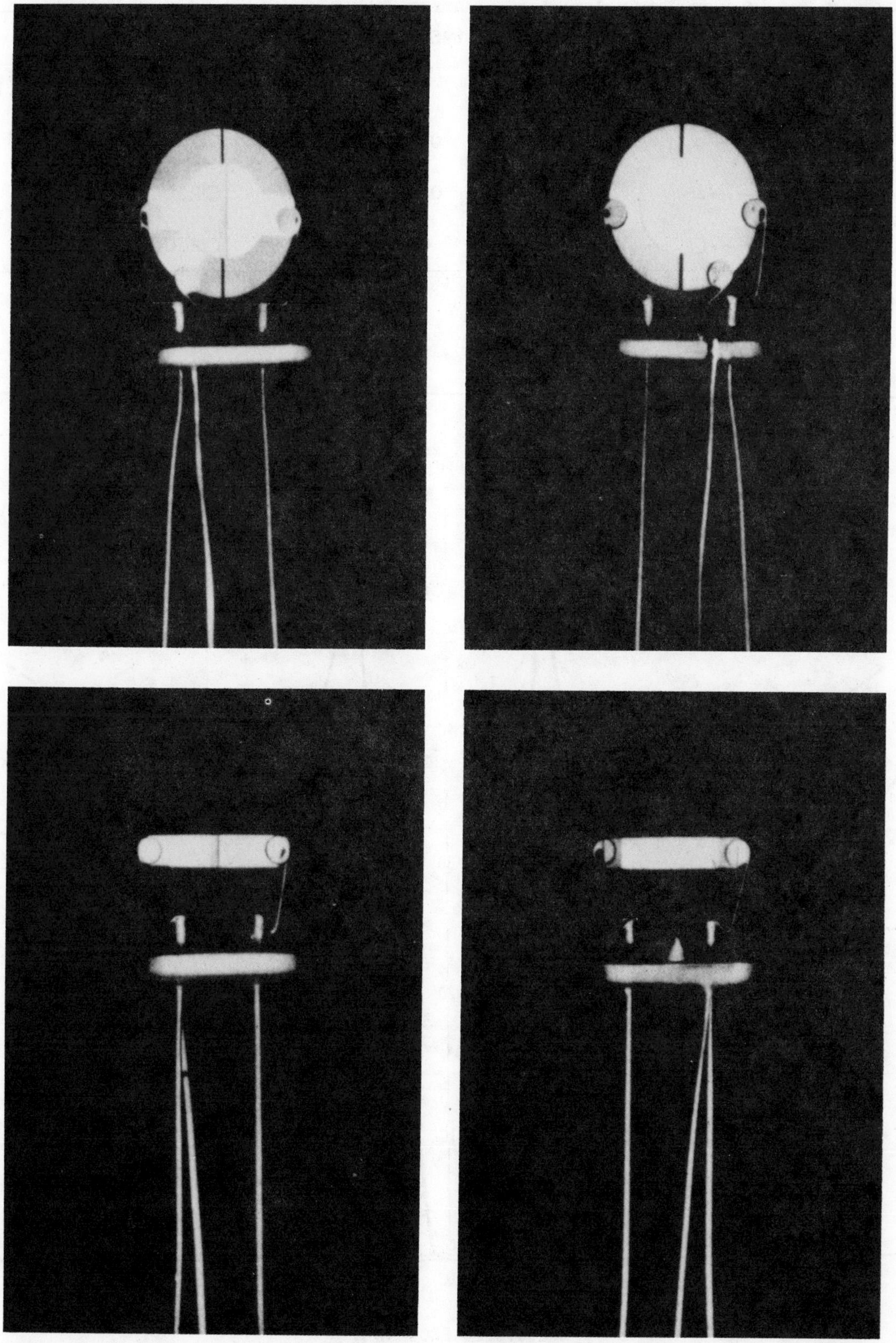

Fig. 7 Front and rear view of simplest electromechanical filter comprising one crystal.

	LINE	TERMINATION	INSERTION LOSS	DIMENSION IN mm
A	●———●	1.5 kΩ	1 ~ 2 db	RECTANGULAR 8 × 0.157 × 1.8
B	×- - - - - - -×	1 kΩ	0.5 ~ 1.5 db	CIRCULAR DIA. 12
C	△———△	2.5 kΩ	0.5 ~ 1.5 db	CIRCULAR DIA. 9
D	⊡- - - - - -⊡	5 kΩ	0.5 ~ 1 db	CIRCULAR DIA. 8

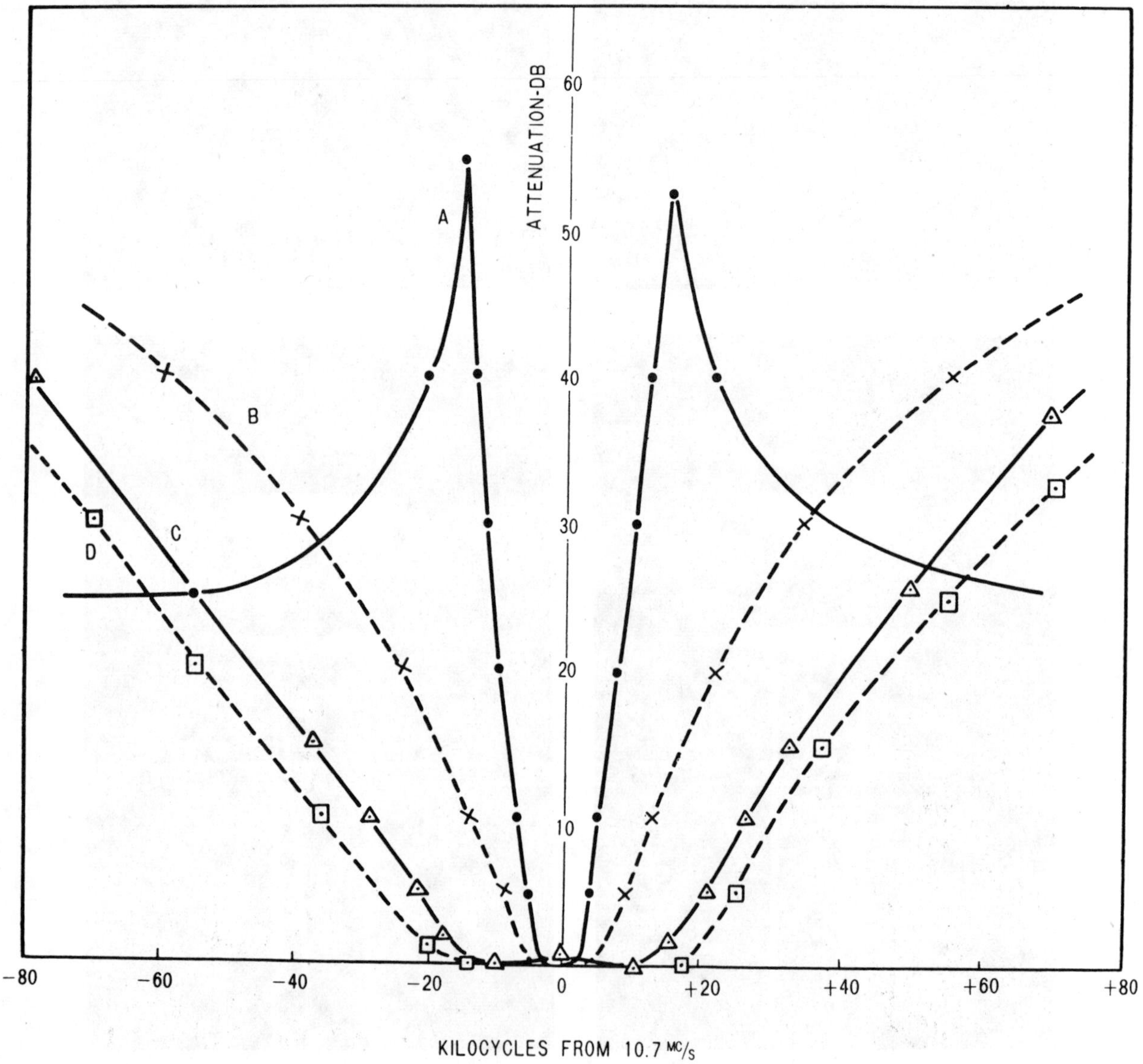

Fig. 8 Attenuation characteristics of one section electromechanical filters.

LINE	SECTION	INSERTION LOSS	TERMINATION
●———●	2 SECTION	2.3 db	2.5 kΩ
×- - - -×	3 SECTION	5.0 db	2.5 kΩ

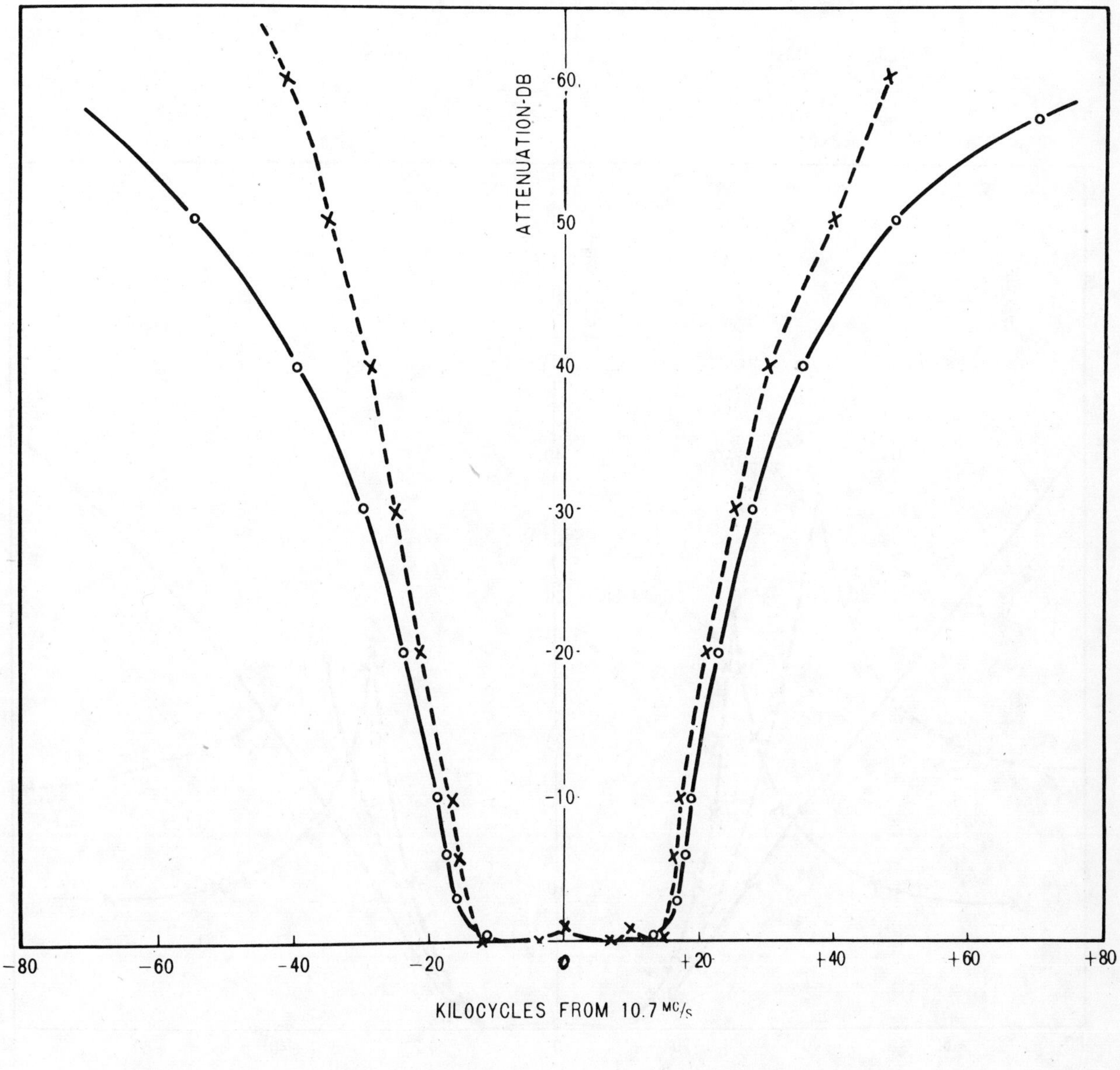

Fig. 9 Attenuation characteristics of two and three sections cascade connected electromechanical filter.

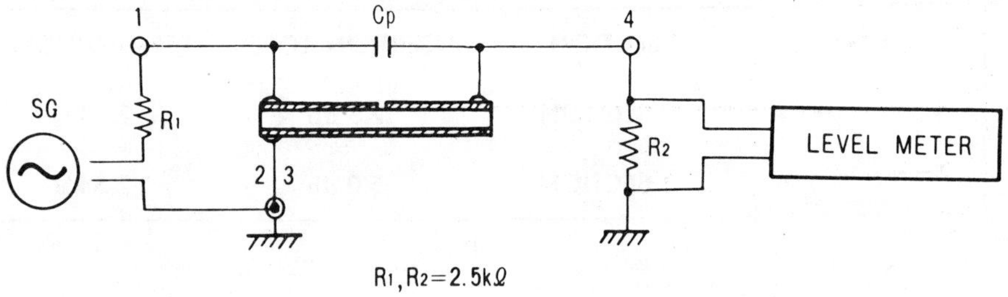

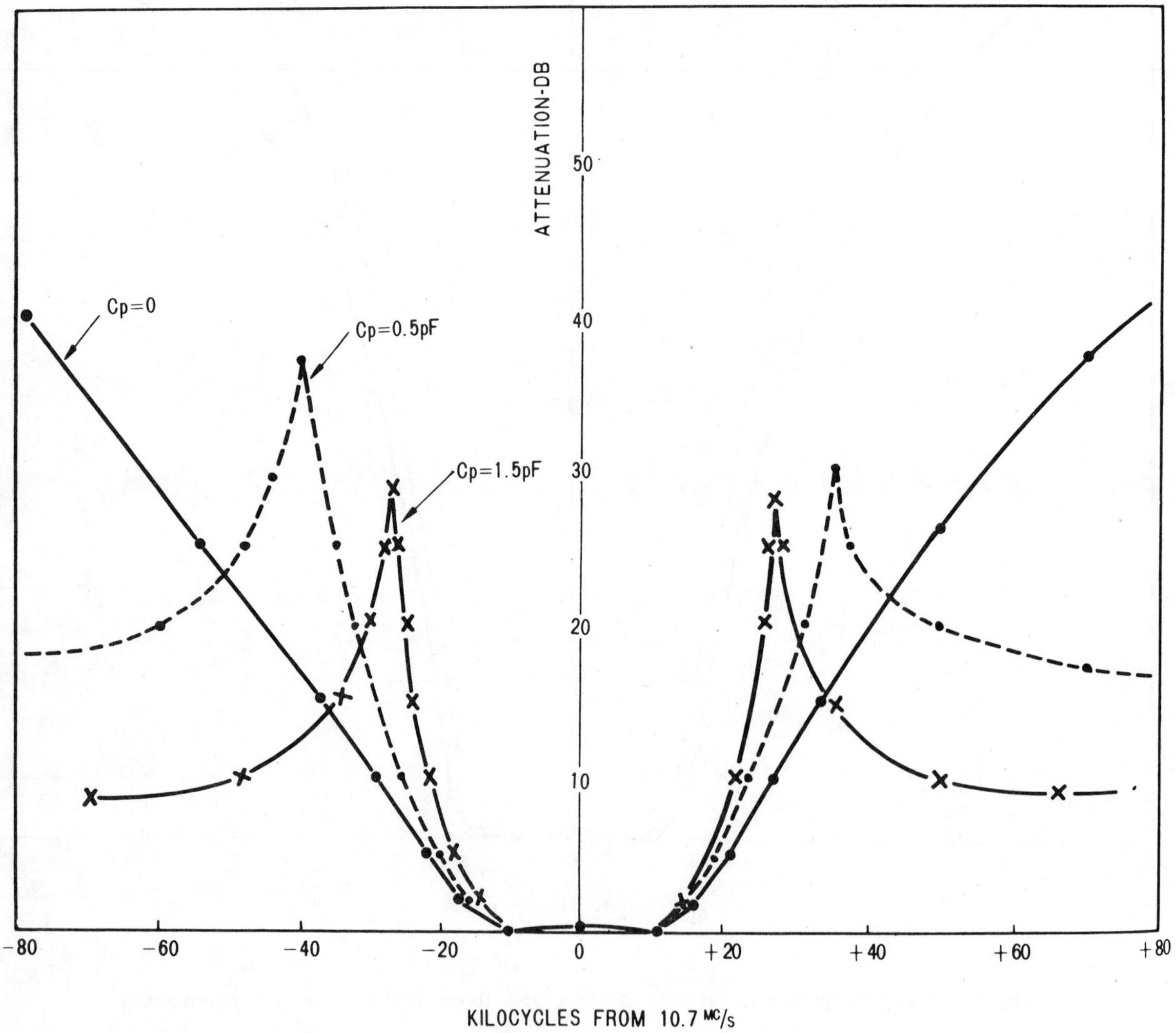

Fig. 10 Polarized band pass filters.

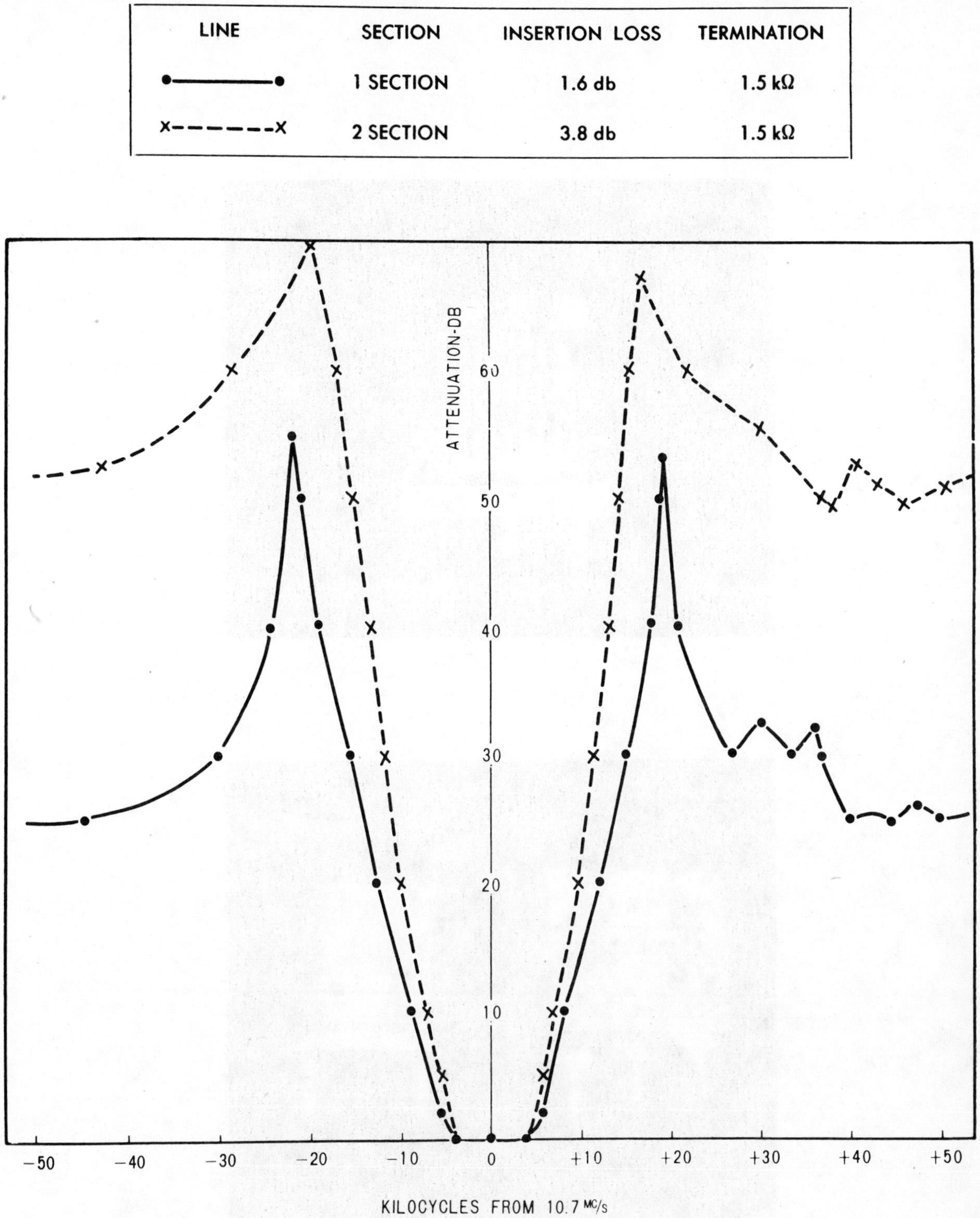

Fig. 11 Narrow band pass filters.

Fig. 12 Interior and exterior view of prototype electromechanical filter in comparison with HC-6/U holder.

ANALYSIS OF PIEZOELECTRIC RESONATORS VIBRATING IN TRAPPED-ENERGY MODES

Morio Onoe and Hiromichi Jumonji, Members

Institute of Industrial Science,
University of Tokyo

SUMMARY

One of the most important problems in piezoelectric resonators for filter application utilizing thickness vibration is to suppress unwanted responses. At frequencies above approximately 10 Mc the method of suppression based on the principle of trapped energy is very effective.

The wave propagating along the surface of a plate has a cutoff frequency below which its amplitude attenuates exponentially with distance. When a structure is devised such that its natural frequency under the electrode is made to fall below the cutoff frequency by increasing the thickness of the electrode or other means, the vibrational energy concentrates only in the vicinity of the electrode so that the effects of its configuration (the unwanted modes resulting from that configuration) are reduced.

The present article is an analysis of such trapped-energy resonators. First, with regard to the case of a finite plate, we calculate the frequency spectrum of each mode in the range above and below the cutoff frequency. It is shown that the wanted mode can be separated from unwanted modes by employing the effect of trapped energy. We also calculate the equivalent inductance and show that the equivalent inductance for the wanted mode is remarkably small compared with that of other modes so that the wanted mode can be preferentially excited. In particular, we clarify the existence of antisymmetric modes and show that multimode filters can be constructed by employing these modes and further that a multisection filter can be obtained using a single piezoelectric plate.

Next, we calculate the frequency spectrum in the case where two trapped-energy resonators are arranged on the same plate and analyze the degree of coupling between them. From this it is possible to determine the required distance between resonators when each resonator is to be used independently. It also shows that mechanical filters can be constructed by effectively utilizing the coupling between resonators.

1. Introduction [1,2]

Demands for filters having sharp response at high frequencies have increased in recent years. This is due to the widespread adoption of SSB in HF radio networks, closer channel spacing in the VHF band and the increased upper frequency limits of wire carrier systems. The midfrequencies demanded of filters have increased from several Mc to several tens of Mc and even higher than 100 Mc.

The characteristics of filters are influenced by the resonators used. Since resonators utilizing mechanical vibration have higher Q than ordinary LC resonator circuits, they are suitable for constructing filters with steep-skirted response. They have other advantageous features: their temperature characteristics are good, their operating characteristics are stable with respect to aging, and they can be made small in size. Crystal filters, ceramic filters and mechanical filters belong to this class. Crystal filters have been used exclusively at high frequencies where the utilization of thickness vibration is required to obtain the desired vibration modes. In filter resonators using thickness vibration it is important to suppress modes other than the main or unwanted mode. At frequencies below 10 Mc this is usually accomplished by the careful selection of the shape and dimensions. At frequencies above 10 Mc it is generally effective to reduce the electrode areas [3].

Reprinted with permission from *Electron. Commun. Japan*, no. 9, pp. 84-93, Sept. 1965. Reprinted with the consent of Scripta Publishing Co., 1511 K St. N.W., Washington DC 20005.

Curran and others showed that it is possible to obtain thickness vibration with suppression of unwanted modes when a small electrode is provided on a thin piezoelectric ceramic plate. They also showed that more than one pair of electrodes could be applied to the same plate with little mutual acoustical coupling, thereby permitting the construction of a multisection filter using a single plate. They named this the "Uni-wafer filter" [4].

In this way, ceramic filters using thickness vibration have been made available. Recently, similar filters have been constructed by using crystals [5, 6]. Shockley and others established that the above phenomena are attributable to the trapping of the vibrational energy of a transverse wave at frequencies below the cutoff frequency of the plate [5]. However, their analysis is confined to the case where resonators exist independently on an infinite plate.

The present article is an analysis of such trapped-energy resonators. First, we determine the effects of configuration and dimensions on the resonant frequency and investigate the frequency spectra of unwanted modes above and below the cutoff frequency. We then calculate the equivalent-circuit constants and show the effect of trapped energy. We also clarify the existence of antisymmetric modes and show that it is possible to produce a multimode filter resembling the so-called crystal mechanical filter [7] by combining these modes with symmetric modes.

Lastly, we analyze the case where two resonators are arranged on the same plate and determine the degree of coupling between them. This furnishes basic data which can be used whether each resonator is to function independently or it is desired to construct a mechanical filter utilizing the coupling between resonators.

2. Trapped-Energy Resonator

Let us consider a wave propagating along the z-axis of a plate having the cross section illustrated in Fig. 1. For simplicity, we shall assume that its dimension in the direction of the x-axis is infinite. The direction of wave displacement is considered to be parallel to the x-axis; that is, we will consider the SH wave only. This corresponds to the case of crystal plates R_1 and R_2 (AT and BT), or a piezoelectric ceramic plate with polarization axis taken in the x-axis direction. The displacement U which satisfies the boundary condition that the upper and lower surfaces are free assumes the following form [8, 9]

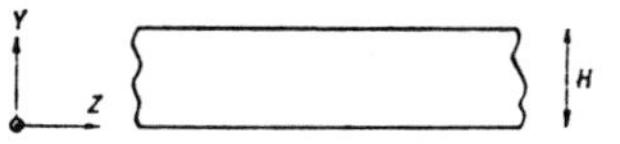

Fig. 1. Dimensions of a plate in which SH waves propagate along z-axis

$$U \propto \cos\left(\frac{n\pi y}{H}\right)\exp j(\omega t \pm \beta z) \qquad (1)$$

where H is the thickness of the plate and ω is the angular frequency. Quantity β is the propagation constant for the z-axis direction and is given by

$$\beta = \frac{n\pi}{H}\sqrt{\left(\frac{f}{nf_0'}\right)^2 - 1} \qquad (2)$$

where f_0' is the natural frequency of thickness-shear vibration of an infinitely wide plate and n denotes the order of the harmonic. Here, we consider the case of $n = 1$, namely, the principal mode, which is most important in practical application. The case of $n > 1$ may be handled similarly. As clear from Eq. (2), if $f > f_0'$, β is a real number, and if $f < f_0'$, it is a pure imaginary number. Thus, f_0' is the cutoff frequency of this wave. Propagation occurs at frequencies higher than this but at frequncies lower than this the amplitude attenuates exponentially and the wave ceases.

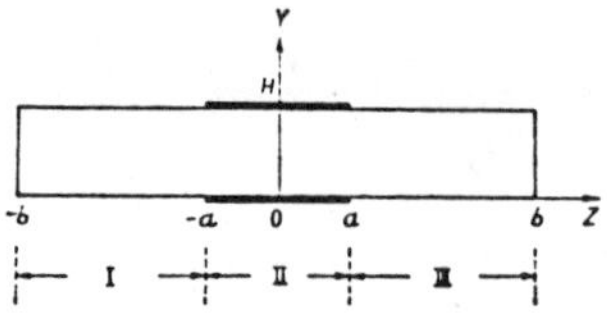

Fig. 2. A trapped energy resonator on a finite plate

Now, electrodes are applied to both sides of the plate (region II in Fig. 2) by electroplating or other means. Let us assume that the natural frequency of thickness-shear vibration in this area has become f_0, falling below f_0' due to the mass loading of the electrodes. In order for the frequencies of regions I and III to lie below the cutoff frequency at an intermediate frequency between these two ($f_0' > f > f_0$), the vibration energy produced in region II is trapped inside. Since the vibration amplitudes in regions I and III attenuate exponentially, they have practically no effect on the end surfaces.

It is also possible to obtain more than one resonator on the same plate without mutual coupling. This is the principle of the trapped-energy resonator and the Uni-wafer filter. In designing such devices, however, it is necessary to know the effects of the distance from the end surfaces or adjacent resonators. In this article, the effects of the end surfaces are analyzed in Sections 3 and 4 and the coupling between adjacent resonators is discussed in Section 5.

In the case when thickness longitudinal vibration is excited in a piezoelectric ceramic plate with the polarization axis perpendicular to the plate surface, or in the case when the plate dimension is finite in the x-axis direction or the electrode areas are finite, it is necessary to consider the propagation of the longitudinal wave (P wave) and the transverse wave (SV wave) whose displacement is perpendicular to the plate. Analysis in such cases becomes extremely complicated and will not be discussed here. However, in the case of either wave (P or SV) a cutoff frequency exists in the same way as in the case of the SH wave and the phenomenon of trapped energy may occur. Hence, a qualitative evaluation is possible.

3. Frequency Spectrum of a Trapped-Energy Resonator in a Finite Plate

The dimensions and symbols for the various parts of a resonator are shown in Fig. 2. Let us determine the natural frequency of the resonator. In consideration of Eq. (1), the displacement is given as follows [10]:

$$U=u\cdot\cos\left(\frac{\pi y}{H}\right)\cdot\exp(j\omega t) \tag{3}$$

where u is a function of z. Employing k and k' to redefine the propagation constant β so that it is always a real number, we obtain

$$u_{\mathrm{I}}=B{}^{\mathrm{ch}}_{\mathrm{sh}}k'z-C{}^{\mathrm{sh}}_{\mathrm{ch}}k'z,\ (-b\leq z\leq -a) \tag{4}$$

$$u_{\mathrm{II}}=A{}^{\cos}_{\sin}kz \qquad (-a\leq z\leq a) \tag{5}$$

$$u_{\mathrm{III}}=B{}^{\mathrm{ch}}_{\mathrm{sh}}k'z+C{}^{\mathrm{sh}}_{\mathrm{ch}}k'z\ \ (a\leq z\leq b) \tag{6}$$

where

$$k=\frac{\pi}{H}\sqrt{\left(\frac{f}{f_0}\right)^2-1} \tag{7}$$

$$k'=\frac{\pi}{H}\sqrt{1-\left(\frac{f}{f_0'}\right)^2} \tag{8}$$

The upper symbol in each double expression gives the symmetric mode in which the displacement becomes an even function with respect to the origin, and the lower symbol gives the antisymmetric mode in which the displacement becomes an odd function. From the boundary condition that the displacement and force are continuous at $z=\pm a$ and the ends are free at $z=\pm b$, the following frequency equation can be obtained (see Appendix I):

$$\mathrm{th}\,k'(b-a)=\frac{k}{k'}{}^{\ \tan}_{-\cot}ka \tag{9}$$

In particular, when the plate is infinitely large ($b\rightarrow\infty$), the following form is obtained:

$$\frac{k}{k'}{}^{\ \tan}_{-\cot}ka=1 \tag{10}$$

Figure 3 gives the result of calculations by Eq. (9). Symbol s denotes the symmetric mode and symbol a denotes the antisymmetric mode. The subscripts, denoting the orders of the modes are employed in a manner explained below.

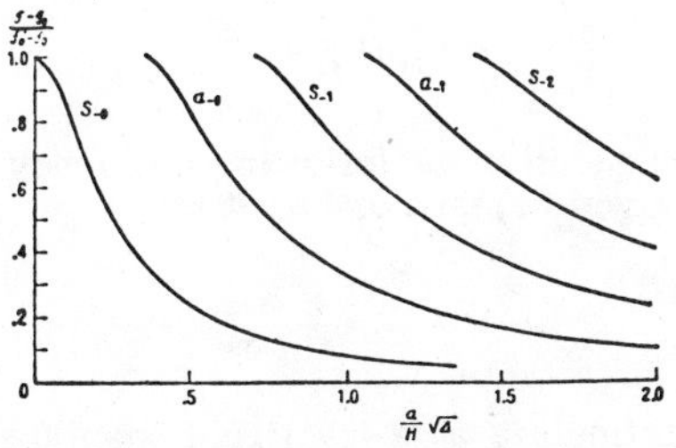

Fig. 3. Resonant frequencies of trapped modes in an infinite plate as functions of resonator parameters. (s : symmetric mode, a : antisymmetric mode, Δ : measure, defined in Eq. (11), of mass loading due to electrodes.)

In the figure

$$\Delta=\frac{f_0'-f_0}{f_0'} \tag{11}$$

is the relative decrease in natural frequency in region II. It is seen from the figure that as the electrode areas are increased and the natural frequency is lowered, higher-order modes appear. This is not desirable from the standpoint of suppressing unwanted modes. Hence, in designing the resonator it is necessary to select a range $\left(\frac{a}{H}\sqrt{\Delta}<0.305\right)$ such that only the lowest-order mode, s_{-0}, may exist. However, since the a_{-0} mode is also used in the multimode filters to be discussed later, the dimensions of the electrodes and the decrease in natural frequency should be selected in the range $\left(0.305<\frac{a}{H}\sqrt{\Delta}<0.710\right)$. Symmetric modes of the infinite plate have already been obtained by Shockley and others [5]. Antisymmetric modes have been obtained for the first time here. However, as explained above, these modes are of importance in controlling the decrease of the natural frequency and in application to multimode filters. Before proceeding to calculation of the frequency spectrum of a finite plate by the use of Eq. (9), we shall analyze the case in which the frequency has become higher than the cutoff frequency ($f > f_0'$). In this case, the wave propagates freely in regions I and III and behaves in the same way as in an ordinary resonator where the mass of the electrodes is taken into consideration. The displacements in the respective regions can be obtained by replacing the hyperbolic functions with the corresponding trigonometric functions in Eqs. (4) — (6) and using k" given below in place of k' defined by Eq. (8):

$$k''=\frac{\pi}{H}\sqrt{\left(\frac{f}{f_0'}\right)^2-1} \qquad (12)$$

Consequently, the following frequency equation is given in place of Eq. (9):

$$\tan k''(b-a)=\frac{k}{k''}\begin{matrix}-\tan\\ \cot\end{matrix}ka \qquad (13)$$

In particular, at a = 0 (i.e., when the effects of the electrodes are neglected) the above equation becomes the following familiar equation for thickness-shear vibration:

$$k''b=\begin{matrix}m\pi\\ \frac{2m+1}{2}\pi\end{matrix} \quad (m=0,1\cdots\cdots) \qquad (14)$$

At a = b (i.e., when the electrodes cover the entire plate surfaces) Eq. (13) assumes a form similar to the above equation:

$$kb=\begin{matrix}m\pi\\ \frac{2m+1}{2}\pi\end{matrix} \quad (m=0,1,\cdots\cdots) \qquad (15)$$

This is obvious when we consider that Eq. (14) refers to the natural frequency f_0' and Eq. (15) to f_0. The subscript affixed to the mode symbol coincides with m. The minus sign is used to denote the trapped-energy mode. We calculated the frequency spectrum of a trapped-energy resonator in a finite plate as a function of the ratio of the sides b/H by the use of Eqs. (9) and (13). Figure 4 shows a case where the s_{-0} mode exists as the only trapped-energy mode. Figure 5 gives an instance where the a_{-0} mode coexists. The resonant frequency of the trapped energy mode rapidly approaches the resonant frequency of the infinite plate given by Eq. (10) when b increases to a certain magnitude. The dotted lines in Fig. 4 show the frequency spectra where the effects of the electrodes in Eq. (14) are neglected. In this case, the unwanted mode approaches the main mode s_0 as b increases. On the other hand, in the case of the trapped-energy resonator, since all unwanted modes exist above f_0', it is possible to secure a frequency range such that there are no unwanted modes between it and the trapped energy mode s_{-0}.

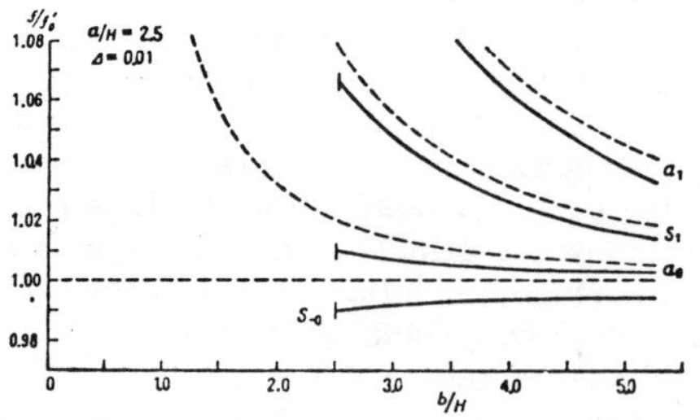

Fig. 4. Frequency spectrum of a trapped energy resonator on a finite plate. Resonator parameters are so chosen that only the first trapped mode, s_{-0}, exists. Dotted lines represent frequency spectrum of a conventional resonator which has the same parameters except the electrodes, of which mass is negligible.

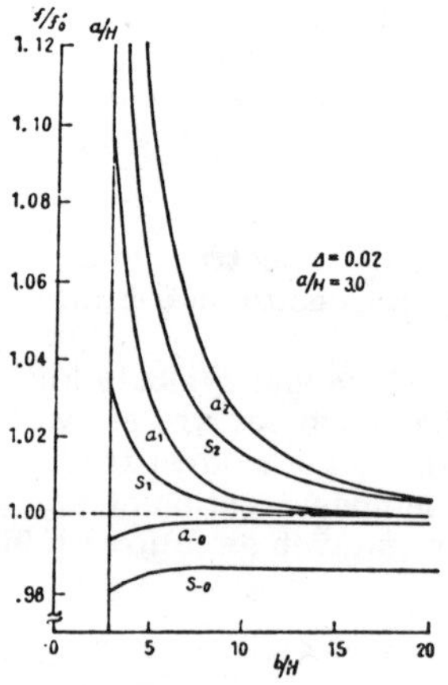

Fig. 5. Frequency spectrum of a trapped energy resonator on a finite plate. Resonator parameters are so chosen that both the first symmetric and antisymmetric trapped modes, s_{-0} and a_{-0}, exist.

Consequently, filters using such resonators are free from disturbance due to unwanted modes in the vicinity of the passband. In using these resonators as oscillators, it is possible to obtain a smooth temperature characteristic since there is no jump due to unwanted modes [11]. Although the unwanted modes are separated by wide intervals above f_0' in the trapped-energy resonator, their existence is not desirable in a resonator for filter application in securing a guaranteed attenuation throughout the reject band. Fortunately, however, the unwanted modes become extremely weak compared with the main mode in a trapped-energy structure. This will be explained by the analysis of the equivalent circuit constants in the next section.

4. Equivalent Circuit Constants of a Trapped-Energy Resonator in a Finite Plate

The equivalent circuit for a trapped-energy resonator can be given by Fig. 6 in the same way as for an ordinary resonator. In a crystal or piezoelectric ceramic for filter application where the stabilized coupling coefficient is comparatively small the equivalent parallel capacitance C_0 may be regarded as an electrostatic parallel capacitance. Since the resonant frequency of the series arm has already been obtained, all constants can be determined when the equivalent inductance and Q are given. In the case of a trapped-energy mode, the effect of the support on the Q is small. In the case of a piezoelectric ceramic, it may be considered that the elasticity loss of its material appears directly. In the case of a crystal, since the loss of the material itself is very small, the Q is influenced by the loss caused by the elasticity loss of the electrode material as well as by a very small energy leakage taking place to the surroundings. Consequently, the principal task of the analysis is to obtain the equivalent inductance.

The equivalent inductance can be obtained from the time-averaged kinetic energy that gives the displacement [12]. The piezoelectric effect is assumed to be the same as that of crystal plates R_1 and R_2 (AT and BT).

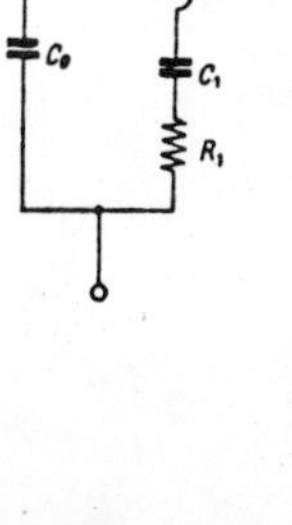

Fig. 6. Equivalent circuit of a piezoelectric resonator

First, with regard to the trapped-energy mode below the cutoff frequency f_0', the inductance per unit length in the x-axis direction is given by the following equation (see Appendix II):

$$\frac{L}{L_{00}} = \frac{1}{2} \frac{(ka)^2}{\begin{pmatrix} \sin^2 ka \\ (1-\cos ka)^2 \end{pmatrix}} \left[(1 \mp 2\Delta)\left(1 \pm \frac{\sin 2ka}{2ka}\right) + \begin{pmatrix} \cos^2 ka \\ \sin^2 ka \end{pmatrix} \left\{ \frac{b-a}{a} \cdot \frac{1}{\mathrm{ch}^2 k'(b-a)} + \frac{\mathrm{th}\, k'(b-a)}{k'a} \right\} \right] \tag{16}$$

However, in the case of a single electrode, no antisymmetric vibration can be excited as the charges cancel out each other. Hence, the values given refer to the case where the electrode is divided in half and so connected that the electrical change can be applied to them equally. In the case where a thin electrode covers the entire surface of an infinite plate, L_{00} is the inductance per unit area divided by the area of the electrode under consideration, 2a, as shown by the following equation:

$$L_{00} = \frac{\rho}{16\, e_{26}^2} \cdot \frac{H^3}{a} \tag{17}$$

Here, ρ is the specific gravity, and e_{26} is the piezoelectric constant. For $b \to \infty$, the terms within braces in Eq. (16) become 1/ka. This gives the inductance of the trapped-energy resonator in an infinite plate. Figure 7 gives an example of its calculated values.

Next, when the frequency is higher than cutoff, the terms within braces in Eq. (16) become as follows.

$$\{\ \} \to \left\{ \frac{b-a}{a} \cdot \frac{1}{\cos^2 k''(b-a)} + \frac{\tan k''(b-a)}{k''a} \right\} \tag{18}$$

In particular, as $\Delta \to 0$ (i.e., in an ordinary resonator in which the electrode is sufficiently thin and has no effect upon the mode), by substituting $k = k'' \to 0$ in Eqs. (16) and

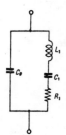

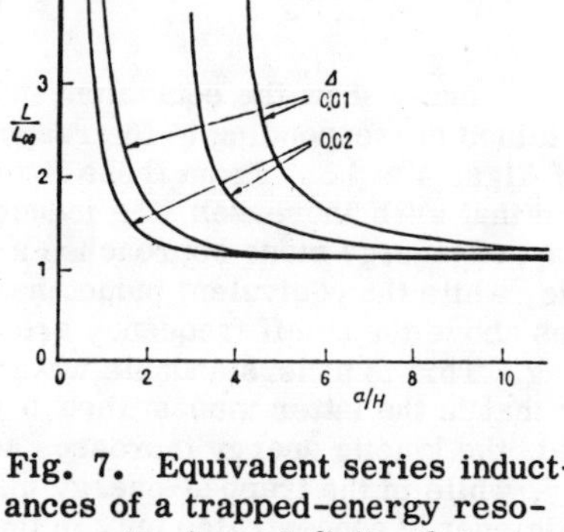

Fig. 7. Equivalent series inductances of a trapped-energy resonator in an infinite plate

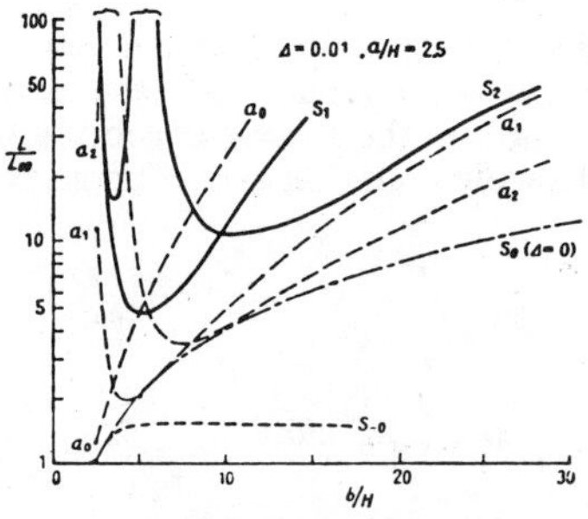

Fig. 8. Equivalent series inductances of a trapped-energy resonator in a finite plate whose frequency spectrum was shown in Fig. 4. Dotted lines show the inductance of a conventional resonator.

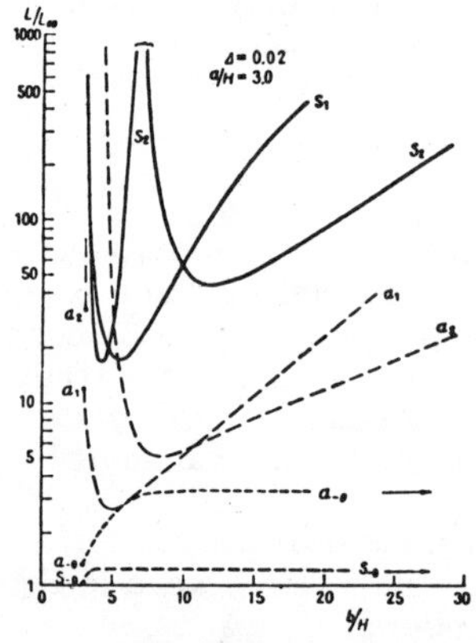

Fig. 9. Equivalent series inductances of a trapped energy resonator in a finite plate whose frequency spectrum was shown in Fig. 5

(18) with respect to the s_0 mode, the following equation is obtained:

$$\frac{L}{L_{\infty}} = \frac{b}{a} \tag{19}$$

Figures 8 and 9 show the equivalent inductances obtained corresponding to the respective modes of Figs. 4 and 5. From these figures it is found that as b increases, the inductance of the trapped-energy mode approaches a certain value, while the equivalent inductances of the modes above the cutoff frequency are all increasing. This is understandable when we consider that in the latter modes when b is increased, the kinetic energy increases accordingly, while in the trapped-energy mode the kinetic energy concentrates only in the vicinity of the electrode and is not much influenced by b. The broken lines in Fig. 8 show the inductances of the s_0 mode of an ordinary resonator ($\Delta = 0$) by Eq. (19), namely where the electrode is thin. It is found that, even if the area of the electrode is the same, when b has a certain magnitude the inductance is much smaller in the energy-trapped type.

Needless to say, the inductances of modes above the cutoff frequency are not much different from the case of ($\Delta \to 0$). Consequently, relative to the main mode, the unwanted modes of the trapped-energy resonator appear to be very weak. As clear from Fig. 9, the inductance of the a_{-0} mode is 2 to 3 times that of the s_{-0} mode, hence it can be preferentially excited. Consequently, it is possible to construct a multimode filter resembling a CM filter [7] by combining both. Namely, we consider a 3-terminal element in which the electrode on one side is divided in the center as shown in Fig. 10 (a). The equivalent circuit obtained by the theory of the divided electrode oscillator [13] will be as shown in Fig. 10 (b). Here, L_s and L_a are respectively the inductances of modes s_{-0} and a_{-0} by Eq. (16). This circuit can be further transformed into the lattice circuit of (c) and when the resonance and antiresonance points of each arm are properly arranged, it is possible to construct a bandpass filter by this alone.

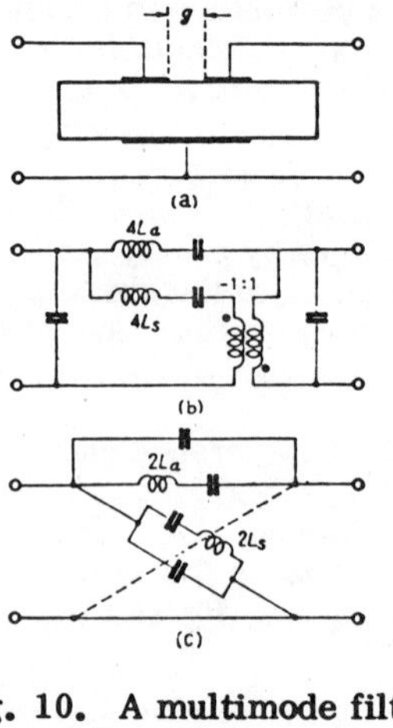

Fig. 10. A multimode filter. A trapped energy resonator with split electrodes (a) can vibrate in both the symmetric and the antisymmetric modes. Hence its characteristics are represented by an equivalent circuit (b), which can be transformed into a lattice bandpass filter (c).

The distance between the resonance points can be varied greatly by a/H and the quantity Δ as shown in Fig. 3. On the other hand, the distance between the resonance point and the antiresonance point in each arm is influenced by the electromechanical coupling coefficient. For filter application it is necessary to design so that the antiresonance point of the arm to which L_s belongs comes near the resonance point of the arm to which L_a belongs. Consequently, the pass band of the filter becomes about twice the distance between the resonance and antiresonance points of each arm. In order that the filter characteristic may become nearly symmetrical with respect to the mid-frequency, it is necessary that L_a and L_s be approximately equal. Adjustment for this can be made by utilizing the fact that when the width g of the dividing line of the electrode shown in Fig. 10 (a) is increased, L_s sharply increases compared with L_a. This is because the piezoelectrically induced change becomes maximum in the center portion in the case of the s_{-0} mode but becomes zero in the case of the a_{-0} mode.

Compared with conventional CM filters, this type of filter is less susceptible to disturbance by unwanted modes, since the trapped-energy mode is used. Also, in this type of filter, multiple sections can be provided on the same plate. These are their advantageous features. Since connection of each electrode can be made by thin plating, it is possible to construct a multisection filter by the use of a piezoelectric plate with electrode etching similar to that of a printed-circuit board. A detailed report on this type of filter together with experimental data will be given on another occasion. In conventional Uniwafer filters many resonators are provided on the same plate. But since the mode of each resonator is single and independent, in order to obtain a filter with a large guaranteed attenuation it is necessary to use at least one external differential transformer [6].

5. Coupling Between Trapped-Energy Resonators Arranged in Parallel

Now let us analyze the coupling of two trapped-energy resonators of identical shape arranged in parallel as shown in Fig. 11. The natural frequency of the plate is the same as in the preceding section. In regions I, III and V, this frequency is f_0' and in regions II and IV under electrodes it is f_0, with f_0 lower than f_0'.

When the dimensions of each part are taken as shown in the figure, the displacement can be expressed by Eq. (3) also. However, u

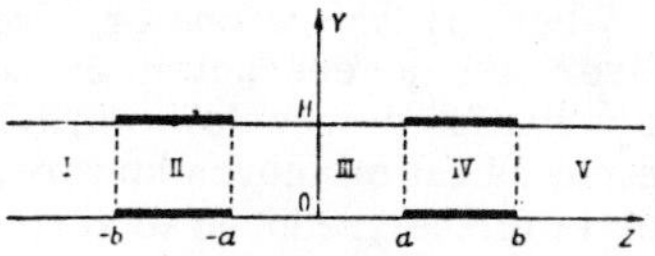

Fig. 11. Two trapped energy resonators in parallel on the same plate

becomes as follows, corresponding to each region:

$$u_{I} = \pm A \exp(+k'z) \qquad (z<-b) \tag{20}$$

$$u_{II} = \mp B \sin kz \pm C \cos kz \qquad (-b<z<-a) \tag{21}$$

$$u_{III} = D \begin{matrix}\mathrm{ch}\\ \mathrm{sh}\end{matrix} k'z \qquad (-a<z<+a) \tag{22}$$

$$u_{IV} = B \sin kz + C \cos kz \qquad (a<z<b) \tag{23}$$

$$u_{V} = A \exp(-k'z) \qquad (b<z) \tag{24}$$

Quantities k and k' are the same as defined by Eqs. (7) and (8). From the condition that the displacement and force are continuous at each boundary of the regions, the following frequency equation can be obtained (see Appendix III):

$$k(b-a) = \tan^{-1}\frac{k'}{k} + \tan^{-1}\left(\frac{k'}{k} \begin{matrix}\mathrm{th}\\ \mathrm{cth}\end{matrix} k'a\right) \tag{25}$$

Figure 12 gives the resonant frequencies of the two resonators calculated with the distance between them taken as a parameter. The dashed curves in the center correspond to the case where the distance is infinite ($a \to \infty$), namely, where the resonators exist independently. It coincides with the s_{-0} mode

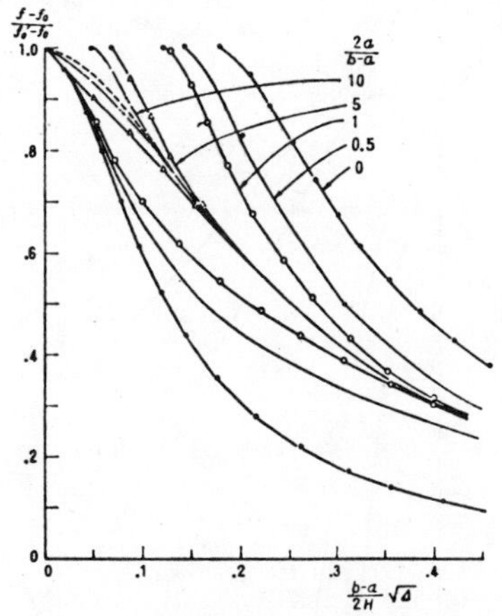

Fig. 12. Resonant frequencies of two trapped-energy resonators on the same plate as functions of resonator parameters

in Fig. 3. When the two resonators are brought closer, the degeneracy is broken, the symmetric vibration moves downward, and the antisymmetric vibration moves upward. As the distance is further reduced to zero ($a \to 0$) we again have the case of an independent resonator. In this case, when the horizontal scale is taken into consideration, the symmetric vibration coincides with the s_{-0} mode of Fig. 3 and the antisymmetric vibration coincides with the a_{-0} mode of the same figure. These are shown by the outermost curves with solid dots. The antisymmetric mode does not exist at frequencies above the cutoff frequency f_0'. This is obvious when we consider that the plate is infinitely large.

From the standpoint that the degeneracy of the resonant frequency has been broken due to the coupling of the two identical resonators by reducing their spacing, the distance between the upper and lower resonant frequencies expressed with respect to frequency is nothing but the coupling coefficient Γ between the resonators. In Fig. 13 this coefficient is calculated for a far wider range. Since the natural frequency drop Δ is at most of the order of 10^{-2}, when (Γ/D) is selected at 10^{-2}, the coupling coefficient becomes 10^{-4}, and it may well be considered that there is no mutual interference in ordinary application. Conversely, since the maximum (Γ/D) is about 0.5, assuming $\Delta = 1\%$, the specific bandwidtb as a mechanical filter becomes below 0.5%. Consequently, although wide-band application may not be feasible, it is possible to obtain a mechanical filter using thickness vibration for narrow-band application. In this case, the use of a piezoelectric material is not

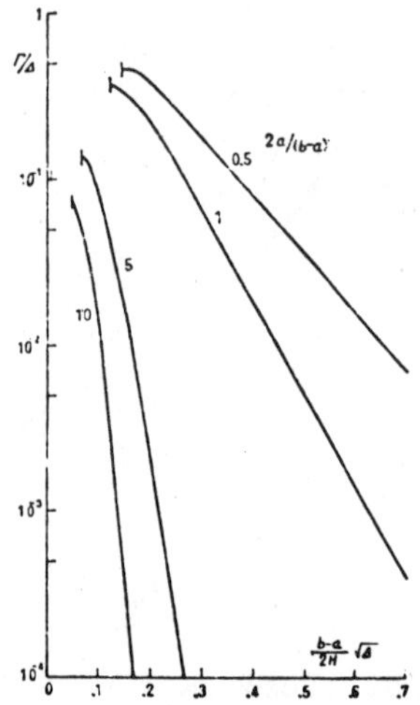

Fig. 13. Coupling coefficient (Γ) between two trapped-energy resonators on the same plate

mandatory but will be more convenient since the transducers at both terminals can be constructed simultaneously. It is not necessary that the electrodes in the midsection be connected to each other, unlike the case of the multimode filter discussed in the preceding section. The electrodes are not required to have conductivity. They are required only to serve the function of lowering the natural frequency of the plate. However, it is convenient for measurement and adjustment of the equivalent constants that each section be excited piezoelectrically [14].

Details of the mechanical filter using thickness vibration will be reported at a later date. We only mention the possibility here.

6. Conclusion

We have analyzed the piezoelectric resonator applying the new principle of trapped energy. First, for the case of a finite plate, we calculated the frequency spectrum and equivalent inductance of each mode above and below the cutoff frequency. We demonstrated that the main mode can be isolated from unwanted modes by virtue of the effect of trapped energy and that the inductance of the main mode is far smaller than that of the other modes, hence the main mode can be excited strongly. In particular, we demonstrated the existence of antisymmetric modes. We showed that it is possible to construct a multimode filter employing these antisymmetric modes and also that it is possible to construct a multisection filter using only one piezoelectric plate. Next, we calculated the frequency spectrum for the case in which two trapped-energy resonators are arranged on the same plate and analyzed the coupling between these two. It was seen from this that it is possible to determine the required spacing between resonators when they are to be operated independently and that, conversely, by effectively using the coupling between these resonators it is possible to obtain a mechanical filter with thickness vibration, although its bandwidth is not very wide.

The foregoing analyses are limited to the case of thickness-shear vibration wherein the length of the plate in the direction of displacement is infinite. We will report other more complex cases and a proposed filter together with experimental results in the future.

Submitted February 27, 1965

NOTES AND REFERENCES

1. A subsidy from the 1964 Scientific Experimental Research Fund was given for the present study.
2. The present work was reported at the Supersonic Research Committee meeting (Feb. 1, 1965) as well as on the following occasions: Papers of Barium Titanate Application Research Committee XIII-70-437 (Sept. 1964), XIII-70-438 (Sept. 1964), and XIV-71-450 (Nov. 1964). Crystal Subcommittee of the Research Committee on Standards 301 and 303; Papers of Joint Subcommittee, 48 (Sept. 1964), 49 (Sept. 1964), and 53 (Dec. 1964).
3. R. Bechmann: Proc. IRE, 49, p. 523 (1961).
4. D.R. Curran and D.J. Koneval: Proc. Natl. Elec. Conf., 17, p. 514 (1961).
5. W. Shockley, D.R. Curran and D.J. Koneval: Proc. 17th Freq. Control Symp., p. 88 (1963).
6. D.R. Curran et al.: WESCON Rec. 18 (1964).
7. Nakazawa: Proc. 16th Freq. Control Symposium, p. 373 (1962); Electronics (Japan), 8, p. 1407 (Dec. 1963).
8. Sato: Bull. Earthquake Rec. Inst., 29, p. 223 (1951).
9. Onoe: Symp. of Joint Conv. on Nondestructive Testing, S 8-1, 10, p. 135 (1961).
10. Equation (3) assumes that the electrode thickness is negligible compared with the plate thickness and does not affect the displacement distribution in the y-direction. Actually, since f_0 and f_0' differ only by several %, this assumption holds. However, when the difference is large, it is necessary to allow for other modes ($n \neq 1$) in the boundaries. Even in such case, however, so long as no zero-order mode ($n = 0$) appears, the energy can still be trapped since all other modes are cutoff modes.
11. Onoe: Analysis and detection of unwanted modes in quartz crystal units for oscillators, Jour. I.E.C.E., Japan, 47, 1, p. 38 (Jan. 1964); available in English in E.C.J., same date, p. 82.
12. Koga: Piezoelectricity and High Frequency, Ohm Pub. Co. (1937).
13. Onoe: Jour. I.E.C.E., Japan, 37, 2, p. 113 (Feb. 1954).
14. Onoe: A method for measuring equivalent circuit constants of electromechanical filters, Jour. I.E.C.E., Japan, 43, 8, p. 884 (Aug. 1960).

APPENDIX

I. Derivation of Eq. (9)

In Eqs. (4) — (6) the symmetric and antisymmetric vibrations are written separately. Hence, it suffices to consider the boundary condition in the range of $0 < z$. Of the stresses acting on the plane perpendicular to z, only the following component is not identically zero:

$$T_{zz} \propto \frac{\partial U}{\partial z} \tag{A 1}$$

Consequently, for the boundary condition at $z = a$ we have

$$(u_{\mathrm{II}})_{z=a} = (u_{\mathrm{III}})_{z=a} \tag{A 2}$$

$$\left(\frac{\partial u_{\mathrm{II}}}{\partial z}\right)_{z=a} = \left(\frac{\partial u_{\mathrm{III}}}{\partial z}\right)_{z=a} \tag{A 3}$$

Also, at $z = b$

$$\left(\frac{\partial u_{\mathrm{III}}}{\partial z}\right)_{z=b} = 0 \tag{A 4}$$

Equations (A 2) — (A 4) are simultaneous linear equations in A, B and C and can be arranged in the following form:

$$\begin{bmatrix} {\cos \atop \sin} ka & {\mathrm{ch} \atop \mathrm{sh}} k'a & {\mathrm{sh} \atop \mathrm{ch}} k'a \\ {-k \sin \atop k \cos} ka & k' {\mathrm{sh} \atop \mathrm{ch}} k'a & k' {\mathrm{ch} \atop \mathrm{sh}} k'a \\ 0 & k' {\mathrm{sh} \atop \mathrm{ch}} k'b & k' {\mathrm{ch} \atop \mathrm{sh}} k'b \end{bmatrix} \begin{bmatrix} A \\ -B \\ -C \end{bmatrix} \equiv 0 \tag{A 5}$$

In order that there exist a solution such that (A, B, C) are not identically zero, it is necessary that the determinant consisting of the coefficients on the left in the above equation be zero. This is applied in Eq. (9). When the resonant frequency that satisfies Eq. (9) is substituted in any two of Eqs. (A 2) – (A 4), (A : B : C) is established and the displacement distribution can be obtained.

II. Derivation of Eq. (18)

The time average K of the kinetic energy per unit length in the x-direction can be written in the following form:

$$K=\frac{1}{2}\omega^2\int_0^H \rho[(N_I+N_{II})+q\,N_{II}]\cos^2\left(\frac{\pi y}{H}\right)dy \qquad \text{(A 6)}$$

where N is the integral of the following form obtained for each region:

$$N=\int u^2 dz \qquad \text{(A 7)}$$

In region II, since the mass of the electrode is finite, the kinetic energy increase slightly, being q times greater than in the case where the mass is neglected. On the other hand, the increase of the elastic energy due to the existence of the electrode can be neglected. This is because, in thickness vibration, the distortion of the surface due to the presence of an electrode is very small. Consequently, according to Rayleigh's theorem (the time average of the kinetic energy and that of the elastic energy are equal at resonance), the following relation between q and the natural frequency drop Δ defined in Eq. (11) is obtained and is substituted in Eq. (A 6):

$$q \doteq 1+2\Delta \qquad \text{(A 8)}$$

On the other hand, the effective value I of the current flowing in the electrode per unit length in the x-direction is given by

$$I=\frac{1}{\sqrt{2}}\cdot e_{26}\cdot\omega\cdot 2\cdot\int_0^a\left(\frac{1}{H}\cdot\int_0^H\frac{\partial u_{II}}{\partial y}\cdot\cos\frac{\pi y}{H}\cdot dy\right)dz \qquad \text{(A 9)}$$

Setting both energies equal, the inductance L is given by the following equation:

$$L=\frac{K}{I^2} \qquad \text{(A 10)}$$

When Eqs. (A 6) and (A 9) are integrated, Eq. (16) is obtained.

III. Derivation of Eq. (25)

In Eqs. (20) – (24) the symmetric and the antisymmetric vibrations are written separately. Hence, it suffices to consider the boundary condition in the range of 0 < z. When the boundary condition is inserted with reference to Appendix I, it corresponds to (A 5) and the following equation is obtained:

$$\begin{pmatrix} \exp(-k'b) & \sin kb & \cos kb & 0 \\ -k'\exp(-k'b) & k\cos kb & -k\sin kb & 0 \\ 0 & \sin ka & \cos ka & {}^{\mathrm{ch}}_{\mathrm{sh}}\,k'a \\ 0 & k\cos ka & -k\sin ka & k'\,{}^{\mathrm{sh}}_{\mathrm{ch}}\,k'a \end{pmatrix}\begin{pmatrix} A \\ -B \\ -C \\ D \end{pmatrix}=0 \qquad \text{(A 11)}$$

From the determinant of this coefficient, Eq. (25) is obtained.

HIGH FREQUENCY MONOLITHIC CRYSTAL FILTERS WITH POSSIBLE APPLICATION TO SINGLE FREQUENCY AND SINGLE SIDE BAND USE

R. A. Sykes and W. D. Beaver

Bell Telephone Laboratories, Inc.
Allentown, Pennsylvania

Summary

A discussion of the various crystal filter networks is presented as they apply to single frequency and voice frequency channel filters. Particular emphasis is placed on the divided electrode type crystal unit at high frequency and its lattice equivalent. Previous attempts to use multiresonant high frequency crystal units with divided platings in conventional filter configuration have been limited by the range of resonance frequency placement and resultant image impedances of the filters. Proper choice of resonator electrode disposition gives more latitude in the placement of the critical resonant frequencies. The unique placement of critical frequencies in the lattice equivalents permits an improved characteristic resistance over the transmission band. No additional components are required to provide essentially flat transmission bands.

From a user's point of view, the performance characteristics of monolithic crystal filters are very simply stated in the form $\frac{f_B - f_A}{f_m} = \frac{10^{-6}}{2} \times R_0$; that is, the fractional bandwidth in Hz/MHz is a direct multiple of the termination resistance. A specific case is shown where a single crystal plate may be used to form a monolithic filter that may be terminated in 100-ohm circuits with a resulting bandwidth of .005% of the mean frequency. This may be used for a single frequency selector such as carrier and pilot filters in the range from 6 to 25 MHz over normal central office temperature ranges. Bandwidths as high as .07% may easily be realized which can allow the use of single side band channel filters at carrier frequencies as low as 6 MHz. Some examples of transmission characteristics for monolithic crystal filters are shown for narrow band and wide band cases at approximately 10 MHz.

Other conventional crystal filters may be realized by the addition of inductors and/or capacitors within the limits that the equivalent lattice of crystal units imposes.

Finally, it is shown that the general case, for monolithic filters of this type, may be considered as a mechanical filter formed by elastically coupled resonators and may be dealt with by conventional transmission methods.

Reprinted with permission from *Proc. 20th Ann. Symp. Frequency Control*, Apr. 19-21, 1966, pp. 288-308, sponsored by the U.S. Army Electronics Command, Ft. Monmouth, NJ.

Discussion

Before we consider the use of "monolithic" crystal filters for high frequency band selection, we will briefly review how crystal units with divided and undivided electrodes have been used previously.[1,2] In Figure 1, the original four-crystal lattice, as conceived by Mason for telephone multiplex, is shown at (a), in which equal crystal units were used in the series arms and in the lattice arms, respectively. He showed that the four crystal units can be replaced by two with divided electrodes and connected to form a lattice structure, as indicated in (b). This divided plate structure resulted in the first practical application of crystal units to frequency selection, and formed the basis of the original crystal channel filters used in carrier telephone transmission systems. As the needs developed for higher frequency channel filters utilizing smaller size quartz plates, the structure shown at (c) was developed. This permitted the use of undivided electrodes, but still retained all the advantages of the original crystal lattice; namely, that of a band-pass structure having complete flexibility of loss peak placement with approximately equal impedance crystal units. It was found that the divided electrode crystal unit could be connected as shown at (d), an unbalanced structure, and still retain the transmission properties of the lattice structure. The important advantage is the freedom to place the loss peaks in the vicinity of the transmission band. The use of crystal units with undivided electrodes in a ladder structure, as is well known, limits the placement of the loss peaks to regions close to the pass band.[3] All four of the circuits shown in Figure 1 have approximately equivalent transmission characteristics, and the type of circuit used would be dependent on the frequency range and crystal unit construction available. A single section of the structure shown in Figure 1(d) has been used in a number of applications for carrier and pilot selection at both low and high frequencies.

In Figure 2(a) is shown one such application of a single divided plate crystal unit that was developed in the early 1930's for the selection of a pilot frequency at 3096 kHz. The usual circuit schematic and equivalent lattice are shown at (b) and (c). While the insertion loss characteristic (d) definitely shows two pass bands as well as poor loss discrimination on the upper side of the band, due to unwanted responses, it did have sufficient loss characteristics below the pilot frequency to satisfy the requirement. It has long been known that these two pass bands result from the antisymmetric and symmetric modes of the high frequency thickness shear. When this type of design is used employing present high frequency crystal unit principles (e), improved loss characteristics may be obtained as shown in (f). It can be seen that the same two transmission bands result from antisymmetric and symmetric modes produced by the divided electrode. A number of attempts have been made to fully utilize the two frequencies resulting from these two modes. For example, adding series inductance to make these two bands overlap usually results in a good transmission band, but too wide for most applications. Only small variations in these resonant frequencies are possible because they are determined by the dimensions of the crystal plate. The design of crystal filters employing this principle has been reported by Y. Nakazawa.[4]

During the course of some experiments to determine the dependence of the separation in frequency between these two modes on the individual resonator design, it was found that the symmetric and antisymmetric modes of a quartz crystal unit with divided electrodes could be considered as two individual coupled resonators, and that in fact, the coupling between them would be a direct function of each resonator design as well as the separation between them. A summary of this investigation is shown in Figure 3, in which the solid line shows the fractional frequency difference between the resonant frequencies of the symmetric and the antisymmetric modes, as a function of separation of an optimally designed pair of resonators. In accordance with trapped energy principles, larger or smaller area electrodes may be used, provided appropriate changes in the electrode thickness are made. The two dotted lines for $\frac{r}{t} = .6$ and $\frac{r}{t} = 20$ are practical limits which may be used. It may be well to point out at this time that further discussion of symmetric and antisymmetric modes is not as important to consider as that of simply coupled resonators. For example, let us recast the statement. Figure 3 shows the characteristics that may be obtained with two coupled resonators formed on a common quartz plate and how coupling between them varies with separation. The equivalent circuit of this four-terminal network is a lattice network composed of two equivalent crystal units whose resonant frequencies are determined by the frequency of the individual resonators offset by the coupling between them. The difference frequency between the equivalent crystal units shown in the lattice network can be approximated by a logarithmic function, as shown in Figure 3. Specific values for K_1 and K_2, as well as their dependence on propagation along X or Z' of the crystal plate are shown in the Appendix. The inductance of the equivalent crystal is also obtained by previously established methods involving the area of each individual electrode, as well as the piezoelectric coupling constant. Its approximate value, following trapped energy optimum design, is shown in Figure 3. This now gives us the degree of flexibility that is required for high frequency crystal filter design; namely, complete freedom in the placement of the resonant frequencies. For example, a conventional narrow band filter would be one in which the second resonance were made coincident with the antiresonance of the first mode and a standard series of filters could be developed. The addition of film type capacitors might be necessary.

If we add no additional components to this structure, the reactance frequency characteristic of the two arms Z_A and Z_B in the lattice equivalent will be as shown in Figure 4(a). Upon applying the usual equations to determine its transmission properties, we find that two pass bands occur, $(f_B - f_A)$ and $(f_4 - f_3)$. Between the two resonant frequencies, a mid series constant-K type impedance is observed, and between the antiresonant frequencies a mid-shunt constant-K type impedance is observed. If each band is somewhat less than the total frequency difference between resonance and antiresonance, a high ratio of midband resistances is found. If we terminate this network to obtain the transmission properties for the lower transmission band, the reflection loss at the higher transmission band is so high that the band is hardly noticed, as shown in Figure 4(b). The termination resistance for the lower transmission band is proportional to the inductance of the resonator and the bandwidth desired. Using the value for inductance given for dual resonators on Figure 3, an extremely simple design criteria may be established. The bandwidth in Hz of a monolithic crystal filter composed of two resonators equals one half the product of termination impedance in ohms and the midband frequency in MHz. For example, a 100-ohm filter at 10 MHz will have a bandwidth of 500 cycles. A 1,000-ohm filter will have a bandwidth of 5,000 cycles, or more generally, 100-ohm filters at any frequency will have a bandwidth of .005%. These should be useful for the selection of any pilot or carrier frequencies that are generated by a common base precision oscillator or clock. It is understood, of course, that different impedances and bandwidths assumed would call for an appropriate design of electrodes and separations.

Some specific monolithic crystal filters have been produced and will be used in the L4 Coaxial Carrier Telephone Transmission System. Figure 5 shows the measured loss characteristic of a filter for the selection of a pilot frequency at 11.648 MHz. A particularly flat loss (Scale B) exists over a 1.5 kHz band, which is important since this filter is used in a pilot regulator. Also due to the high Q of the resonators only .5 db loss occurs at the mid-band. The loss characteristic shown over a wide frequency range (Scale A) shows transmission loss peaks which are due to the small distributed capacity between input and output connections. Figure 6 shows the circuit of a conventionally designed crystal filter for the same application. Its characteristics are nearly identical with those of the monolithic filter, with about 1 additional db insertion loss due to the transformers. The size of the overall filter is much larger because of the additional components required.

As mentioned earlier, it was indicated that a simple design relationship exists for monolithic filters. The bandwidth is proportional to the product of frequency and termination resistance. Figure 7 shows the measured transmission characteristics of a filter nearly 5 kHz wide at approximately 10 MHz.

Since the monolithic filter composed of two resonators does in fact have an equivalent lattice network of two independent crystal resonators, it is obvious that other types of conventional crystal filters may be obtained by the addition of other components such as placing them in series, shunt, bridged across, or sharing the common connection to ground, as shown in Figure 8. There are many networks of this type. To name a few, an inductor may be added in series or shunt to widen the band, a capacitor may be bridged across to control the placement of attenuation peaks, and a capacitor or inductor may be used in the common arm to increase or decrease the frequency of one of the equivalent lattice crystal resonators.

The discussion this far has been concerned with two coupled resonators and their equivalent lattice network, but it is obvious that these are simply mechanically coupled acoustic resonators. If we visualize an equivalent circuit on this basis, that shown in Figure 9 appears to meet the test for validity. Reducing it to its equivalent lattice by the open and short circuit bisection theorem gives us the identical equivalent lattice shown previously. It is also consistent with the theories developed by Mason for coupled modes in crystal units. In addition one might consider the use of coupled shunt antiresonant circuits in the equivalent ladder network instead of coupled series resonant circuits. The general case for coupled multiple resonators on a common piezoelectric substrate is shown in Figure 10. The antiresonant form is shown here to better illustrate the mechanical nature of this structure. It also permits easier analysis for conditions that must prevail for utilizing the low impedance lower band or the high impedance upper band. Either the equivalent lattice or the ladder network may be used for design purposes.

Let us carry this one step further and assume that we have a high Q material, not necessarily piezoelectric, and produce a series of resonators following the trapped energy principle, we arrive at the general case shown in Figure 11. This is a mechanical filter comprised of a series of individual shear type resonators coupled by the elastic constants of the material. There is considerable flexibility here, as in the case of the monolithic crystal filter, in that the impedance as well as the resonant frequency of each resonator may be varied as well as the coupling between them. This is a structure that may now be designed by computer methods that will yield minimal loss variation in the transmission band and maximum loss outside the band for a given number of resonant elements. This structure may be driven by piezoelectric or any other form of mechanical transducers having a bandwidth equal to or greater than that of the mechanical filter, or can be made of a piezoelectric material with only the end resonators acting as not only transducers, but part of the mechanical filter.

In summary, Figure 12 shows results that have been obtained with a two-resonator model of the monolithic crystal filter. This is a composite of several different actual filters in which the center frequency and bandwidths have been normalized to result in a single characteristic. By the introduction of capacity between input and output electrodes which may be of the film type, deposited on the same quartz plate, loss peaks may be introduced close to the transmission band as shown. A two resonator model at 11.6 MHz is shown in Figure 13, together with a conventional crystal filter, each of which have similar transmission characteristics. It is obvious that higher selectivities may be obtained by the use of these filters in series or by the addition of resonators on the same quartz plate. The dashed curve of Figure 12 shows the measured loss characteristic of a three resonator monolithic crystal filter made by R. L. Reynolds. This filter had a bandwidth of 4 kHz at a nominal frequency of 11.5 MHz. The dotted curve shows the measured results that have been obtained with a six resonator model. This filter had a bandwidth of 10.5 kHz at a nominal frequency of 9.5 MHz, a physical model of which is shown in Figure 14. All characteristics shown in Figure 12 have been normalized for comparison purposes.

References

1. "Electro-Mechanical Transducers and Wave Filters," by W. P. Mason, D. Van Nostrand Co., New York, 1942.

2. "Quartz Crystals for Electrical Circuits," by R. A. Heising, D. Van Nostrand Co., New York, 1946.

3. "New Approach to the Design of High Frequency Crystal Filters," by R. A. Sykes, 1958, IRE National Convention Record, Vol. 6, Part 2, pp. 18-29.

4. Proceedings of the 16th Annual Symposium on Frequency Control, April 6, 1962, "High Frequency Crystal Electro-Mechanical Filters," Y. Nakazawa, Toyo Communications Equipment Co., Ltd., Japan, p. 373.

Appendix

The equation for fractional frequency separation shown in Figure 3 may be written in the form

$$\frac{f_B - f_A}{f_O} = \left[\frac{3\, C_{11}}{8\rho\, f_O^2\, \phi_x^2} \right] \exp\,(-\,\xi d) \tag{1}$$

where f_O is the frequency of the uncoupled resonators, C_{11} is the elastic coefficient along the x-crystallographic axis, ρ is the density of quartz, d is the electrode separation, ξ is the propagation constant for thickness shear. ξ is given approximately by the equation

$$\xi = \frac{2.298}{t}\,\Delta^{\frac{1}{2}} \tag{2}$$

where t is the thickness of the crystal plate and Δ is the plate-back, which is given by the equation

$$\Delta = \frac{f_s - f_O}{f_s} \tag{3}$$

where f_s is the frequency of the unelectroded crystal blank. ϕ_x is the effective length of the motional volume along the x-crystallographic axis and is given by the expression

$$\phi_x = d + 2r + \frac{0.418t}{\Delta^{\frac{1}{2}}} \tag{4}$$

where Δ is the plate-back and t the crystal plate thickness. When the two resonators are coupled along the Z' crystallographic axis it is necessary to substitute C_{55} for C_{11} in equation (1), 2.88 for 2.298 in equation (2), and ϕ_z for ϕ_x as well as .33 for .418 in equation (4).

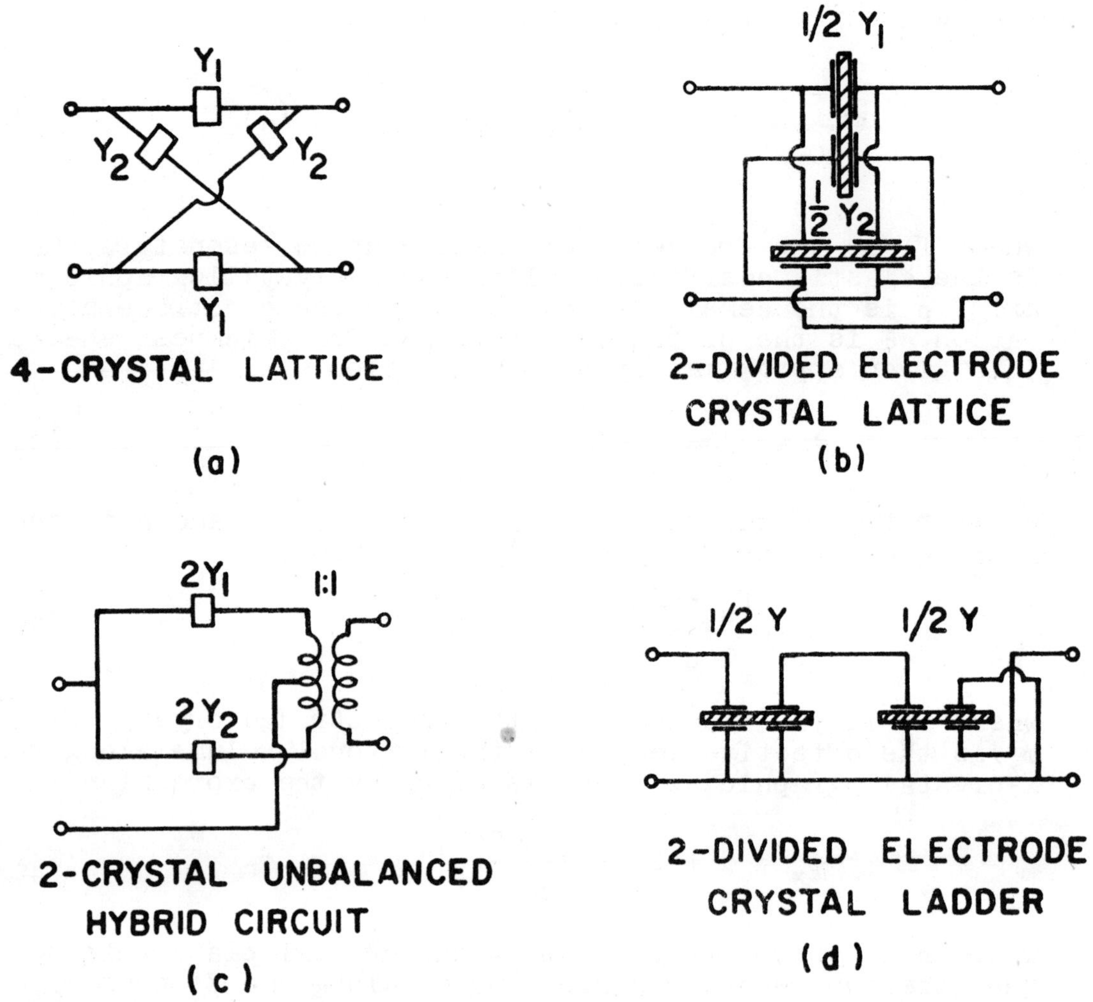

FIG. 1 – CRYSTAL UNIT NETWORKS HAVING ESSENTIALLY THE SAME LATTICE EQUIVALENT

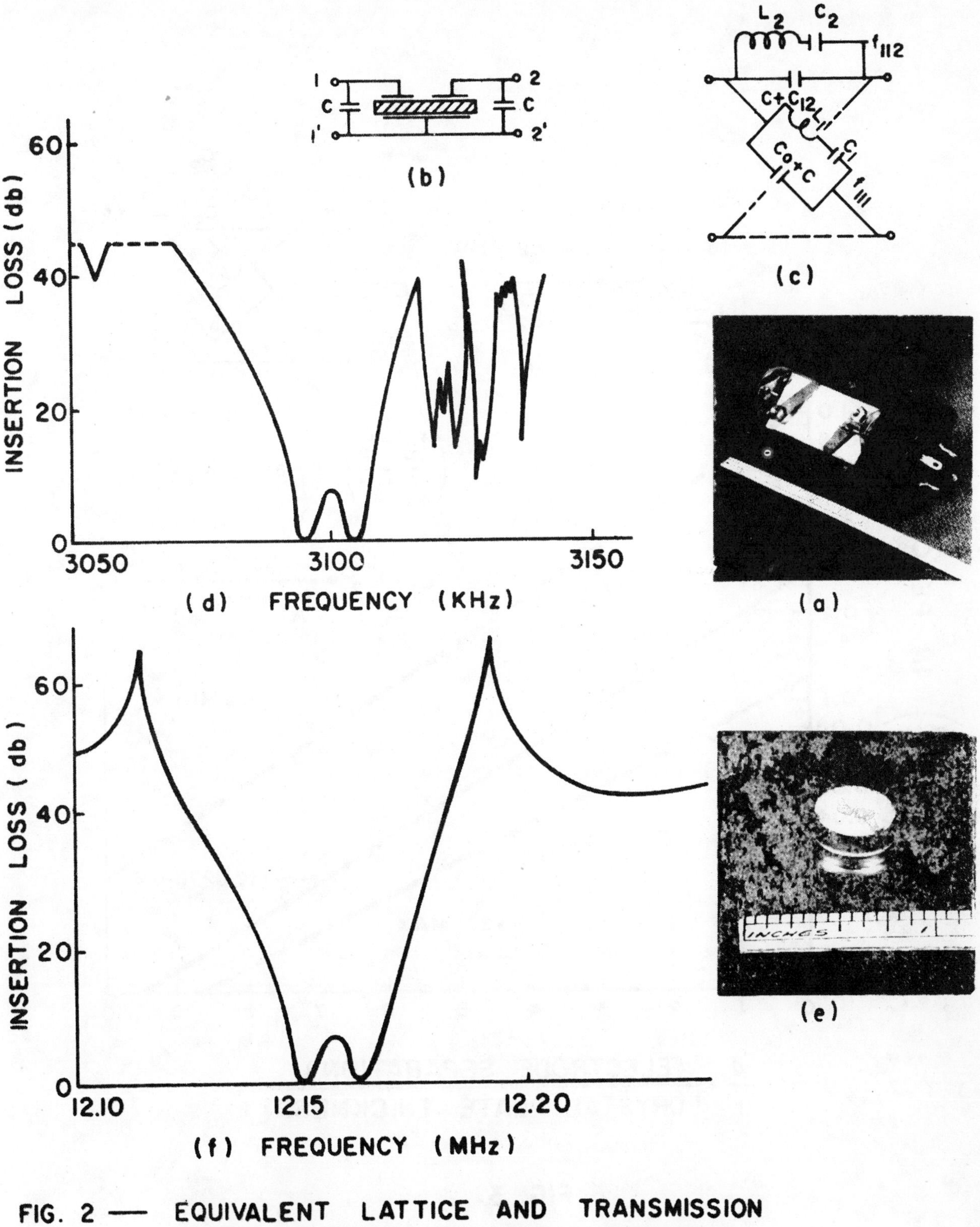

FIG. 2 — EQUIVALENT LATTICE AND TRANSMISSION CHARACTERISTICS OF HIGH FREQUENCY THICKNESS SHEAR CRYSTAL FILTERS WITH CLOSE SPACED ELECTRODES.

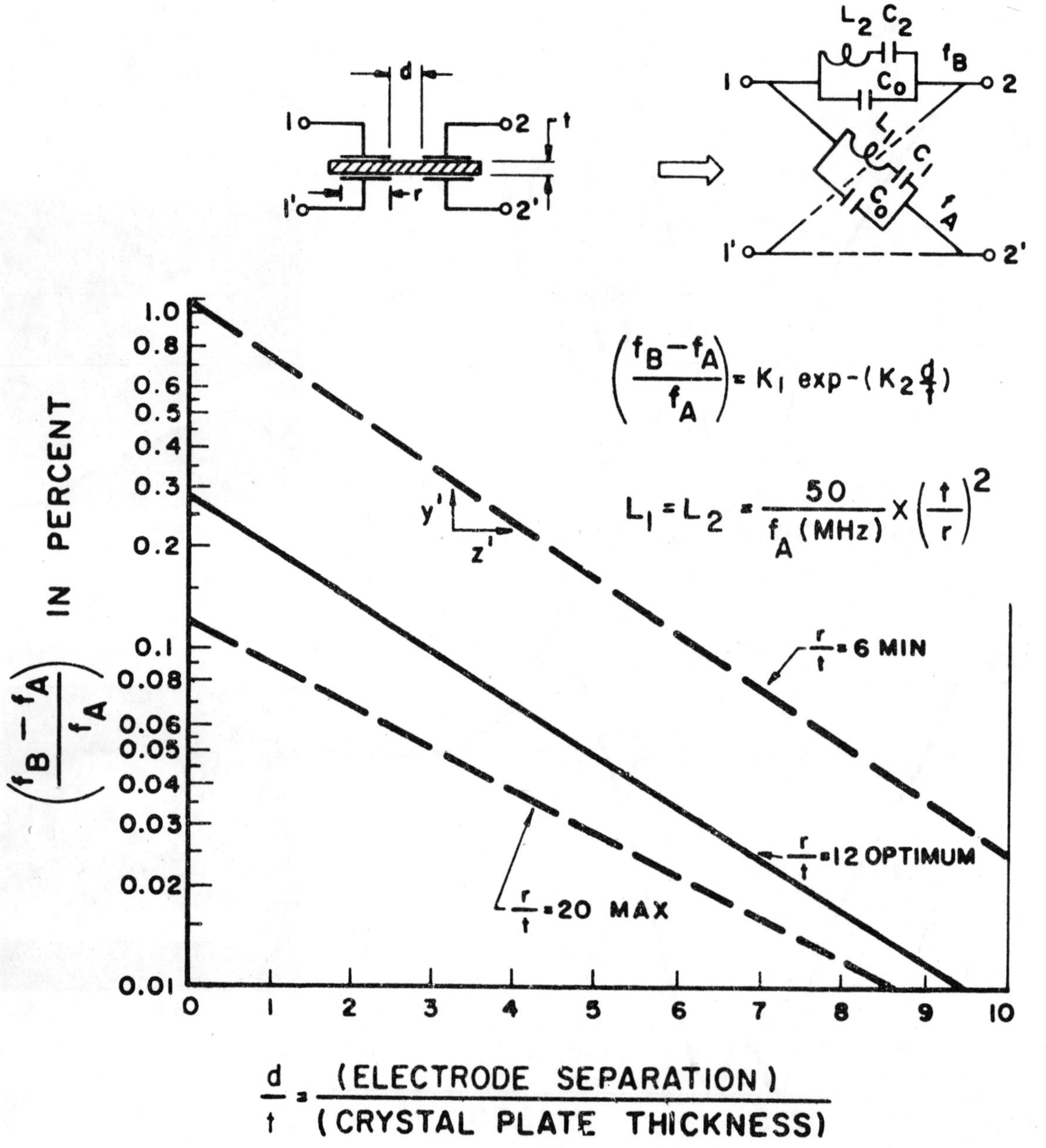

FIG. 3

PERCENT FREQUENCY DIFFERENCE BETWEEN EQUIVALENT LATTICE CRYSTAL UNITS AS A FUNCTION OF RESONATOR SEPARATION.

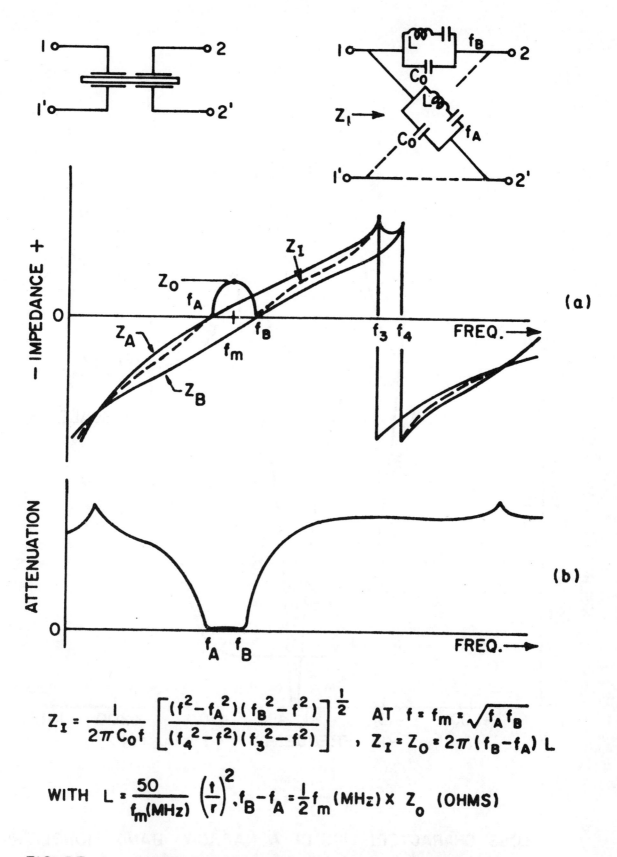

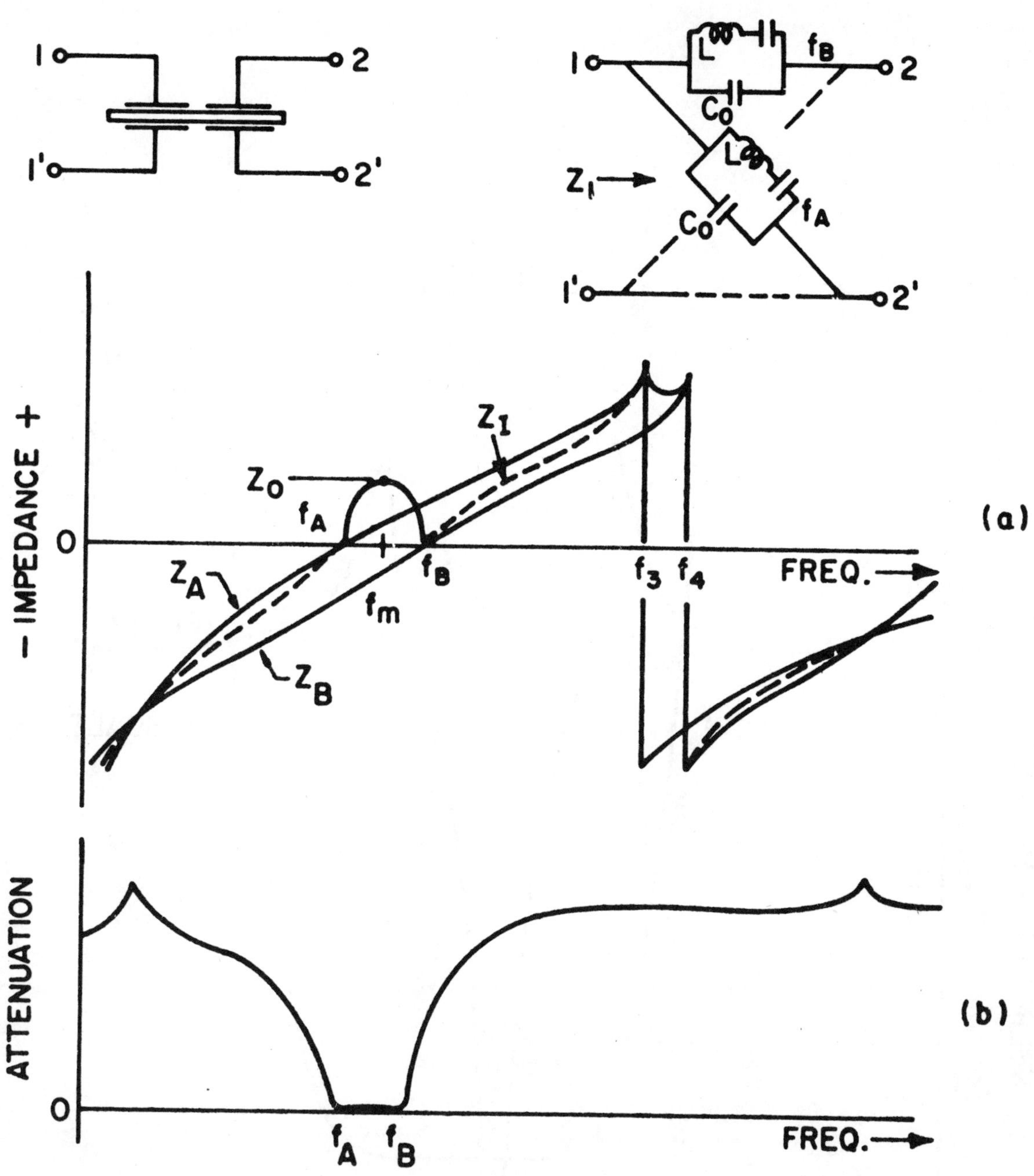

$$Z_I = \frac{1}{2\pi C_0 f}\left[\frac{(f^2-f_A^2)(f_B^2-f^2)}{(f_4^2-f^2)(f_3^2-f^2)}\right]^{\frac{1}{2}}$$

AT $f = f_m = \sqrt{f_A f_B}$, $Z_I = Z_O = 2\pi (f_B - f_A) L$

WITH $L = \frac{50}{f_m(\text{MHz})}\left(\frac{t}{r}\right)^2$, $f_B - f_A = \frac{1}{2} f_m(\text{MHz}) \times Z_o$ (OHMS)

FIGURE 4 — CHARACTERISTICS OF DUAL RESONATOR MONOLITHIC CRYSTAL FILTER.

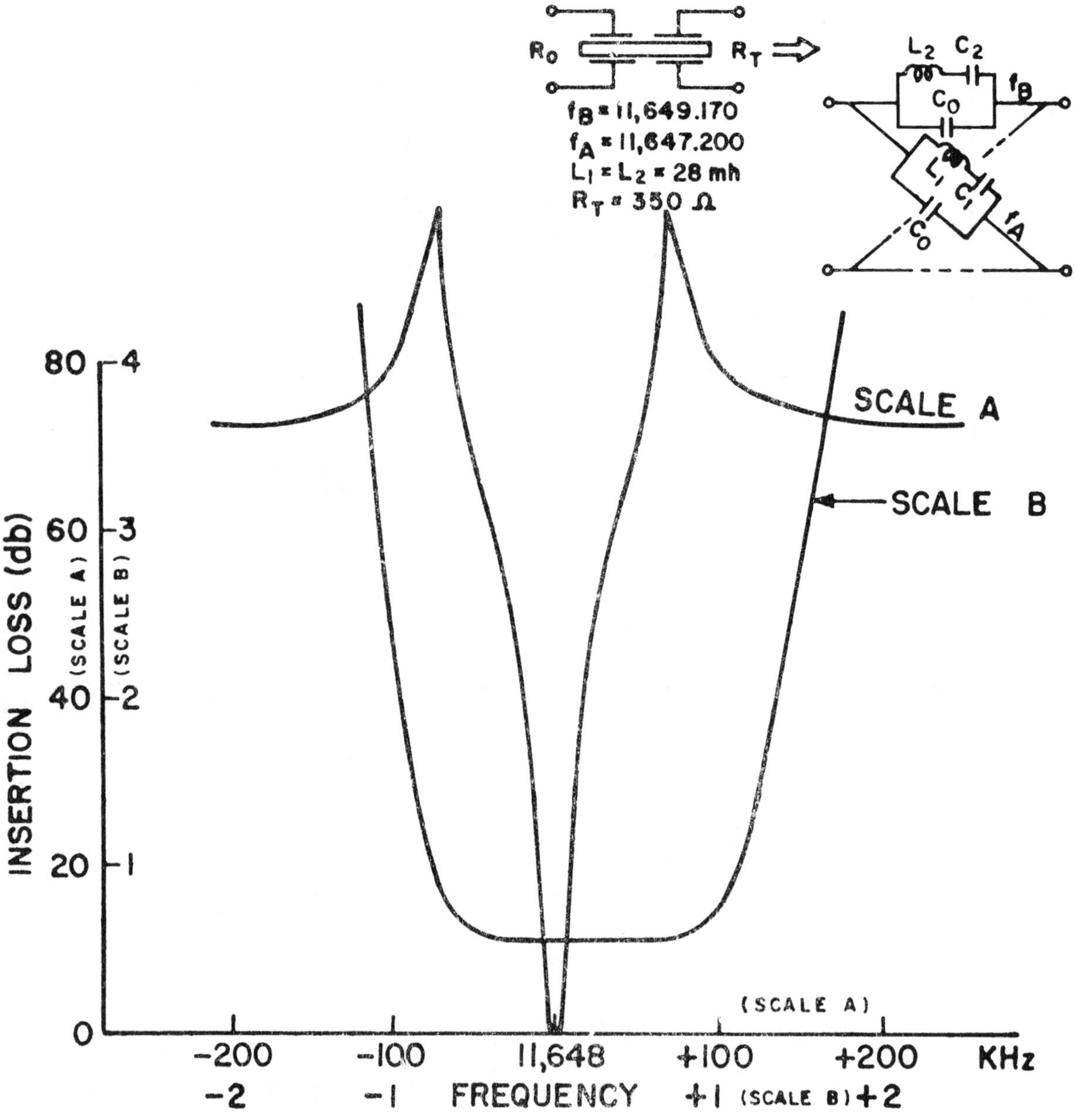

FIG. 5

LOSS CHARACTERISTICS OF A NARROW BAND MONOLITHIC FILTER.

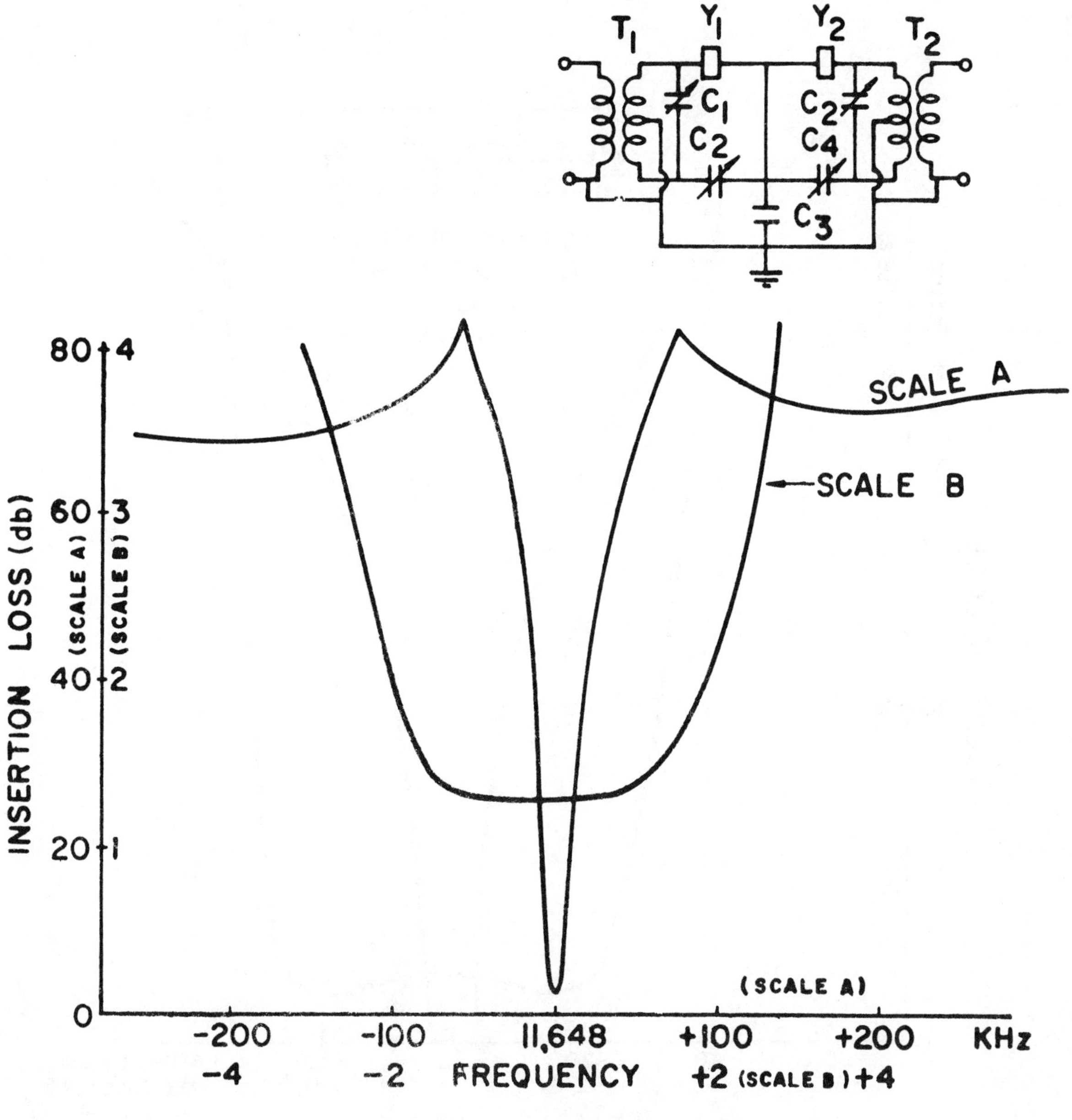

FIG. 6

— LOSS CHARACTERISTICS OF AN 11.648 MHz L-4 PILOT FILTER.

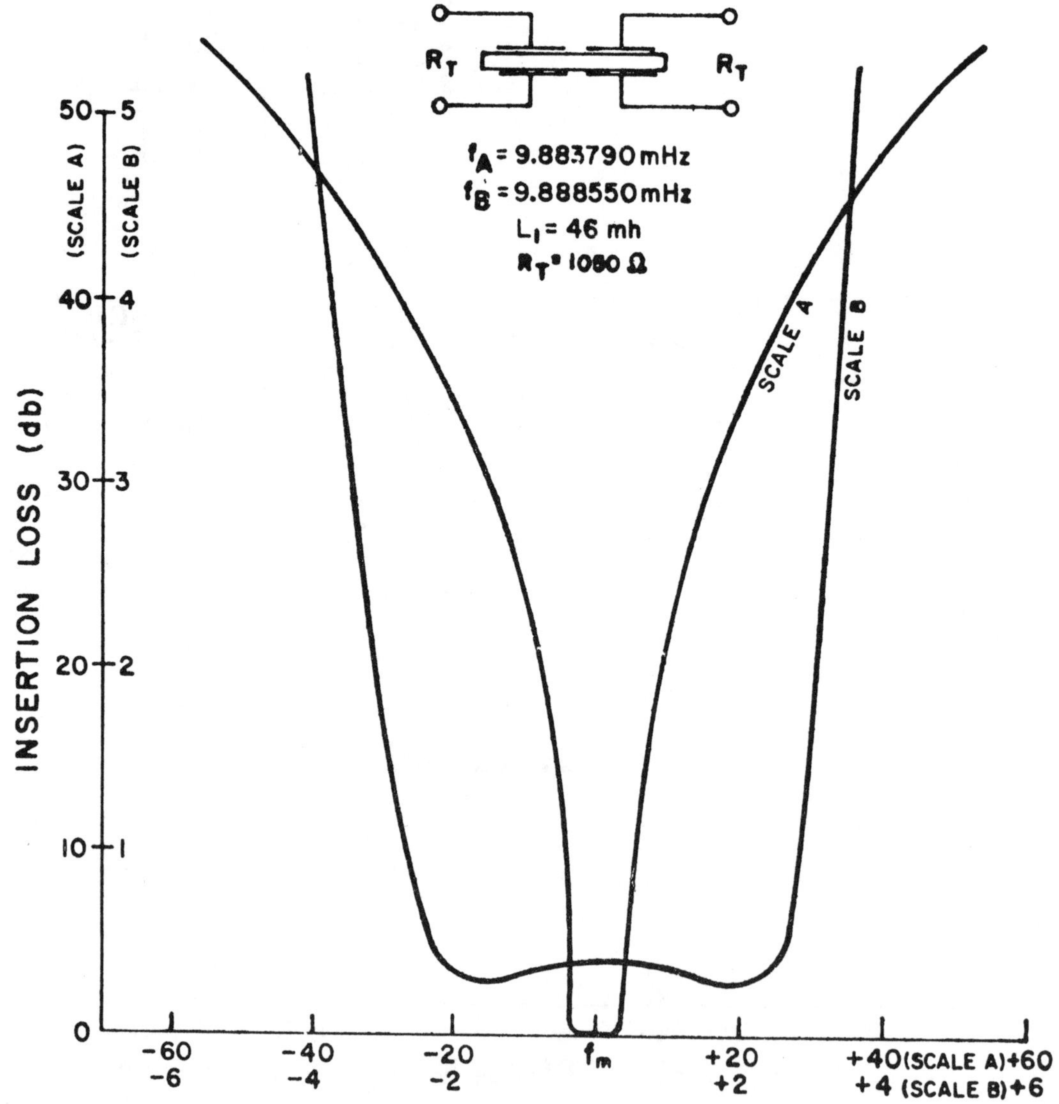

FIG. 7 — LOSS CHARACTERISTICS OF A MONOLITHIC CRYSTAL FILTER OF 4.3 KHz BANDWIDTH.

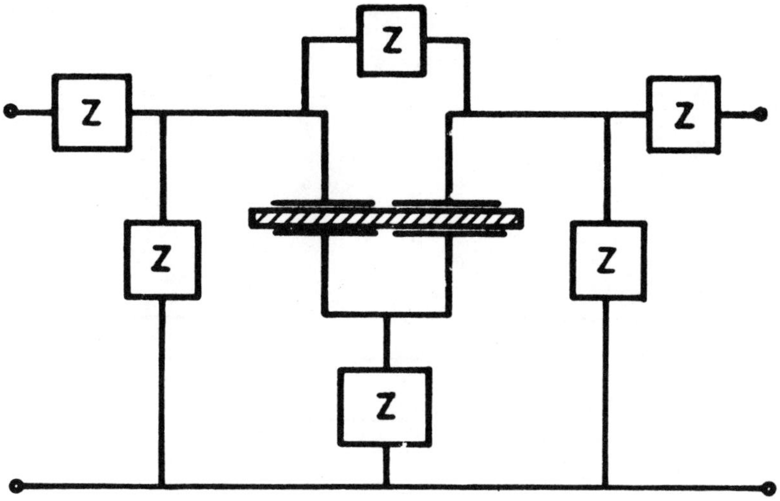

FIG. 8 — ADDITION OF GENERAL IMPEDANCES TO PRODUCE CONVENTIONAL CRYSTAL FILTER STRUCTURES REQUIRING A CRYSTAL UNIT IN EACH ARM OF THE EQUIVALENT LATTICE

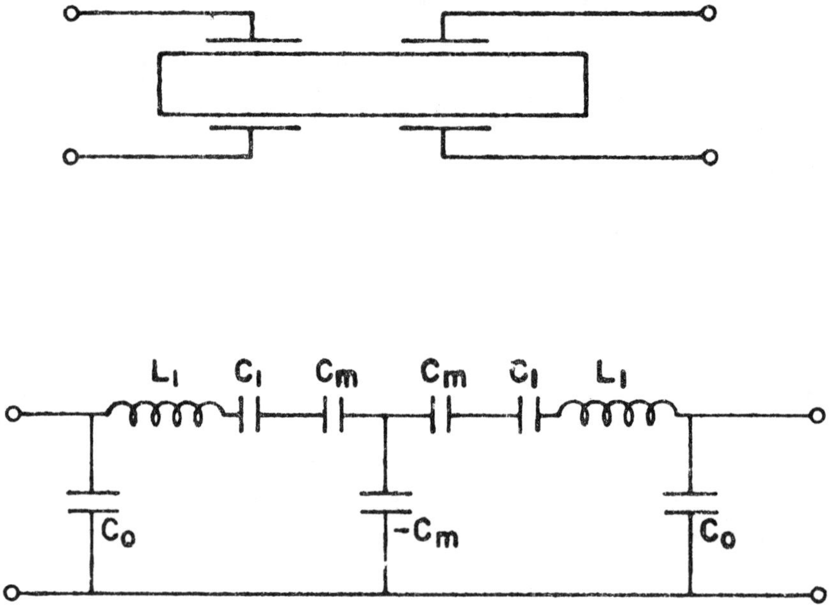

FIG. 9 – EQUIVALENT LADDER STRUCTURE OF A MONOLITHIC CRYSTAL FILTER.

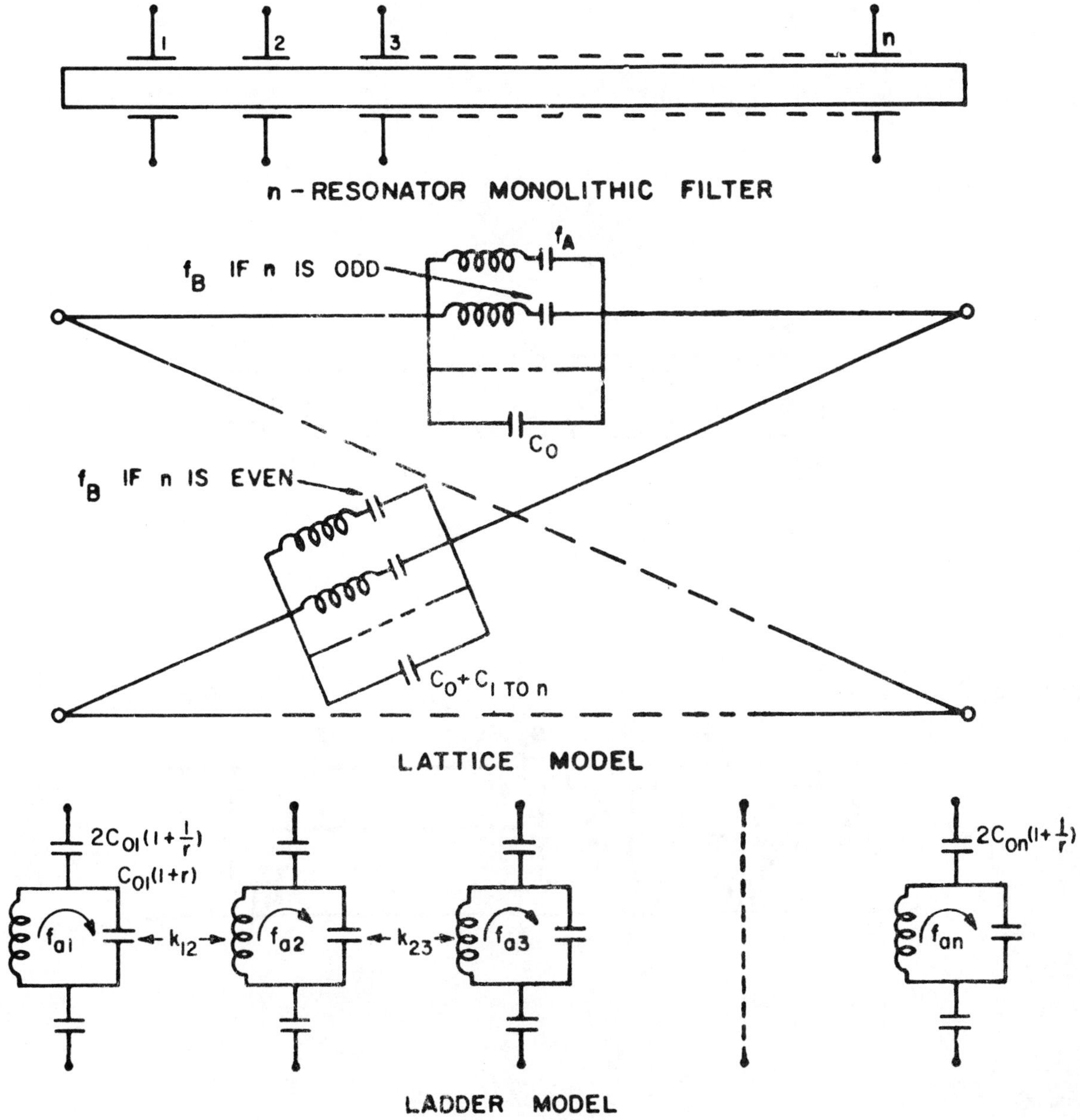

FIG. 10 — GENERALIZED LATTICE AND LADDER EQUIVALENT OF MULTIRESONATOR CRYSTAL FILTERS.

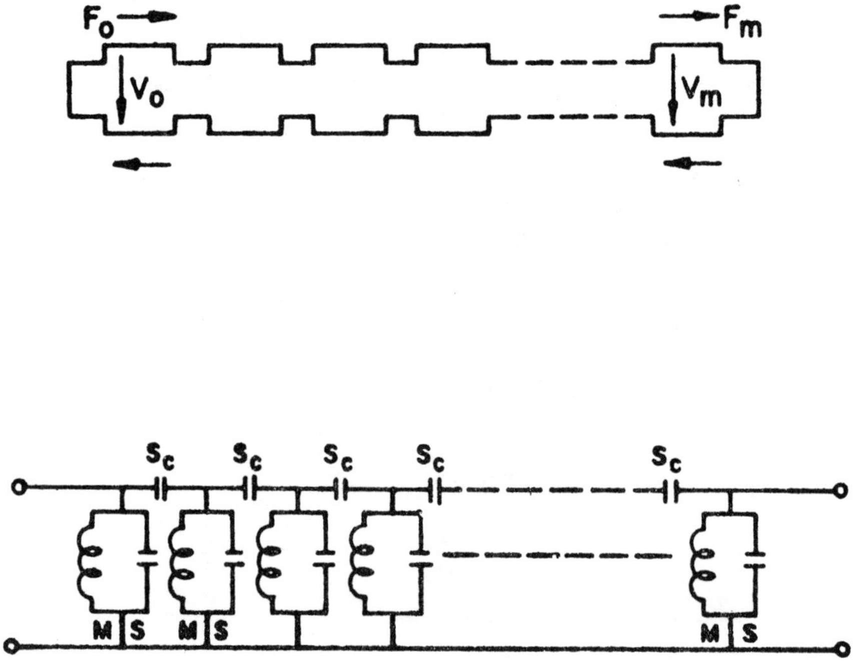

FIG. 11 — GENERALIZED MECHANICAL EQUIVALENT FOR COUPLED SHEAR MODE RESONATORS.

FIG. 12 - NORMALIZED TRANSMISSION CHARACTERISTICS OF MONOLITHIC CRYSTAL FILTERS.

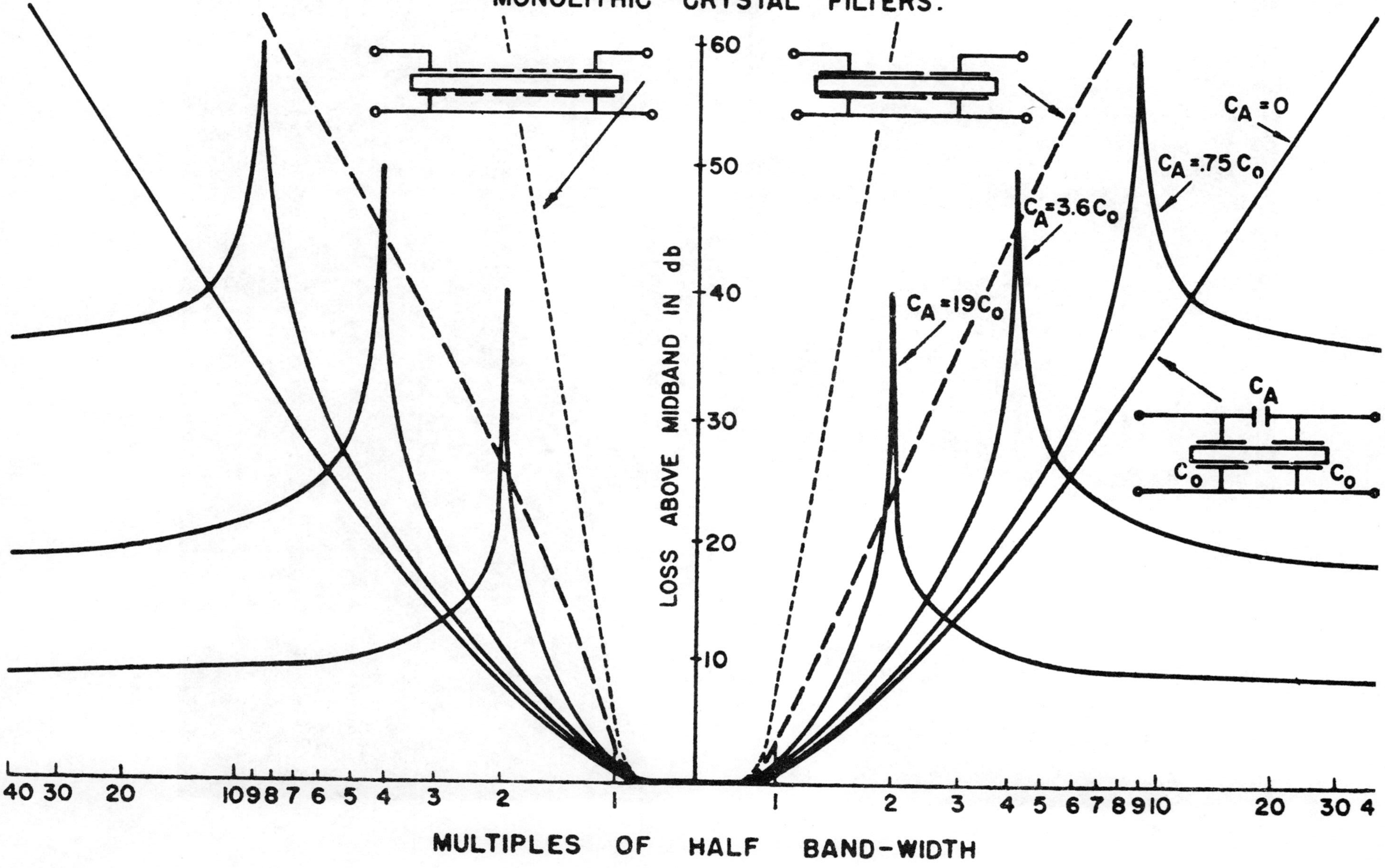

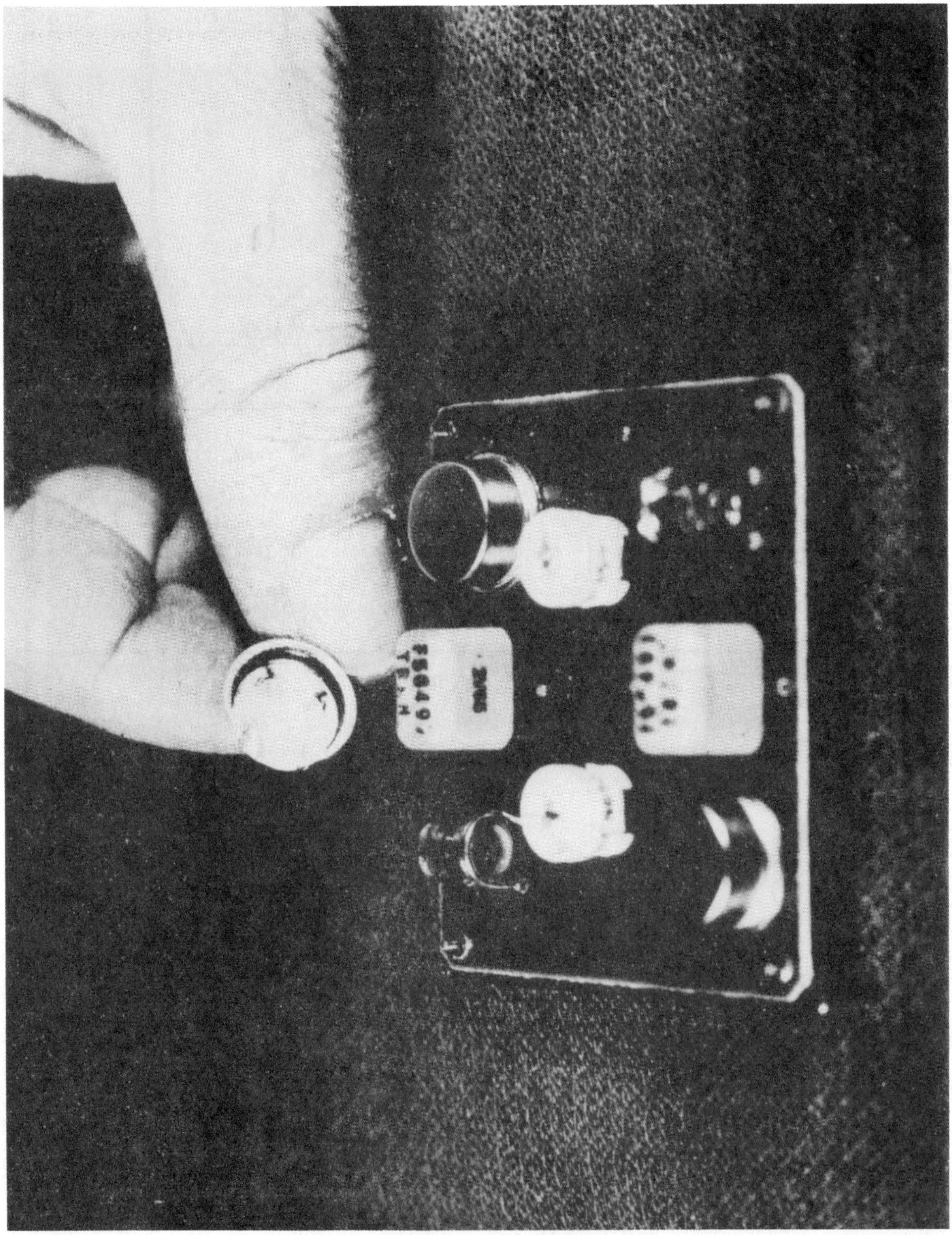

FIGURE 13

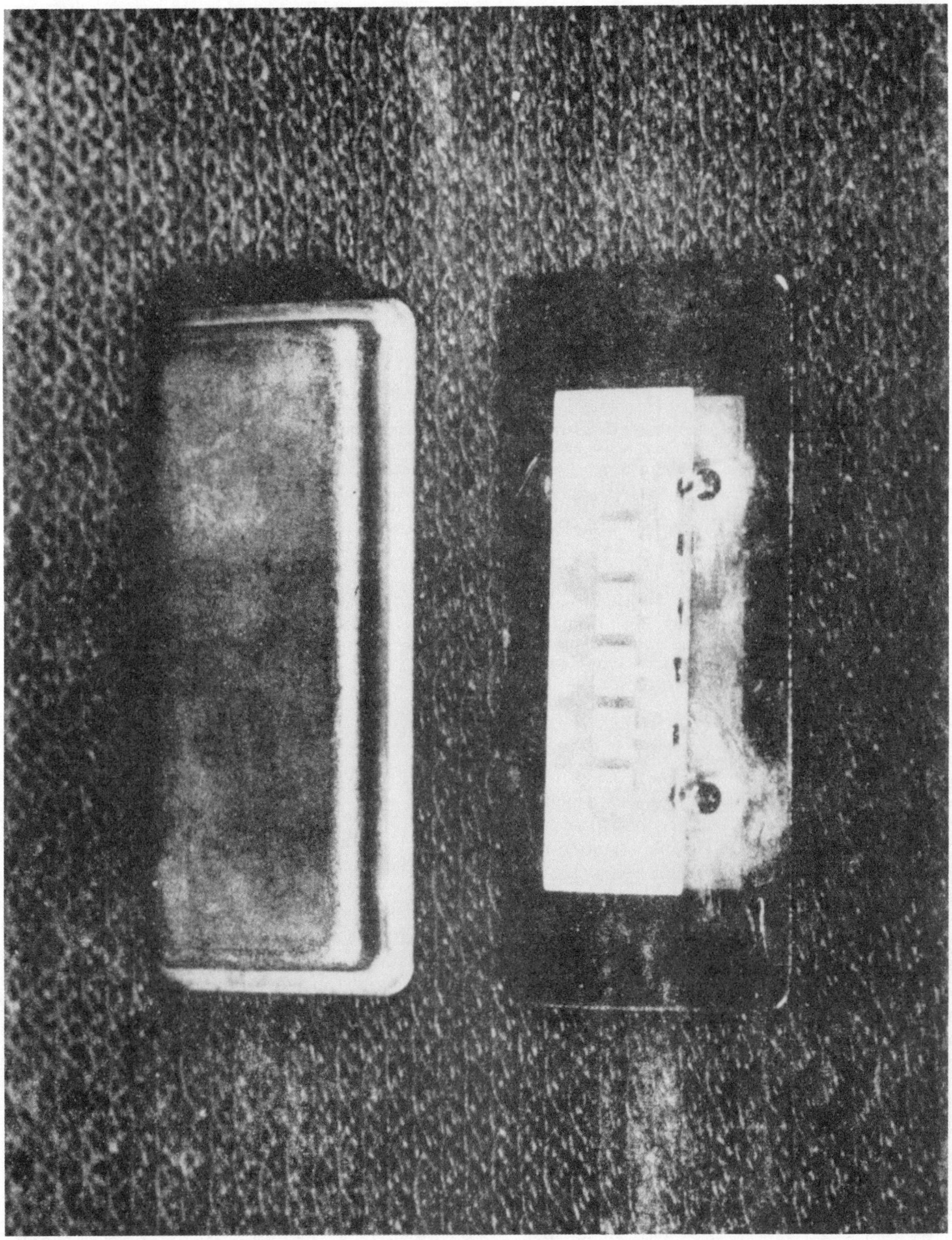

FIGURE 14

Received 14 September 1966

Trapped-Energy Modes in Quartz Filter Crystals

W. Shockley,* D. R. Curran, and D. J. Koneval

Clevite Corporation, Cleveland, Ohio 44108

The behavior of thickness/shear-mode quartz resonators is explained both mathematically and experimentally in terms of lateral standing-wave trapped-energy modes. A cutoff phenomenon, similar to that occurring in optical total internal reflection, results in energy trapping or the restriction of vibratory energy almost exclusively to the electroded region of a resonator, with an energy distribution decreasing exponentially with distance from the electrode edge. Consideration of boundary conditions at the electrode edges of an idealized two-dimensional model yields an expression for the eigenfrequencies of symmetric inharmonic overtone modes (lateral standing waves) that can exist in the electroded region of fundamental- and harmonic-mode resonators. Design formulas for the suppression of these unwanted modes are included. They show that electrode thickness and lateral dimensions can be traded off against one another to obtain a wide range of motional parameters. Previously, unwanted responses were eliminated by restricting the electrode diameter to empirically established values. The present understanding has led to a reduction in resonant resistance and an increase in motional capacitance by at least a factor of four with hf and vhf fundamental- and harmonic-mode filter crystals. Experimental data that verify results predicted by energy-trapping theory are given, and application to design of single- and multi-electrode resonators is considered. Specific examples in the 10- to 180-MHz frequency range are given.

INTRODUCTION

In almost all filter-crystal applications, the reduction or elimination of unwanted spurious responses is of paramount importance, and considerable attention has been devoted to this subject in recent years. In the hf (high-frequency) and vhf (very-high-frequency) ranges, where AT-cut crystals are used almost exclusively, most of this work has centered around the limitation of electrode area to reduce spurious responses. This effect was first observed in the late 1930's by Bechmann,[1] who later noted the progressive disappearance of inharmonic overtone responses with reduction of electrode diameter and suggested design criteria on electrode and wafer diameter in terms of wafer thickness.[2] In related work, Gerber[3] cited the effect of nonuniform wafer thickness on the generation of spurious modes. All of these effects can be explained in terms of a cutoff phenomenon and the resulting vibratory-energy trapping.

Energy trapping and the need for electrodes of finite thickness to obtain a strong resonance was described by the authors in 1963 (Ref. 4). At that time, Guttwein[5] noted that excessive electrode thickness introduced spurious responses. This was subsequently explained both mathematically and experimentally in terms of lateral standing-wave trapped-energy modes.[6,7]

Work described in part in these three papers is presented here, including expressions for the eigenfrequencies of harmonic and symmetric inharmonic overtone modes and design formulas for suppressing unwanted modes. These resonator-design expressions show that electrode thickness and lateral dimensions can be traded

* Present address: Alexander M. Poniatoff Professor of Engineering Sciences, Stanford Univ., Stanford, Calif. 94305, and Exec. Consultant, Bell Telephone Labs., Inc., Murray Hill, N. J. 07971.

[1] R. Bechmann, U. S. Pat. No. 2,249,933 (22 July 1941).

[2] R. Bechmann, "Quartz AT-Type Filter Crystals for the Frequency Range 0.7 to 60 MHz," Proc. IRE **49**, No. 2, 523–524 (1961).

[3] E. A. Gerber, "VHF Crystal Grinding," Electronics **27**, No. 3, 161–163 (Mar. 1954).

[4] W. Shockley, D. R. Curran, and D. J. Koneval, "Energy Trapping and Related Studies of Multiple Electrode Filter Crystals," Proc. Ann. Freq. Control Symp., 17th, 27 May **1963**, 88–126 (1963).

[5] G. K. Guttwein, "Status of Quartz Crystal Research and Development," Proc. Ann. Freq. Control Symp., 17th, 27 May **1963**, 190–214 (1963).

[6] D. R. Curran and D. J. Koneval, "Energy Trapping and the Design of Single- and Multi-Electrode Filter Crystals," Proc. Ann. Freq. Control Symp., 18th, 4 May **1964**, 93–119 (1964).

[7] D. R. Curran and D. J. Koneval, "Factors in the Design of VHF Filter Crystals," Proc. Ann. Freq. Control Symp., 19th, 21 April **1965**, 213–268 (1965).

Reprinted with permission from *J. Acoust. Soc. Amer.*, vol. 41, pt. 2, pp. 981–993, Apr. 1967.

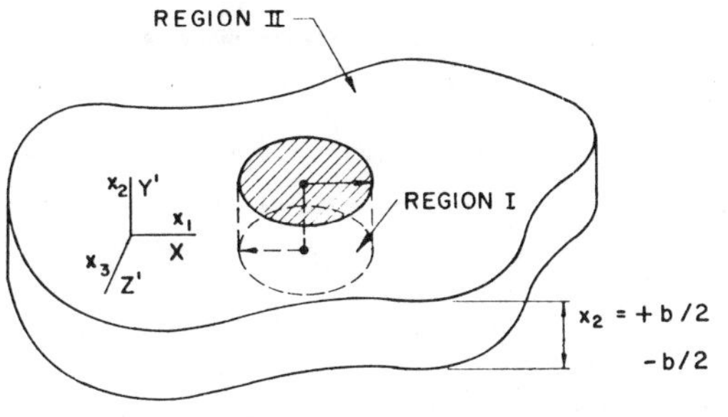

FIG. 1. AT-cut quartz wafer.

off against one another or with wafer dimensions to obtain a wide range of motional parameters in the vhf range or the reduction in resonator dimensions desired in the hf range. The dramatic effects that can be achieved with large-area thin electrodes have been demonstrated for the fundamental-mode and harmonic-mode resonators,[6–8] where a factor of four reduction in resonant resistance and increase in motional capacitance was obtained.

It is also interesting to note that pioneering work on energy trapping was done by Mortley.[9,10] However, this work did not become generally known, nor was its significance recognized, until late 1966 (Ref. 11).

I. ENERGY TRAPPING

An electric field applied in the thickness direction of an AT-cut quartz wafer will produce a thickness/shear distortion as shown in Fig. 1, with principal particle displacement in the X direction. Using Mindlin's notation[12] and sinusoidal excitation, the resulting waves propagating in the X direction are called thickness/shear TS_1 and in the Z' direction thickness/twist TT_3. Further, the plane-wave thickness/shear resonance in the thickness or Y' direction is a cutoff frequency for waves propagating in the plane of the wafer. Thickness/shear and thickness/twist waves with frequencies below cutoff cannot propagate in any direction in the plane of the wafer. The same situation also occurs in the fully electroded wafer, except that the cutoff frequency ω_e is lower in value because of electrode mass-loading and a small electroelastic effect. This gives a mechanism for energy trapping, which is essential to the performance of all high-frequency quartz resonators.

[8] T. J. Lukaszek, "Improvements of Quartz Filter Crystals," Proc. Ann. Freq. Control Symp., 19th, 21 April **1965**, 269–296 (1965).

[9] W. S. Mortley, "F.M.Q.," Wireless World **57**, 399–403 (Oct. 1951).

[10] W. S. Mortley, "Frequency Modulated Quartz Oscillators for Broadcasting Equipment," Proc. IEE **104B**, 239–249 (Dec. 1956).

[11] W. S. Mortley, "Priority in Energy Trapping," Phys. Today **19**, No. 12, 11–12 (Dec. 1966).

[12] R. D. Mindlin and D. C. Gazis, "Strong Resonances of Rectangular AT-Cut Quartz Plates," Proc. U. S. Natl. Congr. Appl. Mech., 4th, **1963**, 305–310 (1963).

If electrodes of finite thickness are applied to a limited portion of an AT-cut wafer, then separate cutoff frequencies exist for the electroded and surrounding regions. These cutoff frequencies, designated ω_e and ω_s respectively, divide the spectrum of interest in three ranges. Below ω_e, TT_3 and TS_1 waves cannot propagate in either region and standing-wave resonances cannot occur. Between ω_e and ω_s, waves can propagate in the "e" region but not in the "s" region, and total internal reflection occurs at the boundary between the regions. This is illustrated in Fig. 2(a). Above ω_s, waves can propagate in both regions [Fig. 2(b)] so that vibratory energy generated in region "e" will propagate away and therefore cannot contribute to a localized standing-wave response.

Excitation at frequencies between ω_e and ω_s will produce in the "e" region trapped waves that cannot escape into the "s" region. Because of boundary conditions, a portion of this vibratory energy fringes out into the cutoff region, but tails off exponentially with distance away from the electrode. This effect is similar to that occurring in optical total internal reflection in which attenuated electromagnetic waves are present in the medium of lower index of refraction. These exponential tails are represented in Fig. 2(c). Thus, trapped-energy-mode responses usually are acoustically isolated from other portions of the wafer. As a result, whole networks of acoustically independent resonators can be realized on a single wafer. Conversely, a controlled degree of coupling can be achieved to obtain coupled-resonator filter responses.[13]

At specific excitation frequencies, standing waves will occur in the "e" region, resulting in a trapped-energy-mode response, as indicated in Fig. 2(c). Its resonance or eigenfrequency is dependent on the relative values of ω_s and ω_e and on the lateral dimensions of the "e" region. It is obvious that a whole series of such responses could occur. These are called the inharmonic-overtone series; and the lowest of these is normally considered to be the fundamental thickness/shear response of the resonator. It should be noted that the series ends at ω_s, because at ω_s standing-wave amplitudes approach zero as the waves escape into the "s" region.

Using an idealized two-dimensional model, which, among other things, neglects effects of coupling to flexural or face shear modes, depending on direction, an expression is derived for these eigenfrequencies as a function of ω_s, ω_e, electrode length $2a$, and wafer thickness b for both fundamental and overtone modes. A more rigorous mathematical treatment that includes coupling to flexural modes is given by Mindin.[14]

[13] Y. Nakazawa, "High Frequency Crystal Electromechanical Filter," Proc. Ann. Freq. Control Symp., 16th, 27 Apr. **1962**, 373–390 (1962).

[14] R. D. Mindlin and P. C. Y. Lee, "Thickness–Shear and Flexural Vibrations of Partially Plated, Crystal Plates," Intern. J. Solids & Structures **2**, 125 (1966).

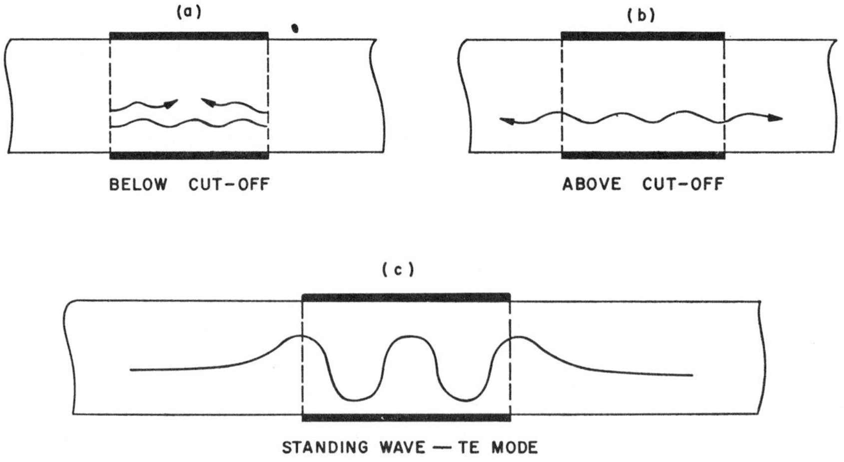

FIG. 2. Energy trapping.

A. Inharmonic-Overtone Series

Consider an idealized two-dimensional wafer of the form shown in Fig. 3, with thickness b in the x_2 (Y') direction and of infinite extent in the lateral x_3 (Z') direction. Solutions of the wave equation for particle displacement u (in the x_1 direction) for thickness/twist modes propagating in the x_3 direction are of the form

$$u = U \sin\eta x_2 \exp j(\zeta x_3 - \omega t). \quad (1)$$

To satisfy the zero-stress boundary condition at the major faces ($\partial u/\partial x_2 = 0$ at $x_2 = \pm b/2$), displacement u can have nonvanishing solutions only for

$$\eta = p\pi/b, \quad (2)$$

where $p = 1, 3, 5, \cdots$ is the order of the harmonic overtone—e.g., fundamental, 3rd harmonic, 5th harmonic, etc. Substitution of Eq. 1 into the wave equation

$$(1/\rho)(c'_{55}\partial^2 u/\partial x_3^2 + c'_{66}\partial^2 u/\partial x_2^2) = \partial^2 u/\partial t^2 \quad (3)$$

gives the expression relating the propagation constants

$$\omega^2 = (1/\rho)(c'_{55}\zeta^2 + c'_{66}\eta^2), \quad (4)$$

and with Eq. 2

$$\zeta = (c'_{66}/c'_{55})^{\frac{1}{2}}[(\omega/v)^2 - (p\pi/b)^2]^{\frac{1}{2}}, \quad (5)$$

where $v = (c'_{66}/\rho)^{\frac{1}{2}}$ is the velocity for propagation of shear waves and c'_{55} and c'_{66} are elastic constants for the **AT** cut. The elastic crosscoupling constant c'_{56} is very small for the **AT** cut and has been taken as zero in this treatment.

Stress waves can propagate freely for all real values of ζ, but reduce to nonpropagating vibrations that decay exponentially with distance for imaginary values of ζ. The angular frequency $\omega = \pi p v/b$, below which wave propagation cannot occur, is called the cutoff frequency, and is the thickness/shear resonant frequency for plane waves in the x_2 direction. Referring again to Fig. 3, the electroded "e" and the unelectroded surrounding "s" regions of the wafer will have cutoff frequencies that differ slightly because of the mass-loading and electroelastic effects of the electrodes.[15] The cutoff frequencies for fundamental-mode propagation in the "e" and "s" regions, respectively, are given by $\omega_e = \pi v_e/b$ and $\omega_s = \pi v_s/b$. Cutoff frequencies for the pth harmonic mode are given in turn by $p\omega_e$ and $p\omega_s$. For each mode, resonant responses associated with limited electrode areas can occur only at frequencies between the mode's respective "e" and "s" cutoff frequencies. These mode-resonant frequencies, or eigenfrequencies, can be determined by application of boundary conditions at the edges of the electroded region $x_3 = \pm a$.

The independence of cutoff frequency upon propagation direction is easily seen from the form of Eqs. 1 and 4. The lowest propagating frequency, for which $\zeta \to 0$ as $\omega > \omega_s$ approaches ω_s, corresponds to a disturbance in Eq. 1 that is independent of x_3 and, hence, uniform over the entire "s" region. This same unique uniform vibration of the plate as a whole will evidently occur no matter what propagation direction is chosen because the wavelength in that direction becomes infinite and the corresponding ζ becomes zero.

Solutions of the wave equation for the electroded and unelectroded regions are of the form of Eq. 1. Continuity of particle displacement and shear stress across the interfaces at $x_3 = \pm a$ impose four boundary conditions on these expressions. As a result, nonvanishing standing-wave (resonant) solutions that satisfy the following equation for symmetric TT_3 modes can occur only at specific frequencies between ω_e and ω_s:

$$\tan\zeta_e a = \gamma_s/\zeta_e, \quad (6)$$

where

$$\zeta_e = (\pi/b)(c_{66}'/c_{55}')^{\frac{1}{2}}[(\omega/\omega_e)^2 - p^2]^{\frac{1}{2}}, \quad (7)$$

and

$$\gamma_s = -j\zeta_s = (\pi/b)(c'_{66}/c'_{55})^{\frac{1}{2}}[p^2 - (\omega/\omega_s)^2]^{\frac{1}{2}}, \quad (8)$$

[15] This effect can be approximated by postulating a slight difference in a material property such as density to give the observed difference in cutoff frequencies in two homogeneous regions of a wafer of uniform thickness. Mathematically, we represent it by a higher effective velocity v_s for the "s" region than v_e for the "e" region.

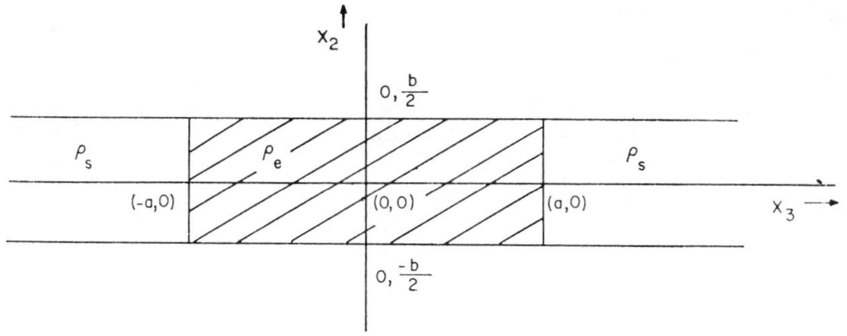

FIG. 3. Plate with trapped-energy structure.

are obtained from Eq. 5. Substitution of Eqs. 7 and 8 in Eq. 6 yields

$$\tan\frac{\pi a}{b}\left(\frac{c'_{66}}{c'_{55}}\right)^{\frac{1}{2}}\left[\left(\frac{\omega}{\omega_e}\right)^2-p^2\right]^{\frac{1}{2}}=\frac{\omega_e}{\omega_s}\left[\frac{p^2\omega_s^2-\omega^2}{\omega^2-p^2\omega_e^2}\right]^{\frac{1}{2}}. \quad (9)$$

The solutions of this equation $\omega=\omega_{te}$, which are the eigenfrequencies of all nonvanishing mode resonances, can be expressed more conveniently as a fraction of the frequency-lowering

$$\psi=(\omega_{te}-p\omega_e)/(p\omega_s-p\omega_e), \quad (10)$$

giving

$$\tan\frac{p\pi a}{b}\left(\frac{c'_{66}}{c'_{55}}\right)^{\frac{1}{2}}\left\{\left[\psi\left(\frac{\omega_e}{\omega_s}-1\right)+1\right]^2-1\right\}^{\frac{1}{2}}$$

$$=\left\{p^2-\left[p\psi\left(1-\frac{\omega_e}{\omega_s}\right)+p\frac{\omega_e}{\omega_s}\right]^2\right\}^{\frac{1}{2}}\Big/,$$

$$\left\{\left[p\psi\left(\frac{\omega_s}{\omega_e}-1\right)+p\right]^2-p^2\right\}^{\frac{1}{2}}, \quad (11)$$

where it can be noted that the order of the harmonic p will cancel out of the righthand side of this expression. Noting that this expression is of the form $\tan\alpha=\beta$ and can have one solution or many, depending on the magnitude of α, where $\alpha=n\pi+\tan^{-1}\beta$, Eq. 11 can be manipulated to obtain a useful expression relating mode frequencies with resonator parameters.

$$\frac{pa}{b}=\Omega_0\left(\frac{c'_{55}}{c'_{66}}\right)^{\frac{1}{2}}\left(n+\frac{1}{\pi}\tan^{-1}\theta\right)\Big/$$

$$\{\psi(1-\Omega_0)[2\Omega_0+\psi(1-\Omega_0)]\}^{\frac{1}{2}}, \quad (12)$$

where

$$\theta^2=(1-\psi)\Omega_0^2[1+\Omega_0+(1-\Omega_0)\psi]/\psi[2\Omega_0+\psi(1-\Omega_0)],$$

and

$$\Omega_0\equiv\omega_e/\omega_s.$$

Here $p=1, 3, 5$, etc., is the order of the harmonic mode and $n=0, 1, 2, 3, 4$, etc., give the symmetric inharmonic-mode series for each value of p. It should be noted that fundamental- and basic harmonic-mode series are given by $n=0$, $p=1, 3, 5$, etc.

Equation 12 relates the eigenfrequencies of all possible two-dimensional symmetric TT_3 modes to the resonator parameters a/b and $\Omega_0=\omega_e/\omega_s$; from this, the mode series for each harmonic ($p=1, 3, 5$, etc.) can be presented graphically in terms of these parameters. If, however, these modes are plotted in normalized fashion as

$$\psi_{pn}=(\omega_{te}-p\omega_e)/p(\omega_s-\omega_e) \quad (13)$$

with $(pa/b)[(\omega_s-\omega_e)/\omega_e]^{\frac{1}{2}}$ as the resonator-parameter variable, then the series for each harmonic coincides and the complete solution can be presented graphically in the simple fashion of Fig. 4. Criteria for the suppression of a portion of all of each inharmonic-overtone series can be obtained from Eq. 12 or Fig. 4.

B. Trapped-Energy Modes

In any piezoelectric element, a strong resonant response can be observed electrically only for those modes of vibration that have a high-mode Q_m and/or large electromechanical coupling. In quartz with moderate to weak coupling, strong resonant responses are dependent on high-mode Q_m. For example, using a half-lattice bridge, the response of a mode with $Q_m=1000$ would be 40 dB below that for a comparable mode with $Q_m=100\,000$. However, even the weaker of these responses (which in many cases would be considered to be negligible as compared to the stronger) at any given time must retain approximately 99.9% of its incident energy as coherent vibrations.

An electroded area of limited extent on an infinite quartz wafer, or even on a finite wafer that in turn is mounted on low-Q supports, can have a high-mode Q_m and a substantial resonant response only if a very large fraction of its vibratory energy is restricted to the electroded and surrounding regions. The application of energy trapping to the case at hand to explain the generation or suppression of specific modes is obvious, and can be considered for the idealized two-dimensional wafer in terms of a single internal reflection.

Consider the internal reflection of an incident thickness/twist wave at the boundary of the electroded region. If the reflected wave is equal in amplitude to the incident wave, then in the usual case total internal reflection occurs and energy has not been lost to a refracted wave. Assuming that this occurs at each of the edges of the electrode, then the resulting mode, whether fundamental or inharmonic overtone, will have a strong response and a high-mode Q_m. On the other hand, if the reflected-wave amplitude is less than unity, then the response of the resultant mode must be relatively weak and have a low-mode Q_m.

Wave amplitudes were obtained on the abovementioned basis for an internally incident thickness/twist wave, using wavefunctions in the convenient exponential form

$$u_e = [A_e \exp(j\zeta_e x_3) + B_e \exp(-j\zeta_e x_3)]$$
(INCIDENT) (REFLECTED)

$$\times \sin(\eta x_2) \exp(-j\omega t), \quad (14)$$

$$u_s = A_s \sin(\eta x_2) \exp(-j\omega t + j\zeta_s x_3), \quad (15)$$

where again "e" and "s" refer to electroded and surrounding regions of the wafer. The usual boundary conditions of continuity of stress and displacement at the electrode edge $x_3 = a$ gives

$$A_e \exp(j\zeta_e a) + B_e \exp(-j\zeta_e a) = A_s \exp(j\zeta_s a), \quad (16)$$

and

$$\zeta_e[A_e \exp(j\zeta_e a) - B_e \exp(-j\zeta_e a)] = \zeta_s A_s \exp(j\zeta_s a). \quad (17)$$

Values of the desired ratio B_e/A_e can be calculated as a function of frequency, if values of ζ_e and ζ_s are substituted in Eqs. 16 and 17. However, conclusions can be drawn from the form of the resulting B_e/A_e equation, the knowledge of the frequency regions in which ζ_e and ζ_s are real and imaginary, and the relative magnitudes of these propagation constants.

For $\omega_e < \omega_s < \omega$, both ζ_e and ζ_s are real and $\zeta_e > \zeta_s$. This gives

$$B_e/A_e = [(\zeta_e - \zeta_s)/(\zeta_e + \zeta_s)] \exp(2j\zeta_e a), \quad (18)$$

and

$$|B_e/A_e| < 1. \quad (19)$$

For $\omega_e < \omega < \omega_s$, ζ_e is real and $\zeta_s = j\gamma$ is imaginary. Then

$$B_e/A_e = \exp 2j(\zeta_e a - \theta), \quad (20)$$

where $\theta = \tan^{-1}(\gamma_s/\zeta_e)$, and

$$|B_e/A_e| = 1. \quad (21)$$

For $\omega < \omega_e < \omega_s$, both ζ_e and ζ_s are imaginary with $\zeta_e < \zeta_s$, giving

$$B_e/A_e = -[(\gamma_s - \gamma_e)/(\gamma_s + \gamma_e)] \exp(-2\gamma_e a), \quad (22)$$

and

$$|B_e/A_e| < 1. \quad (23)$$

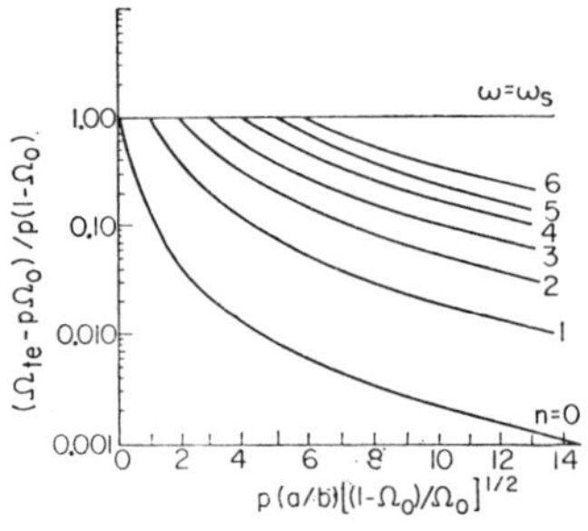

FIG. 4. Normalized eigenfrequencies Ω_{te} for harmonic (p) and inharmonic (n) series as a function of resonator parameters.

From Eqs. 19, 21, and 23, it can be concluded that modes of vibration associated principally with the electroded region can have strong responses and high Q's only when their mode resonance f_{te} falls between f_e and f_s. In addition to the fundamental or harmonic trapped-energy modes, which obviously satisfy this condition, one or more inharmonic-overtone modes as plotted in Fig. 4 could also have their mode resonance $\omega_{te}/2\pi$ fall below f_s. These inharmonics, if unwanted, would usually be called spurious responses. Referring again to Fig. 4, the modes that can be excited for any set of resonator parameters can be identified, as can the conditions for which any group of inharmonics will be suppressed—i.e., when their eigenfrequencies are equal to or greater than ω_s. Resonator parameters $2a/b$ and Ω_0 for the suppression of the nth and all higher symmetric inharmonic modes are given by

$$2a/b \leq (M_n/p)[\Omega_0/(1-\Omega_0)]^{\frac{1}{2}}, \quad (24)$$

where the parameter M_n has theoretical values as follows:

$n=1$, $M_n = 2.17$;
$n=2$, $M_n = 4.35$;
$n=3$, $M_n = 6.53$;
$n=4$, $M_n = 8.72$;
$n=5$, $M_n = 10.9$;
$n=6$, $M_n = 13.1$; etc.

Equation 6 gives only the symmetric solutions while the antisymmetric solutions are given by $\tan\zeta_e a = -\zeta_e/\gamma_s$. The lowest antisymmetric solution has $M'_n = 1.08$.

If the 1st antisymmetric inharmonic mode is not excited because of symmetry conditions, then Eq. 24 can be used as a design criterion for the suppression of all remaining inharmonic modes by setting $M = 2.17$. This, in effect, gives a design criterion that includes both electrode dimension and electrode mass-loading for the fundamental and also harmonic-overtone modes for the Z' directions. On the basis of these equations, it is probable that Bechmann's original empirical results, which recommended approximately a $1/p^{\frac{1}{2}}$ relationship for the electrode diameters, were obtained with progressively thinner electrodes and smaller degrees of frequency-lowering for the higher harmonics, e.g., with frequency-lowering divided by p.

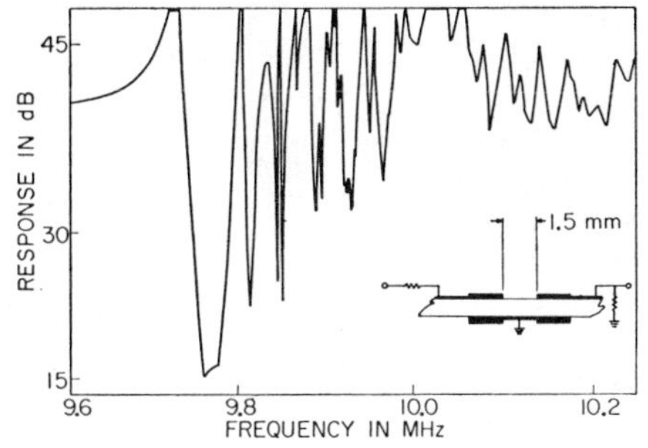

FIG. 5. Effect of interresonator coupling on 10-MHz AT-cut quartz multiresonator wafer. 2.54-mm-diam electrodes.

It should be noted that, in the original theoretical analysis of an ideal isotropic plate,[6] an error regarding identification of the symmetric and antisymmetric modes was made. This has been pointed out by Horton.[16] The error has been corrected in the analysis given here on the propagation of TT_3 modes in an ideal anisotropic plate.

II. EXPERIMENTAL VERIFICATION

A. Experimental Verification of Exponential-Energy Distribution

Experiments, on interresonator coupling and on interaction between a resonator area and the edge of its wafer, for values of $\omega<\omega_s$, have been used to verify the predicted exponential-energy distribution in the region of the wafer outside of the electroded area. Interresonator coupling measurements, which are difficult to perform accurately, confirmed the exponential distribution, but gave only fair correlation with theory on attenuation magnitudes. Figure 5 gives the frequency response of one unit that exhibits a degree of interresonator coupling. The multiresonator wafer has two 2.54-mm-diam resonators separated by 1.5 mm. More recently, this coupling has been utilized in the design of mechanically coupled filters.[17–19] On the other hand, measurements of the interaction between a resonator area and the wafer edge were in excellent agreement with theory both in functional dependence and in magnitude.

Observations of the interaction between dot resonator and wafer edge can be made in terms of dot resonator Q_m as a function of distance to the edge or in terms of position and magnitude of spurious responses as a function of distance. Initial investigation quickly

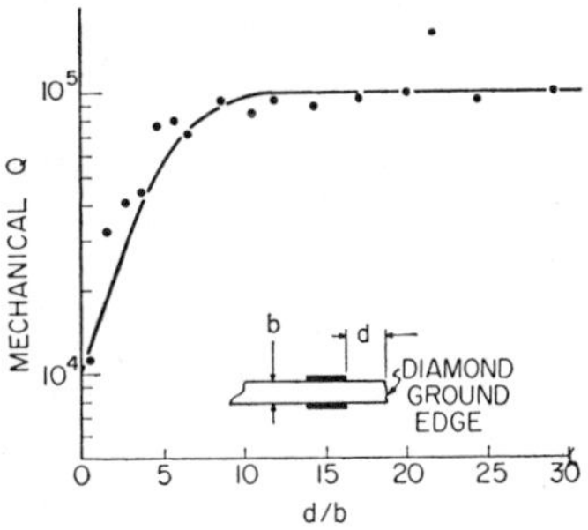

FIG. 6. Mechanical Q as a function of distance to diamond-ground edge of wafer. $Q=(\delta_1e^{-2\gamma d}+\delta_2)^{-1}$, where $\delta_1=1.0\times10^{-4}$, $\delta_2=1.0\times10^{-5}$. Experiment: $\gamma b=0.31$. Theory (——): $\gamma b=0.25$. Z' direction. 2nd run. 10-MHz AT cut.

established that spurious responses were a poor indicator for dot resonators with Ω_0 around 0.99, because spurious responses do not appear to be generated to any appreciable extent for electrode-to-edge separations greater than 1 or 2 wafer thicknesses. On the other hand, Q_m is a very sensitive indicator, gives quantitative data, and can easily be correlated with theory.

A functional dependence can be readily derived for Q_m versus distance to an absorbing or scattering interface. The energy absorbed or lost per unit time at such an interface or edge is given by $P_\delta=\omega\delta E_i$, where δ is a dissipation factor and E_i is the incident energy. For a dot resonator electrically excited at resonance with $\Omega_0<1$, an exponential-energy distribution would be observed as a function of distance d out from the edge of its electroded area. The energy distribution can be expressed as

$$E=E_0e^{-2\gamma d}, \qquad (25)$$

where E_0 is proportional to u_1^2 and γ is the attenuation constant given by Eq. 8. More-accurate values of the attenuation constant are given by Mindlin and Gazis[12] for both the Z' and X directions. From this, an expression for the fraction of the total stored energy of the dot resonator, E_T, which is loss per unit time (e.g., per cycle) as a result of the interface or edge, can be written using a second proportionality constant:

$$P_{\delta 1}=\omega\delta_1E_Te^{-2\gamma d}. \qquad (26)$$

With the acknowledgment that the wafer as a whole, in conjunction with its environment (air), is not quite a lossless medium, a second energy-dissipation factor $P_{\delta 2}=\omega\delta_2E_T$ should be included, giving

$$1/Q_m=(P_{\delta 1}+P_{\delta 2})/\omega E_T=\delta_1e^{-2\gamma d}+\delta_2. \qquad (27)$$

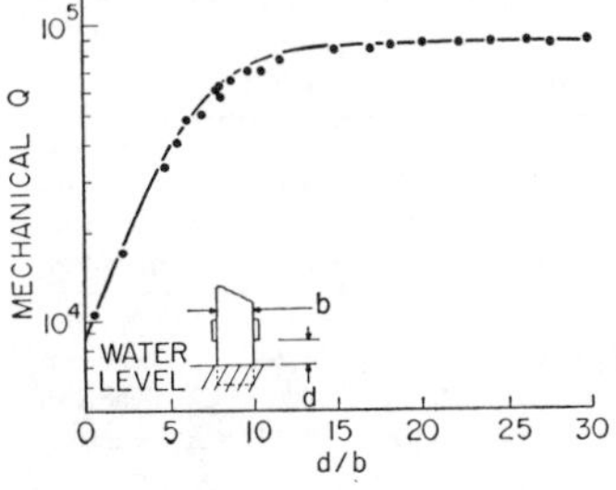

FIG. 7. Mechanical Q as a function of distance to attenuating edge (water). $Q=(\delta_1e^{-2\gamma d}+\delta_2)^{-1}$, where $\delta_1=1.1\times10^{-4}$, $\delta_2=1.16\times10^{-5}$. Experiment: $\gamma b=0.18$. Theory (——): $\gamma b=0.20$. X direction. 4th run. 10-MHz AT cut.

[16] W. H. Horton, Proc. IEEE (to be published).

[17] H. Mailer and D. R. Beuerle, "Incorporation of Multi-Resonator Crystals into Filters for Quantity Production," Proc. Ann. Freq. Control Symp., 20th, 20 Apr. **1966**, 309–342 (1966).

[18] M. Onoe, H. Jamonii, and N. Kobori, "High Frequency Crystal Filters Employing Multiple Mode Resonators Vibrating in Trapped Energy Modes," Proc. Ann. Freq. Control Symp., 20th, 20 Apr. **1966**, 266–287 (1966).

[19] R. A. Sykes and W. D. Beaver, "High Frequency Monolithic Crystal Filter and Possible Application to Single Frequency and Single Side Band Use," Proc. Ann. Freq. Control Symp., 20th, 20 Apr. **1966**, 288–308 (1966).

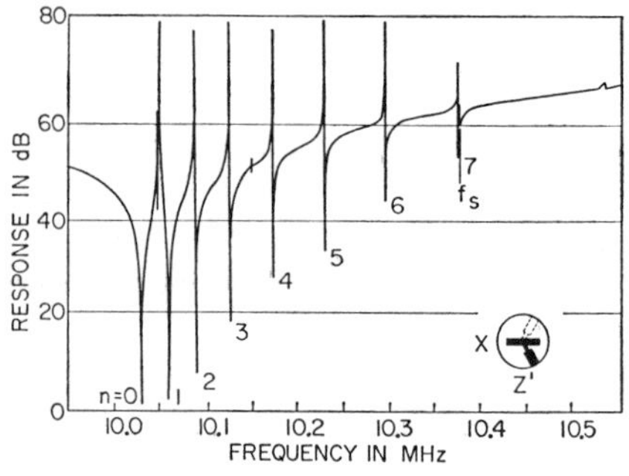

(a) Electrode length 100*b* (16.5 mm); width 10*b*.

FIG. 8. Resonator response ES-R5 ($\Omega=0.966$) with rectangular electrode in X direction. $\Omega_0=0.966$.

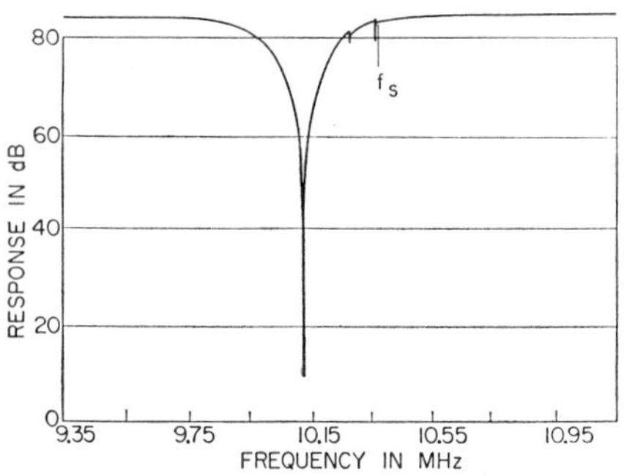

(b) Electrode length 10*b* (1.65 mm); width 10*b*.

In the experiments described, δ_2 was evaluated for each dot resonator from the large wafer Q_m value ($d>35b$) and δ_1 from the extrapolated and/or measured values of Q_m for $d=0$. Empirical values of γ were obtained as best fits to the Q_m versus d data. Theoretical values of γ were calculated using Mindlin and Gazis[12] expression and air-gap measurements of Ω_0.

The problem of varying the distance between the electrode and an absorbing edge was approached in two ways. In the first, the distance between electrode and wafer edges was reduced stepwise by physically removing the quartz, i.e., by grinding or sawing. The second method used a liquid as the absorbing medium, and a screw-feed mechanism to lower the crystal into the liquid. The latter technique has the obvious advantage of being nondestructive. In addition, the data were more consistent and not subject to the erratic deviations observed in the results obtained by either grinding or sawing. In both cases, the experimental results are in good agreement with theory. The results are given in Table I, and representative curves for each method in Figs. 6 and 7. These data verify the exponential decay of vibratory energy in the region outside the electroded area of a resonator.

B. Experimental Verification of Inharmonic-Overtone Series

The principal verification of energy-trapping theory has been in inharmonic-mode frequency measurements at the fundamental, 3rd-, and 5th-harmonic modes at 10, 30, and 50 MHz for both the TS_1 and TT_3 modes. These data were obtained using long narrow rectangular electrodes, as an approach to a two-dimensional configuration. In each case, the electrode was oriented in the appropriate crystallographic direction, X for TS_1 and Z' for TT_3, and half-lattice-bridge response curves were plotted for each electrode length as electrode length was reduced stepwise by photoetching. Photoetching techniques were used so that electrode thickness and bond characteristics could both remain invariant. Initial electrode dimensions for the fundamental, 3rd-, and 5th-harmonic modes, respectively, were 100*b* by 10*b*, 56*b* by 5.2*b*, and 29*b* by 2.3*b*, with the exception of one 5th-harmonic-mode unit, where dimensions of 56*b* by 6.2*b* were used to ease measurement problems.

TABLE I. Experimental Q_m parameters from Eq. 27.

Method	δ_1	δ_2	γb exptl	γb theoret	
Grinding	1.6×10^{-4}	8.9×10^{-6}	0.23	0.17	X direction
Grinding	2.0×10^{-4}	9.6×10^{-6}	0.18	0.18	X direction
Grinding	1.3×10^{-4}	8.9×10^{-6}	0.29	0.23	Z' direction
Sawing	1.0×10^{-4}	1.0×10^{-5}	0.31	0.25	Z' direction
Liquid					
Water	1.7×10^{-4}	1.37×10^{-5}	0.20	0.20	X direction
Water	1.1×10^{-4}	1.16×10^{-5}	0.18	0.20	X direction
Transformer oil	2.4×10^{-4}	1.01×10^{-5}	0.22	0.20	X direction

A classic example of these measurements is shown for ES-R5, a fundamental-mode unit with substantial mass loading ($\Omega_0=\omega_e/\omega_s=0.966$) and electrode length in the X direction. Its response plot is shown in Fig. 8(a) for a 100*b* electrode. The numbered responses are the symmetric inharmonic-overtone series with $n=0$ as the fundamental thickness/shear response. Inharmonic overtones through the 7th also can be identified. In each run, as the electrode length was reduced, the whole inharmonic-overtone series shifted up in frequency toward the cutoff ω_s. As each response approached ω_s, it decreased in amplitude and then vanished. Finally, only the fundamental response ($n=0$) is left, as shown in Fig. 8(b).

Eigenfrequencies for the inharmonic modes for each experiment in the X and Z' directions were recorded. The latter were compared with theory by plotting the experimental points on the theoretical-mode frequency plots of Fig. 4. This was done for three series of experiments, i.e., Z' orientation for the fundamental, 3rd-, and 5th-harmonic modes, with the results shown in Figs. 9(a)–(c). In every case, very good agreement is observed between theory and experiment. The experimental results obtained with the length of the rectangular electrodes oriented in the X direction were also compared with theory for the thickness/twist (TT_3) modes. In this case, however, theoretical and experimental values of the eigenfrequencies were equated at the point $\omega=\omega_s$ for $n=1$. See Figs. 9(d)–(f). It is interesting to note that the theory developed for thickness/twist modes appears to give the correct functional de-

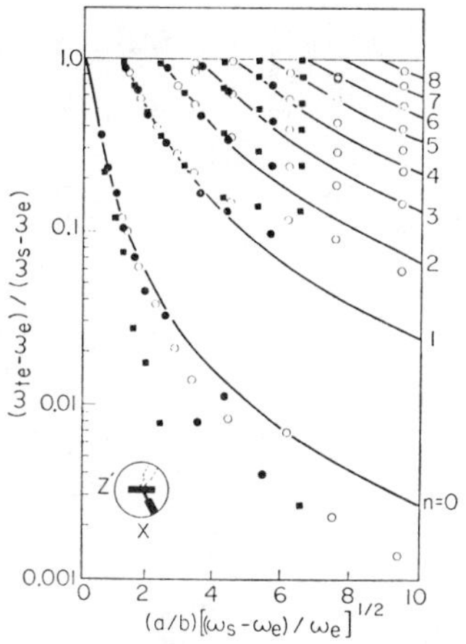

(a) 10-MHz fundamental-mode AT-cut quartz resonator. 25.4-mm-diam wafer. Results for ES-R1 (●; $\Omega_0=0.988$), ES-R3 (■; $\Omega_0=0.984$), and ES-R6 (○; $\Omega_0=0.967$) rectangular electrodes compared with theoretical curves for TT_3 modes. Width: $10b$. Length: $10b\leq 2a\leq 100b$. Z' orientation.

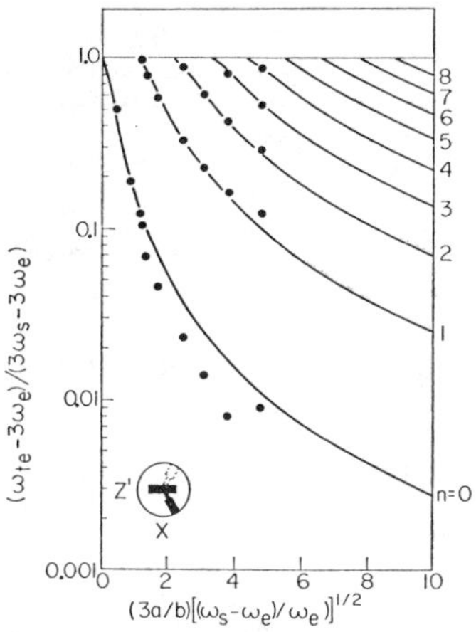

(b) 30-MHz 3rd-harmonic-mode AT-cut quartz resonator. 14-mm-diam wafer. Results for ES-R11 (●; $\Omega_0=0.997$) rectangular electrodes compared with theoretical curves for TT_3 modes. Width: $5.2b$. Length: $4.8b\leq 2a\leq 56b$. Z' orientation.

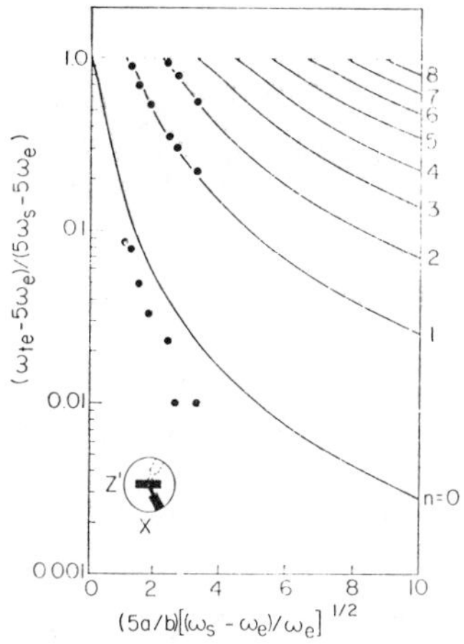

(c) 50-MHz 5th-harmonic-mode AT-cut quartz resonator. 14-mm-diam wafer. Results for ES-R13b (●; $\Omega_0=0.999$) rectangular electrodes compared with theoretical curves for TT_3 modes. Width: $6.2b$. Length: $12\leq 2a\leq 56$. Z' orientation.

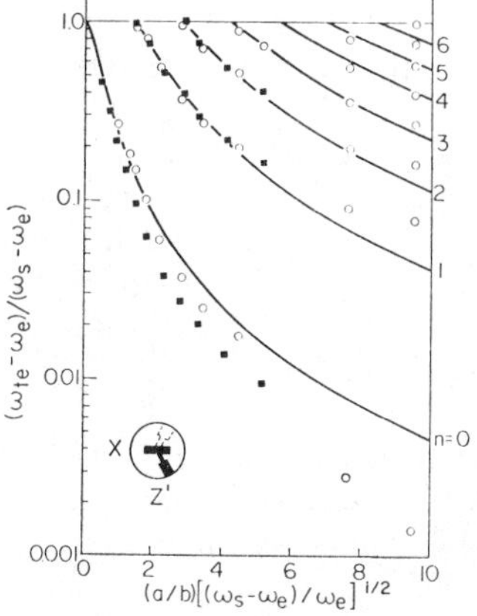

(d) 10-MHz fundamental-mode AT-cut quartz resonator. 25.4-mm-diam wafer. Results for ES-R5 (○; $\Omega_0=0.966$) and ES-R8 (■; $\Omega_0=0.990$) rectangular electrodes compared with theoretical curves for TT_3 modes adjusted to fit experimental data at $\omega_{te}=\omega_s$ for $n=1$. Width: $10b$. Length: $10b\leq 2a\leq 100b$. X orientation.

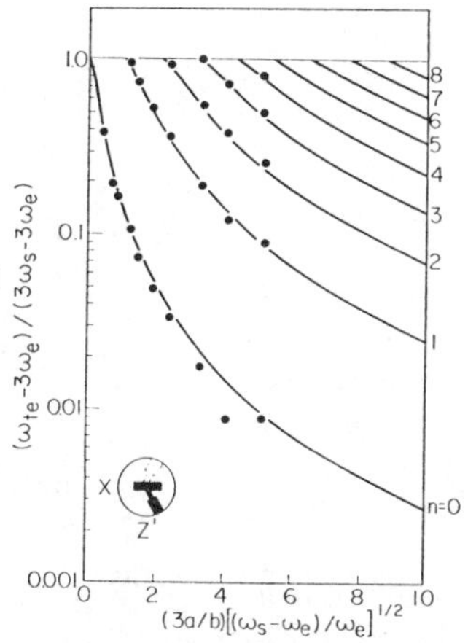

(e) 30-MHz 3rd-harmonic-mode AT-cut quartz resonator. 14-mm-diam wafer. Results for ES-R10 (●; $\Omega_0=0.996$) rectangular electrodes compared with theoretical curves for TT_3 modes adjusted to fit experimental data at $\omega_{te}=\omega_s$ for $n=1$. Width: $5.2b$. Length: $4.8b\leq 2a\leq 56b$. X orientation.

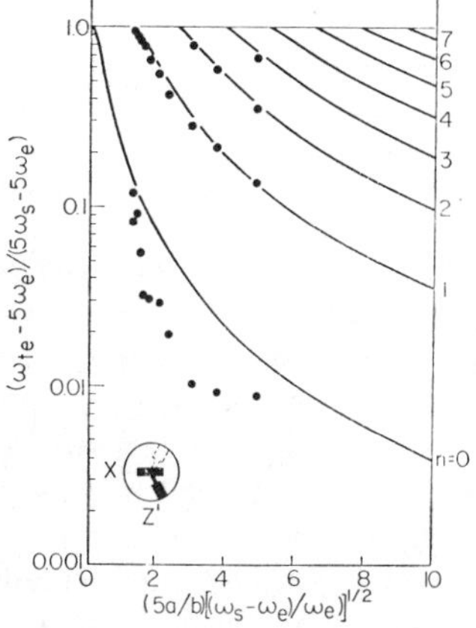

(f) 50-MHz 5th-harmonic-mode AT-cut quartz resonator. 14-mm-diam wafer. Results for ES-R12 (●; $\Omega_0=0.996$) rectangular electrodes compared with theoretical curves for TT_3 modes adjusted to fit experimental data at $\omega_{te}=\omega_s$ for $n=1$. Width: $2.3b$. Length: $7.7b\leq 2a\leq 29b$. X orientation.

FIG. 9. Mode frequencies ω_{te} as a function of electrode parameters. ——: Theoretical for two-dimensional symmetric thickness/twist modes.

pendence for the eigenfrequencies of thickness/shear inharmonic overtone modes and differs only in the numerical value of the constant M_n in Eq. 24. It should also be noted that excellent agreement between theory and experiment was observed when coupling between thickness/shear and /flexure modes was considered by Mindlin and Lee.[14]

III. GENERALIZED RESONATOR-DESIGN CRITERIA

The experiments described in the preceeding Section serve not only to verify energy-trapping theory but also to improve on theory by obtaining design criteria with greater accuracy. From Figs. 9(a)–(f), again assuming that antisymmetric modes can be avoided with symmetric excitation, a generalized design cri-

TABLE II. Design-criterion constant M for suppression of symmetric inharmonic-overtone modes in AT-cut filter crystals.

p	Empirical X	Empirical Z'	Theoretical X^a	Theoretical Z'
1	2.8	2.4	2.75	2.17
3	2.4	2.4		2.17
5	2.6	2.4		2.17

[a] Ref. 14.

terion of the form of Eq. 24 for filter crystals without inharmonic-overtone responses can be written with both theoretical and the more-accurate empirical values of M. Again, $2a/b$ is the electrode dimension in wafer thicknesses and $\Omega_0=\omega_e/\omega_s$.

$$2a/b \leq M/p[\Omega_0/(1-\Omega_0)]^{\frac{1}{2}}, \tag{28}$$

where M is given by Table II.

From these criteria, it is evident that resonators that have a range of motional parameters can be designed, while maintaining suppression of inharmonic overtones associated with the electrode region, by trading off electrode dimensions and frequency-lowering. This has been demonstrated with excellent results for both fundamental and harmonic-overtone-mode resonators.

A. Results on vhf Filter Crystals

The flexibility that can be achieved in the synthesis of crystal-filter networks depends primarily on the extent of the range of motional-parameter values available in filter crystals. At higher frequencies, the goals have been first to obtain resonators with clean responses and usable values of motional parameters, and then to try to extend these to include desired ranges of values. With the techniques outlined in this paper, both goals are directly achievable.

In realizing these goals, resonators have been designed with progressively larger area electrodes. As required by the design criteria—i.e., Eq. 28—the electrode mass lowering of these units is reduced as the lateral electrode dimensions are increased. In addition, it is necessary to consider the distribution of vibratory energy in the unelectroded surrounding regions of the wafer to obtain a strong response and, further, to consider the acoustic transmission-line characteristics of the resonator to obtain reasonable suppression of inharmonic responses associated with the whole wafer or larger regions thereof.

These techniques have been used in the design of fundamental-, 3rd-, 5th-, and 7th-harmonic-mode resonators in the 10- to 140-MHz frequency range. A factor of 4 improvement in motional parameters over the previous state of the art has been realized for 3rd- and 5th-harmonic-mode resonators and fundamental-, 3rd-, 5th-, and 7th-harmonic-mode units with electrode diameters up to $48b$, $30b$, $20b$, and $20b$, respectively, are currently under investigation. The motional parameters of a group of 60-MHz 3rd- and 5th-harmonic-mode resonators are given in Table III. The data on the 3rd-harmonic-mode units are the parameters of the individual resonators of a multiresonator wafer (2 resonators/wafer). Each resonator has an electrode diameter of $20b$ and evaporated aluminum electrodes that provide the frequency-lowering calculated from energy-trapping theory, $\Omega_0=0.998$. The frequency-response characteristics of one of the units, obtained using a half-lattice bridge terminated in the characteristic impedance of the resonators, are given in Fig. 10(a). In Table IV are given the parameters of a group of 90- and 100-MHz 5th-harmonic-mode resonators. The latter, as in the case of the 60-MHz 3rd-harmonic-mode units, are the parameters of the individual resonators comprising a multi-

TABLE III. Resonator parameters for 60-MHz, 3rd- and 5th-harmonic-mode AT-cut filter crystals.

Unit	Frequency (kHz)	R (Ω)	C_0 (pF)	C_1 (10^{-4} pF)	Q_m
3rd-Harmonic Mode					
60-3-1	59 914.000	59	1.57	5.04	89 300
	59 919.469	59	1.67	5.41	83 300
60-3-2	59 757.455	77	1.64	5.65	61 300
	59 752.615	75	1.59	5.48	64 900
60-3-3	59 763.648	67	1.66	5.50	72 500
	59 766.922	79	1.56	4.47	75 800
60-3-4	59 757.125	69	1.62	5.39	72 500
	59 761.888	77	1.60	5.11	67 600
60-3-5	59 987.000	73	1.62	6.03	102 000
	59 999.922	41	1.67	6.45	100 000
5th-Harmonic Mode					
80-60-1	59 916.022	96	1.50	1.93	144 000
-2	59 954.880	128	1.50	2.10	98 800
-3	59 928.325	106	1.50	1.84	136 000
-4	59 911.174	100	1.50	2.00	133 000
-5	59 916.048	98	1.70	2.09	130 000
-6	59 931.292	91	1.68	2.13	137 000

TABLE IV. Resonator parameters for 90- and 100-MHz, 5th-harmonic-mode AT-cut filter crystals.

Unit	Frequency (kHz)	R (Ω)	C_0 (pF)	C_1 (10^{-4} pF)	Q_m
80 90-1	88 933.681	150	1.30	1.46	81 700
-2	88 892.897	146	1.30	1.34	91 600
-3	89 051.061	136	1.29	1.34	98 800
-4	88 908.851	136	1.31	1.47	89 600
-5	88 888.361	160	1.30	1.49	68 700
-6	88 885.660	137	1.30	1.53	85 500
100-5-1	100 079.611	136	1.28	1.58	74 000
	100 068.641	130	1.30	1.64	74 600
100-5-2	100 359.221	121	1.30	1.68	78 400
	100 339.724	128	1.31	1.54	80 500
100-5-3	100 400.592	124	1.30	1.64	77 900
	100 383.962	105	1.33	1.76	85 900
100-5-4	100 016.618	107	1.30	1.88	76 000
	100 114.536	112	1.34	1.48	95 900
100-5-5	100 111.777	129	1.28	1.65	75 000
	100 114.718	116	1.34	1.56	87 800

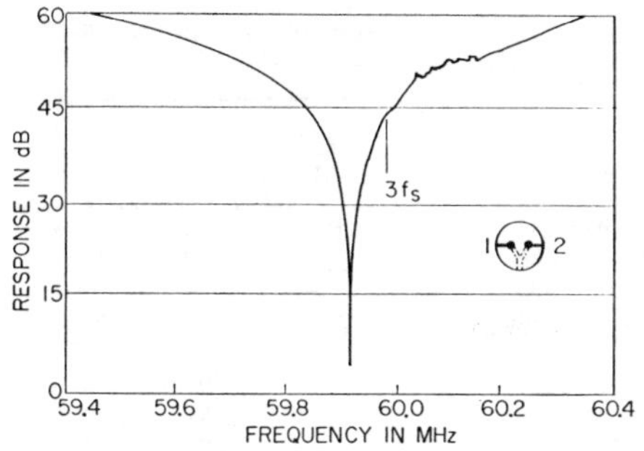

(a) Frequency response of Resonator No. 2 on multiresonator wafer 60-3-1. 60-MHz 3rd-harmonic resonator. 14-mm AT-cut quartz wafer. 1.6-mm-diam electrodes. $2a/b=20$. $f_r=59\,919.469$ kHz. $R=59\,\Omega$. $C_0=1.67$ pF; $C_1=5.41\times10^{-4}$ pF. $Q=83\,300$.

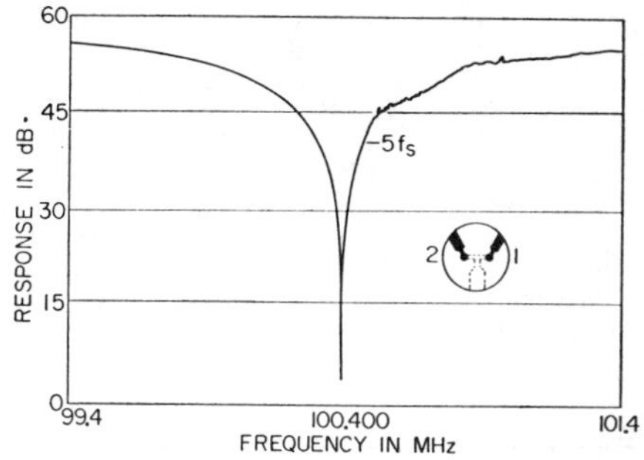

(b) Frequency response of Resonator No. 1 on multiresonator wafer 100-5-3. 100-MHz 5th-harmonic resonator. 14-mm AT-cut quartz wafer. 1.3-mm-diam electrodes. $2a/b=15$. $f_r=100\,400.592$ kHz. $R=124\,\Omega$. $C_0=1.30$ pF; $C_1=1.64\times10^{-4}$ pF. $Q_m=77\,900$.

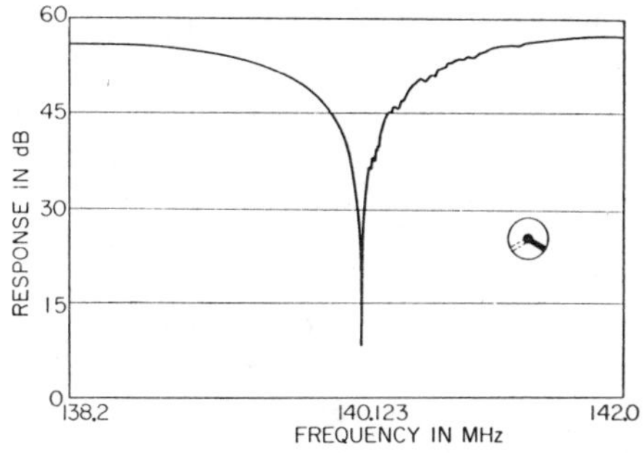

(c) Frequency response of Resonator 140-7-7. 140-MHz 7th-harmonic resonator. 7-mm AT-cut quartz wafer. 1.6-mm-diam electrode. $2a/b=20$. $f=140\,123.174$ kHz. $R=228\,\Omega$. $C_0=1.76$ pF; $C_1=1.18\times10^{-4}$ pF. $Q_m=42\,400$.

FIG. 10. Frequency-response characteristics of single-electrode resonators.

resonator wafer (2 resonator/wafer). The 90- and 100-MHz resonators had an electrode diameter of $15b$, evaporated aluminum electrodes, and frequency lowering $\Omega_0=0.999$. The frequency response of the 100-MHz resonator 100-5-3 is given in Fig. 10(b). The improvement represented by these units can be seen best by comparing the average motional parameter of each group with the parameters of prior state-of-the-art resonators, as given by Bechmann.[20] As shown in Table V, the crystals designed using energy-trapping theory offer a factor of 4 improvement in motional parameters. It is important to note that this extension of the range of motional parameters was realized without sacrificing unwanted mode suppression.

TABLE V. Comparison of resonator parameters for 3rd- and 5th-harmonic-mode filter crystals.

Frequency	Crystal parameters	Prior art[a]	Trapped-energy crystals
60 MHz (3rd harmonic)	$2a/b$	10	20
	$C_1(10^{-4}$ pF)	1.50	5.45
	R (Ω)	160	65
	C_0/C_1	3500	2970
	Q_m	101 000	78 900
60 MHz (5th harmonic)	$2a/b$	7.5	15
	$C_1(10^{-4}$ pF)	0.60	2.02
	R (Ω)	400	103
	C_0/C_1	25 000	7800
	Q_m	112 000	130 000
100 MHz (5th harmonic)	$2a/b$	7.5	15
	$C_1(10^{-4}$ pF)	0.40	1.64
	R (Ω)	700	125
	C_0/C_1	25 000	8000
	Q_m	66 000	78 700

[a] Ref. 20.

Energy-trapping design principles have been applied to higher harmonics. For example, 100-MHz 5th- and 7th-harmonic-mode resonators have been designed with an electrode diameter of $20b$ and frequency-lowering of $\Omega_0=0.9994$ and 0.9997, respectively. The frequency response and motional parameters of a 140-MHz 7th-harmonic-mode resonator are given in Fig. 10(c).

Energy-trapping theory has therefore significantly extended the range of motional-parameter values available in filter crystals and in so doing has given filter designers greater freedom and flexibility in the design of crystal-filter networks.

B. Multielectrode Resonators and Multiresonator Wafers

The same energy-trapping techniques that are used to design single resonators can be used also to design

[20] R. Bechmann, "Crystals for Filter Application, Part 3," Frequency 1, No. 7, 18–21 (1963).

TABLE VI. Equivalent-circuit parameter values for 3rd-harmonic mode multielectrode resonators.

Unit	Frequency (kHz)	R (Ω)	C_0 (pF)	C_1 (10^{-3} pF)	Q_{app}
ME-30-11	30 104.00	64	3.08	1.10	75 100
-12	30 088.13	69	3.44	1.54	50 000
-13	30 098.46	60	3.43	1.60	55 200
-14	30 169.26	42	3.20	1.21	104 000
-15	30 144.84	45	3.32	1.31	89 200
ME-30-16	30 112.83	58	3.34	1.23	74 000
-17	30 130.18	60	3.36	1.42	62 100
-18	30 160.10	40	3.82	1.73	76 300
-19	30 104.66	51	3.80	1.38	75 100
-20	30 193.02	49	4.06	1.81	59 100
ME-60-1	59 794.26	70	2.39	0.732	52 000
-2	59 834.29	66	1.80	0.624	64 500
-3	59 817.64	49	2.25	0.798	68 000
-4	59 807.83	61	2.06	0.796	54 900
ME-60-5	59 767.47	61	2.78	0.974	44 800
-6	59 787.13	40	2.68	0.890	74 600
-7	59 800.19	54	2.83	0.902	54 900
-8	59 865.91	42	2.69	0.792	80 000

multielectrode resonators or multiresonator wafers. The individual electrode areas are designed to give spurious free resonators and these, in turn, are separated sufficiently far from another so that interaction does not occur, or, in the case of coupled mode resonators, resonators can be spaced such that a desired degree of coupling exists. These design goals can be met by using Eqs. 25 and 28.

Multielectrode resonators provide a second method of obtaining filter crystals that have a range of motional parameters. In this case, the desired resonator parameters are obtained by electrically connecting in parallel an appropriate number of acoustically independent resonators that have the same or different parameter values, where the resonators and all interconnections are contained on a single AT-cut quartz wafer. The individual resonators are tuned to the same frequency and the resulting multielectrode resonator has the frequency response and motional parameters equivalent to those of a parallel combination of conventional single-electrode resonators. An example is given in Fig. 11(a) of a 10-MHz fundamental-mode multielectrode resonator that has 4 resonators connected in parallel.

Harmonic-mode multielectrode resonators have been fabricated at 30 and 60 MHz, using 3rd-harmonic-mode resonators, at 60 and 90 MHz, using 5th-harmonic-mode resonators, and at 140 MHz, using 7th-harmonic-mode resonators. The characteristics of the 3rd-harmonic-mode units are given in Table VI and representative frequency-response curves in Figs. 11(b) and (c). For these units, 4 dot resonators, with an electrode diameter/thickness ratio of 10, were appropriately spaced on a 14-mm-diam polished-quartz blank and electrically connected in parallel. Units ME-30-16 through ME-30-20 had four elliptical electrodes connected in parallel. The electrode diameter/thickness ratio was 10 in the Z' direction and 13 in the X direction.

TABLE VII. Equivalent-circuit parameters of 60- and 90-MHz 5th-harmonic-mode multielectrode resonators.

Unit	Frequency (kHz)	R (Ω)	C_0 (pF)	C_1 (10^{-4} pF)	Q_{app}
80-6- 7	59 945.931	59	2.94	4.19	107 000
8	59 921.646	59	2.68	3.78	119 000
9	59 958.023	65	2.96	4.39	93 000
10	59 939.257	51	2.92	4.25	123 000
80-9- 7	88 815.093	101	2.32	2.97	59 700
8	88 880.912	106	2.34	2.97	56 900
9	88 886.967	90	2.34	2.94	67 700
10	89 209.937	86	2.24	2.92	71 000

It should also be noted that, with one or two exceptions, the desired 40-dB suppression of spurious responses was obtained. The 5th-harmonic, multielectroded resonators at 60 and 90 MHz consisted of 2 dot resonators connected in parallel on either a 12- or 14-mm-diam polished-quartz wafer. On the basis of the good results obtained with large-area single-electrode resonators, an electrode diameter/thickness ratio of 15 was selected for these units. This gives a factor of 4 increase in motional parameters of a single resonator over the maximum values specified from the Bechmann criterion, and, hence, a factor of 8 increase would be expected for the resulting multielectrode units. Characteristics of the finished units mounted in HC-6/U holders are given in Table VII. The 7th-harmonic multielectrode resonators at 140 MHz had two resonators with an electrode diameter of $10b$ connected in parallel on a 14-mm-diam quartz blank. Resonator parameters are given in Table VIII and the frequency response of a typical unit in Fig. 11(d). The 9th-harmonic response of the same unit is given in Fig. 11(e). It is interesting to compare these results with those given in Fig. 10(c) for a single-electrode resonator. The latter, also a 7th-harmonic-mode resonator at 140 MHz, has an electrode diameter of $20b$ and, hence, comparable parameter values.

The average resonator-parameter values for the multielectrode resonators are compared in Table IX with prior art values obtained from Bechmann.[20] Several important factors should be considered in a critical examination of Tables VI–IX and in any comparison between single- and multi-electrode resonators. It is apparent that motional capacitance adds up in the

TABLE VIII. Equivalent-circuit parameters of 140-MHz 7th-harmonic-mode multielectrode resonators.

Unit	Frequency (kHz)	R (Ω)	C_0 (pF)	C_1 (10^{-5} pF)	Q_{app}
140-7-1	139 923.485	194	1.57	8.41	69 900
-2	140 192.017	190	1.54	8.51	70 200
-3	139 885.643	235	1.48	8.32	58 200
-4	140 046.446	264	1.52	8.27	52 100
-5	140 230.710	236	1.45	8.72	53 400
-6	140 320.306	220	1.46	7.50	68 800

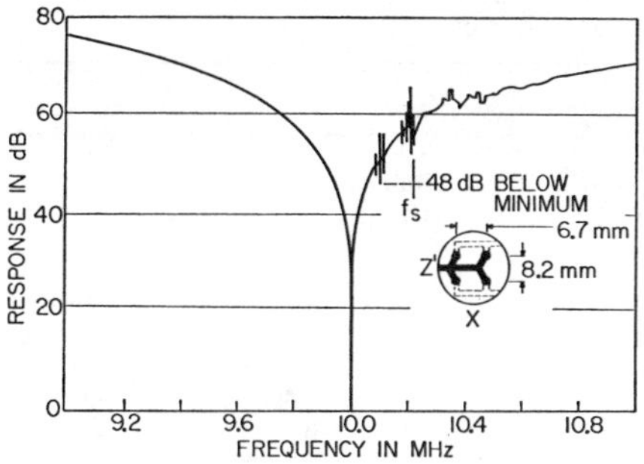

(a) Multielectrode resonator ME-10-2 ($\Omega_0=0.979$; fundamental mode). 25.4-mm-diam **AT**-cut quartz wafer. 2.5-mm-diam electrodes. $2a/b=15$. $f_r=10\,000.761$ kHz. $R=11\ \Omega$. $C_0=7.4$ pF; $C_1=30.2\times10^{-3}$ pF. $Q_m=47\,800$.

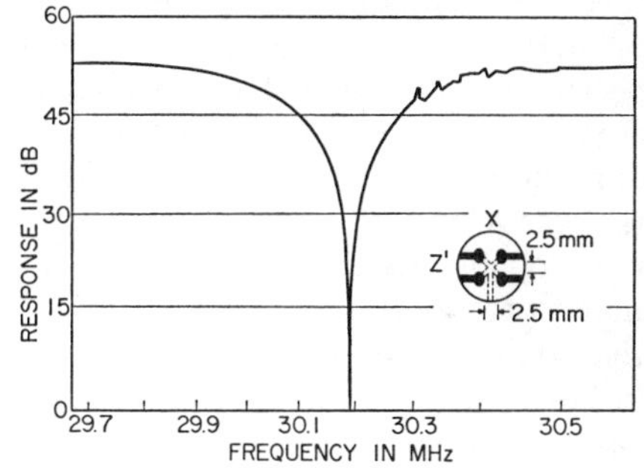

(b) Multielectrode resonator ME-30-20. 3rd-harmonic mode. 14-mm-diam **AT**-cut quartz wafer. Elliptical electrodes. $2a/b=10$ in Z' direction. $2a/b=13$ in X direction. $f=30\,193.02$ kHz. $R=49\ \Omega$. $C_0=4.06$ pF; $C_1=1.81\times10^{-3}$ pF. $Q_m=59\,100$.

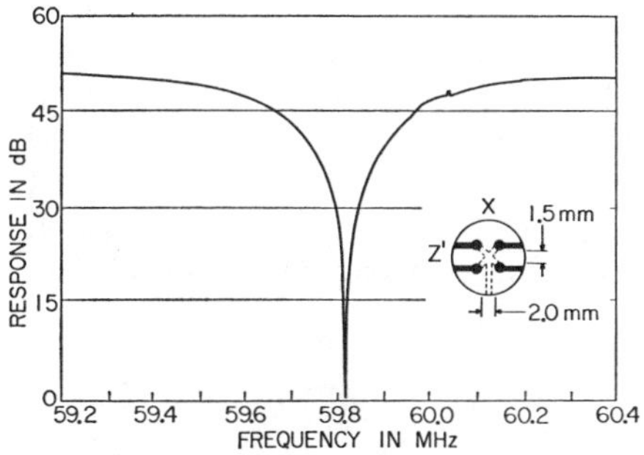

(c) Multielectrode resonator ME-60-3. 3rd-harmonic mode. 14-mm-diam **AT**-cut quartz wafer. 0.86-mm-diam electrodes. $2a/b=10$. $f=59\,817.64$ kHz. $R=49\ \Omega$. $C_0=2.25$ pF; $C_1=7.98\times10^{-4}$ pF. $Q_m=68\,000$.

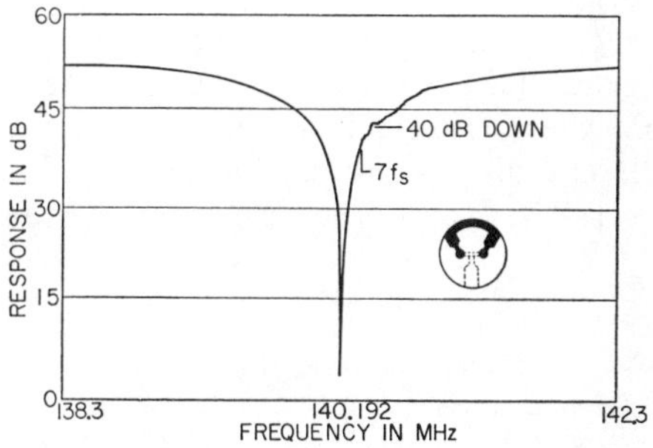

(d) Multielectrode resonator 140-7-2. 7th-harmonic mode. Two resonators in parallel. 14-mm **AT**-cut quartz wafer. 0.9-mm-diam electrodes. $2a/b=11$. $f_r=140\,192.017$ kHz. $R=190\ \Omega$. $C_0=1.54$ pF; $C_1=8.51\times10^{-5}$ pF. $Q_m=70\,200$.

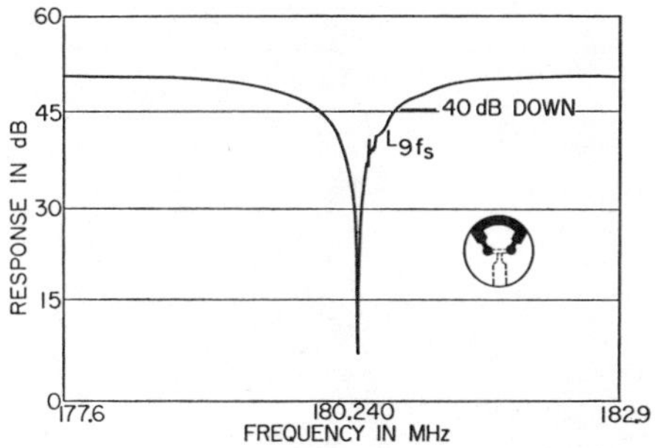

(e) 7th-harmonic multielectrode resonator 140-7-2 at 9th-harmonic mode. Two resonators in parallel. 14-mm **AT**-cut quartz wafer. 0.9-mm-diam electrodes. $2a/b=11$. $f_r=180.240$ MHz. $R=280\ \Omega$.

FIG. 11. Characteristic frequency responses of multielectrode resonators.

predicted fashion and that the capacitance ratio is slightly better (smaller) for the multielectrode resonator. It is also apparent that resonant resistance does not usually decrease by the appropriate factor. Resonant resistance actually behaves in the expected manner for a parallel combination of resonant circuits that are stagger-tuned one from another by very slight amounts. Thus, the observed resonant resistance is a measure not so much of a loss factor as of a deviation from perfect tuning. Similarly, the apparent Q (Q_{app}), calculated from the observed resistance is not a good measure of the Q_m of the individual resonators. It should be noted again that the performance of filter networks is dependent on the loss factors of the individual circuit components and therefore depends on the true Q_m of the individual resonators. If the individual resonators are tuned within the design-tolerance limits set for the filter network, then the resultant filter performance will equal that predicted from the true Q_m values for the individual resonators and will not be affected by any apparent Q values observed for the multielectrode combination of resonators. True Q_m values should not be affected by the parallel interconnection of resonators, and therefore the multielectrode resonators have really met all motional parameter goals. It should be noted, however, that multielectrode resonators involve more operations in fabrication than their large-area electrode counterparts and therefore would not be favored where either is applicable. The multielectrode resonators do offer higher upper limits on motional parameters, however.

TABLE IX. Resonator parameters for harmonic-mode multielectrode AT-cut quartz filter crystals.

	Multielectrode		Prior art[a]	
3rd-Harmonic-Mode Resonators				
	30 MHz	60 MHz	30 MHz	60 MHz
Number of resonators	4	4	1	1
$2a/b$	10	10	10	10
R (Ω)	54	55	80	140
C_1(10^{-4} pF)	14.3	8.1	3.0	1.4
C_0/C_1	2460	3000	3500	3500
Q_{app}	72 000	61 700	200 000	110 000
5th-Harmonic-Mode Resonators				
	60 MHz	90 MHz	60 MHz	90 MHz
Number of resonators	2	2	1	1
$2a/b$	15	14	7.5	7.5
R (Ω)	59	96	400	600
C_1(10^{-4} pF)	4.15	2.95	0.6	0.4
C_0/C_1	6930	7830	25 000	25 000
Q_{app}	111 000	63 800	110 000	75 000

[a] Ref. 20.

IV. CONCLUSIONS

Energy-trapping theory has been applied to a simplified model of an AT-cut quartz wafer to derive an expression relating the eigenfrequencies of each harmonic mode p and its symmetric inharmonic overtones n, as functions of the cutoff frequencies ω_e and ω_s and electrode dimensions. This expression has been shown to be in reasonable agreement with experimental data obtained for the TS_1 and TT_3 modes in the crystallographic X and Z' directions, respectively, for the fundamental, 3rd, and 5th harmonics and the inharmonic overtones of each. From these data, the Bechmann criteria for suppression of inharmonic-overtone modes have been generalized to include harmonic-mode crystals.

These criteria place interrelated upper bounds on lateral dimensions and thickness of electrodes on filter crystals. It has been demonstrated that electrode area and thickness can be traded off against one another to obtain an extended range of motional parameters. Increases in motional-parameter values by approximately a factor of 4 have been attained at frequencies from 10 to 90 MHz using this technique, and additional increases have been achieved using multielectrode filter crystals.

ACKNOWLEDGMENT

Most of the work described here has been supported by the U. S. Army Electronics Command, Fort Monmouth, New Jersey. The authors should like to acknowledge helpful conversations and communications with Dr. R. Bechmann, of that laboratory, and with Professor R. D. Mindlin, of Columbia University, and the assistance of their colleagues, K. A. Pim and R. B. McEntee.

Section II-B
Analysis of Acoustically Coupled Systems

The following series of papers are analytical in nature. They add to the understanding of monolithic crystal filters and they also provide some useful design information.

Byrne, Lloyd, and Spencer describe the operation of resonators on AT-cut quartz, and they put particular emphasis on the various possible modes of vibration. The importance of understanding these various modes increases in proportion to the number of resonators that are to be placed on one crystal plate. If a multiresonator filter is to be built in a manner where no more than two resonators are acoustically coupled, a detailed study of anharmonic modes will not be necessary. If, on the other hand, the filter is to be built with many resonators on a single crystal plate, then such a study is essential to achieving good stopband attenuation.

Beaver's paper illustrates the difficulty of analyzing a multiresonator system from a purely physical point of view. The results that he obtains are useful for a filter designer since they provide a relationship between electrode separation, plateback, and the degree of interresonator coupling. However, this information loses some of its usefulness as extra plating has to be placed between resonators in order to adjust the coupling.

Finally, Ashida analyzes a multiresonator system by a transmission matrix approach. This permits him to avoid carrying all the physical parameters along in his analysis, thus reducing its complexity. He also derives relationships between the degree of coupling and the mass loading on the resonators. The calculated values are, however, for the limited case of equal electrode widths and equal mass loadings on the electrodes.

Received 18 August 1967

Thickness–Shear Vibration in Rectangular *AT*-Cut Quartz Plates with Partial Electrodes

R. J. Byrne, P. Lloyd, and W. J. Spencer

Bell Telephone Laboratories, Inc., Allentown, Pennsylvania 18103

The thickness–shear vibrations in *AT*-cut quartz plates with partial electrodes including the effects of a pair of free edges are discussed in a qualitative manner. The frequencies and mode shapes as a function of plate and electrode dimensions are determined from an analysis similar to that of Mindlin and Lee. These results are compared with measured values obtained from rectangular quartz plates with rectangular electrodes. The ratio of capacitances of thickness–shear vibrations are calculated from an expression given by Lewis. These calculations indicate that the ratio of electrode length to plate length is equally as importamt in the suppression of anharmonic thickness–shear vibrations in finite quartz plates as is the ratio of electrode length to plate thickness (Bechmann's number).

INTRODUCTION

The metallic electrodes on *AT*-cut quartz resonators that are used principally for excitation of thickness–shear vibrations may also be used to confine the deformation associated with these vibrations to the electroded area and to suppress unwanted modes of vibration. The suppression of unwanted modes in *AT*-cut plates was first achieved by Bechmann[1] and the effect correctly related to a cutoff frequency for thickness–shear modes by Mortley.[2] Similar effects had been known for the propagation of electromagnetic waves in wave guides for some time. Mortley's work remained relatively unknown[3] and the effect was later recognized independently by Shockley, Curran, and Koneval.[4] They treated qualitatively the case of a finite electrode on an infinite *AT*-cut quartz plate. The effect of the mass loading and length of the electrode on the propagation constant for thickness–shear modes in all rotated *y*-cut[5] quartz plates was treated rigorously by Mindlin and Lee[6] for the case of a finite electrode strip on an infinite plate.

The role of a finite boundary on the generation of unwanted modes in *AT*-cut quartz plates with partial electrodes has been experimentally investigated by LaBrie.[7] The approach used was similar to that suggested by Vormer[8] for the suppression of overtones in length–extensional resonators and did not include the effects of the electrode mass loading on the thickness–shear mode shapes. The shape of the thickness–shear displacement, which determines the phase and magnitude of the charge due to the piezoelectric effect induced on the electrode and thus the relative strength of any particular resonance, can be determined for finite plates with partial electrodes by extending the approach used by Mindlin and Lee.

In this paper, the frequency, mode shape, and the ratio of capacitances for a partial electrode on an finite *AT*-cut quartz strip are calculated. These are compared with the measured values obtained from rectangular *AT*- cut plates with partial rectangular electrodes. The effect of the free boundary generates resonances that do not exist in the infinite plate. These modes are particularily important when the electrode area is 5% or more of the quartz plate area. It is possible, by proper choice of the electrode mass and length, as well as the plate length, to suppress certain of the anharmonic overtones of thickness–shear by electrical charge cancellation. The plate and electrode dimensions necessary to charge cancel a particular mode are determined from a

[1] R. Bechmann, U. S. Patent No. 2,249,933 (1941).

[2] W. S. Mortley, Wireless World **57**, 399–403 (1951); Proc. Inst. Elec. Engrs. (London) **104B**, 239–249 (1957).

[3] W. S. Mortley, Phys. Today **19**, 11–12 (1966).

[4] W. Shockley, D. R. Curran, and D. J. Koneval, Proc. 17th Ann. Symp. on Freq. Control, pp. 88–126 (May 1963).

[5] W. P. Mason, *Piezoelectric Crystals and Their Applications to Ultrasonics* (D. Van Nostrand Co., Inc., Princeton, N. J. 1950).

[6] R. D. Mindlin and P. C. Y. Lee, Int. J. Solids Struct. **2**, 125–139 (1966).

[7] L. J. LaBrie, Fall Meeting of IRE Profess. Group on Vehicular Commun., 1952.

[8] J. J. Vormer, Proc. IRE **39**, 1086 (1951).

Reprinted with permission from *J. Acoust. Soc. Amer.*, vol. 43, pp. 232–238, Feb. 1968.

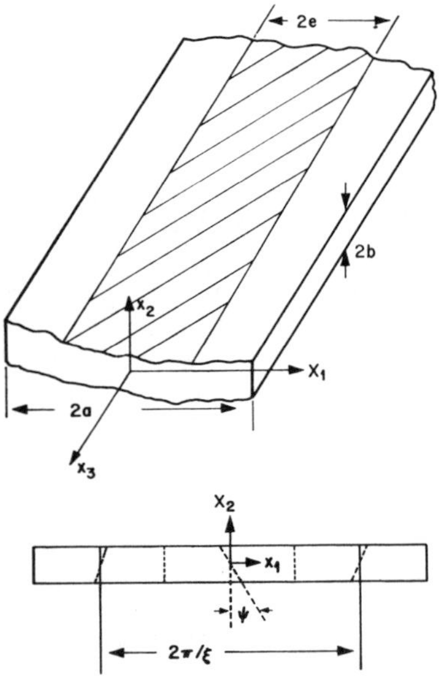

FIG. 1. Plate and electrode geometry for a partially electroded infinite *AT*-cut quartz strip, showing rotation ψ due to thickness–shear displacement only.

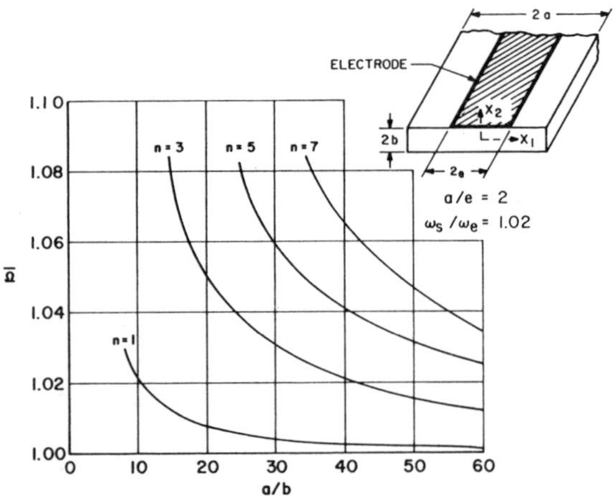

FIG. 2. Frequency of the thickness–shear modes $\bar{\Omega}$ as a function of width-to-thickness ratio a/b for a partially plated *AT*-cut quartz strip.

technique developed by Lewis,[9] using Mindlin's approximate theory for thickness–shear modes. The present analysis considers only the thickness–shear modes. The effects of flexural and twist modes on resonator design and performance are qualitatively discussed.

I. APPROXIMATE THEORY FOR THICKNESS MODES

The frequency of most of the strong resonances near the lowest thickness–shear frequency in *AT*-cut quartz plates is given in some detail by the approximate theory developed by Mindlin and his co-workers.[10] Mindlin has reduced the three-dimensional equations of elasticity for plates to two dimensions by expanding in a power series about the thickness dimension. Generally this series expansion is truncated after the linear term. The resulting equations may be further simplified so that only the thickness–shear motion remains with suitable correction terms for the motions that have been neglected. The validity of this approximation has been discussed by Mindlin and Forray.[11]

The thickness–shear motion in this approximation is represented as a rotation of the plate normal by the angle ψ with a wavelength $\lambda=2\pi/\xi$ in the x_1 direction as shown in Fig. 1. In this case, the plate is assumed to be infinite in the x_3 direction, of thickness $2b$ and width $2a$, with an infinite electrode strip of width $2e$ parallel with the strip edges. Normally, there is a displacement in the x_2 direction associated with the rotation ψ, the magnitude of which depends on the amount of coupled flexure that is always present in the thickness–shear mode in finite plates. This has been neglected and a correction term included in the approximate equations.

The equations and boundary conditions governing the vibration of the strip shown in Fig. 1 have been given by Tiersten.[12] The equation for ψ in the unelectroded region is

$$\psi_{s,11}-(\gamma_{11}+\kappa_6^2c_{66})^{-1}[1-(\rho\omega^2b^2/3\kappa_6^2c_{66})]\times(3\kappa_6^2c_{66}/b^2)\psi_s=0,$$

where γ_{11} is the Voigt stretch modulus[6] in the x_1 direction, $\kappa_6^2=\pi^2/12$, and c_{66} is the shear modulus for the *AT*-cut quartz plate. This may be written as

$$\psi_{s,11}-(K^2/b^2)(1-\omega^2/\omega_s^2)\psi_s=0, \tag{1}$$

where

$$K^2=3\kappa_6^2c_{66}/(\gamma_{11}+\kappa_6^2c_{66})$$

and

$$\omega_s^2=3\kappa_6^2c_{66}/\rho b^2.$$

In the latter equation, ω_s is the thickness-shear frequency of the infinite plate (in the x_1 direction) without an electrode. In the electroded region, the equation governing the rotation is

$$\psi_{e,11}-\frac{K^2}{b^2}(1-\omega^2/\omega_e^2)\psi_e=\frac{3\kappa_6e_{26}V}{2b^3}, \tag{2}$$

where V is the voltage applied to the electrodes, e_{26} is the piezoelectric constant, and ω_e is the frequency of the infinite quartz plate completely covered with an electrode of thickness b' and density ρ'. The frequency ω_e is related to ω_s by[6]

$$\omega_e^2(1+R)=\omega_s^2,$$

where

$$R=2\rho'b'/\rho b$$

is the ratio of the mass per unit area of the electrode to the mass per unit area of the quartz plate.

There is, in addition, an equation for the electric displacement given by

$$D_2^{(0)}=\kappa_6e_{26}\psi_e-\epsilon_{22}V/2b. \tag{3}$$

[9] J. A. Lewis, Bell System Tech. J. **40**, 1259–1280 (1961).
[10] R. D. Mindlin, Quart. Appl. Math. **19**, 51–61 (1961).
[11] R. D. Mindlin and M. Forray, J. Appl. Phys. **25**, 12–20 (1954).
[12] H. F. Tiersten (unpublished notes).

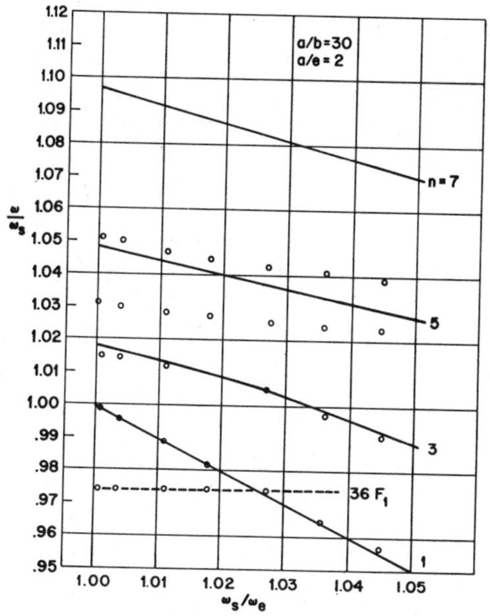

FIG. 3. Measured (circles) and calculated frequencies as a function of mass loading of the electrode for a partially plated AT-cut strip with $a/b=30$ and $a/e=2$.

The boundary conditions at the plate edges are

$$\psi_s(\pm a)=0 \tag{4}$$

and, at the electrode edge,

$$\psi_s(\pm e)=\psi_e(\pm e)$$

and

$$\psi_{s,1}(\pm e)=\psi_{s,1}(\pm e). \tag{5}$$

There are both symmetric and antisymmetric solutions to Eqs. 1 and 2; however, since the electrodes are symmetric with respect to the plate center, only the symmetric solutions produce strong responses. The functions

$$\psi_e=A\cos\xi_e x_1-3\kappa_6 e_{26}V/2b^3\xi_e^2 \tag{6}$$

and

$$\psi_s=B\sin\xi_s x_1+C\cos\xi_s x_1 \tag{7}$$

are solutions of Eqs. 1 and 2, if

$$\xi_e^2=K^2/b^2[(\omega/\omega_e)^2-1]$$

and

$$\xi_s^2=K^2/b^2[(\omega/\omega_s)^2-1].$$

From application of Eqs. 4 and 5,

$$B\sin\xi_s a+C\cos\xi_s a=0,$$

$$A\xi_e\sin\xi_e e+B\xi_s\cos\xi_s e-C\xi_s\sin\xi_s e=0, \tag{8}$$

and

$$A\cos\xi_e e-B\sin\xi_s e-C\cos\xi_s e=3\kappa_6 e_{26}V/2b^3\xi_e^2.$$

Resonance occurs when the determinant of the coefficients of A, B, and C in Eq. 8 goes to zero, or when

$$\tan\xi_e e=\xi_s/\xi_e\cot\xi_s(a-e). \tag{9}$$

The roots of interest in Eq. 9 all lie in the range $\omega>\omega_e$. When $\omega_s>\omega>\omega_e$, ξ_s is imaginary and Eq. 9 becomes

$$\tan\xi_e e=\xi_s'/\xi_e\coth\xi_s'(a-e), \tag{10}$$

where $\xi_s'=i\xi_s$. When $\omega>\omega_s$, both ξ_e and ξ_s are real and the resonant frequencies may be calculated from Eq. 9.

The roots of Eqs. 9 and 10 are shown in Fig. 2 for the first four symmetric modes of a strip in which the electrode covers one-half the strip and $\omega_s/\omega_e=1.02$. The ordinate is $\bar{\Omega}=\omega/\omega_e$ and the abscissa is the strip width-to-thickness ratio a/b. The parameter n corresponds to the number of antinodes of ψ along x_1. The frequency versus length-to-thickness ratio is very similar to that for the plate without an electrode. The major difference is that below $\bar{\Omega}=1.02$, where ξ_s is imaginary and the displacement ψ_s exponentially decays in the region outside the electrode strip. This means that most of the energy associated with a mode of frequency below 1.02 is primarily trapped or bound to the electroded portion of the plate. Above $\bar{\Omega}=1.02$, both ξ_s and ξ_e are real, and the energy of the mode spreads over the entire plate. Since only the deformation under the electrodes contributes to the electrical charge on the electrode, the trapped modes generally give rise to a stronger response than the untrapped modes.

The change in frequency of these same modes for a fixed plate size, $a/b=30$, as a function of the electrode thickness is shown as the solid lines in Fig. 3. At small values of ω_s/ω_e, only the first mode is trapped. When $\omega_s/\omega_e=1.035$, the third mode also becomes trapped. In the case of the electrode strip on an infinite plate, treated by Mindlin and Lee,[6] the thickness–shear resonances do not exist when their frequency is above ω_s. The additional modes whose frequencies are higher than ω_s in Fig. 3 are due to the free edges of the plate. The frequency and particularily the mode shapes depend on the electrode mass and length, as well as the plate length.

The circles in Fig. 3 represent measured resonances in square AT-cut quartz plates with $a/b=c/b=30$, where $2c$ is the plate width along x_3. The plate thickness was 0.5 mm making the fundamental thickness–shear frequency about 3.3 MHz. The electrodes were rectangular with $a/e=2$ and $c/e=3$. The electrodes were initially vacuum-deposited gold or silver. The electrode mass was increased in steps by electroplating nickel over the original vacuum-deposited films. The plates were supported by nickel ribbons soldered to tabs that were brought out to the plate edges normal to the x_3 direction. The frequency and relative amplitude of the resonances were measured in the usual scanning bridge.[13]

The experimental points fit the first and third modes quite well, since both modes are principally thickness–shear. Two sets of points fall on either side of the $n=5$ curve. The upper set of points is principally the fifth

[13] R. D. Mindlin and W. J. Spencer, Proc. 21st Ann. Freq. Control Symp., 3–27 (April 1967).

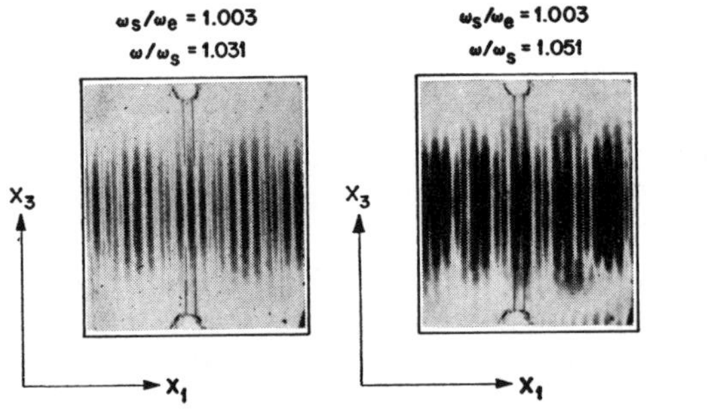

FIG. 4. X-ray diffraction topographs of the fifth anharmonic thickness–shear mode ($\omega/\omega_s = 1.051$) and coupled 38th flexural mode ($\omega/\omega_s = 1.031$). These modes correspond to points at $\omega_s/\omega_e = 1.003$ in Fig. 3.

FIG. 5. The variation of the rotation ψ for the third anharmonic thickness–shear mode as a function of x_1 for various values of electrode mass loading.

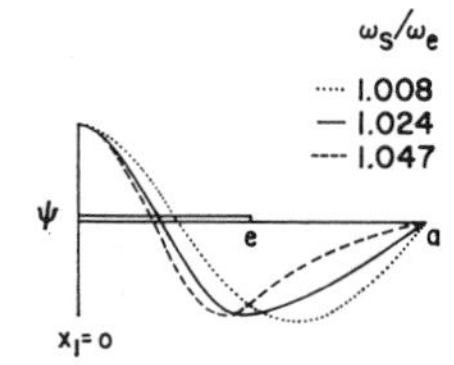

anharmonic thickness–shear mode for small values of ω_s/ω_e, while the lower curve is principally the 38th overtone of flexure. The two modes are coupled and have nearly interchanged their identities for values of ω_s/ω_e above 1.04. The 36th flexural mode is also shown as a dotted line. Neither of these flexural modes is included in the preceeding analysis.

All of the modes were identified by x-ray diffraction topography.[14] This method gives a quick, sensitive, and permanent record of the rotation ψ associated with each resonance. The topographs of two of the modes for $\omega_s/\omega_e = 1.003$, and $\omega = 1.031\omega_s$ and $1.051\omega_s$ are shown in Fig. 4. The dark areas correspond to regions of high diffracted x-ray intensity due to positions of maximum ψ. The topograph at $\omega/\omega_s = 1.05$ shows five major dark regions due to the five antimodes of ψ. Superimposed on this pattern is a banded structure due to the flexural component of the fifth thickness–shear mode. The displacements do not extend over the entire width of the plate because of the finite electrode used to excite the vibration. Fine dark lines are also visible that are due to the edge of the plated electrodes, whereas the larger strained areas at the plate edge are due to the plate support. The mode at $\omega/\omega_s = 1.03$ still has five principal dark regions; however, they are not as pronounced as in the higher frequency mode. The apparent variation in the flexural wave length along x_1 is due to the varying amount of thickness–shear at each point. Note the change from a minimum to a maximum in ψ at the center of the topograph.

At higher length-to-thickness ratios, the simplified theory used here would fit more of the thickness–shear modes since the coupling with flexure would be lower. The fit to the experimental results at the plate dimensions used here could be improved by including the flexural branch in the analysis. The same analysis can be applied to the thickness–twist modes which have displacement similar to the thickness–shear modes but with phase changes along the x_3-direction. The constants in Eqs. 1–10 must be changed, but the form of the solution is similar. These calculations are not presented here.

[14] W. J. Spencer, J. Acoust. Soc. Am. **39**, 929–935 (1966).

It should be noted, however, since the value of K^2 in Eq. 2 is higher for the twist modes than for the shear, the third twist mode will become trapped with less electrode mass than the third shear mode for a square plate with a square electrode.

II. MODE SHAPES

The frequencies of the thickness–shear modes in quartz plates with partial electrodes decrease in an approximately linear manner with the mass of the electrode. Even at the critical frequency $\omega = \omega_s$, there is a smooth uniform change in the frequency with changes in electrode thickness. There is, however, a change in the form of the displacement of the thickness–shear modes as their wave numbers change from real to imaginary in the unelectroded portion of the plate.

The displacement of the various modes may be determined as a function of frequency and x_1 from Eqs. 6 and 7. The constants A, B, and C may be determined from Eq. 8. Since there is no loss included in the analysis, the amplitudes are unbounded at resonance. The constants A, B, and C are

$$A = (3\xi_s\kappa_6 e_{26}V/2b^3\xi_e^2\Delta)\cos\xi_s(a-e),$$

$$B = 3\kappa_6(e_{26}V\xi_s/2b^3\xi_e^2\Delta)\sin\xi_e e\cos\xi_s a,$$

and

$$C = (3\kappa_6 e_{26}V\xi_s/2b^3\xi_e^2\Delta)\sin\xi_e e\sin\xi_s a,$$

where Δ is the determinant of the coefficients given by Eq. 9. This term along with the other frequency-independent terms may be lumped together as part of an arbitrary constant. The remaining terms can then be used to determine the displacements of any of the thickness–shear modes as a function of plate and electrode dimensions and frequency. These calculations do not easily provide information as to the amplitude of one mode with respect to another. It is possible to determine this information in terms of the equivalent electrical parameters that are discussed in the next section.

The change in the displacement of the third anharmonic thickness–shear mode as a function of electrode thickness is shown in Fig. 5. Only half of the displacement is shown, since the mode is symmetric about $x_1 = 0$. The electrode is one-half the length of the plate, and the plate length-to-thickness ratio is 30. When the electrode is thin, $\omega_s/\omega_e = 1.008$, and the mode shape is nearly what would be expected of a plate without an electrode. At $\omega_s/\omega_e = 1.035$, ξ_s becomes imaginary, and the displacement begins to decay in the unelectroded portion of the plate. For $\omega_s/\omega_e \gg 1$, the mode depends primarily on the length of the electrode alone.

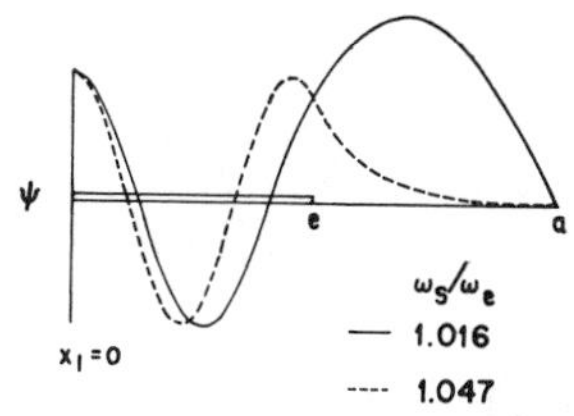

FIG. 6. The variation of the rotation ψ for the fifth anharmonic thickness–shear mode as a function of x_1 for different values of electrode mass loading.

The effect of the electrode on the mode shape is even more apparent for the higher overtones. Figure 6 shows the fifth anharmonic mode in this case when $a/e=2$ and $a/b=60$, with two different electrode masses. When $\omega_s/\omega_e=1.016$, ξ_s is real and the displacement is sinusoidal over the entire plate. The antinode that is outside the electroded region is wider, since $\xi_s<\xi_e$, and higher in amplitude than the antinodes under the electrode. When the mode is trapped, $\omega_s/\omega_e=1.047$, this is no longer the case. The larger amplitude in the surrounding region, due to the absence of the loading of the electrode, does not contribute to the charge on the electrode and tends to make the electrical response of this mode appear even weaker, than if the displacement amplitudes were uniform along x_1.

The increase in amplitude in the unelectroded portion of the plate has been observed experimentally. Figure 7 shows an x-ray diffraction topograph of the fifth anharmonic mode that is not trapped. The electrode edges are apparent as dark lines normal to the x_1 axis. The electrodes cover half of the plate. The antinodes of ψ due to the thickness–shear displacements near the plate edge are much darker than the central antinodes indicating a larger rotation amplitude in this unelectroded region.

III. MODE SUPPRESSION

The displacement associated with the resonant modes discussed in the preceding section determine to a large extent the electrical charge on the plate electrodes in a given piezoelectric material, and thus, the relative strength of the particular mode. The strength of a mode is usually expressed in terms of its equivalent electrical parameters. An equivalent circuit for a lossless piezoelectric resonator is shown in Fig. 8. The C_n and L_n represent the equivalent motional parameters near the nth mode. C_0 is the static capacitance.

The impedance of the circuit in Fig. 8 goes to zero at each resonant frequency ω_n. If each of the modes is assumed to have equal values of $Q=\omega L_n/R_n$, where Q is large but finite, then the relative strength of each mode measured in any of the usual methods is related to L_n or to the ratio of C_n/C_0. The larger the value of C_n/C_0, the lower the impedance of the particular resonant mode and the stronger this mode will appear. Similarly the smaller the value of L_n, the lower the impedance of the nth mode.

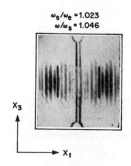

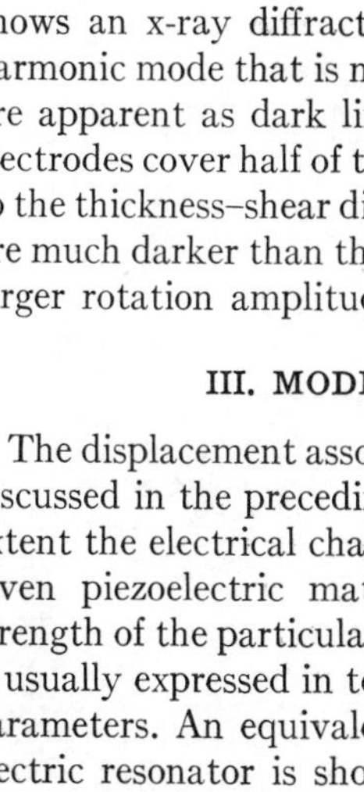

FIG. 7. X-ray diffraction topograph of the fifth anharmonic thickness–shear mode showing the higher amplitude of the rotation ψ in the unelectroded portion of the quartz plate.

The equivalent inductance L of the lowest thickness–shear mode for a given area A in an infinite plate may be shown to be[15]

$$L=\rho b^3/2Ae^2_{26},$$

or for the AT-cut quartz plate

$$L=0.82/Af^3, \tag{11}$$

where A is in square centimeters and f is in megahertz. The result of Eq. 11 does not include any affect due to changes in displacement in the plane of the plate. A more realistic value for the equivalent inductance of a finite electrode on an infinite plate is

$$L=1.8/Af^3. \tag{12}$$

For thickness overtones of the lowest thickness–shear mode, Eq. 12 should be multiplied by m^3, where m is the order of the overtone.

In the case of a finite plate completely covered with a thin electrode, the equivalent inductance may be obtained from Eq. 3. The charge on the electrode is given by

$$q=\int_{-e}^{e} D_2^{(0)}dx_1.$$

Since $I=\dot{q}=YV$ then, using Eq. 6, the admittance Y becomes

$$Y=i\omega\epsilon_{22}e/b-2i\omega(\kappa_6e_{26}/\xi_e)A\sin\xi_e. \tag{13}$$

As $\omega\to 0$, the admittance becomes

$$Y_0=i\omega\epsilon_{22}e/b,$$

which is the static capacitance per unit width of the electroded plate. The inductance at the resonant frequencies ω_n for any series resonant circuit is given approximately by

$$dZ/d\omega\,|_{Z=0}=2L,$$

where $Z=Y^{-1}$ is the impedance of the resonator that, in this analysis, is purely reactive. Then by differentiation of Eq. 13, setting $\sin\xi_e=0$ and $\cos\xi_e=1$, the inductance of the nth anharmonic overtone of thickness–shear is

$$L_n=n^2L. \tag{14}$$

In addition,

$$C_n=C/n^2,$$

where L is given by Eq. 12. The inductance of the an-

[15] R. A. Sykes, W. L. Smith, and W. J. Spencer, IEEE Intern. Conv. Record, 78–93 (1967).

harmonic modes that are well trapped should also be given to a good approximation by Eq. 14.

When the mode is not trapped or when the electrode only partially covers the plate, the inductance or the ratio of capacitances may be strongly affected by the free plate edges. An equation similar to Eq. 13 could be obtained for the AT-cut quartz plate with a partial electrode, and the equivalent electrical parameters predicted to the extent that the preceding analysis is valid. An alternative approach is to use the results obtained by Lewis[9] to calculate C_n for a piezoelectric body of arbitrary shape, when the displacements u_i are known. The expression given by Lewis is

$$C_n=\left(\int_A [D_i^n]n_i dA\right)^2 \Big/ \left(\rho\omega_n^2\int_V u_i^{(n)}u_i^{(n)}dV\right), \quad (15)$$

where u_i^n and D_i^n are the particle displacements and the electric displacements due to the nth mode. This expression does not include the mass loading of the electrodes. Including the effect of the electrode with mass per unit area ρ_s, Eq. 15 becomes[16]

$$C_n=\left(\int_A [D_i^n]n_i dA\right)^2 \Big/ \left(\int_V \rho\omega_n^2 u_i^n u_i^n dV+\int_A \rho_s\omega_n^2 u_i^n u_i^n dV\right). \quad (16)$$

This expression may now be used to compute the motional capacitance as a function of the plate and electrode dimensions including the mass of the electrodes with the displacements obtained in the first section.

The values of C_n/C_0 for the first four symmetric modes have been calculated as a function of electrode thickness ω_s/ω_e from Eq. 16 for the geometry in Fig. 1 with $a/b=30$ and $a/e=2$. The results are plotted in Fig. 9. There is very little effect of electrode thickness on C_1/C_0, except when the electrode is extremely thin. The capacitance ratios of the other modes, however, are quite strongly affected by changing the electrode mass for a particular electrode-length to plate-length ratio. The ratio C_3/C_0 starts at a relatively high value for very thin electrodes, goes to zero at about $\omega_s/\omega_e=1.02$, and returns to nearly its initial value as ω_s/ω_e increases. If this mode were measured as a function of increasing the electrode thickness, it would initially appear as a very strong mode, nearly disappear for $\omega_s/\omega_e=1.02$, and then become a strong mode again for values of $\omega_s/\omega_e>1.04$.

The curves in Fig. 3 show that ξ_s becomes imaginary for this mode at a frequency ratio of about 1.035. The theory of "energy trapping" based on a finite electrode on an infinite plate would predict that the $n=3$ mode would not be present for $\omega_s/\omega_e<1.035$ and would occur as a strong mode for values larger than 1.035. In reality, for the case of finite plates, all of these modes are present, as well as many others, and their relative amplitudes depend on the plate length as well as the electrode dimensions.

[16] P. Lloyd, Bell System Tech. J. 46, 1881–1900 (1967).

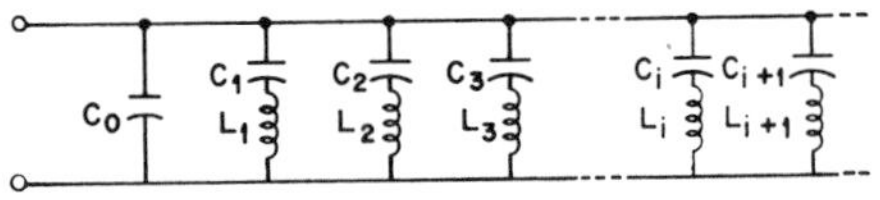

FIG. 8. Equivalent electrical circuit of a piezoelectric crystal.

In Fig. 9, C_3 goes to zero at $\omega_s/\omega_e=1.02$ due to charge cancellation. That is, at this particular ratio of plate and electrode dimensions, should the plate be undergoing free vibration with its electrodes short circuited, the positive and negative charge induced on the electrode due to reversals in the rotation ψ would exactly cancel, and no current would flow in the external circuit. In terms of the equivalent circuit values, $C_n\rightarrow 0$, $L_n\rightarrow\infty$, and $Y\rightarrow 0$.

A similar effect takes place at $\omega_s/\omega_e=1.045$ for C_7. In general, as may be seen from Eq. 16, there can be $(n-1)/2$ values of ω_s/ω_e for which $C_n=0$ for a given mode. The magnitude of C_n for other values of the electrode thickness will depend on the ratio of the plate-to-electrode length. In fact, as the fraction of the plate covered by the conducting electrode becomes smaller, the lower will be the values of C_n, when ξ_s is real for the nth mode.

The values of C_1 and C_3 for a plate with $a/b=90$ and $a/e=3$ have been calculated as a function of electrode thickness from Eq. 16 and are shown in Fig. 10. In this case, even for very thin electrodes, ξ_s is imaginary

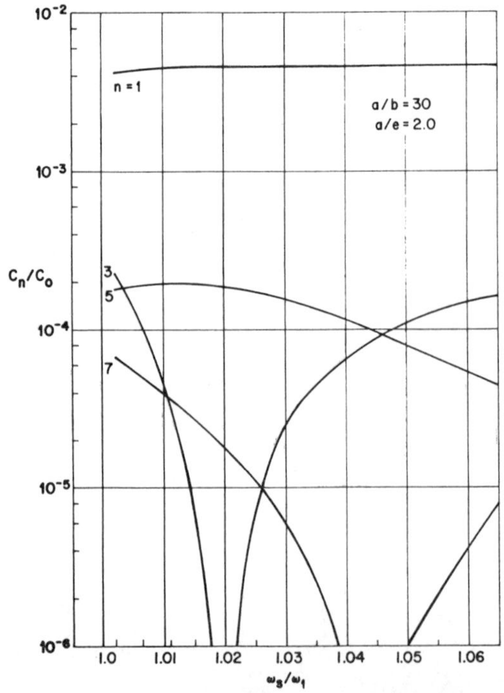

FIG. 9. Ratio of capacitances C_n/C_0 as a function of mass loading calculated from Eq. 16 for an infinite AT-cut quartz strip with an electrode covering half the plate surfaces.

for both the $n=1$ and $n=3$ modes, so that the value of C_3 increases as ω_s/ω_e is increased and approaches a value of $\frac{1}{9}C_1$ as predicted by Eq. 14. The other impedances were too high to be measured by present techniques. The experimental points plotted in Fig. 10 were obtained by plating successively thicker gold films by vapor deposition on several different quartz plates. To ensure that the films were of uniform thickness, the electrodes were evaporated from a single filament with a filament-to-plate distance approximately 10 times the electrode dimensions. Attempts to produce the same results by successive electroplatings were unsuccessful. The values of C_1 were relatively insensitive to the method of plating the film, but C_3 was much lower for the electroplated films. The electroplating is believed to be thicker near the film edge, which would affect the mode shape and hence the value of C_n. Usually, in the design of this type of resonator, it is desirable to make C_1 as large as possible and the other C_n as small as possible. Electroplated films appear to be better for this sort of suppression than are vapor-deposited films.

Similar experiments have been performed on plates having the dimensions used to compute the curves in Fig. 9. The impedances of the modes in the plates that can be fabricated are too high to measure with current techniques. However, the mode strengths qualitatively follow the results shown in Fig. 9.

The results given here, of course, are not complete. They include only the effects of a single mode and only one of the plate dimensions. From the topographs shown in Figs. 4 and 6, the flexural components are certainly present and will affect the values of C_n particularly for the larger values of n. The third boundary will also affect the equivalent electrical parameters.

IV. EFFECT OF OTHER MODES

The frequency of the fifth anharmonic mode in Fig. 3 was higher than predicted by the simple theory in Sec. I due to coupling with the 38th overture of flexure. At low length-to-thickness ratios, the coupling between flexure and shear will be the stronger, and the effect on both the frequency and equivalent electrical parameters greater. Although the results shown here indicate only the effect on the higher order thickness–shear modes, the fundamental mode will also be affected, particularly if the plate dimensions are chosen to coincide with a flexure or flexure-twist overtone.[13] The coupled flexural mode will produce only small changes in frequency but can radically affect the electrical impedance of the thickness–shear mode. This may show up as a change in the series resonant resistance as a function of temperature or as erratic behavior when the amplitude of the driven mode is changed.

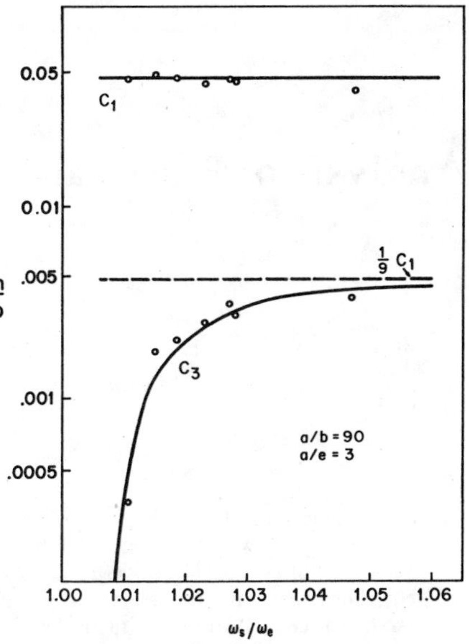

FIG. 10. Experimental and calculated ratio of capacitances for a square AT-cut quartz plate with $a/b=90$ and $a/e=3$.

Some of these effects could be analyzed by including the flexural branch in the theory. The case of a finite rectangular electrode on an infinite plate could probably be handled although the number of boundary conditions to be satisfied becomes rather large. A more interesting and somewhat simpler problem would be to analyze the electrode strip on a plate including edges normal to both the x_1 and x_3 axis.

Recent work has shown that both the thickness–shear and flexural modes and their twist overtones can be predicted very well as a function of both plate dimensions by a relatively simple theory.[13] The next step would be to include the effect of an electrode strip covering the plates completely along the x_3 direction and partially along the x_1 direction. This geometry has the advantage that all the twist overtones of thickness–shear and flexure should be suppressed by charge cancellation. The twist overtones of both shear and flexure depend on x_3 as follows:

$$\omega = f(x_1)\cos\zeta x_3,$$

where $\zeta = n\pi/2c$, $n=0, 2, 4\cdots$. For all n with the exception of $n=0$, the charge due to this type of displacement should cancel, if the electrodes extend completely to the plate edge in the x_3 direction. The rules discussed in the previous section could then be applied to suppress the anharmonic overtones of thickness–shear. This should permit the fabrication of a singly resonant plate with extremely low impedance since the electrode and plate length could be extended in the x_3 direction without introducing additional modes.

Received 12 December 1967

Analysis of Elastically Coupled Piezoelectric Resonators*

W. D. Beaver

Collins Radio Company, Newport Beach, California 92663

This paper presents an analysis of an elastically coupled piezoelectric resonator system consisting of a linear array of two or more counter electrode pairs placed on the major surfaces of a quartz crystal plate. Each pair of counter electrodes establishes a resonator having a transverse thickness mode of motion. The equations describing this system are derived and applied to an array of n coupled resonators and specific numerical computations are made for the cases where n is equal to 2, 3, and 6. These results are plotted and show the dependence of the effective interresonator coupling on the dimensions of the electrode configuration.

LIST OF SYMBOLS

c_{pq}	components of elastic moduli (p, q = 1 to 6)
c_{11}	(AT-cut quartz) $8.674(10^{10})$ N/m^2
c_{12}	(AT-cut quartz) -8.2246 (10^{9})
c_{22}	(AT-cut quartz) 1.2966 (10^{11})
c_{55}	(AT-cut quartz) 6.8827 (10^{10})
c_{66}	(AT-cut quartz) 2.8988 (10^{10})
$D^{(0)}_i$, $D^{(1)}_i$, $D^{(2)}_i$	two-dimensional components of electric displacement (i, j = 1, 2, 3)
e_{ip}	components of piezoelectric moduli
e_{11}	(AT-cut quartz) 0.171 C/m^2
e_{12}	(AT-cut quartz) -0.15233
e_{25}	(AT-cut quartz) 0.10718
e_{26}	(AT-cut quartz) -0.09595
E_i	electric-field vectors
f	frequency, hertz
f_{up}	frequency of unelectroded crystal plate
f_{ep}	frequency of electroded resonator
f_1, f_2	in-phase and out-of-phase resonant frequencies of a two-resonator system
h	half-thickness of crystal plate
R	mass loading which is equal to the ratio of combined mass per unit area of both electrodes to mass per unit area of the plate
$S_q^{(0)}$, $S^{(1)}_q$	two-dimensional components of strain (q = 1, 2, 3, 4, 5, 6)
$T^{(0)}_p$, $T^{(1)}_p$	two-dimensional components of stress (p = 1, 2, 3, 4, 5, 6)
$u^{(0)}_i$, $u^{(1)}_i$	two-dimensional components of acoustic displacement
x_k	Cartesian coordinate (k = 1, 2, 3)
δ_{ij}	Kronecker delta
ϵ_{ij}	components of the dielectric constants
ζ_e, ζ_u	wavenumber for the thickness–twist mode in the electroded and unelectroded regions of the crystal plate
κ_4, κ_6	shear correction factors (Ref. 4)
ξ_e, ξ_u	wavenumbers for the thickness–shear mode in the electroded and unelectroded retions of the crystal plate
ρ	density of quartz
$\phi^{(0)}$, $\phi^{(1)}$	components of electrical potential
$\Phi^{(0)}$	driving surface potential on electrodes
ω	angular frequency
ω_e, ω_u	frequency of infinite electroded and unelectroded plates

Notation: (1) Subscript e or u designates variables and parameters as being characteristic of either the electroded or unelectroded regions of the crystal plate. (2) Commas denote differentation with respect to the independent variables; e.g., $u_{k,i} = \partial u_k / \partial x_i$. (3) The usual summation convention is used—i.e., repeated indices indicate a sum.

* This work was performed while the author was employed by Bell Telephone Labs., Inc., Allentown, Pa. It is part of a dissertation presented to the graduate faculty of Lehigh Univ. in partial fulfillment of the requirements for the degree of PhD in Electrical Engineering.

Reprinted with permission from *J. Acoust. Soc. Amer.*, vol. 43, pp. 972–981, May 1968.

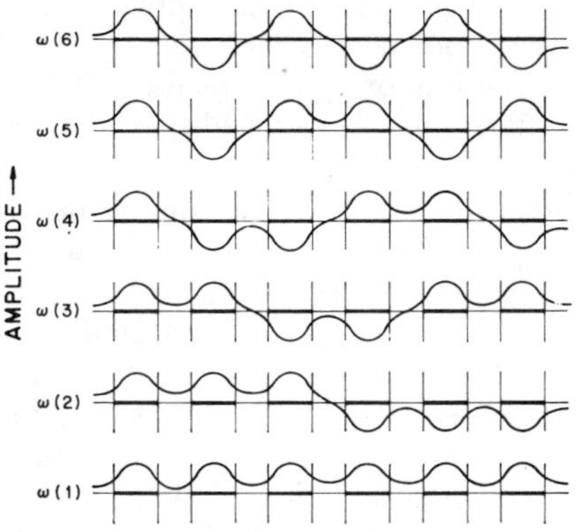

FIG. 1. Representation of the amplitude of the wavefunctions of the acoustic displacement for a six-coupled resonator system.

INTRODUCTION

THIS paper concerns the analysis of a system of elastically coupled piezoelectric resonators where both the resonators and the coupling medium are embodied in a single plate of quartz (the analysis should hold equally well for similar piezoelectric materials of the trigonal crystal class *D*3). Each resonator is established by a pair of electrodes mounted directly opposite each other on the two major surfaces of the crystal plate. The resonators can have either a thickness–shear, or thickness–twist mechanical mode of motion. The inverse piezoelectric effect is employed to establish a standing wave of the appropriate thickness mode within the input resonator. The interresonator coupling is achieved by controlled leakage of acoustical energy from resonator to resonator via the crystalline medium. Owing to the presence of the metallic electrode, the cutoff frequency of the thickness mode is lower in an electroded portion of the crystal plate than in the unelectroded portion, and therefore the standing wave set up in an electroded portion is nonpropagating in the unelectroded portion. In the unelectroded portion there is an exponential decay of the amplitude of acoustical displacement. The rate of this decay is a function of the difference between the resonant frequency of the thickness mode in the electroded and unelectroded portions of the plate and of the distance from the edge of the electrode. The method of controlling the coupling[1] between resonators is to control the separation between electrodes and the difference between the resonant frequencies of thickness vibration for the electroded and unelectroded regions. This frequency difference is related to the mass loading of the electrode. The coupling between resonators may be adjusted to obtain the desired transmission of acoustic energy from input to output resonator. The direct piezoelectric effect is then employed to transform the acoustical wave in the output resonator into an electrical signal.

[1] R. A. Sykes and W. D. Beaver, "High Frequency Monolithic Filter," Trans. Ann. Freq. Control Symp., 20th (Apr. 1966).

The physical system described above is that of the monolithic crystal filter,[1] which is an electric wave filter now finding extensive use in communications.

In this paper, the approximate equations that describe the system are presented. Solutions to these equations are found which satisfy prescribed boundary conditions for a general system consisting of n coupled resonators. Specific calculations are given for systems consisting of two, three, and six coupled resonators. These illustrate the dependence of the inter resonator coupling upon the electrode configuration and the mass loading.

I. COUPLED RESONATORS

The resultant motion of each component of a system with two or more coupled components for the case of small displacements from the equilibrium positions can be considered as a linear superposition of simple harmonic oscillations. These oscillations will be at characteristic frequencies known as coupled or "normal" frequencies. For a nondegenerate, nondissipative coupled system of n oscillators each having one degree of freedom, there will exist n normal frequencies. The above description fits very closely the physical system being analyzed, where the 1 deg of freedom is the fundamental mode of vibration. The thickness–shear

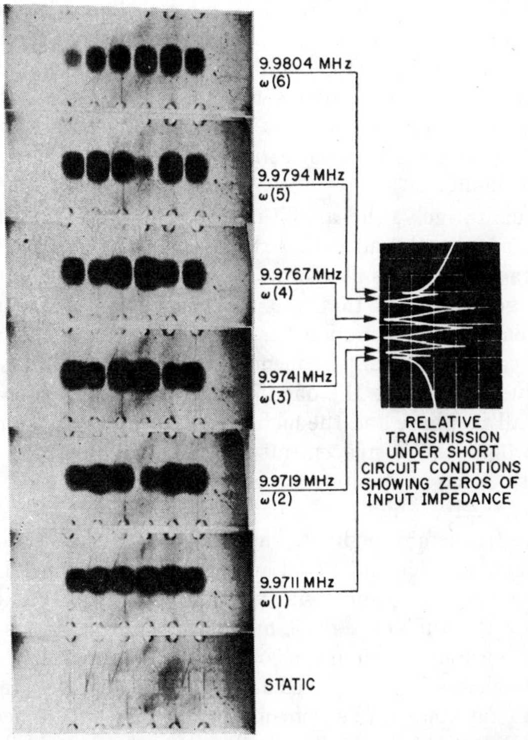

FIG. 2. X-ray topographs of the six normal resonances having the thickness–twist mode of the six coupled resonator system in *AT*-cut quartz plate. Maximum darkening in the topographs corresponds to the maximum amplitude of the acoustical displacement. The light areas are quiescent or nodal regions. The dark lines in these light regions are defects in the quartz.

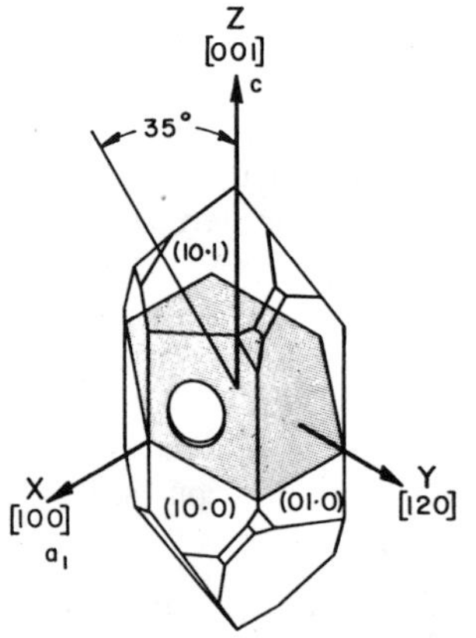

FIG. 3. Orientation of AT-cut quartz.

and thickness–twist modes in rotated Y-cut quartz plates have very high Q resonances; the particle motion is essentially one-dimensional with displacement in the x crystallographic direction. Also, for each coupled resonator a normal resonant frequency is found. Thus a system composed of n coupled resonators has n normal resonant frequencies distributed about the frequency of the fundamental thickness resonance. The normal modes are separated by frequency differences which decrease as the mass loading increases or the interresonator spacing increases, so that in the limits when the interresonator spacing and/or the mass loading become very great the normal frequencies coincide as is demonstrated in later sections. For an AT-cut quartz plate with mass loading of 0.02, the interresonator spacing would have to be greater than 35 crystal-plate thicknesses at 10 MHz to decrease the coupling to the extent that there would be less than 1-Hz frequency difference between the normal resonant modes.

Qualitatively, the amplitudes of the wavefunctions representing the acoustical displacements in a system of six resonators is illustrated in Fig. 1. In keeping with the general rule[2] that the more nodes or zeros an oscillatory wavefunction has, the higher is its energy and natural resonant frequency, we see that the lowest-frequency mode is the one in which the displacements are all in phase and the highest-frequency mode is one in which the displacements are sequentially out of phase. Each of the four wavefunctions in between has, respectively, a higher number of nodes than its nearest lower-frequency neighbor. This correspondence between frequency order and nodal arrangement is verified by the x-ray topographs[3] shown in Fig. 2. This Figure shows the thickness–twist mode shapes of a six-resonator system driven in each of its six normal resonances (with electrode pairs placed on an AT-cut quartz plate). The x-ray topographs were obtained using Bragg reflection from the crystallographic plane having Miller indices ($2\bar{1}.0$). The darkening is approximately proportional to the tranverse displacement of the plane for the thickness-twist or –shear modes. Therefore, the maximum darkening corresponds to the maximum displacement.

The orientation of the AT-cut quartz plate relative to the crystallographic directions are shown in Fig. 3. The X, Y, and Z axes are crystallographic axes corresponding to the [100], [120], and [001] directions, respectively. The numbers in brackets are the Miller indices of the associated planes. Thus, the coordinate variables x_1, x_2, and x_3, that are used in the analysis, are aligned such that x_1 coincides with X, but x_2, and x_3 are rotated 35° from the Y and Z axes, respectively.

The form of the mechanical displacement during resonance is shown for the thickness–twist mode in Fig. 4 and for the thickness–shear mode in Fig. 5. Both of these Figures show the in-phase and out-of-phase conditions of two coupled resonators. They also show that resonators coupled via the thickness–twist mode are arrayed along the x_3 axis and resonators coupled via the thickness–shear mode are arrayed along the x_1 axis. Thickness–twist waves in a plate are those in which the displacement is parallel to the middle plane of the plate and normal to the direction of propagation, and thickness–shear waves in a plate are those in which the displacement is parallel to both the middle plane of the plate and the direction of propagation.

II. ANALYSIS OF ELASTICALLY COUPLED PIEZOELECTRIC RESONATORS

The objectives of the present analysis are to obtain the normal resonant frequencies for a particular resonator configuration, or, by supplying a set of normal

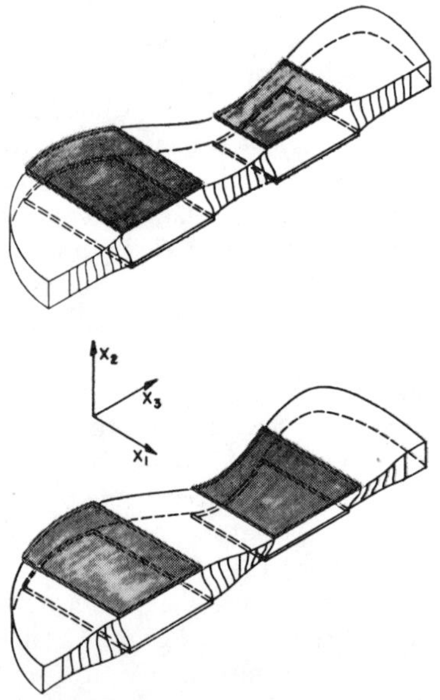

FIG. 4. Thickness–twist deformation for (a) in-phase and (b) out-of-phase coupling of two resonators.

[2] J. C. Slater, *Quantum Theory of Atomic Structure* (McGraw-Hill Book Co., New York, 1960), Vol. I.

[3] W. J. Spencer, "Modes in Circular Quartz Plates," IEEE Trans. Sonics Ultrasonics **SU-12**, No. 1, 1–5 (1965).

frequencies and the ratios of the interresonator couplings, to obtain the actual interresonator dimensions. The analysis is performed separately for the thickness–twist mode in which the resonators are coupled along the x_3 axis and the thickness–shear mode in which the x_1 axis is the coupling direction. The equations of the approximate theory in Refs. 4 and 5 serve as the starting point for this analysis.

We assume the crystal plate and electrode geometry shown in Fig. 6, and neglect edge effects, which are small, by assuming an infinite plate. The coordinate axes are listed for coupling via the thickness–shear, or thickness–twist mode.

The stress–strain relations for a rotated Y-cut quartz plate when referred to the rectangular Cartesian coordinate system, x_1, x_2, x_3, with x_1 a digonal axis and $x_2=0$ the middle plane of the plate in Fig. 6, exhibit effective monoclinic symmetry.[6] In Refs. 4 and 5, the electric and acoustic displacements, D and u, are expanded in a power series of the thickness variable as follows:

$$D_j=\sum_{n=0}^{g} x^n{}_2 D^{(n)}{}_j+\delta_{2j}x^{g+1}{}_2 D^{(g+1)}{}_2, \quad (1)$$

and

$$u_j=\sum_{n=0}^{g} x^n{}_2 u^{(n)}{}_j=u^{(0)}{}_j+x_2u^{(1)}{}_j +x^2{}_2u^{(2)}{}_j+\cdots+x^g{}_2u^{(g)}{}_j, \quad (2)$$

where g is made equal to unity for this analysis. The bracketed superscripts in Eqs. 1 and 2 denote the order of the coefficients in these expansions. The stress components are given by the equations

$$T^{(0)}{}_p=2h\bar{c}_{pq}S^{(0)}{}_q-\bar{e}_{ip}E^{(0)}{}_i \quad (3)$$

and

$$T^{(1)}{}_r=\tfrac{2}{3}h^3\gamma_{rs}S^{(1)}{}_s-\Psi_{ir}E^{)1)}{}_i. \quad (4)$$

The electric potentials are defined by the equations

$$\phi^{(n)}\equiv\int_{-h}^{h} x^n{}_2\phi dx_2 \quad (5)$$

and

$$\Phi^{(n)}=[x^n{}_2\phi]^h{}_{-h}. \quad (6)$$

The stress equations of motion, along with the electric displacement equations for a plate with traction-free

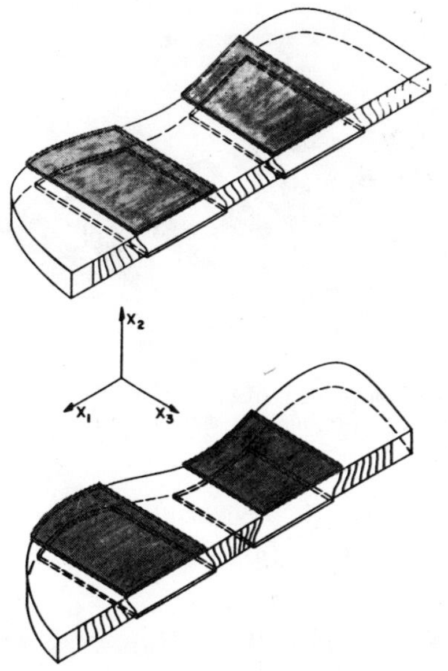

FIG. 5. Thickness–shear deformation for (a) in-phase and (b) out-of-phase coupling of two resonators.

surfaces, may be written as follows:

$$\begin{aligned}
T^{(0)}{}_{1,1}+T^{(0)}{}_{5,3}&=2h\rho\ddot{u}^{(0)}{}_1,\\
T^{(0)}{}_{6,1}+T^{(0)}{}_{4,3}&=2h\rho\ddot{u}^{(0)}{}_2,\\
T^{(0)}{}_{5,1}+T^{(0)}{}_{3,3}&=2h\rho\ddot{u}^{(0)}{}_3,\\
T^{(1)}{}_{1,1}+T^{(1)}{}_{5,3}-T^{(0)}{}_6&=\tfrac{2}{3}\rho h^3\ddot{u}^{(1)}{}_1,\\
T^{(1)}{}_{5,1}+T^{(1)}{}_{3,3}-T^{(0)}{}_4&=\tfrac{2}{3}\rho h^3\ddot{u}^{(1)}{}_3,\\
D^{(0)}{}_{1,1}+D^{(0)}{}_{2,2}+D^{(0)}{}_{3,3}+D^{(1)}{}_2&=0,
\end{aligned} \quad (7)$$

and

$$D^{(1)}{}_{1,1}+D^{(1)}{}_{2,2}+D^{(1)}{}_{3,3}+2D^{(2)}{}_2=0.$$

The constitutive relations for the stress and electric displacements in terms of the electric potentials, ϕ, and the acoustical displacements, u, are listed as follows[7]:

$$\begin{aligned}
T^{(0)}{}_1&=2h[\bar{c}_{11}u^{(0)}{}_{1,1}+\bar{c}_{14}\kappa_4(u^{(0)}{}_{2,3}+u^{(1)}{}_3)+\bar{c}_{13}u^{(0)}{}_{3,3}]+\bar{e}_{11}\phi,^{(0)}{}_1,\\
T^{(0)}{}_3&=2h[\bar{c}_{13}u^{(0)}{}_{1,1}+\bar{c}_{34}\kappa_4(u^{(0)}{}_{2,3}+u^{(1)}{}_3)+\bar{c}_{33}u^{(0)}{}_{3,3}]+\bar{e}_{13}\phi,^{(0)}{}_1,\\
T^{(0)}{}_4&=2h\kappa_4[\bar{c}_{14}u^{(0)}{}_{1,1}+\bar{c}_{44}\kappa_4(u^{(0)}{}_{2,3}+u^{(1)}{}_3)+\bar{c}_{34}u^{(0)}{}_{3,3}]+\bar{e}_{14}\kappa_4\phi,^{(0)}{}_1,\\
T^{(0)}{}_5&=2h[\bar{c}_{56}\kappa_6(u^{(0)}{}_{2,1}+u^{(1)}{}_1)+\bar{c}_{55}(u^{(0)}{}_{1,3}+u^{(0)}{}_{3,1})]+\bar{e}_{25}\Phi^{(0)}+\bar{e}_{35}\phi,^{(0)}{}_3,\\
T^{(0)}{}_6&=2h\kappa_6[\bar{c}_{66}\kappa_6(u^{(0)}{}_{2,1}+u^{(1)}{}_1)+\bar{c}_{56}(u^{(0)}{}_{1,3}+u^{(0)}{}_{3,1})]+\bar{e}_{26}\kappa_6\Phi^{(0)}+\bar{e}_{36}\kappa_6\phi,^{(0)}{}_3,\\
T^{(1)}{}_1&=\tfrac{2}{3}h^3(\gamma_{11}u^{(1)}{}_{1,1}+\gamma_{13}u^{(1)}{}_{3,3})+\Psi_{11}\phi,^{(1)}{}_1,\\
T^{(1)}{}_3&=\tfrac{2}{3}h^3(\gamma_{13}u^{(1)}{}_{1,1}+\gamma_{33}u^{(1)}{}_{3,3})+\Psi_{13}\phi,^{(1)}{}_1,\\
T^{(1)}{}_5&=\tfrac{2}{3}h^3\gamma_{55}(u^{(1)}{}_{1,3}+u^{(1)}{}_{3,1})+\Psi_{25}(\Phi^{(1)}-\phi^{(0)})+\Psi_{35}\phi,^{(1)}{}_3,
\end{aligned} \quad (8)$$

[4] H. F. Tiersten and R. D. Mindlin, "Forced Vibrations of Piezoelectric Crystal Plates," Quart. Appl. Math. **20**, No. 2, 107–119 (1962).

[5] R. D. Mindlin and P. C. Y. Lee, "Thickness Shear and Flexural Vibration of Partially Plated, Crystal Plates," Intern. J. Solids and Structures **2**, 125–139 (1966).

[6] W. P. Mason, *Piezoelectric Crystals and Their Application to Ultrasonics* (D. Van Nostrand Co., Inc., New York, 1950).

[7] These relations were independently obtained by H. F. Tiersten and cross checked with these equations. It should also be pointed out that R. D. Mindlin derived a similar set of equations for a purely elastic system which were published in *An Introduction to the Mathematical Theory of the Vibrations of Elastic Plates* (U. S. Army Signal Corps. Eng. Lab., Fort Monmouth, N. J., 1955), 99–100. Signal Corps Contract DA-36-039 SC-56772.

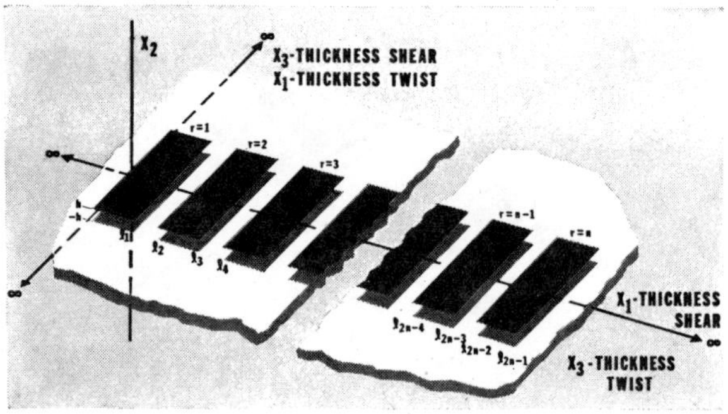

FIG. 6. The electrode configuration and location of coordinate axes of a general n resonator system.

and

$$D^{(0)}{}_1=\bar{e}_{11}u^{(0)}{}_{1,1}+\bar{e}_{14}\kappa_4(u_{2,3}+u^{(1)}{}_3)+\bar{e}_{13}u^{(0)}{}_{3,3}-\frac{\bar{\epsilon}_{11}}{2h}\phi,^{(0)}{}_1,$$

$$D^{(0)}{}_2=\bar{e}_{26}\kappa_6(u^{(0)}{}_{2,1}+u^{(1)}{}_1)+\bar{e}_{25}(u^{(0)}{}_{1,3}+u^{(0)}{}_{3,1})+\frac{\bar{\epsilon}_{22}}{4h}\left(\frac{11}{2}\Phi^{(0)}-\frac{15}{h^2}\phi^{(1)}\right)-\frac{\bar{\epsilon}_{23}}{2h}\phi,^{(0)}{}_3,$$

$$D^{(0)}{}_3=\bar{e}_{36}\kappa_6(u^{(0)}{}_{2,1}+u^{(1)}{}_1)+\bar{e}_{35}(u^{(0)}{}_{1,3}+u^{(0)}{}_{3,1})-\frac{1}{2h}[\bar{\epsilon}_{23}\Phi^{(0)}+\bar{\epsilon}_{33}\phi,^{(0)}{}_3]-\frac{15\bar{\epsilon}_{23}}{4h^3}\phi^{(1)}+\frac{15\bar{\epsilon}_2}{8h}g^{(0)},$$

$$D^{(1)}{}_1=\Psi_{11}u^{(1)}{}_{1,1}+\Psi_{13}u^{(1)}{}_{3,3}-\frac{3\zeta_{11}}{2h^3}\phi,^{(1)}{}_1, \tag{9}$$

$$D^{(1)}{}_2=\Psi_{25}(u^{(1)}{}_{1,3}+u^{(1)}{}_{3,1})-\frac{3}{2h^3}[\zeta_{22}(\Phi^{(1)}-\phi^{(0)})+\zeta_{23}\phi,^{(1)}{}_3],$$

$$D^{(1)}{}_3=\Psi_{35}(u^{(1)}{}_{1,3}+u^{(1)}{}_{3,1})-\frac{3}{2h^3}[\zeta_{23}(\Phi^{(1)}-\phi^{(0)})+\zeta_{33}\phi,^{(1)}{}_3],$$

and

$$D^{(2)}{}_2=\frac{15}{4h^3}\bar{\epsilon}_{22}\left(\frac{3}{h^2}\phi^{(1)}-\Phi^{(0)}\right)+\frac{15}{8h^3}\bar{\epsilon}_{23}\phi,^{(0)}{}_3.$$

In these equations,

$$\begin{aligned}
\bar{c}_{pq}&=c_{pq}-c_{p2}c_{2q}/c_{22},\\
\bar{e}_{ip}&=e_{ip}-e_{i2}c_{p2}/c_{22},\\
\bar{\epsilon}_{ij}&=(9/4)\epsilon_{ij}+e_{i2}e_{j2}/c_{22},\\
\gamma_{rs}&=c_{rs}-c_{rw}c_{vs}/c_{vw},\\
\Psi_{ir}&=e_{ir}-e_{iv}c_{rw}/c_{vw},\\
\zeta_{ij}&=\epsilon_{ij}+e_{iv}e_{jw}/cv_{w},
\end{aligned} \tag{10}$$

and

$$p,\ q=1,\ 2,\ 3,\ 4,\ 5,\ 6;\quad i,\ j=1,\ 2,\ 3;$$
$$r,\ s=1,\ 3,\ 5;\qquad v,\ w=2,\ 4,\ 6.$$

III. THICKNESS–TWIST MODE

In the first calculation, we wish to consider the thickness–twist mode with the resonators coupling along the x_3 axis. We will therefore assume the following acoustical displacements:

$$\begin{aligned}
u^{(0)}{}_1&=u^{(0)}\exp(j\omega t),\\
u^{(1)}{}_1&=u^{(1)}\exp(j\omega t),\\
\text{and}\quad u^{(0)}{}_2&=u^{(0)}{}_3=u^{(1)}{}_3=0,
\end{aligned} \tag{11}$$

where $u^{(0)}$ and $u^{(1)}$ are assumed to be functions of x_3 alone.

The nonzero components are the thickness–twist, $u^{(1)}{}_1$, and the face shear, $u^{(0)}{}_1$.

The potentials $\phi^{(0)}$ and $\Phi^{(1)}$ may be taken as zero, leaving the function $\phi^{(1)}$. The driving potential is given by

$$\Phi^{(0)}=V\exp(j\omega t). \tag{12}$$

Also, $\phi^{(1)}$ is assumed to depend only on x_3. With these simplifications, the Equations of set, Eqs. 8 and 9, reduce to

$$T^{(0)}{}_5=2h(\bar{e}_{56}\kappa_6u^{(1)}{}_1+\bar{e}_{55}u^{(0)}{}_{1,3})+\bar{e}_{25}\Phi^{(0)},$$

$$T^{(0)}{}_6=2h\kappa_6(\bar{e}_{66}\kappa_6u^{(1)}{}_1+\bar{e}_{56}u^{(0)}{}_{1,3})+\bar{e}_{26}\kappa_6\Phi^{(0)},$$

$$T^{(1)}{}_5=\tfrac{2}{3}h^3\gamma_{55}u^{(1)}{}_{1,3}+\Psi_{35}\phi,^{1)}{}_3,$$

$$D^{(0)}{}_2=\bar{e}_{26}\kappa_6u^{(1)}{}_1+\bar{e}_{25}u^{(0)}{}_{1,3}+\frac{\bar{\epsilon}_{22}}{4h}\left(\frac{11}{2}\Phi^{(0)}-\frac{15}{h^2}\phi^{(1)}\right),$$

$$D^{(0)}{}_3=\bar{e}_{36}\kappa_6u^{(1)}{}_1+\bar{e}_{35}u^{(0)}{}_{1,3}-\frac{\bar{e}_{23}}{2h}\Phi^{(0)}-\frac{15\bar{\epsilon}_{23}}{4h^3}\phi^{(1)}+\frac{15\bar{\epsilon}_{22}}{8h}\Phi^{(0)}, \tag{13}$$

$$D^{(1)}{}_2=\Psi_{25}u^{(1)}{}_{1,3}-\frac{3\zeta_{23}}{2h^3}\phi,^{(1)}{}_3,$$

$$D^{(1)}{}_3=\Psi_{35}u^{(1)}{}_{1,3}-\frac{3\zeta_{33}}{2h^3}\phi,^{(1)}{}_3,$$

and

$$D^{(2)}{}_2=\frac{15}{4h^3}\bar{\epsilon}_{22}\left(\frac{3}{h^2}\phi^{(1)}-\Phi^{(0)}\right).$$

The equations of motion and electric potential equation are obtained by substituting the above results into Eqs. 7. This gives

$$2h(\bar{c}_{56}\kappa_6u^{(1)}{}_{1,3}+\bar{c}_{55}u^{(0)}{}_{1,33})=2h\rho\ddot{u}^{(0)}{}_1, \tag{14a}$$

$$\tfrac{2}{3}h^3\gamma_{55}u^{(1)}{}_{1,33}+\Psi_{35}\phi,^{(1)}{}_{33}-2h\kappa_6(\bar{c}_{66}\kappa_6u^{(1)}{}_1+\bar{c}_{56}u^{(0)}{}_{1,3})+\bar{e}_{26}\kappa_6\Phi^{(0)}=\tfrac{2}{3}h^3\rho\ddot{u}^{(1)}{}_1, \tag{14b}$$

and

$$\Psi_{35}u^{(1)}{}_{1,33}-\frac{3\zeta_{33}}{2h^3}\phi,^{(1)}{}_{33}-\frac{15}{2h^3}\bar{\epsilon}_{22}\left(\frac{3}{h^2}\phi^{(1)}-\Phi^{(0)}\right)=0. \tag{14c}$$

A further approximation is made by noting that $\bar{c}_{56} \ll \bar{c}_{55}$, and therefore the terms multiplied by $\bar{c}_{56}$ may be neglected. With this assumption, Eqs. 14 reduce to

$$\bar{c}_{55} u^{(0)}{}_{1,33} + \rho\omega^2 u^{(0)}{}_1 = 0, \qquad (15a)$$

$$\tfrac{2}{3}h^3\gamma_{55} u^{(1)}{}_{1,33} + \Psi_{35}\phi,^{(1)}{}_{33} - 2h\kappa^2{}_6\bar{c}_{66} u^{(1)}{}_1 + \tfrac{2}{3}h^3\rho\omega^2 u^{(1)}{}_1 + \bar{c}_{26}\kappa_6\Phi^{(0)} = 0, \qquad (15b)$$

and

$$\Psi_{35} u^{(1)}{}_{1,33} - \frac{3}{2h^3}\zeta_{33}\phi,^{(1)}{}_{33} - \frac{15}{2h^3}\bar{\epsilon}_{22}\left(\frac{3}{h^2}\phi^{(1)} - \Phi^{(0)}\right) = 0. \qquad (15c)$$

The equations of motion given by Eqs. 15(a) and 15(b) are decoupled as far as the mechanical displacements are concerned. The mechanical displacement $u^{(1)}{}_1$ and the electric potential are still coupled in Eqs. 15(b) and 15(c). However, since the piezoelectric coupling in quartz is small, the contribution from the electric potential may be neglected in Eq. 15(b). Also, there is very little contribution from the face-shear deformation at the high frequencies we are considering and the contribution may thus be neglected in calculations concerning this system. The equation

$$u^{(1)}{}_{1,33} + \frac{3\kappa^2{}_6\bar{c}_{66}}{\gamma_{55}h^2}\left[\frac{\omega^2}{(3\kappa^3{}_6\bar{c}_{66}/\rho h^2)} - 1\right] u^{(1)}{}_1 + \frac{3\bar{e}_{26}\kappa_6\Phi^{(0)}}{2h^3\gamma_{55}} = 0 \qquad (16)$$

and Eq. 15(c) are therefore the equations describing the vibration of the system.

Once the functional form of $u_1^{(1)}$ has been determined from Eq. 16 it may then be substituted into Eq. 15(c) so that $\phi^{(1)}$ may be determined.

Equations 15(c) and 16 can be applied to the analysis of the system. However, before proceeding further we must consider the additional corrections due to the presence of the electrodes.

It was shown in Refs. 5 and 8 for a quartz crystal plate, with counter electrodes on each major surface, of thickness h', and density ρ', that an alternating voltage, impressed across the electrodes, is mechanically equivalent to a couple, C per unit area, distributed uniformly over the electroded portion. Also in Ref. 9, it was shown that the elastic constant $\bar{c}_{66}$ is stiffened by the piezoelectric effect, so that our effective elastic modulus becomes

$$\bar{c}^*{}_{66} = \bar{c}_{66}(1 + \bar{e}^2{}_{26}/\bar{c}_{66}\bar{\epsilon}_{22}) = \bar{c}_{66}(1 + p^2). \qquad (17)$$

Incorporating these two conditions and neglecting the stiffnesses of the electrodes, Eq. 16 for the electroded regions of the plate becomes

$$u^{(1)}{}_{1e,33} + \frac{3\kappa^2{}_{6e}\bar{c}^*{}_{66}}{\gamma_{55}h^2}\left[\frac{\omega^2}{[3\kappa^2{}_{6e}\bar{c}^*{}_{66}/(1+3R)\rho h^2]} - 1\right] u^{(1)}{}_{1e} + \frac{3\bar{e}_{26}\kappa_{6e}\Phi^{(0)}}{2h^2\gamma_{55}} = 0, \qquad (18a)$$

where

$$\kappa^2{}_{6e} = \frac{\alpha^2{}_e}{3}(1+p^2)\frac{(1+3R)}{(1+R)^2} \simeq \kappa^2{}_{6u}\frac{(1+3R)}{(1+R)^2}$$

and α_e is the first root of the equation

$$\tan\alpha_e = \alpha_e(1 + \bar{c}_{66}\bar{\epsilon}_{22}/\bar{e}^2{}_{26}). \qquad (19)$$

The parameter R is called the "mass loading" and is the ratio of the combined mass per unit area of both electrodes to the mass per unit area of the plate, or

$$R = 2\rho' h'/\rho h.$$

In the unelectroded portion, $\alpha_u = \pi/2$. Since there are no driving potentials, Eq. 16 becomes

$$u^{(1)}{}_{1u,33} + \frac{3\kappa^2{}_{6u}\bar{c}^*{}_{66}}{\gamma_{55}h^2}\left[\frac{\omega^2}{(3\kappa^2{}_{6u}\bar{c}^*{}_{66}/\rho h^2)} - 1\right] u^{(1)}{}_{1u} = 0. \qquad (18b)$$

Defining ω_u and ω_e as the resonant frequencies of the thickness–twist mode in the infinite unelectroded and electroded plates, respectively, or ω_e by

$$\omega^2{}_e = 3\kappa^2{}_{6e}\bar{c}^*{}_{66}/(1+3R)\rho h^2 \qquad (20a)$$

and ω_u by

$$\omega_u{}^2 = 3\kappa^2{}_{6u}\bar{c}_{66}/\rho h^2, \qquad (20b)$$

we note that

$$\omega_u/\omega_e \doteq 1 + R. \qquad (21)$$

Thus, in Eqs. 15(c), 18(a), and 18(b), we have a mathematical characterization of the piezoelectric system where the x_3 axis is the axis along which the inter-resonator coupling takes place.

The boundary conditions to be imposed are that the stress $(T^{(1)}{}_5)$ and displacement $(u^{(1)}{}_1)$ be continuous at the boundaries between the electroded and unelectroded regions of the quartz plate. These boundary conditions may be expressed formally by the relations

$$u^{(1)}{}_{1e,3} = u^{(1)}{}_{1u,3} \qquad (22)$$

and

$$u^{(1)}{}_{1e} = u^{(1)}{}_{1u}. \qquad (23)$$

For the frequencies of interest, which are in the vicinity of ω_e, the term in brackets in Eq. 18(b) will be negative. Rewriting Eqs. 18(a) and 18(b) as follows:

$$u^{(1)}{}_{1e,33} + \zeta^2{}_e u^{(1)}{}_{1e} + \bar{V} = 0 \qquad (24a)$$

and

$$u^{(1)}{}_{1u,33} - \zeta^2{}_u u^{(1)}{}_{1u} = 0, \qquad (24b)$$

[8] R. D. Mindlin, J. Appl. Phys. 23, 83–88 (1952).

[9] R. D. Mindlin and H. Deresiewicz, J. Appl. Phys. 25, 21–24 (1954).

we obtain the form of the solution directly as

$$u^{(1)}{}_{1e}=A\cos(\zeta_e x_3)+B\sin(\zeta_e x_3)-\bar{V}/\zeta^2{}_e \quad (25a)$$

and

$$u^{(1)}{}_{1u}=C\exp(\zeta_u x_3)+D\exp(-\zeta_u x_3), \quad (25b)$$

where the A, B, C, and D are arbitrary constants of integration, and ζ_e and ζ_u are given by

$$\zeta^2{}_e=(\omega^2{}_e/\gamma_{55})\rho(1+3R)[(\omega^2/\omega^2{}_e)-1] \quad (26a)$$

and

$$\zeta_u{}^2=(\omega^2{}_u\rho/\gamma_{55})[1-\omega^2/\omega^2{}_e(1+R)^2]. \quad (26b)$$

Since the acoustical displacement must remain finite in the regions where $x_3<0$ and for values of x_3 beyond the last resonator, the solution for $u^{(1)}{}_{1u}$ in these regions must have the respective forms

$$u^{(1)}{}_{1u}=\exp(\zeta_u x_3) \quad (27)$$

and

$$u^{(1)}{}_{1u}=\exp(-\zeta_u x_3). \quad (28)$$

We use the following notation to designate the constants of integration for a system containing n coupled resonators. The solution just above is written

$$u^{(1)}{}_{1u0}=C_1\exp(\zeta_u x_3) \quad (29)$$

and

$$u^{(1)}{}_{1un}=C_{4n}\exp(-\zeta_u x_3). \quad (30)$$

The solution for a typical resonator designated by subscript r is

$$u^{(1)}{}_{1er}=C_{4r-2}\cos\zeta_e x_3+C_{4r-1}\sin\zeta_e x_3-\bar{V}_r/\zeta^2{}_e. \quad (31)$$

The solution in the unelectroded region between resonator r and $r+1$ becomes

$$u^{(1)}{}_{1ur}=C_{4r}\exp(\zeta_u x_3)+C_{4r-1}\exp(-\zeta_u x_3). \quad (32)$$

IV. TWO-RESONATOR EXAMPLE

Consider the simplest case of two coupled resonators. The solutions in the different regions are represented by the following set of equations:

$$u^{(1)}{}_{1u0}=C_1\exp(\zeta_u x_3), \quad (33a)$$

$$u^{(1)}{}_{1e1}=C_2\cos\zeta_e x_3+C_3\sin\zeta_e x_3-\bar{V}_1/\zeta^2{}_e, \quad (33b)$$

$$u^{(1)}{}_{1u1}=C_4\exp(\zeta_u x_3)+C_5\exp(-\zeta_u x_3), \quad (33c)$$

$$u^{(1)}{}_{1e2}=C_6\cos\zeta_e x_3+C_7\sin\zeta_e x_3-\bar{V}_2/\zeta^2{}_e, \quad (33d)$$

and

$$u^{(1)}{}_{1u2}=C_8\exp(-\zeta_u x_3). \quad (33e)$$

The boundary conditions establish the equations

$$\begin{aligned} &u^{(1)}{}_{1u0}=u^{(1)}{}_{1e1}\text{ and }u^{(1)}{}_{1u0,3}=u^{(1)}{}_{1e1,3}\text{ at }x_3=0;\\ &u^{(1)}{}_{1e1}=u^{(1)}{}_{1u1}\text{ and }u^{(1)}{}_{1e1,3}=u^{(1)}{}_{1u1,3}\text{ at }x_3=l_1;\\ &u^{(1)}{}_{1u1}=u^{(1)}{}_{1e2}\text{ and }u^{(1)}{}_{1u1,3}=u^{(1)}{}_{1e2,3}\text{ at }x_3=l_2; \end{aligned} \quad (34)$$

and

$$u^{(1)}{}_{1e2}=u^{(1)}{}_{1u2}\text{ and }u^{(1)}{}_{1e2,3}=u^{(1)}{}_{1u2,3}\text{ at }x_3=l_3.$$

We obtain the following set of equations from Eqs. 34;

$$C_1-C_2=-\bar{V}_1/\zeta^2{}_e, \quad (35a)$$

$$\zeta_u C_1-\zeta_e C_3=0, \quad (35b)$$

$$[C_2\cos\zeta_e l_1+C_3\sin\zeta_e l_1]-C_4\exp(\zeta_u l_1)-C_5\exp(-\zeta_u l_1)=\bar{V}_1/\zeta^2{}_e, \quad (35c)$$

$$[-C_2\zeta_e\sin\zeta_e l_1+C_3\zeta_e\cos\zeta_e l_1]-C_4\zeta_u\exp(\zeta_u l_1)+C_5\zeta_u\exp(-\zeta_u l_1)=0, \quad (35d)$$

$$C_4\exp(\zeta_u l_2)+C_5\exp(-\zeta_u l_2)-[C_6\cos\zeta_e l_2+C_7\sin\zeta_e l_2]=-\bar{V}_2/\zeta^2{}_e, \quad (35e)$$

$$C_4\zeta_u\exp(\zeta_u l_2)-C_5\zeta_u\exp(-\zeta_u l_2)+[C_6\zeta_e\sin\zeta_e l_2-C_7\zeta_e\cos\zeta_e l_2]=0, \quad (35f)$$

$$[C_6\cos\zeta_e l_3+C_7\sin\zeta_e l_3]-C_8\exp(-\zeta_u l_3)=\bar{V}_2/\zeta^2{}_e, \quad (35g)$$

and

$$[-C_6\zeta_e\sin\zeta_e l_3+C_7\zeta_e\cos\zeta_e l_3]+C_8\zeta_u\exp(-\zeta_u l_3)=0. \quad (35h)$$

These eight equations may be simplified in form and arranged as follows:

$$\begin{aligned} &C_1A_{11}+C_2A_{12} &&=-\bar{V}_1/\zeta^2{}_e,\\ &C_1A_{21}+C_3A_{23} &&=0,\\ &C_2A_{32}+C_3A_{33}+C_4A_{34}+C_5A_{35} &&=\bar{V}_1/\zeta^2{}_e,\\ &C_2A_{42}+C_3A_{43}+C_4A_{44}+C_5A_{45} &&=0,\\ &C_4A_{54}+C_5A_{55}+C_6A_{56}+C_7A_{57} &&=-\bar{V}_2/\zeta^2{}_e,\\ &C_4A_{64}+C_5A_{65}+C_6A_{66}+C_7A_{67} &&=0,\\ &C_6A_{76}+C_7A_{77}+C_8A_{78} &&=\bar{V}_2/\zeta^2{}_e,\\ &C_6A_{86}+C_7A_{87}+C_8A_{88} &&=0, \end{aligned} \quad (36)$$

where the A_{ij} are the appropriate coefficients of the C's in Eqs. 35. Equations 36 may be abbreviated as

$$A_{ij}C_j = V_i. \tag{37}$$

The normal resonant frequencies are the values of ω which make the determinant of A_{ij} equal to zero. In the two-resonator case there will be an infinite set of values for ω which make $|A_{ij}|$ equal to zero; but only the first two normal frequencies will be of interest.

The determinant itself for the two-resonator system is as follows:

$$\begin{vmatrix} A_{11} & A_{12} & 0 & 0 & 0 & 0 & 0 & 0 \\ A_{21} & 0 & A_{23} & 0 & 0 & 0 & 0 & 0 \\ 0 & \mathbf{A}_{32} & \mathbf{A}_{33} & \mathbf{A}_{43} & \mathbf{A}_{35} & \mathbf{0} & \mathbf{0} & 0 \\ 0 & \mathbf{A}_{42} & \mathbf{A}_{43} & \mathbf{A}_{44} & \mathbf{A}_{45} & \mathbf{0} & \mathbf{0} & 0 \\ 0 & \mathbf{0} & \mathbf{0} & \mathbf{A}_{54} & \mathbf{A}_{55} & \mathbf{A}_{56} & \mathbf{A}_{57} & 0 \\ 0 & \mathbf{0} & \mathbf{0} & \mathbf{A}_{64} & \mathbf{A}_{65} & \mathbf{A}_{66} & \mathbf{A}_{67} & 0 \\ 0 & 0 & 0 & 0 & 0 & A_{76} & A_{77} & A_{78} \\ 0 & 0 & 0 & 0 & 0 & A_{86} & A_{87} & A_{88} \end{vmatrix} = 0. \tag{38}$$

Its general form is similar to that of any system containing n coupled resonators in a linear array. This form of determinant lends itself readily to the pivotal-reduction technique[10] for reducing determinants and a very simple computer program can be written to find the normal frequencies. The central field, in boldface type, is the "basic block" of the determinant. For a system having n coupled resonators the "basic block" is repeated sequentially along the diagonal of the determinant for $n-1$ times differing only in the dimensional value of length present in the exponential or sinusoidal arguments. This basic block can be put into the following general form:

$$\begin{aligned}
A_{4s-1,4s-2} &= \cos\zeta_e l_{2s-1}, \\
A_{4s-1,4s-1} &= \sin\zeta_e l_{2s-1}, \\
A_{4s-1,4s} &= -\exp(\zeta_u l_{2s-1}), \\
A_{4s-1,4s+1} &= -\exp(-\zeta_u l_{2s-1}), \\
A_{4s+1,4s} &= \exp(\zeta_u l_{2s}), \\
A_{4s+1,4s+1} &= \exp(-\zeta_u l_{2s}), \\
A_{4s+1,4s+2} &= -\cos\zeta_e l_{2s}, \\
A_{4s+1,4s+3} &= -\sin\zeta_e l_{2s}, \\
A_{4s,4s-2} &= -\zeta_e \sin\zeta_e l_{2s-1}, \\
A_{4s,4s-1} &= \zeta_e \cos\zeta_e l_{2s-1}, \\
A_{4s,4s} &= -\zeta_u \exp(\zeta_u l_{2s-1}), \\
A_{4s,4s+1} &= \zeta_u \exp(-\zeta_u l_{2s-1}), \\
A_{4s+2,4s} &= \zeta_u \exp(\zeta_u l_{2s}), \\
A_{4s+2,4s+1} &= -\zeta_u \exp(-\zeta_u l_{2s}), \\
A_{4s+2,4s+2} &= \zeta_e \sin\zeta_e l_{2s},
\end{aligned} \tag{39}$$

and

$$A_{4s+2,4s+3} = -\zeta_e \cos\zeta_e l_{2s}.$$

The range of s is from 1 to $n-1$.

The first and last groups of elements on the diagonal, which are the ones outside the central field in the determinant in Eq. 38, will have the same form for filters of all orders. These are

$$\begin{aligned}
&A_{11} = 1, \qquad A_{21} = \zeta_u, \\
&A_{12} = -1, \qquad A_{23} = -\zeta_e, \\
&A_{n-1,n-2} = \cos\zeta_e l_{2n-1}, \\
&A_{n-1,n-2} = \sin\zeta_e l_{2n-1}, \\
&A_{n-1,n} = -\exp(-\zeta_u l_{2n-1}), \\
&A_{n,n-2} = -\zeta_e \sin\zeta_e l_{2n-1}, \\
&A_{n,n-1} = \zeta_e \cos\zeta e l_{2n-1},
\end{aligned} \tag{40}$$

and

$$A_{n,n} = \zeta_u \exp(-\zeta_u l_{2n-1}),$$

where n is the number of coupled resonators. All elements of the determinant not specified above are zero. With these general expressions, we can generate the elements of the determinant for a system containing any number coupled resonators and calculate the normal frequencies for various arrangements of the electrode configuration and plate back.

V. THICKNESS–SHEAR MODE

A similar derivation may be made when employing the thickness–shear mode of coupling along the x_1 axis where the acoustical displacement depends on the x_1 coordinate variable. The equations which are obtained in this case are the following[11]:

$$u^{(1)}{}_{1u,11} - \xi^2{}_u u^{(1)}{}_{1u} = 0, \tag{41a}$$

$$u^{(1)}{}_{1e,11} + \xi^2{}_e u^{(1)}{}_{1e} + 3\bar{e}_{26}\kappa_{6e}\Phi^{(0)}/2h^2\gamma_{11} = 0, \tag{41b}$$

and

$$\tfrac{2}{3}h^3\Psi_{11}u^{(1)}{}_{1e,11} - \zeta_{11}\phi,^{(1)}{}_{11} + (15/h^2)\bar{\epsilon}_{22}\phi^{(1)} - 5\bar{\epsilon}_{22}\Phi^{(0)} = 0, \tag{41c}$$

where

$$\xi^2{}_u = \frac{\omega^2{}_u\rho}{\gamma_{11} + \kappa^2{}_6\bar{e}_{66}}\left[1 - \frac{\omega^2}{\omega^2{}_e(1+R)^2}\right] \tag{42a}$$

and

$$\xi^2{}_e = \frac{\omega^2{}_e\rho(1+3R)(1+p^2)}{\gamma_{11} + \kappa^2{}_{6e}\bar{e}^*{}_{66}}\left[\frac{\omega^2}{\omega_e{}^2} - 1\right]. \tag{42b}$$

The boundary conditions for the thickness–shear mode in an electroded plate are expressed by the

---

[10] J. B. Scarborough, *Numerical Mathematical Analysis* (Johns Hopkins Press, Baltimore, Md., 1966), 261–263.

[11] The method of derivation of these equations was indicated in private communication from H. F. Tiersten. However, a similar set of equations without the piezoelectric contributions were derived by Mindlin and Lee and are given in Ref. 5.

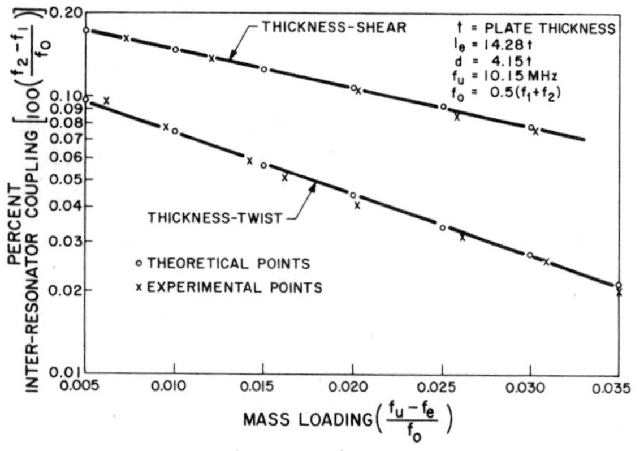

FIG. 7. The effect of mass loading on the frequency difference between the two normal resonances of a two resonator system. Comparison of theoretical and experimental results.

relations

$$u^{(1)}{}_{1e}=u^{(1)}{}_{1u}, \tag{43a}$$

and

$$u^{(1)}{}_{1e,1}=u^{(1)}{}_{1u,1}. \tag{43b}$$

VI. NUMERICAL RESULTS

A computer program was written to compute the frequencies that make the determinant in Eq. 38 equal to zero. Such frequencies are the normal frequencies of the homogeneous boundary value problem. Calculations were made by applying this analysis to some particular examples. The results for two-, three-, and six resonator systems are discussed, starting with the two-resonator system.

The frequency separation between the two normal resonances is calculated as a function of the mass loading and interresonator separation. Some of the results of these calculations are shown in Figs. 7 and 8. The ordinate of the semilog plot is the percent frequency separation relative to the average of the two frequencies. The abscissa in Fig. 7 is mass loading where the interresonator spacing is held constant at five plate thicknesses; and the abscissa in Fig. 8 is interresonator spacing in units of crystal plate thicknesses where the mass loading is held constant at 0.02. The electrode length along the coupling axis for these calculations is 15 plate thicknesses. The dots in the Figures are the theoretical points and the crosses are experimental data.

The plots show substantial agreement between the theoretical and experimental findings. Also, one finds that the decoupling of the resonators is achieved much more effectively along the x_3 axis than along the x_1 axis. Since increasing the mass loading beyond 0.025 generally begins to degrade the resonator Q, increased interresonator spacing is preferable to increased mass loading for decoupling the resonators.

By increasing the value of n to 3, we can calculate the normal frequencies for the three-resonator system and show how these frequencies vary with electrode geometry. The results of two calculations are shown in Fig. 9. Again, the percent frequency difference is plotted as ordinate and electrode separation as abscissa.

In one calculation, the results of which are shown in Fig. 9 as solid lines, the individual electrode length l_e was held constant at $16t$ and the interresonator spacing d was increased. The other calculation, the results of which are shown in Fig. 9 as broken lines, was made holding the length of the entire electrode configuration l_t constant and increasing the interresonator spacing d by decreasing the individual electrode lengths. In both calculations the mass loading was held at 0.021. This shows the effect of the over-all configuration length. The decoupling of resonators is more readily achieved when the length of the electrode configuration is increased along with the interresonator spacing.

The last calculation to be discussed is that for the normal resonances for a six-resonator system. The results for coupling of the thickness–twist mode and the thickness–shear mode are shown in Figs. 10 and 11, respectively. In both Figures, the percent frequency differences between resonances $\omega(2)-\omega(1)$ through $\omega(6)-\omega(1)$ are plotted as a function of mass loading.

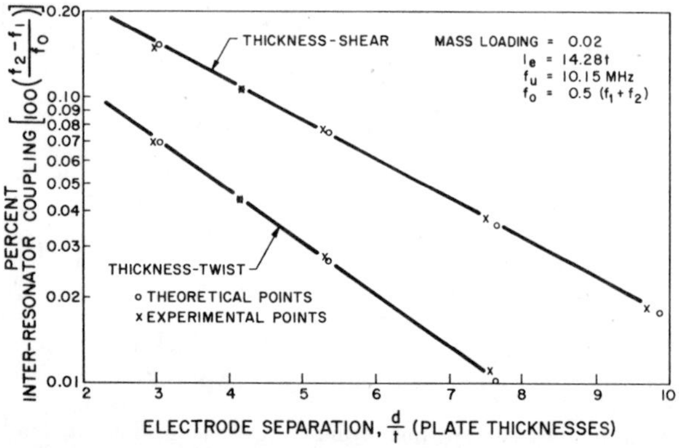

FIG. 8. The effect of electrode separation on the frequency difference between the two normal resonances of a two resonator system. Comparison of theoretical and experimental results.

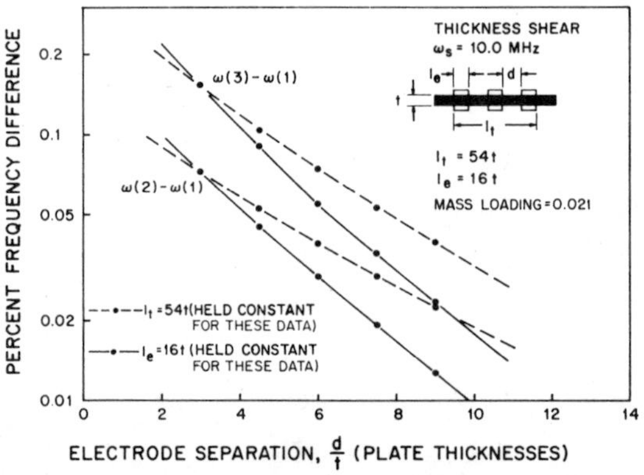

FIG. 9. The effect of electrode separation on the calculated frequency difference between the three normal resonances of a three-resonator system.

VII. SUMMARY

The mathematical model used in this analysis has proved to be an accurate representation of the piezoelectric system it was meant to describe. It has therefore become a tool that may be employed to determine the normal frequencies of such a system. An important application may be found in the fact that the "monolithic crystal filter"[1] is such a piezoelectric system. The transmission properties of this device depend on the electrode geometry such that, by proper control of interresonator coupling and termination, a single plate of quartz can be made to perform a bandpass-filter function. This analysis may be employed in the design of such filters to compute the cutoff frequencies of the passband and the zeros of the short-circuit impedance.

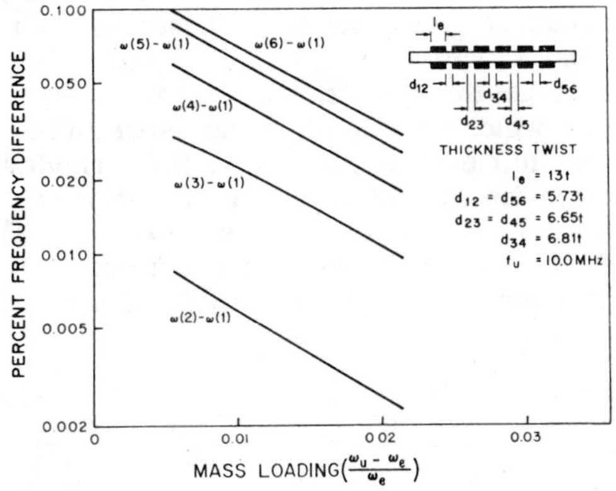

FIG. 10. The effect of mass loading on the calculated frequency difference between the six normal resonances of a six resonator system having the x_3 axis as the coupling axis.

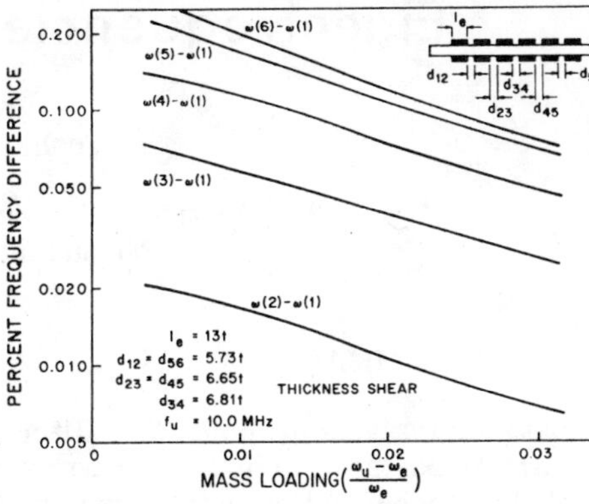

FIG. 11. The effect of mass loading on the calculated frequency difference between the six normal resonances of a six-resonator system having the x_1 axis as the coupling axis.

ACKNOWLEDGMENTS

In the course of developing this analysis, the author has become indebted to many workers for their helpful discussions and encouragement. He would like to thank expecially W. P. Mason, W. J. Spencer, and H. F. Tiersten, who has influenced much of this investigation by his contribution to the general theory of plate vibrations, and through many private discussions. He also wishes to thank R. N. Thurston for his critical review of the initial draft. Finally, the author wishes to thank R. A. Sykes for his invaluable direction and advice.

Eigenfrequencies of Monolithic Filters

Toshio Ashida, Member

The Electrical Communication Laboratory, N. T. T.,
Musashino, Tokyo, Japan 180

SUMMARY

In a theoretical analysis of monolithic filters it is of primary importance to obtain the numerical values of the mechanical eigenfrequencies of the vibrating system. In this paper a new approximate formula is presented by which the eigenfrequencies are easily calculated from the given electrode configuration and mass loading. Using this formula, specific calculations were made for the cases where there are 2, 3 and 4 electrodes, and the results show good agreement with exact computations.

1. Introduction

Monolithic filters [1, 2] have been used as intermediate-frequency filters and have been applied in other fields of communication as SSB, channel and pilot filters. Monolithic filters are compact in size, easy to fabricate and can be used in the circuit without matching transformers.

A monolithic filter is a thin piezoelectric plate with more than two pairs of electrodes plated on both surfaces. The electrode-plated region works as a mechanical resonator. If the electrodes are properly spaced, therefore, the resonators are elastically coupled. This coupled vibratory system is made use of in monolithic filters. The eigenfrequencies of a coupled vibratory system give the zeros of the transmission function of a filter that characterize the system. Furthermore, the eigenfrequency distribution with respect to the center frequency f_{oo} is not symmetric, unlike in a coupled resonant system of electrical circuits. The frequency intervals are wider on the higher-frequency side than those on the lower-frequency side. The eigenfrequencies of the monolithic filter are a fundamental factor in its analysis and design.

The eigenfrequencies of a monolithic filter have numerically been calculated by solving a transcendental equation derived from the equation of motion under proper boundary conditions by means of a digital computer [3, 4]. However, the transcendental equation contains design parameters, such as the electrode dimensions and the mass loading of the electrodes, which forms an implicit function. Therefore, a large number of calculations is repeatedly required to find the roots. This is the reason why the design of monolithic filters is complex. In the present paper, an electrical equivalent circuit of the monolithic filter is first developed to overcome this problem. The circuit is capable of expressing the asymmetrical eigenfrequency distribution. Then, based on the equivalent circuit developed, the simple formulas for obtaining the approximate eigenfrequencies are derived, which contain the parameters in the form of an explicit function. Finally, the formulas are applied to the case of crystal plates on which several pairs of electrodes are provided. The results are compared with the exact ones. It is also confirmed by experiment that the approximate formulas are practically sufficient.

2. Notations

A_i, B_i, C_i, D_i : elements of the transmission matrix, Eq. (41)

c_{ij} : elastic stiffness

d_i : interval between the i-th and (i+1)-th electrodes

f_e : cutoff frequency of the electrode-plated region, Eq. (10)

f_s : cutoff frequency of the non-electrode region, Eq. (9)

f_i : i-th eigenfrequency, (the number is subscripted from the lower eigenfrequency up)

f_{oo} : eigenfrequency of the resonator with a pair of electrodes of width l_o

$[F_i^e], [E_i]$: equivalent transmission matrices of the electrode-plated region

$[F_i^s], [S_i]$: equivalent transmission matrices of the non-electrode region

$[F_0^s], [F_0^s]^{-1}$: equivalent transmission matrix of a non-electrode plate of infinite length, and its inverse matrix

G_i : image gyrator constant, Eq. (39)

g_{ij} : ratio of the two image gyrator constants Eq. (46)

H : plate thickness

k_e, k_s : propagation constants of the electrode-plated region and the non-electrode region, Eq. (7)

K_{ij} : frequency interval between two eigenfrequencies f_i and f_j, Eq. (53)

Reprinted with permission from *Electron. Commun. Japan*, vol. 54, pt. A, pp. 41-49, June 1971. Reprinted with the consent of Scripta Publishing Co., 1511 K St., N.W., Washington, DC 20005.

l_i : width of the i-th electrode
M, N: functions given by Eq. (48)
p_i : energy-trapping factor, Eq. (26)
q_i : ratio of the electrode width to the interval between the electrodes
R : effect of mass loading due to the electrode plating, to be called a frequency-lowering factor
T_s : shear stress
$\dot{u}_1$: displacement velocity in the x-direction
W_i : impedance of the series resonator circuit, Eq. (33)
z_e, z_s : characteristic impedances of the electrode-plated region and the non-electrode region, Eq. (6)
α_i : given by Eq. (35)
θ_i, Θ_i : phases of the electrode-plated region and the non-electrode region, Eq. (8)
ρ : plate density
ϕ : a variable with respect to the frequency given by Eq. (18)
ϕ_{i0} : resonant frequency of the plate provided with a pair of electrodes of width l_i given by Eq. (25)
ϕ_{00} : resonant frequency of the plate if the width of all electrodes is equal to l_0
ω : angular frequency
ϵ : minute change of ϕ, given by Eq. (27)
Δ_{ij} : frequency interval between two frequencies ϕ_i and ϕ_j, given by Eq. (51)
δ_i : minute change of frequency f_i, given by Eq. (52)
γ_{55} : a term with respect to crystal orthotropy given by Eq. (11)

3. Transmission Matrix of the Elastic Vibrations

Thickness-shear modes associated with an AT-cut quartz crystal plate or a piezoelectric ceramic plate have been analyzed, including their piezoelectric effect [5]. This type of vibration can be expressed by transmission matrices similar to those in electrical transmission lines [6, 7]. The treatment can be applied to a monolithic filter as shown in Fig. 1, where n resonators with plated-on electrodes are coupled by n-1 couplers. A resonator with a pair of electrodes of width l_i is expressed by a matrix as

$$\begin{pmatrix} T_s \\ \dot{u}_1 \end{pmatrix}_{2i-1} = \begin{pmatrix} \cos\theta_i & , \ jz_e \sin\theta_i \\ j\sin\theta_i/z_e, & \cos\theta_i \end{pmatrix} \begin{pmatrix} T_s \\ \dot{u}_1 \end{pmatrix}_{2i} \tag{1}$$

A coupler of width d_i which couples the resonators is then given as

$$\begin{pmatrix} T_s \\ \dot{u}_i \end{pmatrix}_{2i} = \begin{pmatrix} \cosh\Theta_i & , \ -jz_s \sinh\Theta_i \\ j\sinh\Theta_i/z_s, & \cosh\Theta_i \end{pmatrix} \begin{pmatrix} T_s \\ \dot{u}_1 \end{pmatrix}_{2i+1} \tag{2}$$

Therefore, the transmission matrices $[F_i^e]$ and $[F_i^s]$ given by Eqs. (1) and (2) are expressed as

$$[F_i^e] = \begin{pmatrix} \cos\theta_i & , \ j\sin\theta_i \\ j\sin\theta_i, & \cos\theta_i \end{pmatrix} \tag{3}$$

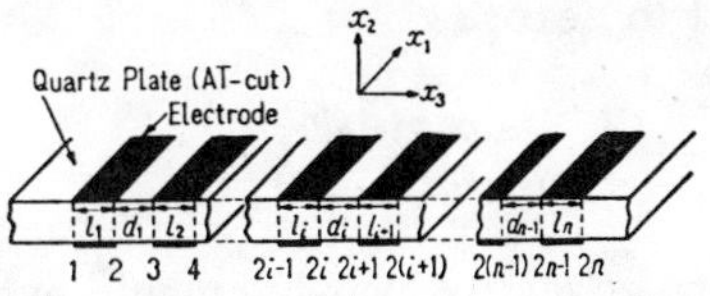

Fig. 1. Monolithic structure of mechanically n-coupled resonator system.

$$[F_i^s] = \begin{pmatrix} \cosh\Theta_i & , \ -j(k_s/k_e)\sinh\Theta_i \\ j(k_e/k_s)\sinh\Theta_i, & \cosh\Theta_i \end{pmatrix} \tag{4}$$

which are normalized with respect to the characteristic acoustical impedance z_e of the electrode-plated region. Furthermore, both ends of the plate can be expressed by the two-terminal impedance $-jz_s$, in which the plate length is assumed to be infinite. This matrix $[F_0^s]$ is then given as

$$[F_0^s] = \begin{pmatrix} 1, & -jk_s/k_e \\ 0, & 1 \end{pmatrix} \tag{5}$$

The parameters that appeared in Eqs. (1) to (5) are as follows:

$$z_e = c_{55}k_e/\omega, \quad z_s = c_{55}k_s/\omega \tag{6}$$

$$\left.\begin{aligned} k_e &= (\pi/\sqrt{\gamma_{55}}\,H)\sqrt{(f/f_e)^2 - 1} \\ k_s &= (\pi/\sqrt{\gamma_{55}}\,H)\sqrt{1 - (f/f_s)^2} \end{aligned}\right\} \tag{7}$$

$$\theta_i = k_e l_i, \quad \Theta_i = k_s d_i \tag{8}$$

$$f_s = (1/2H)\sqrt{c_{66}/\rho} \tag{9}$$

$$R = (f_s - f_e)/f_s \tag{10}$$

$$\gamma_{55} = c_{55}/c_{66} \tag{11}$$

Therefore, the transmission matrix [F] of the whole monolithic filter with n electrodes is

$$\begin{aligned} [F] = {} & [F_0^s][F_1^e][F_1^s]\cdots[F_i^e][F_i^s] \\ & \cdots[F_{n-1}^s][F_n^e][F_0^s] \end{aligned} \tag{12}$$

The eigenfrequency equation is given by setting the elements of the first row and the second column in the matrix [F] equal to zero. We define new matrices $[E_i]$ and $[S_i]$ as follows:

$$\begin{aligned} [E_i] &\equiv [F_0^s][F_i^e][F_0^s] \\ &= \begin{pmatrix} \cos\theta_i + (k_s/k_e)\sin\theta_i, & 2j\{\sin(\theta_i/2) - (k_s/k_e) \cdot \cos(\theta_i/2)\}\{\cos(\theta_i/2) + (k_s/k_e)\sin(\theta_i/2)\} \\ j\sin\theta_i & , \ \cos\theta_i + (k_s/k_e)\sin\theta_i \end{pmatrix} \end{aligned} \tag{13}$$

$$\begin{aligned} [S_i] &\equiv [F_0^s]^{-1}[F_i^s][F_0^s]^{-1} \\ &= \begin{pmatrix} e^{-\Theta_i} & , \ 2j(k_s/k_e)e^{-\Theta_i} \\ j(k_e/k_s)\sinh\Theta_i, & e^{-\Theta_i} \end{pmatrix} \end{aligned} \tag{14}$$

where $[F_0^s]^{-1}$ is the inverse matrix of $[F_0^s]$. Equation (12) is then written as

$$[F] = [E_1][S_1]\cdots[E_i][S_i]\cdots[S_{n-1}][E_n] \tag{15}$$

As defined in Eq. (13), the matrix $[E_i]$ means the transmission matrix of a crystal plate of infinite length with electrodes of width l_i. Therefore, eigenfrequency equations are given by the following two relations which are obtained by setting the elements of the first row and the second column in $[E_i]$ equal to zero;

$$\tan(\theta_i/2) = k_s/k_e \tag{16}$$

$$\tan(\theta_i/2) = -k_e/k_s \tag{17}$$

They indicate so-called non-harmonic overtone modes of the thickness modes. Equation (16) shows the eigenfrequencies of the symmetrical modes while Eq. (17) shows those of the oblique symmetrical modes [8]. Monolithic filters make use of the first, symmetrical modes.

Finally, in Eq. (15), the crystal plate with n mutually coupled electrodes can be expressed by the two elements, the first one $[E_i]$ indicating each pair of the electrodes singly plated on the plate of infinite length, and the second $[S_i]$ the coupling between them.

4. Narrowband Approximation

Following Belevitch [9], one can introduce the following frequency variable ϕ:

$$(f-f_e)/(f_s-f_e) = \cos^2\phi \tag{18}$$

From Eqs. (10) and (18), one obtains

$$f/f_s = 1 - R\sin^2\phi \tag{19}$$

$$f/f_e = 1 + R\cos^2\phi \tag{20}$$

for a light mass loading. Substitution of the equations above into Eq. (17) leads to

$$k_e = (\pi/\sqrt{\gamma_{55}}\,H)\sqrt{2R}\cos\phi \tag{21}$$

$$k_s = (\pi/\sqrt{\gamma_{55}}\,H)\sqrt{2R}\sin\phi \tag{22}$$

$$k_s/k_e \doteqdot \tan\phi \tag{23}$$

Equation (19) is exact, while Eqs. (20) to (23) are approximate, the square term with R being assumed to be much smaller than unity.

ϕ_{io} is assigned to the resonant frequency of the symmetric mode of the quartz crystal plate of infinite length with a pair of the electrodes of width l_i. θ_i is assigned to the phase angle at that frequency. Substitution of Eq. (23) into Eq. (16) leads to

$$\phi_{io} = \theta_{io}/2 \tag{24}$$

which, by the use of Eqs. (8) and (21), is finally given as

$$\phi_{io} = p_i \cos\phi_{io} \tag{25}$$

p_i in the equation is a parameter named "energy-trapping factor" by Onoe. The factor relates to two parameters, the electrode dimension l_i/H and the mass loading R, which contribute to the trapping effect of elastic energy. That is,

$$p_i = (\pi/\sqrt{2}\sqrt{\gamma_{55}})(l_i/H)\sqrt{R} \tag{26}$$

The relation between p_i and ϕ_{io} that satisfies Eq. (25) is to be used in the approximate calculation that follows. The relation is tabulated in Table 1.

In the following, we confine ourselves to the narrow-band characteristic, where the frequency variable ϕ can be written as

$$\phi = \phi_{io} + \epsilon \tag{27}$$

ϵ is a small change of the frequency. On the other hand, θ_i is expressed

$$\theta_i/2 = \phi_{io} - \epsilon\, p_i \sin\phi_{io} \tag{28}$$

for which substitution of Eq. (27) into Eqs. (8) and (21), and a Taylor expansion with respect to the first order of ϵ were made. Similarly, from Eqs. (8), (22) and (25), one obtains

$$\Theta_i = 2\,q_i\phi_{io}(\tan\phi_{io} + \epsilon) \tag{29}$$

Table 1. Relation between the resonant frequency ϕ_{oo} of the plate with a pair of electrodes and the trapping factor p_o

p_o	ϕ_{oo}	$\phi_{oo}\tan\phi_{oo}$	$\dfrac{\sin 2\phi_{oo}}{1+\phi_{oo}\tan\phi_{oo}}$	$\dfrac{\sin^2 2\phi_{oo}}{1+\phi_{oo}\tan\phi_{oo}}$
0.1	0.099505	0.0099341	0.19575	0.038701
0.2	0.19616	0.038982	0.36800	0.14070
0.3	0.28767	0.085116	0.50144	0.27285
0.4	0.37256	0.14560	0.59188	0.40133
0.5	0.45018	0.21757	0.64354	0.50425
0.6	0.52053	0.29841	0.66462	0.57353
0.7	0.58399	0.38595	0.66378	0.61065
0.8	0.64113	0.47848	0.64841	0.62161
0.9	0.69262	0.57470	0.62414	0.61343
1.0	0.73909	0.67361	0.59495	0.59240
1.1	0.78113	0.77449	0.56352	0.56350
1.2	0.81929	0.87679	0.53160	0.53038
1.3	0.85403	0.98012	0.50027	0.49556
1.4	0.88577	1.0842	0.47017	0.46073
1.5	0.91486	1.1887	0.44166	0.42694
1.6	0.94160	1.2936	0.41489	0.39481
1.7	0.96625	1.3987	0.38992	0.36469
1.8	0.98906	1.5039	0.36670	0.33670
1.9	1.0102	1.6092	0.34517	0.31087
2.0	1.0299	1.7145	0.32523	0.28713
2.1	1.0482	1.8197	0.30678	0.26537
2.2	1.0653	1.9249	0.28970	0.24548
2.3	1.0813	2.0300	0.27390	0.22730
2.4	1.0964	2.1349	0.25925	0.21071
2.5	1.1105	2.2398	0.24568	0.19555
2.6	1.1238	2.3446	0.23308	0.18170
2.7	1.1364	2.4492	0.22138	0.16904
2.8	1.1482	2.5537	0.21049	0.15746
2.9	1.1595	2.6581	0.20036	0.14685
3.0	1.1701	2.7624	0.19091	0.13713
3.1	1.1802	2.8665	0.18210	0.12821
3.2	1.1898	2.9706	0.17386	0.12002
3.3	1.1990	3.0745	0.16615	0.11248
3.4	1.2077	3.1783	0.15893	0.10554
3.5	1.2160	3.2820	0.15216	0.09914
3.6	1.2239	3.3856	0.14581	0.09324
3.7	1.2315	3.4890	0.13983	0.08778
3.8	1.2387	3.5924	0.13421	0.08272
3.9	1.2457	3.6957	0.12892	0.07804
4.0	1.2524	3.7989	0.12392	0.07370

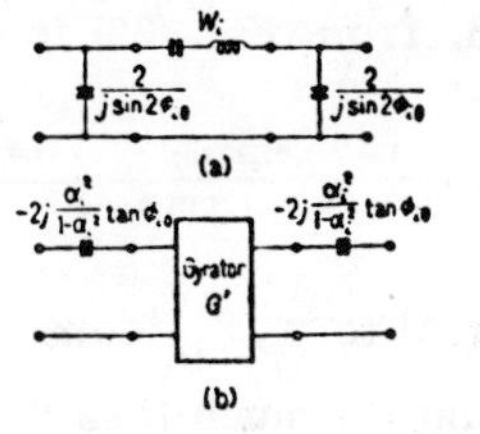

Fig. 2. Equivalent circuits of transmission matrices $[E_i]$ in Eq. 31 (a) and $[S_i]$ in Eq. 32 (b).

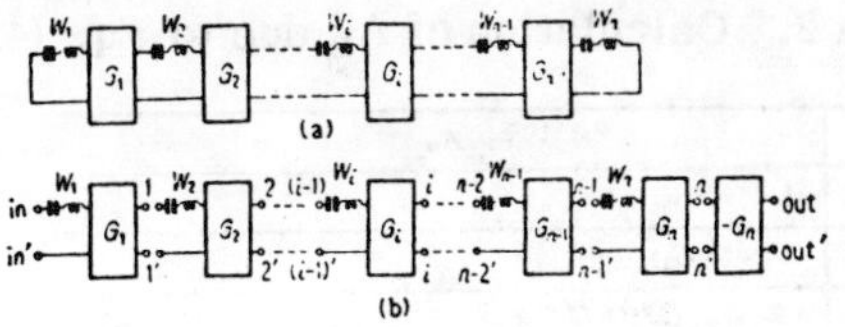

Fig. 3. Equivalent circuit of monolithic filter (a) and its representation by four-terminal networks (b).

where

$$q_i = d_i/l_i \tag{30}$$

By substituting them into the elements in $[E_i]$ and $[S_i]$ of Eqs. (13) and (14), one obtains the approximate equations as follows:

$$[E_i] = \begin{pmatrix} 1 & , 0 \\ j\sin 2\phi_{i0}/2, & 1 \end{pmatrix} \begin{pmatrix} 1, & W_i \\ 0, & 1 \end{pmatrix} \cdot \begin{pmatrix} 1 & , 0 \\ j\sin 2\phi_{i0}/2, & 1 \end{pmatrix} \tag{31}$$

$$[S_i] = \begin{pmatrix} 1, & -2j\alpha_i^2 \tan\phi_{i0}/1-\alpha_i^2 \\ 0, & 1 \end{pmatrix} \cdot \begin{pmatrix} 0 & , jG' \\ j/G', & 0 \end{pmatrix} \begin{pmatrix} 1, & -2j\alpha_i^2 \tan\phi_{i0}/1-\alpha_i^2 \\ 0, & 1 \end{pmatrix} \tag{32}$$

where

$$W_i = -2j\epsilon(1+\phi_{i0}\tan\phi_{i0})/\cos^2\phi_{i0} \tag{33}$$

$$G' = 2\alpha_i \tan\phi_{i0}/1-\alpha_i^2 \tag{34}$$

$$\alpha_i = e^{-\theta_i} \tag{35}$$

ϵ is much smaller than unity. (See Appendix.) Equations (31) and (32) are equivalent to the electrical circuits shown in Fig. 2. As shown in Fig. 2(a), the equivalent circuit for $[E_i]$ consists of a series resonant circuit (W_i) shunted by the impedance $2/j \sin 2\phi_0$ in both its arms. If the following relation is satisfied:

$$2/\sin 2\phi_{i0} > 2\epsilon(1+\phi_{i0}\tan\phi_{i0})/\cos^2\phi_{i0} \tag{36}$$

the shunt terms can be neglected. That is, the transmission matrix $[E_i]$ of the electrode-plated region is approximately expressed by the series resonant circuit of W_i. On the other hand, as shown in Fig. 2(b), $[S_i]$ consists of a gyrator (G_i') with the impedance $-2j\alpha_i^2 \tan\phi_{i0}/1-\alpha_i^2$ in its arms inserted in series. The impedance can be omitted if it is smaller than W_i such that

$$2\alpha_i^2 \tan\phi_{i0}/1-\alpha_i^2 < 2\epsilon(1+\phi_{i0}\tan\phi_{i0})/\cos^2\phi_{i0} \tag{37}$$

$[S_i]$ is then equivalent to an image gyrator G_i'.

Furthermore, if

$$1 > \alpha_i \tag{38}$$

G_i' is approximately equal to, then

$$G_i = 2\alpha_i \tan\phi_{i0} = 2(1-2q_i\phi_{i0}\epsilon)\tan\phi_{i0} \cdot e^{-2q_i\phi_{i0}\tan\phi_{i0}} \tag{39}$$

Finally, the electrode-plated region of the monolithic filter can be expressed in narrowband approximation by a series resonant circuit W_i and the coupling region by the image gyrator G_i. The frequency dependence of the circuit constants is given by Eqs. (33) and (39). It is noted that here the gyrator constant depends on the frequency, while that in the usual electric circuits is constant and independent of frequency. An expression of the asymmetricity of the eigenfrequency distribution which is difficult to understand in electrically coupled resonant circuits is possible by this frequency dependence characteristic. This is explained in more detail in the following Chapter.

5. Fundamental Equations for the Eigenfrequencies

Making use of the previous results, one can now form the equivalent electrical circuit of a monolithic filter with n pairs of electrodes as shown in Fig. 3(a). The engenfrequencies are the frequencies of the free vibrations of this coupled system. One can now obtain the fundamental equations governing the free vibrations of the circuit. First of all, the circuit is divided into a series of four-terminal networks, each of which consists of a series resonant circuit and an image gyrator. The terminals of the networks are numbered as shown in Fig. 3(b). It is noted that for the n-th (last) electrode-plated region, two gyrators G_n and $-G_n$ are assumed because of the generalization of the circuit. Now, the i-th network is expressed as

$$\begin{pmatrix} V_{i-1} \\ I_{i-1} \end{pmatrix} = \begin{pmatrix} jW_i/G_i, & jG_i \\ j/G_i & , 0 \end{pmatrix} \begin{pmatrix} V_i \\ I_i \end{pmatrix} \tag{40}$$

If the networks are connected in cascade up to the

Table 2. Calculation of A_n due to Eq. (42)

n	A_n
0	1
1	$W_1 \times (j/G_1)$
2	$(W_1W_2+G_1^2)(1/G_1G_2)$
3	$(W_1W_2W_3+W_1G_2^2+W_3G_1^2)(-j/G_1G_2G_3)$
4	$(W_1W_2W_3W_4+W_1W_2G_3^2+W_3W_4G_1^2$ $+W_4W_1G_2^2+G_1^2G_3^2)(1/G_1G_2G_3G_4)$
5	$(W_1W_2W_3W_4W_5+W_1W_2W_3G_4^2+W_3W_4W_5G_1^2$ $+W_4W_5W_1G_2^2+W_5W_1W_2G_3^2+W_1G_2^2G_4^2+W_3G_4^2G_1^2$ $+W_5G_1^2G_3^2)(j/G_1G_2G_3G_4G_5)$
6	$(W_1W_2W_3W_4W_5W_6+W_1W_2W_3W_4G_5^2+W_3W_4W_5W_6G_1^2$ $+W_4W_5W_6W_1G_2^2+W_5W_6W_1W_2G_3^2+W_6W_1W_2W_3G_4^2$ $+W_1W_2G_3^2G_5^2+W_3W_4G_5^2G_1^2+W_4W_1G_2^2G_5^2$ $+W_6W_1G_2^2G_4^2+W_3W_6G_4^2G_1^2+W_5W_6G_1^2G_3^2$ $+G_1^2G_3^2G_5^2)(-1/G_1G_2G_3G_4G_5G_6)$

i-th network, the matrix is given as

$$\begin{pmatrix} V_{in} \\ I_{in} \end{pmatrix} = \begin{pmatrix} A_i, & B_i \\ C_i, & D_i \end{pmatrix} \begin{pmatrix} V_i \\ I_i \end{pmatrix} \tag{41}$$

where A_i, B_i, C_i and D_i are the elements of the transmission matrix. Combining this with Eq. (40), one obtains the asymptote equations as follows:

$$\left.\begin{aligned} A_i &= j(A_{i-1}W_i/G_i + B_{i-1}/G_i) \\ B_i &= jA_{i-1}G_i \end{aligned}\right\} \tag{42}$$

where

$$A_0 = 1, \quad B_0 = 0$$

Therefore, the matrix for the whole circuit is

$$\begin{pmatrix} V_{in} \\ I_{in} \end{pmatrix} = \begin{pmatrix} A_n, & B_n \\ C_n, & D_n \end{pmatrix} \begin{pmatrix} 0 & , -jG_n \\ -j/G_n, & 0 \end{pmatrix} \begin{pmatrix} V_{out} \\ I_{out} \end{pmatrix} \tag{43}$$

Since the matrix element of the first row and the second column is $-A_nG_n$, the final secular equation for the free vibrations is

$$A_n = 0 \tag{44}$$

The results of calculation of A_n are shown in Table 2 for n = 0 to 6. As seen in the Table, A_n is of the n-th order with respect to (W_i/G_i). Since W_i and G_i are given by Eqs. (33) and (39), respectively, W_i/G_i is given by a fractional equation of the first order with respect to ϵ, such as

$$\frac{W_i}{G_i} = -j\frac{2\epsilon(1+\phi_{i0}\tan\phi_{i0})}{(1-2q_i\phi_{i0}\epsilon)\sin 2\phi_{i0}\cdot e^{-2q_i\phi_{i0}\tan\phi_{i0}}} \tag{45}$$

A quantity frequently involved in obtaining the results in Table 2 is the gyrator proportionality factor g_{ij}, which, from Eq. (39) is expressed as

$$g_{ij} \equiv G_i/G_j = \frac{(1-2q_i\phi_{i0}\epsilon)\tan\phi_{i0}\cdot e^{-2q_i\phi_{i0}\tan\phi_{i0}}}{(1-2q_j\phi_{j0}\epsilon)\tan\phi_{j0}\cdot e^{-2q_j\phi_{j0}\tan\phi_{j0}}} \tag{46}$$

of p_i is nearly equal to p_j, ϕ_{io} comes close to ϕ_{jo}. Therefore, for small ϵ much less than unity, the gyrator proportionality factor becomes

$$g_{ij} = e^{2\phi_{i0}\tan\phi_{i0}\cdot(q_j - q_i)} \tag{47}$$

Since g_{ij} excludes ϵ, A_n is the n-th order equation with respect to ϵ. There exist n roots for ϵ that satisfy Eq. (44). That is, there are n eigenfrequencies corresponding to ϵs.

6. Calculated Examples

In the previous analysis, the electrode width, the electrode interval and the mass loading were taken to be arbitrary. However, it is very complicated to deal with a monolithic filter of many electrodes unless f_s given by Eq. (9), R by Eq. (10) and f_e are all taken to be equal for each section, respectively. At the same time, it is also convenient for adjustment of the resonant frequencies as well as improvement of the adjustment accuracy in practical fabrication that not only R but also the electrode width, respectively, are taken to be equal.

Therefore, in the following we confine ourselves to the case where the width of all electrodes is l_o and R is equal for all electrode-plated regions. ϕ_{oo} is assigned to the resonant frequency of the plate with a pair of electrodes of width l_o, which is given by Eq. (25). W_o is the impedance of the series resonance given by Eq. (33). For convenience of the calculation, two functions M and N are defined as

$$\left.\begin{aligned} M &= \sin 2\phi_{00}/2(1+\phi_{00}\tan\phi_{00}) \\ N &= e^{-2q_i\phi_{00}\tan\phi_{00}} \end{aligned}\right\} \tag{48}$$

In Figs. 4 to 7, the solid lines indicate the exact results [4, 7] calculated on the basis of Eq. (12) by means of a computer. The approximate results based on the present analysis are shown by the symbols O and ◉ in Figs. 4 to 6, and by the dashed lines in Fig. 7.

6.1 n = 2 (Two Pairs of Electrodes)

From Table 2, the fundamental equation for the eigenfrequencies is

$$(W_o/G_1)^2 + 1 = 0 \tag{49}$$

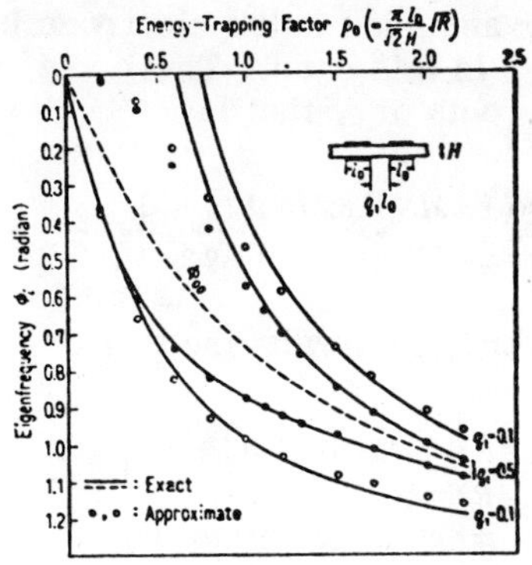

Fig. 4. The relationship between eigenfrequencies and energy-trapping factor in coupled 2-resonator system. Circles are approximate values calculated from Eq. (50) and lines are exact values.

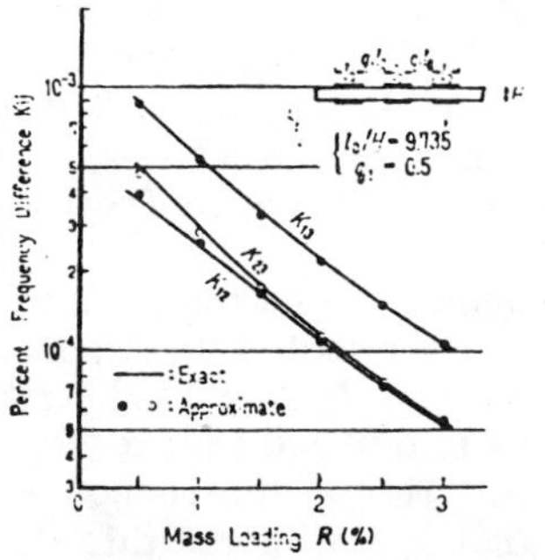

Fig. 6. Frequency difference between eigenfrequencies of coupled 3-resonator system as a function of mass loading

By solving the equation after substitution of Eq. (45), one obtains two roots of ϵ;

$$\epsilon_{\frac{1}{2}} = \pm MN - (MN)^2 2 q_1 \phi_{oo} \tag{50}$$

The corresponding eigenfrequencies are $\phi_1 = \phi_{oo} + \epsilon_1$, and $\phi_2 = \phi_{oo} + \epsilon_2$. The frequency interval Δ_{12} is

$$\Delta_{12} \equiv \phi_1 - \phi_2 = 2MN \tag{51}$$

On the other hand, to restore the original function $f_i = f_{oo} + \delta_i$ to the eigenvalue $\phi_i = \phi_{oo} + \epsilon_i$, the following relation reduced from Eq. (18) is utilized;

$$\delta_i / f_{oo} = -(R \sin 2\phi_{oo})\epsilon_i \tag{52}$$

of the eigenvalue difference K_{ij} in the frequency dimension is defined as

$$K_{ij} \equiv (f_j - f_i) \Big/ \sqrt{f_i f_j} = (\delta_j - \delta_i)/f_{oo} \tag{53}$$

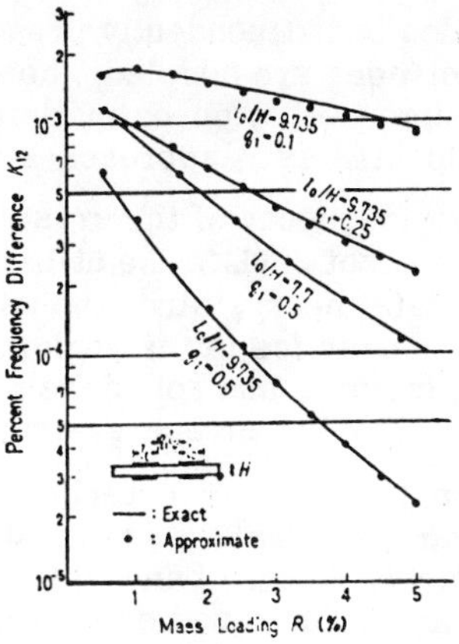

Fig. 5. Frequency difference between two eigenfrequencies of coupled 2-resonator system as a function of mass loading.

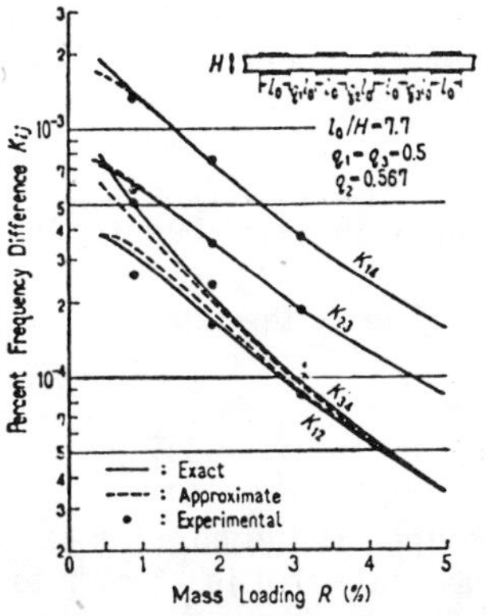

Fig. 7. Frequency difference between eigenfrequencies of coupled 4-resonator system as a function of mass loading.

K_{12} is given for two pairs of electrodes by Eqs. (51) and (52) as

$$K_{12} = R \sin 2\phi_{oo} \cdot 2MN \tag{54}$$

This is equal to the coefficient of the elastic coupling between the resonators. This coincidence occurs only for the case of two pairs of electrodes. The values necessary for the calculation of Eqs. (50) to (54) are tabulated in Table 1.

Figures 4 and 5 show the results for some concrete examples. The relation between the eigenfrequencies ϕ_i calculated on the basis of Eq. (50) and the energy-trapping factor p_o is shown in Fig. 4. In the figure, the ordinate for ϕ_i is increasing downward, which is, as usual, upward in the original frequency dimension f_i. The dashed line labelled ϕ_{oo} is for the case that one of two pairs of electrodes is independently present. If two pairs of electrodes are coupled, the eigenvalue separates into two. Approximation of the treatment deteriorates as ϕ_i approaches zero, that is, the cutoff frequency of the crystal plate. However, this does not obviate the utility of the present approximate theory, since monolithic filters are used in practice in the region of larger energy-trapping factors, and not in the region near $\phi_i = 0$. Even for larger energy-trapping factors the approximation also deteriorates if the electrode interval is as narrow as $q_1 = 0.1$, that is, when a relatively greater band width is required. However, an improved expression can be given if the eigenvalue difference or the coupling factor is employed instead.

Figure 5 shows the relation between the coupling coefficient due to Eq. (54) and the mass loading, if the electrode dimension is taken as a parameter. The approximation is satisfactory, including the case of $q_1 = 0.1$. It is quite convenient for the design of monolithic filters that the coupling coefficient can now be calculated simply in good approximation.

6.2 n = 3 (Three Pairs of Electrodes)

As shown in Table 2, the fundamental equations for n = 3 are

$$W_0 = 0, \quad (W_0/G_i) = \pm\sqrt{1+g_{ji}^2} \tag{55}$$

where the subscript i is the number assigned to the electrode interval, that is, 1 or 2 for the present case. The number j is correspondingly 2 or 1. g_{ji} is calculated by means of Eq. (47) for a given electrode dimension and mass loading. As such, one obtains three values of ϵ

$$\left.\begin{aligned} \epsilon_{\substack{1\\3}} &= \pm\sqrt{1+g_{ji}^2}\,MN - (\sqrt{1+g_{ji}^2}\,MN)^2\, 2q_i\phi_{00} \\ \epsilon_2 &= 0 \end{aligned}\right\} \tag{56}$$

According to the definition of Eq. (53), the eigenfrequency intervals are

$$\left.\begin{aligned} K_{13} &= R\sin 2\phi_{00}\cdot\sqrt{1+g_{ji}^2}\cdot 2MN \\ K_{\substack{12\\23}} &= R\sin 2\phi_{00}\cdot\{\sqrt{1+g_{ji}^2}\,MN \\ &\quad \mp(\sqrt{1+g_{ji}^2}\,MN)^2\, 2q_i\phi_{00}\} \end{aligned}\right\} \tag{57}$$

The numerical results are shown in Fig. 6. It is noted that the eigenfrequency distribution is asymmetrical as $K_{12} > K_{23}$, while $K_{12} = K_{23}$ in usual electrical circuits.

6.3 n = 4 (Four Pairs of Electrodes)

There are four fundamental equations for n = 4.

$$W_0/G_i = \pm j a_{\substack{1\\2}}$$

where

$$a_{\substack{1\\2}} = \{\sqrt{g_{2i}^2 + (g_{1i}+g_{3i})^2} \mp \sqrt{g_{2i}^2 + (g_{1i}-g_{3i})^2}\}/2 \tag{58}$$

i is the number assigned to the electrode intervals, that is 1, 2 or 3 in this case. There are correspondingly four roots of ϵ, that is

$$\left.\begin{aligned} \epsilon_{\substack{1\\4}} &= \pm a_2 MN - (a_2 MN)^2\, 2q_i\phi_{00} \\ \epsilon_{\substack{2\\3}} &= \pm a_1 MN - (a_1 MN)^2\, 2q_i\phi_{00} \end{aligned}\right\} \tag{59}$$

The eigenfrequency intervals are then

$$\left.\begin{aligned} K_{\substack{14\\23}} &= R\sin 2\phi_{00}\cdot a_{\substack{2\\1}}\cdot 2MN \\ K_{\substack{12\\34}} &= R\sin 2\phi_{00}\cdot\{(a_2-a_1)MN \\ &\quad \mp(a_2^2-a_1^2)(MN)^2\, 2q_i\phi_{00}\} \end{aligned}\right\} \tag{60}$$

The calculated results are shown in Fig. 7. The mark ⊙ in the figure indicates the value experimentally obtained.* In this case, as in the previous case of n = 3, the frequency interval K_{34} on the higher-frequency side is wider than K_{12} on the lower-frequency side. The relation $K_{12} = K_{34}$ is always retained in electrically coupled resonant circuits. That is, there is always present a symmetric relation between f_1 and f_4, and f_2 and f_3 with respect to the central frequency f_{00}. The asymmetric distribution of the eigenfrequencies in the monolithic filters is due to the fact that the image gyrator constants given by Eq. (39) consist of the term of ϵ dependent on the frequency.

The eigenfrequencies are important in giving the zeros of the transmission function of a filter. As discussed, the eigenfrequency distribution in monolithic filters is quite different from that in electrical filters. This is the reason why the design and the analysis of monolithic filters are so complex. As seen in Figs. 6 and 7, the approximate formulas developed are practical enough and capable of expressing the asymmetry of the eigenfrequency distribution. The experimental data also proved the validity.

7. Conclusion

In the present paper a monolithic filter has been analyzed as a coupled vibratory system. First, the system is expressed in terms of a sequence of transmission matrices. It is found that the electrode-plated regions are expressed in terms of an equivalent series resonant circuit and the non-electrode regions in terms of an image gyrator. On the basis of the equivalent circuit derived, the approximate formulas capable of calculating the eigenfrequencies are given when the electrode width, the electrode interval

---

*The material used in the experiment is an AT-cut quartz crystal of thickness 0.3 mm. Four pairs of electrodes of 3×3 mm^2 were formed in the direction of the electrical axis at intervals of 1.5—1.7—1.5 mm. The thickness-shear modes are utilized. Four eigenfrequencies were measured for each pair of shorted electrodes.

and the mass loading are known. Some calculated examples of monolithic filters of practical dimensions are shown for n = 2, 3 and 4, where the electrode width and the mass loading, respectively of each element are taken to be equal.

On the other hand, the asymmetry of the frequency distribution, unlike in electrically coupled resonant circuits, has been pointed out for the first time. This is also confirmed by experiment. The formulas derived provide a practical enough approximation, including the asymmetric characteristic of the eigenfrequency distribution.

The design of a monolithic filter making use of the formulas derived in the present analysis is to be reported in due course.

Acknowledgement. The author expresses his gratitude to Mr. Hanazawa, Head of the Circuit Component Division, NTT, for his guidance and to Prof. Onoe, Tokyo University, for his discussions. Messrs. Sawamoto and Tsukamoto are also acknowledged for their valuable comments and cooperation in the numerical calculation.

REFERENCES

1. Onoe, Jumonji and Kobori: High-frequency crystal filters employing multiple-mode resonators vibrating in trapped energy modes, Proc. of 20th Ann. Symp. on Frequency Control, p. 266 (April 1966).

2. R. A. Sykes and W. D. Beaver: High-frequency monolithic crystal filters with possible application to single-frequency and single-side band use, Proc. of 20th Ann. Symp. on Frequency Control, p. 288 (April 1966).

3. W. D. Beaver: Analysis of elastically coupled piezoelectric resonators, J. Acoust. Soc. Am., 43, 5, p. 972 (1968).

4. P. Schnabel: Frequency equations for n mechanically coupled piezoelectric resonators, Acustica, 21, p. 351 (1969).

5. Onoe: Thickness-twist vibrations of a piezoelectric plate, Trans. I. E. C. E., Japan, 52-A, 10, p. 403 (Oct. 1969): available in English in ECJ, same date, p. 38.

6. Jumonji and Onoe: Analysis of a trapped-energy filter with multiple modes by means of equivalent electrical transmission line theory, Joint Conv. Record of Electrical and Electronics Engrs. of Japan, 1233, (1969).

7. Onoe and Okada: Analysis of contoured piezoelectric resonators vibrating in thickness-twist modes, Proc. of 23rd Ann. Symp. on Frequency Control, p. 26 (1969).

8. Onoe and Jumonji: Analysis of a trapped-energy piezoelectric resonator, Jour, I. E. C. E., Japan, 48, 9, p.1574 (Sept. 1965): available in English in ECJ, same date, p. 84.

9. V. Belevitch and Y. Kamp: Theory of monolithic crystal filters using thickness-twist vibrations, Philips Res. Repts., 24, p. 331 (1969).

Submitted September 30, 1970

APPENDIX

Derivation of Eqs. (31) and (32)

Substituting Eqs. (23), (27) and (28) into each element in the matrix $[E_i]$ of Eq. (13), one obtains

$$[E_i]=\begin{pmatrix} 1+2\varepsilon\tan\phi_{i0}\cdot(1+\phi_{i0}\tan\phi_{i0}), & -2j\varepsilon(1+\phi_{i0}\tan\phi_{i0})\cdot(1+\varepsilon\tan\phi_{i0})/\cos^2\phi_{i0} \\ j\sin 2\phi_{i0}\{1+\varepsilon\phi_{i0}\cdot(\tan^2\phi_{i0}-1)\}, & 1+2\varepsilon\tan\phi_{i0}\cdot(1-\phi_{i0}\tan\phi_{i0}) \end{pmatrix} \quad \text{(A. 1)}$$

where a Taylor expansion of the first order with respect to ε is performed. Furthermore, neglecting the first-order term which is much smaller than unity, one obtains

$$[E_i]=\begin{pmatrix} 1, & -2j\varepsilon(1+\phi_{i0}\tan\phi_{i0})/\cos^2\phi_{i0} \\ j\sin 2\phi_{i0}, & 1 \end{pmatrix} \quad \text{(A. 2)}$$

The equation is reduced to Eq. (31). Then, Eq. (31) is synthesized to obtain

$$[E_i]=\begin{pmatrix} 1+2\varepsilon\tan\phi_{i0}\cdot(1+\phi_{i0}\tan\phi_{i0}), & -2j\varepsilon(1+\phi_{i0}\cdot\tan\phi_{i0})/\cos^2\phi_{i0} \\ j\sin 2\phi_{i0}\{1+\varepsilon\tan\phi_{i0}\cdot(1+\phi_{i0}\tan\phi_{i0})\}, & 1+2\varepsilon\tan\phi_{i0}\cdot(1-\phi_{i0}\tan\phi_{i0}) \end{pmatrix} \quad \text{(A. 3)}$$

By comparing each element in Eq. (A. 2) with that in Eq. (A. 3), one finds that the diagonal elements are equal to one another, and others also coincide if the first-order term with respect to ε is neglected.

Similarly, substituting Eqs. (23) and (27) into $[S_i]$ of Eq. (14), one obtains

$$[S_i]=\begin{pmatrix} \alpha_i, & 2j\alpha_i\tan(\phi_{i0}+\varepsilon) \\ j\{(1-\alpha_i^2)/2\alpha_i\}\cdot\cot(\phi_{i0}+\varepsilon), & \alpha_i \end{pmatrix} \quad \text{(A. 4)}$$

This is reduced to

$$[S_i]=\begin{pmatrix} 1, & -2j\{\alpha_i^2/(1-\alpha_i^2)\}\tan(\phi_{i0}+\varepsilon) \\ 0, & 1 \end{pmatrix}\cdot\begin{pmatrix} 0, & jG' \\ j/G', & 0 \end{pmatrix}\begin{pmatrix} 1, & -2j\{\alpha_i^2/(1-\alpha_i^2)\}\tan(\phi_{i0}+\varepsilon) \\ 0, & 1 \end{pmatrix} \quad \text{(A. 5)}$$

without any restriction. By neglecting the ε term one finally obtains Eq. (32).

Section II-C
Crystal Filters Containing Acoustically Coupled Resonators

This group of papers describes some of the different types of filters that have been designed with acoustically coupled resonators, and they also illustrate the different design techniques that have been used.

The first monolithic telephone channel filter, as described by Sykes, Smith, and Spencer, was a tandem connection of eight resonators. Some of the design philosophy was based on image parameter theory, but the actual design itself was performed via Dishal's [1] equations derived from insertion loss theory for a tandem connection of resonators. An alternate approach is to make use of the large body of knowledge that has been built up on inductor–capacitor filters [2] and to relate this information to the design of crystal filters via electrically equivalent circuits for the crystal resonators. This is the technique that was used by Sheahan in which he describes the polylithic approach to channel filter design. Rennick expands on this equivalent circuit approach to filter design. The use of equivalent circuits lends itself to designing filter structures that are very difficult to treat by other means. An example of this can be seen in the capacitor bridging approach as described by Lee. Capacitor bridging and its usefulness for achieving finite frequency attenuation poles has been known for a number of years. Its usefulness was, however, limited until a method was found for using it in exact insertion loss filter designs such as is described by Lee.

Simpson, Finch, and Weeman describe some unusual filter structures where they combine inductors and capacitors with the crystal filters to achieve complex band-elimination filters that would have been difficult to achieve by any other means.

Smythe gives some examples of how acoustic coupling can be used for high-frequency crystal filters that employ the overtone modes of the crystal. Filters have been built at frequencies as high as 350 MHz by this means.

Finally, Watanabe and Tsukamoto describe the design of an experimental 14-resonator telephone channel filter that was intended for an out-of-band signaling FDM system. Resonator Q values of 3×10^5 with a resulting flat passband response were obtained with this filter. These were obtained by optimizing the plateback and the resonator inductances for the highest Q value.

References

[1] M. Dishal "Design of dissipative band-pass filters producing desired exact amplitude-frequency characteristics," *Proc. IRE*, vol. 37, pp. 1050–1069, Sept. 1949.

[2] H. J. Orchard and G. C. Temes, "Filter design using transformed variables," *IEEE Trans. Circuit Theory*, vol. CT-15, pp. 385–408, Dec. 1968.

MONOLITHIC CRYSTAL FILTERS

R. A. Sykes, Fellow IEEE, W. L. Smith, Senior Member, IEEE, and W. J. Spencer, Member, IEEE
Bell Telephone Laboratories
Allentown, Pa.

Summary

High frequency bandpass filters having bandwidths of 0.1% or less and high stop-band discrimination have traditionally been realized by combining quartz crystal units with balanced transformers and additional electrical components. Recent developments in the theory of vibrations of partially electroded, finite quartz plates lead to the concept of controlled mechanical coupling between individual resonators fabricated on a single plate. With proper control of inter-resonator coupling, the realization of highly selective filter functions with a single quartz device have been achieved.

An analysis of the thickness shear quartz resonator is presented, deriving the interdependence of electrode geometry and mass loading required to obtain relative freedom from inharmonic modes. Then, the resonances resulting from a system consisting of two or more of these singly-resonant regions in close geometric proximity are calculated. From the frequency separation of the resonances of a pair of coupled, singly-resonant regions, the equivalent coefficient of coupling is obtained as a function of electrode geometry and mass loading.

The synthesis of a desired filter function is then carried out in terms of an equivalent all-pole network. The required coefficients of coupling between adjacent resonators are obtained in the conventional manner for the desired inband ripple and from these coefficients, the filter geometry can be calculated.

Examples are given of monolithic filters of various complexities, ranging from the simplest case of the two-pole filter to a structure of ten coupled resonators. Measured transmission characteristics are shown to be in quite good agreement with theoretical characteristics.

Additional modes can be generated due to a finite plate geometry and large area close spaced electrodes. These modes will produce distortion near the band edge as well as sharp unwanted transmission regions above the passband. Some of these limitations are discussed.

It should also be recognized that the impedance level of the device is determined within fairly narrow limits by the requirement that essentially singly-resonant elements be employed in the structure. Even with near optimum resonator geometries, spurious transmission levels can be expected within about 0.5% to 1.5% above the band, at discrimination levels ranging from 40 to 50 db which may be a severe limitation in some cases.

Within these limitations, the monolithic crystal filter has shown the capability of producing highly selective high frequency filters having nearly ideal performance in the near vicinity of the passband. They provide the inherent advantages of simplicity, small size, and economy which should prove to be of significant importance in communications systems of many types.

I. Introduction

Wave filters used to select different frequency bands of the electromagnetic spectrum are an integral part of radio communication equipment. Since their introduction there has been considerable progress in the art and design of radio frequency filters. Initially they were composed of large size air core inductors and the filter comprised the major volume of any communication equipment. The introduction of ferrite cores provided the first reduction in the size of filters but, compared with that resulting from the use of multipurpose miniature vacuum tubes, the filter still was a large portion of the final equipment. The quality factor (Q) of air or ferrite coils has always been limited to the 300-500 range and hence for single and double sideband type filters the maximum frequency, for reasonable selectivity for the bandwidths required, is about 500 kHz and this was the principal reason for standardizing on 465 kHz for Intermediate Frequency filters. The introduction of quartz crystal units (with Q value ranging from 10^4 to 10^6) in filters, and later the resonant disc mechanical filter,[1] served to greatly improve the selectivity and reduce their size. Crystal filters still required the use of inductors at low frequencies to obtain sufficient bandwidth and at high frequencies for hybrid balance. Here again the advancement in semiconductors as active elements for gain left the major volume of communication equipment occupied by the filter. The latest improvement in filters, the monolithic crystal filter,[2,3,4] in which the entire filter is composed of a series of coupled trapped energy resonators on the same quartz plate has again reduced the size and complexity of the radio frequency filter. The use of thin film circuitry for passive components and beam lead semiconductors is now at such a stage that it could probably be deposited on the same substrate as the monolithic crystal filter.

This paper will describe the operation of monolithic filters, from the standpoint of the coupled vibrations of the quartz plate and its equivalent electrical circuit. The details of the plate vibration are in section III and of the design of the filters from its equivalent electrical circuit in section IV. The other sections

Reprinted from *1967 IEEE Int. Conv. Rec.*, pt. II, Mar. 20–23, 1967, pp. 78–93.

show the performance that has been obtained on some experimental models (V) together with a general discussion of the principal advantages and uses of the new device with some of the inherent limitations (VI).

II. General Considerations

The various types of filters that have been used in most electronic communication equipment may be broadly described as 1) electrical, 2) mechanical, 3) crystal. The network configurations that have been used for filters in most cases fall into one of the 3 categories shown in Figure 1, 2 and 3 which are the basic ladder, lattice and bridge-T type of networks.

The design of transmission networks with selective properties has been achieved by several methods. The earliest was that of the image parameter theory and is still used to a large extent for determining possible configurations and feasibility. To obtain a final design, more modern methods such as insertion loss techniques are now available. These procedures will lead one to the design of filters having maximum loss outside the transmission band with a specified minimum or no ripple within the transmission band. Control may also be exercised over phase linearity.

The number of elements in these types of networks and sections is dependent upon the degree of selectivity required. The elements Z_1, Z_2, etc., are shown in a generalized form and are usually comprised of inductors and capacitors or combinations of the two. In Figures 1 and 2, where only a capacitor or an inductor is used for each element, low pass and high pass filters result. When the elements comprise both inductors and capacitors in resonant or antiresonant combinations, bandpass characteristics can result. In Figure 1, frequencies of high loss may be obtained with antiresonant elements in the series branches or resonant elements in the shunt branches. In Figure 2, they are obtained by balance between the elements in the series and lattice arms. In Figure 3, frequencies of high loss may be obtained by cancellation through the bridge arm.

What has been said above with reference to the use of inductors and capacitors in electromagnetic filters also applies generally to the design of similar structures involving mechanical elements. In mechanical filters there is a correspondence of mass to inductance, stiffness to capacitance and mechanical resonance to electrical resonance. Here also the degree of selectivity depends on the number of resonators used in the filter. A transducer may be used to convert from electrical to mechanical energy and vice versa as shown in Figure 4. In some electromechanical filters the transducer may be designed to improve the selectivity. The overall input and output device is dealt with in electrical terms, but the mechanical filter is usually designed separately with transducers to convert from electrical to mechanical energy. An example of a highly selective mechanical filter is one using a series of elastically coupled resonant discs.[5] Many mechanical filters have been designed using resonators in the flexural, extensional and shear modes, but very few have had widespread use in comparison with the disc type mechanical filter.

Some of the electrical elements in Figures 1, 2, and 3 may be replaced by piezoelectric resonators and this network has often been called a "crystal filter." The use of piezoelectric materials allows the use of mechanical resonators to be connected electrically to form an electronic filter. The quartz resonators are combined with electrical elements to produce the desired degree of frequency selectivity. The high mechanical Q and low temperature-coefficient of frequency found in quartz has made possible the design of small highly selective filters that have found wide application in many radio communication systems and particularly as narrow band and voice frequency filters in telephone multiplex systems.

Each of the various filters; electrical, mechanical or crystal makes use of individual components coupled electrically or mechanically to produce the desired frequency selectivity. Various attempts have been made to produce transducer driven electromechanical filters and, except for the coupled resonant disc type, they have found only limited application due to their complexity and cost.

The recent increase in the understanding of the propagation of shear waves in quartz plates with partial electrodes has made possible integrated electromechanical filters for frequencies above 1 MHz. These monolithic crystal filters use a series of thickness-shear resonators in a single quartz plate. The impedance and coupling of these individual resonators may be controlled by the size and separation of the electrodes on the quartz plate. Using modern evaporation and photo-etch techniques it is possible to design a complete filter on a single quartz wafer with various bandwidths and termination impedances. The degree of selectivity depends on the number of resonators and since they may be added by simply evaporating additional pairs of electrodes, highly selective, economical filters may be obtained.

Figure 5A schematically shows two such resonators on a quartz plate whose area is large compared to the electrode area. These two sets of electrodes are used to form coupled resonators separated by a distance "d". These mechanical resonators are coupled through the elastic properties of the crystal and would result in a mechanical filter even if no piezoelectric effect were present. Since the material is piezoelectric, it is no longer necessary to consider the transducer separate from the mechanical filter. An equivalent electric circuit for this electromechanical filter is immediately evident and is shown in

Figure 5B. When more than two resonators are used, the same equivalent lattice network results but with an increased number of resonators alternately in the lattice and series arms. As will be shown later, this simple lattice equivalent will not completely specify the high loss region but is a good approximation for the low loss transmission region. Figures 6A, 6B and 6C show how this equivalent lattice network may be used to exhibit selective properties. This approach follows the conventional design of crystal filters that has been used for many years. The equation relating termination resistance (R_o) to bandwidth (Δf) and resonator inductance (L) is given by

$$R_o = 2\pi \, \Delta f L \, . \qquad (1)$$

When the zeros and poles for the reactances in the equivalent lattice network are as shown in Figure 6D, the image impedance and insertion loss characteristics are as shown in Figures 6E and 6F. The equation relating the characteristic impedance to resonator properties is:

$$Z_I = \frac{1}{2\pi f C_o} \left[\frac{(f^2 - f_A^2)\,(f_B^2 - f^2)}{(f_{A1}^2 - f^2)\,(f_{B1}^2 - f^2)} \right]^{1/2} \qquad (2)$$

It will be noticed that while a second transmission band exists above the first, its characteristic resistance is high in comparison with that for the first. When terminated in a resistance (R_o) appropriate for the first band the loss at the second band will be high due to reflection. A small amount of capacitance bridged from input to output will serve to reduce the width of the second band, increase its characteristic resistance and also move the two loss peaks nearer the transmission band. If we assume the bandwidth Δf is 1/3 or less of the total frequency difference from the zero to the pole ($f_{A1}-f_A$) of the resonator, equation 2 reduces to a good approximation to equation 1. Because of the improved termination conditions the bandwidth, for a small inband ripple (< .25 db), will be more than double that for the conventional design. Conversely, for a given practical bandwidth using the same resonators, the termination resistance will be one half that required by the conventional design.

Since the maximum separation between the resonant and antiresonant frequencies of quartz resonators is the order of 0.3% it would appear that monolithic filters of this type would find their best use above 5 MHz for such applications as single frequency, single sideband and narrow band FM. For bandwidths greater than 0.1% one can resort to the same technique now employed in conventionally designed crystal filters to widen the band but still retain the improved resistance characteristics. The addition of series or shunt inductors will separate the critical frequencies and the design approach shown in Figure 7 may be used. For example, the separation of the extreme antiresonant frequencies with a coil shunting the crystal resonator may be as high as 7.5% for AT type resonators. This should result in usable bandwidths between the resonant frequencies as high as 1% to 2%. Here again equation (1), to a good approximation, defines the termination conditions. Relatively low Q inductors may be used. If high Q inductors are used and there is coincidence of poles and zeros except at the cut-off frequencies, then bandwidths as high as 10% may be obtained.

The above use of image parameter theory and equivalent lattice networks shows the advantage of this method to arrive at a simple general analysis and to determine initial feasibility. The critical frequencies of the lattice are the coupled frequencies and not the natural frequencies of the individual resonators. The final design approach then will be based on an equivalent ladder network where the natural frequencies of the resonators are used together with the coupling as derived from the physical size and disposition of the electrodes forming the individual resonators.

With this general approach the design of monolithic crystal filters will be dependent upon the characteristics obtainable from trapped energy resonators and the geometric relationships determining their reactances and the coupling between them. Insertion loss design techniques may then be used to obtain the specific transmission characteristics desired.

III. Analysis of Single and Coupled Mode Thickness - Shear Resonators

If one were able to solve all the equations of linear piezoelasticity it would be possible to predict the bandwidth and termination impedance of a monolithic crystal filter as a function of the plate and electrode dimensions. The approximate equations of Mindlin[5] for transverse thickness modes in rotated y-cut quartz plates are adequate to determine the resonant frequencies and uncoupled equivalent electrical parameters as a function of plate and electrode geometries and are completely sufficient to understand the physics of the device. Although this problem may be formulated rather completely, its solution is presently beyond the capability of even the largest computers. Fortunately the transmission characteristics may be obtained more simply from the equivalent network approach as will be shown in section IV.

This section will briefly review the work on plate vibrations leading to the monolithic filter, present the pertinent solution of Mindlin's equations which determine the modes of vibration of the mechanical part of the filter as well as give the electrical properties of the uncoupled resonators. The theory* used here to describe the

*H. F. Tiersten has influenced much of this investigation by his contribution to the general theory of plate vibrations.

vibration of coupled thickness-shear resonators has been formulated by K. Haruta and is discussed in greater detail by Beaver.[6]

The monolithic filter is based on the coupling of acoustic vibrations trapped or bound to the electroded regions of AT cut quartz resonators. The trapping is related to a cut-off frequency for thickness-shear modes similar to that used for electromagnetic waves in microwave cavities. Bechmann[7] experimentally observed a relation between the mode spectrum of thickness-shear vibrators and the electrode size. Mortley[8] correctly associated this phenomena with a cut-off frequency for acoustic modes similar to that found in waveguide resonators. Mindlin and Lee[9] have shown how this effect is related to elastic waves propagated in a partially electroded plate.

The principal mode in AT cut quartz plates is the thickness-shear mode. In an infinite plate the frequency of this mode is given by

$$\omega = \frac{m\pi}{2b}\sqrt{\frac{c_{66}}{\rho}} \tag{3}$$

where 2b is the plate thickness, c_{66} the shear elastic constant, ρ the density of quartz and m the order of the harmonic thickness overtone. Generally only the odd modes may be driven piezoelectrically due to electrode symmetry. These modes have a displacement u parallel to the diagonal axis in the quartz plate (this corresponds to the x-axis in IRE[10] notation) which may be piezoelectrically excited by applying an electric field normal to the plate surface. Since the piezoelectric coupling in quartz is very low, the modes in quartz plates are given very closely by solution of only the mechanical equations neglecting electrical effects. Of course, to predict the equivalent electrical parameters it is necessary to include the piezoelectric effects.

When one attempts to find the modes of vibration in a finite plate, it becomes quickly apparent that the spectrum is exceedingly complex. It is possible to predict the frequency of many of the modes near the lowest thickness-shear mode in quartz plates by an approximate method developed by Mindlin.[5] Even here to include the flexure, face-shear and extensional modes presents a rather formidable problem. By considering only the thickness-shear mode near the frequency given by equation (3) Mindlin's equations reduce to[11]

$$\frac{\partial^2\psi}{\partial x_1^2} + \frac{K^2}{b^2}\left(\frac{\omega^2}{\bar{\omega}^2} - 1\right)\psi = 0 \tag{4}$$

for rotated y-cut quartz plates. Where K depends only on the elastic constants, b is one-half the plate thickness, $\bar{\omega}$ is given by equation (3) and u_1 the displacement in the x_1 direction is $u_1 = x_2\psi$. In the Mindlin approximation, ψ is the rotation of the plate normal about the x_3 - axis. It is also the S_6 shear strain when thickness-shear alone is considered.

For an infinite strip[12] bounded by the planes $x_1 = \pm a$, the boundary conditions are that $\psi(\pm a) = 0$. If we consider only the symmetric solutions of equation (4) then

$$\psi = A\cos\xi x_1 \tag{5}$$

if

$$-\xi^2 + \frac{K^2}{b^2}\left(\frac{\omega^2}{\bar{\omega}^2} - 1\right) = 0. \tag{6}$$

Since $\psi(\pm a) = 0$, then $\xi = \frac{n\pi}{2a}$ where n = 1, 3, 5...

These solutions correspond to an infinite series of thickness-shear modes all with a half wavelength through the plate thickness and 1, 3, 5 ..... half wavelengths along the plate length. The frequency of these modes is given by

$$\omega = \bar{\omega}\left[\frac{n\pi^2}{2K}\left(\frac{b}{a}\right)^2 + 1\right]^{1/2} \tag{7}$$

For AT cut quartz $K \sim .8$ and many of these modes lie within a few percent of $\bar{\omega}$ for length-to-thickness (a/b) ratios above thirty. These modes are generally called anharmonic thickness-shear modes as opposed to the harmonic overtones given by n = 1, 3, 5 ..... in equation (3). There is an infinite series of these anharmonic modes associated with each harmonic overtone of thickness shear. The frequency of the infinite plate $\bar{\omega}$ is a cut-off frequency for these thickness-shear modes. For $\omega > \bar{\omega}$, ξ is real and the modes exist as real propagating modes. When $\omega < \bar{\omega}$, ξ is imaginary and the solutions of equation (4) are no longer propagating modes but are exponentially decaying everywhere. This cut-off frequency has particularly important implications in partially electroded plates.

The anharmonic thickness-shear modes may be greatly suppressed by using a partial electrode on the quartz plate. In the case of an infinite electrode strip normal to the x-axis on an infinite AT quartz plate as shown in Figure 8, equation (4) becomes

$$\frac{\partial^2\psi_p}{\partial x_1^2} + \frac{\bar{K}^2}{b^2}\left(\frac{\omega^2}{\bar{\omega}_p^2} - 1\right)\psi_p = 0 \tag{8}$$

for the electroded portion of the plate. The dependence of $\bar{K}$ and $\bar{\omega}_p$ on electrode thickness is given by Mindlin and Lee.[10] In most practical applications $\bar{\omega} \geq \omega \geq .97\bar{\omega}$ and $\bar{K} \approx K$. Then at $x_1 = (\pm e/2)$ the electrode edge

$$\psi\left(\pm\frac{e}{2}\right) = \psi_p\left(\pm\frac{e}{2}\right) \tag{9a}$$

and

$$\left.\frac{\partial\psi}{\partial x_1}\right|_{\pm\frac{e}{2}} = \left.\frac{\partial\psi_p}{\partial x_1}\right|_{\pm\frac{e}{2}} \tag{9b}$$

These approximations are only good for $\bar{\omega}_p/\bar{\omega} \approx 1$. Assuming

$$\psi = A \cos \xi x_1 \tag{10a}$$

and

$$\psi_p = Be^{-\xi_p x} \tag{10b}$$

where

$$\xi_p^2 + \frac{K^2}{b^2}\left(\frac{\omega^2}{\bar{\omega}_p^2} - 1\right) = 0, \tag{11}$$

and ξ is given by equation (6). Then substituting (10) into (9) and simplifying one obtains

$$\tan \xi_p \frac{e}{2} = \xi/\xi_p. \tag{12}$$

The roots of equation (12) give the resonant frequency ω as a function of the material constants of the quartz and the geometry of the plate and electrode. There is at least one solution of equation (12) for which ξ_p is real and ξ is imaginary to the approximation assumed for this solution. This root corresponds to the fundamental thickness-shear resonance of the infinite electrode strip on an infinite quartz plate. The frequency of this mode is very close to $\bar{\omega}$ and the displacement is sinusoidal in the electroded region and exponentially decaying into the surrounding region. In this case most of the mechanical energy associated with this mode is trapped or bound to the electroded region. Essentially the same thing happens if the quartz plate is finite. This exponential character of the fundamental thickness-shear mode in the unelectroded region permits the finite plate to be supported with a low Q mounting with very little loss.

There may, of course, be more than one value of ω for which $\bar{\xi}$ is real and ξ is imaginary. When this occurs, there are additional modes trapped and these produce relatively strong responses in addition to the fundamental thickness-shear mode. It is possible to determine the value of $\bar{\omega}_p$, the cut-off frequency of the electroded portion of the plate, e the electrode length and 2b the plate thickness for which only the fundamental mode is trapped.

The value of ω for which ξ becomes real from equation (4) is $\omega = \bar{\omega}$. Substitution of this value in equation (12) and simplifying then

$$e/2b = \frac{r\pi}{\sqrt{2}K}\left(\frac{\bar{\omega}}{\bar{\omega}-\bar{\omega}_p}\right)^{1/2} \tag{13}$$

When r = 0, the fundamental thickness-shear mode is trapped which occurs independently of the plate and electrode dimensions. When r = 1 the next higher mode becomes trapped. Since the assumed solution in equation (10) included only the symmetric modes, this value of r gives the relation between the electrode length-to-plate thickness e/2b and electrode thickness $(\bar{\omega}-\bar{\omega}_p/\omega)^{1/2}$ for which the next symmetric mode is trapped. This mode is higher in frequency than the fundamental and will have approximately three half wavelengths of displacement ψ along x_1.

Equation (13) may be written as

$$e/2b \leqq 2.8\left(\frac{\bar{\omega}}{\bar{\omega}-\bar{\omega}_p}\right)^{1/2} \tag{14}$$

for AT cut quartz when r = 1. This equation gives the maximum electrode length along x_1 to trap only the fundamental thickness-shear mode. Although the thickness-twist modes are not included in this discussion, it is possible to derive a similar equation for the electrode length along x_3 (the z′ axis) to trap only the fundamental mode. In this case

$$e/2b \leqq 2.2\left(\frac{\bar{\omega}}{\bar{\omega}-\bar{\omega}_p}\right)^{1/2} \quad \text{(thickness-twist).} \tag{15}$$

Equations (14) and (15) may be used to design singly resonant electroded areas on a quartz plate. These regions will be singly resonant only if the plate is very large with respect to the electrode area. However, in most practical cases, it is possible to significantly suppress all the anharmonic overtones of the thickness-shear mode.

When two or more of these singly resonant electroded areas are placed on the same substrate, the modes associated with each region immediately couple through the exponential character of the displacement in the unelectroded region.[6] If these two regions are sufficiently separated, the coupling is quite small. When two of these regions are brought together, two frequencies are observed for the system in accordance with established theory of coupled resonators.

The effect of electrode size and separation on the frequency of such a coupled system may be determined from equations (4) and (8) with the proper solutions for the geometry shown in Figure 9. The origin of the coordinate system is assumed to be at the plate center and equidistant between the two electrodes. The displacements are different for the region between the two electroded regions depending on whether the modes trapped under the electrodes are in phase (symmetric case) or out of phase (antisymmetric case). The displacements for the two cases are:

Electroded Region,

$$\psi = B \cos (\xi x_1 + c) \tag{16a}$$

Central Region,

$$\psi = A \begin{matrix} \cosh \xi x_1 \\ \sinh \xi x_1 \end{matrix} \quad \begin{matrix} \text{(symmetric)} \\ \text{(antisymmetric)} \end{matrix} \tag{16b}$$

Surrounding Region

$$\psi = De^{-\xi x_1} \qquad (16c)$$

The boundary conditions at the electrode edge are the same as previously given in equation (9) for the single electrode. Substitution of (16) into (9) at $x_1 = d$ and d+e gives

$$\tanh \xi d = \frac{\alpha^{-1} \tan \bar{\xi} e - 1}{\alpha \tan \bar{\xi} e + 1} \quad \text{(symmetric)}$$

and (17)

$$\coth \xi d = \frac{\alpha^{-1} \tan \bar{\xi} e - 1}{\alpha \tan \bar{\xi} e + 1} \quad \text{(antisymmetric)}$$

where

$$\alpha = \frac{\left(1 - \frac{\omega^2}{\bar{\omega}^2}\right)^{1/2}}{\left(\frac{\omega^2}{\omega_p^2} - 1\right)^{1/2}} .$$

The roots of equation (17) give the frequencies of the symmetric and antisymmetric resonant modes of two coupled resonators as a function of the plate and electrode dimensions as shown in Figure 10. The upper branch is the antisymmetric mode and the lower branch the symmetric mode. As $d \to \infty$ these two modes become degenerate.

The maximum separation of the two frequencies is determined by the frequency difference between the fundamental thickness-shear mode and the first antisymmetric anharmonic thickness-shear mode. This separation is a function of the total electrode length of the two sections. This plot was made for $\bar{\omega}/\bar{\omega}_p = 1.02$ and $e/2b = 12.5$ for the thickness-shear modes. A similar plot for thickness-twist modes, that is electrodes separated along the x_3 axis, shows that the frequency difference between the two modes is about half that shown in Figure 10 for equivalent electrode separation. This is due to the different values of the elastic constants for shear and for twist modes. By making the split between the electrodes at an arbitrary angle with respect to either the x_1 or x_3 axis any frequency separation between that found for the thickness-shear and thickness-twist modes may be obtained.

The roots of equation (17) may be plotted in several ways. One may wish to know the frequency separation of the two modes as a function of electrode thickness for a particular separation and electrode length, or the separation as a function of electrode length for a particular separation and electrode thickness. The relation of the individual electrode length and thickness for the maximum suppression of unwanted modes must also be considered, and as shown in Section V where particular filters are discussed, the separation as well as the total electrode length must be considered in some cases.

The results plotted in Figure 10 include only the effects due to thickness-shear modes and neglect not only the contributions of all other modes but also the effect of the x_3 dimension of the plate. Even with this highly simplified model, the frequency separation predicted by equation (17) fits the measured values quite well. Figure 11 shows the normalized frequency difference between the symmetric and antisymmetric mode as a function of electrode thickness for a rectangular AT cut quartz plate with rectangular electrodes. The measurements were made as part of a complete study of two section monolithic filters by E. C. Thompson. The fundamental frequency of the plates was 8.0 MHz. The plates were 15 mm square and the electrodes covered one third of the plate along x_3. The solid line was computed from equation (17) and the circles are experimentally measured frequencies of the coupled modes.

The coupled frequencies given by equation (17) determine the bandwidth of the monolithic filter. Physically these two frequencies correspond to the shear displacements under each electrode being in and out of phase at resonance. At the lower edge of the filter band the displacements, which are sinusoidal under the electrodes and exponential between them, are in phase. At the upper edge of the band displacements under each electrode are 180° out of phase. This may be experimentally verified by taking x-ray diffraction topographs[13] of a monolithic filter at these two resonant frequencies. These two modes are shown in Figure 12. The dark regions in the photographs correspond to the maximum shear displacement. At the lower frequency f_s the two dark regions under each electrode are joined by a continuous grey in the region between the two electrodes. At the higher frequency f_a there is a white line between the two dark regions where the shear displacement is zero and the displacement in the two electroded regions is 180° out of phase.

Although the analysis here was for only two coupled regions, it may be extended in a straightforward manner to include any number of coupled regions. Experimentally the first multisection models were fabricated by R. L. Reynolds.

In addition to the coupled frequencies given by equation (15), it is necessary to know the reactance of each uncoupled resonator. This reactance could be calculated by including the piezoelectric constants in the previous analysis and determining the reactance of the thickness-shear resonator as a function of frequency. A simpler procedure is to determine the displacement current due only to the thickness-shear mode in an infinite AT cut plate and from this calculate the reactance by introducing in an ad hoc manner from the shape of the thickness-shear mode due to a finite electrode length in both the x_1 and the x_3 direction.

Tiersten[14] has shown that the displacement

current D_2 as a function of plate thickness alone in an infinite AT cut quartz plate is given by

$$D_2 = e_{26}S_6 + \epsilon_{22}E_2 \tag{18}$$

where e_{26} is a component of the piezoelectric tensor S_6 the shear strain, ϵ_{22} a component of the dielectric tensor and E_2 the electric field normal to the plate. Using Tiersten's solution equation (18) may be put in the form

$$D_2 = -\frac{\epsilon_{22}\varphi_o\eta\left(1 - \frac{1}{k_{26}^2}\right)e^{j\omega t}}{\tan\eta b - \frac{\eta b}{k_{26}^2}} \tag{19}$$

where k_{26} is the electromechanical factor, φ_o the voltage associated with E_2 and η the wave number for the thickness-shear modes. The current i through the quartz plate is proportional to $\partial D_2/\partial t$ and by differentiating equation (19) the impedance Z may be written as

$$\varphi_o e^{i\omega t} = Z = \frac{\tan\eta b - \frac{\eta b}{k_{26}^2}}{Ai\omega\epsilon_{22}\eta\left(1 - \frac{1}{k_{26}^2}\right)} \tag{20}$$

where A is the electrode area. Equation (20) corresponds to a pure reactance since no loss was included in the initial equations. In the case of a series resonant circuit it may be shown that $dX/d\omega|_{x=o} = 2L$. Then by differentiating equation (20) and setting $\tan\eta b = \eta b/b_{26}^2$ the equivalent inductance at resonance is given by

$$L = \frac{b^3\eta^3}{2A\epsilon_{22}\omega^2 b_{26}^2} \tag{21}$$

Equation (21) assumes no x_1 or x_3 dependence of the thickness-shear mode. This may be included to a good approximation as a cosine function of both dimensions. Integration of this dependence over the electrode area A adds a constant factor of .405 to the denominator of equation (21). Substituting in the proper values of k_{26} and ϵ_{22} for AT-cut plates gives

$$L = \frac{1.8m^3}{A(\text{cm}^2)\, f^3(\text{MHz})}\text{, in henries} \tag{22}$$

where m is the harmonic overtone of the thickness shear mode.

Equations (17) and (22) may be used to calculate the coupled resonant frequencies and the reactance of each electrode as a function of the plate and electrode geometry. These values may now be used to compute the characteristics of a filter using these coupled resonators by the methods of section IV. Although both equations are approximations, they have been found quite satisfactory for calculating most filter characteristics.

IV. Equivalent Network Considerations

The preceding section has given the equations required to determine the frequency of multiple resonators on a single quartz substrate. If two such resonators are constructed in close juxtaposition, it was shown that two characteristic frequencies for the system will exist, as would be expected for a pair of coupled resonators. In terms of the geometric and elastic constants of the system, these two frequencies are given as the first symmetric and first antisymmetric solutions of equation (17).

The corresponding coupling coefficient, then, may be defined in terms of the frequency separation of the two characteristic frequencies. If two meshes, each having the same self-resonant frequency, ω_o, are coupled, such as those shown in Figure 13, it is evident that the entire network will possess two resonances. In the case of small coupling, the separation between these frequencies can be shown to be

$$\frac{\omega_b - \omega_a}{\omega_o} = \frac{\sqrt{C_1C_2}}{C_{12}} = k_{12} \tag{23}$$

Consequently, the equivalent coupling coefficient for two adjacent trapped-energy resonators can be given as

$$k = \frac{\omega_a - \omega_s}{\sqrt{\omega_a\omega_s}} \tag{24}$$

where ω_a is the first antisymmetric solution of (17) and ω_s is the first symmetric solution. It is therefore possible to define the coefficient of coupling, as well as the equivalent electrical parameters, of the monolithic filter structure in terms of the geometry of the system and the thickness of the deposited electrodes.

The required values for these parameters may best be determined from a consideration of the equivalent electrical network. It has been shown by Green[15] and Dishall[16], for example, that all-pole bandpass filter structures providing either Butterworth (maximally flat) or Chebycheff (equal ripple) characteristics can be realized through the use of ladder structures consisting of circuits self resonant at the midband frequency with nearest-neighbor coupling as shown in Figure 4. The amplitude response functions are (Butterworth)

$$\left(\frac{V_p}{V}\right)^2 = 1 + \left(\frac{\Delta\omega}{\Delta\omega_{3\ db}}\right)^{2n} \tag{25}$$

and (Chebycheff)

$$\left(\frac{V_p}{V}\right)^2 = 1 + \left[\left(\frac{V_p}{V_v}\right)^2 - 1\right] \cosh^2 \left(n \cosh^{-1} \frac{\Delta\omega}{\Delta\omega_v}\right) \quad (26)$$

where V_p = peak output voltage in the passband

V = output voltage at a frequency separation $\Delta\omega/2$ from the center of the passband

V_v = valley output voltage in passband

n = number of resonators in the filter

$\Delta\omega_{3\ db}$ = frequency separation of the 3 db cut-off points of the filter

$\Delta\omega_v$ = frequency separation of the points on the skirts where the output voltage equals V_v

The conditions which must be imposed on interresonator coupling coefficients and termination resistances in order to satisfy these response functions are the following:

Butterworth:

$$Q_1 = \frac{\omega L_1}{R_1} = Q_n = \frac{\omega L_n}{R_n}$$

$$Q_1 = \left[\frac{\omega_o}{\Delta\omega_{3\ db}}\right] q_1 = \left[\frac{\omega_o}{\Delta\omega_{3\ db}}\right] 2 \sin \frac{\pi}{2n} \quad (27)$$

$$K^2_{r,r+1} = \left(\frac{\Delta\omega_{3\ db}}{\omega_o}\right)^2 k^2_{r,r+1} =$$

$$\left(\frac{\Delta\omega_{3\ db}}{\omega_o}\right)^2 \frac{0.25}{\left[\sin (2r-1) \frac{\pi}{2n}\right]\left[\sin (2r+1) \frac{\pi}{2n}\right]} \quad (28)$$

Chebycheff:

$$Q_1 = \frac{\omega L_1}{R_1} = Q_n = \frac{\omega L_n}{R_n}$$

$$Q_1 = \left[\frac{\omega_o}{\Delta\omega_{3\ db}}\right] q_1 = \left[\frac{\omega_o}{\Delta\omega_{3\ db}}\right] \frac{2 \sin \frac{\pi}{2n}}{S_n} \quad (29)$$

$$K^2_{r,r+1} = \left(\frac{\Delta\omega_{3\ db}}{\omega_o}\right)^2 k^2_{r,r+1} =$$

$$\left(\frac{\Delta\omega_{3\ db}}{\omega_o}\right)^2 \frac{\left[\sin^2 \frac{r\pi}{2n} \cos^2 \frac{r\pi}{2n} + \frac{1}{4} S_n^2\right]}{\left[\sin (2r-1) \frac{\pi}{2n}\right]\left[\sin (2r+1) \frac{\pi}{2n}\right]} \quad (30)$$

In these expressions, R_1 and R_n are the input and output termination resistances, L_1 and L_n are the self-inductances of the input and output meshes, and $K_{r,r+1}$ is the coefficient of coupling between rth and (r+1) th meshes. q_1, q_n, and $k_{r,r+1}$ are the normalized quality factors and coupling coefficients, and

$$S_n = \sinh \left\{\frac{1}{n} \sinh^{-1} \left[\left(\frac{V_p}{V_v}\right)^2 - 1\right]^{-1/2}\right\} \quad (31)$$

Values of the normalized quality factors and normalized coupling factors have been calculated by Dishall[17] and others, for various passband ripple values and for filters comprised of two to seven resonators. Tabulations of these values are readily available, (Reference Data for Radio Engineers, published by IT&T) which provide a simple method for the design of all-pole filters which will meet most communications systems requirements.

It is evident then, that the techniques for the design of coupled-resonator wave filters exist. It is only necessary to establish the correspondence between the acoustically coupled thickness-shear resonators and an equivalent electrical network, in order to utilize these design procedures. The preceding discussion has derived the necessary relationships between electrode geometry and mass loading to insure that the electroded regions will be essentially singly resonant. In addition, the acoustic coupling coefficients between adjacent resonators have been derived in terms of geometric and mass-loading factors. An equivalent ladder circuit which has been found satisfactory to describe the transmission characteristics of a filter comprised of a number of adjacent, acoustically coupled resonators is shown in Figure 15. No attempt is made to provide a rigorous derivation of this network from the piezoelectric and mechanical properties of the device; however, experimental measurements of the response functions of many filters consisting of up to ten coupled resonators have verified the equivalence over a narrow band of frequencies in the vicinity of the pass band. Qualitatively, the capacitance tee-sections at input and output represent the piezoelectric coupling, the antiresonant circuits represent the individual resonators, and the series inductors represent the interresonator coupling. C_o and C_o' are the static capacitance associated with the input and output resonators, including wiring capacitance and capacitance due to the plated electrodes.

Remembering that we are dealing with a quartz device having very high Q and inherently limited to passbands of no more than a few percent under optimum conditions, we may transform the network of Figure 15 into that shown in Figure 16 by simply transforming each component through the capacitive transformer. In this circuit, L_1, L_2, L_3, etc., represent the equivalent inductances of the respective resonators, and the interresonator coupling factors are represented by

$$K_{r,r+1} = \frac{\sqrt{C_r C_{r+1}}}{C_{r,r+1}} \quad (32)$$

The shunt capacitances C_o and C_o' appear across the termination resistors, so that in effect we may consider the filter to be terminated by the parallel combination of C_o and R_1. Since our filter consists of AT-cut quartz piezoelectric resonators, we know that the ratio of shunt capacitance C_o to resonator series capacitance C_1 will remain essentially constant; also, while the Q factors for the input and output resonators will in actuality differ for different ripple factors, the approximate relationship will be

$$Q_1 = \frac{\omega L_1}{R_1} \simeq \frac{\omega_o}{(\Delta\omega)_{3\ \mathrm{db}}} \tag{33}$$

so that the termination resistance will be

$$R_1 \simeq \frac{(\Delta\omega)_{3\ \mathrm{db}}}{\omega_o} (\omega_o L_1) \tag{34}$$

If r is the ratio of capacities for AT-cut quartz, then

$$\frac{C_o}{C_1} = r; \left(\frac{1}{\omega_o C_1}\right) = r\left(\frac{1}{\omega_o C_o}\right) = (\omega_o L) \tag{35}$$

so that

$$\left(\frac{1}{\omega C_o}\right) = \frac{1}{r}\left(\omega_o L\right) \tag{36}$$

and we see that the ratio of shunt reactance to termination resistance is dependent upon bandwidth:

$$\frac{\left(\frac{1}{\omega_o C_o}\right)}{R_1} = \frac{1}{r} \frac{\omega_o}{(\Delta\omega)_{3\ \mathrm{db}}} \tag{37}$$

Since $r \approx 200$ for AT-cut resonators of this type, the presence of C_o can be neglected for bandwidths of the order of 0.1% or less. For wider bands, however, the effect of C_o may be determined by considering the transformation shown in Figure 17. The parallel combination may be considered as an equivalent series network, from which it is apparent that the effect of C_o is to reduce the effective termination from the value of R_1 to the new value R_1', and to introduce an effective series capacitance C_o' which will result in an increase in the effective mesh resonant frequency of the input and output meshes. Adjusting the values of R_1, R_n and the equivalent frequency of the input and output resonators accordingly will result in restoration of the desired conditions. This technique may be used to obtain satisfactory results for bandwidths approaching

$$\frac{(\Delta\omega)_{3\ \mathrm{db}}}{\omega_o} = \frac{1}{2r} \tag{38}$$

at which value the required effective termination resistance is

$$\begin{aligned} R_1' &= r\left[\frac{(\Delta\omega)_{3\ \mathrm{db}}}{\omega_o}\right]\left(\frac{1}{\omega C_o}\right) \\ &= \frac{1}{2}\left(\frac{1}{\omega C_o}\right) \end{aligned} \tag{39}$$

This condition will be satisfied if R_1 in Figure 17 is

$$R_1 = \frac{1}{\omega C_o} \tag{40}$$

The resulting shift in frequency of the terminal resonators is that produced by a series reactance

$$\frac{1}{\omega C_o'} = \frac{1}{2}\left(\frac{1}{\omega C_o}\right) \tag{41}$$

so that the equivalent series resonant frequencies of the input and output resonators would be adjusted to a lower angular frequency

$$\omega' = \omega_o \left(1 - \frac{1}{6r}\right) \tag{42}$$

For bandwidths greater than $(\Delta\omega)_{3\ \mathrm{bd}}/\omega_o = 1/2r$, it becomes necessary to add shunt inductors across the terminations, as shown in Figure 18. In this way, passbands of greater than 0.5% may be obtained. Obviously, additional capacitance may also be added to permit the required inductor to be incorporated in an impedance-matching network, if desired. The addition of these inductors will result in two other passbands, one above and one below the desired band, as pointed out earlier. The difference in characteristic impedance between the desired and undesired bands is so great that the observed insertion loss is modified only slightly.

V. Experimental Results

Several filters designed in accordance with the preceding techniques have been fabricated, and the results obtained are in good agreement with the predicted behavior. Some examples of these designs are described below.

One of these experiments consisted of the design of a 3-pole filter at a center frequency of 15 MHz, having a 3 db bandwidth of 3.1 kHz and an inband ripple of 0.1 db. Dishall's tabulated values for the normalized coupling coefficients and quality factors are

$$q_1 = q_3 = 1.43$$

$$k_{12} = k_{23} = 0.665$$

Multiplying these factors by the fractional bandwidth factor yields

$$Q_1 = Q_3 = 6.9 \times 10^3$$

$$K_{12} = K_{23} = 1.375 \times 10^{-4}$$

0.100" square electrodes were chosen for the resonators, giving an equivalent inductance of 12 mH for each resonator. The calculated terminations, then, are found to be

$$R_1 = R_3 = 164 \text{ ohms}$$

The electrode thickness and resonator separation were chosen in accordance with the equations in Section III to give the specified interresonator coupling coefficients.

The response obtained from this filter is shown by the oscilloscope traces in Figure 19. The measured 3 db bandwidth was found to be 3.3 kHz rather than the 3.1 kHz objective, and the distortion across the flat band region was found to be approximately 0.2 db, rather than the 0.1 db theoretical value. Both of these effects are believed to be due to the fact that the actual quality factors of the unloaded resonators were somewhat less than the 200,000 value required for this design. Figure 20 shows the response function of the experimental filter in comparison with the theoretical loss curves computed for a 3-pole Chebycheff filter with 0.1 db distortion. The experimental model shows non-symmetrical skirts, due to the presence of a small residual bridging capacitance which results in an attenuation peak about 35 kHz below the passband. This enhances the attenuation slope below the band and reduces it above the band for a 3-pole structure. The degradation of selectivity near the band edge and the rounding of the passband curve are the result of inadequate resonator Q which was not taken into account in the design.

Another experimental design was that of a 6-pole filter at 10 MHz, having a 3 db bandwidth of 5 kHz, and 1 db inband ripple. Figure 21 shows the observed transmission characteristic. Once again, there is an apparent rounding of the passband corners, due probably to internal dissipation of the resonators, which was not taken into account in the design. Figure 22 shows the experimental characteristic superimposed on the calculated Chebyscheff characteristic; the deviations at the corners of the passband are apparent, and there is a slight skewing of the attenuation curves in the transition region due to the residual stray feed-through of the circuit which introduced a peak below the band.

A model of a 10-pole Chebyscheff filter was constructed based on the preceding design equations, using a ripple factor of 0.1 db. Center frequency was 10 MHz, and 3 db bandwidth was 4.5 kHz. Electrode geometry was chosen to give approximately 38 mH inductance for each resonator. The oscilloscope traces in Figure 23 show the attenuation curves which were obtained experimentally for this design. The degradation due to resonator dissipation is obviously more severe in this case, as evidenced by the rounding of the passband characteristic. The superimposed ripple is approximately 0.1 db, however, indicating that the actual values of coupling coefficients are quite close to the intended values. In spite of this limitation, however, the selectivity achieved is good. The 60 db/ 3 db bandwidth ratio is found to be 1.48, which is adequate for most applications in communications systems. The 2 db distortion across the 4 kHz band would be acceptable for most point-to-point systems, whether single sideband AM or narrow band FM. For the most critical applications, such as telephone multiplex systems, it would be necessary to provide equalization across the band, or to alter the design to take into account the actual dissipation factors of the resonators. The design and evaluation of both the 6- and 10-pole models were carried out by W. D. Beaver.

The attenuation characteristics obtained from experimental monolithic filters of varying degrees of complexity are shown in Figure 24, plotted against a normalized frequency scale. The 2-pole and 6-pole models were designed for a 1 db ripple factor, while the 3-pole and 10-pole models were based on a 0.1 db ripple. The results obtained with these models demonstrate adequately the feasibility of realizing simple, inexpensive, integrated wave-filter structures having high selectivity, by making use of the controlled mechanical coupling between thickness shear mode resonators on a single quartz plate. The design techniques described, and the equivalent networks presented, have proven sufficiently accurate to permit the design and realization of satisfactory filters in the frequency range from about 3 MHz to above 35 MHz. Investigations are continuing to extend the frequency range, as well as determine design methods for wider bandwidths.

VI. Discussion

From the material presented in this paper it appears obvious that the principal advantages of the monolithic filter are its unique simplicity, small size and low cost.

For double and single sideband or narrow band FM, station to station radio or wire transmission, it should find wide application. The cost factor that limited high selectivity applications to the more sophisticated transmission systems should now be low enough to re-examine the more competitive mass production market. Its small size should open new applications that have previously been closed to high selectivity due to size and weight of present filters.

A monolithic filter in conjunction with an additional quartz plate comprising four uncoupled resonators can be used for the more stringent requirements of telephone multiplex channel filters.[18]

There are a number of limitations to its use regarding bandwidth, terminal impedance and transmission of other than the wanted frequencies, dependent upon the specific design. There are, in addition, many problems that remain to be solved to make full use of its capabilities.

Since overtones of flexure and face shear modes are propagated throughout the plate and may be reflected from the boundary, they will cause additional narrow transmission regions. In addition, the anharmonic and twist modes of the thickness shear that are not trapped may be generated due to a finite boundary. While all these boundary generated modes may be reduced in level by a suitable absorbing medium[19], the Q of the useful main shear mode may be degraded depending upon its coupling to these modes. This is particularly true in the lower frequency region (1 to 5 MHz). It may, in some cases, be preferable to use two filters in tandem with noncoincidence of unwanted modes in the manner now standard practice with conventional crystal filters.

In cases where the widest band possible is desired with low impedance, it is necessary to use a small separation of the resonant regions. Under these conditions the combined array of resonators acts to generate more modes along the axis of the wave being propagated than are required for proper operation of the filter. This results in poor selectivity on the upper side of the band and will in some cases generate additional transmission regions.

It is obvious from the equations presented that the bandwidth of the filter is limited by the piezoelectric coupling coefficient and that to obtain wider bands it is necessary to introduce inductance in the case of quartz filters. This may be overcome by using materials other than quartz with a higher coefficient or the use of ceramic transducers as driving elements. If high coefficient transducers are used, then the filter itself may be made of any high Q, low temperature coefficient material.

While there are many problems to be solved for some specialized applications, it appears that within the limitations presented herein this device should find many uses in future communication systems.

References

1. Adler, R., "Compact Electromechanical Filter," Electronics, Vol. 20, pp. 100-105, April 1947.

2. Onoe, M., Junouji, H., and Kobori, K., Proc. Nat. Conv. Inst. Elec. Comm. Eng., Japan #116 (1965).

3. Sykes, R. A. and Beaver, W. D., Proc. 20 Annual FCS, Atlantic City, April 1966.

4. Mailer, H. and Bunerle, D. R., Multi-Resonator Crystals for Filters, Proc. 20 Annual FCS, Atlantic City, April 1966.

5. Mindlin, R. D., Quartz Appl. Math. 19, 51-61, 1961.

6. Beaver, W. D., PhD Thesis, Lehigh University, 1967.

7. Bechmann, R., U.S. Patent No. 2,249,933 (1941).

8. Mortley, W. S., Wireless World, 57, pp. 399-403, October 1951.

9. Mindlin, R. D. and Lee, P. C. Y., Int. J. Solids Structures 2, 125, 1966.

10. IEEE Standard #176 Piezoelectric Crystals 37, 1378, 1949.

11. Spencer, W. J., (to be published Jour. Acous. Soc. Am.).

12. Mindlin, R. D. and Forray, M., Jour. App. Physics 25, 12, 1954.

13. Spencer, W. J., IEEE Trans. Sonics and Ultrasonics SU12, 1, 1965.

14. Tiersten, H. F., J. Acous. Soc. Am. 35, 53, 1963.

15. Greene, E., "Exact Amplitude-Frequency Characteristics of Ladder Networks," Marconi Review, Vol. 16, No. 108, pp. 25-68; 1953.

16. Dishall, M., "Design of Dissipative Bandpass Filters Producing Exact Amplitude-Frequency Characteristics," Proc. IRE, Vol. 37, pp. 1050-1069, September 1949.

17. Dishall, M., "Two New Equations for the Design of Filters," Electrical Communications, Vol. 30, pp. 324-337, December 1952.

18. McLean, D., "Physical Realization of Miniature Bandpass Filters with Single Frequency or Single Sideband Characteristics," to be presented at 21st Annual Frequency Control Symposium, Atlantic City, April 1967.

19. Sykes, R. A., "High Frequency Plated Quartz Crystal Units," 1946 Winter Tech. Meeting, IRE,; Proc. IRE, January 1948.

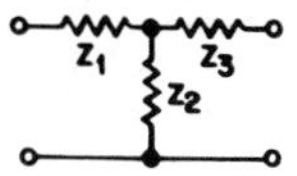

Figure 1. Generalized network for frequency selection.

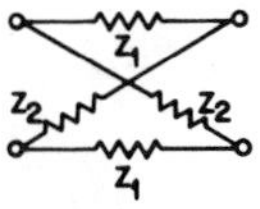

Figure 2. Generalized network for frequency selection.

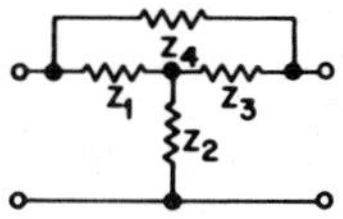

Figure 3. Generalized network for frequency selection.

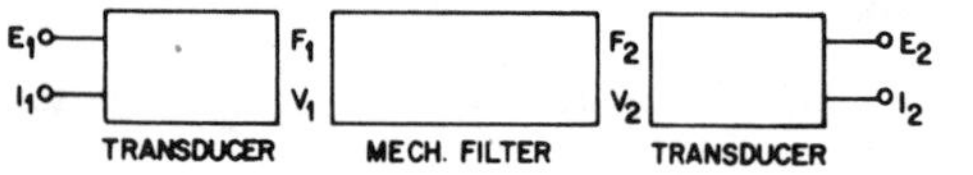

Figure 4. Generalized electromechanical filter.

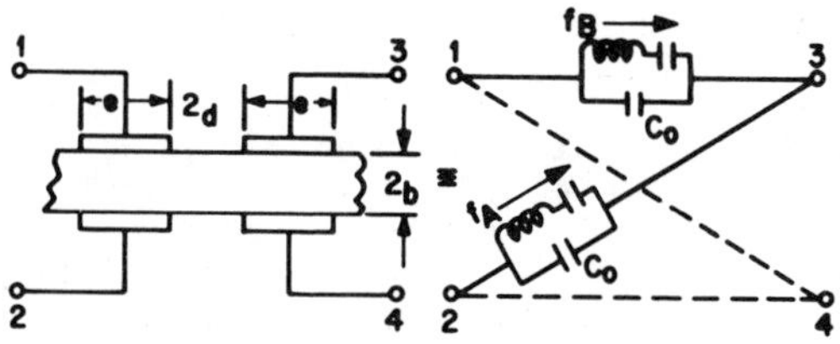

Figure 5. Coupled thickness shear resonators with equivalent lattice network.

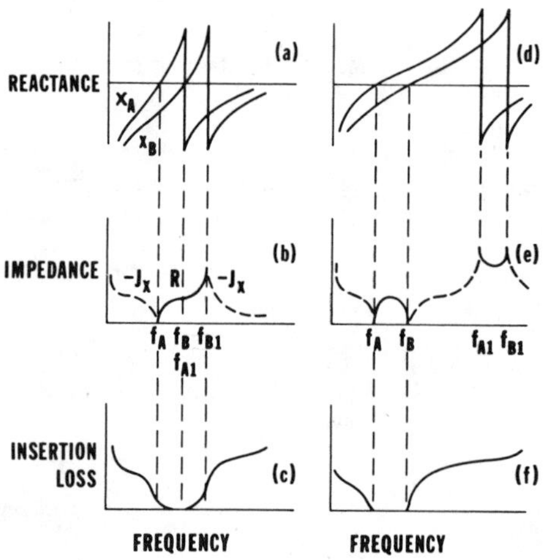

Figure 6. Disposition of poles and zeros for coupled resonators with resulting impedance and loss characteristics.

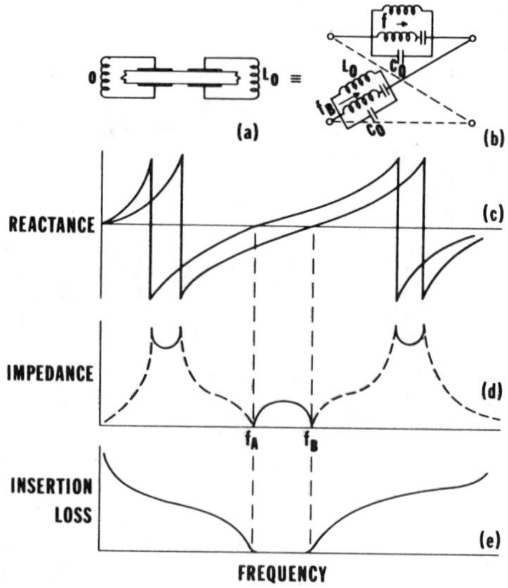

Figure 7. Use of inductance to increase the bandwidth of coupled quartz resonators.

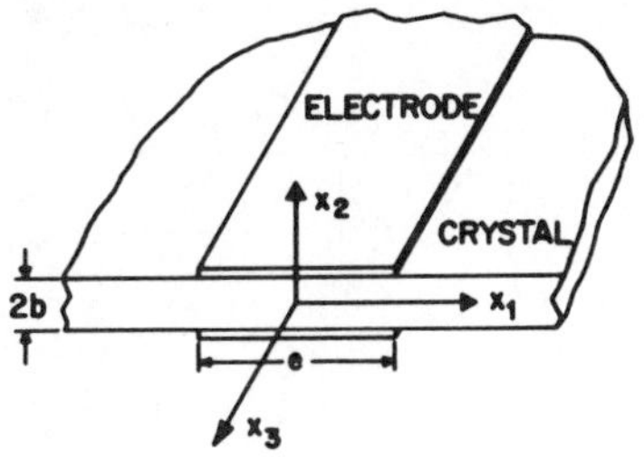

Figure 8. Electrode disposition illustrating trapped energy concept.

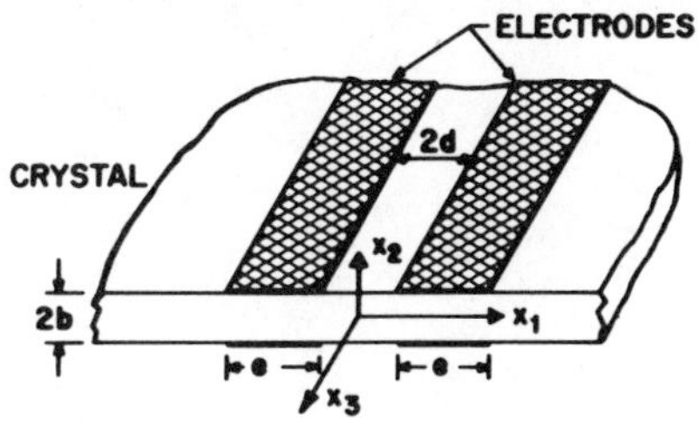

Figure 9. Electrode disposition illustrating coupled resonators.

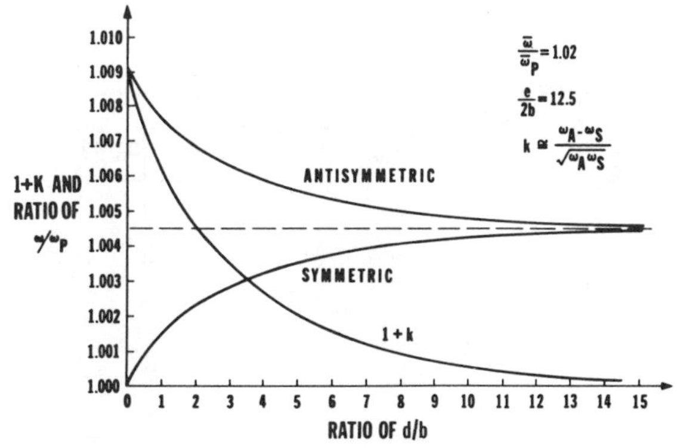

Figure 10. Dependence of coupled modes and coefficient k on resonator spacing.

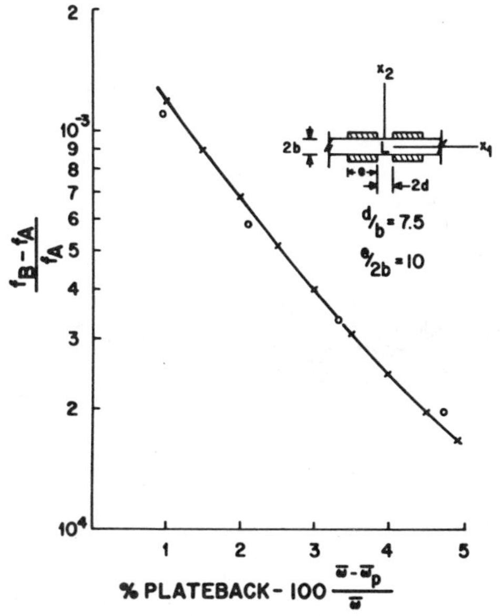

Figure 11. Dependence of coupled modes on electrode thickness.

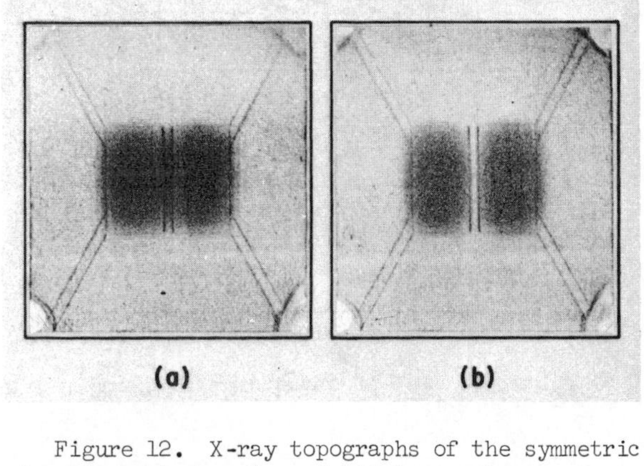

Figure 12. X-ray topographs of the symmetric and antisymmetric modes of thickness shear.

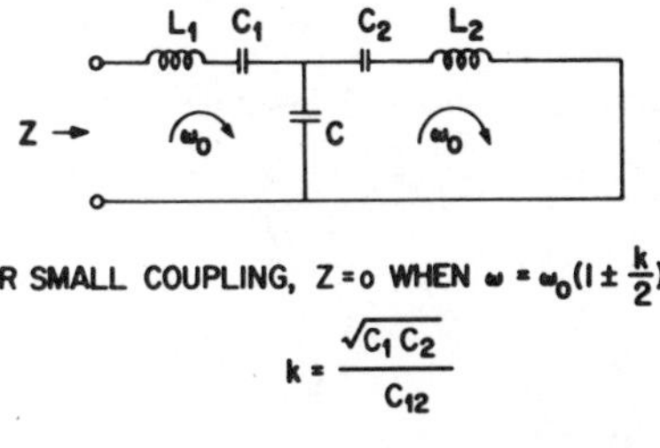

Figure 13. Network consisting of two coupled meshes.

$$Q_1 = \frac{\omega_0 L_1}{R_1} \quad Q_n = \frac{\omega_0 L_n}{R_n}$$

$$K_{r,r+1} = \frac{\sqrt{C_r C_{r+1}}}{C_{r,r+1}}$$

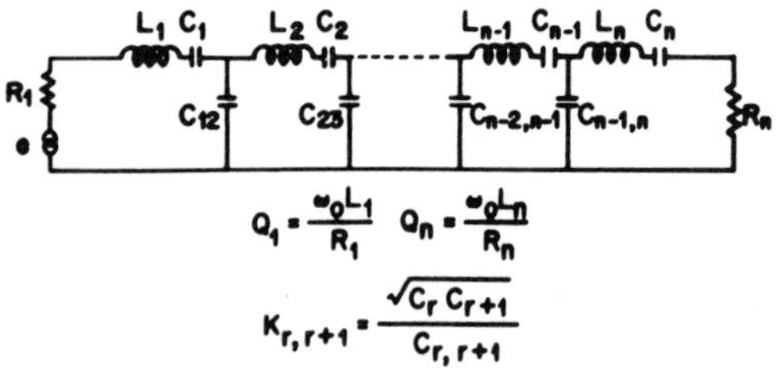

Figure 14. All-pole filter having n coupled resonators.

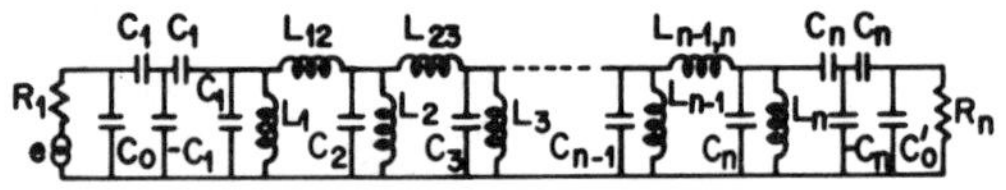

Figure 15. Equivalent network for monolithic filter having n coupled resonators.

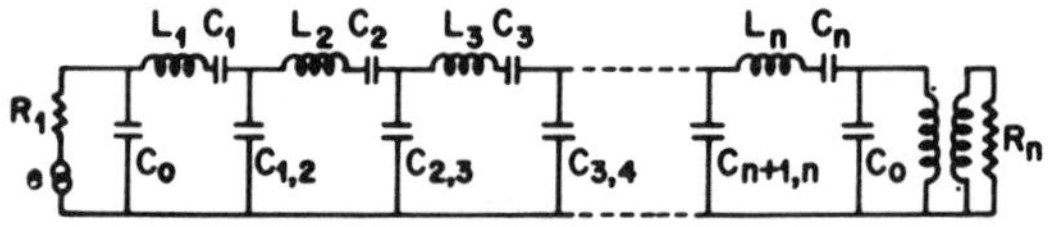

Figure 16. Transformed equivalent filter network.

$$R_1' = R_1 \left[\frac{\left(\frac{1}{r}\right)^2 \left(\frac{\omega_0}{\Delta \omega_3}\right)^2}{1 + \left(\frac{1}{r}\right)^2 \left(\frac{\omega_0}{\Delta \omega_3}\right)^2} \right]$$

$$C_0' = C_0 \left[1 + \left(\frac{1}{r}\right)^2 \left(\frac{\omega_0}{\Delta \omega_3}\right)^2 \right]$$

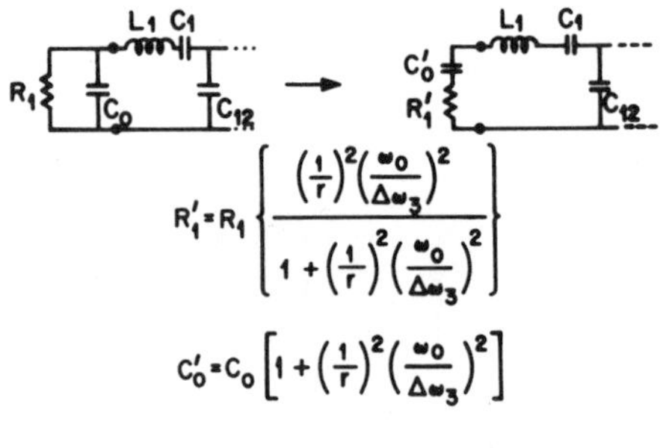

Figure 17. Compensation for shunt capacitance.

$$\omega_0 L_T = \frac{1}{\omega_0 C_0}$$

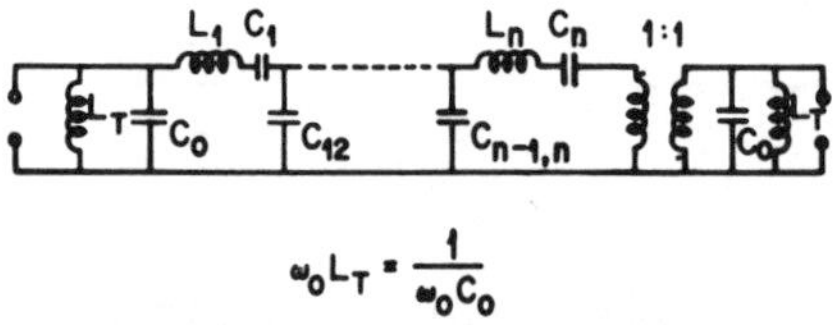

Figure 18. Use of shunt inductors to reduce the effects of shunt capacitance on wide band filters.

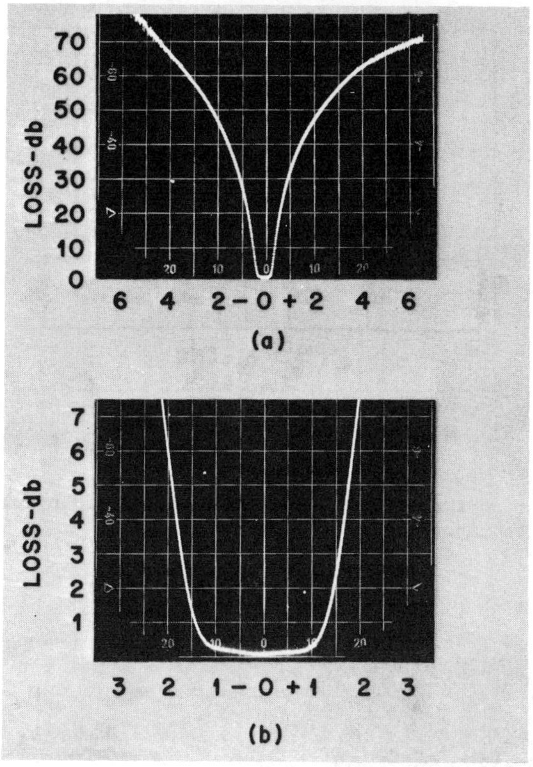

Figure 19. Measured response of 3-pole 15 MHz monolithic filter.

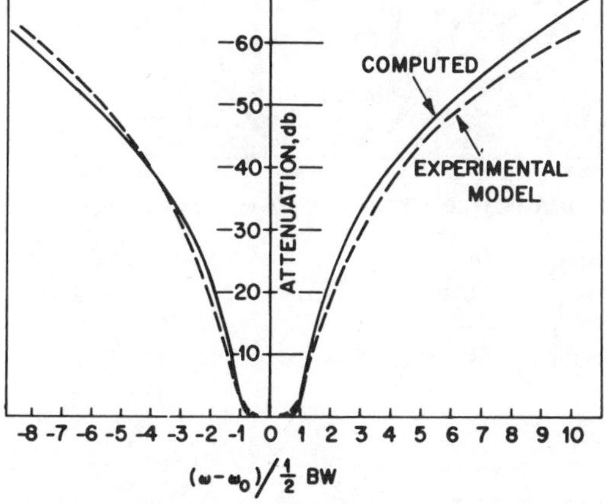

Figure 20. Comparison of measured and calculated response for 3-pole filter of Figure 19.

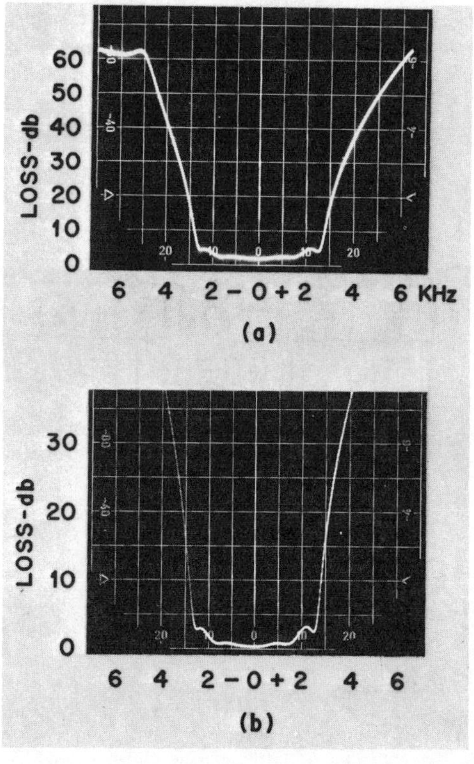

Figure 21. Loss characteristic of 6-pole monolithic filter.

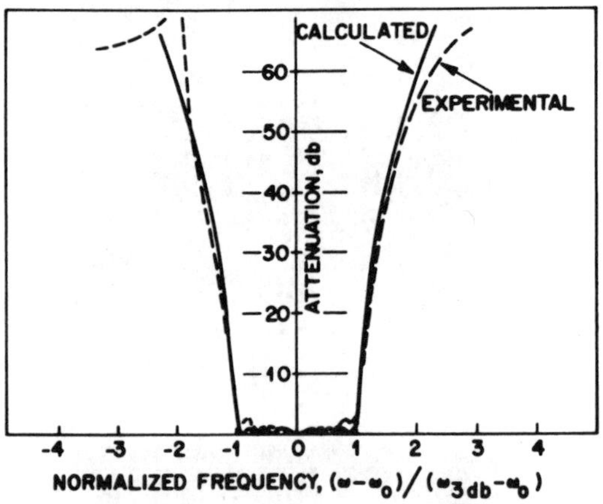

Figure 22. Comparison of measured and calculated response for 6-pole filter of Figure 21.

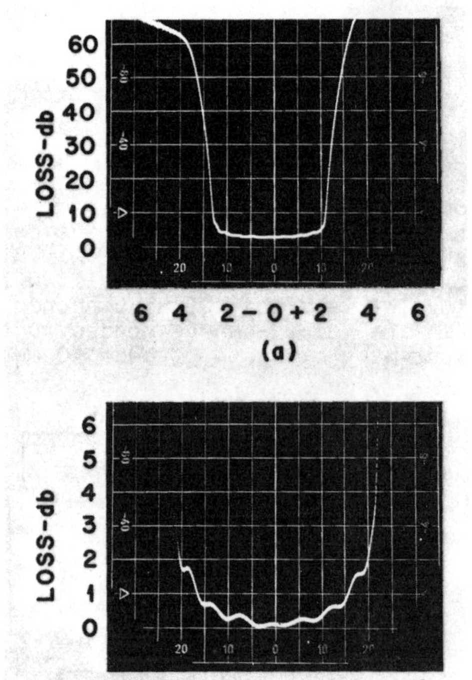

Figure 23. Loss characteristic of 10-pole monolithic filter.

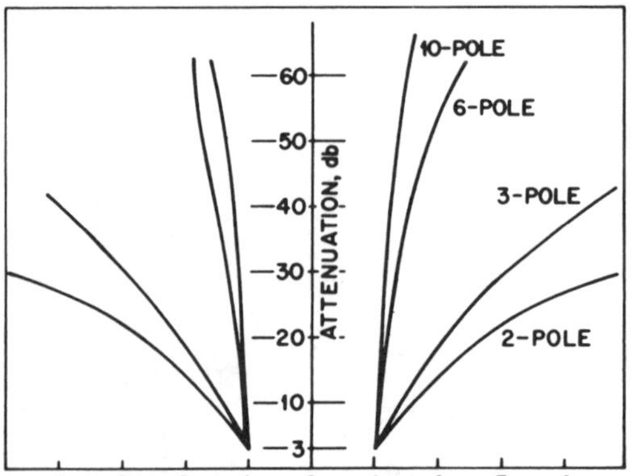

Figure 24. Comparison of loss characteristics of experimental 2, 3, 6 and 10-pole monolithic filters.

HF AND VHF INDUCTORLESS FILTERS FOR MICROELECTRONIC SYSTEMS

R.C. Smythe

Piezo Technology Inc.
Orlando, Florida

Summary

Integrated crystal filters utilize acoustical coupling to obtain a monolithic structure electrically equivalent to a conventional crystal filter section composed of discrete resonators and balanced, tuned transformer. Tandem monolithic sections are used to obtain highest stopband performance. Performance of monolithic and tandem monolithic integrated crystal filters at center frequencies up to 350 MHz is illustrated by examples. Applications include spectrum clean-up, carrier channel and other single-sideband filters, IF channel filters, and simulated front-end filtering using parametric up-conversion.

Key Words

Filters, crystal filters, miniaturization, integration, quartz, piezoelectric devices.

1. Introduction

This paper discusses current progress concerning a new kind of crystal filter, the integrated crystal filter, which promises to make important contributions to the solution of the frequency selectivity problem in microelectronics. The term "integrated crystal filter" is used to include both monolithic and tandem monolithic crystal filters. Conventional crystal filters consist of a number of individual piezoelectric resonators (usually quartz) interconnected in a suitable manner, usually with the aid of one or more balanced transformers. For frequencies above 1 MHz, thickness-shear-mode resonators are invariably used. It has been known for many years that by dividing the electrodes of a thickness-shear resonator it is possible to form a two-pole bandpass filter of sorts. Until recent years, this device was very imperfectly understood and found little application. Recent contributions to acoustic wave theory by R.D. Mindlin and others[1,2] as well as the earlier theory of W.S. Mortley[3] have provided the background for the understanding of this filter[4,5] which is now seen to consist of two acoustically-coupled resonators. The recognition of the acoustical coupling mechanism led to the postulation of multi-resonator monolithic structures, and both two-resonator and multi-resonator devices are currently undergoing intensive development.[6,7,8]

In the next section of the paper, the principle of operation of the monolithic crystal filter is briefly reviewed and requirements for inductorless operation are presented. Cascading considerations are then discussed. If, for example, an eight-pole response is desired, it may be realized as a single eight-resonator monolithic unit, as two four-resonator monolithic units in tandem connection, or as four two-resonator units in tandem. In each case, insertion loss design techniques can be used. Due to fabrication problems and unwanted mode considerations the full monolithic approach is not always the most successful and tandem operation frequently offers the best performance at the lowest cost. On the other hand, for fractional bandwidths too large to be realized by inductorless operation, the monolithic units can be used with simple, low-Q inductors at input and output, while the tandem configurations require additional inductors of higher Q at each interconnection point.

Initial developments of integrated crystal filters were limited to frequencies of 5 to 20 MHz. We have extended the practical frequency range to 300 MHz and beyond. The third portion of the paper presents performance data on a variety of integrated crystal filters at center frequencies from 10 MHz to 350 MHz and comments on some important applications.

Reprinted from *Proc. 1969 Electron. Components Conf.*, Apr. 30–May 2, 1969, pp. 115–119.

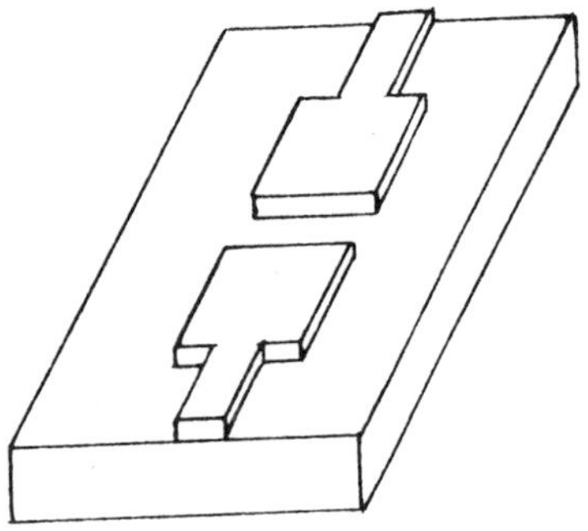

Fig. 1. Two-Resonator Monolithic Crystal Filter. (Electrodes on rear face not shown.)

2. Design Considerations

2.1 Acoustical Coupling

Figure 1 is a sketch of a two-resonator acoustically-coupled monolithic crystal filter, drawn to illustrate the principle of operation rather than actual details of construction. In current practice the plate is AT-cut quartz, but other piezoelectric materials can be used. The electrodes are thin films, typically of aluminum, silver, or gold, of carefully controlled thickness. Rectangular electrodes are shown, but circular and semi-circular electrodes are also in use. The piezoelectric plate can be considered as the analogue of a parallel-plate waveguide with a cut-off frequency inversely proportional to its thickness. Regions of lower cut-off frequency are formed by opposing pairs of electrodes. In a region formed by an electrode pair, standing waves can occur at one or more frequencies above the cut-off frequency of the electroded region and below the cut-off frequency of the surrounding plate. By properly dimensioning the electrodes, only one such resonant frequency will be obtained.[3,4,9] Two such resonators placed in close proximity will be mechanically coupled by acoustic energy which spreads beyond the electrode edges. From the modal point of view, the single resonant frequency splits into a pair of frequencies at which standing waves occur under the two electrode pairs. Input and output electroacoustical transducer action is provided by the direct and converse piezoelectric effects. Thus a two-resonator acoustically-coupled crystal filter is formed.

Now if a third electrode pair is added between the input and output electrode pairs it can be seen that a three-resonator filter has been formed. By extension, multi-resonator monolithic filters can be obtained.[4,5]

2.2 Inductorless Operation

Of particular interest for microelectronic applications are the conditions under which inductorless operation is possible. Using AT-cut quartz, the maximum inductorless 3 db bandwidth is given approximately by

$$BW_{max} = .003\ f_o/N^2 \qquad (1)$$

where f_o is the center frequency and N is the overtone (N=1,3,5,...). When the maximum inductorless bandwidth is desired, N should of course be chosen as small as possible. The fundamental frequency of the quartz plate is approximately f_o/N and this frequency is inversely proportional to plate thickness. Therefore the minimum value of N will be determined by the highest practical fundamental frequency....that is, by the thinnest practical blank. The bandwidth limitation of (1) is shown in figure 2 for maximum fundamental frequencies, $\hat{f}$, of 30, 40 and 60 MHz. This latter figure has been achieved only in a developmental facility and is well beyond the state of the art for production applications.

In many cases, particularly for center frequencies in the VHF, the bandwidths obtainable without the use of inductors are impractically narrow and "spreading" coils must be employed to resonate excess shunt capacitance and obtain useful bandwidths. This is not a consequence of the use of acoustical coupling--conventional discrete resonator crystals suffer the same limitation - but is characteristic of the piezoelectric material used. Eventually inductorless operation for bandwidths much greater than those indicated by (1) should be made possible through the use of other piezoelectric materials.

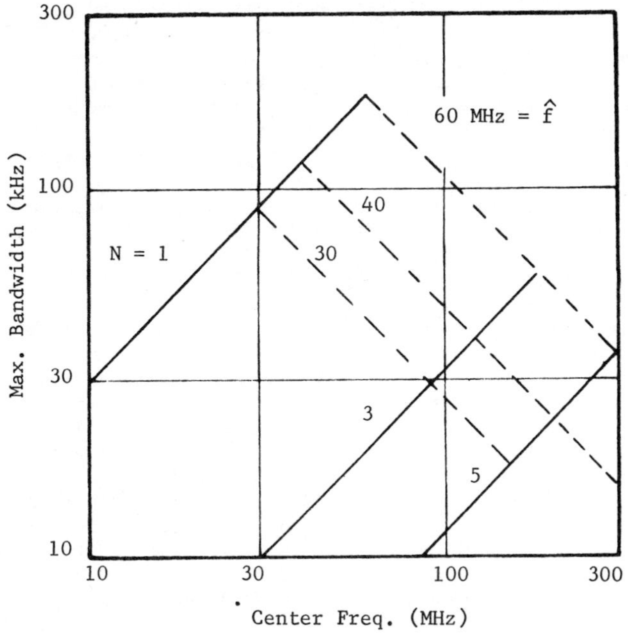

Fig. 2. Maximum Inductorless Bandwidth Using AT-cut Plates.

2.3 Cascading Considerations

In practical filter design it is necessary to distinguish between, and give separate consideration to, the number of resonators (number of poles) and the number of sections into which the filter is subdivided. The choice of the number of poles does not concern us here. To achieve high levels of stopband attenuation, for almost all types of filter construction, it is usually expedient to subdivide the filter into two or more sections, each having more modest attenuation requirements. (This is not intended to imply that the electrical design is carried out on a section-by-section basis. What is meant is that, having obtained an electrical design by insertion loss synthesis, the resultant network is then manipulated so that it can be subdivided into an appropriate number of sections.)

Although the resonators of a monolithic crystal filter are arrayed in tandem, it is in effect a single-section filter. We see that in addition to the desired acoustic path along the resonator array there exist a multiplicity of acoustic paths from input to output. (These paths are of importance only in the upper stopband and transition region, where the frequency is above the cut-off frequency of the plate. Their existence is evidenced by so-called unwanted modes of vibration which occur in the upper stopband.)

To achieve sectionalization it is necessary to use two or more plates, with appropriate tandem connections. A filter of this type is called a tandem monolithic crystal filter to distinguish it from the single-plate, or monolithic crystal filter. As stated earlier the term integrated crystal filter is used to embrace both monolithics and tandem monolithics.

It seems likely that both monolithics and tandem monolithics will find their sphere of usefulness. At least for the near term, the stopband performance advantage of the tandem monolithic integrated crystal filter is considerable, however. Furthermore, although the tandem approach requires two or more quartz plates, as opposed to one for the pure monolithic, the former can be much smaller than the latter. Since large blanks are more costly to manufacture than small ones, economics frequently favors tandem construction.

3. Examples and Applications

The principle conventional roles of bandpass crystal filters are as IF filters in AM, FM, and SSB communications receivers, as sideband filters in SSB transmitters, and as channel filters, pilot filters, etc., in frequency-division multiplex wire transmission systems. On occasion they have found use in providing receiver front-end selectivity. Integrated crystal filters are being employed in all these ways. In addition, new uses are being found.

One of the most important new applications for integrated crystal filters is in frequency synthesis, where it is desired to "clean-up" the output of a mixer or harmonic generator. Figure 3 shows the response of a typical two-resonator monolithic integrated crystal filter used in such an application at 180 MHz. In this instance, it was desired to select the 18th harmonic of a stable 10 MHz reference and suppress the other harmonics. In the complete system this filter is one member of a bank of filters used to select the desired harmonic of 10 MHz.

The importance of integrated crystal filters in the VHF region is illustrated by the previous example and the following one. Figure 4 shows a typical response of a four-resonator tandem monolithic integrated crystal filter with a center frequency of 125 MHz. This example serves to illustrate another application of increasing importance: in a communioations receiver, the advantages of a front-end filter can be obtained, together with tunability, by the use of a parametric up-converter followed by an appropriate filter. In general the center frequency of the filter is required to be at least four times the highest signal frequency. Then in order to cover the HF range up to 30 MHz or thereabouts, a filter in the 120 MHz range is needed. Since a narrow bandwidth is also needed, the resultant Q requirements dictate that a crystal filter be used.

Figure 5 shows the response of a two-resonator monolithic integrated crystal filter at a center frequency of 350 MHz. This filter is also used in a spectrum clean-up application.

In these examples the emphasis has been on the VHF since this represents an important area of performance not readily obtainable by conventional techniques as well as an area of interesting new applications. Integrated crystal filters also are finding very wide application at frequencies as low as 2 to 3 MHz. An important example is the carrier channel filter being developed at Bell Telephone Laboratories in the 8-10 MHz region .[5,6] For application to two-way radios and paging receivers, 10.7 MHz and 21.4 MHz integrated crystal filters have great potential.[8] An eight-pole 10.7 MHz tandem monolithic integrated crystal filter is currently available in a flatpack with overall dimension 1.75 x 0.6 x 0.2 in. With a 6 db bandwidth of 15 kHz, the 80 db bandwidth is less than 40 kHz. Stopband attenuation well above 100 db is typical. By comparison a conventional eight-pole crystal filter having the same bandwidth and center frequency measures 2.38 x 1.00 x 0.75 in.

Figure 6 shows the attenuation characteristic of a 23 MHz four-pole tandem monolithic integrated crystal filter. Three db bandwidth is 32 kHz, 50 db bandwidth is 120 kHz, and unwanted modes are attenuated 60 db. A filter of this type could be packaged in a flatpack measuring approximately 0.9 x 0.6 x 0.2 in.

References

(1)R.D. Mindlin and P.Y. Lee, "Thickness-Shear and Flexural Vibrations of Partially Plates Crystal Plates," Int. J. Solids and Structures, vol. 2, pp. 125-139, 1966.

(2)R.D. Mindlin, "Thickness-Twist Vibrations of an Infinite, Monoclinic, Crystal Plate," ONR Contract Nonr-266(09) Tech. Rpt. 52; USAERDL Contract DA-36-039-AMC-0065(E) Interim Tech. Rpt. 1; Columbia U.; June, 1964.

(3)W.S. Mortley, "Frequency-Modulated Quartz Oscillators for Broadcast Equipment," Proc. IEE, vol. 104, pt. B, pp. 239-249; 1957.

(4)M. Onoe and H. Jumonji, "Analysis of Piezoelectric Resonators Vibrating in Trapped-Energy Modes," J. Inst. Elec. Comm. Eng'rs. Japan; vol. 48, No. 9, Sept. 1965.

(5)R.A. Sykes and W.D. Beaver, "High Frequency Monolithic Crystal Filters with Possible Application to Single Frequency and Single Sideband Use," Proc. 20th Freq. Cont. Symp., pp. 288-308; 1966.

(6)R.A. Sykes, W.L. Smith, and W.J. Spencer, "Monolithic Crystal Filters," IEEE conv. Rec. vol. 15, pt. 11, pp. 78-93; 1967.

(7)M. Onoe, H. Jumonji, and N. Kobori, "High Frequency Crystal Filters Employing Multiple Mode Resonators Vibrating in Trapped Energy Modes," Proc. 20th Freq. Cont. Symp., pp. 266-287; 1966.

(8)H. Yoda, Y. Nakazawa, S. Okano, and N. Kobori, "High Frequency Crystal Mechanical Filters," Proc. 22nd Freq. Cont. Symp., pp. 188-205; 1968.

(9)W.Shockley, D.R. Curran, and D. J. Koneval, "Energy Trapping and Related Studies of Multiple Electrode Filter Crystals," Proc. 17th Freq. Cont. Symp., pp. 88-126; 1963.

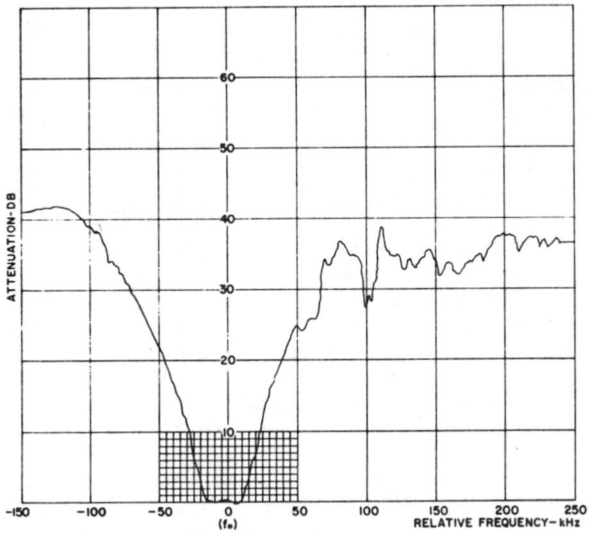

FIGURE 3.
ATTENUATION CHARACTERISTIC OF TWO-RESONATOR
MONOLITHIC INTEGRATED CRYSTAL FILTER
CENTER FREQUENCY 180 MHz

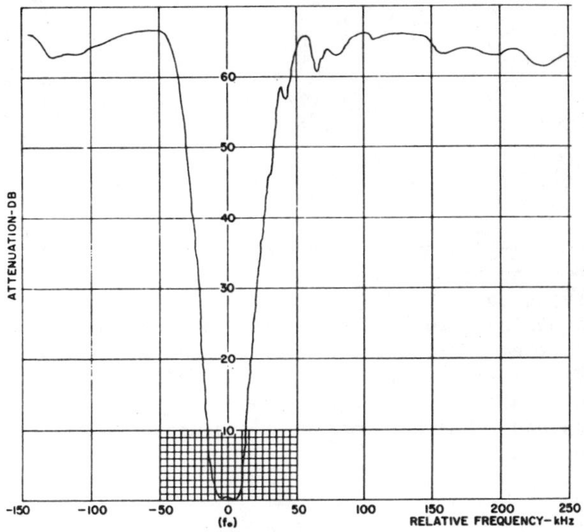

FIGURE 4.
ATTENUATION CHARACTERISTIC OF FOUR-RESONATOR
TANDEM MONOLITHIC INTEGRATED CRYSTAL FILTER
CENTER FREQUENCY 125 MHz

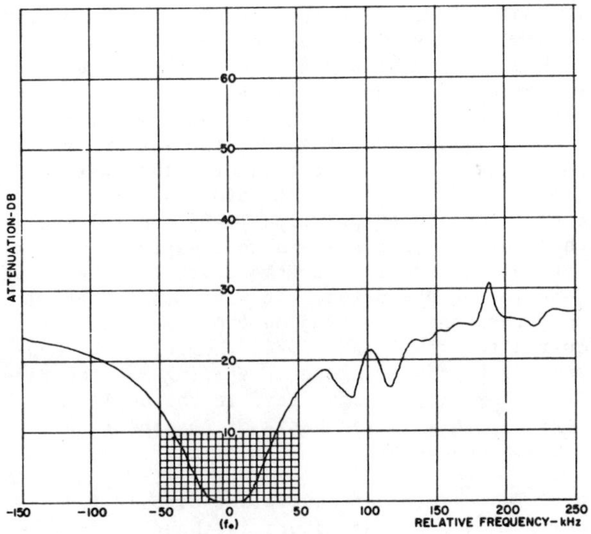

FIGURE 5.
ATTENUATION CHARACTERISTIC OF TWO-RESONATOR
MONOLITHIC INTEGRATED CRYSTAL FILTER
CENTER FREQUENCY 350 MHz

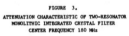

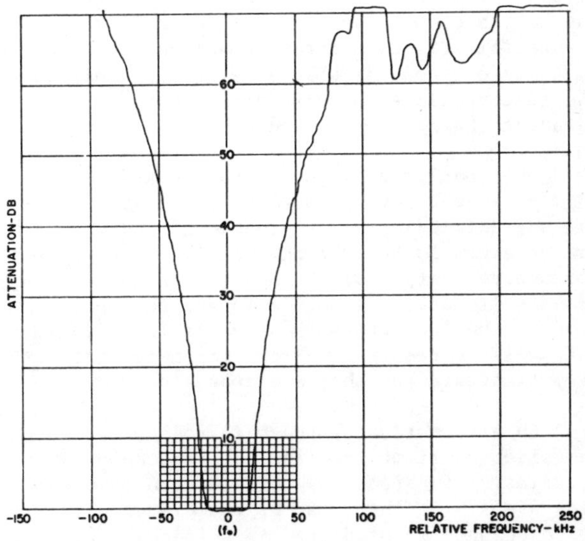

FIGURE 6.
ATTENUATION CHARACTERISTIC OF FOUR-RESONATOR
TANDEM MONOLITHIC INTEGRATED CRYSTAL FILTER
CENTER FREQUENCY 23 MHz

SINGLE SIDEBAND FILTERS FOR SHORT HAUL SYSTEMS

Desmond F. Sheahan
Lenkurt Electric Company, Inc.
San Carlos, California, USA

Summary

The discovery of the theory of acoustical coupling in AT cut-quartz resonators has created a great deal of interest in the telephone industry. Most of this interest has been directed towards the realization of miniature channel filters for single sideband FDM systems. This paper will describe a particular design approach that has been used successfully to realize such filters.

Introduction

Frequency division multiplex single sideband systems generally operate by modulating 3.4 kHz wide voice channels into the frequency range of 60-108 kHz to form groups of 12 channels. Each voice channel is translated in frequency by using it to amplitude modulate the appropriate carrier frequency. The modulation process will produce two sidebands of the voice channel, one on either side of the carrier. A channel filter must therefore be used to select just one of these sidebands before transmission. The filter will also be required to isolate the adjacent channels from each other.

The modulation into the 60-108 kHz frequency range can be performed in one step without any difficulty. The problem arises when one attempts to build a channel filter at this frequency. A typical channel filter requires a transfer function of degree 16 and for operation in the 100 kHz range very stable components will be required. Such components are not easy to obtain and they are expensive.

An alternative approach is to separate the sideband filtering from the frequency translating functions by means of two modulation steps. In the first of these steps, the voice channel is modulated to a frequency where the single sideband filtering can be accomplished most economically. The second step modulates the filtered single sideband to the assigned slot in the 60-108 kHz range. Once this two step modulation philosophy is adopted many filtering options become attractive. Some that have been studied are LC filters at 20 kHz and gyrator filters at the same frequency, mechanical filters at 50 kHz, and 8 MHz crystal filters that utilize the principle of acoustic coupling. In this paper we will only concern ourselves with the latter crystal filters.

Acoustically Coupled Resonators

The low temperature coefficient AT cut in quartz has the property that an electric field in the thickness direction of such an AT plate will create a mechanical shear strain about a plane midway between the two faces of the plate. An alternating electric field will likewise create a mechanical shear vibration. The vibrating piece of quartz will have a resonant frequency that is determined by its density and elasticity. If the frequency of the alternating electric field is the same as the mechanically resonant frequency of the quartz plate, the plate will appear to be an electric resonant circuit with an equivalent inductance that is inversely proportional to the electrode area. The equivalent circuit of such a resonant plate is shown in Figure 1.

Such single resonators have been known and used for many years. What was not known until recent years, however, was the fact that when the mass of metal in the electrodes of a plated crystal is large enough the mechanical vibration will be confined to the area underneath the electrodes and the vibrating energy will decay exponentially with distance away from them.[1] In such a situation, the energy is said to be trapped in the resonator. This was an important discovery as it meant that resonators could be designed principally by controlling their electrode areas and what happened at the external connections to the crystal plate could be relatively ignored. It also meant that with more than one resonator on the same piece of quartz the coupling between them could be controlled by adjusting their spacing and the rate of exponential energy decay in the unplated region. This is the basis of the theory of coupled resonators.

The electrical equivalent circuit for a pair of coupler resonators is shown in Figure 2. This figure also shows the relationships between the circuit elements and the resonator dimensions. In the equivalent circuit L_1C_1 and L_2C_2 are the motional parameters of the resonators, C_4 and C_5 are the parallel plate capacitances formed by the dielectric material of the quartz. C_3 which represents the coupling[2,3] has a negative sign to account for the phase change that is caused by the interaction of the direct and the converse piezoelectric effects in the quartz. If the coupled resonators are never to be used with

Reprinted from *Memorias Conf. Int. IEEE Sobre Sistemas, Redes y Computadores*, vol. II, Jan. 19–21, 1971, pp. 744–748.

bridging circuit elements between their input and output terminals the negative sign on the coupling capacitor can be disregarded.

Filter Design With Acoustically Coupled Resonators

The very nature of the electrical equivalent circuit in Figure 2 suggests that a suitable bandpass filter structure for acoustically coupled resonators would be that shown in Figure 3. This filter is drawn as an eight resonator design because that is the degree of complexity required for use as a telephone channel filter. This type of filter is also the one that has received the greatest attention in the literature. It can be realized in three different ways that use varying degrees of acoustical coupling. It can be realized as a monolithic filter where all the resonators are on the same piece of quartz and each resonator is acoustically coupled to the adjacent one.[4,5] Alternatively it can be realized by two pieces of quartz, each of which contains more than two acoustically coupled resonators and the coupling between the plates is accomplished by a single capacitor. This is called the Bilithic form.[6] It can also be realized by four separate coupled resonator pairs and three capacitors.[7] This is the polylithic form and it is the one that will be described in this paper.

The most common method of designing the filter in Figure 3 has been the coupled resonator theory of Dishal.[8] This design method is intended specifically for this type of all pole transfer function filter. The design information is obtained in the form of coupling coefficients and loaded Q's for each of the resonators. An alternate design method due to Orchard which is more consistent with insertion loss filter design techniques is shown in Figure 4. In this method the filter is first designed as a low pass filter and it is then transformed into a bandpass one. Gyrators are then inserted and finally each gyrator is replaced with its narrow band equivalent of a capacitive T. This design method allows the use of low pass filter design tables that are readily available and it also presents the design information in a form that is suitable for use with network analysis computer programs.

The filter in Figure 3 has a symmetrical stopband response. Such a symmetrical stopband response, however, is not always the most desirable one for single sideband multiplex systems. A more preferable stopband attenuation would be one that has a steeper attenuation slope on the unwanted sideband side of the carrier. A greater attenuation slope could of course be obtained if an additional coupled resonator section were added to the filter in Figure 3. This increase in the degree of the transfer function would increase the attenuation equally as much in both stopbands. It would also increase the attenuation at the passband corner frequencies and would worsen the group delay distortion.

The desired effect can be achieved without any increase in the degree of the transfer function if we introduce attenuation holes at finite frequencies in either stopband. A convenient method of achieving such attenuation holes becomes apparent if we consider the equivalent circuit for a single resonator as shown in Figure 1. The series resonance in this equivalent circuit can be used to produce an attenuation pole if it is placed in the shunt arm of a ladder filter and the attenuation pole thus produced will be in the lower stopband.

A particular filter design which meets the desired requirements is shown in Figure 5. It can be realized with three coupled resonator pairs, two single resonators, and two capacitors. The filter is designed as an equi-ripple passband filter with two coincident attenuation holes in its lower stopband. The actual synthesis itself is performed with a computer program based on the procedures of Orchard and Temes.[9] It is then rearranged in the form shown in Figure 5. In order for this filter to be easily realizable in polylithic form, it is important that the inductances of the single resonators not be very much smaller than that of the coupled resonators. This is to ensure that the same size of crystal blank can be used for both.

Assembly Method

A common assembly technique is used for the polylithic filters in Figures 3 and 5. The crystal blanks are all 0.5" dia and they have the appropriate single resonator or double resonator pattern evaporated onto them. The crystals are mounted in ceramic rings that are 0.75" dia and while in these rings they are tuned by the addition of some more gold. For assembly, the rings are stacked on top of each other and wired together with the necessary capacitors. The completed filter is sealed in an evacuated cold welded can that is 1" high and 1" in diameter. NPO ceramic chip capacitors are used.

Measured Results

The measured responses obtained with these filters are shown in Figure 6. They agree very closely with those calculated for resonator Q's equal to 200,000. The measured temperature performance of the filters is a frequency shift in the whole response that is typically less than ±40 Hz for temperature changes of ±25°C.

Spurious responses in both filters were all measured to be more than 70 dB below the passband level. The relative freedom of the filter stopbands from spurious responses is one advantage of the polylithic form of realization. This type of filter is not susceptible to the spurious mode problems that were experienced with monolithic filters. [4]

Conclusions

This paper has described polylithic crystal filters that are suitable for use in single sideband FDM systems. When the equivalent circuit for acoustically coupled resonators is used these filters can be designed with the well known insertion loss techniques. The filters actually realized by this polylithic technique offer substantial size and cost savings over their LC counterparts and they are ideally suited for use in short haul FDM systems.

Acknowledgments

The author would like to thank Prof. H. J. Orchard of UCLA for his consultations during the development of these filters, also Mr. G. C. Callander for his expertise in building them, and finally the management of Lenkurt Electric Co., Inc. for permission to publish this paper.

References

1. Shockley W., Curran D., & Koneval D. "Energy Trapping and Related Studies of Multiple Electrode Filter Crystals," Proc. 17th Annual Symposium on Frequency Control, Atlantic City, New Jersey, May, 1963, pp. 88-126.

2. Sykes R. H. & Beaver W. D., "High Frequency Monolithic Crystal Filters with Possible Application to Single Frequency and Single Sideband Use," Proc. 20th Annual Symposium on Frequency Control, April, 1966, pp. 288-308.

3. Sheahan, D. F., "An Improved Resonance Equation for AT-Cut Quartz Crystals," Proc. IEEE Vol. 58, No. 2, February, 1970, pp. 260-261.

4. Beaver W. D., "Theory and Design of the Monolithic Crystal Filter," Proc. 21st Annual Symposium on Frequency Control, April, 1967, pp. 179-199.

5. Sykes R. A., Smith W. L., and Spencer W. J., "Monolithic Crystal Filters," 1967 IEEE International Convention Record, Vol. 15, Part 11, pp. 78-93.

6. Werner J. F., Dyer A. J., and Birch J., "The Development of High Performance Filters Using Acoustically Coupled Resonators on AT Cut Quartz Crystals," Proc. 23rd Annual Symposium on Frequency Control, May, 1969, pp. 65-75.

7. Yoda H., Nakazawa Y., and Kobori N., "High Frequency Crystal Mechanical Filters," Proc. 22nd Annual Symposium on Frequency Control, April, 1968, pp. 188-206.

8. Dishal M., "Design of Dissipative Band Pass Filters Producing Desired Exact Amplitude-Frequency Characteristics," Proc. I.R.E., Sept., 1949, pp. 1050-1069.

9. Orchard, H. J. and Temes G. C., "Filter Design Using Transformed Variables," IEEE Trans on Circuit Theory, CT-151 No. 4, Dec., 1968, pp. 385-408.

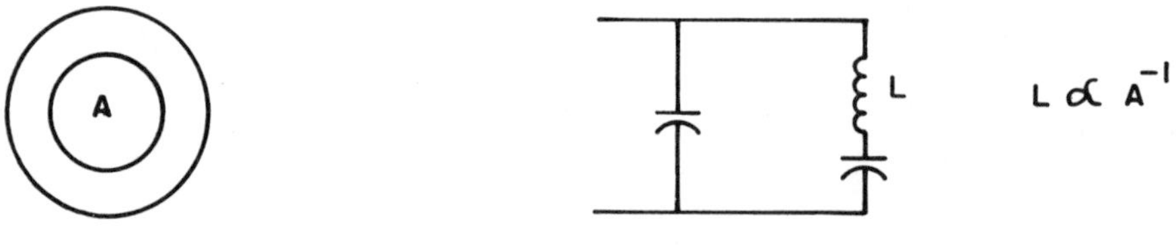

$$L \propto A^{-1}$$

FIG. 1 SINGLE RESONATOR AND ITS EQUIVALENT CIRCUIT.

$$L_1, L_2 \propto A^{-1}$$
$$C_4, C_5 \propto A$$
$$-C_3 \propto A e^{d\Delta^{\frac{1}{2}}}$$

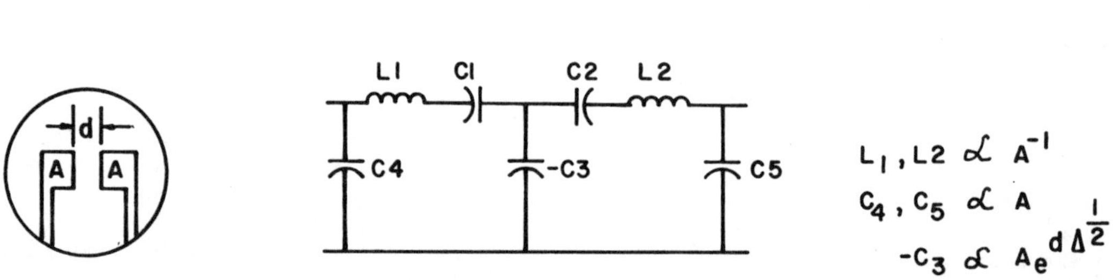

$$\Delta = \text{PLATEBACK} = \frac{f_{\text{CRYSTAL}} - f_{\text{RESONATOR}}}{f_{\text{RESONATOR}}}$$

FIG.2 COUPLED PAIR OF RESONATORS

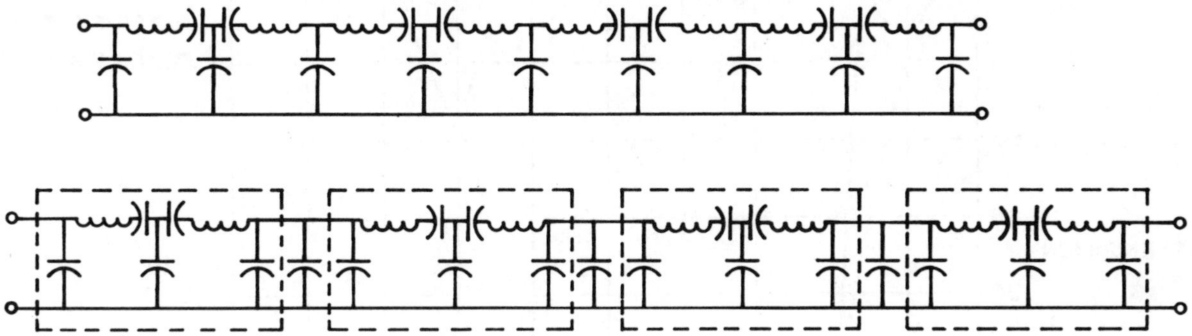

FIG.3 ALL POLE POLYLITHIC FILTER

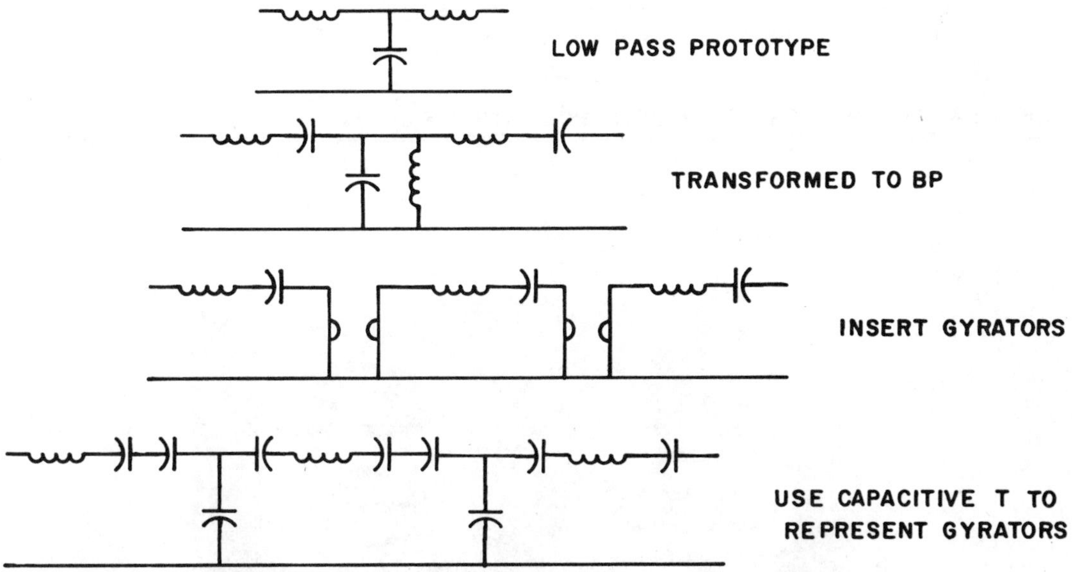

FIG.4 DESIGN TECHNIQUE FOR ALL POLE FILTER.

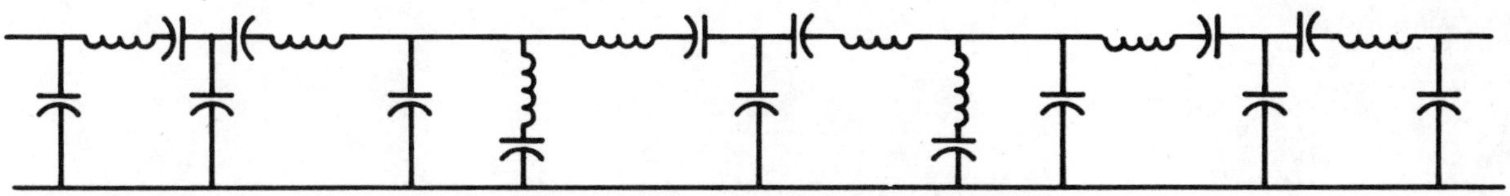

FIG. 5 FILTER WITH FINITE FREQUENCY ATTENUATION POLES.

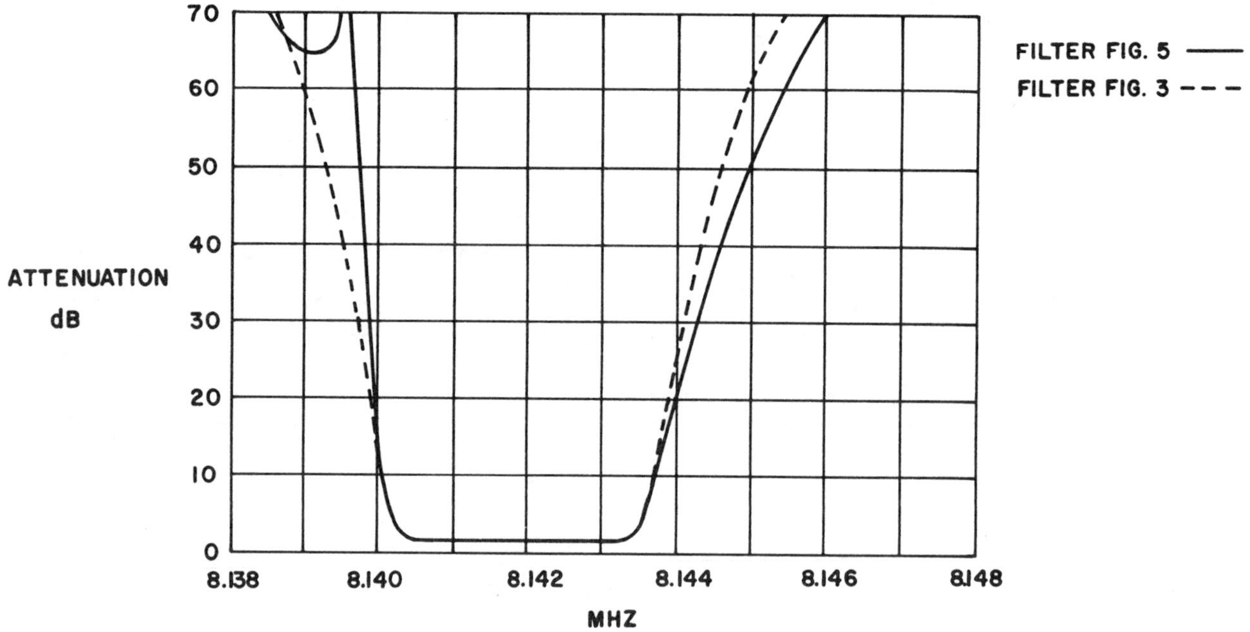

FIG.6 MEASURED RESPONSES OF BOTH POLYLITHIC FILTERS.

FIG. 7 PHOTOGRAPH OF BOTH POLYLITHIC FILTERS. FIG.3 FILTER ON LEFT, FIG.5 FILTER ON RIGHT.

COMPOSITE FILTER STRUCTURES INCORPORATING MONOLITHIC CRYSTAL FILTERS AND LC NETWORKS

H. A. Simpson, E. D. Finch, Jr., and R. K. Weeman
Bell Telephone Laboratories, Incorporated
North Andover, Massachusetts

Summary

The monolithic crystal filter may be considered from either of two points of view: 1) as a complete, self-contained filter entity; or 2) as a two-port device which may be incorporated as a building block into some larger frequency selective network. The present paper considers the monolithic crystal filter from the latter point of view, with application to problems in bandpass and band-elimination filter design. For the bandpass case, this paper is principally concerned with the problem of realizing Chebyshev passband behavior from a cascade of two or more symmetrical monolithic crystal filters (MCFs) coupled by LC networks. For band-elimination filters, the theory of operation of both inductor-derived and capacitor-derived monolithic crystal filter structures is presented. Examples of practical designs for both bandpass and band-elimination filters are used to illustrate design techniques based on closed-form solutions.

Building Blocks for Composite Multi-resonator Crystal Filters

The building blocks considered in this paper are illustrated in Fig. 1. They include the following:

1. Symmetrical, coupled two-resonator crystal devices (MCF_S) having the equivalent circuit shown in the figure.
2. Narrowband LC networks of two types:
   a) Impedance inverters having ±90° phase shift; and
   b) Impedance matching networks of the "L" or "Pi" configuration.
3. Two-terminal, single-element reactances (inductors and capacitors).

The monolithic crystal filter devices considered in this paper are restricted to two-resonator symmetrical devices. These MCFs represent complete filters in their own right and can be specified by their own filter parameters. In the composite structure, however, this identity is lost, and the MCF becomes merely a device for realizing a desired performance - in this instance, a narrowband crystal filter having Chebyshev behavior in the passband. As will be seen later, Butterworth parameters permit convenient description of the individual two-resonator coupled devices.

The impedance inverter networks shown in Fig. 1 are composed of positive and negative capacitors. These are realized in practice as LC coupling networks, based on a narrowband approximation in which the positive reactance of an inductor is used to realize the reactance of a negative capacitor. As a matter of interest as will be shown later, the equivalent circuit of the MCF includes an inverter having the required positive and negative capacitances provided by the acoustic coupling between resonators.

Impedance matching networks are required at the input and output ports of the composite filter to provide an impedance match to the source and load impedances. This is necessary because the source and load impedances are generally fixed at a value different from the optimum impedance level of the composite filter.

The discrete components shown hardly need explanation. They are listed here, however, as two-terminal building blocks required for realization of inductor-derived and capacitor-derived monolithic band-elimination filters.

Some of the features of the composite filter approach to narrow bandpass filter design may be summarized as follows:

1. It allows the use of symmetrical MCF designs.
2. The coupling and impedance matching networks can be used to provide additional discrimination against overtone passbands.
3. Additional degrees of freedom are available in the coupling networks for adjustment, thus allowing requirements on the MCFs to be relaxed.
4. For small volume production, high-order crystal filters may be produced quickly - as contrasted to development times required for single plate designs.

Types of Monolithic Crystal Filters

Figure 2 shows the wide variety of MCFs that have been developed for high frequency filter applications to radio and wire systems. Since 1966, over 100 designs have been developed. Of these, approximately 12 designs of the type shown at the top of the display and identified as multi-order MCFs are for application to a telephone channel bank.

The two lower levels in the display show what are commonly called second-order or two-resonator MCFs, although they have fourth-order transfer functions. Such MCFs find application, individually, as single-tone pick-off filters or as components in a composite filter structure. The frequency ranges indicated for each header size are only approximate.

Design Parameters of Two-resonator Devices

The mathematical analyses to follow employ the equivalent electrical network as shown in Fig. 3 for the MCF. The element values of the components comprising this equivalent network completely determine the mathematical model that is used to represent the coupled resonator device. These element values are given on the left-hand side of the figure in terms of the network design parameters: R_o, the impedance

Reprinted with permission from *Proc. Twenty-Fifth Ann. Symp. on Frequency Control*, Apr. 26-28, 1971, pp. 287-296, sponsored by the U.S. Army Electronics Command, Ft. Monmouth, NJ.

level of the filter; q_1, the normalized loaded Q of the resonators; $Bw_{3\ dB}$, the 3-dB bandwidth of the two-resonator device operated as an individual filter; f_o, the center frequency of the filter; and a device parameter n, the order of overtone operation (n=1 for the fundamental mode). The right-hand side of Fig. 3 relates the network and device design parameters. The device parameters are: A, the electrode area; t, the plate thickness; ϵ, the material permittivity; and K_{12}, the actual coupling between resonators in terms of the frequencies of symmetric (fs) and anti-symmetric (fa) thickness shear displacements.[1]

The network representing the acoustic coupling between resonators is electrically an impedance inverter that has the property of presenting an impedance $Z = \frac{k^2}{Z_{load}}$ at the input port when the output port is terminated in Z_{load}. The inversion constant, $k^2 = \left(\frac{1}{\omega Cm}\right)^2$, is approximately constant over the narrowband passband of the filter. Furthermore, since the z parameters Z_{11} and Z_{22} of an impedance inverter are zero, mesh resonant frequencies are unaltered when meshes are coupled by such a network.

Electrical Matching Networks

Figure 4 illustrates three LC narrow-band impedance matching networks that have been used extensively to provide a narrowband impedance match between the MCF and the source and load impedances. Back-to-back connection of such sections provides electrical coupling networks for coupling two-resonator MCFs. The networks shown in low-pass, high-pass, and bandpass forms all match R_1 to R_2 at the frequency $\omega = \omega_o$. In all cases, R_2 must be greater than R_1. The three-element bandpass section requires an additional parameter $\Delta\omega$, the 3-dB bandwidth of the section, to uniquely specify all elements. This bandpass section is particularly useful with overtone mode MCFs where the increased discrimination of the section aids in the elimination of the fundamental passband. The low-pass and bandpass-type sections provide for a broadband absorption of the shunt capacitance C_o of the MCF by supplying capacitance abutting the MCF. The remainder of this paper will consider only the low-pass-type section, although similar results have been achieved with the other sections.

Single-frequency Resistive Matching at Junctions

A possible approach to the design of a composite four-resonator crystal filter is indicated in Fig. 5. In this instance, a resistive matching technique is used at selected junctions of the network. With a load impedance of R_1, each MCF is designed at an impedance level of R_o. Considering the coupling network in the center of the schematic as two low-pass matching sections operating back-to-back, the impedance level R_x becomes the only free parameter.

The band-widening and band-shifting properties of this network are illustrated in Fig. 6. In this figure, the insertion losses of the network, for specific values of the dimensionless parameter R_x/R_o are plotted. The specific examples selected show the response of two identical Butterworth MCFs coupled by a Pi-network at a center frequency of 8.448 MHz. Individually, each MCF was designed for a 3-dB bandwidth of 1400 Hz. For the case $R_x/R_o = 0.5$, an equiripple passband having a 3-dB bandwidth of approximately 1600 Hz and 1-dB ripple is obtained. If the losses of the Butterworth filters were merely additive, a bandwidth of only 1000 Hz would result. This increase in bandwidth using an LC coupling section is approximately 60 percent. For R_x/R_o above and below 0.5, large loss bumps appear above and below the passband center frequency.

Computer Optimized Results[3]

The technique of impedance matching previously shown is not, in general, capable of producing sufficiently flat passband performance for critical applications where return loss specifications might be in excess of 26 dB. Departing from the previous concept of resistive impedance matching, additional degrees of freedom result if the elements of the input matching and coupling networks are allowed to assume new design values. Typical results that can be obtained by computer-based optimization are illustrated in Fig. 7. The parameters of the MCFs were fixed, and the overall network was constrained to a symmetrical structure. This results in a total of four free variable elements. In the computer optimization routine, the flatness of the passband was optimized over a selected band.

Although four return loss peaks were obtained, the return loss characteristic was not equi-minima, nor was the insertion loss (not observable in the scale shown) equi-ripple. The deviations from the desired four-resonator Chebyshev response were only slight and negligible for all practical purposes. A closed-form solution, giving exact Chevyshev response, however, was found to exist.

Modified Impedance Matching Approach

The first impedance matching approach to composite filter design, illustrated in Fig. 5, failed to result in four-resonator Chebyshev performance. Figure 8 illustrates a modified impedance matching approach that results in exact Chebyshev response.

The matching sections are designed to present an impedance αR_o to the MCFs. The MCFs, in turn, are coupled by an LC network, the narrowband equivalent of an impedance inverter, having an inversion constant of $(R_o/\alpha K)^2$. The constant, α, is defined in terms of the coefficients K_{12} and Q_1, which are the normalized coupling and loaded resonator Q specified by Dishal[2] for a four-resonator Chebyshev filter. K is related to the desired reflection coefficient of the composite network at center frequency. The reflection coefficient, ρ, is related to A_{max}, the maximum allowable insertion loss ripple.

Decomposition of a General N-resonator Filter into a Composite Structure

Figure 9 illustrates the steps that may be taken to decompose an n-resonator (n even), equal inductance bandpass filter into an equivalent structure for realization as a composite crystal filter. The elements in the top schematic may be computed simply from considerations outlined by Dishal.[2] The desired coupled resonator form is that shown in the top figure. Within the structure, a number of capacitive-tee impedance inverters are evident. The actual value of source and load resistance, R, may be realized as a physical resistance or, preferably, as a resistance presented by one of the narrowband matching sections of Fig. 4. The low-pass and bandpass forms of these matching sections permit broadband absorption of the shunt capacitance C_o, which has therefore been neglected in Fig. 9. The first pair of coupled resonators may be realized as a symmetrical Butterworth MCF shown schematically in the lower figure. The 3-dB bandwidth of the Butterworth filter is related to the ripple bandwidth of the composite filter and to the coupling coefficient, K_{12}, of the n-resonator Chebyshev filter as shown. The second capacitive-tee impedance inverter is realized by the narrowband equivalent Pi-network. Following this realization procedure throughout the filter, all elements may be realized as symmetrical two-resonator MCFs or as narrowband equivalents of impedance inverters.

Figure 10 summarizes the number of building blocks and the number of different building block designs required for an n-resonator composite crystal filter. The "rounding" indicated in Fig. 10 infers that improper fractions resulting from these computations must be rounded to the next whole integer. The result of the application of the decomposition technique to the realization of a 42.496-MHz eight-resonator Chebyshev composite filter is shown in Fig. 11. As determined by the chart of Fig. 10, two different symmetrical MCF designs and two different coupling network designs (narrowband impedance inverter equivalents) are required. The schematic of Fig. 11 identifies the MCFs and components. The wideband insertion loss, passband ripple, and inband return loss are seen to correspond to the exact Chebyshev filter case. In this instance, the impedance matching networks required to match R_o of the filter to specific source and load impedances are not shown.

Band-elimination Filters

Two types of monolithic crystal band-elimination filters have been developed. These filters are derived directly from monolithic crystal bandpass filters by the simple addition of inductive or capacitive elements across the ungrounded electrodes of single MCFs. The filters have been designated as inductor- or capacitor-derived types, depending on the type of bridging element.

Both the inductor-derived and capacitor-derived band-elimination filters (BEFs) are shown in Fig. 12. For the inductor - derived type, input and output ports are coupled by an inductor having a reactance equal in magnitude to that of one of the acoustic coupling capacitors of the MCF at the rejection frequency. This rejection frequency is identical to the passband center frequency of the MCF when used in its bandpass mode. For the capacitor-derived filter, a phase reversal of the monolithic filter is required. This is achieved by the connection indicated in the schematic. The phase reversal achieved is clearly evident on comparison of the two impedance inverters representing the acoustic coupling in the equivalent networks. Band-elimination characteristics are then obtained when C_X is made equal to one of the acoustic coupling capacitors, C_m.

The band-eliminating properties of these filters may be studied in a variety of ways, but perhaps the easiest approach is the application of Bartlett's Bisection Theorem to derive the equivalent lattice network. For the inductor-derived case, the sequence of networks shown at the top of Fig. 13 illustrates the method and verifies the existence of zero transmission at the rejection frequency, f_o, from the properties of a balanced lattice. The broadband equivalent lattice network for the inductor-derived filter is shown in the sequence at the bottom of Fig. 13.

For practical applications, the inductor-derived form of this BEF has superior passband performance as compared to its capacitor-derived counterpart. This is clear when it is observed, as in Fig. 14, that the BEF is essentially comprised of two filters connected in parallel: 1) a low-pass filter (C2, L_X, C2), and 2) a monolithic crystal filter. The rejection properties at f_o have already been explained. In the passband, the impedances of the individual resonators comprising the MCF are extremely high, and the resulting network reduces to a simple low-pass filter (C2, L_X, C2). These elements, and the associated source and load impedances, determine the cut-off frequency of the low-pass filter which limits the highest frequency that may be passed in a particular design. Furthermore, this impedance level cannot be selected arbitrarily in practical designs because of limitations in broadband transformer capabilities. Some compromise and iteration may be necessary at this stage to achieve a satisfactory compromise between the low-pass filter design and the design of the MCF. One must also realize that vibrational modes other than the main mode exist in the MCF and produce passband irregularities. Design considerations related to spurious mode suppression confine the MCF parameters to a limited range. The design procedure outlined in Fig. 14 has been found useful. Some iteration, however, is generally required before a satisfactory overall design is obtained.

An example of the application of these design principles to the realization of a 42.88-MHz pilot-blocking filter is shown in Fig. 15. Autotransformers are incorporated for broadband impedance matching to the 75-ohm source and load. Two inductor-derived BEF sections are required to meet the stopband requirements, and a four-section low-pass filter is chosen to permit compensation for the distributed capacitance of the transformers and for the static capacitances of the MCFs. To achieve the passband indicated as flat to within ±0.05 dB from 1 to 75 MHz, an additional amplitude equalizer is necessary. A fundamental mode MCF was required to meet passband specifications.

Conclusion

In this paper, the authors have treated the MCF as a device specifically designed as a component or building block of some larger frequence selective network. Applications to both bandpass and band-elimination filters have been cited. The theory of operation has been discussed and pertinent design techniques have been presented. Certain advantages in the building block approach have been pointed out.

References

1. Sykes, R. A., Smith, W. L., and Spencer, W. J., "Monolithic Crystal Filters," 1967, IEEE International Convention Record, pp. 78-93.

2. Dishal, M., "Two New Equations for the Design of Filters," Electrical Communication, Vol. 30, December 1952, pp. 324-337.

3. Garrison, J. L., Georgiades, A. N., and Simpson, H. A., "The Application of Monolithic Crystal Filters to Frequency Selective Networks," Digest of Technical Papers, 1970 International Symposium on Circuit Theory, 1970, pp. 177-178.

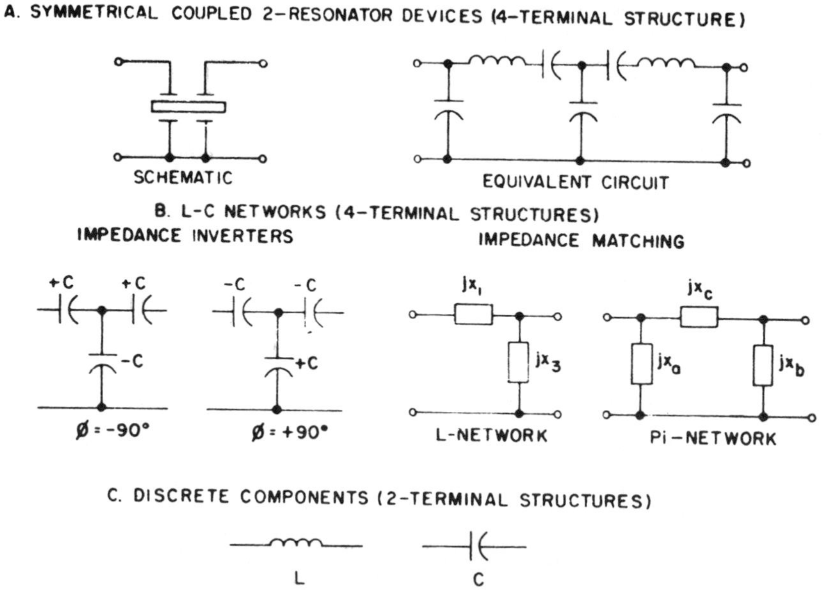

BUILDING BLOCKS FOR COMPOSITE MULTI-RESONATOR CRYSTAL FILTERS
FIGURE 1

FIGURE 2

Family of Monolithic Crystal Filters

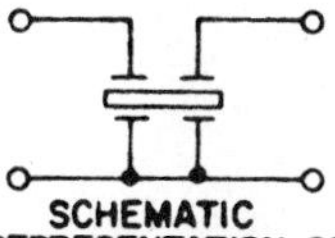

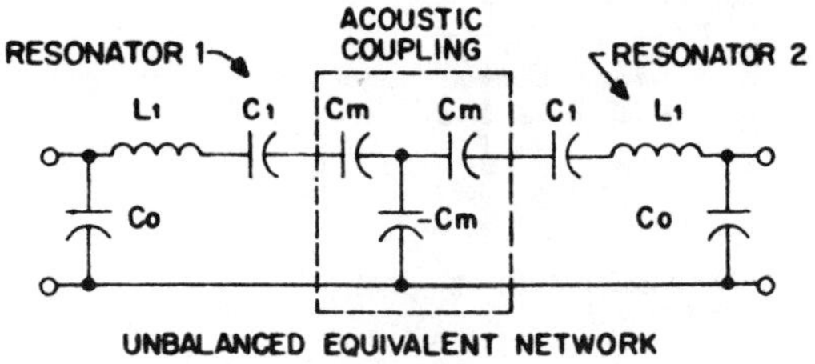

NETWORK DESIGN PARAMETERS

$$L_1 = \frac{R_{oq1}}{2\pi Bw_{3dB}} \quad (H)$$

$$C_1 = \frac{1}{\omega_o^2 L_1} \quad (F)$$

$$C_m = \frac{C_1 f_o}{k_{12} Bw_{3dB}} \quad (F)$$

$$C_o = r n^2 C_1 \quad (F)$$

$$r = 250$$

DEVICE DESIGN PARAMETERS

$$L_1 = \frac{1.8 n^3}{A(Cm^2) f_o^3 (MHz)} \quad (H)$$

$$C_1 = \frac{1}{4\pi^2 f_o^2 L_1} \quad (F)$$

$$K_{12} = \frac{f_a - f_s}{\sqrt{f_a f_s}}\ ;\ C_m = \frac{C_1}{K_{12}} \quad (F)$$

$$C_o = \frac{\epsilon A}{t} \quad (F)$$

$$t = \frac{0.1665 n}{f_o (MHz)} \quad (cm)$$

EQUIVALENT NETWORK AND DESIGN PARAMETERS: 2-RESONATOR MCF
FIGURE 3

NETWORK	ELEMENT VALUES		
	L_1	C_1	C_2
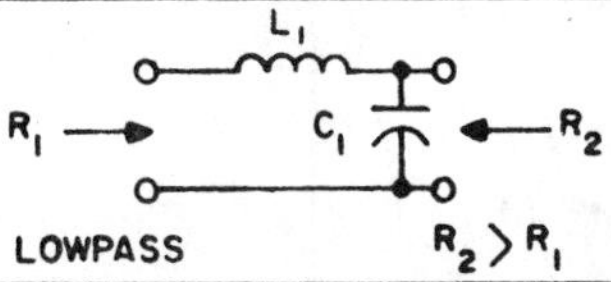 LOWPASS $R_2 > R_1$	$\frac{1}{\omega_o}\sqrt{R_1(R_2-R_1)}$	$\frac{1}{\omega_o R_2}\sqrt{\frac{R_2}{R_1}-1}$	—
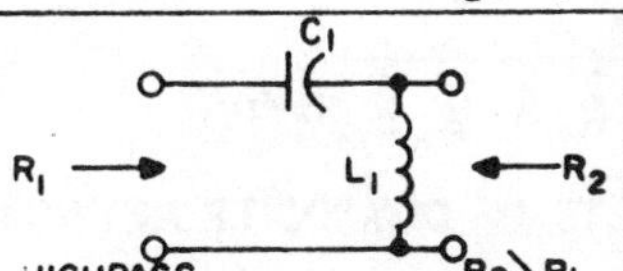 HIGHPASS $R_2 > R_1$	$\frac{R_2}{\omega_o}\sqrt{\frac{R_1}{R_2-R_1}}$	$\frac{1}{\omega_o\sqrt{R_1(R_2-R_1)}}$	—
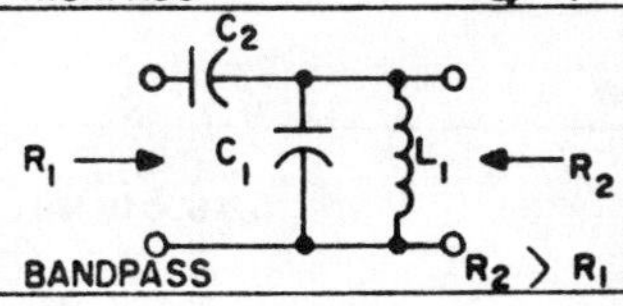 BANDPASS $R_2 > R_1$	$\frac{R_2 \Delta\omega}{2\omega_o^2}$	$\frac{1}{\omega_o^2 L_1} - C_2\left(1-\frac{R_1}{R_2}\right)$	$\frac{1}{\omega_o\sqrt{R_1(R_2-R_1)}}$

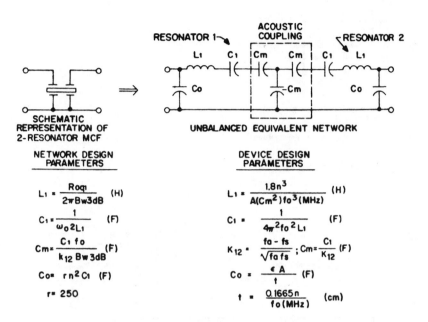

NARROW BAND IMPEDANCE MATCHING NETWORKS
FIGURE 4

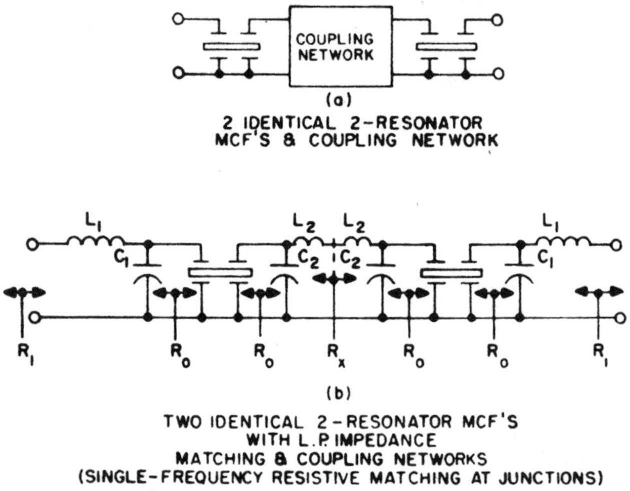

FOUR-RESONATOR MONOLITHIC CRYSTAL FILTERS
FIGURE 5

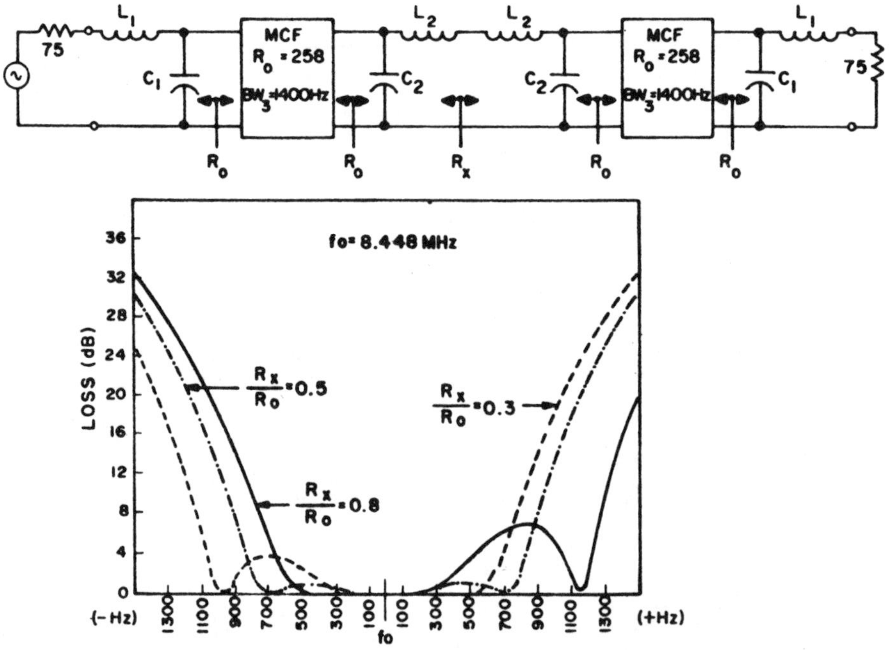

BAND-WIDENING & BAND-SHIFTING EFFECTS OF COMPOSITE NETWORK
FIGURE 6

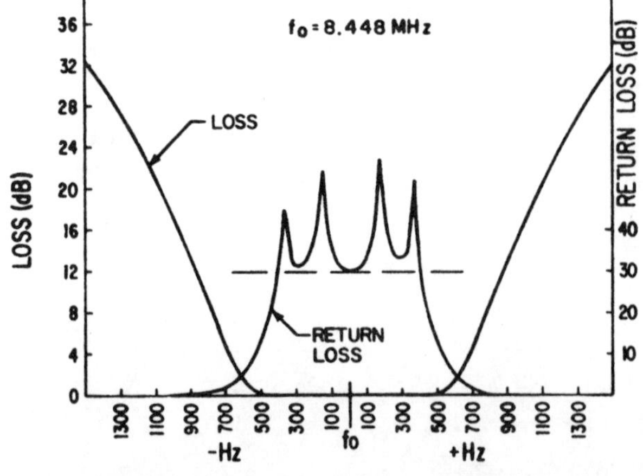

COMPUTER OPTIMIZED RESPONSE OF COMPOSITE 4-RESONATOR FILTER
FIGURE 7

$$\alpha = \frac{1}{K_{12}Q_1}$$

$$K = \sqrt{\frac{1+\rho}{1-\rho}}$$

$$\rho = \sqrt{1-10^{-A_{MAX}/10}}$$

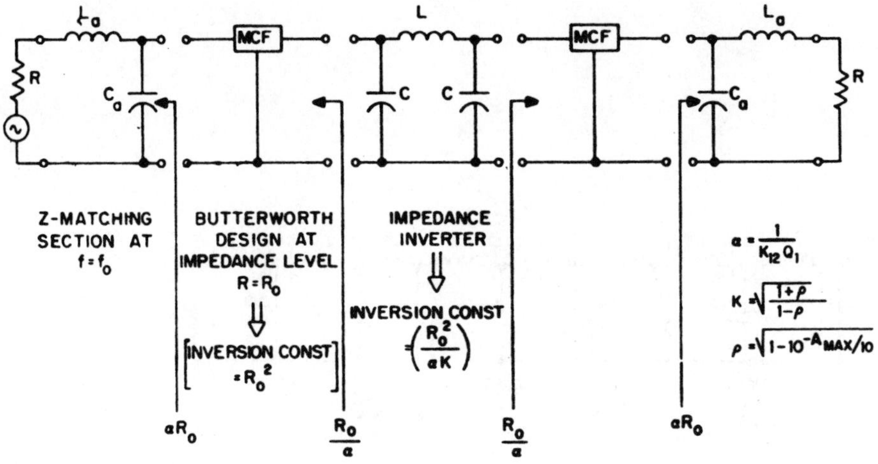

MODIFIED IMPEDANCE MATCHING REALIZATION
OF EXACT 4-RESONATOR CHEBYSHEV RESPONSE
FIGURE 8

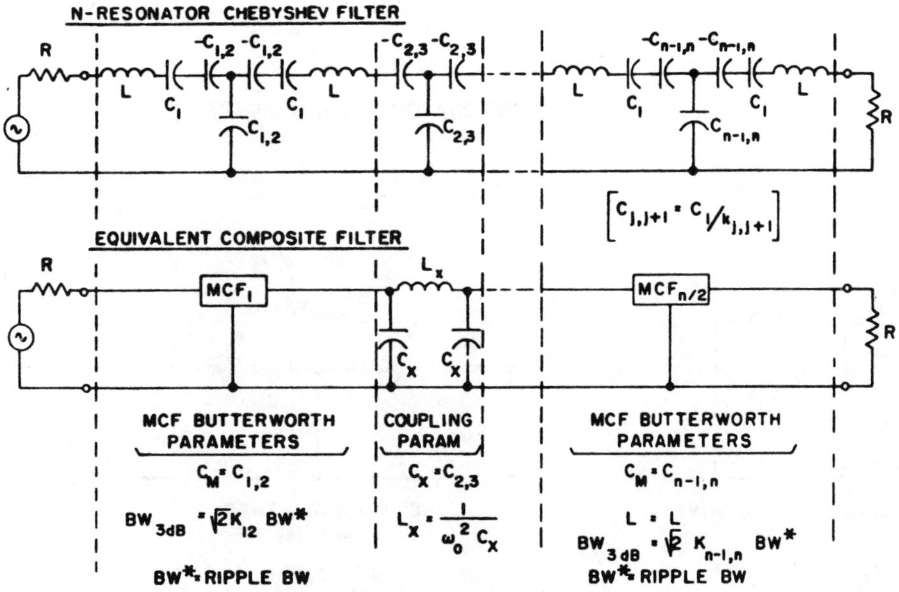

DECOMPOSITION OF n–RESONATOR FILTER TO COMPOSITE STRUCTURE
FIGURE 9

NO. OF RESONATORS	NO. OF MCF'S	NO. OF COUPLING NETS	NO. OF MCF DESIGNS	NO. OF COUPLING NET. DESIGNS
4	2	1	1	1
6	3	2	2	1
8	4	3	2	2
10	5	4	3	2
(n)	(n/2)	(n/2 –1)	(n/4)*	$\left(\frac{n/2-1}{2}\right)$*

* ROUNDED

BUILDING BLOCKS REQUIRED FOR N-RESONATOR COMPOSITE FILTERS
FIGURE 10

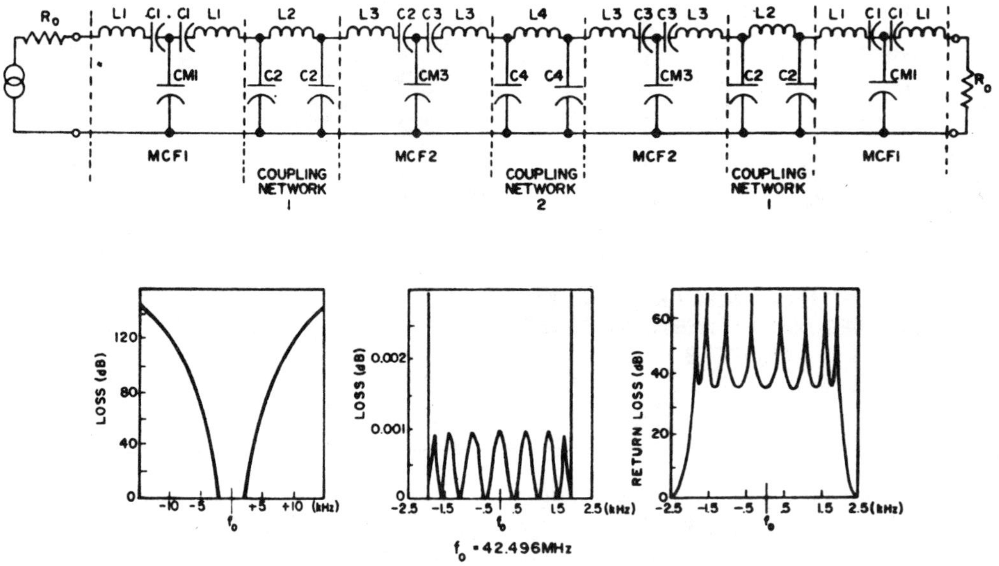

COMPOSITE EIGHT-RESONATOR CRYSTAL FILTER
FIGURE 11

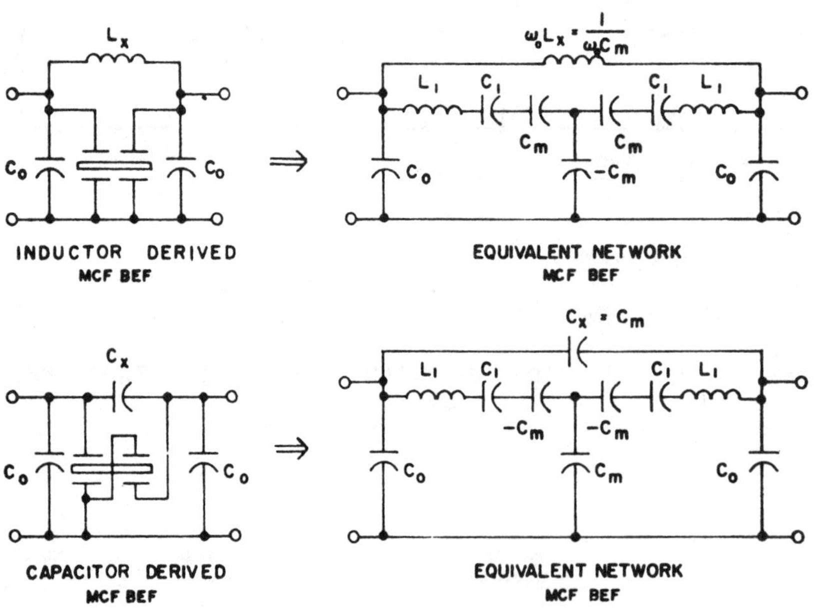

EQUIVALENT NETWORKS: INDUCTOR & CAPACITOR-DERIVED MCF BEF'S
FIGURE 12

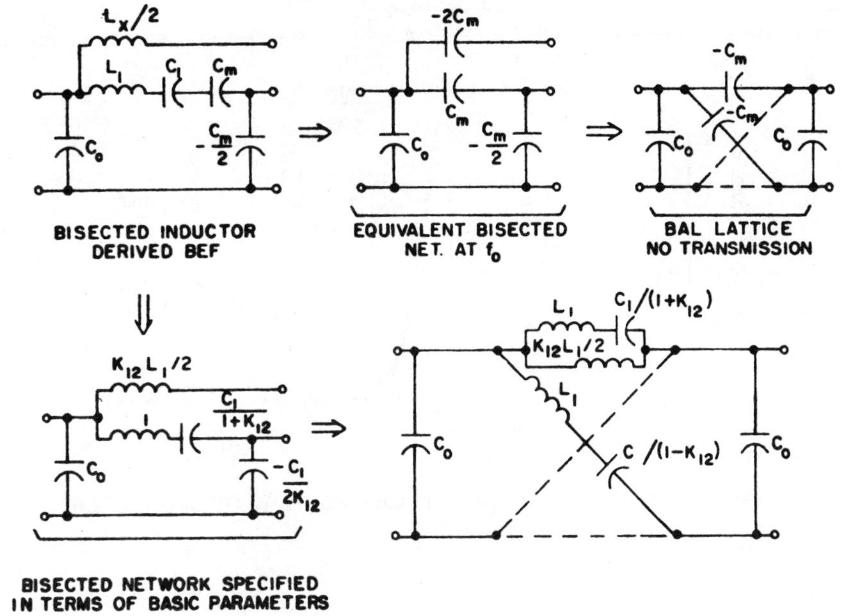

LATTICE EQUIVALENTS OF INDUCTOR-DERIVED MONOLITHIC B.E.F.
FIGURE 13

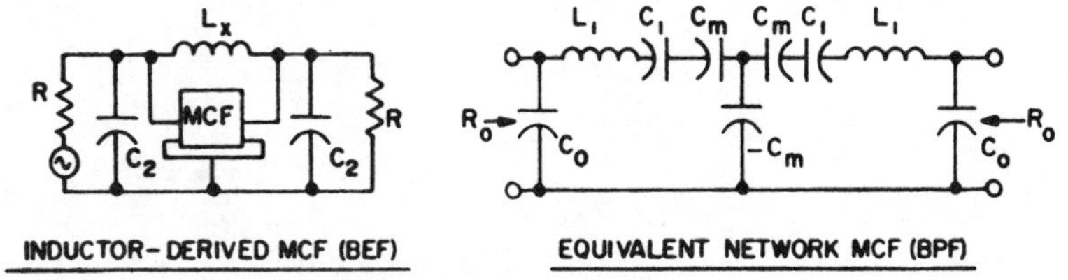

DESIGN PROCEDURE

1. SELECT SUITABLE IMPEDANCE LEVEL R
2. SELECT SUITABLE f_c FOR L.P. FILTER
3. COMPUTE $L_x = 2R/\omega_c$
4. COMPUTE $C_m \; 1/(\omega_o^2 L_x)$ ω_o = SUPPRESSION PEAK FREQUENCY
5. COMPUTE $R_o = 1/(\omega_o k_{12} q_1 C_m)$ $k_{12} = 1/\sqrt{2}$; $q_1 = \sqrt{2}$ FOR BUTTERWORTH
6. COMPUTE $L_1 = R_o \, q_1 / (2\pi BW_{3dB})$
7. COMPUTE $C_1 = 1/(\omega_o^2 L_1)$
8. COMPUTE $C_3 = 1/(\omega_c R)$; $C_3 > C_o$; $C_o = n^2 r \, C_1$, $C_3 = C_o + C_2$

DESIGN PROCEDURES- MCF BAND-ELIMINATION FILTERS
FIGURE 14

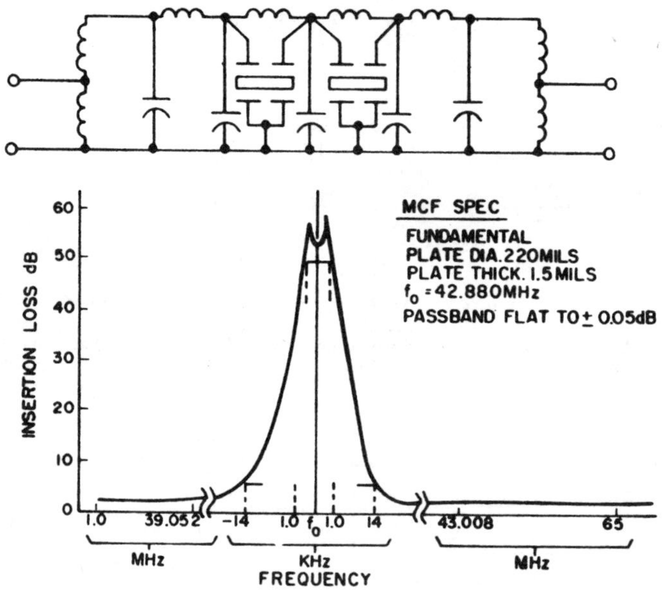

INSERTION LOSS OF A 42.880 MHz PILOT BLOCKING FILTER
FIGURE 15

An Equivalent Circuit Approach to the Design and Analysis of Monolithic Crystal Filters

R. C. RENNICK

Abstract—A general n port equivalent network is developed for the Monolithic Crystal structure, where n is the number of electrode pairs on the quartz substrate. This network is reduced to a two-port network in a form suitable for the design of n pole crystal filters. An analysis of the two-port network yields filter design parameters in terms of well-known coupling coefficients and dissipations. The analysis is extended to include impedance transformations within the Monolithic Crystal structure. The practical aspects of measuring the filter design parameters during fabrication, and the analysis of existing designs, are discussed. The results are presented in terms of various short and open circuit frequencies which are easily measured and calculated. The design of a six-pole filter is presented to illustrate the methods described.

INTRODUCTION

The monolithic crystal filter (MCF) [1, 2, 3, 4] is a piezoelectric device designed to perform an electrical filter function and consists of a quartz substrate upon which metallic electrodes are vacuum deposited. Fig. 1 shows a typical 8 pole (16th order) MCF and the passband characteristic realized.

A complete understanding of this device requires both an acoustical and electrical analysis. This paper is concerned with the electrical analysis and design methods. Toward that end, an equivalent circuit is developed and analyzed and an example employing the design methods is also given.

The MCF shown in Fig. 1 consists of eight electrode pairs deposited on a quartz substrate. The filter characteristic realized with this MCF is determined by the acoustical parameters of the system, which in turn are determined by the physical dimensions of the quartz plate and electrodes, and by the spacing between these electrodes. The acoustical design of an MCF uses as input information electrical design parameters [5] (such as the various short circuit frequenceis of the system, impedance levels Q, etc.) which must be realized in order to perform the filter function. This paper will be concerned with the electrical equivalent circuit of the MCF used as a tool for providing the electrical design parameters and analyzing the various measurements made on MCF's such as resonator frequencies, coupled frequencies, and filter performance.

Manuscript received October 30, 1972.
The author is with the Bell Laboratories, Inc., Allentown, Pa. 18103.

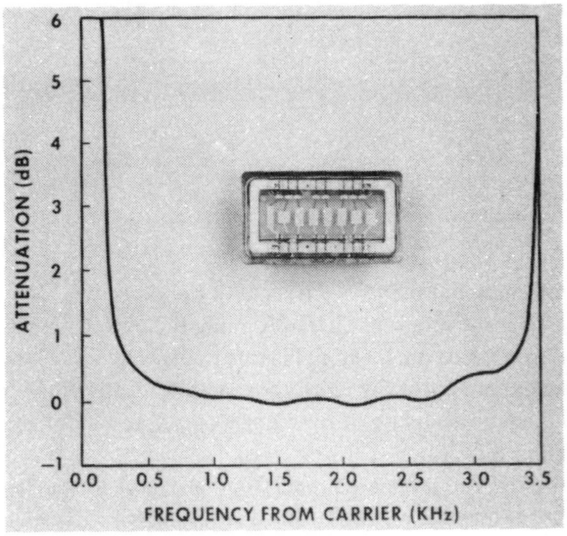

Fig. 1. Typical 8 Pole MCF.

EQUIVALENT CIRCUIT OF MCF

An intuitive look at the MCF suggests a set of coupled resonators as the equivalent circuit and hence that the electrical filter design be based on the vast amount of knowledge available in the field of coupled resonators. This information has been tabulated in many places in the form of low-pass element values or coupling coefficients and dissipations [6].

Consider first the equivalent circuit of a single quartz resonator as shown in Fig. 2(a). The inertia and compliance of the system are represented by the motional inductance L and motional capacitance C, respectively. The finite Q is represented by the conductance G and the parallel plate capacitor, due to the dielectric properties of quartz and the electrode pair, is represented by the static capacitance C_0. The inverter tee of capacitance represents the piezoelectric coupling, and since this will be involved throughout the analysis it will be worthwhile to discuss its characteristics. If this network is terminated in an impedance Z, as in Fig. 2(b), and we look at the input impedance we find that:

$$Z_{in} = 1/\omega^2 C^2 Z. \tag{1}$$

Thus if we terminate the network in a short (open) circuit it reflects an open (short) circuit at the input. In the vicinity of resonance ($\omega_0^2 = 1/LC$), equation (1) reduces to:

$$Z_{in} \approx L/CZ. \tag{2}$$

Reprinted from *IEEE Trans. Sonics Ultrason.*, vol. SU-20, pp. 347–354, Oct. 1973.

(a) EQUIVALENT CIRCUIT OF QUARTZ CRYSTAL

(b) CAPACITIVE TEE TRANSFORMATION

$$Z_{in} = \frac{1}{\omega^2 C^2 Z}$$

$$Z_{in} \cong \frac{L}{CZ} \quad (\omega = \omega_0)$$

(c) TRANSFORMED QUARTZ CRYSTAL EQUIVALENT

$$Q = \omega_0 C/G \qquad Q = \omega_0 L/R$$

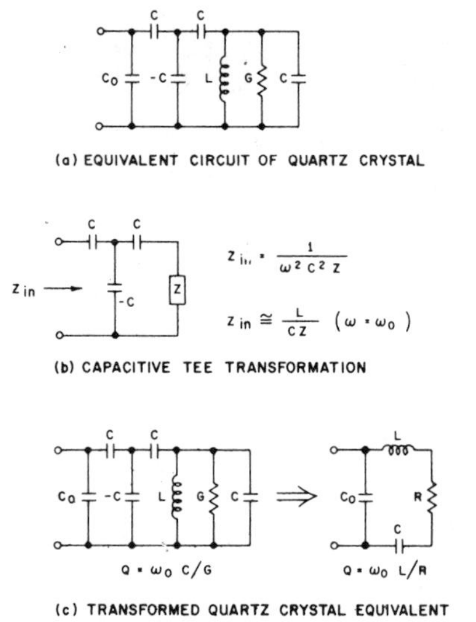

Fig. 2. (a) Equivalent circuit of quartz crystal. (b) Capacitance tee transformation. (c) Transformed quartz crystal equivalent.

Therefore, in the vicinity of resonance the capacitive tee (inverter) has the property of transforming the terminating impedance to its dual. Applying this transformation to the equivalent circuit of Fig. 2(a), for example, results in the more familiar equivalent circuit for a single crystal resonator as shown in Fig. 2(c).

In the MCF, energy is coupled from resonator to resonator by the mechanical motion of the quartz, and since the electrical equivalent of mechanical motion is inductive the coupling element in the equivalent circuit is assumed to take this form. The resulting equivalent circuit of the MCF is shown in Fig. 3(a). For simplicity the equivalent of a four resonator or four pole structure is shown. This is complex enough to involve all of the peculiarities of MCF's. The analysis which follows can easily be extended to any number of resonators, and the results will be presented in general form.

It is advantageous, from the standpoint of analysis, to redraw the equivalent circuit in the form of Fig. 3(b). A coupling coefficient $k_{l,m}$ is introduced, as in the treatment of coupled circuits by Dishal [7]. The equivalence of element values in the networks of Fig. 3(a) and (b) are also given in Fig. 3.

The circuit of Fig. 3 is obviously an n-port device, however we are only interested in the two port characteristics. In general the internal resonators may be terminated in any impedance, as long as each node is tuned to the filter center frequency (f_0) when all other nodes are shorted, and the Q is not degraded beyond that required of the filter design. The simplest and most reliable termination is a short-circuit across the internal resonator terminals. As was pointed out earlier this short-circuit will be transformed to an open circuit at the associated node thus reducing the equivalent circuit to that of Fig. 4(a). The nodal equivalent circuit of Fig. 4(a) can be transformed to a mesh circuit by simply transforming the elements through either tee of capacitors associated with the external resonators using (2). The circuit is shown in Fig. 4(b) where the resulting two capacitive tees in tandem are represented by an ideal transformer having a turns ratio of 1:−1.

(a)

$$L_{1,m} = L/k_{1,m}$$
$$L_1 = L/(1-k_{12}) \qquad L_2 = L/[1-(k_{12}+k_{23})]$$
$$L_4 = L/(1-k_{34}) \qquad L_3 = L/[1-(k_{23}+k_{34})]$$

(b)

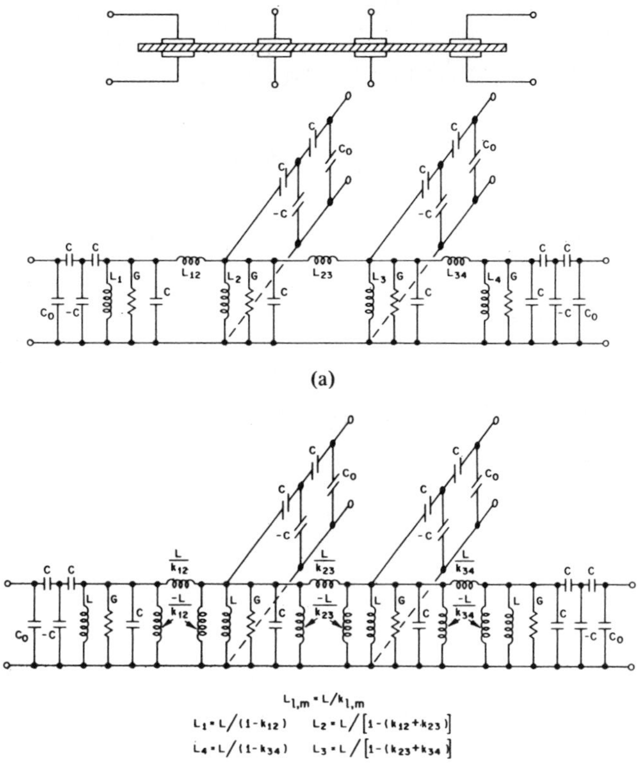

Fig. 3. Equivalent circuit for 4 pole MCF with open internal resonators.

(a) NODAL EQUIVALENT CIRCUIT

(b) MESH EQUIVALENT CIRCUIT

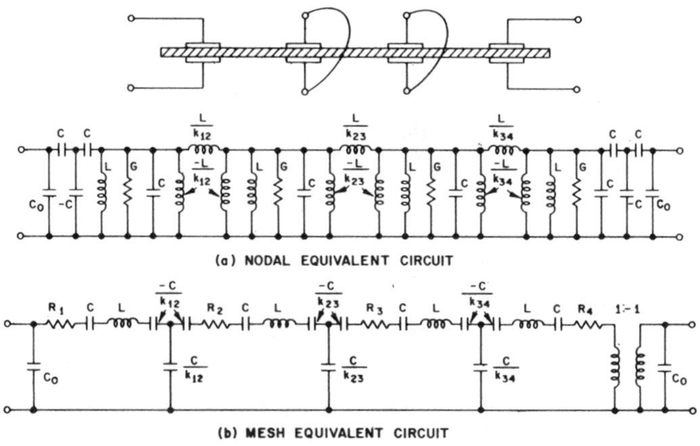

Fig. 4. Equivalent circuit for 4 pole MCF with shorted internal resonators.

This simply results in an additional 180° of phase shift which must be kept in mind but which does not affect the analysis. In Fig. 5(b) the transformer has been dropped and a source and termination have been added to complete the filter structure.

The shunt capacitor (C_0) which terminates the filter can of course cause some in-band distortion depending upon the ratio of the reactance of this capacitance to the terminating resistance. This distortion will be considered later, but at this point it is important to realize that this effect can be essentially eliminated by tuning it out with an inductor, or performing a parallel-to-series transformation of C_0 with the

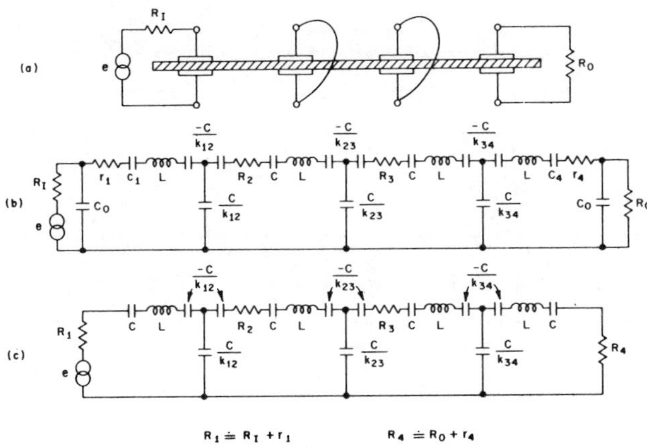

Fig. 5. Parallel to series transformation of terminal and static capacitance.

termination resistance and absorbing the resulting series capacitance in the end resonators, which amounts to a slightly different tuning frequency for these resonators. In either case the circuit may be reduced to that shown in Fig. 5(c) where the resistance of the end resonators has been absorbed into the termination.

FILTER-DESIGN PARAMETERS

The mesh equations for the circuit of Fig. 5(c) may now be written as

$$e = \left[j\left(\omega L - \frac{1}{\omega C}\right) + R_1\right] I_1 + \left[j\frac{k_{12}}{\omega C}\right] I_2$$

$$0 = \left[j\frac{k_{12}}{\omega C}\right] I_1 + \left[j\left(\omega L - \frac{1}{\omega C}\right) + R_2\right] I_2 + \left[j\frac{k_{23}}{\omega C}\right] I_3$$

$$0 = \left[j\frac{k_{23}}{\omega C}\right] I_2 + \left[j\left(\omega L - \frac{1}{\omega C}\right) + R_3\right] I_3 + \left[j\frac{k_{34}}{\omega C}\right] I_4$$

$$0 = \left[j\frac{k_{34}}{\omega C}\right] I_3 + \left[j\left(\omega L - \frac{1}{\omega C}\right) + R_4\right] I_4.$$

Multiplying these equations by $(-j\omega_0 C)$, or normalizing with respect to the mesh impedance level, gives

$$\frac{e}{j\omega_0 L} = \left[\left(\frac{\omega}{\omega_0} - \frac{\omega_0}{\omega}\right) - j\omega_0 R_1 C\right] I_1 + \left[k_{12}\frac{\omega_0}{\omega}\right] I_2$$

$$0 = \left[k_{12}\frac{\omega_0}{\omega}\right] I_1 + \left[\left(\frac{\omega}{\omega_0} - \frac{\omega_0}{\omega}\right) - j\omega_0 R_2 C\right] I_2 + \left[k_{23}\frac{\omega_0}{\omega}\right] I_3$$

$$0 = \left[k_{23}\frac{\omega_0}{\omega}\right] I_2 + \left[\left(\frac{\omega}{\omega_0} - \frac{\omega_0}{\omega}\right) - j\omega_0 R_3 C\right] I_3 + \left[k_{34}\frac{\omega_0}{\omega}\right] I_4$$

$$0 = \left[k_{34}\frac{\omega_0}{\omega}\right] I_3 + \left[\left(\frac{\omega}{\omega_0} - \frac{\omega_0}{\omega}\right) - j\omega_0 R_4 C\right] I_4$$

where

$$\omega_0^2 = 1/LC.$$

These equations may now be normalized with respect to the relative bandwidth (BW/f_0) in which case they reduce to the following matrix equation.

$$\begin{bmatrix} \dfrac{e}{j2\pi \cdot \mathrm{BW} \cdot L} \\ 0 \\ 0 \\ 0 \end{bmatrix} = \begin{bmatrix} (F - jD_1) & K_{12} & 0 & 0 \\ K_{12} & (F - jD_2) & K_{23} & 0 \\ 0 & K_{23} & (F - jD_3) & K_{34} \\ 0 & 0 & K_{34} & (F - jD_4) \end{bmatrix} \begin{bmatrix} I_1 \\ I_2 \\ I_3 \\ I_4 \end{bmatrix}$$

where

- f_0 center frequency of the filter
- BW bandwidth
- δ $(f - f_0)$ = frequency deviation from f_0 (Band-pass to low-pass transformation).
- F $(2\delta/\mathrm{BW})$ = fractional frequency deviation
- d_n $\omega_0 R_n C = R_n/\omega_0 L$ = dissipation of the nth mesh
- D_n $(d_n \cdot f_0/\mathrm{BW})$ = normalized dissipation of the nth mesh
- K_{mn} $(k_{mn} \cdot f_0/\mathrm{BW})$ = normalized coupling between meshes m and n.

The bandpass-to-low-pass transformation $\delta = f - f_0$ and the narrow-band approximation $1 >> \delta/f_0$ have been applied to these equations.

From the matrix equation we may calculate any of the filter characteristics within the bounds of the narrow band approximations, for example the transmission charactersitic for any number of resonators,

$$\frac{e_{\mathrm{out}}}{e_{\mathrm{in}}} = \frac{(D_N - D_{Nr}) K_{12} K_{13} \cdots K_{N-1,N}}{\text{Determinant}}$$

where N is the number of resonators, and D_{Nr} is the dissipation in the Nth resonator.

The electrical design of the filter then requires that the coupling coefficients and dissipations be realized. The normalized coupling coefficients and dissipations required for many designs are tabulated in filter handbooks [6], or they may be calculated.

IMPEDANCE TRANSFORMATION

In the previous analysis the impedance level of all resonators were considered to be equal. There are many filter designs where the impedance level of the resonators are either not allowed or not desired to be equal. For example many designs require unequal input and output dissipations where the design may require equal terminations. Or in the design of higher ordered filters it may be desirable to provide two impedance transformations in order that the internal resonator electrodes be smaller to minimize the size of the quartz substrate, or to avoid unwanted responses.

These situations can be analyzed in the same manner as before if we redraw the equivalent circuit as in Fig. 6(a). The different impedance of each electrode is a function of the electrode size and is reflected in the values of L_n and C_n. The internal resonators are assumed to have been shorted. If, as before, we transform the elements through the capacitive tee at the input terminal pair we arrive at the mesh equivalent circuit of Fig. 6(b). The transformer represents the two capacitive tees in tandem and now has a turns ration of $1 : -(L_4/L_1)$. Ignoring again the 180° phase shift due to the transformer, the mesh equations may be written,

$$e = \left[j\left(\omega L_1 - \frac{1}{\omega C_1}\right) + R_1\right] I_1 + \left[j\frac{k_{12}}{\omega C_1} \cdot \frac{L_1}{\sqrt{L_1 L_2}}\right] I_2$$

$$0 = \left[j\frac{k_{12}}{\omega C_1} \cdot \frac{L_1}{\sqrt{L_1 L_2}}\right] I_1 + \left[j\left(\omega L_1 \cdot \frac{C_2}{C_1} - \frac{1}{\omega C_1} \cdot \frac{L_1}{L_2}\right) + \frac{C_2 L_1}{C_1 L_2} \cdot R_2\right] I_2 + \left[j\frac{k_{23}}{\omega C_1} \cdot \frac{L_1}{\sqrt{L_2 L_3}}\right] I_3$$

$$0 = \left[j\frac{k_{23}}{\omega C_1} \cdot \frac{L_1}{\sqrt{L_2 L_3}}\right] I_2 + \left[j\left(\omega L_1 \cdot \frac{C_3}{C_1} - \frac{1}{\omega C_1} \cdot \frac{L_1}{L_3}\right) + \frac{C_3 L_1}{C_1 L_3} \cdot R_3\right] I_3 + \left[j\frac{k_{34}}{\omega C_1} \cdot \frac{L_1}{\sqrt{L_3 L_4}}\right] I_4$$

$$0 = \left[j\frac{k_{34}}{\omega C_1} \cdot \frac{L_1}{\sqrt{L_3 L_4}}\right] I_3 + \left[j\left(\omega L_1 \cdot \frac{C_4}{C_1} - \frac{1}{\omega C_1} \cdot \frac{L_1}{L_4}\right) + \frac{C_4 L_1}{C_1 L_4} \cdot R_4\right] I_4.$$

Multiplying the nth equation by $(-j\omega_0 C_1 L_n/L_1)$, they become

$$\frac{e}{j\omega_0 L_1} = \left[\left(\frac{\omega}{\omega_0} - \frac{\omega_0}{\omega}\right) - jd_1\right] I_1 + \left[k_{12}\frac{\omega_0}{\omega}\sqrt{\frac{L_1}{L_2}}\right] I_2$$

$$0 = \left[k_{12}\frac{\omega_0}{\omega}\sqrt{\frac{L_2}{L_1}}\right] I_1 + \left[\left(\frac{\omega}{\omega_0} - \frac{\omega_0}{\omega}\right) - jd_2\right] I_2 + \left[k_{23}\frac{\omega_0}{\omega}\sqrt{\frac{L_2}{L_3}}\right] I_3$$

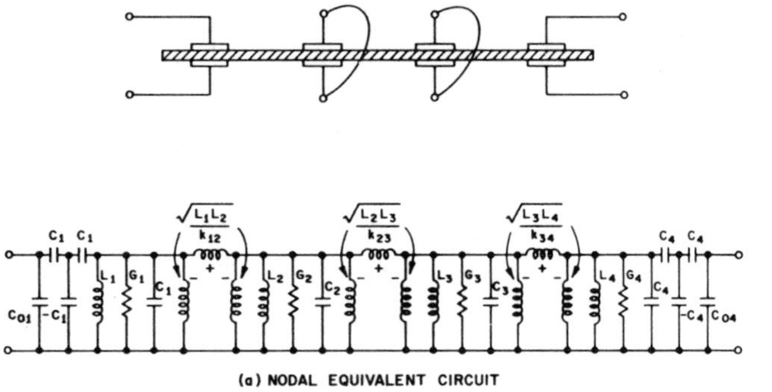

(a)

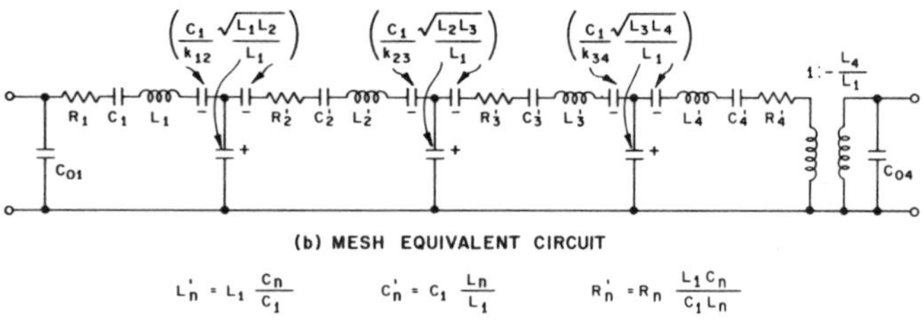

(b)

Fig. 6. Equivalent circuit for 4 pole MCF involving an impedance transformation.

$$0 = \left[k_{23}\frac{\omega_0}{\omega}\sqrt{\frac{L_3}{L_2}}\right] I_2 + \left[\left(\frac{\omega}{\omega_0} - \frac{\omega_0}{\omega}\right) - jd_3\right] I_3 + \left[k_{34}\frac{\omega_0}{\omega}\sqrt{\frac{L_3}{L_4}}\right] I_4$$

$$0 = \left[k_{34}\frac{\omega_0}{\omega}\sqrt{\frac{L_4}{L_3}}\right] I_3 + \left[\left(\frac{\omega}{\omega_0} - \frac{\omega_0}{\omega}\right) - jd_4\right] I_4.$$

Again normalizing with respect to the relative bandwidth and applying the narrow-band approximation as before these equations may be written in matrix form.

$$\begin{bmatrix} \dfrac{e}{j2\pi \cdot \mathrm{BW} \cdot L_1} \\ 0 \\ 0 \\ 0 \end{bmatrix} = \begin{bmatrix} (F - jD_1) & K_{12}\sqrt{\dfrac{L_1}{L_2}} & 0 & 0 \\ K_{12}\sqrt{\dfrac{L_2}{L_1}} & (F - jD_2) & K_{23}\sqrt{\dfrac{L_2}{L_3}} & 0 \\ 0 & K_{23}\sqrt{\dfrac{L_3}{L_2}} & (F - jD_3) & K_{34}\sqrt{\dfrac{L_3}{L_4}} \\ 0 & 0 & K_{34}\sqrt{\dfrac{L_4}{L_3}} & (F - jD_4) \end{bmatrix} \begin{bmatrix} I_1 \\ I_2 \\ I_3 \\ I_4 \end{bmatrix}$$

From this matrix equation we may calculate any of the filter characteristics. For example the transmission characteristic,

$$\frac{e_{\text{out}}}{e_{\text{in}}} = \frac{(D_4 - D_{4r})\sqrt{L_4/L_1}\,K_{12}K_{34}K_{34}}{\text{Determinant}}.$$

In general for any number of resonators,

$$\frac{e_{\text{out}}}{e_{\text{in}}} = \frac{(D_N - D_{Nr})\sqrt{L_nL_1}\,K_{12}K_{23}K_{34}\cdots K_{N-1,N}}{\text{Determinant}}.$$

This of course reduces to the same expression as for the case of equal inductance since the value of the determinant is the same for a given design and is determined by the required location of the poles that realize the design. From the foregoing analysis we can see that impedance transformation is achieved by simply changing the size of the resonator electrodes. The internal electrodes may be of any size, within the acoustical design requirement, as long as the required dissipation is realized.

The electrical design of an MCF then requires that each resonator be adjusted to the proper frequency (f_0), and that the design coupling coefficients and dissipations be realized. Where terminations (R) are prescribed the inductance of the external resonators (L_1 or L_N) are determined by the required dissipations (D_1 or D_N).

$$L = \frac{R}{2\pi \cdot \text{BW} \cdot D}.$$

The internal resonator inductances may have any value.

ANALYSIS OF MCF MEASUREMENTS

Consider the following simplified process for making an MCF having some specified frequency, bandwidth, and ripple requirements. The coupling between resonators determines the bandwidth and influences ripple, and the individual resonator frequencies determine the frequency of the filter and influences the ripple. The coupling between resonators is a function of the electrode separation and the mass loading of the electrodes on the quartz. The individual resonator frequency is a function of the mass loading. The mass loading required to achieve the desired frequency is in turn a function of the quartz plate dimensions. The formidable task of determining the quartz plate and electrode geometries which will simultaneously satisfy these conditions when the resonators are plated to frequency (mass loaded) is accomplished through an acoustical analysis of the structure [5]. The question here is: To what frequencies should the individual resonators be plated, and what are the coupling coefficients realized?

If the MCF were as ideal as the equivalent circuit of Fig. 5(b) we would simply require that each mesh be resonant at f_0 when all other meshes were open circuited, since that was assumed in the preceding analysis. In the MCF we cannot isolate the resonator to be measured since by its nature it is coupled to the other resonators on the plate which will influence the measurement of the resonator in question. Sim-

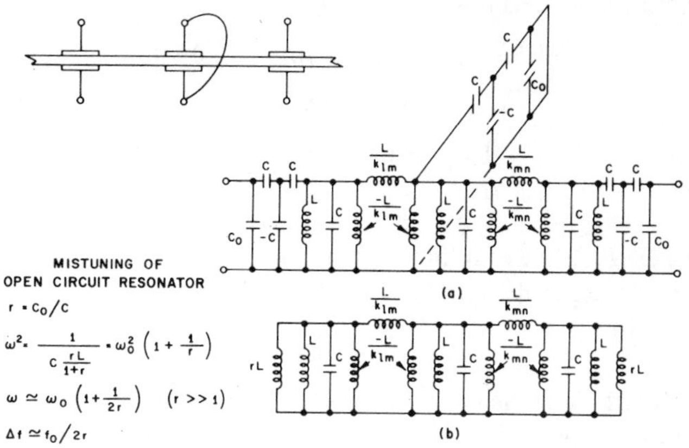

Fig. 7. Mistuning due to adjacent open-circuit resonators.

ilarly the coupling coefficient between any two meshes of Fig. 5(b) could be measured by opening all meshes except the two in question and finding the two short-circuit frequencies. The normalized coupling coefficient is then:

$$K_{mn} = \frac{f_2 - f_1}{\text{BW}}.$$

Consider Fig. 7(a) which shows the equivalent circuit of three adjacent resonators, in an N resonator array, where the center resonator is being measured. We can essentially remove the other resonators if we place a shunt inductor across these electrodes which resonates with C_0 at the frequency of interest (f_0). The capacitive tee will reflect this antiresonance as a series resonance or short-circuit across these nodes, thus we should measure a resonance at f_0 across the terminals of the resonator in question. This method is undesirable in production since we will make these measurements in a vapor plating machine. The simplest and probably the most reliable method is to leave the resonators not being measured open circuited. This terminates these resonators in the capacitor C_0 which is transformed to the nodes as a shunt inductor having a value rL as in Fig. 7(b). This mistunes all of the other resonators and we can calculate the resonant frequency we should measure at the resonator in question. This frequency will be referred to as the tuning frequency.

The tuning frequency of any resonator occurs at a zero of the input impedance to the resonator in question and this input impedance may be expressed as:

$$Z = H\frac{(\text{short circuit natural frequencies})}{(\text{open circuit natural frequencies})}$$

In our measurement there will be N zeros of this input impedance, but ($N - 1$) of these will be far removed from the other, due to the mistuning of the other resonators. The frequency of interest is the lowest of the N short-circuit natural frequencies under the condition that the resonator in question is shorted and all others are open and thus mistuned. The short-circuit natural frequencies of the system are simply the zeros of the determinant of the system. If the equivalent circuit is analyzed in the same manner as before under the condition that the resonant frequencies are in error, the resulting determinant is

$$\begin{vmatrix} F - E_1 & K_{12} & 0 & 0 \\ K_{12} & F - E_2 & K_{23} & 0 \\ 0 & K_{23} & F - E_3 & K_{34} \\ 0 & 0 & K_{34} & F - E_4 \end{vmatrix}$$

where E_n is the error in the nth resonator frequency expressed in half-bandwidths, and can be expressed as

$$E_n = (f_n - f_0)\, 2/\text{BW}.$$

In this determinant the dissipations have been omitted under the assumption that the resonator Q is large, which is the case for the MCF.

The value of E_n is easily determined for any calculation. For example if we were to determine the tuning frequency of resonator three in an N resonator system.

$$E_3 = 0$$

$$E_n = f_0/r_n\, \text{BW}, \qquad n \neq 3.$$

The zeros of the determinant are then calculated using a computer. The lowest zero is then the tuning frequency of the third resonator expressed in terms of F. The frequency which should be measured at this resonator is determined by

$$f_n = f_0 + \frac{E_n \cdot \text{BW}}{2}.$$

In a similar manner we may determine the apparent coupling between resonators. In this case we would short two adjacent resonators, open all others, and determine the two short-circuit frequencies f_2 and f_1. The apparent coupling is then given by

$$K_{mn} = (f_2 - f_1)/\text{BW}.$$

The tuning frequencies and apparent coupling coefficients as calculated above are those that should be measured on a filter, where all of the frequencies and couplings are correct to begin with. If this is not the case, a perturbation technique on the method described will give the corrected frequencies and couplings.

The short-circuit natural frequencies of the system are useful in the acoustical analysis of the filter. These are easily determined in the same manner as before. All of the resonators are shorted (E_1 thru $E_N = 0$) and the N zeros of the determinant are calculated.

All of the short-circuit frequencies used to determine the resonator frequencies, couplings, etc. which have been calculated may also be measured by providing the shorted and opened terminals as required, and making a measurement which will indicate the series resonant frequencies at one of the shorted terminals.

It was mentioned earlier that the shunt capacitance of the input and output resonators may in some cases, depending upon the filter bandwidth and terminating impedance, be compensated for by purposely mistuning these resonators. The end resonators have as a termination the shunt combination of R and C_0 as in Fig. 5(a). If a parallel-to series

$$R_S = R_1 \Big/ \left[1 + \left(\frac{R_1}{X_0}\right)^2\right]$$

$$C_S = C_0 \left[1 + \left(\frac{X_0}{R_1}\right)^2\right]$$

RESONATOR MISTUNING

$$C_1 = \frac{C C_S}{C_S - C}$$

$$\omega_1^2 = \frac{1}{L C_1} = \frac{C_S - C}{L C C_S}$$

$$\omega_1^2 = \omega_0^2 \left(1 - \frac{C}{C_S}\right) \qquad (\omega_0^2 = 1/LC)$$

$$\omega_1 \cong \omega_0 \left(1 - \frac{C}{2 C_S}\right) \qquad \left(\frac{C}{C_S} \ll 1\right)$$

$$\Delta f_1 \cong -f_0 \frac{C}{2 C_S}$$

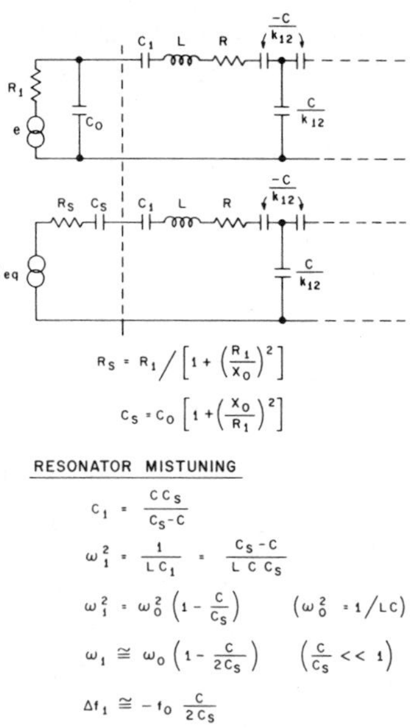

Fig. 8. Mistuning of terminating resonators.

transformation[1] is performed as in Fig. 8 the result is a series resistor and capacitor the values of which are given in the figure. Evaluation of the effective series capacitance at f_0 results in a series capacitor which can be absorbed in the end resonators by mistuning them from f_0. The mistuning required is given by

$$\Delta f \cong -f_0 \frac{C}{2 C_S}$$

where

$$C = \frac{1}{\omega_0^2 L}.$$

This method of compensating for the shunt capacitance results in very little in-band distortion for relative bandwidths up to 0.05 percent. For bandwidths wider than this the shunt capacitance should be tuned with an inductor unless the distortion can be tolerated.

If the end resonators are mistuned as described, the mistuning must be considered when the tuning frequencies and coupling frequencies are determined. When the end resonators are shorted E_1 or E_N must be set to the mistuning

$$E_1 \text{ or } E_N = -\frac{C}{C_S} \frac{f_0}{\text{BW}}.$$

When E_1 or E_N is open-circuited:

$$E_1 \text{ or } E_N = \left(\frac{1}{r_{1,N}} - \frac{C}{C_S}\right) \frac{f_0}{\text{BW}}.$$

All of the preceding calculations for the electrical design or analysis of an MCF are done by computer. Starting with a set of coupling coefficients and dissipations, either from tables or calculated, design parameters for the MCF are easily determined.

[1] Only allowed in the narrow-band case.

EXAMPLE DESIGN

Consider the design of a 6 pole, 0.1 dB Chebyshev filter, centered at 8 MHz having a 3-dB bandwidth of 4 KHz. The filter is to be terminated at both ends with 500 Ω.

Therefore, the specifications for the design are:

$$f_0 = 8 \text{ MHz}$$

$$\text{Design BW} = 4 \text{ KHz}$$

$$\text{Ripple} = 0.1 \text{ dB}$$

$$R_1 = R_6 = 500\ \Omega.$$

The required dissipations and coupling coefficients are [6, p. 348],

$$D_1 = D_6 = 0.7833$$

$$K_{12} = K_{56} = 0.7145$$

$$K_{23} = K_{45} = 0.5385$$

$$K_{34} = 0.5180$$

We may now calculate the electrical design parameters

$$L_1 = L_6 = R/2\pi \text{BW}\, D_{1,6} = 25.4 \text{ mH}$$

The internal resonator inductances may be different but for ease of manufacture there is an advantage if they are the same.

$$L = 25.4 \text{ mH}$$

$$C = 1/\omega_0^2 L = .0156 \text{ pF}$$

$$r = C_0/C = 250 \text{ (including mounting capacitance)}$$

$$C_0 = rC = 3.9 \text{ pF.}$$

Assuming an additional 1 pF due to wiring in the filter the shunt capacitance terminating the filter will be

$$C_{01} = C_{06} = 4.9 \text{ pF.}$$

We may now calculate the equivalent series capacitance of the termination.

$$C_{S1} = C_{S6} = 4.9\left[1 + \left(\frac{X_{01,06}}{R}\right)^2\right] = 328 \text{ pF } (\omega = \omega_0).$$

The values of E_n to be inserted in the determinant for calculating the tuning frequencies, coupled frequencies, and short-circuit natural frequencies are now easily determined.

$$E_1 = E_6 = -\frac{C}{C_S}\frac{f_0}{\text{BW}} = -0.0950 \quad \text{(terminals shorted)}$$

$$E_1 = E_6 = \left(\frac{1}{r} - \frac{C}{C_S}\right)\frac{f_0}{\text{BW}}$$

$$= 7.8958 \quad \text{(terminals open)}$$

$$E_2 \text{ through } E_5 = 0 \quad \text{(terminals shorted)}$$

$$E_2 \text{ through } E_5 = \frac{1}{r}\frac{f_0}{\text{BW}} = 7.9908 \quad \text{(terminals open).}$$

When the appropriate values are inserted into the determinant and the zeros are found the results are as follows.

Tuning Frequencies:

$$f_1 = f_6 = 7.999684 \text{ MHz}$$

$$f_2 = f_5 = 7.999800 \text{ MHz}$$

$$f_3 = f_4 = 7.999861 \text{ MHz.}$$

Coupling Frequencies:

$$f_{12} = f_{56} = 8001295 - 7998442 = 2853 \text{ Hz}$$

$$f_{23} = f_{45} = 8000974 - 7998832 = 2142 \text{ Hz}$$

$$f_{34} = 8000958 - 7998896 = 2062 \text{ Hz.}$$

where the subscript refers to the shorted electrodes.

Short-Circuit Natural Frequencies:

7.997877 MHz

7.998262 MHz

7.999308 MHz

8.000532 MHz

8.001589 MHz

8.002052 MHz.

These frequencies and the resonator inductance are now used as input data for the acoustical analysis which will determine the proper mask for realizing the design when each resonator is adjusted to the frequency calculated in the preceding.

DESIGN ANALYSIS

The design process may be reversed to analyze an MCF. The resonator frequencies are measured when all other resonators are open circuited to determine the tuning frequencies. The coupled frequencies, and therefore the coupling coefficients, are determined by shorting adjacent resonators while opening all others, and measuring the difference of the two resonant frequencies.

With these coupling coefficients the determinant is evaluated to determine the frequencies at which the resonators should be tuned. These frequencies are then compared with the measured frequencies to determine the mistuning of each resonator. The transmission characteristic of the filter can then be calculated using these tuning errors, the measured couplings and the measured Q of the resonator.

This method of analysis is used for the evaluation of filter designs without having to fabricate the devices. It is also used for determining the variation in filter performance due to errors in the adjustment of design parameters.

CONCLUSIONS

The equivalent circuit along with the resulting design and analysis techniques discussed have been used successfully for the electrical design of various monolithic crystal filters ranging from two to ten resonators.

The calculations have all been computerized so that starting from a set of coupling coefficients and dissipations the design parameters are determined in a matter of minutes. The design of filters involving impedance transformations is equally as simple.

The MCF is an acoustical device and the electrical design must of course comply with the acoustical design, such as electrode size and achievable coupling, and the propagation of unwanted modes. Some practical consideration must also be given to the size of quartz plate and the difficulty of plating where impedance transformations require various electrode sizes.

REFERENCES

[1] Sykes, R. A., and Beaver, W. D., "High Frequency Monolithic Crystal Filters with Application to Single Frequency and Single Sideband Use," Prec. 20th Ann. Symp. Freq. Control; April 1966.

[2] Sykes, R. A., Smith, W. L., and Spencer, W. J., "Monolithic Crystal Filters," 1967 IEEE International Convention Record, pt. 11, pp. 78–93.

[3] Sykes, R. A., and Smith, W. L., "A Monolithic Crystal Filter," Bell Labs Record, pp. 52–54, Feb. 1968.

[4] Spencer, W. J., "Monolithic Crystal Filters," presented at Third Asilomar Conference, Dec. 1969.

[5] Haruta, K., and Lloyd, P., "Monolithic Crystal Filter," to be published.

[6] Zverev, A. I., "Handbook of Filter Synthesis," John Wiley and Sons, Inc., 1967.

[7] Dishal, M., "Design of Dissipative Band-Pass Filters Producing Desired Exact Amplitude–Frequency Characteristics," Proc. of IRE, pp. 1050–1069, Sept. 1949.

POLYLITHIC CRYSTAL FILTERS WITH LOSS POLES AT FINITE FREQUENCIES*

Man Shek Lee
GTE Lenkurt, Inc.
San Carlos, CA 94070

ABSTRACT

This paper describes a design investigation of polylithic crystal filters with finite frequency loss poles, and in particular, loss poles in both the lower and upper stopbands obtained by bridging capacitors across the coupled resonators. Practical design considerations and experimental results are also reported.

INTRODUCTION

Polylithic crystal filters (filters consisting of more than one crystal wafer where at least one wafer contains acoustically coupled resonators) have been in successful production for almost 5 years. The physical properties of the quartz resonators make them suitable for operation between 1-100 MHz and they are especially suitable for use in narrow band filters. One of the largest applications of polylithic filters is in the communications industry where they are used in single-sideband frequency-division-multiplexing systems.[1-4] They are superior for this type of application because of their small size, excellent frequency stability characteristics and low manufacturing cost. They also require no power.

Most of the polylithic filters now in production are of the "all-pole" type or those with loss poles in the lower stopband only. Each of these loss poles is produced by a single crystal resonator in a shunt branch of a ladder structure. Occasions arise where more system attenuation is required in the upper stopband. One method of accomplishing this is to use a lowpass filter following baseband demodulation. Another method would be to increase the degree of the single-sideband filter by increasing the number of loss poles at infinite frequency. However, both of these methods are costly and the latter would also increase the dissipation and the delay distortion in the passband, which is undesirable for data transmission. These factors lead to the interest in the investigation of crystal filters with finite loss poles, especially loss poles in the upper stopband. There are several means of obtaining finite loss poles in a crystal filter[5-8] but the technique of bridging a capacitor across a coupled resonator is especially suitable for polylithic crystal filters. Very narrowband filters, such as pilot-pick-off filters, have been made using this technique. Most filters of this type have been designed by trial and error by perturbing an exact non-bridged design. This paper will describe some results on an exact synthesis of a complex channel filter using network transformations that were developed for this purpose.

DESIGN METHOD

Fig. 1 shows the equivalent circuit of a coupled crystal resonator[2] with an external capacitor, C_b, bridged across the two non-grounded terminals. In this circuit, C_o represents the stray capacitances. L_a, L_d and C_a, C_d are the motional inductances and capacitances, respectively. The coupling capacitor, C_c, may be positive or negative depending on the connection of the resonator electrodes.

Fig. 2 shows an equivalent ladder circuit[9] of the bridged network in Fig. 1. The circuits are equivalent if, and only if, the following constraint on the elements of the network in Fig. 2 is satisfied:

$$L_1(C_3+C_4)+L_2(C_4+C_5)=0 \quad (1)$$

For ease of manufacturing, the value of L_a is required to be the same as that of L_d. This imposes a further constraint on the elements:

$$L_1C_3^2=L_2C_5^2 \quad (2)$$

It is clear from eqn. (1) that, for positive inductances, some of the series capacitors in Fig. 2 must be negative and thus one of the loss poles is necessarily in

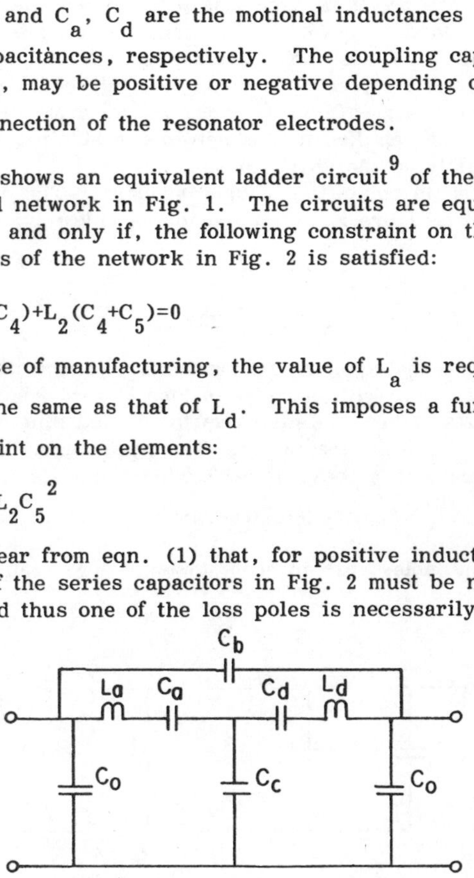

Fig. 1. Equivalent circuit of coupled resonator with bridging capacitor.

*Part of the material presented in this paper was first presented at the Crystal and Mechanical Filters Workshop in Rancho Santa Fe, California in October 1974.

Reprinted from *Proc. 1975 IEEE Int. Symp. on Circuits and Syst.*, Apr. 21-23, 1975, pp. 297-300.

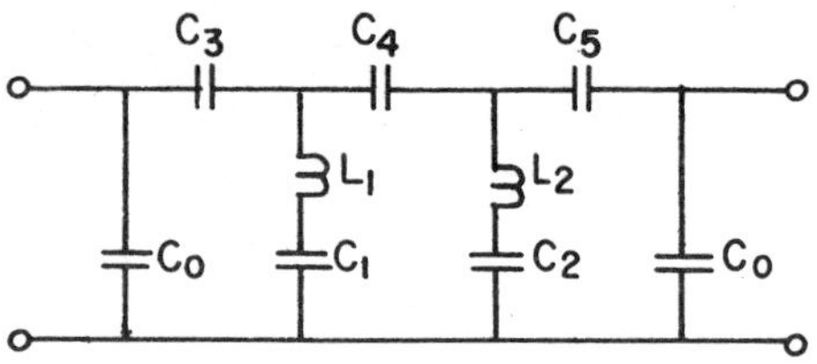

Fig. 2. Equivalent ladder circuit of Fig.1.

the upper stopband. If the above two constraints are satisfied, the element values of the circuit in Fig. 1 are given directly by

$$C_b=1/(1/C_3+1/C_4+1/C_5) \qquad (3)$$

$$L_a=L_d=2/(1/L_1+1/L_2)$$

$$C_c=(1+L_1/L_2)^2/((L_1/L_2)^2(1/C_2+1/C_5)$$
$$+(L_1/L_2)(1/C_3+1/C_5-1/C_1-1/C_2)+1/C_1+1/C_3)$$

$$C_a=\tfrac{1}{2}(1+L_2/L_1)^2C_5/((L_2/L_1)(C_5/C_1-C_3/C_1)$$
$$-(1+L_2/L_1)(C_4/C_2+1))$$

$$C_d=\tfrac{1}{2}(1+L_1/L_2)^2C_3/((L_1/L_2)(C_3/C_2-C_5/C_2)$$
$$-(1+L_1/L_2)(C_4/C_1+1))$$

To obtain a polylithic bandpass filter with finite loss poles, a ladder network without any bridging is first designed. The loss poles are realized by shunt branches and the sections to be bridged should have the form shown in Fig. 2. Norton transformations (Fig. 3) are applied to this network to obtain a structure satisfying the above constraints. Eqn. (3) is then used to obtain the bridged sections. The detailed design procedure is illustrated in the following example.

DESIGN EXAMPLE

A 20th degree equal-ripple-passband polylithic filter was designed, with one pole of loss at zero frequency, five poles of loss at finite frequencies and nine poles of loss at infinite frequency. The preliminary design was done with a conventional ladder synthesis program.[10] The ladder structure is shown in Fig. 4.

Four loss poles, two in each stopband, are realized

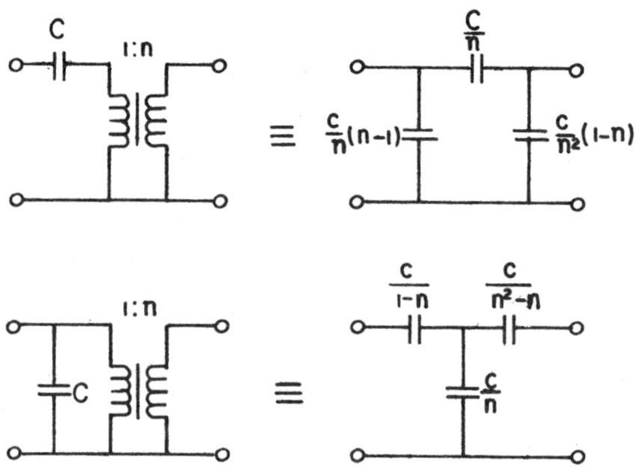

Fig. 3. Impedance transformations used in crystal filter design.

by two bridged sections. Each bridged section realizes two loss poles, one in the lower stopband and the other in the upper stopband. To obtain a physically symmetrical filter, the two finite poles in the upper stopband are at the same frequency. The same is true for the two lowest finite frequency loss poles in the lower stopband. The remaining loss pole, which is the one closest to the lower cutoff frequency, is to be realized by a single shunt resonator. The single resonator is put at the center of the filter with one bridged-coupled resonator on each side. Since the filter is symmetrical, it can be developed by concentrating the design on a half section taken about the center components.

Impedance transformations are made as shown in Fig. 5 to obtain elements in the ladder structure such as to satisfy the constraints in eqns. (1) and (2). Note that, as shown in the figure, the transformations are not unique. By suitably choosing the transformer turns ratios, a physically realizable filter is possible. Eqn. (3) is applied to obtain the bridged sections. Finally, impedance transformations are made on the end resonators (those without bridging capacitors) to equalize the inductances. The final configuation of the filter is shown in Fig. 6.

The calculated response of this filter is shown in Fig. 7. Note that there is substantially more attenuation in the lower stopband than in the upper stopband. The pole frequency locations have been optimized to give the best physical element values. The pole locations have a considerable effect on the impedance levels

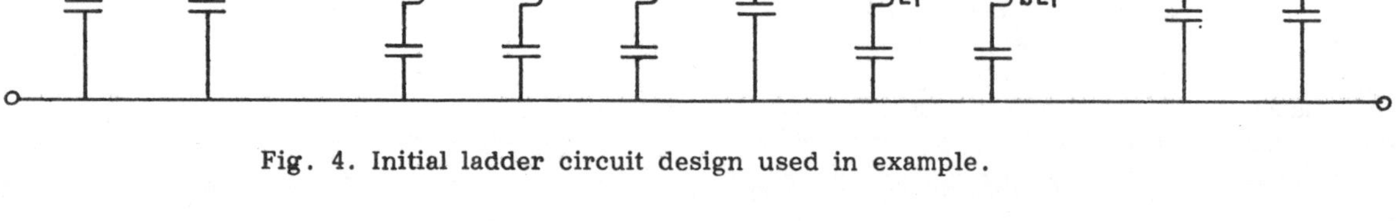

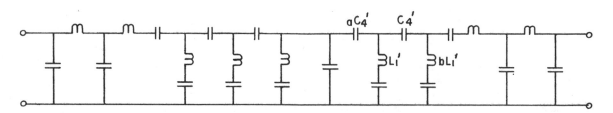

Fig. 4. Initial ladder circuit design used in example.

$$n = \frac{(a+1)^2 + b}{a((a+1)\pm\sqrt{b})}$$

$$m = \frac{1}{1+a(1-n)}$$

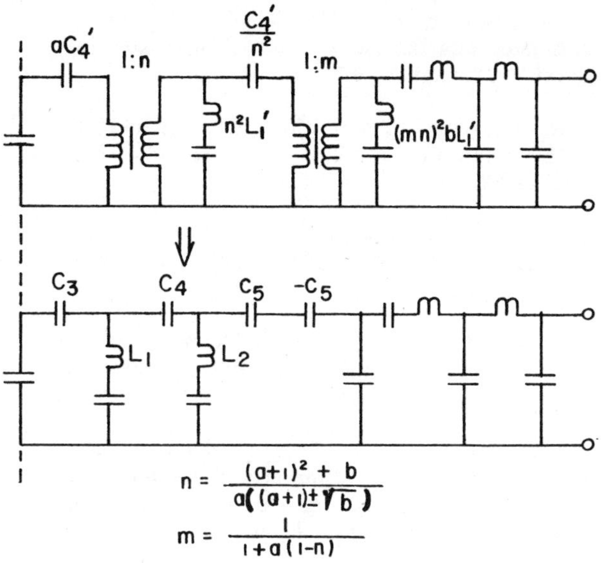

Fig. 5. Transformation for the bridged resonator section.

in the filter. There are constraints on the inductance values that depend on the physical size of the quartz blank and the frequency of operation. For the 8 MHz frequency range where this is used, the lower limit is approximately 5 mH and the upper limit is approximately 50 mH. The sections which are bridged by capacitors may also be selected to obtain the best element values. An acceptable design should have an inductance ratio of less than 2 : 1 in the coupled resonators. It is this that imposes the most critical constraint in the design of crystal filters.

EXPERIMENTAL RESULTS

The above example was built with one single resonator and four coupled resonators. NPO ceramic chip capacitors were used. The crystal resonators were tuned by evaporating more metal on the electrodes to to decrease the frequency to $\pm$ 10 Hz of the desired value. Tuning was done before mounting the capacitors. No tuning was needed on the bridging capacitors. The capacitor tolerance used was $\pm$ 2%. The filter response was relatively insensitive to the bridging capacitors and it was found that there was very little change in the passband and stopband responses when the capacitors were changed by 10% from their nominal

values. This can be explained by looking at the sensitivity for a stopband loss pole with respect to its corresponding bridging capacitor. The sensitivity is given by

$$S_{Cb}^{\omega_0} = \pm 1/(2\,\omega_0^2 LC_b C_c((1/C_a - 1/C_d)^2 + 4(1/C_c - 1/C_b)/C_c)^{\frac{1}{2}})$$

(4)

where the positive sign is used for the upper stopband pole and the negative sign is used for the lower stopband pole. Note also that C_c is negative and $L_a = L_d = L$. Eqn. (4) shows that the pole frequencies move closer to the passband as the the value of the bridging capacitor, C_b, is increased. The sensitivity in this example is in the order of 10^{-4}. From this it is not surprising that the bridging capacitors are the least sensitive elements in the filter.

The resistive terminations have to be adjusted slightly to compensate for the uncertainty in the impedance of the experimental model of the filter. The designed passband ripple is 0.13 dB. and the measured passband ripple on two samples is about 0.5 dB. Stopband attenuation is well above 70 dB and spurious responses are not observable. The measured response is shown in Fig. 8.

CONCLUSION

An equivalent network for a bridged-coupled resonator is described in this paper. This facilitates the synthesis of high-grade polylithic filters suitable for use in SSB FDM systems. It was found, both by experiment and by calculation, that the frequency response of a polylithic filter is very insensitive to the bridging capacitors. This, in conjunction with the well-known good temperature and aging characteristics of the AT-cut quartz crystal, allows the production of good quality crystal filters with finite loss poles. This can be done with little modification to the present technology used in polylithic crystal filter production.

Since polylithic filters are mass production items, every aspect of design and production must be given very careful consideration before going into the production stage. There are many areas in the field of crystal filters that are worthy of further investigation by either university or industrial researchers. Some of these will now be mentioned. Since very tight impedance ratios are required within a polylithic filter,

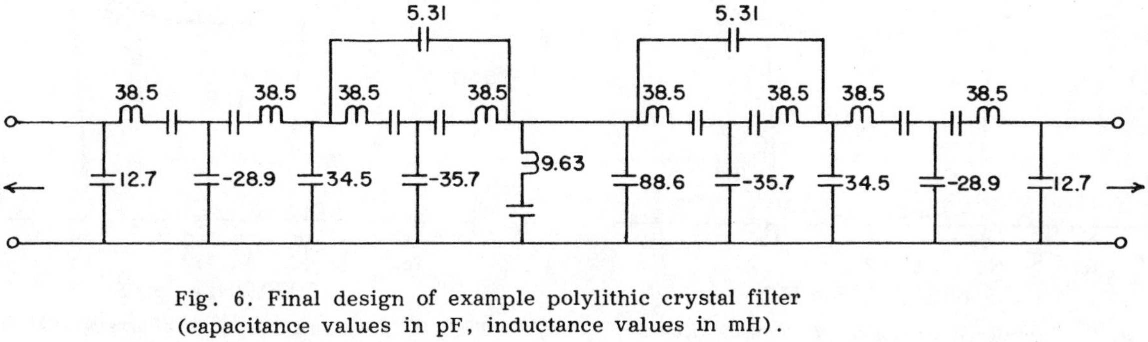

Fig. 6. Final design of example polylithic crystal filter (capacitance values in pF, inductance values in mH).

equal-ripple responses in the passband and stopbands are normally traded-off for better element values. It would be useful to find the approximation that gives the best set of element values. The Q of a crystal resonator is about 200,000 , which is much lower than the Q of the intrinsic crystal. This should be able to be improved. This in itself is a large topic worthy of much investigation. Another possibility is to predistort the filter to compensate for the finite Q characteristics. Another area of interest is the application of multiple bridgings (by bridging a capacitor across two or more resonators to produce multiple loss poles) to crystal filters. This technique has been applied successfully to mechanical filters and in a similar fashion to microwave filters.

ACKNOWLEDGEMENT

I would like to thank C. E. Schmidt and D. F. Sheahan of GTE Lenkurt for their constant support and assistance. I would also like to thank Prof. H. J. Orchard of UCLA for his valuable comments.

REFERENCES

1. D. F. Sheahan, "Single sideband filter for short-haul systems," Proc. Mexico, Int., 1971, pp.744-748.

2. D. F. Sheahan and C. E. Schmidt, "Coupled resonator quartz crystal filters," WESCON, Session 8, August 1971.

3. S.W. Anderes, "A polylithic filter channel bank," IEEE Trans. Comm., Vol. COM-20, Feb. 1972, pp. 48-52.

4. D. F. Sheahan, "Channel bank filtering at GTE Lenkurt," Proc. IEEE ISCAS,April 1974, pp.115-119.

5. H. Yoda, Y. Nakazawa and N. Kobori, "High frequncy crystal monolithic (HCM) filters," Proc. 23rd Annual Sym. Freq. Cont., 1969, pp.76-92.

6. A. A. Simpson, E. D. Finch, Jr. and R. K. Weeman, "Composite filter structures incorporating monolithic crystal filters and LC networks," Proc. 25th Annual Sym. Freq, Cont., 1971, pp.278-296.

7. Y. Masuda, I. Kawakami and Kobayashi, "Monolithic crystal filter with attenuation poles utilizing 2 dimensional arrangement of electrode," Proc. 27th Annual Sym. Freq. Cont., 1973. pp.227-232.

8. D. F. Sheahan and R. A. Johnson, "Crystal and mechanical filters," IEEE Trans. Circuits and Systems, vol. CAS-22, Feb. 1975, pp. 69-89.

9. M. S. Lee, "Equivalent network for bridged crystal filters," Electronics Letters, Vol. 10, No. 24, Nov.1974, pp. 507-508.

10. H. J. Orchard and G. C. Temes, "Filter design using transformed variables," IEEE Trans. Circuit Theory, Vol. CT-15, Dec. 1968, pp.385-408.

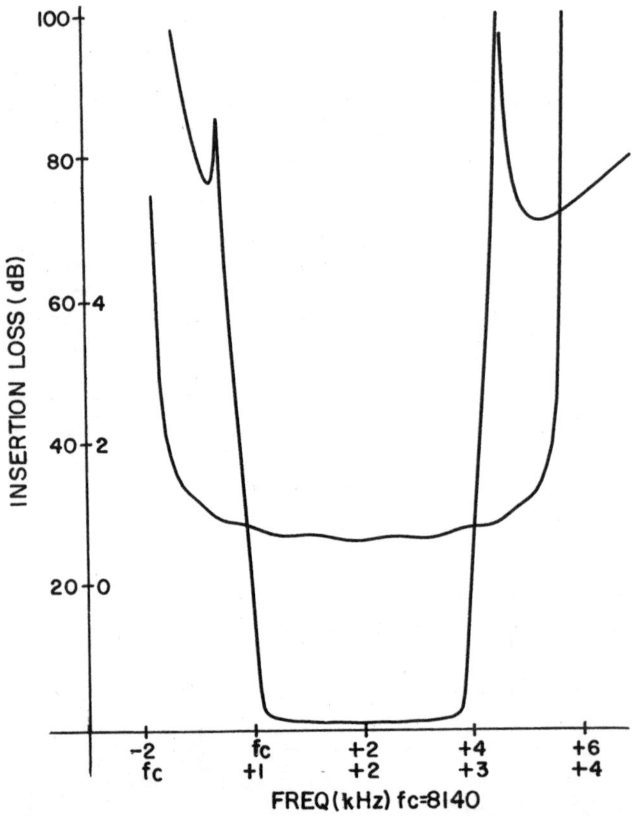

Fig. 7. Calculated response of example filter.

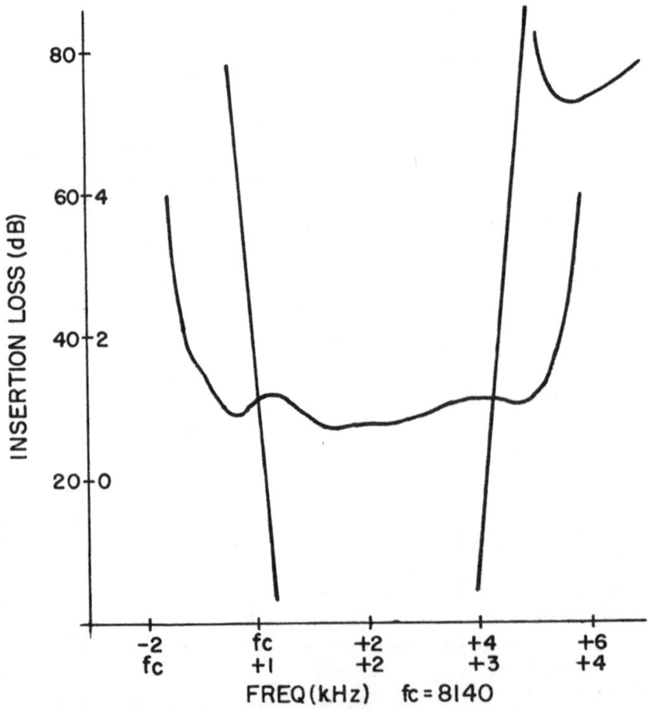

Fig. 8. Measured response of example filter.

High-Performance Monolithic Crystal Filters with Stripe Electrodes

Noboru Watanabe and Kazuo Tsukamoto, Members

Musashino Electrical Communication Laboratory, N.T.T., Musachino, Japan 180

SUMMARY

This paper describes a method of designing high-performance monolithic crystal filters (MCF) with stripe electrodes which are used to adjust coupling between resonator electrodes. A loop frequency of resonator electrodes is set to a value derived by the perturbation method. The relation between the loop frequency and the frequency observable at the resonator terminal is examined. An experimental 14-resonator tandem MCF has a Q-value of above 3×10^5 and meets with the CCITT 1/10 specifications.

1. Introduction

The monolithic crystal filters (MCF) operating in energy-trapped, thickness-vibrational modes have high Q-values and are stable in operation [1, 2]. The MCF's have been used extensively as narrow-band FM IF filters and pilot-signal extraction filters. Eight-resonator tandem MCF's which meet with the CCITT 1/5 specifications are expected to be used as A6-band channel filters [3]. In Japan, the out-of-band carrier transmission systems have been adopted which require high-performance channel filters. Such filters can be realized by use of stripe electrodes for adjustment of coupling between resonator electrodes [4].

This paper presents an MCF design method which is based on an equivalent circuit of stripe electrodes. The relation between design values and observable values is clarified and the causes of high loss at low frequencies are examined. An experimental 14-resonator tandem MCF can realize the Q value of 3×10^5 and meets with the CCITT 1/10 specifications.

2. Design of MCF with Stripe Electrodes

2.1 Equivalent circuit of stripe electrodes

The MCF can be constructed by depositing electrodes on both surfaces of crystal plate [see Fig. 1(a)]. To improve the MCF performance, stripe electrodes are deposited on both surfaces or one surface of the crystal plate and their edges are trimmed by a laser to realize a desired coupling between resonator electrodes. Figures 3 and 4 show the relations of the stripe electrode width to the eigenfrequency and coupling coefficients, respectively. In Fig. 3, a slanty symmetric mode F_a is independent of the stripe electrode width but a symmetric mode F_s is considerably dependent on it; the location of the stripe electrode corresponds to nodes of oscillation in the former and to valleys in the latter.

Figure 2 also shows the calculation model of the short-circuit stripe electrodes, where the electrode film thickness in case of one surface deposited is assumed as half the actual thickness. The distributed-parameter circuit for trapped-energy thickness modes on piezoelectric plate was derived by Nakamura et al. [6] and Jumonji et al. [7], and the equivalent circuit for the MCF on crystal AT plate was proposed by Ashida [5]. The dotted and solid lines in Figs. 3 and 4 show the calculated results on the above equivalent circuit of the stripe electrodes [5]. The case of one surface deposited shows better agreement between the experimental and calculated results than the case of both surfaces deposited. Accordingly, the MCF was designed based on the result of short-circuit terminal model with one surface deposited.

2.2 Design method

A narrow-band MCF design is based on the concept that the resonator electrodes in the MCF

Reprinted with permission from *Electron. Commun. Japan*, vol. 57, part A, pp. 53-60, 1975. Reprinted with the consent of Scripta Publishing Co., 1511 K St., N.W., Washington, DC 20005.

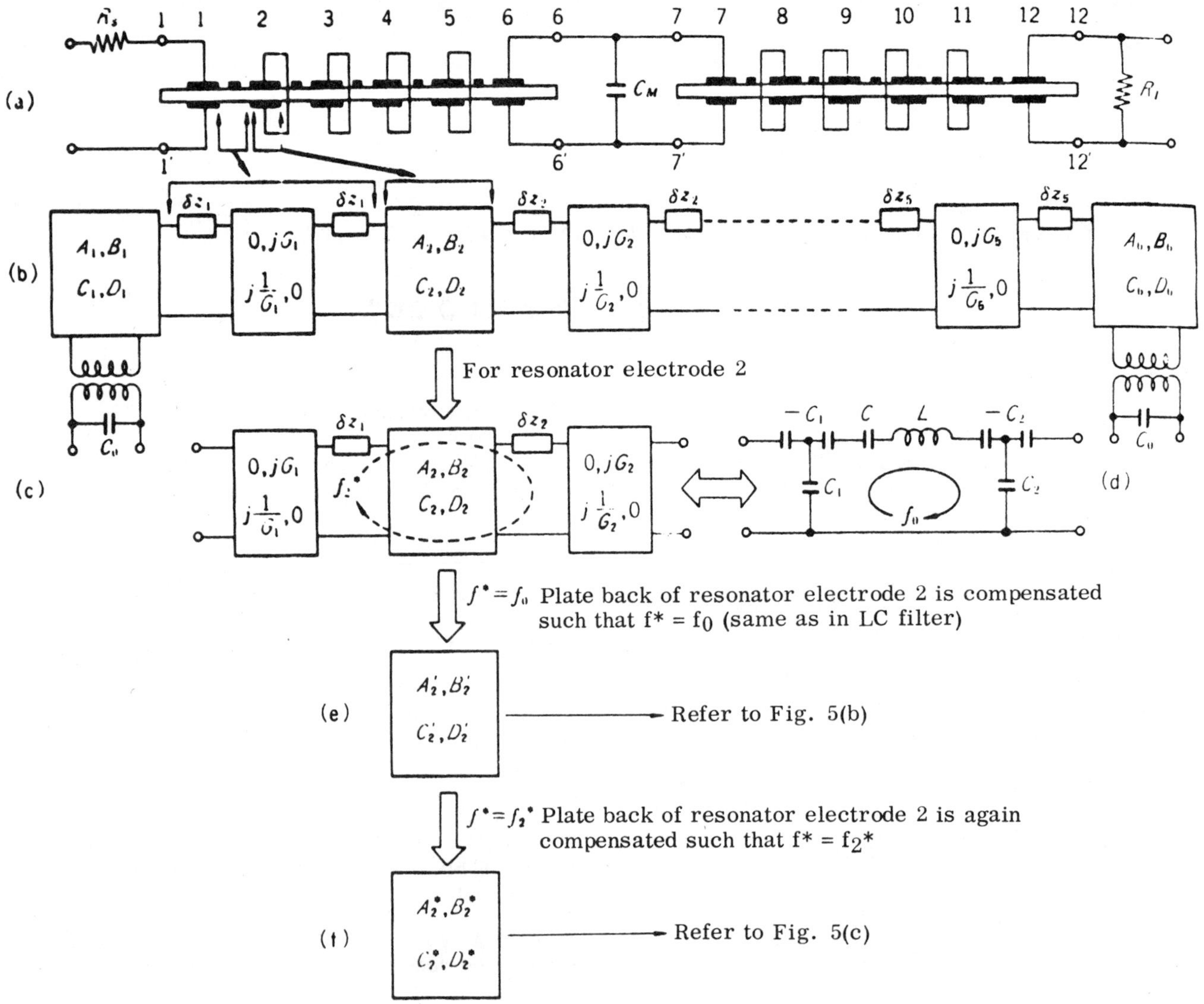

Fig. 1. Equivalent circuit of monolithic filter and its design procedure.

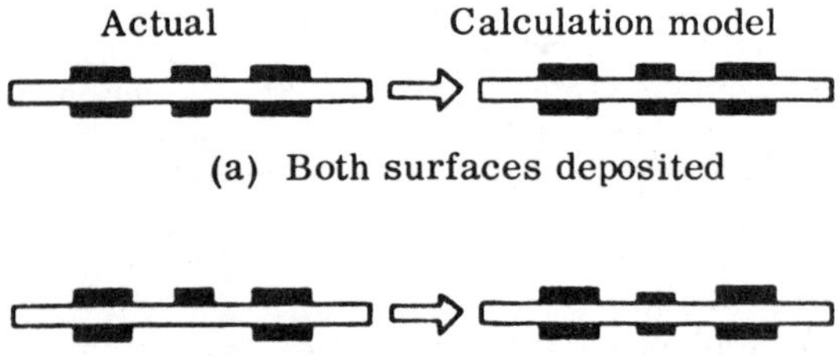

Fig. 2. Two kinds of stripe electrodes deposited for adjusting coupling coefficients between adjacent resonators and their calculation models.

lumped-parameter equivalent circuit are approximated by a series resonant circuit and the coupling portion between resonator electrodes by a virtual gyrator [5]. Figure 1(b) shows an equivalent circuit of the left half of the 12-element tandem MCF shown in Fig. 1(a), where R_0 is the plate back of resonator electrodes, f_0 is the resonant frequency; the resonator electrode is represented by the matrix of A_i, B_i, C_i and D_i and the coupling portion is approximated by a matrix of a virtual gyrator G_i and series element δz_i. As shown in Fig. 5(a), the MCF with stripe electrodes exhibit the loss characteristics with about 1.7-dB ripple at high frequencies and with a slight increase at low frequencies.

In Fig. 1(b) we consider the loop frequency f_i^* for a loop containing a resonator surrounded by two gyrators. The second resonator shown in Fig. 1(c) can be transformed into a lumped-parameter circuit shown in Fig. 1(d) if the plate back of resonator electrode 2 is compensated such that the loop frequency f_2^* coincides with f_0. Performing this operation to loops of all resonator

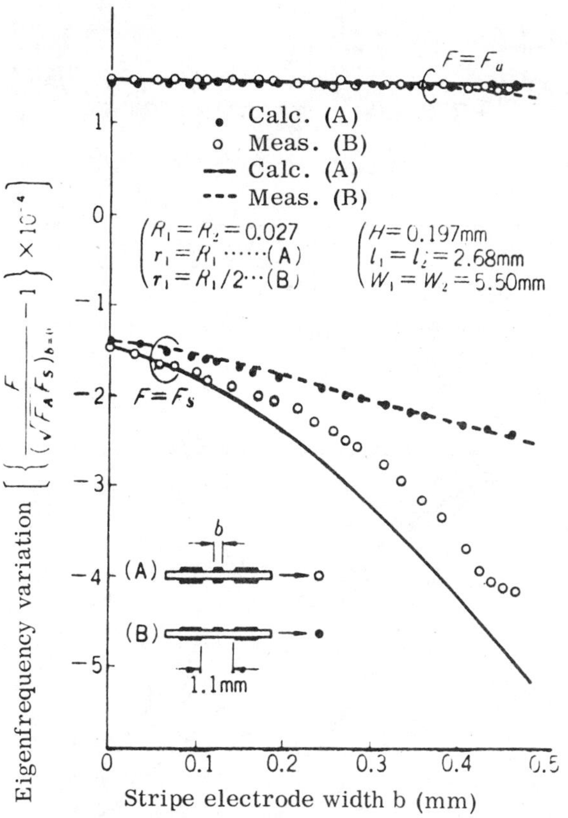

Fig. 3. Relation between the stripe electrode width and eigenfrequencies of two coupled resonators.

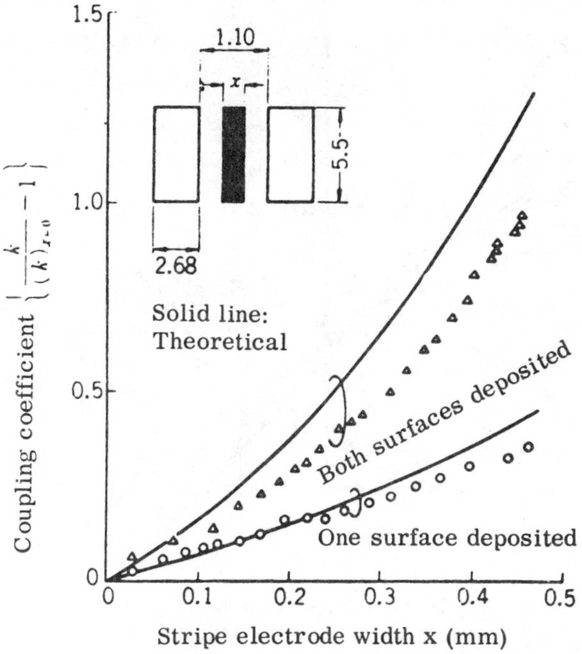

Fig. 4. Relation between stripe electrode width and coupling coefficient of resonators.

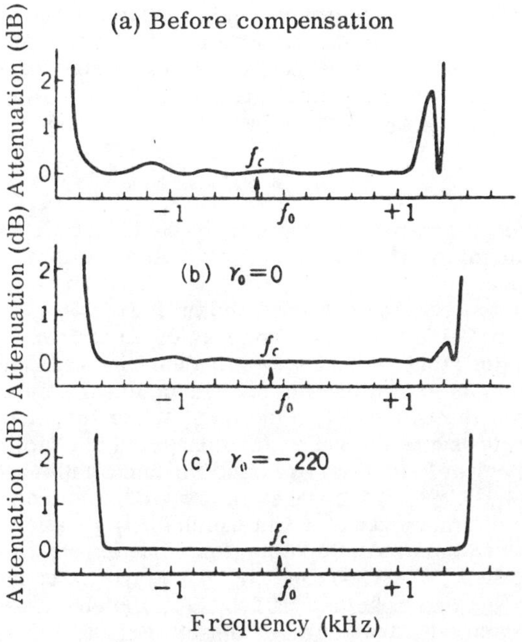

Fig. 5. Calculated loss characteristics of monolithic crystal filter with γ_0 in Eq. (1) as a parameter.

electrodes, the MCF frequency setting can be made to correspond to the frequency setting of LC filters. Figure 1(e) illustrates an equivalent circuit of this state and Fig. 5(b) shows the MCF transmission characteristics, revealing an improvement of loss characteristics at low frequencies and reduced ripple (about 0.5 dB) at high frequencies [5].

To improve the loss characteristics further, the loop frequency f_i^* is set to

$$f_i^* = f_0\{1 - \frac{1}{2}(k_i^2 + k_{i+1}^2)\gamma_0\} \qquad (1)$$

where k_i is a coupling coefficient between ith and (i+1)st resonators electrodes and can be obtained from the calculation, and γ_0 represents a negative capacitance ratio and is a parameter in the perturbation method. The MCF characteristics can be obtained by changing γ_0 to set f_i^* according to Eq. (1) and the optimum value of γ_0 for the optimum MCF characteristics can be found by comparing these characteristics with ideal characteristics. Figure 1(f) shows this operation while Fig. 5(c) illustrates the MCF transmission characteristics thus designed; a ripple at high frequencies disappears, and the loss characteristics at low frequencies are ideal.

3. Discussion on Observable Frequency

In this section, we discuss the relation between the loop frequency f_i^* and the observable frequencies which can be measured at the resonator electrode terminals. An open-circuit state of the resonator electrode terminals signifies an

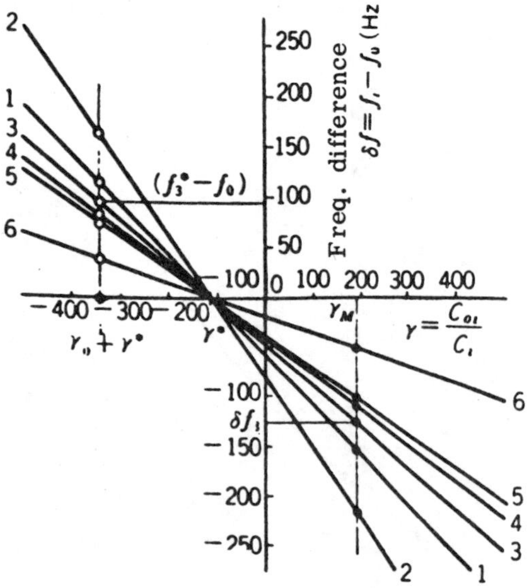

Fig. 6. Observation frequencies of the individual resonators as a function of capacitance ratio of resonators.

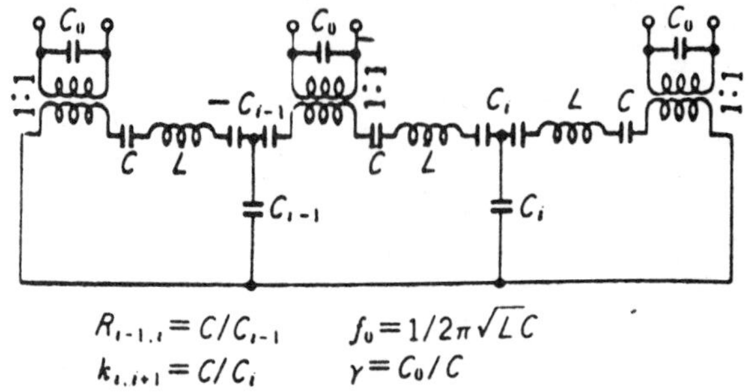

(a) LC equivalent circuit

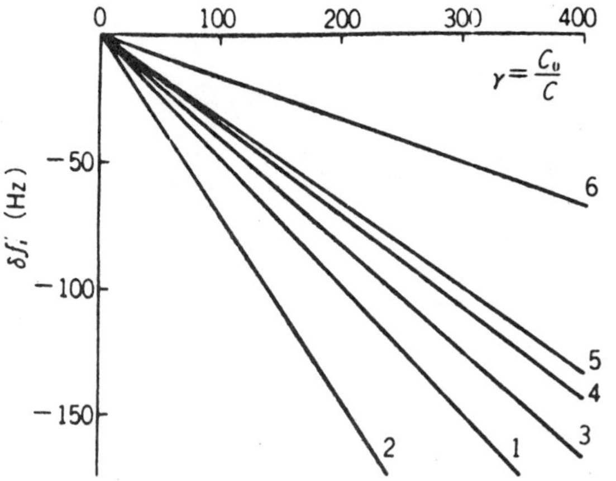

(b) Approximate calculation

Fig. 7. Approximate calculation of the observation frequencies.

addition of a capacitance C_0 in parallel to the resonator in the equivalent circuit. Letting an equivalent series capacitance of the resonator be C and a capacitance ratio be γ, the parallel capacitance C_0 is given by

$$C_0 = \gamma C \tag{2}$$

Since C can be calculated from the electrode dimensions and the plate back, the parallel capacitance can be regarded as a function of γ. For the 12-element MCF shown in Fig. 5(c), the relation between f_i and γ is obtained from the distributed constant equivalent circuit. The result is shown in Fig. 6. The numeral 1 represents the resonator in the input converter, and 6 represents the resonator connected to coupling capacitor. In a narrow band for bandwidth ratio of 4×10^{-4}, f_i increases in proportion to γ and the optimum value of γ at which all f_i coincide with f_0 obtains. Denoting this value as γ^*, we obtain $\gamma^* \doteqdot -120$ from Fig. 6. Letting a capacitance ratio of actual MCF be γ_M, the observable frequencies are shown by dots in Fig. 6, to which the resonator resonant frequencies are adjusted in the MCF design.

In the LC equivalent circuit [see Fig. 7(a)] where an adjacent resonator alone is considered, f_i is given by

$$f_i = f_0 \{ 1 - \frac{1}{2} (k_i^2 + k_{i+1}^2) \gamma \} \tag{3}$$

Figure 7(b) shows the relation between γ and δf_i which is calculated from Eq. (3).

Figure 6 agrees approximately with Fig. 7(b). Accordingly, the observable frequency f_i can be approximately expressed by substituting $\gamma_M - \gamma^*$ into γ in Eq. (3):

$$f_i = f_0 \{ 1 - \frac{1}{2} (k_i^2 + k_{i+1}^2)(\gamma_M - \gamma^*) \} \tag{4}$$

Since Eq. (1) is in the same form as Eq. (4), the perturbation parameter γ_0 varies in proportion to a capacitance ratio and f_i^* coincides with f_i for $\gamma = \gamma_0 + \gamma^*$ as shown by circles in Fig. 6.

Figure 6 differs from Fig. 7(b) by γ^* in terms of γ, signifying the existence of parallel capacitance $\gamma^* C$ even in short-circuited state of the MCF resonator electrode terminals. This is due to a difference between lumped-constant and distributed-constant equivalent circuits.

The observable frequency depends on $\gamma_M - \gamma^*$ as seen in Eq. (4). Figure 8 shows the MCF filter characteristics with a parameter of $\Delta\gamma$, i.e., the duration of γ from the optimum value

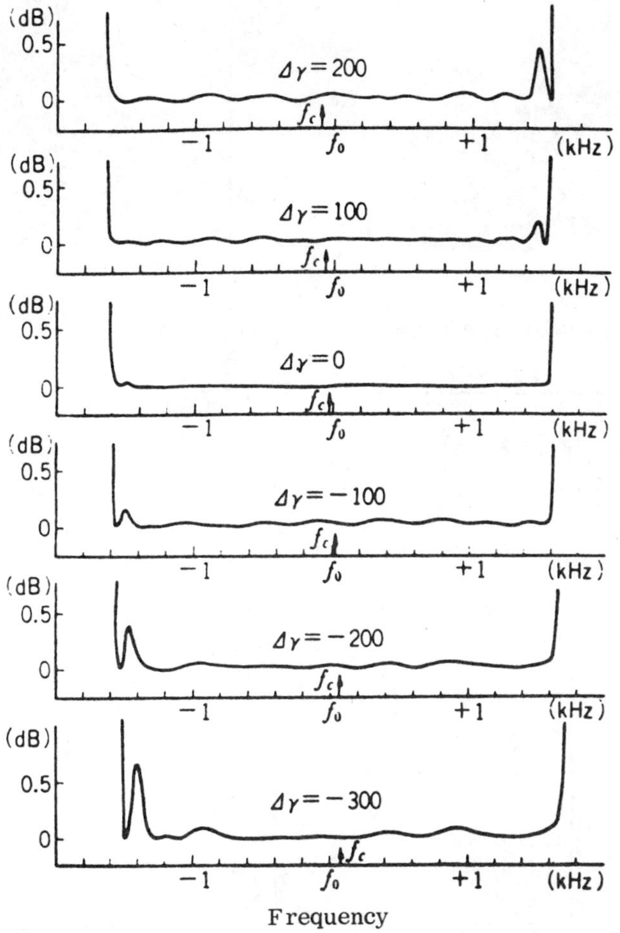

Fig. 8. Calculated inband-ripple characteristics with $\Delta\gamma$ as a parameter, deviation from the optimum capacitance ratio.

of $\gamma_M - \gamma^*$. The ideal characteristics for $\Delta\gamma = 0$ reveal a ripple at high frequencies in case of positive value of $\Delta\gamma$ and at low frequencies in case of negative value. In the conventional MCF design, the observable frequencies for all resonators were set to a constant (i.e., $\gamma_M - \gamma^* = 0$ and $\Delta\gamma = -310$ in case of Fig. 6), resulting in a large ripple at low frequencies as seen in Fig. 8. In actual MCF's, the Q-value is limited and the ripple results in a loss increase at low frequencies.

4. Q-Values

The previous discussion has assumed no loss and infinite Q-values. In practical narrow-band filters, the loss increases with an increase of the number of resonators. The CCITT requires the Q-values of above 3×10^5 for the channel filters.

Mindlin et al. [8] discussed the dependency of the electrode dimensions on the Q-values of rectangular-electrode resonators on crystal AT plates. The Q-value is periodic in the direction of X-axis (electrical axie) of crystal AT plate. The Q-value reaches the maximum for the electrode in which the mode conversion from the thickness-shear vibration to bending vibration does not occur. By contrast, the Q-value becomes minimum when this mode conversion occurs. This was confirmed experimentally by Sasaki et al. [9].

Figures 9 and 10 show the relation between electrode length and Q-value of the MCF with a lapped crystal plate; nichrome and gold were deposited by evaporation on a #4000 lapped crystal AT plate ($x \times y' \times z' = 11.2 \times 0.197 \times 35.0$ mm^3). The Q-value was measured by removing the electrode by several microns using a laser. The Q-values in these figures represent those in air. In vacuum, these values double. The arrows in the figures represent the electrode dimensions which maximize Q-values according to Mindlin:

$$L_x = 1.634 \times n \times H \quad (n : \text{integer}) \qquad (5)$$

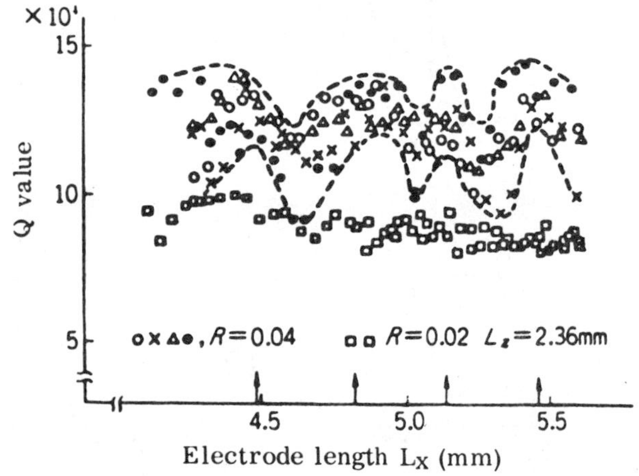

Fig. 9. Effect of electrode length (L_X) on Q value.

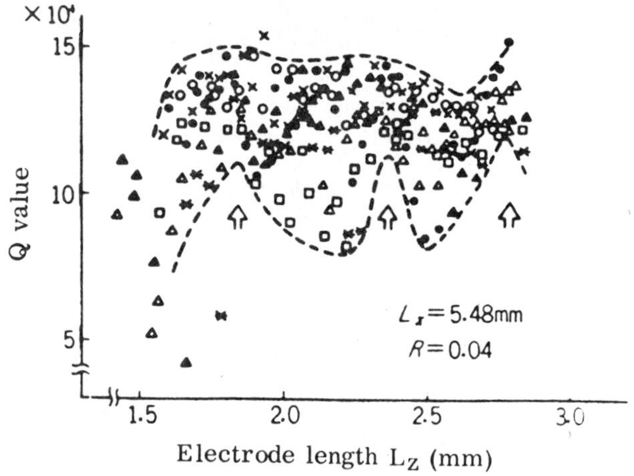

Fig. 10. Effect of electrode width (L_Z) on Q value.

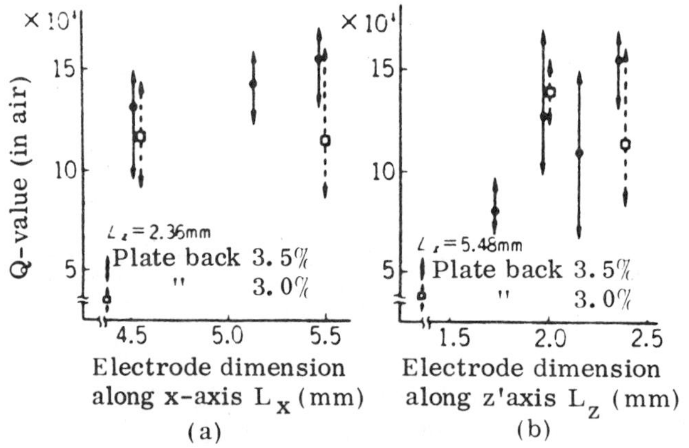

Fig. 11. Effect of electrode dimension on the variation of mechanical Q values.

where H is a crystal plate thickness. For R = = 0.04, Q-value exhibits periodicity, but for R = 0.02, the Q-value is small and does not exhibit periodicity. The plate back is about 4% and the optimum value of L_X is about 5.48 mm which corresponds to n = 17 in Eq. (5). In Fig. 10, the variation of Q value is reduced at L_Z = = 1.84, 2.36 and 2.78 mm.

In actual MCF with multielectrodes, it is necessary to reduce the variation of Q-values as small as possible. Figure 11 shows the effect of the electrode dimension on the variation of the Q-value of the MCF with 6 resonator electrodes. The arrow represents the variation of Q-value and the dots signify the mean values. In Fig. 11(a), L_X does not affect the variation of Q-value, while Fig. 11(b) shows a large dependency of L_X on the Q-value. In short, L_X affects the absolute value of Q and L_Z influences the variation of Q.

5. Trial MCF

We have constructed a 14-resonator tandem MCF. The crystal AT plate has the same dimension as described before. The parallelism deviation was less than 0.1%. The resonator dimension was $L_Z \times L_X = 2.36 \times 5.48$ mm^2 and 2.36×5.14 mm^2. The plate back R was 0.037. Figure 12 shows a trial MCF. The electrode sizes L_X on two crystal plates differ, which enables one to suppress in-harmonic overtones of thickness-shear modes. Figure 13 shows the filter characteristics of the trial MCF which conforms with the CCITT 1/10 specifications.

Figure 14 shows the spurious characteristics of 14-resonator tandem MCF in which the dimensions of all resonators on two crystal plates are identical [Fig. 14(a)]; a spurious ($TS\text{-}S_1$) of TS mode occurs near f_c + 100 kHz. This spurious

Fig. 12. 14-resonator tandem MCF.

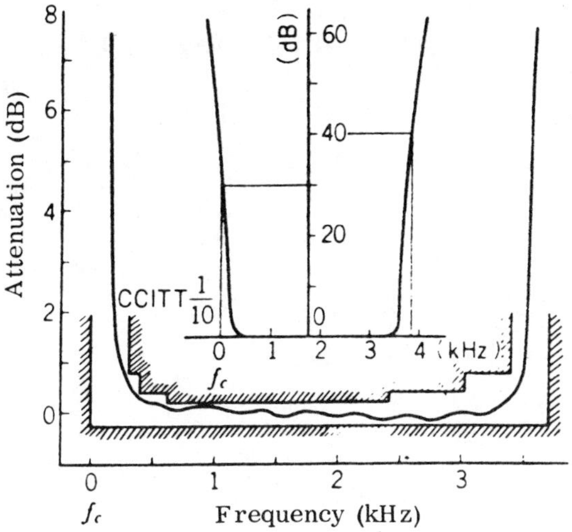

Fig. 13. Filter characteristics of 14-resonator tandem MCF.

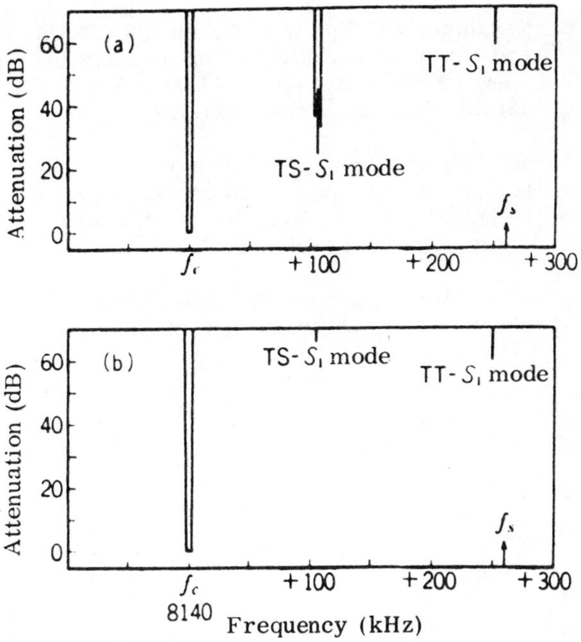

Fig. 14. Measured spurious characteristics of 14-resonator tandem MCF.

characteristic can be suppressed by use of resonators with different dimensions [Fig. 14(b)]. The TT-S_1 mode signifies a spurious characteristic along the L_x direction.

6. Conclusion

This paper has described a design method of high-performance monolithic crystal filters with stripe electrodes. An equivalent circuit for stripe electrodes was derived and a calculation model for energy-trapped mode resonators with short-circuited terminals was proposed. A new MCF design method was proposed in which a loop frequency of resonator electrodes including stripe electrodes, was set to a specified value. The frequency observable at the electrical terminals of each resonator electrode was examined.

A trial 14-resonator tandem MCF which can realize the Q-value of above 5×10^5 was constructed. It conforms with the CCITT 1/10 specifications.

Acknowledgment. The authors would like to thank Mr. Takahara, director of data communication section, Mr. Hanawa, director of electronics device section, Mr. Kawamata, Mr. Kaneoya, Mr. Sawamoto, Mr. Jumonji and Mr. Ashida, all of the E.C.L., for their guidance and encouragement in this research.

REFERENCES

1. Onoe and Jumonji: Analysis of piezoelectric resonators vibrating in trapped-energy modes, Jour. I.E.C.E., Japan, 48, 9, p. 1574 (Sept. 1965); available in English in E.C.J., same date, p. 84.

2. R.A. Sykes and W.D. Beaver: High frequency monolithic crystal filters with possible application to single frequency and single side band use, Proc. of 20th Ann. Symp. on Frequency Control, p. 288 (April 1966).

3. R.J. Byrne: Monolithic crystal filters, Proc. of 23rd Ann. Symp. on Frequency Control, p. 84 (1969).

4. J.L. Hokanson: Laser machining thin film electrode arrays on quartz crystal substrates, Ibid., p. 163.

5. Ashida and Tsukamoto: Design of monolithic filters and their transmission characteristics, Data of the 25th Committee of Mechanical Vibration Systems Functional Components (Jan. 1971); Ashida and Tsukamoto: ibid., Data of the 27th same committee (March 1971); Ashida: Design and Characteristic Analysis of monolithic filters, to be published in Trans. I.E.C.E., Japan.

6. Nakamura and Shimizu: Equivalent circuit expression of thickness modes propagating along piezoelectric plates, Trans. I.E.C.E., Japan, 55-A, 2, p. 95 (Feb. 1972).

7. Jumonji and Onoe: Equivalent circuit of trapped-energy mode resonators, Paper of technical Group on Ultrasonics, I.E.C.E., Japan, US 69-33 (Feb. 1970).

8. R.D. Mindlin and P.C.Y. Lee: Thickness shear and flexural vibrations of partially plated, crystal plates, Jour. Solid Structure, 2, p. 125 (1966).

9. Sasaki and Jumonji: Electrode dimension and Q-value of thickness-mode quartz resonators, Paper of Technical Group on Ultrasonics, I.E.C.E., Japan, US 71-39 (Feb. 1972).

Submitted September 19, 1973

Section II-D
Manufacturing Techniques

The next group of papers describes some of the techniques that have been used to economically produce large quantities of modern crystal filters. These papers illustrate two different approaches that have been followed. The monolithic approach as described by Cawley *et al*. was aimed at producing a complete eight-resonator telephone channel filter on a single plate of quartz. Production difficulties dictated that this be modified to the two-plate approach that is described. This monolithic approach does, however, lend itself to a high degree of automation, as can be seen from the descriptions of the processing equipment contained in this paper. Much of this equipment has been designed specifically for this type of filter and it is quite complex. The paper describes the process of using a laser to remove metal from the crystal and thereby adjust the coupling between resonators. It is difficult to maintain adequate coupling tolerances on a large crystal plate, so consequently adjustments have to be made even though there is a resultant penalty in increased processing complexity.

An alternate approach is described by Sheahan, where no more than two resonators are coupled together on a single plate of quartz. This greatly minimizes the coupling accuracy problems to the point where no adjustment is needed. The requirements on the crystal blanks are also simplified so that they can be easily obtained from commercial sources. This approach offers flexibility and lower capital costs at the expense of greater assembly complexity.

Miller, in his paper, describes the precautions that must be taken to guarantee crystal blanks of adequate quality for the monolithic crystal filter. The techniques used are more complex than those normally used in the crystal industry, and they are necessitated by the very tight requirements on crystal orientation and degree of parallelism that the blanks must meet. The study of surface roughness conditions that is also contained in this paper gave new insight into the optimum amount of etching that should be performed on the crystal blanks.

Finally, we come to the reports on the actual manufacturing experiences and the performance statistics of the manufactured product. Some information on these subjects is presented in the paper by Olster *et al*. and also in the paper by Sheahan. From these papers, we see that both processes have been used successfully to produce large quantities of filters that have both the performance characteristics and the long term stability required by the telephone industry.

MANUFACTURE OF MONOLITHIC CRYSTAL FILTERS FOR A-6 CHANNEL BANK

H. F. Cawley, J. D. Jennings, J. I. Pelc, P. R. Perri
and F. E. Snell

Western Electric Company, North Andover, Massachusetts

A. J. Miller

Bell Laboratories, Allentown, Pennsylvania

INTRODUCTION

The A-6 Monolithic Crystal Filter (Figure 1) is an eight pole Chebyshev band-pass filter operating in the 8 mHz frequency range with a 3 dB bandwidth of 3.26 kHz. As Figure 2 indicates, it is composed of; the quartz plate, two lead frames, a ceramic holder, a capacitor, a shorting bar, a header and a cover. The device was designed for the A-6 channel bank, which provides the first and last steps in the overall multiplexing scheme for voice channel transmission. The filter was introduced into manufacture in late 1972. Initial production was limited and build-up slow because prototype facilities and pilot line conditions had to be converted to high volume production. Also, and very important to the success of the project, operator experience had to be accumulated. The operators can effect a considerable influence on the yields at almost every process step, although many facilities have automated closed loop control.

In the interim between initial production and the present, the manufacturing process has undergone constant modification and improvement. Through these changes, and with increased operator efficiency, a reduction of costs has been accomplished. Figure 3 shows the normalized classic learning curve for the A-6 MCF.

This paper will discuss the significant manufacturing process steps, facility improvements, some of the problems encountered, and some innovations that have contributed to lower costs.

The manufacturing process naturally divides into two separate parts - plate fabrication and filter assembly. These parts will be dealt with separately even though some overlap does exist. First the plate fabrication area will be discussed.

PLATE FABRICATION

Our channel bank requirements establish the need for a high volume crystal plate manufacturing process capable of producing equal quantities of twelve codes of AT-oriented crystal plates alike in all respects except the absolute final thickness. The etch end point thicknesses, expressed in frequency, of the twelve codes are in successive increments of 4 kHz and cover a range of 48 kHz.

Figure 4 shows the dimensions and the crystallographic orientation of our unconventionally long and narrow crystal plate. This design provides optimum space for; eight pairs of electrodes with coupling stripes, and sixteen pads for the mechanical and electrical interconnection of the completed crystal filter.

The difficulties entailed in uniform production of twelve frequency variations were insignificant when compared to the challenges presented by two other crystal plate design criteria:

1. Extreme parallelism between each plate's major surfaces;
2. unusually tight control of the AT crystallographic orientation.

The practical range of adjustment of inter-resonator coupling of the original design limits the allowable parallelism deviation over the plate length to 6 kHz, i.e. 6 millionths of an inch.

Redesign of the electrode and coupling stripe arrangement added range for coupling adjustment permitting a change in the parallelism requirement from 6 to 10 kHz. A dramatic yield improvement - to 98% - accompanied the relaxed parallelism tolerance. 10 kHz (10 millionths of an inch) approximates the parallelism capability for double face lapping a crystal plate with this unusual length to width ratio.

The other important design criterion is the AT crystallographic orientation angle.

The completed M.C.F. must include allowances for frequency adjustment, long term frequency stability, and the frequency-temperature characteristic. This latter portion is allocated to the crystal plate as a tolerance on the AT angle. This tolerance is plus or minus 45 seconds of arc.

The orientation is measured in-process mid-way through the plate shaping process while the plate is relatively thick and X-ray measurements are the most economic and accurate. The angular tolerance at this mid-process point is plus or minus 30 seconds. The remainder of the plate machining processes are rigidly controlled to attain plus

Reprinted with permission from *Proc. Twenty-Ninth Ann. Symp. on Frequency Control*, May 28-30, 1975, pp. 113-119, sponsored by the U.S. Army Electronics Command, Ft. Monmouth, NJ.

or minus 45 seconds in the finished plate.

PROCESS STEPS

Relatively unskilled personnel are able to manufacture A-6 Crystal Plates in large volumes with the precision required for plate parallelism, crystallographic orientation, and uniform quantities of twelve different frequencies.

Sophisticated tooling and control systems, justified by high production, are necessities in our manufacturing operation. The most significant process steps are listed in Figure 5.

The graphical presentation of the first several process steps, including wafering to individual plates (.037 inches thick), illustrates the economies derived from quartz crystals (ingots) supplied by our crystal growing facility. The R-face quartz[1] ingots are tailor made to match our crystallographic orientation requirement and to provide economic multiples of individual plates.

It seems appropriate to review the reduction in thickness accomplished in each production step and to discuss the rationale for the step and thickness.

Wafering produces plates of about 37 mil thickness. The final 8 mil thickness is obtained by successive abrasive lappings and one chemical etching step. The first lapping, to 28 mils, removes irregularities left by wafering and provides the degree of smooth texture and planar surface required for X-ray measurement at the X-ray Sorting Step. The planarity achieved by the simultaneous lapping of both major surfaces to .028 inches, is also a prerequisite for attaining the plus or minus 45 second angle tolerance in the finished plate.

Frequent and precise reconditioning of the lap plates and other tooling employed in all the abrasive lapping processes is essential to meet the requirements of AT orientation, plate planarity and parallelism. The dimensional stability exhibited by our 86" diameter master-lap (Figure 6) allows non-skilled operators to easily perform the precise reconditioning.

The X-ray sorting step is a measure of the accuracy of the state of the art for quartz crystal wafering. Currently only about one-half of the plates wafered meet the plus or minus 30 second in-process tolerance. X-ray sorting identifies those plates meeting the angle requirement thereby qualifying for processing without correction, and segregates plates of like orientation error as the first step in repair of defective crystallographic orientation.

ANGLE CORRECTION

Existing angle correction methods were neither sufficiently precise nor economically attractive. Facilities were provided to simultaneously correct large numbers of plates in automatic lapping cycles. Angle correction by lapping instead of grinding is important because its inherent precision allows minimum stock removal while economically providing the required surface texture. Virtually all Angle Corrected Crystal plates meet the orientation requirement.

THINNING

The Thinning Step reduces plate thickness to about 8.5 mils from thicknesses of 26 mils and 28 mils for corrected and uncorrected plates, respectively. Here, a large number of plates are wax-bonded to an ultra-flat mounting plate and stock is removed from one side of the crystal plates by single face lapping. This process contributes a minimum loss of angular orientation.

Although only a small amount of thickness is removed in the last lapping operation, 5 micron aluminum oxide abrasive double-face lapping, this step must provide the required surface finish and parallelism and, together with chemical etching, production of the twelve frequency codes. Other details of final lapping and etching have been reported earlier[2].

CRYSTALLOGRAPHIC MEASUREMENT ERRORS

The last significant information to report from our A-6 Crystal Plate Manufacturing experience is related to misorientations about Y^1 and Z^1 crystal axes. It has been known for some time that errors in orientation about Y^1 and Z^1 adversely affect the indicated AT angle measurement[3]. It has also been known that misorientations occurring exclusively about either (but not both) secondary axis produce relatively insignificant affects[4]. A detailed study[5] of the interrelationship between secondary axis misorientation and indicated AT reading shows that the effect of combined Z^1 and Y^1 rotational errors is substantial, as shown graphically in Figure 7. The dashed lines, indicate the most widely used tolerances (15 or 30 minutes). Note that if both secondary axis rotational errors are 30 minutes, the indicated AT measurement differs from the true reading by about 16 seconds. However, if a 30 minute error occurs exclusively about one secondary axis, the AT measurement error would amount to only about one second.

Following completion of the plate fabrication sequence, the filter assembly process transforms the plate into a Monolithic Crystal Filter. The major process steps are outlined in Figure 8.

FILTER ASSEMBLY

Most of the facilities which are used to manufacture the filter have been modified since production was initiated to provide higher throughput and reliability. A discussion of some of these facilities follows.

BASE PLATERS

The original base plater was a "Bell-Jar" evaporator, incorporating electron beam evaporation sources. It was found that this facility produced product with unacceptable lot-to-lot variations which were at least partly attributible to inconsistent operator performance. The current production facilities (Figure 9) employ automatic closed-loop control over vacuum system cycling, source selection, evaporation rate and mass of deposited film. The lot-to-lot consistency has been greatly improved. In addition, the vacuum system employs a load lock feature which requires that only part of the system need be brought back to atmospheric pressure to load and unload product. This results in dramatically shortened cycle times.

FREQUENCY PLATING

Increased product throughput at frequency plating was provided by a redesign which increased batch size. The larger batch size was made possible by substituting a computer-controlled plating control system (Figure 10) for the earlier hard-wired logic version. Use of the computer permits closed-loop control of plating rate, and allows testing of the product prior to and after adjustment.

The introduction of channel sorting, which will be discussed in further detail later, resulted in greater throughput at frequency plating since in a batch load process total plating time is dependant on the unit that requires the most adjustment. Hence, if preadjustment sorting of frequency is employed, all units in the batch will be adjusted approximately the same amount.

Improved cleaning and preventative maintenance procedures on the frequency plater vacuum systems also contributed greatly to increased throughput as each machine is now able to operate at full capacity a much larger percentage of the time.

LASER: COUPLE ADJUSTMENT

The original laser utilized an X-Y table to move the filter in its test fixture under a fixed laser beam. The system was controlled by hard-wired logic with parameters obtained from an endless loop of paper tape. The current computer-controlled laser (Figure 11) has far greater throughput and reliability. It has a galvonometer-mirror beam positioning system which permits a fixed filter location and faster beam positioning. The computer control permits pre and post-adjustment testing of filters. The flexibility and speed of this system also allows laser fine tuning of the resonator area after the coupling stripes have been adjusted. Laser fine tuning will be discussed in more detail later.

FINAL ADJUSTMENT AND ENCAPSULATION

It was originally believed that successful adjustment of the lower 3 dB frequency would require a continuous vacuum environment for transfer from adjustment to encapsulation. The facility incorporating this concept functioned adequately, but was very expensive, had limited throughput and, because of its complexity, required an undesirable amount of maintenance effort. Experimentation performed in the actual manufacturing environment indicated that transfer from adjustment to encapsulation could be achieved at atmospheric pressure in a dry, clean environment with no adverse effect to the device. This led to development of the present facility (Figure 12) which consists of a separate adjustment module linked via a controlled atmosphere chamber to a cold welding module. The benefits derived from this new facility are vastly increased throughput, reduced facility cost and less maintenance effort.

SHOP INFORMATION SYSTEM (SIS)

The more complex a manufacturing sequence becomes, the more important it is to have real-time information on the performance of the entire sequence and each of its steps. The Western Electric Shop Information System (SIS) has been an important engineering tool in helping to improve throughput and yields. Basically, the system is a computer-based data system that optically reads input data sheets that are tallied by each operator. Reports are generated and distributed to engineering and shop supervision on daily, weekly, and multi-weekly basis. In addition, a CRT terminal allows real-time access to the entire data-base. The reports include first run, repair, or combined yield information by machine, by operator or by process step. Lot histories are also available. Information of interest to shop personnel includeinventory and operator efficiency reports.

CHANNEL SORTING

Originally, the frequency sort for channel determination was performed on the plate after etch. This method depended on the base platers to provide uniform plate-back. However, unit to unit variations in plate-back caused several problems. One was that some units arrived at the frequency plater already too low for the specified channel. Also, units that were very high in frequency coming into the frequency plater required excessive time to plate down, causing a loss in throughput. The excessive plating also caused a decrease in coupling which showed up as a decrease in yield at the coupling adjust operation.

The solution to this problem was to add a semi-automatic channel sort operation to the process after the plate is bonded to the ceramic. The plates are designated for the highest channel possible so that the batch operation of the frequency platers is optimized. An additional advantage obtained by this method of channel sorting is that defective units, no reads, are screened out at the less expensive facility.

PLASMA CLEANING

Lack of consistency in the effectiveness of the rigorous cleaning process required for this device became evident early in the operation of the process line. It was determined that subtle differences in the incoming quality of solvents and other chemicals, coupled with difficulty in determining the end of useful life of the cleaning baths accounted for the inconsistency. Considerable effort was devoted to obtaining a viable alternative to the wet chemical processes which would yield the desired repeatable results. Plasma cleaning, (Figure 13) employing energetic gas as the cleaning medium, was determined to best satisfy these requirements. The Plasma process with the facilities used to implement it is programmable, repeatable, is not product volume sensitive and produces very acceptable results.

LASER FINE TUNING

A new method of fine tuning resonators is now being introduced into the process. Figure 14 shows the flow chart for the new process which uses the laser to vaporize metal from the resonators. The Rough Adjust tuning frequencies obviously have to be modified to bring them below the final desired frequency since laser tuning increases them. The speed of the new computer-controlled laser made this type of tuning possible. Advantages of performing the fine tuning by this method include:

1. Higher through-put per machine.
2. A less expensive facility, as vacuum plater systems are more expensive and less reliable.
3. Eliminates the use of gold for the fine tuning adjustment.

SUMMARY

The manufacture of the A-6 Monolithic Crystal Filter is an accomplished fact. The transformation of the monolithic theory from laboratory models to full scale production has been achieved. To say it was achieved simply and easily would be stretching the truth. The transformation was accomplished only with close cooperation among all individuals involved, and set-backs were encountered along the way. The process is precise and requires constant surveillance.

The Shop Information System provides an up-to-the-minute picture of how well we are implementing the process and indicates those area's where attention is required. By fine tuning our efforts on a day-to-day basis we are able to achieve both high volume and efficiency.

REFERENCES

1. R. L. Barns, Et. Al.
   "Production and Perfection of R-Face Quartz"
   29th Annual Frequency Control Symposium (1975)

2. A. J. Miller
   "Preparation of Quartz Crystal Plates for Monolithic Crystal Filters"
   Proceedings of the 24th Annual Frequency Control Symposium (1970)

3. R. A. Heising
   Quartz Crystal for Electrical Circuits, Chapter III Van Nostrand Co. (Out of print)

4. Roger E. Bennett
   Quartz Resonator Hand Book
   Prepared for the Department of the Army, by Union Thermo Electric Division (1960), Page 145, 146 - AD 251289

5. Private Communication with G. L. Dybwad, Bell Laboratories, Allentown, Pa. and E. V. Conradt, Western Electric Co., North Andover, Mass.

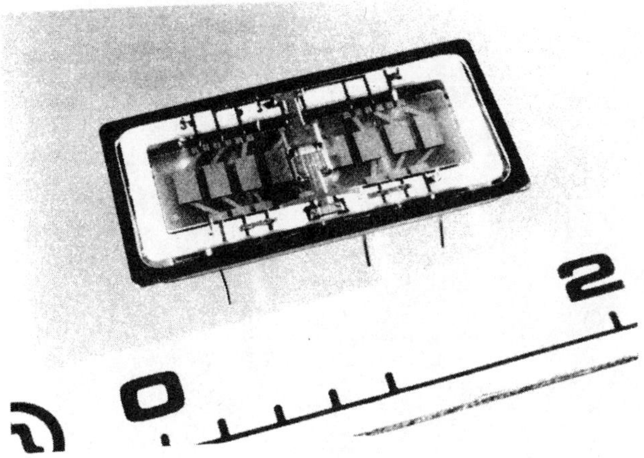

Figure 1 Open A-6 Monolithic Crystal Filter

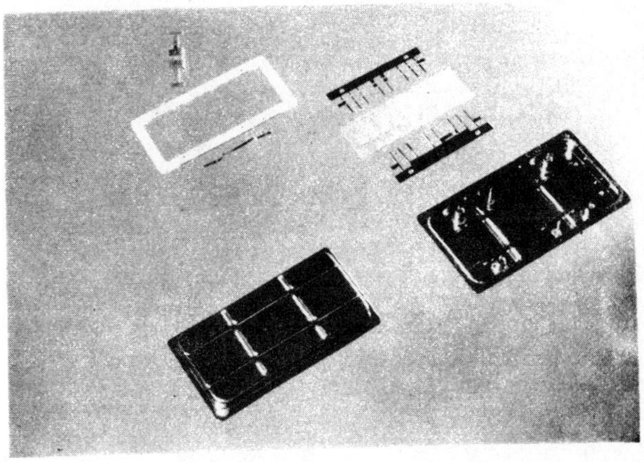

Figure 2 Piece Parts A-6 MCF

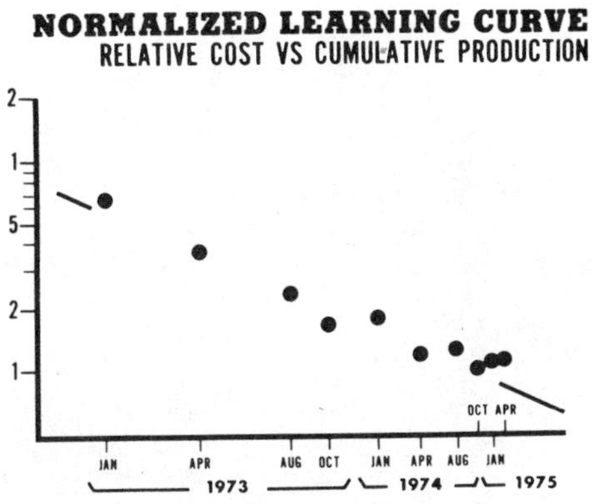

Figure 3 Normalized Learning Curve

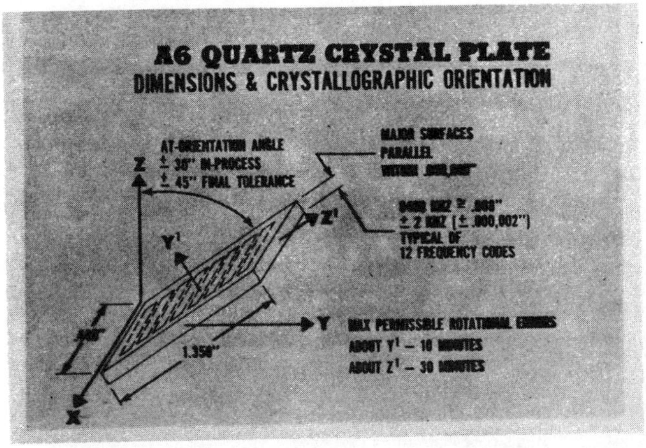

Figure 4

A-6 Quartz Plate Dimensions and Crystallographic Orientation

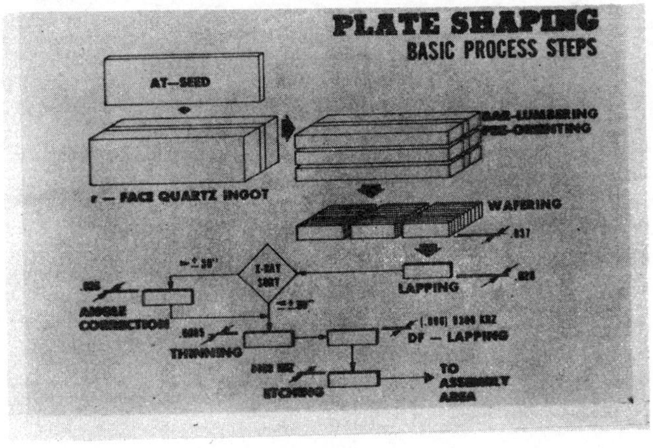

Figure 5 Plate Shaping Process Steps

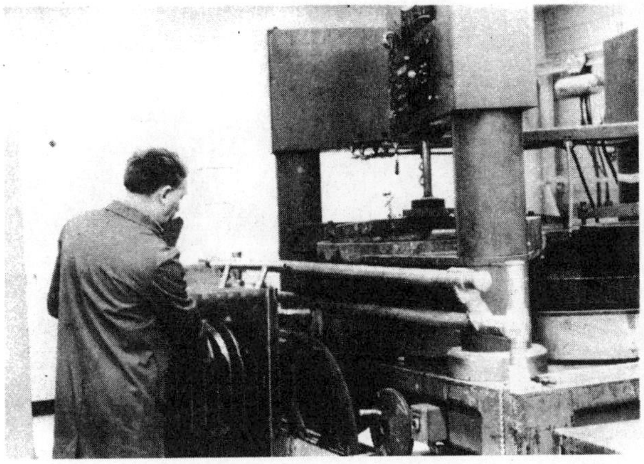

Figure 6 86" Master Lap

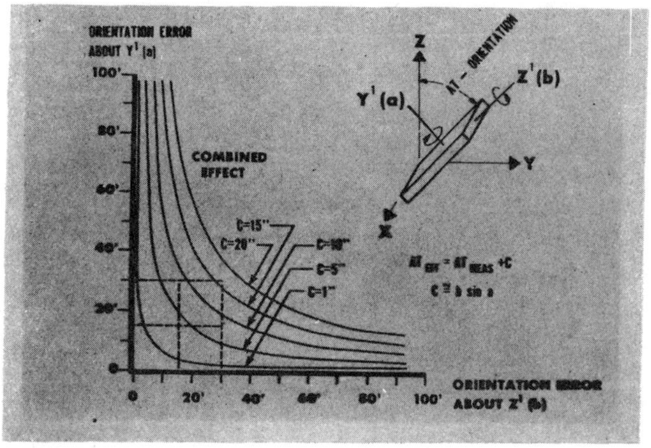

Figure 7

Error Due to Mis-orientation about Z^1 & Y^1

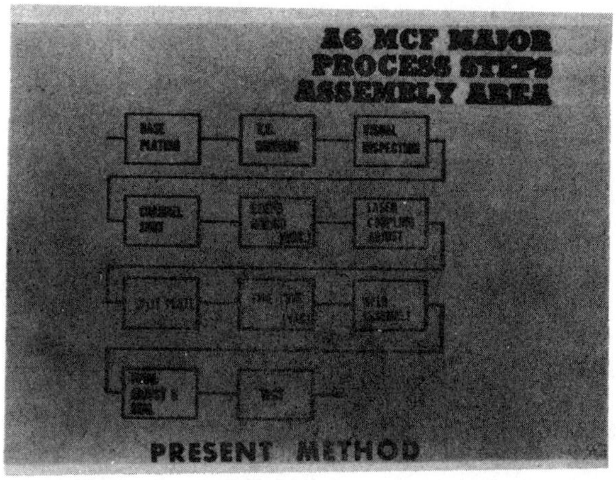

Figure 8 Major Process Steps - Assembly

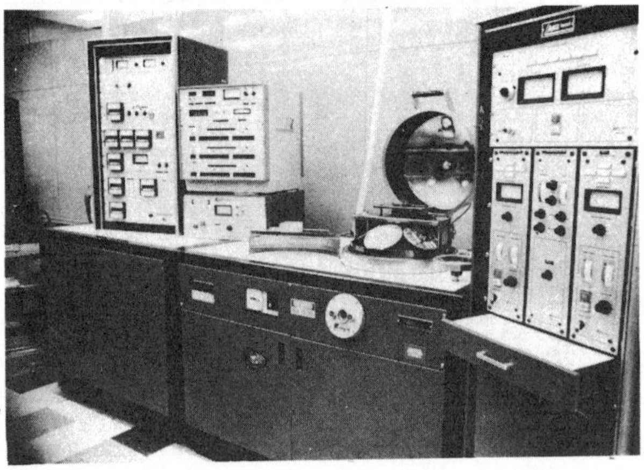

Figure 9 Base Plater

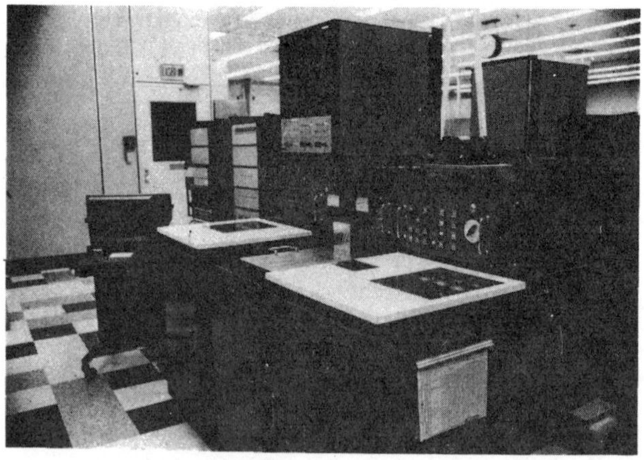

Figure 10 Frequency Plater

Figure 11 Laser

Figure 12 Adjust & Seal Facility

Figure 13 Plasma Cleaner

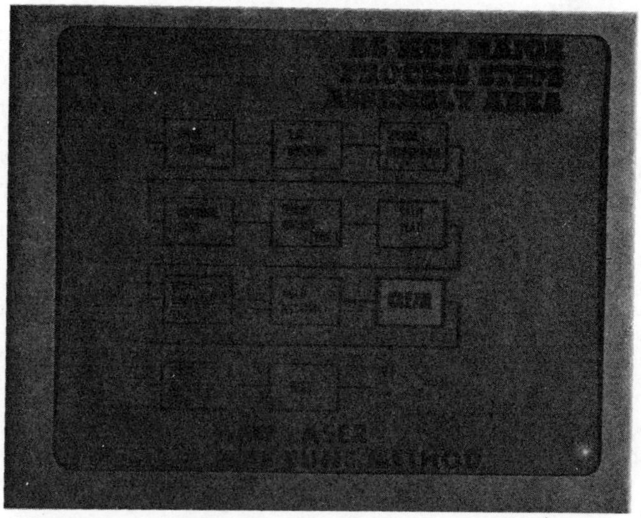

Figure 14 Process Steps - Laser Fine Tuning

PREPARATION OF QUARTZ CRYSTAL PLATES FOR MONOLITHIC CRYSTAL FILTERS

by

A. J. Miller
Bell Telephone Laboratories, Inc.
Allentown, Pennsylvania

Introduction

The previous paper[1] has shown the design and performance of Monolithic Channel Bank Filters (Fig. 1). In my presentation I will concentrate on important details pertinent to processing crystal plates for these filters.

It is apparent that the unusually strict requirements the filter has to meet will in turn place unusually strict requirements on the crystal plate. The principle requirements put on the crystal plate can be summarized as follows:

Of the eight resonators on one plate

A. The frequencies must be almost identical.
B. The frequency variations over the temperature range must be small and uniform.
C. The Q-values must be high and rather identical.
D. The Q-variations over the temperature range must be avoided or minimized.

Quite obviously these requirements can be met only if crystal plates are very parallel, the crystallographic orientation angle is optimum and deviations are kept to minimum, the major surfaces are smooth and uniform in texture, and the effects of unwanted flexure modes are adequately controlled.

Specific Requirements and Process Sequences

Fig. 2 shows the important dimensions of the AT-cut plate and specific requirements. The length along ZZ' is determined by the number of resonators, and the dimensions of electrodes and spacings. For economical reasons the width is kept as small as practical, guided by basic rules to avoid couplings to flexure type modes[2].

The nominal frequency is at 8.14 MHz. The tolerance of the AT-cut orientation is ± 1 minute of arc. Note at this point that the English measurement system is preferred throughout this paper simply because a frequency change of 1 kHz at 8 MHz is identical to a thickness change of 1 μ inch. The frequency variations, or the error of parallelism over the entire plate must be less than 8 kHz or 8 μ inches. This is measured at three points as indicated in the figure.

Figure 3 shows two process sequences, and approximate thicknesses and frequencies at each respective process step. Mainly because of difficulties in preserving the AT-cut orientation within the acoustically active area (roughly the area provided for electrodes) the existing standard process sequence was not satisfactory. In this sequence the thickness was reduced entirely by double face lapping, that is from both sides at the same time. This has led to the modified process sequence. The main difference is that the thickness is now reduced almost entirely by single face grinding, that is from one side only, while crystal plates are mounted firmly onto a backing plate minimizing any angle variation.

By changing from double face lapping to single face grinding we avoid, or reduce, angle variations, but, unfortunately, we pay with parallelism. This means that with single face grinding alone, the yield in obtaining very parallel plates would be rather low. Therefore, we could not neglect double face lapping. In fact it is still the only economical technique capable of producing very parallel crystal plates. Two steps of double face lapping are still needed. The basic step before single face grinding removes irregularities in thickness and surface damage from the cutting step. A minimum of .004 inches was found to be sufficient for that. The final step is to equalize errors in parallelism from the single face grinding, and, of course, to be able to apply radio receiver control during lapping.

For this final lapping step the thickness reduction of ≈.001 inch may be considered a compromise because it is a minimum to equalize errors in parallelism from single face grinding, and a maximum to avoid AT-cut angle variations. With this modified process sequence under control, the only X-ray angle test necessary comes after the basic double face lapping step with the reading made at the center of the plate only.

Orienting-Cutting-Trimming

The R-face cultured quartz (See Fig. 4) recently developed by Western Electric Co.[3] has proven to be satisfactory in both economy and quality. Filters made of natural quartz and R-face quartz were fully identical, within the limits of the present experiment. These two drawings show

Reprinted with permission from *Proc. Twenty-Fourth Ann. Symp. on Frequency Control*, Apr. 27-29, 1970, pp. 93-103, sponsored by the U.S. Army Electronics Command, Ft. Monmouth, NJ.

briefly the principles in orienting, cutting and trimming. Cutting the oriented blocks into crystal plates is done be a multiblade gangsaw. Because abrasives do the actual cutting, the crystal plates have little surface damage. The present yield in meeting the required tolerance of ± 1 minute of arc about the nominal AT-cut angle is greater than 80%. Trimming crystal plates to the final length and width dimension is done after basic double face lapping because the subsequent process steps produce no significant edge damage.

Single Face Grinding

Single face grinding is subdivided into cementing, pressing, calibrating, diamond and abrasive grinding (See Fig. 5). Up to 80 plates may be cemented onto one backing plate. The backing plate is serrated to allow excessive cement and unwanted particles to flow off. Figure 6 shows an alignment fixture placed over a backing plate. This fixture simplifies and speeds up proper locating of crystal plates during cementing. After cementing, the alignment fixture is removed and crystal plates are pressed onto the backing plate during cooling. The minimum pressure per crystal plate is approximately two pounds. Six boron carbide tipped pins screwed into the backing plate have to be adjusted to extend above the top surface so that the grinding action is stopped whenever the thickness of the crystal plates during grinding reaches this preset height. The height is determined by the frequency requirements after single face grinding. Despite the hardness of boron carbide, naturally the grinding action is not stopped at this point, but rather retarded. The retardation, however, is sufficient to allow good thickness control.

The results of one grinding load after single face grinding are shown on Fig. 7. This load consists of 72 crystal plates. The numbers with a plus sign represent frequency differences in kHz from plate to plate as measured at the center of each plate. Differences are recorded in reference to the thickest plate, the one with the lowest frequency which is here 7426 kHz. It can be seen that only a slight wedge, amounting to about 40 kHz is apparent. These results are plotted in the histogram at the lower left hand corner. Note, that the requirement at this point is ± 200 kHz about 7.3 MHz. The other numbers represent the error of parallelism in kHz as measured at three points along the length of each crystal plate. This error may be attributed to uneveness of the cement layer between crystal plate and backing plate. These results are statistically shown in the right hand corner. The requirement here calls for 80 kHz maximum.

Parallelism and Double Face Lapping

The basic requirement that the frequencies of the eight resonators be nearly identical requires that the major surfaces of the plate be very parallel. This is achieved by double face lapping.

Planetary lapping machines can be divided into those having two and those having four motions (See Fig. 8). The two motions are, of course, the rotation of each carrier around its own axis and the counterclockwise revolution of all carriers between two ring-shaped laps. The two additional motions are the clockwise rotation of the upper lap and the counterclockwise rotation of the lower lap. Both types of machines are capable of producing the required parallelism provided the two laps remain very flat and the crystal plates are transposed frequently.

After experimenting with several machines, we chose a four motion machine for our work because (1) the laps maintain their flatness for a longer time, (2) more crystal plates could be loaded into each blue steel carrier (the layout of such a carrier is shown in the lower right figure), (3) there was less breakage and edge damage, and (4) there were significantly less orientation angle changes during lapping. The main reason for these advantages is obviously based to a large extent on the opposing lap rotation which reduces or even equalizes unwanted forces applied to crystal plates and carriers, forces which are in turn applied to the laps. Because of their uniform particle size distribution micrograded aluminum oxide abrasives were found to add to the high degree of parallelism achieved. The histogram in the lower left corner of the figures shows characteristic results of this process. From several machine loads (about 200 plates) over 90% meet the 8 kHz or 8 μ inch parallelism requirement.

AT-Cut Angle Variations

The effect of AT-cut angle variations on frequency-temperature characteristic of filters is shown on results of five crystal plates (See Fig. 9). The frequency of each one of the eight resonators on each plate was individually measured over the temperature range. Subsequently the AT-cut angle was measured at each one of the areas covered by a pair of electrodes. It can be seen that the frequency deviations agree quite well with what one would expect the AT-cut angle to be for such a respective deviation. The parallelism on these plates is better than 6 kHz, and only 2 kHz on plate "E". This plate shows the largest angle deviations of all and represents a worst case condition. In spite of almost perfect parallelism, the AT-cut angle varies 7 minutes of arc from one plate end to the other. On the other side plate "A" may be considered a characteristic example of AT-cut variations obtained from the modified process sequence. The angle varies only 22 seconds of arc. Plate "D" is shown because the angles and frequency slopes measured at areas occupied by the four resonators on one half of the plate were identical, indicating no angle variations. The angles and slopes on the other hand, however, varied with progressively increasing deviation toward the end of the plate. "B" and "C" are of interest because the orientation angles of the resonators do not vary uniformly along the length of the plate but are random. Such variations are generated by deformations of the crystal plate during thickness reduction from double face lapping . A discussion of models which explain why such variation occur is too lengthy a matter to be covered at this time. Suffice to say, the new modified process sequency greatly reduces or even eliminates such variation, as illustrated by the results of plate "A".

The Effect of Mass Loading on the Frequency-Temperature Characteristic

The effect of mass loading (or plate back as it sometimes is called) on the frequency temperature characteristic was experimently evaluated on resonators provided for these filters. The base-electrodes used consist of a combination of metals. A thin chromium film is followed by a layer of copper and a layer of gold. These metals are plated by evaporation. The mass of the base-electrodes was such as to lower the frequency about 1.5% below the frequency of the unplated plate. After evaporation the mass was increased in increments by electroplating nickel or evaporating gold, until a maximum frequency lowering of about 9% had been achieved. After each increment the frequency deviation between -50°C and +110°C was measured. The results for one crystal plate are shown in Fig. 10. The measurement at "zero" mass loading was made while the unplated crystal plate was sandwiched between two metal plates which had the same dimensions as the electrodes subsequently applied by evaporation. This family of frequency-temperature curves is the same as one would obtain by changing the crystal plate's orientation angle in increments up to a maximum of about 12 minutes of arc; keeping the mass loading constant, of course.

The frequency deviations between the lower and upper turnover-points (i.e., points where the tangent has zero slope) have been plotted as a function of mass loading, and are shown in Fig. 11. Six different resonators on six different crystal plates have been studied. On three of these resonators the mass was increased by nickel electroplating only. These results are drawn in solid lines. On the other three resonators the mass was increased entirely by evaporating gold. These results are shown in broken lines. The readings taken between the highest and lowest values have been omitted unintentionally. They do however fall on, or very near to the broken lines shown. Although the frequency deviations show some irregularities as mass increases, especially in the case of nickel electroplating, the relation of mass loading versus frequency-temperature characteristic can clearly be seen. Irregularities are caused probably by slight differences in the mass distribution over the plated area, and by differences in the mass balance between two electrodes of one resonator. No special attention was given to distribution and balancing of mass during this study. In spite of these disturbances, a coarse but useful rule can be established. It can be said that each one percent of mass loading causes the same change in the frequency-temperature characteristic of an AT-cut quartz-resonator as that resulting from a change in the crystallographic orientation angle of about 1-1/3 minutes of arc.

It is interesting to note that the Q-values of nickel electroplated resonators with heavy mass loads are very low and erratic, while the Q-values of resonators with heavy evaporated loads (8% for example) are almost identical to those of moderate mass loads.

Etching

Because of high requirements on Q and aging, much attention was paid to the condition of the two major surfaces. The amount of etching proved to be an important parameter. The minimum amount could not be determined by mechanical surface tests. For example, no differences in surface texture were revealed by stylus type measurements. Using a Talysurf instrument the stylus recordings of lapped, unetched and etched surfaces at a variety of etching stages were rather identical. The minimum etch was finally determined by microscopic tests. Samples were etched sequentially in intervals to increase the frequency a maximum of 1000 kHz. After each etching interval, the surface topology was observed with a scanning electron microscope.[4] Some of these surface samples at several etching stages are shown in Figs. 12, 13, and 14. The viewing angle is 45°, and we look along the X -axis, where minus X is toward the top of the pictures. Magnifications at 2000 and 5000 appeared best to uncover important details. For realistic size comparison the white dust particle at zero etching has about the size of a 3 μm abrasive grain, the abrasive which has been used to lap these surfaces. If one takes complete removal of the layer formed by loose or disturbed matter as measure, he will observe that little has changed after mass was etched away equivalent to 20 kHz. 10 kHz, not included here, indicated hardly any change. Surfaces at 30 kHz (and even at 40 kHz) still show sections which may be unstable with time and hence may contribute to aging of filters. The first indication that the disturbed layer is removed -- that is when the surface really appears crystalline -- may be detected to some degree at 40 kHz, but certainly at 50 and 60 kHz. Thus the minimum etch was put at 50 kHz. At this point I would like to point out that the minimum amount of etching specified in MIL-C-3098 is ≈ 2.5 kHz at 8 MHz for plates lapped with 3.75 μm abrasive.

Because Q does not decrease with longer etching, a maximum amount of etching is not so important. For example, resonators etched about 150 kHz still were at the same Q level as resonators etched about 50 to 60 kHz. On the other hand, resonators etched amounts below 40 kHz, especially between 10 to 30 kHz showed nonuniform, sometimes erratic Q's, and because of that many filters did not perform as required. After the etching and position of electrodes were fully under control, generally no Q differences were observed between polished plates and those lapped with 3 μm, 5 μm or even slightly larger grain sized abrasives. It is strongly believed that by removing the layer of disturbed matter not only the Q but also the aging of filters was advantageously affected.

References

1. R. J. Byrne, Proceedings of the 24th Annual Symposium on Frequency Control, (1970).

2. P. C. Y Lee and W. J. Spencer, The Journal of the Acoustical Society of America, Vol. 45, No. 3, 637-645, March 1969.

3. N. C. Lias and D. W. Rudd, The Western Electric Engineer, Vol. XIII, No. 2, April 1969.

4. P. R. Thornton, Scanning Electron Microscopy, Chapman and Hall Ltd., London, 1968.

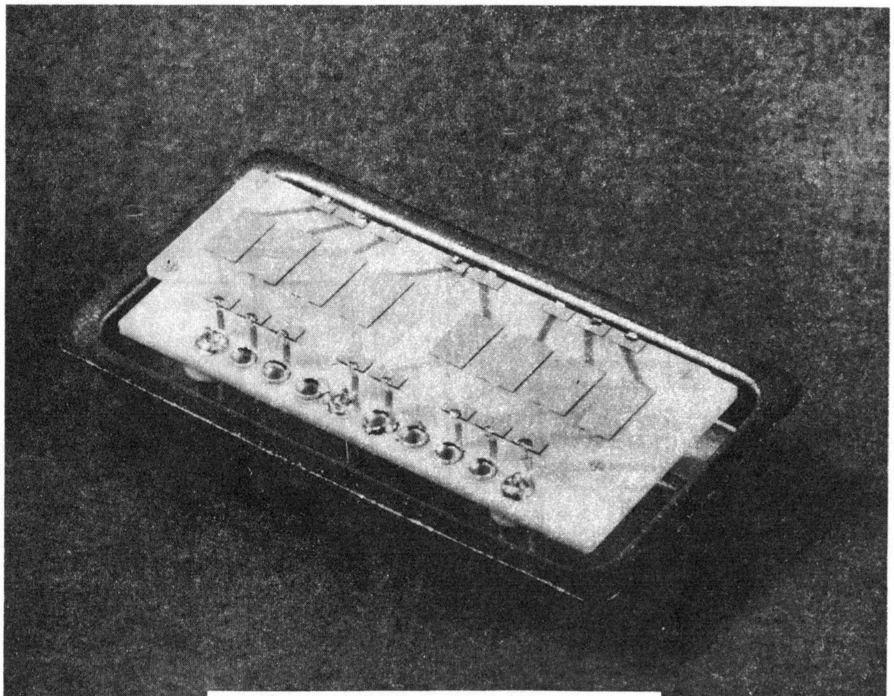

Figure 1 Monolithic Channel Bank Filter.

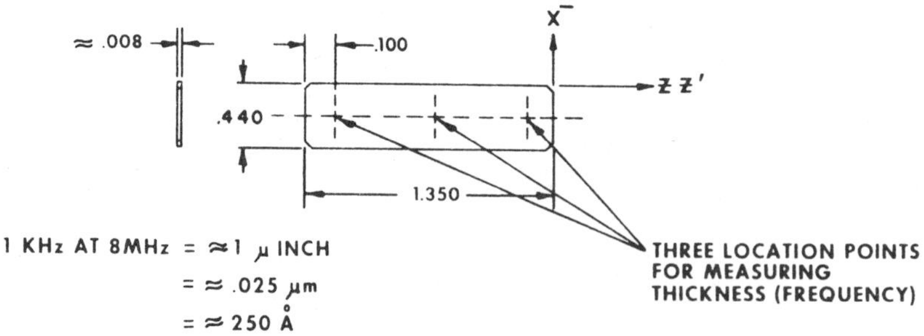

NOMINAL FREQUENCY : 8140 kHz
AT-CUT ORIENTATION : NOMINAL ± 1′
PARALLELISM : 8 kHz MAXIMUM
MEASURED AT THREE POINTS
AS INDICATED
MATERIAL : R-FACE CULTURED QUARTZ
SURFACE CONDITION : LAPPED AND ETCHED
END ABRASIVE : MICRO GRADED 3 μm
ALUMINUM OXIDE

Figure 2 Crystal Plate Dimensions and Specific Requirements.

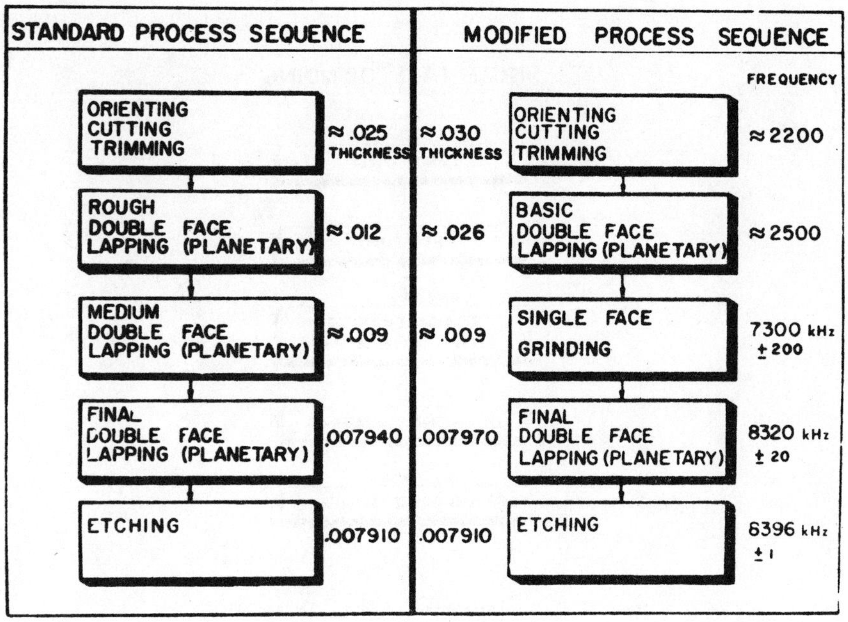

Figure 3 Process Sequences.

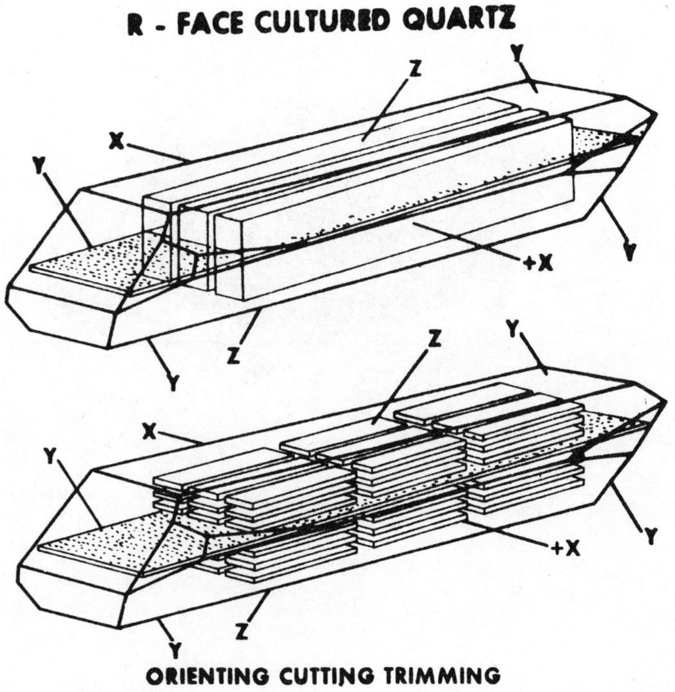

Figure 4 Orienting-Cutting-Trimming.

SINGLE FACE GRINDING

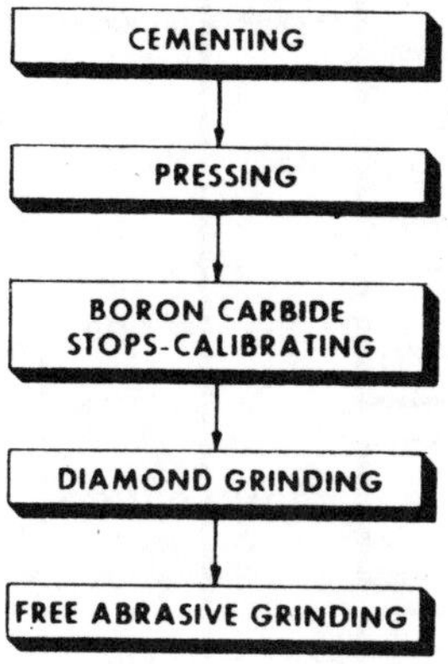

Figure 5 Single Face Grinding - Process Steps.

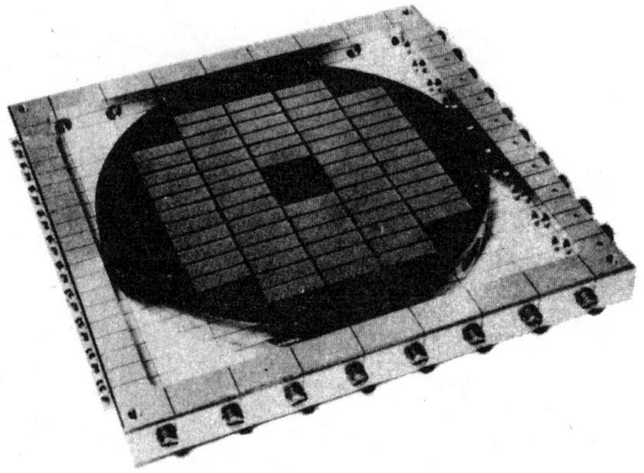

Figure 6 Single Face Grinding - Backing Plate and Alignment Fixture.

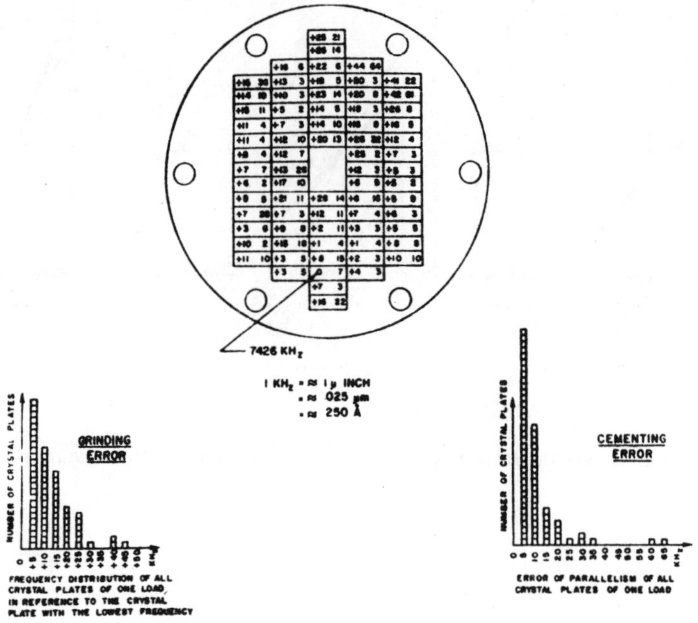

Figure 7 SINGLE FACE GRINDING (CHARACTERISTIC RESULTS)

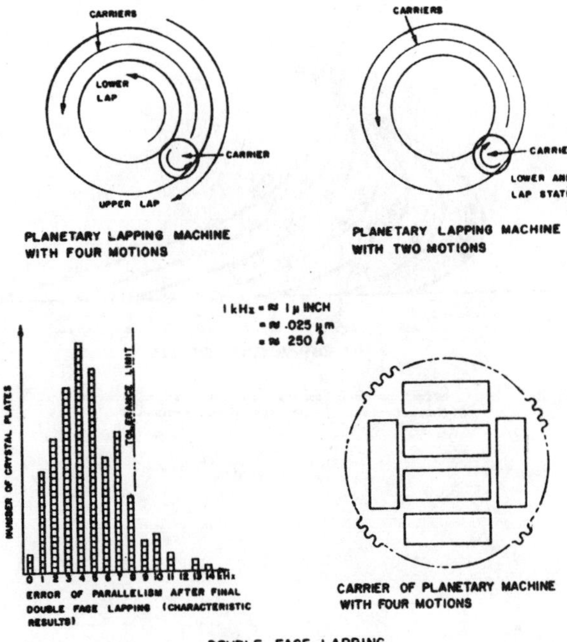

Figure 8 DOUBLE FACE LAPPING

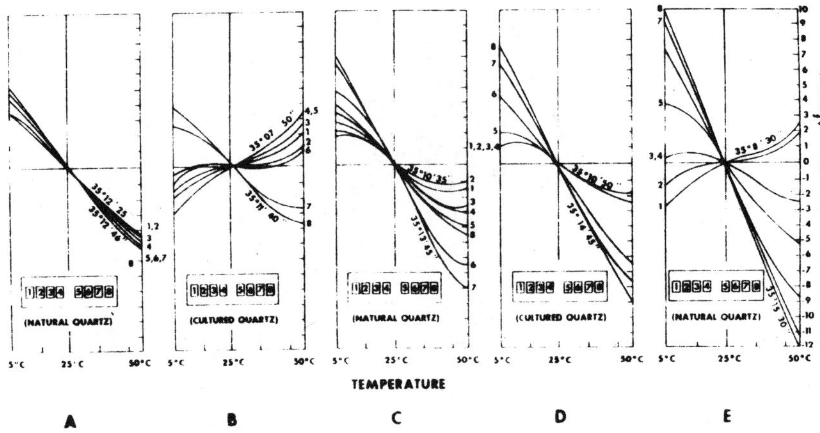

Figure 9

EFFECT OF AT-CUT ANGLE VARIATIONS ON FREQUENCY TEMPERATURE CHARACTERISTICS

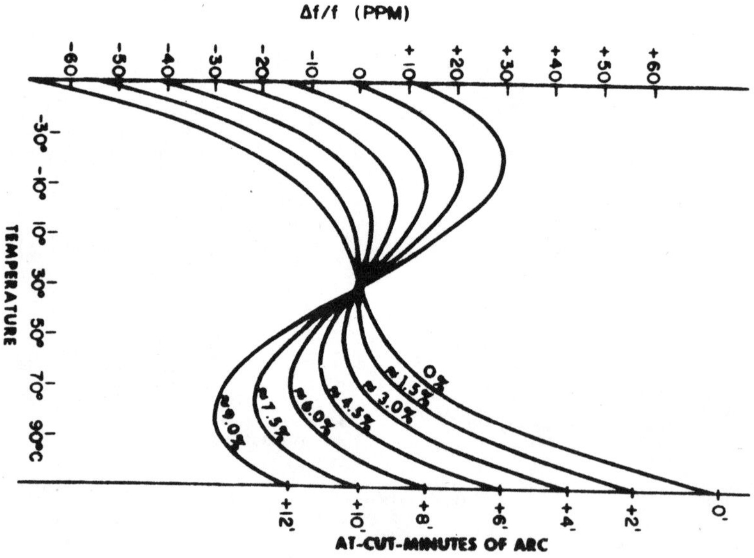

Figure 10 Effect of Mass Loading on Frequency Temperature Characteristics - Results of One Resonator.

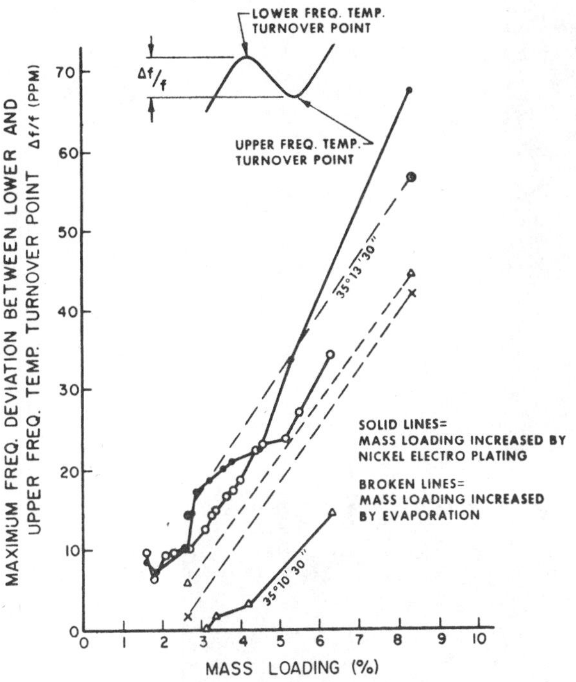

Figure 11 Effect of Mass Loading on Frequency Temperature Characteristics - Frequency Deviation Between Lower and Upper Turnover Point.

Figures 12a, 12b, 12c, 13a, 13b, 13c, 14a, 14b, 14c Etching Sequence - Surface Topology.

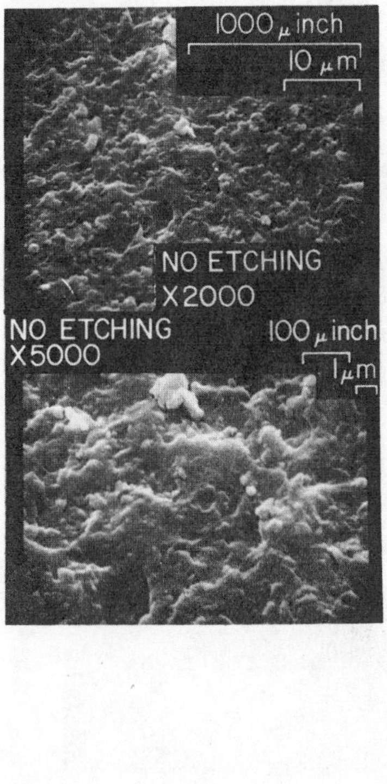

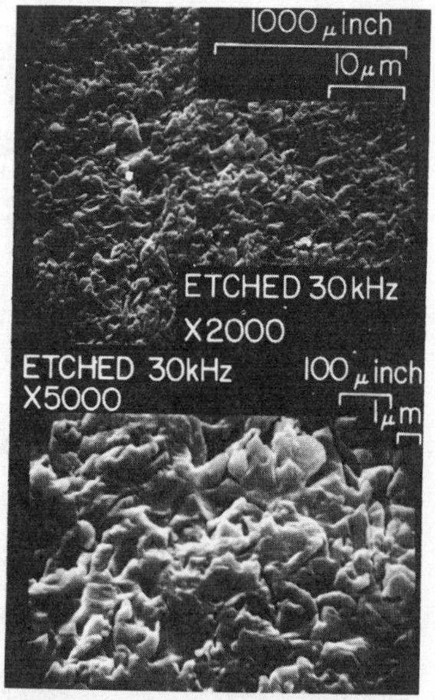

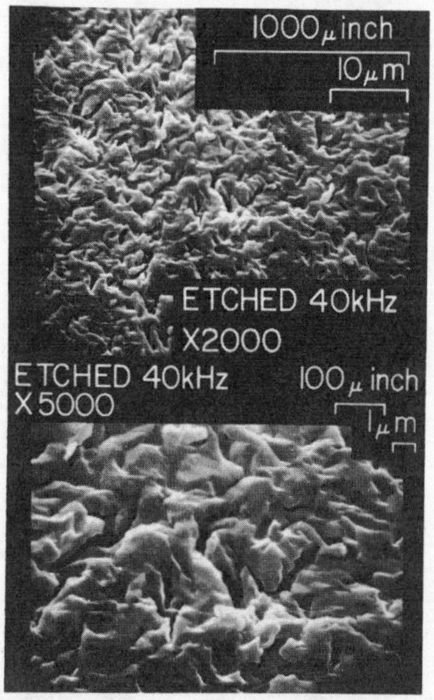

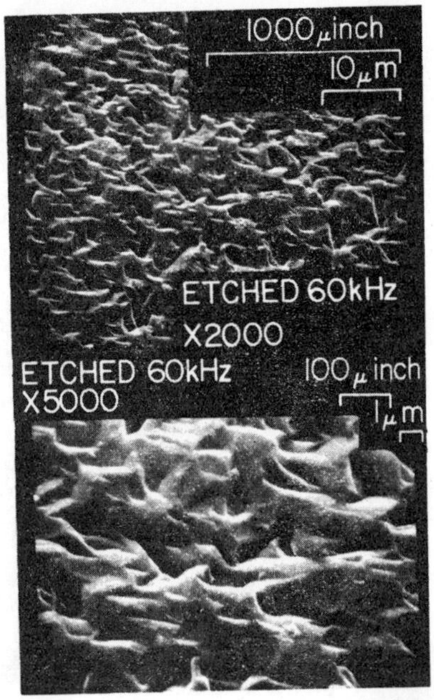
1000 μinch
10 μm
ETCHED 60kHz
X2000
ETCHED 60kHz
X5000
100 μinch
1 μm

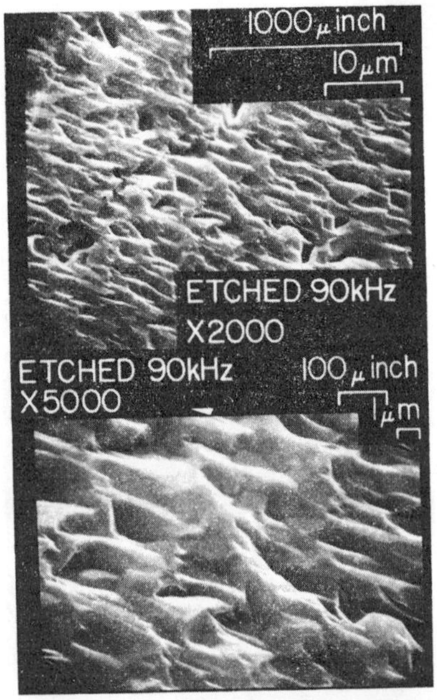
1000 μinch
10 μm
ETCHED 90kHz
X2000
ETCHED 90kHz
X5000
100 μinch
1 μm

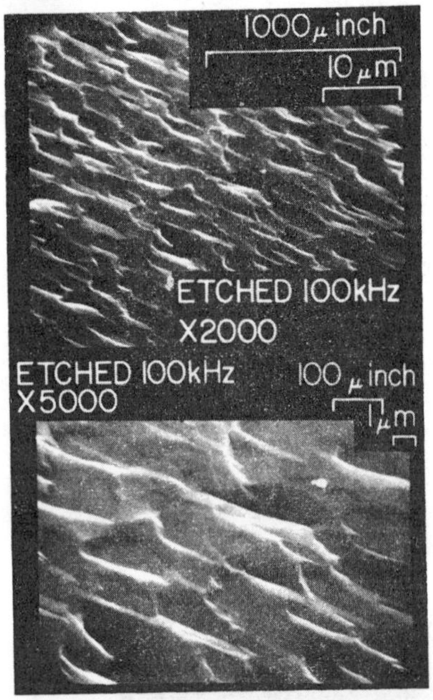
1000 μinch
10 μm
ETCHED 100kHz
X2000
ETCHED 100kHz
X5000
100 μinch
1 μm

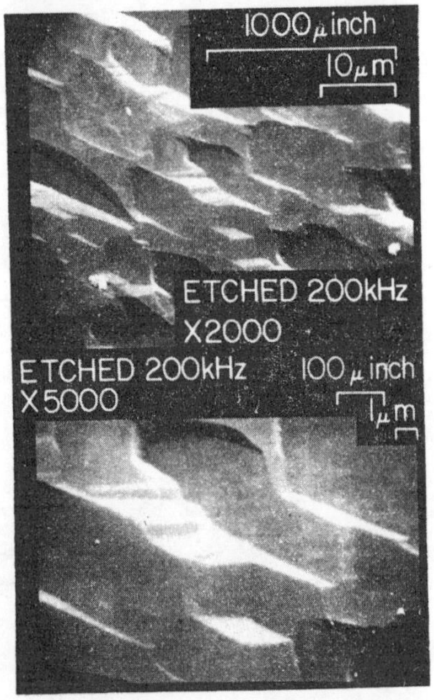
1000 μinch
10 μm
ETCHED 200kHz
X2000
ETCHED 200kHz
X5000
100 μinch
1 μm

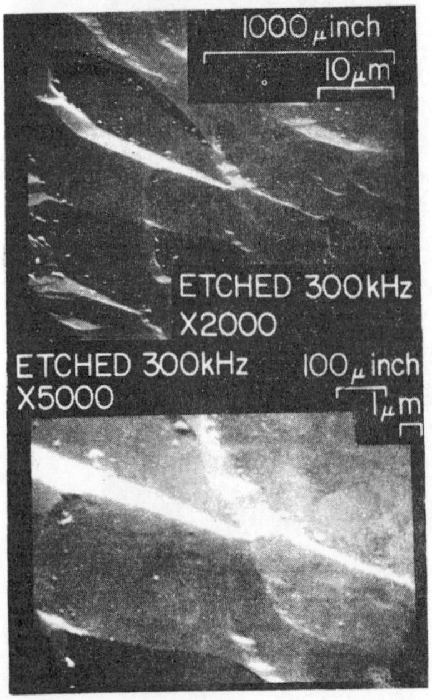
1000 μinch
10 μm
ETCHED 300kHz
X2000
ETCHED 300kHz
X5000
100 μinch
1 μm

POLYLITHIC CRYSTAL FILTERS

Desmond F. Sheahan
GTE Lenkurt Incorporated
San Carlos, California

Summary

This paper will describe the polylithic crystal filter process which uses acoustically coupled resonators on more than one piece of quartz to realize filters for use in frequency division multiplex systems. Several production schemes have been devised in the past few years to realize filters via the acoustical coupling technique, but this paper will concentrate on just one of them. It will give the rationale behind this particular approach and also discuss some of the problems and successes experienced with it as it evolved over the past five years.

Introduction

Our interest in acoustical coupling derived from the significance of Frequency Division Multiplex (FDM) systems to the overall operation of GTE Lenkurt. FDM systems are designed around the channel filters which are the most important part of the system. Traditionally such filters have been built with discrete inductors and capacitors, and they have been bulky and expensive. A research project was begun to find alternate filtering techniques, and at first this concentrated mainly on active filters. Technically suitable active channel filters were designed and built,[1] however when the cost of these filters was examined, and the fact that they consumed power was taken into consideration, it was found that they did not offer any economic advantage. There was therefore no incentive to proceed any further with this approach so it was consequently dropped. At this time we decided to investigate monolithic crystal filters[2] since favorable predictions were being made about the economics of this approach. Upon close examination however the monolithic approach did not prove to be very attractive to us for the following reasons:

1. A very large capital investment was required.
2. The capital investment would be essentially dedicated to one filter type.
3. The quartz blanks were not available from commercial sources and special processing techniques would have to be used to get the quality desired.[3]
4. The all pole 8 resonator monolithic filter design[4] did not have enough stopband attenuation to give the degree of suppression of the unwanted sideband required by our customers, and more than 8 resonators would have too much delay distortion to be equalized readily. A design with finite frequency attenuation poles was therefore required.

The Polylithic Approach

For these reasons it was decided to proceed with a polylithic approach and to design the process to have maximum flexibility to handle many different types of filter topologies.[5] This approach offered the technical advantages of a small stable passive filter without the disadvantages listed above. In order to minimize the capital investment required, it was decided to design the process around 1/2" diameter quartz wafers which are readily available from commercial sources. Another significant feature was to use just one can for enclosing the complete filter rather than have each crystal packaged separately. Because of the fragility of the crystals themselves, a ceramic ring was chosen as a holding fixture for the crystals. This ring serves as a protective device during the manufacturing operations. When the crystal is carefully aligned in the ring and attached to it, all subsequent critical alignment can be done with respect to the ring rather than against the more fragile crystal. Finally the same ring becomes part of the final filter assembly.

Figure 1 shows a photograph of five filters with different topologies that are presently in production, and Figure 2 shows their normalized frequency responses. The two of these filter types that are actually used as channel filters are no. 4 and no. 5, and they contain 4 and 5 rings respectively. Filter no. 4 has an all pole 8 resonator response similar to that obtained with the all monolithic filter.[4] Filter no. 5, on the other hand, has two coincident attenuation poles in the lower stopband and this is what provides the greater attenuation of the unwanted sideband. This extra margin of attenuation means that no special techniques will be required to accurately locate the lower stopband of the filter.[6] Since Filter no. 5 is the channel filter with by far the greatest usage, this is the filter that we mean when we subsequently refer to a channel filter. Figure 3 shows a circuit diagram of the channel filter which consists of 3 dual resonators, 2 single resonators and 2 capacitors.

Manufacturing Process

Figure 4 shows a flow chart of the major processing steps. The process starts with 1/2" diameter AT-cut quartz blanks that have been lapped to within 100 kHz of the desired frequency and are then etched and sorted. Since we have 3 major FDM systems and over 100 different filters at different frequencies in the 8 MHz range, the sorting job is not too difficult as there is potential usage for all of the blanks.

The sorting itself is performed in a precision air gap fixture that is connected to an oscillator. This fixture essentially measures on the 1/4" diameter central area of the 1/2" diameter blank and the sorting is performed to an accuracy of ±1 kHz at 8 MHz.

After the blanks have been sorted to frequency for the particular filter to be built, they are thoroughly cleaned before being loaded into the fixtures through which the electrode patterns will be evaporated. These fixtures are 6" diameter and they consist of chemically milled transmission masks that are mainly copper but which contain a thin lip of nickel to give sharp definition. Each fixture contains 45 crystals and the masks with crystals are held together between two stainless steel plates. Figure 5 shows a picture of these fixtures. These fixtures are loaded 5 at a time into resistively heated evaporators in which the electrode pattern is deposited. The metal system used is a 3-part one of chrome, nickel and gold. The bulk of the metal is gold and the chromium is the adhesion layer for attachment to the quartz whereas the nickel allows

Reprinted with permission from *Proc. Twenty-Ninth Ann. Symp. on Frequency Control*, May 28-30, 1975, pp. 120-127, sponsored by the U.S. Army Electronics Command, Ft. Monmouth, NJ.

tin lead solders to be used in the subsequent attachment process. The baseplating process must be performed slowly to prevent creating any nodules on the metalization as they can have a disastrous effect on the resonator Q values. The vacuum is broken, the fixture turned over and the process repeated in order to plate both sides of the crystal blanks. After baseplating, the plated crystal blanks are checked for frequency to see that they are in a range not greater than 8 kHz above the finally desired frequency. This range was set somewhat arbitrarily considering the spread of frequencies to be expected from a baseplater and the fact that it is desirable to maximize the amount of metal deposited in the baseplater simply because it deals with a large number of blanks simultaneously, whereas the final frequency plating process deals with the crystals one at a time. It must also be kept in mind that the greater the amount of metal that is to be deposited in the final frequency plating process, the better has to be the alignment of this frequency plated metal pattern with the already deposited baseplated pattern.

From the baseplater the crystals are attached to the rings. These are alumina rings 0.75" OD and 0.55" ID. They are used in the as-fired condition so no grinding or lapping is required. The rings are plated with a metal pattern onto which the crystals will be attached, and this will also act as the interconnection points for the filter. The metalization used is a 3-part one of chromium, copper and gold. The chromium is the adhesive layer and the bulk of the metal is copper with a thin protective layer of gold. These rings are plated in a specially designed evaporator and the pattern is formed by a chemically milled metal mask that fits snugly over the ring. The rings are assembled on rods in stacks of 20 high, and the fixturing in the evaporator allows each vertically stacked rod to rotate on its own axis while simultaneously rotating about the 18" diameter of the evaporator. In this manner an evaporator can handle a load of approximately 1000 rings and also get even plating of metal both on the top surface of the rings and all the way around its periphery. A chemically milled lead frame is used to attach the crystal to the ring. This lead frame is chemically milled out of copper and it is supplied in strips. It has a small amount of tin lead solder selectively plated at the points of attachment to the crystal. The procedure is to load the chemically milled strips into a fixture on top of a number of rings. This fixture is then moved under a parallel gap welder which welds the lead frames to the rings with three points of attachment for each ring. Finally the strips with the rings attached are loaded into a punch which separates the rings with the desired portion of the leadframe from the undesired portion.

The next step in the process is to add multilayer NPO ceramic chip capacitors to those rings that will require them in the final filter topology. The capacitors are placed on top of the rings and soldered to the metal pads that were previously formed in the ring evaporator.

Now the crystal can be attached to the ring leadframe combination, and this is done by aligning the crystal and the ring in a fixture and reflowing the small amount of tin lead solder at the tip of the leadframe. This fixturing is designed so that there is a fixed relationship between the center of the crystal and one edge of the ring. Consequently, when we need accurate location of the center of the crystal in the subsequent frequency plating process, we can do so with respect to the edge of the ring and we do not have to contact the more delicate crystal. Figure 6 shows a photograph of the crystal soldered into the ring assembly which we call a resonator assembly. Note the point of attachment of the leadframe to the ring. The 90° bend in the leadframe minimizes the effect of the different coefficients of thermal expansions of the materials, and thus it prevents stressing of the quartz. It was found by experience that this stressing could cause abnormal frequency shifts with temperature and, as expected, these were most severe if the coupling were along the Z axis of the crystal rather than along its X axis. As a result of this, all of our filter designs now use X axis coupling.

The ring crystal assemblies are cleaned thoroughly before being plated to frequency. This is done in the device shown in Figure 7. It consists of two pairs of small chambers connected to a turbomolecular vacuum pump. Two resonators are being loaded into one pair of chambers while the other pair is going through the automatic plating cycle. The use of small chambers in conjunction with a turbomolecular pump minimizes the vacuum pump down cycle because no roughing cycle is used, and plating can commence within 15 seconds after opening the chambers to the pump. Each chamber contains a filament for the gold to be evaporated, a shutter, and a sliding fixture which allows either one of the two resonators on a crystal to be plated. There is also a measuring circuit which is connected to a vector voltmeter, and a programmable frequency synthesizer provides the desired frequency. A hard-wired controller takes the equipment through the processing steps required, and a complete cycle of plating two resonator assemblies to an accuracy of ±20 Hz over a maximum frequency change of 8 kHz takes less than two minutes. The operator has to watch the plating rates in each chamber and keep the filaments supplied with gold, but otherwise the operation is completely automatic. Each crystal ring assembly is next tested and sorted for frequency, Q, and coupling. This is done in an automatic fixture that contains a network analyzer and a frequency synthesizer under the control of a programmable calculator. Note that this is the only measurement that is made of coupling and no adjustment of coupling is performed. Reliance on achieving the desired coupling therefore depends primarily on the accuracy of the blank sorting and on the accuracy of the coupling gap as determined by the baseplater mask. For some of the filters produced, the coupling is not even measured as the general process repeatability is sufficient to give the tolerance required. For the channel filter however, coupling tolerance is important and it will be discussed later.

The sorted resonator assemblies are now ready to be assembled into filters. This is done by stacking them on a header which contains two posts that fit into indents on the rings. A chemically milled piece of gold plated copper is placed between each ring. These pieces of copper provide electrical shielding as they are electrically connected to the grounded posts. They also contain tabs that are connected to the metal patterns on the rings, thereby providing ground points on the filter wherever the particular filter topology requires it. The interconnections between the rings are made via gold plated copper straps that are welded to the rings. Since all of the other connections are made via welding, and since each of the metal shields between the rings is separately welded to the grounding posts, the result is a very rugged and compact filter package.

After the filter has been assembled and welded, all of the weld joints are tested, and the filter is electrically tested for frequency response. This test is done while the filter is open to atmospheric pressure so consequently the resonator Q values will not be as good as when the filter is finally evacuated and sealed. This test however enables any major defect to be spotted and if a resonator assembly needs to be replaced this can be done very easily. After the filter has passed this preliminary test a can is placed over it, a vacuum is drawn and the filter is sealed by cold welding. This completes the manufacturing process.

Before the final electrical test is performed on the filter by a calculator controlled network analyzer and synthesizer combination, the filters are subjected to 12 hours of rapid temperature cycling between $-40^{o}C$ and $+75^{o}C$. This is designed to weed out any potential problems due to weak bonds, seals, etc.

Production Experience

The production experience with this process over the past 4 years and almost one million filters have borne out the fundamental decisions made when the process was being designed. The principal one being the decision not to adjust the inter-resonator coupling but instead to rely on tight control of both blank frequency and the mask dimensions to achieve the coupling tolerance desired. A further advantage was achieved due to the fact that with a polylithic technology, the couplings in a filter alternate between electrical coupling due to capacitors and mechanical coupling due to the so-called acoustical coupling in the quartz. Since the capacitors are of the multilayer ceramic chip variety, they can be trimmed very easily to tolerances of 1% via a sandblasting technique. This means that trade-offs can be made in the filter between tight tolerances on the electrical coupling and looser tolerances on the mechanical coupling. Figure 8 shows the statistical distribution curve of couplings achieved in a typical month's production of filters. The standard deviation of the process was 2.1% of the mean coupling. The actual production tolerance used when sorting resonators for coupling is ±3%. However due to the symmetry of the channel filter design, the 3% tolerance can be exceeded if a high coupling resonator is paired off with a low coupling one. The sorting tolerance for this is ±5%. The ±3% tolerance units can be used in random combinations and we can see from the cumulative normal distribution that this accounts for 84% of the units. If we match a resonator that has a tolerance in the range of +3 to +5% with one in the range of -3 to -5%, the cumulative distribution tells us that this takes care of 98% of the output. Maintaining this distribution curve shown does however require constant vigilance on the part of the production team as it can easily start to drift. The alternative technique[7] of achieving satisfactory coupling tolerances that has been reported for the monolithic crystal filter requires two additional processing steps in the form of an initial frequency adjustment as well as the coupling adjustment itself. Both of these processing steps will require extra handling of the crystals as well as extra processing equipment to perform the operations and it would therefore be unreasonable to expect 100% yield due to this method. Therefore, the fact that some crystals cannot be used due to their being outside acceptable tolerance limits is to us an acceptable alternate process. As mentioned already the principal reliance on achieving the desired coupling tolerances falls on the blank sorting fixtures and on the baseplating masks. A continuous monitoring program therefore had to be set up to watch these two areas and a regular replacement program was established for the masks as the coupling will change when there is excessive metal build-up on them.

The method of assembly just described involves manufacturing each part to a particular tolerance, and it does not allow for any adjustments to be made on the assembled filter. This is in contrast to most of the usual practice with telephone channel filters where with widely different technologies some adjustment is usually performed while observing the filter frequency response.[6] It is useful therefore to consider Figure 9 which shows the statistics of the filter passband for several months of polylithic channel filter production. These statistics have been collected for only the most sensitive part of the passband which is its lower frequency edge. Figure 9 shows the measured passband response of a typical channel filter, also the test limits and the one standard deviation ranges for the lower corner frequencies. The final test of the filter checks the filter stopband as well as its passband and the yield at this test is generally greater than 80%. Of the filters that do fail the test, their components are recycled through the process.

Filter Performance

Figure 10 shows the typical temperature behavior of a channel filter's passband. It shows the excellent stability of the stopband with some minor ripples appearing in the passband. The polylithic process was found to have an unexpected bonus when we consider temperature performance. Each blank in a channel filter has a nominal tolerance of ±1′ but we obtain an averaging effect when we put five of them together in a filter. Therefore the filter has a more stable stopband than would be expected from this tolerance simply because we get random cancellations of temperature drifts. Any drifts that do occur will cause minor ripples in the passband. The mean frequency shift obtained over a temperature range of $+25^{o}C$ to $+60^{o}C$ with a large batch of channel filters was 4.63 Hz with a standard deviation of 1.16 Hz. This is considerably better than had been expected with an AT angle tolerance of ±1′.

The aging achieved with the filters has been quite satisfactory and Figure 11 shows the results obtained with a test group of channel filters that were stored at $+75^{o}C$ for a period of 1 year. If we extrapolate the data in the figure, we see that even with the high temperature accelerated aging environment, the expected drift over 20 years would be an acceptable 12 Hz. More significant however is the fact that after 4 years of production and almost one million filters produced, no significant aging problems have been reported even though the polylithic crystal filter FDM systems are installed around the world and sometimes in hostile environments.

Problem Areas

While the process that has just been described is basically a straightforward one it is nevertheless very sensitive since there are a large number of operations that can go wrong. There are many multi-process steps that do not permit detailed measurement after each step, but one has to wait until the sequence is finished before a measurement can be made. This requires an alert process team to make it operate successfully. Some of the major problem areas are common to the crystal industry in general and they are listed here in order of importance.

1. Breakage of crystals
2. Low Q values
3. Frequency errors
4. Coupling errors
5. Interconnection problems
6. Lack of cleanliness
7. Variation of response with signal level.

In addition to the problem areas listed above, there are the problems with processing equipment. It seems that with complex electro-mechanical processing equipment there is an optimum point beyond which it should not be automated. If this point is exceeded the high probability of an equipment failure due to the added complexity will outweigh any anticipated increase in output due to that complexity. On some occasions we did exceed that limit and we had to remove some of the automatic features to obtain a useful system.

Filter Applications

The technology described is being used in three different FDM systems that have different modulation schemes and the major distinguishing features of these systems are as follows.

1. Single channel per card and all channel filters at 8.14 MHz.[8]
2. 12 Channel filters in 4 kHz increments covering the frequency range of 8.140 to 8.188 MHz.[9] The 12 channels are combined into a group and the entire group is down converted in one step.
3. Single channel filter at 8.140 MHz. An entire supergroup is formed by down converting each channel separately.[10] This system has out-of-band signaling and is CCITT compatible.

The flexibility of system design that is illustrated by the systems listed above is due primarily to performing the channel filtering at 8 MHz with quartz filters. However this wide variety of systems means that a large number of filters with different frequencies and different topologies will be required and it is a feature of the polylithic process that these can all be done on the same processing equipment as shown by the examples in Figures 1 and 2.

Further Developments

So far this paper has only discussed what has been achieved to date with the polylithic technology. Some development work has also been pursued to develop more complex filter topologies. In particular, the technique of capacitor bridging while it is in use on some of the signaling and carrier selection filters has not yet been introduced on a channel filter design. It does look very promising for the channel filter application however and a number of experimental units have been built.[11] Figure 12 shows the frequency response of such a channel filter design. It can be seen that this filter gives more stopband attenuation close in to the passband where it is required and, since it has finite frequency attenuation poles, this extra attenuation is achieved without the penalty of greatly increased delay distortion. The bridging is implemented very easily in the polylithic technology by adding chip capacitors to the spaces provided on the rings. The initial computer studies on this bridged resonator channel filter design show that it is quite promising from a point of view of sensitivity.

The polylithic process described results in a channel filter that is 1" diameter and 0.75" high. The height however is mainly determined by the height of the rings that are used and the ones we use are 0.125" high. Smaller height filters can of course be built just by reducing the height of the rings, and in fact some channel filters have been built this way in cans that are only 0.5" high. To date however there is no system demand for this lower profile filter, but the technology can handle it with very little change whenever it is required.

Conclusions

To sum up, therefore, the polylithic crystal filter process just described has now had 4 years of production experience and almost one million filters have been produced. Like any other new technology, it has had its teething problems but these have been overcome, and the basic philosophy of the process has proven to be a sound one.

References

1) H. J. Orchard & D. F. Sheahan, "Inductorless Bandpass Filters," IEEE Journal of Solid-State Circuits, Vol. SC-5, No. 3, June 1970, pp.108-118.

2) R. A. Sykes, W. C. Smith & W. J. Spencer, "Monolithic Crystal Filters," IEEE International Convention Record, Vol. 15, Part II, 1967, pp. 78-93.

3) A. J. Miller, "Preparation of Quartz Crystal Plates for Monolithic Crystal Filters," Proceedings of the 24th Annual Symposium on Frequency Control, April 1970, pp. 93-103.

4) P. Lloyd, "Monolithic Crystal Filters for Frequency Division Multiplex," Proceedings of the 25th Annual Symposium on Frequency Control, April 1971, pp. 280-286.

5) D. F. Sheahan & C. E. Schmidt, "Coupled Resonator Quartz Crystal Filters," WESCON, Session 8, August 1971.

6) R. P. Grenier, "A Technique for Automatic Monolithic Crystal Filter Frequency Adjustment," Proceedings of the 24th Annual Symposium on Frequency Control, April 1970, pp. 104-110.

7) R. J. Byrne, "Monolithic Crystal Filters," Proceedings of the 24th Annual Symposium on Frequency Control, April 1970, pp. 84-92.

8) J. A. Stewart, "System Considerations for Light Route Multiplex Using Crystal Filters," IEEE International Conference on Communications, Montreal, June 1971, pp. 6.25-6.29.

9) S. W. Anderes, "A Polylithic Filter Channel Bank," IEEE International Conference on Communications, Montreal, June 1971, pp. 6.1-6.6.

10) A. Reading and R. J. Martin, "A New Modulation Concept in Channel Translating Equipment," IEEE International Conference on Communications, Montreal, June 1971, pp. 6.13-6.18.

11) M. S. Lee, "Polylithic Crystal Filters with Attenuation Peaks at Finite Frequencies," IEEE International Symposium on Circuits and Systems, Special Session S1, Newton, Massachusetts, April 1975.

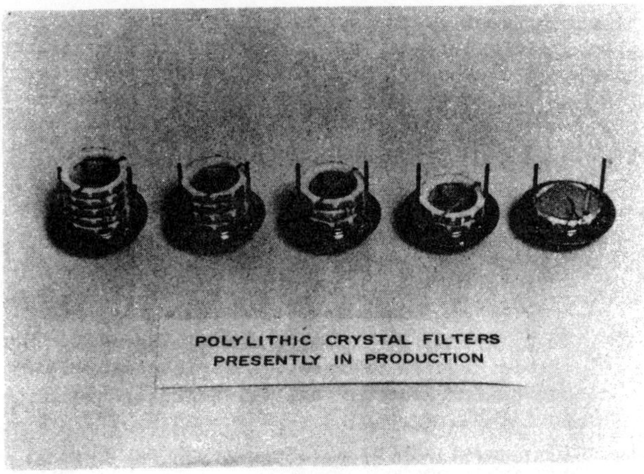

Fig 1 5 different types of polylithic crystal filters presently in production. The polylithic technology can cater to many different filter topologies without any change in the production equipment.

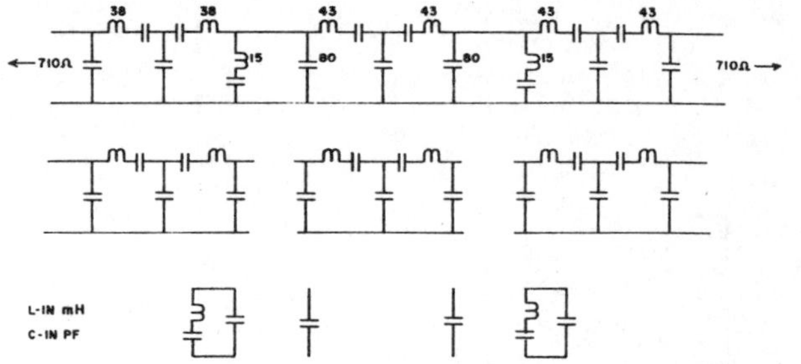

Fig 3 Circuit diagram of the 5 crystal channel filter showing its components. Response No. 5 in Fig.2.

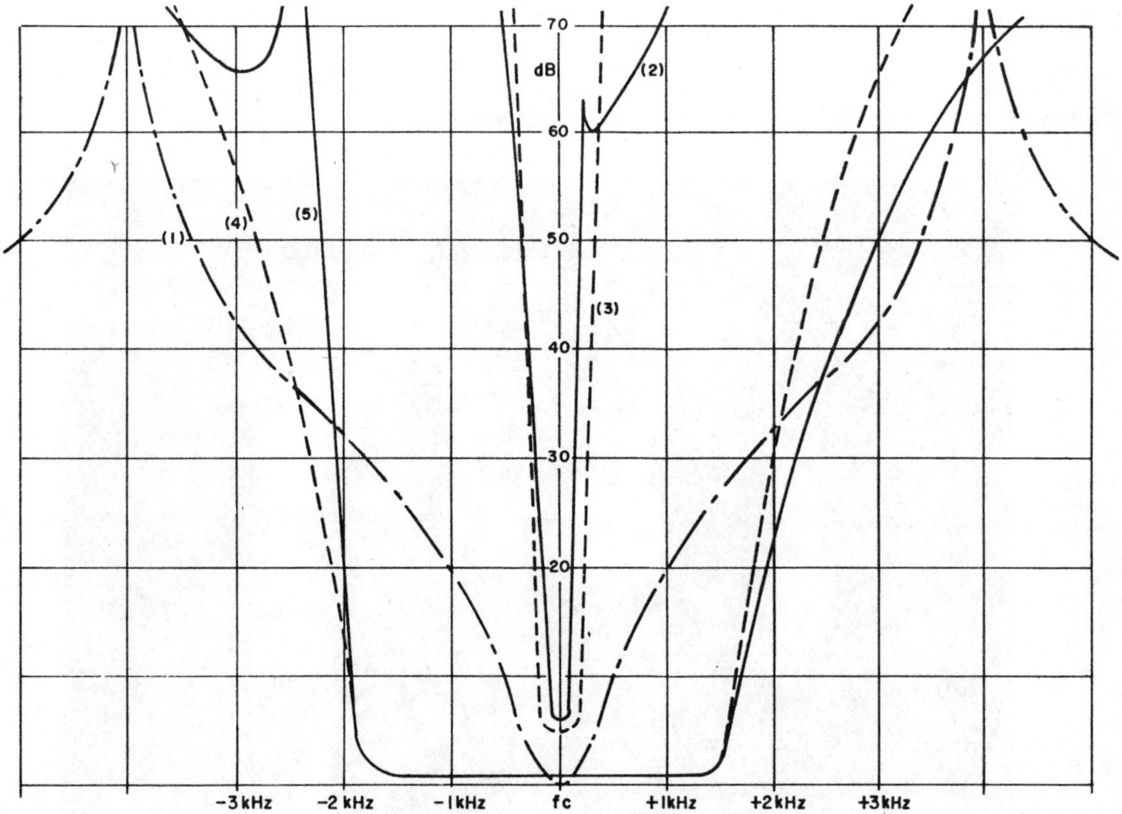

Fig 2 The normalized frequency responses of the 5 filters shown in Figure 1. The number of wafers per filter is shoqn in parentheses. These are also the filter numbers.

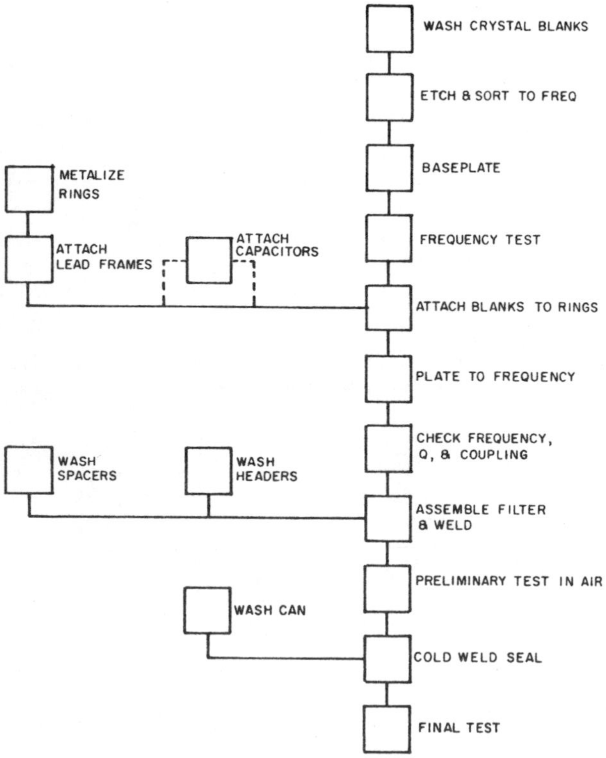

Fig 4 Flow chart of the polylithic crystal filter process.

Fig 5 Loading of the crystals into the evaporation masks prior to baseplating.

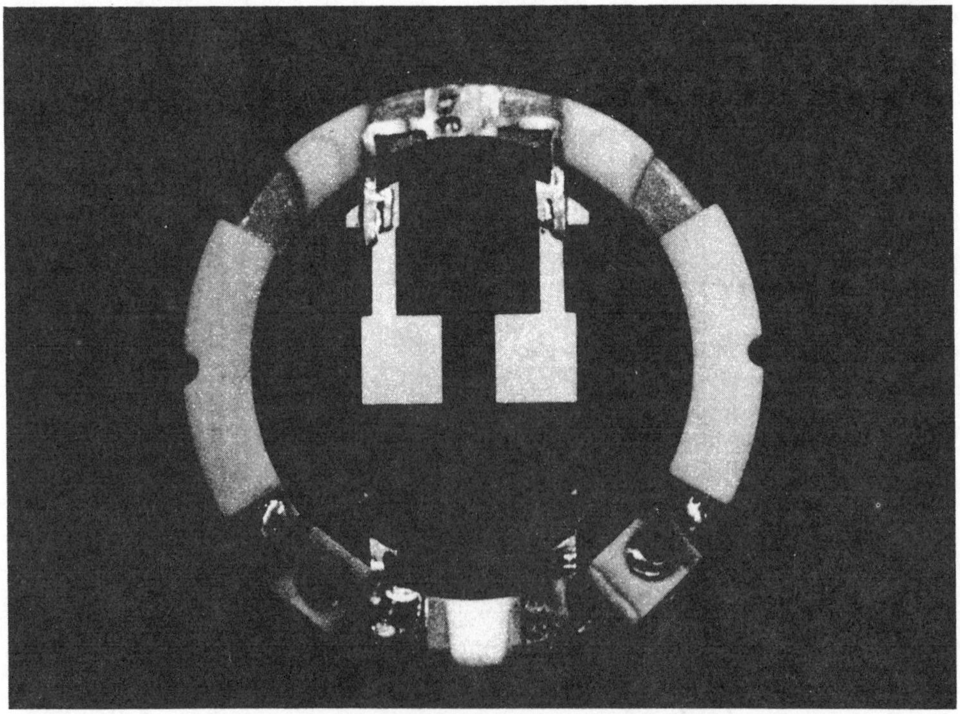

Fig 6 Closeup view of a resonator assembly which is a plated crystal mounted in a ceramic ring onto which are soldered 2 capacitors.

STANDARD DEVIATION σ
= 2.1 % OF MEAN

COUPLING TOLERANCE %

84 %

98 %

CUMULATIVE
DISTRIBUTIONS

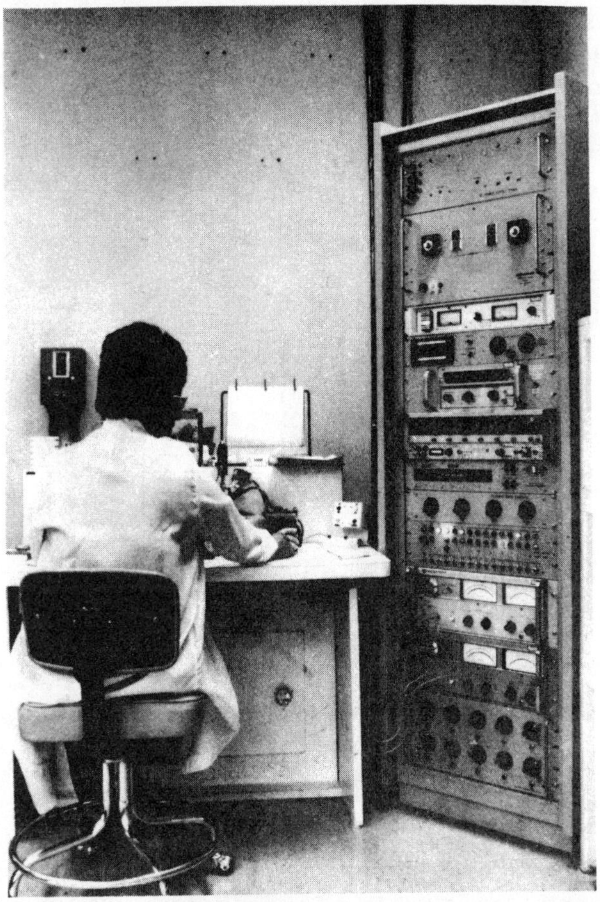

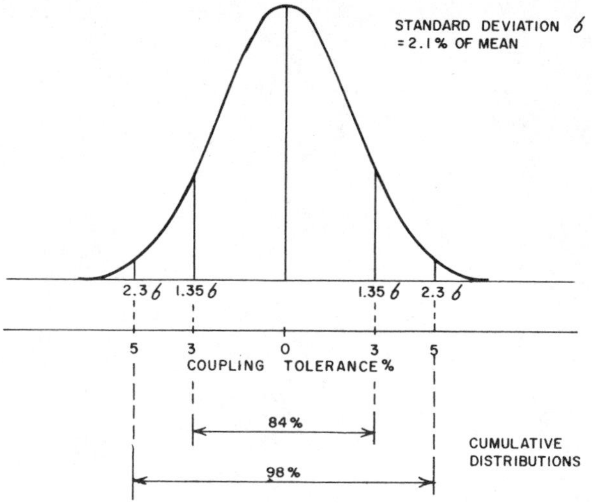

Fig 8 The normal distribution of the inter-resonator
coupling obtained from the polylithic process.
Since the coupling is not adjusted, the blank
frequencies and the mask dimensions have to be
carefully controlled. This is the distribution
of one month's produstion.

Fig 7 The automatic frequency plater. 2 pairs of small
chambers are connected to a turbomolecular pump
in the cabinet jsut in front of the operator.
The controller on the right controls one pair
of chambers at a time.

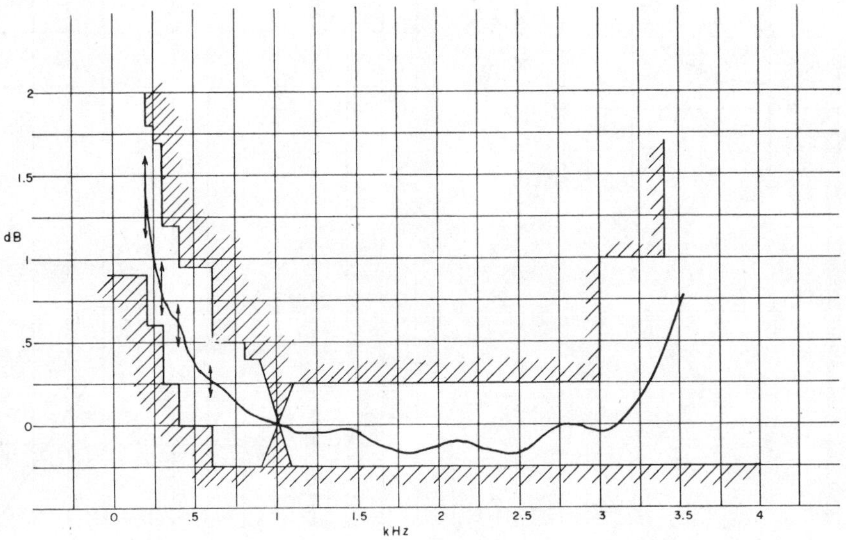

Fig 9 Typical passband response of a channel filter
showing the test limits and the one standard
deviation ranges for the critical lower corner frequencies.

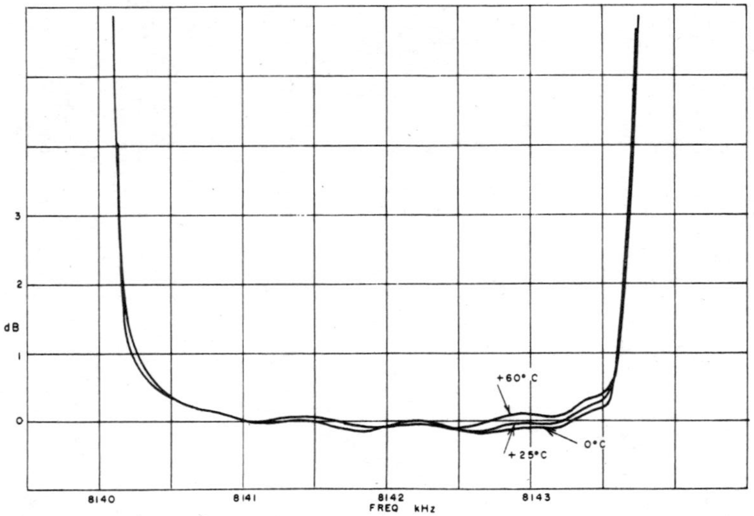

Fig 10 Temperature performance of a typical polylithic channel filter. With 5 crystals in the filter, some of the temperature drifts tend to cancel each other.

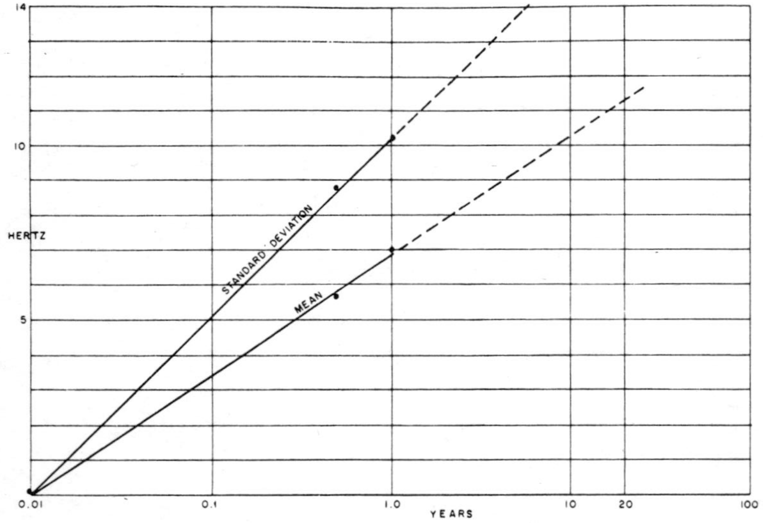

Fig 11 Aging data obtained from a batch of filters that had been stored for one year at+75°C. Normal operating temperature for the filter is +40° C.

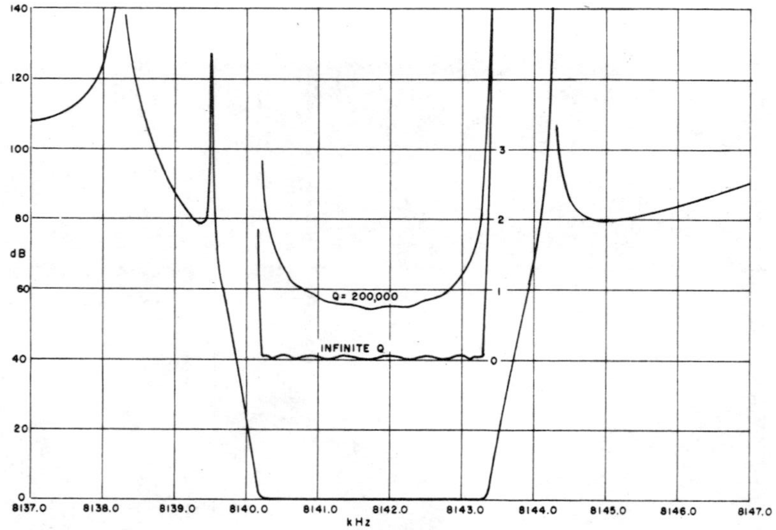

Fig 12 Calculated frequency response of a new polylithic channel filter that has capacitors bridged across the acoustically coupled resonators.

A6 MONOLITHIC CRYSTAL FILTER
DESIGN FOR MANUFACTURE AND DEVICE QUALITY

by

S. H. Olster and I. R. Oak
Western Electric Company, Incorporated
North Andover, Massachusetts 01845

and

G. T. Pearman, R. C. Rennick, and T. R. Meeker
Bell Telephone Laboratories, Incorporated
Allentown, Pennsylvania 18103

Summary

This paper describes the requirements for the monolithic crystal filter (MCF) used in the A6 Channel Bank. The design alternatives considered to satisfy the requirements in manufacture are discussed briefly. Resonator frequency and coupling adjustment are employed in the fabrication in order to maximize the tolerances on the quartz plate, electrode geometry, and plating distribution.

The quality of a shipped A6 MCF is estimated from three types of measurements, i.e., final electrical test results and variations of electrical filter characteristics with temperature and time.

The Final Electrical Test screens A6 MCF's for shipment and is a measure of the overall performance of the manufacturing process. Insertion losses are measured by a computer controlled test set at 10 in-band frequencies and 5 out-of-band frequencies. The test frequencies were selected to optimize both the accuracy of the test in correlating with overall customer requirements and the cost of performing the measurements. In addition, lower and upper 3 dB frequencies are measured. Statistical measures of the Final Electrical Test performance of the manufactured A6 MCF are presented and discussed.

The Temperature Performance Test determines whether the X-ray angle of the AT-cut quartz plate is within the required limits. The temperature performance of the A6 MCF is controlled by requirements on X-ray angle. Temperature performance data are presented to show the crystal orientation tolerances being achieved.

The Time Stability Test determines the expected useful life of the filter. Experimental results of a large scale accelerated aging study of A6 MCF's are discussed.

The shipped filter characteristics, the temperature behavior, and the aging behavior are combined to give an estimate of device life under various operating conditions.

I. Introduction

Western Electric Company is currently manufacturing two filter designs for the A6 Channel Bank; a four pole carrier filter, and an eight pole channel filter. The high production volume and the exacting requirements made it necessary to develop a dedicated facility to manufacture these filters. This development[1] involved all phases of the device manufacture including the growing of r-face cultured quartz,[2] a separate plate preparation facility,[3] and special equipment for the assembly and tuning of the filters.[4-8]

This paper describes the requirements governing the manufacturing process for the channel filter used in the A6 Channel Bank.

The quality of a manufactured device generally depends on its entire history, including the composition and preparation of the starting materials[9] and the processes used in its fabrication.[10] In this paper the quality of a delivered A6 Monolithic Crystal Filter (MCF) is estimated from three types of measurements, i.e., final electrical test results and variations of electrical filter characteristics with temperature and time.

The main role of the Final Electrical Test is to determine if the manufactured filter meets the requirements and is therefore deliverable to the customer. An important secondary role of this test is to indicate the overall performance of the manufacturing process. Highly accurate ($\pm$ 0.02 dB) insertion loss measurements are made on a computer controlled test set at 10 pass-band frequencies and 5 stop-band frequencies on every unit manufactured. These test frequencies were selected to optimize both the accuracy of the test in correlating with overall customer requirements and the cost of performing the measurements. In addition to the insertion loss measurements, the lower and upper 3 dB frequencies are often measured as a sensitive indicator of fabrication process performance.

The main role of the Temperature Test is to determine whether the X-ray angle of the AT-cut quartz plate is within the required limits. Because of the large cost of temperature scanning, the temperature performance of the A6 MCF is controlled by tight requirements on X-ray angle ($\pm$ 45 seconds).

The main role of the Time Stability Test is to determine the end-of-life conditions for the manufactured filter. Experimental results of a large scale accelerated aging study of the A6 MCF and the techniques used to understand end-of-life behavior (long-term extrapolation and activation energy) are discussed.

The shipped filter characteristics, the temperature behavior, and the aging behavior are combined to give an estimate of device life under various operating conditions.

II. A6 Monolithic Crystal Filter Requirements

Overall customer satisfaction in the performance of telephone equipment leads to the channel bank filter requirements shown in Figure 1 for the pass-band and in Figure 2 for the stop-band.[11] All re-

Reprinted with permission from *Proc. Twenty-Ninth Ann. Symp. on Frequency Control*, May 28-30, 1975, pp. 105-112, sponsored by the U.S. Army Electronics Command, Ft. Monmouth, NJ.

quirements are specified with respect to the insertion loss at the individual carrier frequency + 1,000 Hz, which must have an absolute value $\leq$ 1.3 dB to allow for proper gain adjustment in the channel.

In general, variations of the order of $\pm$ 0.30 dB are allowed in the pass-band region. The most critical requirement in the pass-band is the location of the lower 3 dB frequency, which must be located 203 Hz above the carrier within a tolerance of $\pm$ 25 Hz for all causes. This lower 3 dB frequency requirement may be segmented into three parts; a) a manufacturing adjustment tolerance, b) temperature performance (15° - 50°C), and c) aging (20 year life). More will be said with regard to the performance of the filter with respect to these requirements below. In order to meet these objectives, the lower 3 dB frequency is adjusted in manufacture to 1 ppm (9 Hz).

Several filter designs employing the monolithic crystal structure,[(12)] with and without attenuation peaks, were considered to satisfy these requirements. It was determined that a monolithic approach having all resonators acoustically coupled on a single quartz substrate would be the optimum design from the standpoint of cost. This design permits the adjustment of all couplings to very tight tolerances, and also facilitates a Final Filter Adjustment step where the lower 3 dB frequency of the filter is adjusted to within the 1 ppm required after final assembly. These adjustment features permit relatively wide tolerances on the critical parameters of the manufacturing process; i.e., array dimensions, plate-back, resonator frequency adjustment, inductance, and Q. This approach also minimizes quartz usage and piece parts to simplify the assembly.

Due to difficulties in realizing the stop-band requirements, the monolithic approach was modified to a bilithic design shown in Figure 3. This is an 8 pole .1 dB Chebyshev design where capacitive coupling is used between the fourth and fifth resonators. A typical stop-band characteristic for this design is shown in Figure 2, where the stronger high frequency transmission is due to untrapped modes which have been placed in a relatively harmless frequency range by the particular choice of plate-back. This design easily meets the stop-band requirements and thus production testing is not required outside of the immediate vicinity of the pass-band. This paper will not discuss stop-band performance or requirements further. The theoretical pass-band characteristic of this design is shown with respect to the requirements in Figure 1 for the nominal Q realized in manufacture. This design meets the pass-band requirements with a reasonable margin for manufacturing variations.

In order to maintain the features of a monolithic approach in the manufacture of the bilithic design, a single plate is processed until the final tuning step where the plate is split. The capacitor is added at final assembly and therefore is a part of the filter structure during the Final Filter Adjustment step described earlier.

Finally, the performance of the filters must not degrade at temperatures normally found in central office environments (15 degrees Centigrade to 50 degrees Centigrade) and with time (up to 20 years).

III. A6 Monolithic Crystal Filter Manufacturing Process

The system requirements must be converted to controls on the filter parameters in the manufacturing process. The performance of the monolithic crystal filter (as that for any high quality quartz crystal device) is controlled by the crystallographic orientation, crystal quality, thickness, uniformity of thickness, and surface of the quartz plate; by the thickness, length, width, spacing, and material of the electrodes; and by the crystal and electrode strain and contamination introduced in the assembly process. Various tradeoffs between parameter control, adjustment, and repair must be made to optimize cost.[(13)] The A6 manufacturing process is described in greater detail elsewhere.[(10)]

In A6 monolithic crystal filter manufacture, control of the crystallographic orientation replaces production measurement of the filter temperature performance. Interresonator coupling and resonator frequency are adjusted to reduce the need for tight tolerances on metallization and quartz plate uniformity. Time stability is achieved by careful control of filter assembly processes designed to minimize mounting stress and contamination.

IV. Performance Of A6 Monolithic Crystal Manufacturing Process

The performance of a manufacturing process may be measured in many ways -- all of which are ultimately referred to device quality and device cost. In this paper the performance of the manufacturing process for the A6 monolithic crystal filter will be discussed in terms of the quality of a shipped filter estimated from final electrical test results (what kind of a filter population did the shipped unit come from?), from measurements of the dependence of filter parameters on temperature (what kind of control on crystallographic orientation and material was exercised?), and from estimates of the stability of units with time (what kind of control over the assembly process was exercised?).

A. Final Electrical Test

Figure 4 shows a cumulative distribution[(14)] of the lower 3 dB frequencies of a random sample of 99 units as measured in the final electrical test. This kind of analysis has been performed for over 20 groups of units selected during nearly 3 years of production. In Figure 4 the distribution is nearly normal (linear on a probability plot), with one value which does not appear to be a member of the normal distribution. The statistical parameters associated with this distribution show that the requirements ($\pm$ 9 Hz) are consistent with a process two standard deviations wide. In the remainder of this paper the lower 3 dB frequency will be the only filter parameter discussed. Although all requirements are set on insertion losses, it is convenient for discussion to use a frequency to characterize the filter.

Statistical measures of overall filter performance, derived from parameter distributions (as in Figure 4) at the ten pass-band test frequencies, are shown in Figure 5 for a random sample of 180 units. In Figure 5 the test distributions are given as a 50 percentile (median) curve and 2.5 and 97.5 percentile (approximately $\pm 2\sigma$) curves. The crosshatched areas indicate those portions of the test distributions which are out of limits and, therefore, represent nonshippable units. This kind of data gives some assurance that all shipped filters are members of the same group, so that variations in the filter parameters are the result of the many random fluctuations that occur in the manufacturing process and are <u>not</u> a result of a systematic variation caused by a general malfunction at some step in the process.

A second use for this kind of analysis is to suggest reasons for defective devices so that corrective action be taken. For example, in Figure 5 the slightly high insertion losses at the carrier frequency + 1 kHz and the slightly low losses at 2.0 to 3.0 kHz are both likely to be caused by poor Q on a few resonators.

B. Temperature Performance

A second measure of device quality is the variation of filter performance with temperature. In the analysis of filter temperature performance in this paper, all lower 3 dB frequencies are expressed as a shift from that at 25 degrees Centigrade. In the temperature tests made on random samples taken from the manufacturing line, each unit is held at the desired temperature for one-half to three-quarters of an hour before the frequency measurements are made. Temperatures of measurement are typically 0 degrees Centigrade to 75 degrees Centigrade in 5 degree Centigrade increments. Distribution functions are determined at each temperature. A typical distribution of lower 3 dB frequency shifts for 95 filters at 50°C is shown in Figure 6. This distribution is nearly normal, with the exception of one or two values. Statistical parameters from the distribution data on these 95 filters at each temperature are shown in Figure 7. The various percentile curves in Figure 7 all show the typical shape of the AT temperature dependence.[15]

C. Time Stability Or Aging

The final measure of the quality of a quartz monolithic filter to be discussed in this paper is the small long-term drift of the lower 3 dB frequency, commonly called aging. Since this aging occurs very slowly at the operating temperatures of the channel bank, it has been necessary to develop creditable acceleration techniques. In this paper results are reported for acceleration techniques which depend on frequency changes caused by exposure to high temperatures. These frequency changes are subsequently extrapolated to expected changes at the operating temperatures. The first technique used to accelerate the aging of A6 MCF's is commonly known as thermal step stress aging, in which each filter is exposed to successive steps of increasing temperature, each exposure being of a fixed time.[16] In the second method, known as isothermal aging, the temperature is fixed and the frequency is measured as a function of time.[17] Most crystal aging studies have used the isothermal technique. In both techniques the filter is cooled to room temperature for measurement to avoid errors due to temperature variations. This cooling may produce thermal shock effects and frequency errors could result from shifts at room temperature. Thermal shock effects are considered to cause frequency errors of less than 1 part in 10^7 and are therefore ignored in this paper.[17] No appreciable frequency shifts occur during room temperature storage and measurement.

Since aging is likely to be process dependent, it is important to characterize production samples taken periodically from the manufacturing facility. It is for the purpose of evaluating the impact of changes in the MCF manufacturing process on device quality that thermal step stress aging is primarily used.

Figure 8 shows statistical results (median $\pm$ 2 standard deviations) of a thermal step stress study on 68 A6 monolithic crystal filters. The fixed time interval was 16 hours. The median frequency shifts are a few parts in 10^7 for stress temperatures as high as 180°C for 16 hours. The standard deviation increases with temperature. Since half of the units show positive aging and half show negative aging, a complex aging mechanism is indicated by these results. The thermal step stress results indicate that the A6 monolithic crystal filter is quite stable in time, although the suggested complex aging mechanism makes it difficult to assess the acceleration factor needed to use these results for end-of-life estimates. In the event of a process change in the manufacturing facility, a sample from the line could be subjected to the same tests and comparisons with previous results would indicate the effect of the change on aging. The powerful advantage of the step stress technique is that small samples can be used and that the time of measurement is fully determined (and reasonably short) at the beginning of the test.

More direct estimates of end-of-life lower 3 dB frequency may be obtained from isothermal studies. Figure 9 shows the dependence of the median frequency shifts of a group of 9 A6 monolithic crystal filters on time at 80°C. Figure 9 also shows similar results for 10 units at 140°C. Median values increase slightly (1 to 2 parts in 10^7) at each temperature. For 140°C, the median values begin to decrease after about 10 hours. This change in direction of aging also indicates a complex aging mechanism, as suggested above by the thermal step stress results.

The data shown in Figure 9 covers a real time of aging of just over 1,000 hours. For each unit in an aging study, lower 3 dB frequencies are extrapolated to 20 years by a least squares fit to the equation[17]

$$\Delta f = A + B \log (t) \quad .$$

The distribution of these extrapolated values for a group of 20 units aged at 80°C is shown in Figure 10. The frequency shifts for one unit extrapolated to a large positive frequency shift, which is not statistically consistent with the other shifts. Figure 11 shows similar results for units aged at 140°C and this same large positive shift is seen for 2 units.

In one aging study, extrapolated frequency shift distributions like those of Figure 10 for 80°C and of Figure 11 for 140°C were obtained for 90 units at temperatures of 80°C, 100°C, 120°C, 140°C, 160°C and 177°C. The logarithms of the negative median extrapolated frequency shifts are plotted versus temperature in Figure 12.[18] In Figure 12 the temperature scale is linear in the variable [1/(T°C + 273.16)]. The solid line in Figure 12 is a least squares fit to the medians of the frequency shifts extrapolated to 20 years. Actual aging times are 1,500 and 3,400 hours for these samples. The equation of the solid line is

$$DF = DF_{\infty} \exp (- E_a/RT)$$

with $DF_{\infty} = -9.793 \times 10^8$ Hz and $E_a = 0.653$ eV or 15,033 cal.. In this equation $T = T° + 273.16$.

For the purposes of this paper DF_{∞} and E_a are empirical numbers describing the results. Only in the framework of an appropriate model can these results acquire physical significance. Extrapolation to temperatures of 15 to 50°C suggests that the median aging will be less than the measurement error of $\pm$ 1 Hz at the end of 20 years.

The logarithms of the standard deviations of the extrapolated frequency shifts are plotted versus temperature in Figure 13. As in Figure 12, the temperature scale is linear in the variable [1/(T°C +

273.16)]. The solid line is a least squares fit to the standard deviations of the frequency shifts extrapolated to 20 years. The standard deviations are for the same 90 units for aging times of 1,500 and 3,400 hours as described above. The equation of the solid line is

$$\sigma = \sigma_{\infty} \exp(-E_a/RT)$$

with $\sigma_{\infty} = +2.379 \times 10^4$ Hz and $E_a = 0.249$ eV or 5,730 cal.. As above, T in this equation is T°C + 273.16.

From this equation, standard deviations of frequency shifts extrapolated to 20 years may be estimated for operating conditions in the telephone system. At 50°C this estimated standard deviation is about 3.2 Hz, and at 30°C it is about 1.8 Hz.

D. End-Of-Life Performance Estimates

The overall normality of the statistical distributions and the expected statistical independence of the lower 3 dB frequency at the final electrical test and the shifts with temperature and time allow an estimate of combined frequency shifts at end-of-life. This estimate will be temperature dependent since both the frequency and the aging of the frequency depend on temperature. End-of-life (20 years) frequency shift median and standard deviation are found at each temperature as:

$$\text{median (combined)} = \sum_{n=1}^{3} \text{median}_n$$

$$\text{standard deviation(combined)} = \sqrt{\sum_{n=1}^{3} [\text{standard deviation}_n]^2}$$

where n = 1 stands for final test, n = 2 stands for temperature dependence, and n = 3 stands for aging.

These combined frequency shifts (for the data presented in this paper) are plotted versus temperature in Figure 14. Median and median $\pm$ 2 standard deviations are shown, along with the $\pm$ 25 Hz objective. This data suggests that very few filter performance problems will develop within 20 years for all operating temperatures between 15°C and 50°C. Data such as that in Figure 15 is used to optimize the relationship between the cost of achieving demonstrable filter performance and the requirements of the filter. Only in this way can the filter cost be minimized.

V. Conclusions

By statistical analyses of large samples of final electrical test results, temperature performance data, and aging measurements it has been possible to develop some confidence in the overall normality of the parameter (lower 3 dB frequency in this paper) distributions for manufactured A6 monolithic crystal filters. This normality and the expected mutual statistical independences allow a simple end-of-life estimate, which shows that at 20 years the filters will continue to meet their requirements in the thermal environments of the central office. Not considered in this paper are the mechanical reliability of the filter and any non-thermally activated aging mechanisms. Both of these effects appear to be adequately controlled.

VI. Acknowledgments

The scope of this work makes it impossible to properly acknowledge all of the many people at Bell Laboratories and at Western Electric who have made significant contributions.

References

1. W. J. Spencer, Physical Acoustics, Volume IX, Academic Press, New York (1972), Chapter 4, pp. 167-220.

2. N. C. Lias and D. W. Rudd, "Growth of Synthetic Quartz for Use in High Frequency Monolithic Crystal Filters," Western Electric Engineer, Vol. 13, p. 23, 1969.

3. A. J. Miller, "Preparation of Quartz Crystal Plates for Monolithic Crystal Filters," Proc. 24th Annual Symposium on Frequency Control, p. 93, April, 1970.

4. R. J. Byrne, "Monolithic Crystal Filters," Proc. 24th Annual Symposium on Frequency Control, p. 84, April, 1970.

5. R. P. Grenier, "A Technique for Automatic MCF Frequency Adjustment," Proc. 24th Annual Symposium on Frequency Control, p. 104, April, 1970.

6. J. L. Hokanson, "Laser Machining Thin Film Electrode Arrays on Quartz Crystal Substrates," Proc. 23rd Annual Symposium on Frequency Control, May, 1969.

7. P. Lloyd, "Monolithic Crystal Filters for Frequency Division Multiplex," Proc. 25th Annual Symposium on Frequency Control, p. 280, April, 1971.

8. W. C. Morse and R. C. Rennick, "Adjusting Frequency of Monolithic Crystal Filters with an Automatic Vapor Plater," J. of Vac. Sc. and Tech., Vol. 9, No. 1, p. 28, January, 1972.

9. *R. A. Laudise, "Production and Perfection of R-Face Quartz."

10. *J. I. Pelc, et al, "Manufacture of Monolithic Crystal Filters for A6 Channel Bank and Facilities - A6 MCF Assembly."

11. G. W. Bleisch, "The A6 Channel Bank," IEEE International Conf. on Communications (1971), pp. 6-7 to 6-12.

12. G. T. Pearman and R. C. Rennick, "Monolithic Crystal Filters," IEEE Trans. on Sonics and Ultrasonics, SU21, No. 4, October, 1974, pp. 238-243.

13. R. C. Rennick, "Modeling and Tuning Methods for Monolithic Crystal Filters," Proc. 1975 IEEE International Symposium on Circuits and Systems (1975), pp. 309-312.

14. H. Arkin and R. R. Colton, An Outline of Statistical Methods, Fourth Ed., Barnes and Noble, New York (1939).

15. R. E. Bennett, Manufacturing Guide for "AT" Type Units, Dept. of the Army, Union Thermoelectric Company, Niles, Illinois, PB 171839 (1960), pp. 77-97.

16. D. S. Peck and C. H. Zierdt, Jr., "The Reliability of Semiconducting Devices in the Bell System," Proc. IEEE, February, 1974, pp. 185-211.

17. A. W. Warner, D. B. Fraser, and C. D. Stockbridge, "Fundamental Studies of Aging in Quartz Resonators," IEEE Trans. on Sonics and Ultrasonics, 12, June, 1965, pp. 52-59.

18. S. Glasstone, Textbook of Physical Chemistry, Second Ed., D. Van Nostrand Company, New York (1946), pp. 1,087-1,089.

*References 9-10 have been submitted for publication in the 29th Annual Symposium on Frequency Control.

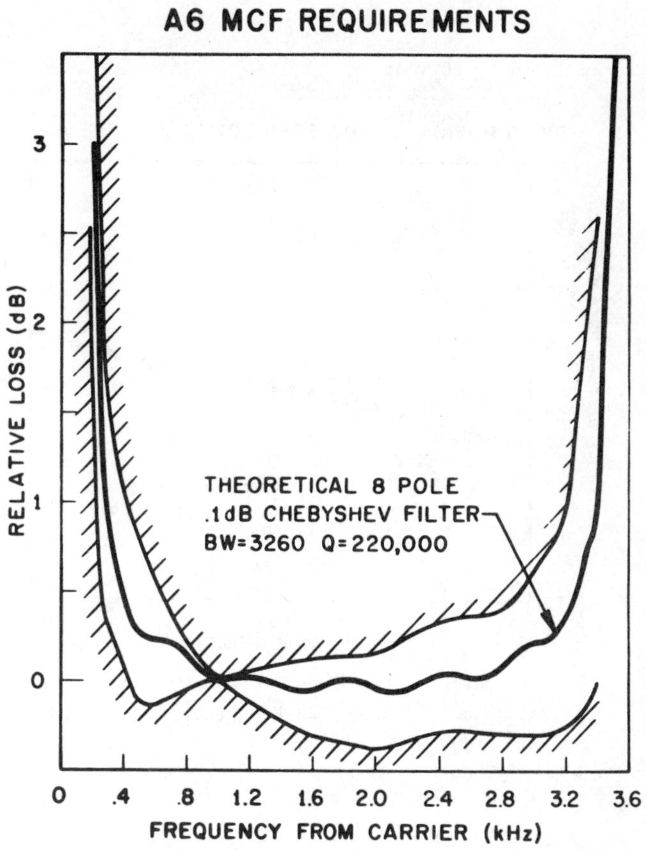

Figure 1 - A6 MCF Requirements - Pass-Band

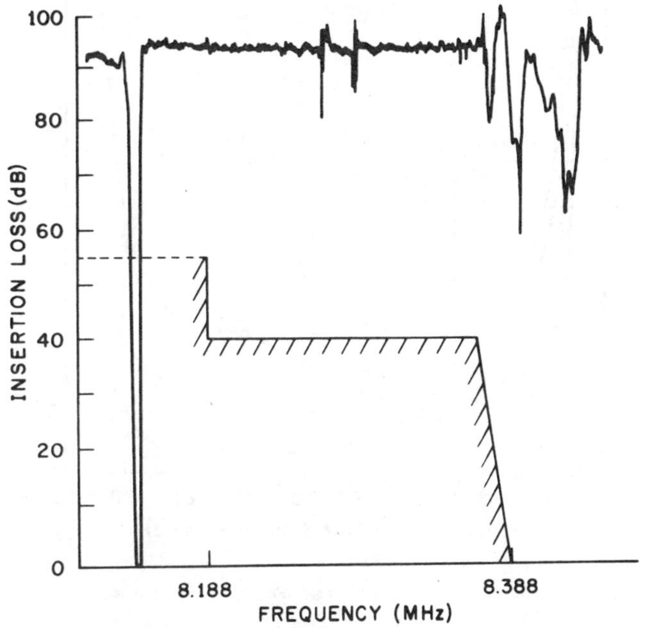

Figure 2 - Stop-Band Requirements
Typical Stop-Band Performance Of A6 Monolithic Crystal Filter

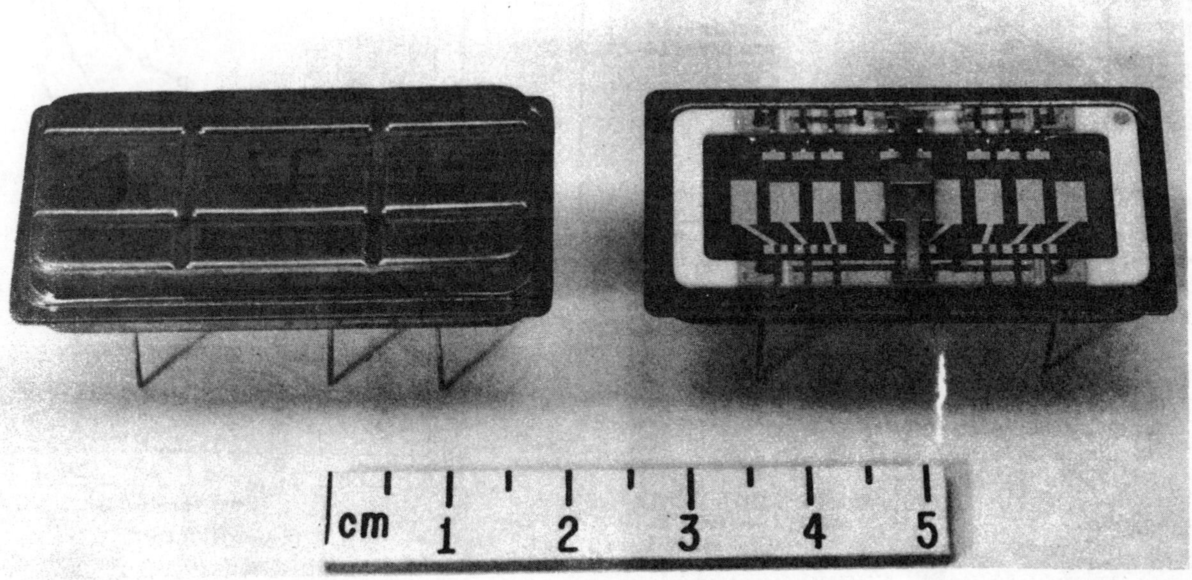

Figure 3 - Bilithic A6 MCF

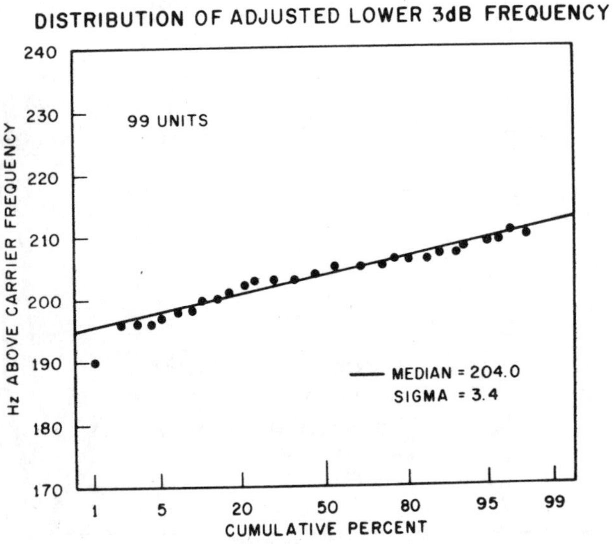

Figure 4

Distribution Of Lower 3 dB Frequency For 99 A6 MCF's At Final Electrical Test

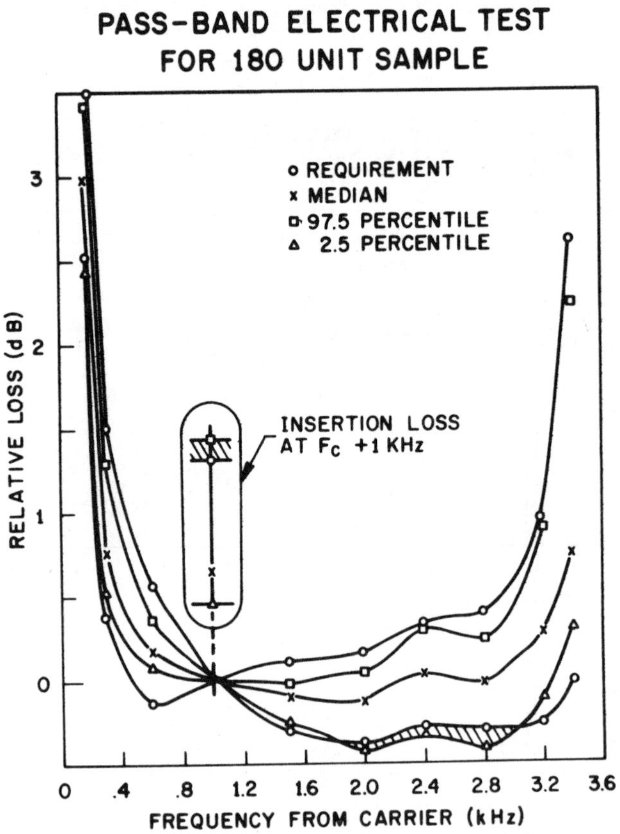

Figure 5

Pass-Band Electrical Test For 180 Unit Sample

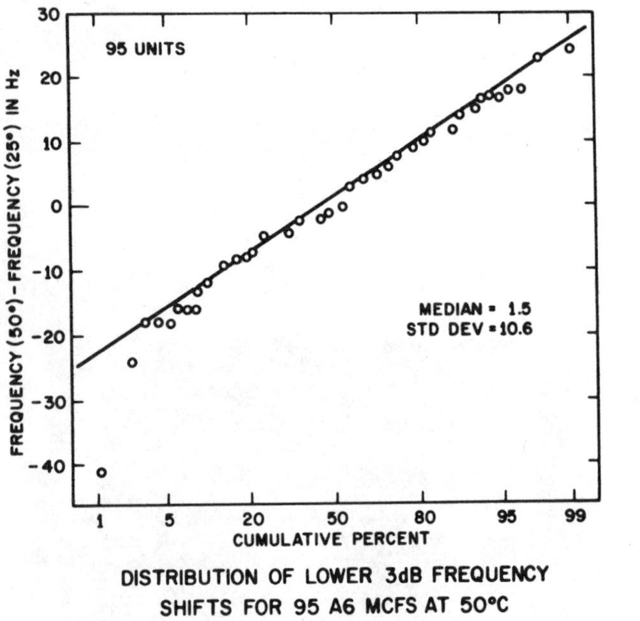

Figure 6

Cumulative Distribution Of Lower 3 dB Frequency Shifts For 95 A6 MCF's At 50°C

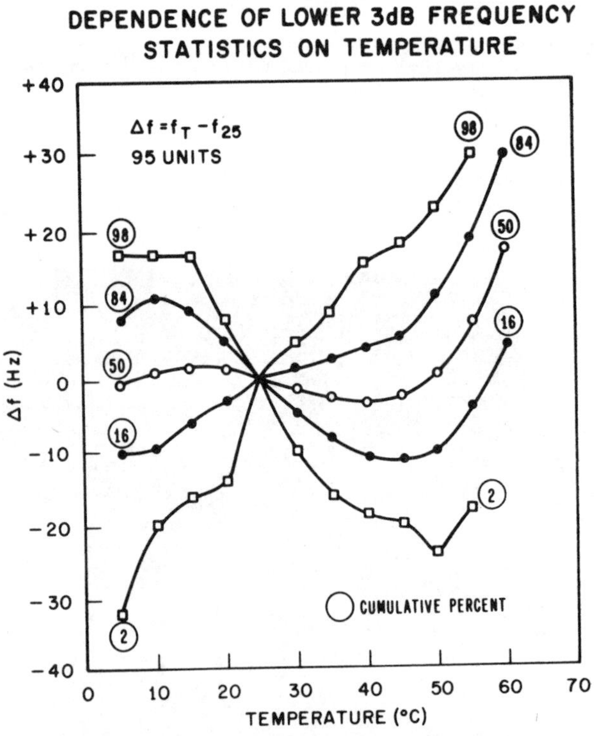

Figure 7

Statistical Temperature Performance Of 95 A6 MCF's

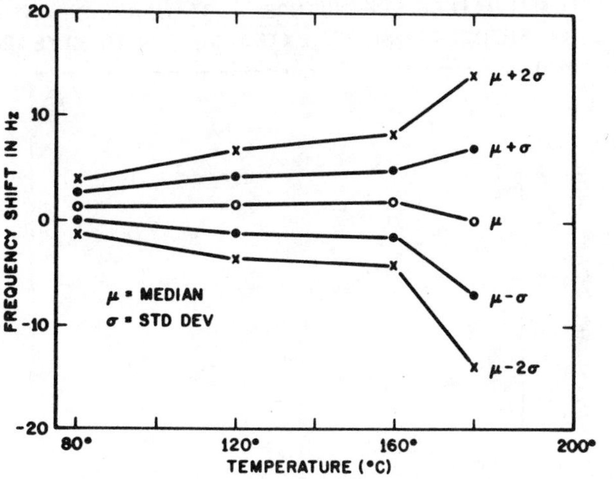

THERMAL STEP STRESS ON 68 A6 MCFS

(16 HOURS PER STEP)

Figure 8

Thermal Step Stress On 68 A6 MCF's (16 Hours Per Step)

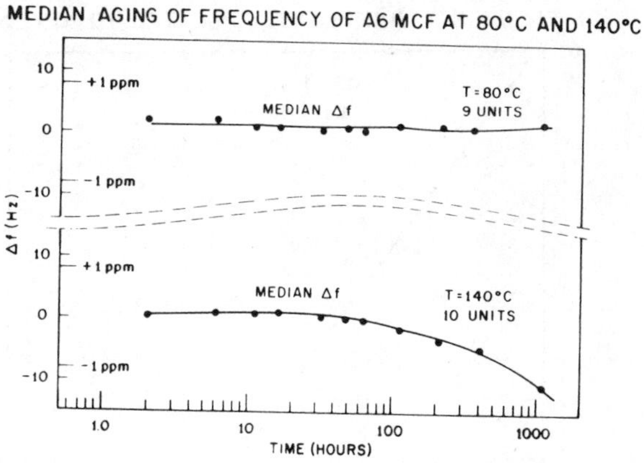

Figure 9

Median Aging Of Lower 3 dB Frequency Of A6 MCF

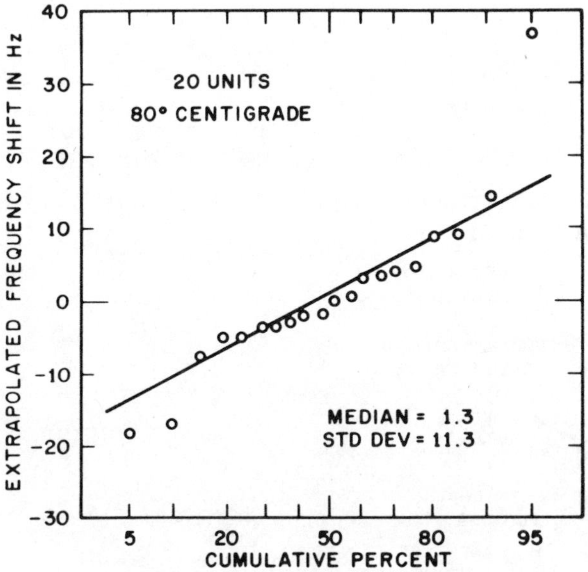

DISTRIBUTION OF FREQUENCY SHIFTS EXTRAPOLATED TO 20 YEARS

Figure 10

Distribution Of Frequency Shifts Extrapolated To 20 Years - 80°C

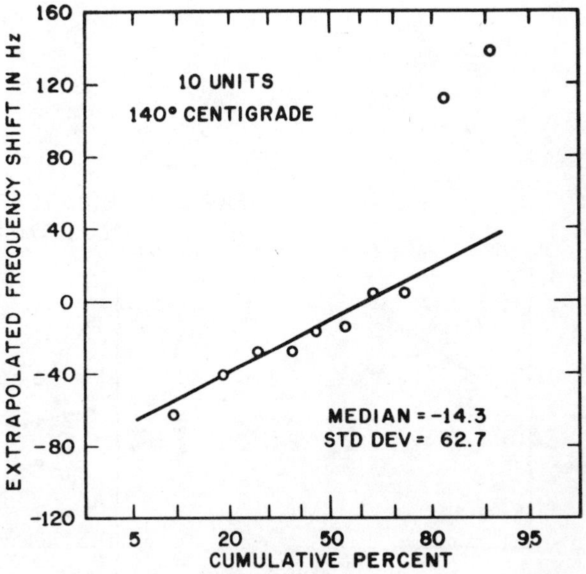

DISTRIBUTION OF FREQUENCY SHIFTS EXTRAPOLATED TO 20 YEARS

Figure 11

Distribution Of Frequency Shifts Extrapolated To 20 Years - 140°C

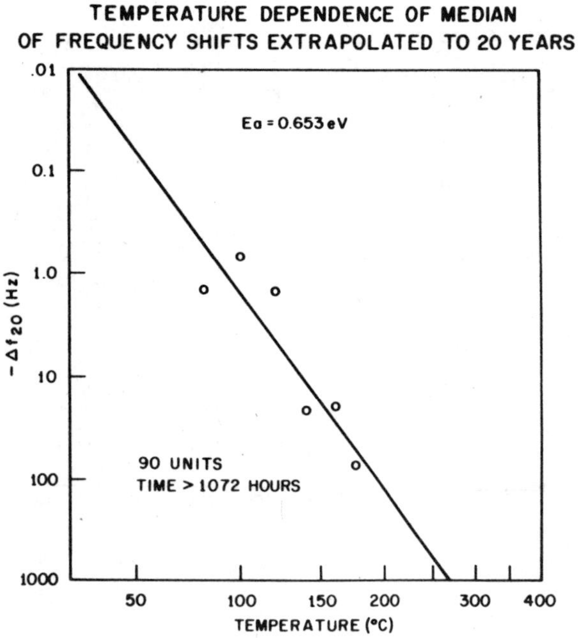

Figure 12

Temperature Dependence Of Median Frequency Shifts Extrapolated To 20 Years

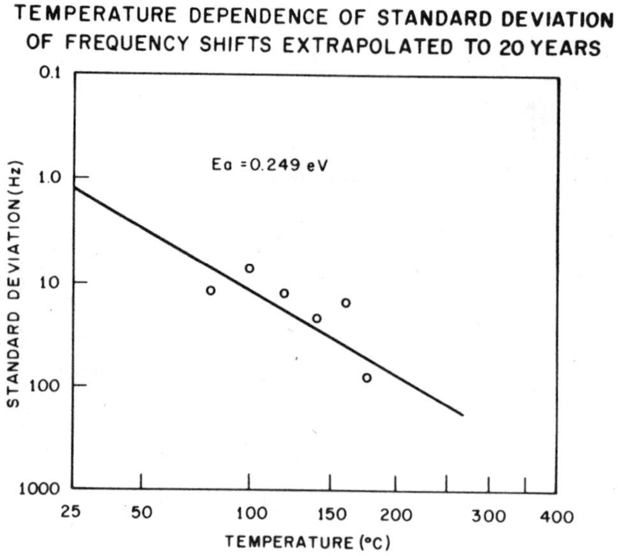

Figure 13

Temperature Dependence Of Standard Deviation Of Frequency Shifts Extrapolated To 20 Years

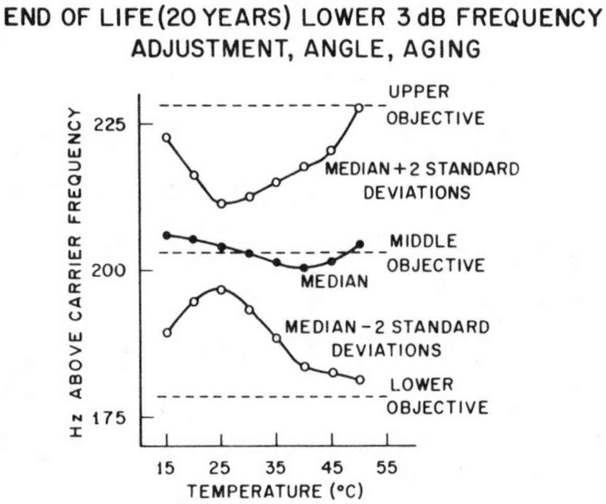

Figure 14 - Lower 3 dB Frequency Extrapolated To 20 Years - Adjustment, Angle, Aging

Section II-E
Materials, Properties, and Acoustically Coupled Filters with Nonquartz Materials

Almost all crystal filters are presently made from quartz because its many advantages far outweigh its major disadvantages, which is limited bandwidth. The search for wider bandwidth materials has been pursued for almost as long as there have been crystal filters. The papers in this section discuss some of the newer materials for which we are starting to find significant areas of usefulness.

Zelenka discusses the bandwidth limitations of quartz, and he shows that the maximum filter bandwidth that can be expected from it is 0.3 percent. He compares this figure with values that he has computed for various piezoelectric ceramics.

Schuessler gives an excellent account of the state of the art in ceramic resonators and ceramic filters in particular. He shows the types of ceramic filters that have found significant areas of usefulness, and he describes the techniques that are used to design them.

Lithium tantalate is a promising new material that is being investigated for use in wide-bandwidth filters. It has several properties similar to quartz, and it can be cut and lapped with the same techniques that are generally used for quartz. The maximum theoretical filter bandwidth that can be obtained with this material is 20 times that of quartz. Uno, in his paper, develops equivalent circuits for use with lithium tantalate, and he does this for a thickness-extensional mode of vibration. He describes the results obtained with a third overtone lithium tantalate filter that operates at a frequency of 200 MHz. Hales and Burgess, on the other hand, concentrate on a thickness-shear mode of vibration, and they use equivalent circuits and design approaches that are virtually identical to those used with quartz. They use this information to build third overtone lithium tantalate filters at 46 MHz.

The influence of material constants on the parameters of high frequency monolithic crystal filters

Jiří Zelenka

The College of Mechanical and Textile Engineering
Liberec, Czechoslovakia

1. Introduction

The trend to miniaturisation and recently also to integration of electronic components and circuits has involved, also in the field of the linear electronic, the creation of suitable selective circuits corresponding to new technological tendencies. Monolithic crystal filters are such components in the field of frequency filters. Nowadays the term "monolithic crystal filter" usually indicates a system of elastically coupled piezoelectric resonators where both the resonators and the coupling medium are embodied in a single plate of piezoelectric material.

Elastically coupled piezoelectric resonators were studied by Beaver /1/, Tiersten /2/, Onoe /3/ et al. They based their research on the theory of the thickness-shear, thickness-twist vibrations of rectangular piezoelectric plates /4/ and on that of coupled thickness-shear and flexural vibrations of partially plated crystal plates /5/ and derived, on certain approximative assumptions, the frequency relation of two or more acoustically coupled resonators. They considered both the thickness-twist and thickness-shear mode of coupling.

The first end n-th resonance frequency given by the frequency relation of the elastically coupled n-resonator structure determines (as was shown by Sykes and Beaver /6/) the bandwidth of the monolithic crystal filter. In order to determine limit para-

Reprinted with permission from *Proc. Summer School on Circuit Theory*, Partizán Tále Hotel, Czechoslovakia, 1971, pp. 12-1–12-11.

meters of the monolitic crystal filter Beaver /1/ studied the influence of the number of coupled resonators and that of the thickness as well of dimensions of the electrodes upon the distance of the resonance frequencies given by the frequency equation of elastically coupled system compound of two to six resonators. Considering the structure created on the AT-cut quartz crystal plate he found that a greater distance between resonances may be obtained with a structure using the thickness-twist mode of coupling.

Referring to the Beaver´s paper /1/ and Sykes´s and Beaver´s paper /6/ we made it our aim to study the two resonator system as a representative of the simplest monolitic filter. Besides a two resonator quartz crystal system we considered also a two resonator system prepared on a piezoelectric plate as a representative of a material with a great electromechanical coupling factor.

2. Frequency equations of an elastically coupled two resonators structure

Let us consider a plate with thickness a, length l and width b prepared from a piezoelectric material and orientated in the orthogonal axes system as shown in Fig. 1. Let us suppose that the plate is from such a material and has such an orientation that there exists a piezoelectric stress constant which makes it posible to excitate, by electric field along X_2 axis, thickness-shear modes of vibration at least in one plane perpendicular to axes X_1 or X_3.

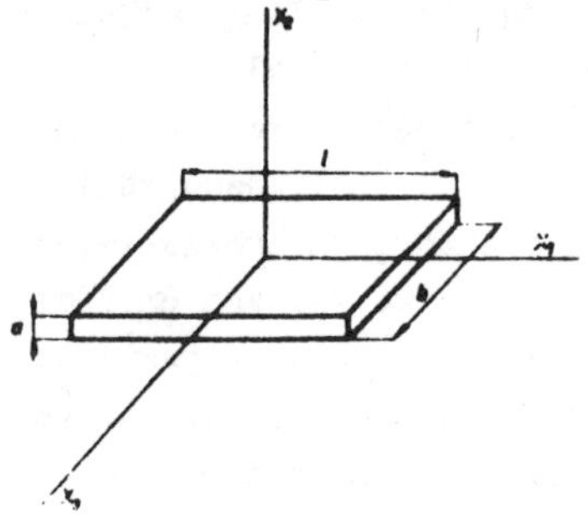

Fig. 1.
Plate in rectangular coordinate axes

This condition is fulfilled for instance by AT-or BT-cut quartz plates where the excitation of the thickness-shear modes of vibration is realized by means of a piezoelectric stress constant e_{26}. The matrix of the elastic stiffnesses C_{pq} for considered quartz plates exhibits monoclinic symmetry, i.e.

$$c_{15} = c_{16} = c_{25} = c_{35} = c_{36} = c_{45} = c_{46} = 0$$

and only piezoelectric stress constants e_{11}, e_{12}, e_{13}, e_{14}, e_{25}, e_{26}, e_{35} and e_{36} are non zero.

In case the piezoceramic plate is orientated in such a way that the X_3 axis is identical with the polarisation direction of the plate it is possible to excitate, by electric field along X_2 axis and by means of the piezoelectric stress constant e_{24}, thickness-shear mode of vibration in the plane perpendicular to X_1 axis. For this piezoceramic plate elastic stifnesses

$$c_{14} = c_{15} = c_{16} = c_{24} = c_{25} = c_{26} = c_{34} = c_{35} = c_{36} = c_{45} =$$

$$= c_{56} = 0$$

are equal to zero and piezoelectric stress constants e_{15}, e_{24}, e_{31}, e_{32} and e_{33} are non zero.

For AT-cut quartz crystal plates the frequency equation of two coupled resonators has been derived by Beaver /1/. He supposed that elastic stiffness c_{56} is negligible in comparison with elastic stiffness c_{55} and so that is possible to neglect the elastic coupling of thickness-shear mode of vibration in the plane perpendicular to X_3 axis and face-shear mode of vibration in the plane of the plate. He neglected also the piezoelectric stress constant e_{25} by means of which face-shear mode of vibration in the plane of the plate is excited by electric field in thickness direction.

The frequency equation derivated by Beaver /1/ was used for calculation of percent inter-resonator coupling given by relative frequency difference between the first two frequencies calculated from the frequency equation. The percent inter-resonator coupling corresponded to the percent band width of the monolitic crystal filter. Calculations were made for a few diferent separations d between electrodes and for two width l_e of electrodes. The results are shown in Fig. 2.

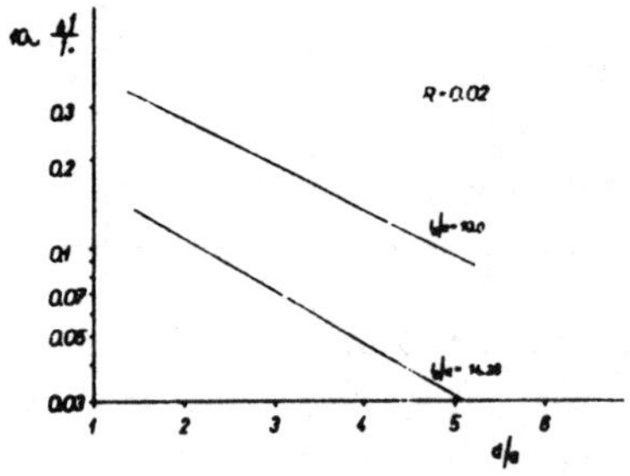

Fig. 2.
The effect of electrode separation on the frequency difference between, the first two frequencies of a AT-cut quartz two resonator system

In case of piezoceramic plates, owing to the fact that by means of electric field applied in thickness direction it is possible to excite thickness-shear mode of vibration in the plane perpendicular tq the X_1 axis, it is necessary to arrange the two resonator system as shown in Fig. 3.

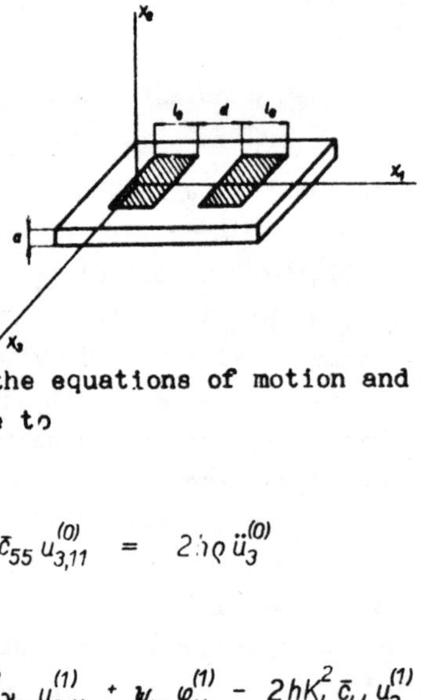

Fig. 3.
The electrode configuration and location of coordinate axes of two resonator system.

Than the equations of motion and electric potential equation reduce to

$$2h\,\bar{c}_{55}\,u^{(0)}_{3,11} = 2h\rho\,\ddot{u}^{(0)}_3 \tag{1}$$

$$\frac{2}{3}h^3\gamma_{55}u^{(1)}_{3,11} + \psi_{15}\,\varphi^{(1)}_{,11} - 2hK_4^2\,\bar{c}_{44}\,u^{(1)}_3 - \bar{e}_{24}K_4\,\phi^{(0)} = \frac{2}{3}h^3\rho\,\ddot{u}^{(1)}_3 \tag{2}$$

$$\Psi_{15} u_{3,11}^{(1)} - \frac{3f}{2h^3} \varphi_{11}^{(1)} - \frac{15}{2h^3} \bar{\varepsilon}_{22} \left(\frac{3}{h^2} \varphi^{(1)} - \phi^{(0)} \right) = 0 \qquad (3)$$

In these equation ρ is density, f is wavenumber, $u_3^{(0)}$ is zero order component of displacement in the X_3 axis direction, $u_3^{(1)}$ is component of rotation, $\varphi^{(1)}$ is the electric-potential resultant, $\phi^{(0)}$ is the inhomogeneous electrical forcing term and for piezoceramic plate of given orientation

$$\bar{c}_{55} = \bar{c}_{44} = c_{44} \; ; \qquad \gamma_{55} = c_{44}$$

$$\bar{e}_{24} = e_{15} ; \qquad \Psi_{15} = e_{15}$$

$$\bar{\varepsilon}_{22} = \frac{9}{4} \varepsilon_{22} \; ; \qquad K_4^2 = \frac{\pi^2}{12}$$

The subscripts 1 following a comma in symbols $u_{3,11}^{(0)}$, $u_{3,11}^{(1)}$ and $\varphi_{,11}^{(1)}$ indicate differentiation with respect to x_1. The symbols $\ddot{u}_3^{(0)}$ and $\ddot{u}_3^{(1)}$ indicate second differentiation of $u_3^{(0)}$ and $u_3^{(1)}$ respectively with respect to time. As the components $u_3^{(0)}$ and $u_3^{(1)}$ are supposed in the form

$$u_3^{(n)} = u^{(n)} e^{j\omega t} \qquad n = 0,1 \qquad (4)$$

are

$$\ddot{u}_3^{(n)} = -\omega^2 u_3^{(n)} \qquad (5)$$

For piezoceramic plate of given orientation is $c_{56} = 0$ and there is, no elastic coupling betwern thicness shear vibration in the plane perpendicular to X_3 axis and face-shear vibration in the plane of the plate. But face-shear mode of vibration is excitated by means of piezoelectric stress constant Ψ_{15}. As $c_{56} = 0$ and driving electrodes are very small in equation (2) the component $\Psi_{15}\ \varphi_{,11}^{(1)}$ may be neglected and equations (1) and (3) reduced to

$$\bar{c}_{55} u_{3,11}^{(0)} + \rho \omega^2 u_3^{(0)} = 0 \tag{6}$$

$$u_{3,11}^{(1)} + \frac{3 K_4^2 \bar{c}_{44}}{\gamma_{55} h^2} \left[\frac{\omega^2}{3 K_4^2 \bar{c}_{44}/\rho h^2} - 1 \right] u_3^{(1)} + \frac{3 \bar{e}_{24} K_4 \phi^{(0)}}{2 h^3 \gamma_{55}} = 0 \tag{7}$$

$$\gamma_{15} u_{3,11}^{(1)} - \frac{3}{2h^3} \int \varphi_{,11}^{(1)} - \frac{15}{2h^3} \bar{\varepsilon}_{22} \left(\frac{3}{h^2} \varphi^{(1)} - \phi^{(0)} \right) = 0 \tag{8}$$

Equations (6) and (8) describe vibration of unplated plates. If the plate is plated by driving electrodes than in the plated regions of the plate we must consider, as was shown by Mindlin and Lee /5/ and Mindlin /7/ the thickness of system 2h and the density ρ'. Instead of elastic stiffness $\bar{c}_{44}$ we must consider elastic stifness $\bar{c}_{44}^{*}$

$$\bar{c}_{44}^{*} = \bar{c}_{44} \left(1 + \frac{\bar{e}_{24}^2}{\bar{c}_{44} \bar{\varepsilon}_{22}} \right) = \bar{c}_{44} (1 + p^2) \tag{9}$$

and equation (7) changes to

$$u_{3e,11}^{(1)} + \frac{3 K_{4e}^2 \bar{c}_{44}^{*}}{\gamma_{55} h^2} \left[\frac{\omega^2}{3 K_{4e}^2 \bar{c}_{44}^{*}/[(1+3R)\rho h^2]} - 1 \right] u_{3e}^{(1)} + \frac{3 \bar{e}_{24} K_{4e} \phi^{(0)}}{2 h^2 \gamma_{55}} = 0 \tag{10}$$

where

$$K_{4e}^2 = \frac{\alpha_e^2}{3} (1 + p^2) \frac{1 + 3R}{(1 + R)^2} \doteq K_4^2 \frac{1 + 3R}{(1 + R)^2} \tag{11}$$

and α_e is first root of the equation

$$tg\, \alpha_e = \alpha_e \left(1 + \frac{\bar{c}_{44} \bar{\varepsilon}_{22}}{\bar{e}_{24}^2} \right)$$

The parameter R is called the "mass loading" and is given by the relation

$$R = 2\varrho' h'/\varrho h$$

For derivation of the frequency equation of two resonator system we must consider the unplated region of the plate ($x_1 < 0$ and $x_1 > 2\,l + d$) where

$$u_3^{(1)} = C_1 e^{f_u x_1}, \qquad u_3^{(1)} = C_8 e^{-f_u x_1}$$

the plated region of the plate ($0 < x_1 < l_e$ and $l_e + d < x_1 <$ $2\,l_e + d$) where

$$u_{3e}^{(1)} = C_{2n} \cos f_e x_1 + C_{2n+1} \sin f_e x_1 - \bar{V}_n / f_e^2, \quad n\ 1,3$$

and the unplated region of the plate between resonator ($l_e < x_1 < l_e + d$)

$$u_3^{(1)} = C_4 e^{f_u x_1} + C_5 e^{-f_u x_1}$$

The wavenumber f_u in the unplated region and the wavenumber f_e in the plated region of the plate are given by relations

$$f_e^2 = \frac{\omega_e^2}{\gamma_{55}} \varrho (1 + 3R)\left(\frac{\omega^2}{\omega_e^2} - 1\right)$$

$$f_u^2 = \frac{\omega_u^2}{\gamma_{55}} \varrho \left[1 - \frac{\omega^2}{\omega_e^2}(1 + R)^2\right]$$

where

$$\omega_e^2 = \frac{3K_{4e}^2 \bar{c}_{44}^*}{(1 + 3R)\varrho h^2}, \qquad \omega_u^2 = \frac{3K_4^2 \bar{c}_{44}}{\varrho h^2}$$

We consider boundary condition

$$\left[u_3^{(1)}\right]_{x_1 = l_n} = \left[u_{3e}^{(1)}\right]_{x_1 = l_n} \quad \text{and} \quad \left[u_{3,1}^{(1)}\right]_{x_1 = l_n} = \left[u_{3e,1}^{(1)}\right]_{x_1 = l_n}$$

Where n = 0, 1, 2, 3 and l_o = 0 and than in the same way as it was shown by Beaver /1/ we obtained the frequency equation

$$/A_{ij}/ = 0$$

where for two resonator system $/A_{ij}/$ is eight row determinant.

It was concidered three different piezoceramic materials described in table 1 and was calculated the relative difference $\Delta f/f_o$ of two lowest frequencies of the structure for l_e = 10 a and for a different distance between electrodes. The results of calculation are given in table 2.

Table 1. Material constants of concidered piezoceramic materials

Piezoceramic material	c_{44} /10^{10} N m^{-2}/	ε_{11} /10^{-9}C $V^{-1}m^{-1}$/	e_{15} /C m^{-2}/	k_{15}
PF 15	5.1	3.1	1.95	0.155
PF 30	6.25	7.45	8.45	0.39
PF 45	8.4	10.65	21.0	0.58

Table 2. Calculated relative difference $\Delta f/f_o$ of two resonator structure realized by l_e = 10 a and R = 0.02 on piezoceramic plate

Piezoceramic material	Δf / fo in percent	
	d = a	d = 2a
PF 15	0.14	0.08
PF 30	0.14	0.08
PF 45	0.135	0.07

It is seen from comperison of results given in table 2 and in Fig. 2 that for the same configuration of electrodes is the relative difference $\Delta f/f_o$ of two lovest frequencies of the structure realized on piezoceramic plate smaller than for that realised on AT-cut quartz plate.

3. Limit bandwidth of the two resonator monolithic filter

As it was shown by Sykes and Beaver /6/ the equivalent electrical circuit of a monolithic filter can be considered in the form of the lattice structure. In the lattice arms there is a series resonance circuit L_1C_1 with the resonance frequency f_1 shunted by the capacities C_{o1} and in the series arms there is a series resonance circuit L_2C_2 with the resonance frequency f_2 shunted by the capacities C_{o2}. In our case of the two resonator monolithic crystal filter the resonance frequencies f_1 and f_2 are identical with two lowest frequencies given by the frequency equation of the elastically coupled two resonator structure. The capacities C_{o1} and C_{o2} correspond approximately to capacities of condensers produced by resonator electrodes and by the part of a plate between electrodes The ratio of the capacities C_1/C_{o1} or C_2/C_{o2} determinates the relative difference of the resonance and antiresonance circuit frequency in the lattice or series arms of the equivalent electric circuit of the monolithic filter. As far as the relative difference $\Delta f/f_o$ of two lowest frequencies of the elastically coupled resonance system is smaller than $C_1/2C_{o1}$ and $C_2/2C_{o2}$ the bandwidth of the monolithic filter is given by the relative difference $\Delta f/f_o$. As far as $\Delta f/f_o \geqslant C_1/2C_{o1}$, $C_2/2C_{o2}$ is needed to compensate the influence of the capacities C_1 and C_{o2} by the suitable induction.

We are interested in the first case when $\Delta f/f_o < C_1/2C_{o1}$, $C_2/2C_{o2}$ and when isn´t needed to consider further auxiliary elements.

For AT-cut quartz resonators the capacitive ratio is given by the relation

$$\frac{C_1}{2C_{01}} \doteq \frac{C_2}{2C_{02}} \leqslant 3.10^{-3}$$

For piezoceramic resonators the capacitive ratio and the distance between resonance and antiresonance frequencies is much more greater. The relative distance between resonance and antiresonance frequencies of the thickness-shear vibration can be expressed by means of the electromechanical coupling faktor k_{15}

by the relation /8/

$$\frac{\Delta f}{f_1} = \frac{k_{15}^2}{1 - 0.810\, k_{15}^2}$$

For instance for the monolithic filter from piezoceramic PF 45 $\Delta f/f_1 = 0.459$.

4. Conclusion

To obtain the information about the influence of material constants on the limit bandwidth of the monolithic filter we consider the two resonator elastically coupled system realised on the AT-cut quartz plate and on the piezoceramic plates from the different piezoceramic material. It was shown that for monolithic quartz crystal filters without further elements the filter bandwidth is limited by the relation C_1/C_{oi} to the value of approximately 3.10^{-3}. For piezoceramic plates is for the same configuration of the electrodes the difference between two lowest frequencies of the two resonator system smaller. As the distance between the resonance and antiresonance frequency is greater for the piezoceramic resonators (generally for the plates from the piezoelectric material with the greater electromechanical coupling factor) the limit bandwidth will be some percent. As a limited bandwidth value of the filter is in this case the difference of the first two frequencies of the two or more elastically coupled resonator system.

References:

/1/ W. D. Beaver: Analysis of Elastically Coupled Piezoelectric Resonators.
J. Acoust. Soc. Amer. 43 (1968), pp 972-981

/2/ H. F. Tiersten: Linear Piezoelectric Plate Vibrations. Plenum Press. New York, 1969.

/3/ M. Onoe, H. Jumonji: Analysis of Piezoelectric Resonators Vibrating in Trapped Energy Modes.

Electronics & Communication Japan 48 (1965), pp. 84-93.

/4/ H. F. Tiersten, R. D. Mindlin: Forced Vibrations of Piezoelectric Crystal Plates. Quart. Appl. Math. 20 (1962), pp 107-119.

/5/ R. D. Mindlin, P.C.Y. Lee: Thickness Shear and Flexural Vibration of Partially Plated, Crystal Plates. Inter. J. Solids and Structures 2 (1966), pp 125-139.

/6/ R. A. Sykes, W. D. Beaver: High Frequency Monolithic Crystal Filters With possible Application to Single Frequency and Single Side Band Use. Proc. of the 20th Annual Symp.

/7/ R. D. Mindlin: Forced Thickness-Shear and Flexural Vibrations of Piezoelectric Crystal Plates. J. Appl. Phys. 23 (1952), pp 83-88.

/8/ M. Onoe, H. Jumonji: Useful Formulus for Piezoelectric Ceramic Resonators and Their Application to Measurement of Parameters. J. Acoust. Soc. Amer 41 (1967), pp. 974-980.

Ceramic Filters and Resonators

HANS H. SCHUESSLER

I. INTRODUCTION

Very soon after the discovery of the piezoelectric effect in quartz crystals and the use of the first quartz resonators for frequency stabilization, numerous attempts were made, using other crystalline substances, particularly to produce resonators with a larger electromechanical coupling and lower impedance level and with simpler technological methods at less expense. These attempts also included electrostrictive ceramic materials[1] on the basis of $BaTiO_3$ soon after their discovery [1]. However, the widespread use of barium titanate and modified barium titanates in resonator and filter applications was prevented by the low Curie point of 120°C, and the dependence on temperature and time of most material constants, which was to high for resonator and filter applications [2]. The decisive breakthrough was achieved with the discovery of polycrystalline materials on the basis of lead titanate zirconate [3] by Jaffe and colleagues. Even at this time quite a number of papers were published about work on the technology of measuring resonators, the design of ceramic resonators, the question of oscillating modes, and including proposals for filter circuits [4]–[8]. The frequency range for the materialization of such resonators was restricted to 1 kHz–1 MHz by the familiar resonator shapes and the materializable dimensions.

The transistors and semiconductor devices available at the beginning of the 1960's for entertainment electronics gave an impetus *inter alia* to investigations for new selective circuits, too. It was thought that a better method of IF selectivity and amplification could be found by the use of piezoelectric ceramic resonators [9], [10]. The aim was to replace the coil circuits incorporated in the AM-IF amplifier by ceramic resonators. However, subsequent experiments did not yield sufficient merits for this concept. First, there was no point in fitting ceramic resonators in combined AM-FM sets for an IF of 460 kHz alone, there being no appropriate ceramic resonators for an IF of 10.7 MHz. Moreover, adequate experience had not been gathered with the lead titanate zirconate ceramics (PZT ceramic) used for the resonators, and the achieved material ratings were dissatisfactory, particularly in regards to their spread, temperature stability, and aging. In view of these circumstances it will also be appreciated that ceramic filters, which were cheaper than coil filters, could not be materialized. However, multicircuit ladder filters with ceramic resonators were fitted in communications equipment, for example, in FM receivers [10], [11].

Recent development trends in the field of monolithic integrated circuits, monolithic crystal and ceramic filters, and the extensive production experience gathered with PZT ceramic in consequence of the widespread use of electromechanical transducers in shear mode in ultrasonic delay lines for color TV sets in Europe led to the resumption of work. In the interim the theoretical work had yielded better insight into the local distribution of the amplitude of oscillation in crystal resonators excited in the thickness and thickness-shear modes [12]. From these considerations possibilities were derived for the materialization of mechanically coupled crystal and ceramic filters, which in the literature are generally termed monolithic crystal or ceramic filters. The proposal of a two-circuit monolithic ceramic filter for 10.7 MHz using thickness resonators then greatly stimulated the discussion about ceramic filters for broadcasting receivers [13].

During recent years a wide range of resonators and filters has been brought on the market which, at least in Japan and Europe, are being widely used in radio and TV sets together with the now implemented monolithic integration of large circuit parts [14]. Here both low-frequency filters for pilot signal selection as well as the AM-IF filters for 455–460 kHz, the TV sound IF filters for 4.5 MHz, and 5.5 MHz and FM-IF filters for 10.7 MHz must be mentioned. Television video IF filters are a still open requirement though a noteworthy alternative solution of the problem has been proposed with the surface wave device. However, agreement has not yet been reached about their merits *vis-à-vis* present filters with printed coils. For this type of filter, hot-pressed niobate ceramics have been proposed as substrate material concurrent to crystalline materials [15].

All in all, the use of ceramic resonators and filters will be restricted to consumer electronics because the most significant merit of the ceramic may be found in its inexpensive production. New crystalline substances like $Bi_{12}GeO_{20}$, $Bi_{12}SiO_{20}$, $LiNbO_3$, and $LiTaO_3$ have been developed to comply with the need for piezoelectric materials featuring better electromechanical coupling than quartz crystals [16].

II. APPLICATION OF CERAMIC ELEMENTS

In communications equipment incorporating analog circuitry, resonators are used for two main tasks: 1) frequency determination in generators, and 2) filtering. For the first task, mechanical resonators may be con-

The author is with the Electronic Component Department, Research Institute of AEG-Telefunken, Ulm, Germany.

[1] In the meantime colloquial usage has developed in such a fashion that the linear effect of the electromechanical and mechanical-electrical interaction of polarized ceramic is called piezoelectric in analogy to the piezoelectric active crystals.

Reprinted from *IEEE Trans. Sonics Ultrason.*, vol. SU-21, pp. 257–268, Oct. 1974.

TABLE I
COMPARISON OF DIFFERENT RESONANCE CIRCUITS

resonance circuits	optimum frequency range f/MHz	quality factor Q/1000	capacitance ratio r	activity Q/r	temperature coefficient of frequency TC(f)/ppm/K
commercial coil circuit active filter	0.01... 0.05	0.1... 1	-	-	20
broadcasting coil circuit	0.1... 100	0.05... 0.20	-	-	50
ceramic resonator	0.01... 10	0.2 ... 2	6...100	20...200	20
composite transducer	0.05... 0.5	1 ...10	20...400	20...200	2
steel resonator	0.05... 0.5	10 ...60	-	-	1
quartz crystal resonator	0.01... 100	30... 1000	300...30000	50... 2000	0.1

sidered only for fixed frequency generators. The requirements specified for these resonators within a wide selectable frequency range, with one production technology as far as possible, are high stability of the resonator frequency as a function to temperature and time, high resonator quality factor and activity to achieve a steep phase as a function of frequency, a resonance resistance matching semiconductor circuit technology, and a relatively restricted pulling range for fine tuning. As yet ceramic resonators have not been widely employed for this task because the quality factor and activity and stability ratings are not adequate and because, on the other hand, special purpose quartz crystal resonators have been developed for purposes such as the clock crystal for 32 kHz and the TV crystal for the PAL color TV standard at 4.433 MHz, which may be inexpensively manufactured and which feature quality factor and stability ratings (Table I) superior to those of possible ceramic resonators by factors 10–100. Though the higher capacitance ration of quartz crystals restricts the pulling range, it is nevertheless wide enough for most applications. A proposal for a multielectrode resonator, allowing pulling of the oscillator frequency in the range by approximately 1% depending on the connection of additional electrodes with capacitors or resistors, has also not found general acceptance [17]. In filter applications, i.e., all types of networks with amplitude and phase response dependent on frequency, the merits in comparison to the coil circuits shown in Table I, may be fully exploited.

For which frequency range can ceramic filters be designed from a rational point of view? The frequency range is limited by the excitable modes of the resonators and by physical size [18]. At low frequencies only resonators for the flexural mode may be taken into consideration. At frequencies below 1 kHz, the resonators are large and very weak, and consequently technically reasonable designs are not conceivable. Longitudinal contour resonators are suitable for the frequency range 100 kHz–1 MHz. For frequencies above 2 MHz, resonators in the thickness-shear mode are used, and for frequencies around 10 MHz thickness resonators. The highest frequencies obtained hitherto for resonators in the basic mode were at approximately 40 MHz [19], [20]. Therefore, the customary PTZ ceramic was replaced by pure titanates or niobates.

Towards lower frequencies the bandwidth is limited by the obtainable resonator quality factor. Filter catalogs [21] indicate that, for bandpass filters of the degree 8–20 with 4 to 10 resonators, the quality factor (Q) standardized for the relative bandwidth (B/f) for these resonators should be

$$q = Q \frac{B}{f} > 10 \tag{1}$$

if the transmission loss does not exceed 2–3 dB and the transmission distortions are not be inadmissibly high [22], [23]. It follows that the lower limit of relative bandwidth is situated between 0.5 and 5% depending on frequency, mode of oscillation, and choice of material. If, in addition, a center frequency stability better than 1/10 of the bandwidth is specified, the stability ratings may be also directly read off from the lower bandwidth limit. At an operating temperature range of 50°C the temperature coefficient $TC\ (f) < 10^{-5} \cdots 10^{-4}$/°C must be obtained, and the aging must be less than $2.10^{-4} \cdots 2.10^{-3}$/decade. For narrower realitive bandwidths, mechanical filters and crystal filters are more suitable. Towards higher bandwidths, the relative bandwidth (B/f) is limited, when additional coils are not used, by the magnitude of the electromechanical coupling (K_{EM}) [9], [16] and, respectively, by the reciprocal of the capacitance ratio (r)

$$\frac{B}{f} \leq K_{EM}^2 = \frac{1}{r} \tag{2}$$

and, in monolithic ceramic filters, by the mechanical coupling between resonators [22]. The ratings obtained are approximately 10% for contour resonators and 5% for resonators in the thickness mode. For wider relative bandwidths only coil filters or active filters come into question.

III. RESONATORS, VIBRATION THEORY AND EQUIVALENT CIRCUITS

In this section the most important basic principles of ceramic resonators shall be summarized in order to obtain a review of the familiar and proven resonators.

A. Vibration Mechanism

The piezoelectric ceramic materials used belong to the perovskites and show specific symmetries, so that, after polarization of the material, we may proceed from multiple crystal though with extensive alignment of the domains. As for piezoelectric crystals, matrices may be established which describe the relationships between electrical field magnitudes and mechanical magnitudes [4], [5]. Here only the d-matrix shall be stated, which describes the electromechanical interaction. Thereby the polarization is situated in the direction 3:

$$d = \begin{pmatrix} 0 & 0 & 0 & 0 & d_{15} & 0 \\ 0 & 0 & 0 & d_{15} & 0 & 0 \\ d_{31} & d_{31} & d_{33} & 0 & 0 & 0 \end{pmatrix}. \quad (3)$$

As may be seen from the arrangement of the matrix, only the direct excitation of some few modes is possible due to the strong symmetries. The axial symmetry in respect to the polarization direction allows solely the transversal (d_{31}) and longitudinal excitation (d_{33}) of compression waves. In addition, there is the possibility of excitation for shear waves (d_{15}) by alternating fields in the two directions perpendicular to the polarization.

B. One-Port Ceramic Resonators

Starting from the equation system described, the equations may be solved for incomplex resonator elements with one-dimensional wave propagation, and from these equations an equivalent circuit diagram is obtained with line components as shown in Fig. 1. In the event that the resonator shall be operated as a one-port, the forces $\boldsymbol{K}_1$ and $\boldsymbol{K}_2$ disappear and the equivalent circuit diagram familiar for crystal resonators may already be stated. The additional series resonant circuits should describe the excitation of spurious waves—harmonics, oscillations at inadequate length-thickness ratio with other frequency-determining dimensions, and parasitic oscillations, which may be excited due to assymmetries in resonators. Apart from thin transversally excited longitudinal resonators, circular disc radial resonators and rectangular almost square resonators, are particularly interesting.

Circular disc resonators in accordance with Fig. 2, which may be quite easily produced, are very frequently used. The resonance mode is a radial symmetrical elonga-

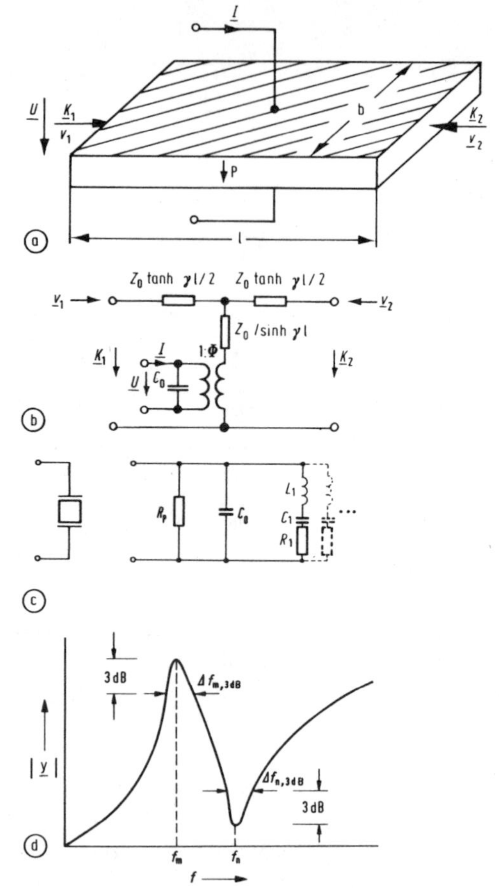

Fig. 1. Three-port equivalent circuit for piezoelectric transversally excited and longitudinally vibrating ceramic resonator. (a) Physical arrangement. (b) Circuit diagram with line elements. (c) Circuit diagram of ceramic resonator. (d) Admittance of ceramic resonator.

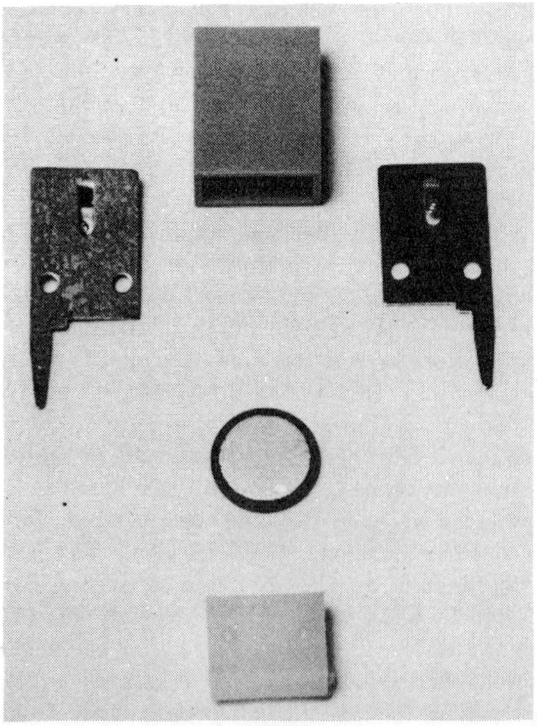

Fig. 2. Circular disk resonator for 455 kHz ceramic disk (5.8 mm diameter) with case and contact parts (products of CRL Elektronische Bauelemente Porz, F.R. Germany and Philips, Netherlands).

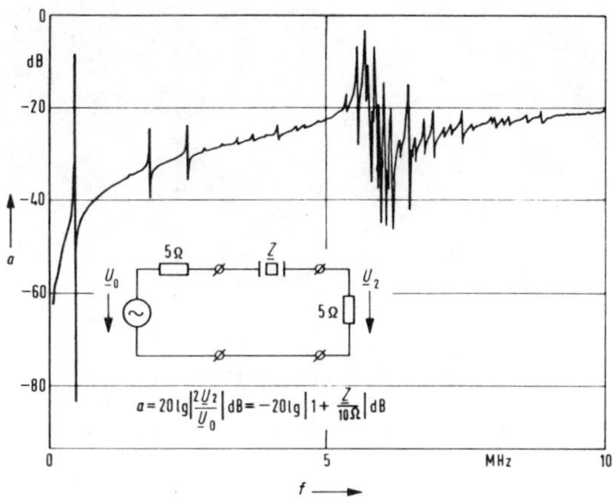

Fig. 3. Resonance frequencies of partly plated circular disk (Fig. 2).

TABLE II
CIRCUIT ELEMENTS OF CIRCULAR DISK RESONATOR (FIG. 2)

C_{NF}	=	182pF	C_0	=	165pF
f_m	=	459.740kHz	C_1	=	16.9pF
f_n	=	483.170kHz	L_1	=	7.09mH
K_{EM}	=	32%	R_1	=	15Ω
r	=	9.8			
Q	=	1370			
TC (f_m)	≤	20ppm/K			

Size: 5.78 mm Ø x 4.5 mm Ø x 0.4 mm
Material: Sonox P6
Producer: CRL Elektronische Bauelemente Porz, F.R. Germany

tion with a node in the center of the resonator, which by polarization in thickness direction of the discs and excitation electrodes on the two top sides, may be excited as lowest mode; the diameter of the disc being the frequency-determining dimension [24], [25]. From Fig. 3 we obtain for a resonator in accordance with Fig. 2 the spurious resonance which interferes with the frequency that could be explained as radial oscillations of higher order at frequencies up to 5 MHz, and as a thickness vibration of the disc at 6 MHz. Since the excitation electrodes of the resonator cover only part of the diameter, all the possible resonance modes are not excited. The much lower activity for all undesired spurious waves is clearly noticeable. From measurements of the input admittance, the equivalent circuit elements may be ascertained as implied in Fig. 1, [4], [22]. In the calculations, approximations are made which give rise to only small error if the quotient Q/r is much larger than 1. The locus of the input admittance then becomes a circle with very good approximation. Correction formulas, which, however, may be disregarded for the accuracies needed here, have been published in the IRE Standards [4] and a number of other papers [26], [27]. A summary of the resonator equivalent circuit elements for the radial resonator in accordance with Fig. 2 is given in Table II. Apart from a radial resonator longitudinally excited, longitudinal resonators of arbitrary cross section and transversally excited longitudinal resonators of square shape are known [28].

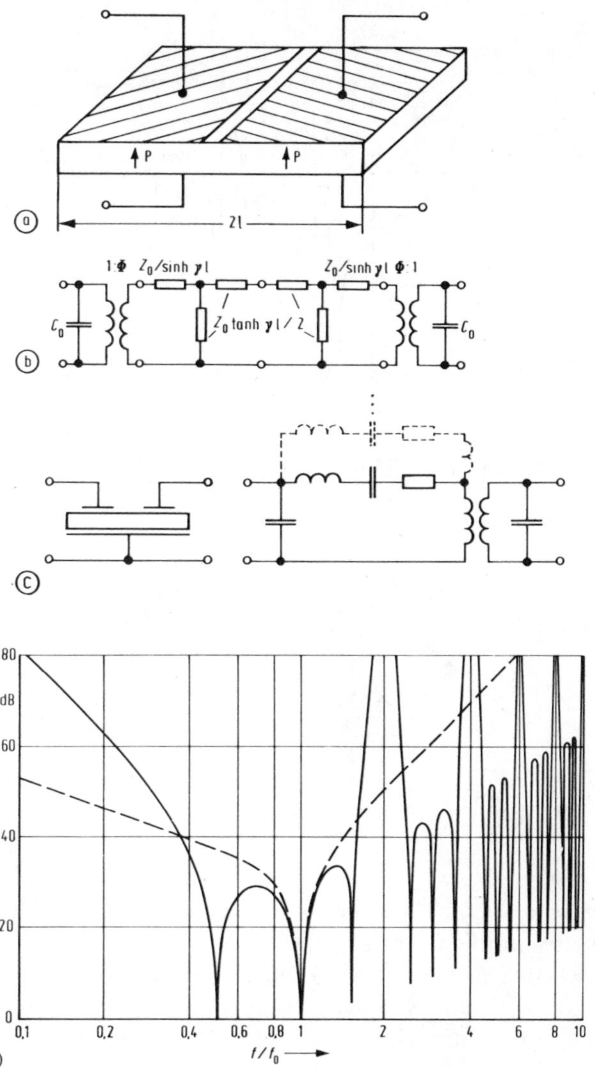

Fig. 4. Two-port ceramic resonator with transversally excited longitudinal vibration. (a) Physical arrangement. (b) Circuit diagram with line elements. (c) Circuit diagram with lumped elements. (d) Calculated transmission characteristic.

C. Two-Port Ceramic Resonators

Concepts have often been proposed with two-port resonators exhibiting the effect of a single resonant circuit [8], [9]. As an example, a two-port resonator is shown in Fig. 4 which may be described as a combination of two one-port resonators. The transmission characteristic may be calculated for such an arrangement if the resonator is sufficiently thin [9].

By measurements on such a resonator, which are reproduced in Fig. 5, the calculated characteristic may be confirmed in principle. In addition, further resonances occur, and due to electrical overcoupling between input and output, the stop-band attenuation is limited to 40 dB, approximately. In supplied resonators, the passband at f_0 is used in each case due to the desired wide bandwidth, although undesired passbands must be accepted below and just above the used frequency [8]. Arrangements such as shown in Fig. 6 that exhibit isolated passband in a wide frequency range but need two reso-

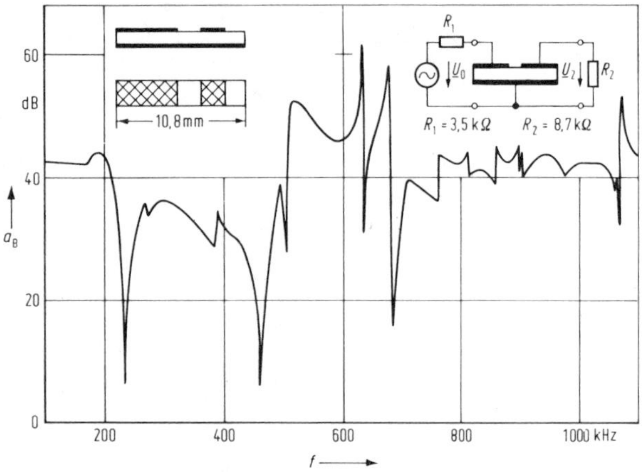

Fig. 5. Measured transmission characteristic of two-port ceramic resonator.

Fig. 6. Two-port ceramic resonator with ceramic disk for 455 kHz (product of CRL Elektronische Bauelemente Porz, F. R. Germany). Diameter of disk: 5.8 mm.

nator discs for the effect of one resonance circuit [9] are more suitable for this purpose.

During recent years very many experiments have been made with almost-square resonators. As a rule, mode 3 of the dilatational modes is used [28]. In Fig. 7 an appropriate arrangement is shown, and in Table III the equivalent ratings are summarized. In comparison with the radial resonator, the merits are low sensitivity towards small faults of contour which allow fine adjustment and the possibility of fixing the mounts near the center of the side surface so that the two-port resonators may be produced in large quantity [29]. Resonators with reasonable dimensions such as discussed are used for frequencies between 100 kHz and 1 MHz.

D. Low-Frequency Ceramic Bending Resonators

Only bending resonators may be considered for frequencies below 100 kHz. According to the matrix (3) direct bending excitation is not possible. Bimorph devices have been made which, however, need two ceramic parts as well as a complicated connection technique. Therefore, the proposal of a resonator resembling a tuning fork in accordance with Fig. 8 which is designed as a three-electrode resonator [30], is notable. Excitation takes place through an assymmetric arrangement of the excitation electrodes and, as shown in Table IV, coupling factors in excess of 10% are obtained with this excitation. Tuning fork resonators as composite resonators will not be discussed here because such resonators are usually considered mechanical filters [31].

Fig. 7. Rectangular shaped two-port ceramic resonators of two producers for 455 kHz (product of CRL Elektronische Bauelemente Porz, F.R. Germany). Edge length of ceramic disk: 5 mm.

TABLE III
CIRCUIT ELEMENTS OF RECTANGULAR SHAPED TWO-PORT CERAMIC RESONATOR (FIG. 7)

C_{NF1}	=	146pF	C_{NF2}	=	156pF	C_{o1}	=	145pF	C_{o2}	=	151pF
f_m	=	448.580kHz				C_1	=	1.2pF			
f_{n1}	=	450.460kHz	f_{n2}	=	455.270kHz	L_1	=	104mH			
K_{EM1}	=	9.1%	K_{EM2}	=	17.2%	R_1	=	405Ω			
r_1	=	120	r_2	=	33.6	ü	=	0.52			
Q	=	715									

$TC\ (f_m) \leq 20$ ppm/K
Size: 4.94mm x 4.93mm x 0.4mm
Material: Sonox P 6
Producer: CRL Elektronische Bauelemente Porz, F.R. Germany

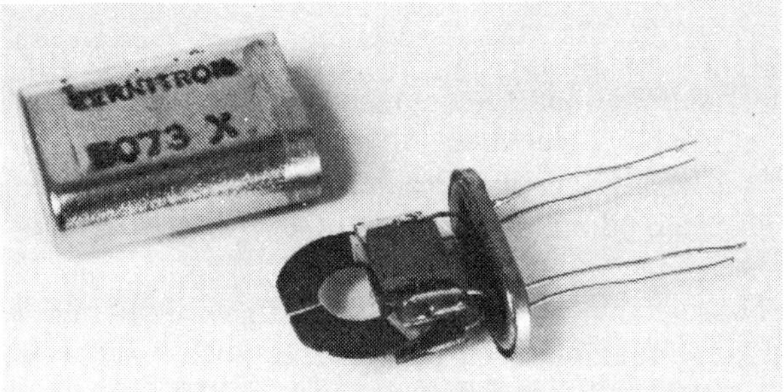

Fig. 8. Low-frequency two-port ceramic resonator for 10 kHz (product of Vernitron, USA). Diameter of disk: 14.6 mm.

E. Two-Port Monolithic Ceramic Filters

Only thickness-shear and thickness resonators come into question for frequencies above 1 MHz. Some proposals have been published for the design of single ceramic resonators in the 10 MHz band [32], [33]. These designs are based on experience gathered with crystal resonators.

The proposal, which is interesting especially for higher frequencies, to permanently mount on a substrate a resonator by means of quarter-wave decoupling films, cannot be realized because materials with appropriate acoustic impedances are not available. An exact analysis of the technical and technological feasibilities shows that the concept of two and multipole monolithic ceramic filters with thickness resonators is the cheapest for broadcast IF filters. Consequently, filters in chain connection are made up of such two-port filters. A photograph of a materialized filter element is shown in Fig. 9. The resonator

TABLE IV

CIRCUIT ELEMENTS OF TWO-PORT LOW-FREQUENCY CERAMIC RESONATOR (FIG. 8)

C_{NF}	=	215pF	C_{NF2}	=	141pF	C_{o1}	=	213pF	C_{o2}	=	138pF
f_m	=	10.160kHz				C_1	=	2.2pF			
f_{n1}	=	10.212kHz	f_{n2}	=	10.251kHz	L_1	=	111.6H			
K_{EM1}	=	10.2%	K_{EM2}	=	13.6%	R_1	=	10.25kΩ			
r_1	=	96.8	r_2	=	54.0	ü	=	0.93			
Q	=	700									

TC (f_m) $\leq$ 30 ppm/K

Size: 14.6mm Ø x 7.0mm Ø x 0.7mm

Material: PZT 6

Producer: Vernitron, USA

Fig. 9. Monolithic filter element with two mechanically coupled resonators for 10.7 MHz (products of Murata, Japan, and CRL Elektronische Bauelemente Porz, F.R. Germany). Ceramic disk length: 6 mm.

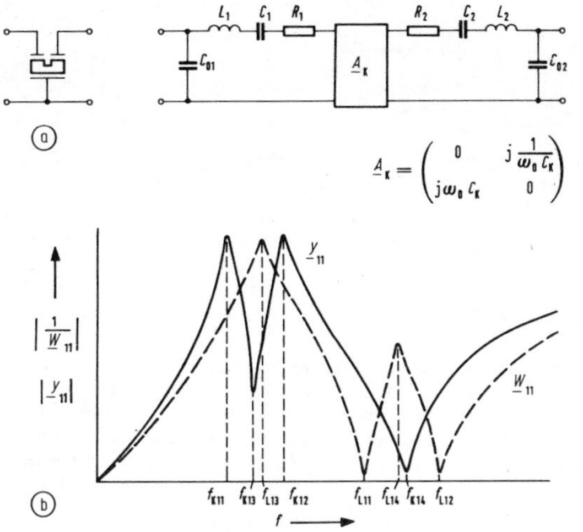

Fig. 10. Monolithic filter element with two mechanically coupled resonators (Fig. 9). (a) Circuit diagram with lumped elements. (b) Admittance at input with output short circuited and open.

TABLE V

CIRCUIT ELEMENTS OF MONOLITHIC FILTER ELEMENT (FIG. 9)

C_{NF1}	=	61pF	C_{NF2}	=	67pF	C_{o1}	=	56.3pF	C_{o2}	=	62.4pF
f_{L13}	=	10,672MHz	f_{L23}	=	10,668kHz	C_1	=	4.2pF	C_2	=	4.45pF
r_1	=	12,4	r_2	=	13.1	L_1	=	51.0μH	L_2	=	47.8μH
Q_1	=	110	Q_2	=	113	R_1	=	30.1Ω	R_2	=	28.1Ω
K_{12}	=	2.53%							C_K	=	175pF

TC (f) $\leq$ 50ppm/K

Size: 5.5mm x 6.0mm x 0.22mm

Material: PZT

Producer: Murata, Japan

areas, which represent the mechanically coupled filter, consist of evaporated silver films separated by a narrow gap.

The thickness tolerance of the disc at 10 MHz together with the tolerances of the acoustic velocity are responsible for the frequency spread of the filter. With lapping accuracies of 1 micron, a frequency spread of 50 kHz must be expected, which, however, may be accepted at filter bandwidths above 200 kHz.

For the measurement and calculation of this two-port filter, the equivalent circuit stated in [34] in modification of the equivalent circuit for single resonators [4] may be used for coupling the two series resonant circuits with a coupler $\boldsymbol{A}_k$. Only a wideband delay element has been disregarded between input and output circuits. From the short-circuit input admittances $\boldsymbol{Y}_{11}$, $\boldsymbol{Y}_{22}$ and the no-load input impedances $\boldsymbol{Z}_{11}$, $\boldsymbol{Z}_{22}$ all the equivalent ratings may be ascertained in one calculation. In Fig. 10 the equivalent circuit and the frequency responses of the admittances and impedances are reproduced. Therefore, the loss angle of the input capacitances C_{01} and C_{02} have been disregarded. Details of the calculation may be obtained from [34]. As an example, Table V shows ratings of monolithic two-port filters which were measured on a filter arrangement in accordance with Fig 9.

The adjustment of the mechanical coupling K between the two resonators and their dependence on shape are interesting for design. Such curves have been published for AT crystal resonators. For the electrode shape shown in Fig. 9 with semicircular electrodes, the coupling was measured as a function of the gap width between the resonators. This dependence is shown in Fig. 11. In the semilogarithmic presentation a straight line is found as was expected according to the theory [12]. The freedom from spurious responses of this filter is surprising. Apart from the weakly excited basic shear mode at 4.8 MHz, only few interfering spurious responses are found below and above the desired passband in accordance with Fig. 12, whereas the desired passband range is completely free from spurious responses. Suitable damping of the ceramic disc outside the resonator surface is important to obtain the proper shape of the transmission characteristic. The transmission characteristic shown in Fig. 12 was measured on an optimally matched two-port filter. Thorough investigations indicate that, apart from the thickness of the applied silver film and the electrode configuration [35], the material data and the polarization treatment also play an important role for the transmission properties. A better insight into the resonating problems of thickness resonators has been published in [36], where all new mathematical methods were used for the solution of boundary value problems.

Two different designs result for a filter frequency of

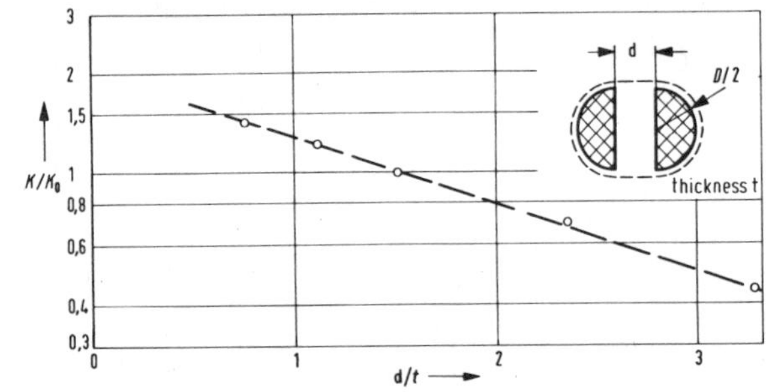

Fig. 11. Mechanical coupling coefficient of monolithic filter element (Fig. 9) as function of distance d between two resonators normalized with thickness t of ceramic plate. (Electrode diameter D = 1.2 mm, thickness of ceramic plate at 10.7 MHz, thickness vibration t = 206 μm, normalizing coupling coefficient k_o = 1.8%).

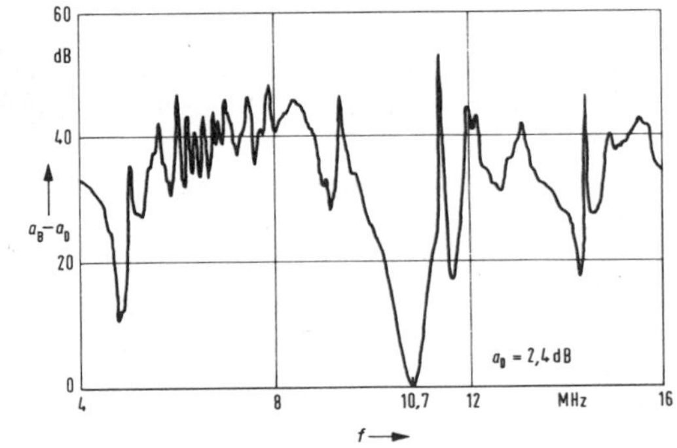

Fig. 12. Transmission characteristic of monolithic filter element (Fig. 9) (coupling coefficient 1.65 %, matching resistors 125Ω, transmission loss a_D = 2.4 dB).

approximately 5 MHz. An arrangement with thickness resonators, which is intended for use as a television sound IF filter, is shown in Fig. 13. Here two ceramic discs with two-port filter units are combined with an additional capacitor to form a four-pole filter. Since the number of resonators required in this design is determined not so much by the desired transmission characteristic as the stopband attenuation at the spurious wave frequencies, a fully monolithic three-pole ceramic filter with thickness-shear resonators in accordance with Fig. 14 could be more suitable because for this filter, too, the tolerance problems, which occur for the coupling capacitor on the cascade circuit of two-port filters, do not apply. However, this three-pole design did not satisfy the spurious wave requirements above the passband.

In addition to the papers about filters designed in accordance with the energy trapping principle, a number of proposals have been made with multimode resonators [37], [38]. However, since in the known cases the difficulty of the adjustment for the various resonances used could not be solved, these designs have not been adopted.

IV. CERAMIC MATERIALS

The development of ceramic piezoelectric materials was concentrated in the USA and Japan. The history of this development has been vividly described in [39]. Hitherto, primarily PZT ceramics were used for resonator and filter applications, the work on these materials having

Fig. 13. Monolithic filter element for television sound IF-filter at 5.5 MHz (product of CRL Elektronische Bauelemente Porz, F.R. Germany). Ceramic disk length: 6 mm.

Fig. 14. Proposal of monolithic filter with three resonators for television sound IF-filter at 5.5 MHz using thickness-shear mode. Ceramic disk length: 6 mm.

been directed during the past years towards a short selection of a combination of material ratings most suitable for filter applications, an improvement of the stability ratings, and the reduction of their sensitivity in production processes.

Starting with work on ceramic whose zirconate to titanate proportions are chosen directly at the phase limit between the tetragonal and rhombohedral crystal structure to achieve a maximum coupling factor [4], there are two combination systems for filter ceramics which exhibit a zero transit of the temperature coefficient of the resonance frequency [41]. The one transit is situated very near the phase limit in the tetragonal system and provides materials with a higher coupling factor though lower resonant quality factor and higher aging. The other transit is situated more in the rhombohedral system with lower ratings for the coupling factor, higher resonant quality factor, but stability ratings which are still inadequate for filter applications [42], [43]. Consequently, work has continued on the doping of PZT ceramic for many years. By the additional of Mn, Fe, Cr, Co, Sb, and Ni or combinations of these additives, it is now possible to achieve a stabilization of both the dielectric constant as well as of the resonance frequency in combination with an improvement in quality factor. The ratings obtainable in production are approximately 1%/decade for the dielectric constant and 1‰/decade for the resonance frequency. A review of the effects of the additives has been given in a number of publications [42], [44], [45]. As an example of a filter material, Fig. 15 shows the four most important resonator equivalent ratings of a radial resonator, namely, resonance frequency f_m, planar coupling factor k_p, mechanical resonance quality factor Q, and capacitance at low frequency C as a function of temperature. In addition to the temperature response, an after effect is noticeable which is not aging and which decays with a time constant of days.

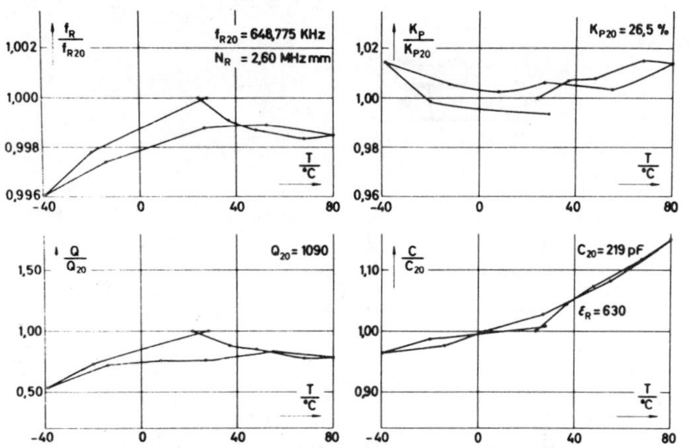

Fig. 15. Temperature stability of the ceramic material Sonox P6 measured on circular radially vibrating disk (fundamental radial mode f_m, planar coupling coefficient K_p, quality factor Q, low-frequency capacitance C) (producer: Rosenthal-Technik Lauf, F.R. Germany).

The loading capacity is also an important parameter for crystal resonators. These ratings are very high for piezoelectric ceramics so that over drive does not occur for normal filter applications [46]. The state of PZT ceramics reached so far is shown in [39] and in earlier papers [47], [48].

Many years ago experiments started on new material systems in order to substantially increase the activity $Q/r \leq 200$ stated in Table I for PZT ceramics. Although ternary systems [49] exhibit higher ratings up to 800 in a restricted temperature range, the ratings are achieved at the expense of temperature stability [50]. The difficulty in the task of developing further materials for ceramic filters consists of bringing a large number of material ratings into a desired framework, whereby tolerances important particularly for production, such as that of frequency constants, must not be ignored, which for PZT ceramic have been brought into the range of parts per thousand in one production batch.

V. FILTER CONFIGURATIONS WITH CERAMIC ELEMENTS

The first designs of ceramic resonators were based on the filter circuits familiar for quartz crystals. A very detailed presentation of the state of the art is given in [51], and it also contains a list of references with more than 150 quotations. On the use of one-port resonators, the familiar bridge circuits are employed. Circuit proposals using push–pull inputs and outputs of differential amplifiers instead of differential transformers approach more the present semiconductor circuitry [52]. Two-port resonators which were not usual as crystal resonators because the problem of electrical overcoupling of input and output was difficult to solve may be designed as ceramic resonators with an electrical decoupling above 40 dB due to the higher dielectric constants. In Fig. 16 a number of circuits are listed for filters in the frequency range below 1 MHz. Due to the low cost of resonators, such complex concepts as ladder filters in accordance with ① may also be tackled [52]. The two bridge concepts ⑤ and ⑥ are well known and, by the inclusion of coils, have the merit of easier matching of impedance ratings and better spurious wave suppression [22], [23]. Capacitively coupled three-electrode resonators allow the flexible design of larger filter complexes in accordance with ②, ③ and ④. The combination of two- and three-electrode resonators also allows the materialization of filters with finite attenuation poles. Over and above these designs, lattice filters have also been proposed [53], although the number of resonators is higher by approximately a factor of 2 than in normal ladder filters. The standardization of resonators, desirable for production reasons, gave rise to the proposal for pulling and shunting single-port resonators with capacitors [52]. However, the additional complexity is much more serious than the more complex production for several types of resonator. For frequencies above 2 MHz, the only circuit proposal which has been adopted calls for the use of two-pole or multipole filter elements in chain connection [34]. In the future, tests must show the extent to which a higher degree of integration than two resonators with mechanical coupling is appropriate. In regard to monolithic crystal filters this question is being discussed [54], [55], [34].

VI. FILTER CALCULATION

In the beginning, the methods of image parameter theory were utilized for the calculation of ceramic filters [56], [57], [51]. Following the publication of a number of filter catalogs [56], [21], the desire grew to also use the published standard designs for ceramic filters. On the other hand, several advances had been made in the rating of effective loss design of crystal filters and mechanical filters [59], [60]. All of these methods used primarily the Norton transformation for a modification of the circuits such that ceramic filters or resonator elements could be incorporated [61]. Moreover, the possibilities were exploited of using narrow-band approximations in the range below 10% due to the narrow relative bandwidth of the filters. Furthermore, a low-pass transformation is possible for frequency-symmetrical filters. The methods adopted shall be described with reference to two examples:

For the AM-IF filter the degree 4 is chosen for the standard receiver. In regard to transmission loss, stability, fitting to the equipment concept, and the suppression of spurious responses, a filter proposal in accordance with Fig. 17 is optimum. Starting from the low pass equivalent circuit, the components of the actual circuit may be calculated directly [22]. The actual circuit contains a number of additional components which may be selected through unspecified additional conditions. Thus, for example, it is possible to make the two ceramic resonators equal and to freely select the impedance level and the capacitance ratio of the ceramic resonators. In [22] the equation system and its solution are stated for this special case. A comparable method of solving the problem has also been attempted for ladder filters [62].

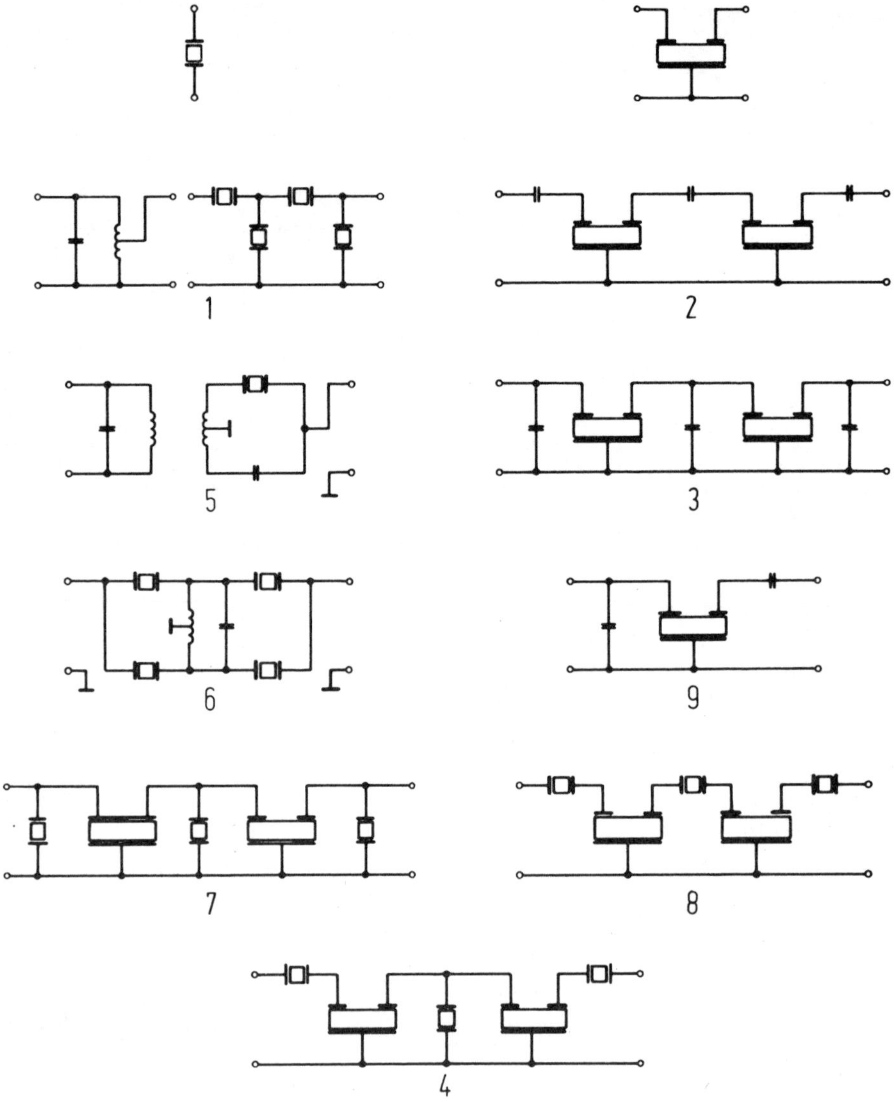

Fig. 16. Filter configurations with single ceramic resonators.

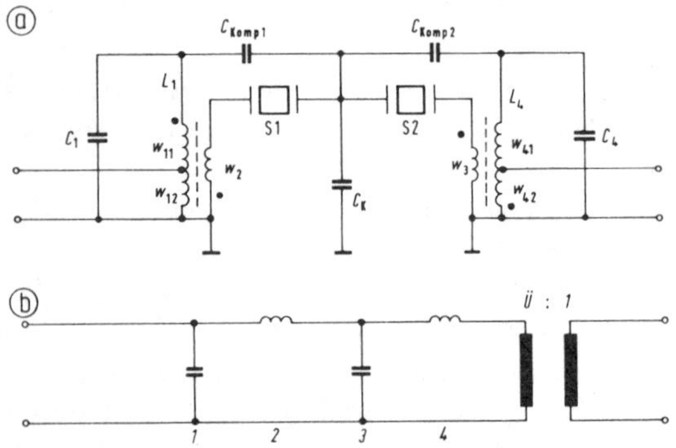

Fig. 17. IF filter circuit with two differential transformers and two ceramic resonators at 460 kHz. (a) Circuit diagram. (b) Equivalent low-pass circuit.

Depending on receiver quality, five to nine circuits are customary for selectivity in the FM-IF filter. For a filter proposal of uneven degree Fig. 18 shows the conversion from a low-pass equivalent circuit to an actual circuit with a coil circuit at the input and a number of two-port ceramic filter elements in chain connection. Here, too, as applied for the AM filter, a number of additional unspecified ratings may be exploited for the technically optimum design of the ceramic filter elements [22]. Details are given in [34] and [63] about the transformations used for both mechanical filters and monolithic crystal filters.

The filter calculation methods implied here allow, starting from the desired transmission characteristic in the first step, the determination of the circuit elements of an equivalent low pass circuit as a ladder network with the usual methods of network synthesis, for example, or also by reference to catalogs and then the statement of materialization with ceramic resonators.

In many cases it is appropriate to continue with an optimization of the circuit with a view to production tolerances and consideration of the real quality factor. For filters of lower degree (4 to 5) a table-top computer with an XY recorder may be used, for example, an interactive optimization with very simple means.

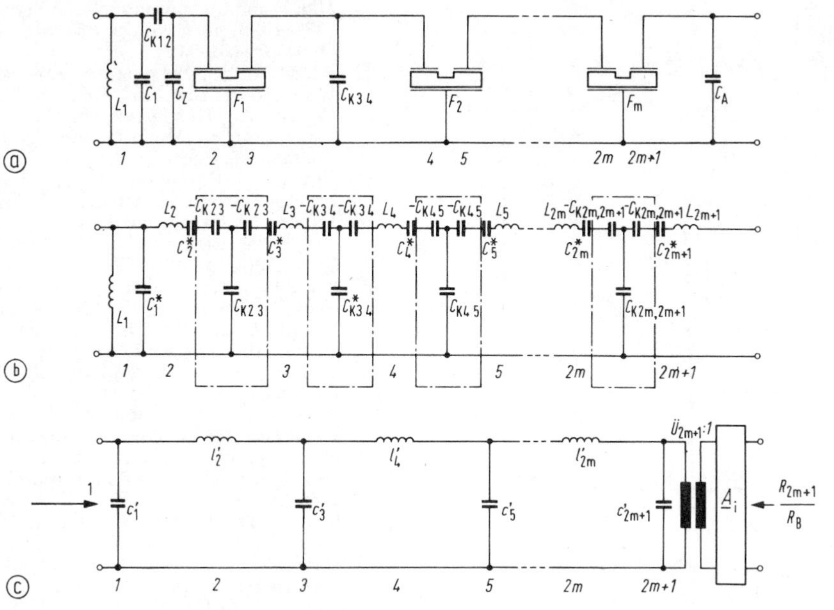

Fig. 18. IF filter circuit with one coil and monolithic filter elements at 10.7 MHz. (a) Circuit diagram. (b) Equivalent circuit with couplers. (c) Equivalent low-pass circuit.

Fig. 19. Photo of two realized IF filters for sound broadcasting together with monolithic amplifier (product of CRL Elektronische Bauelemente Porz, F.R. Germany).

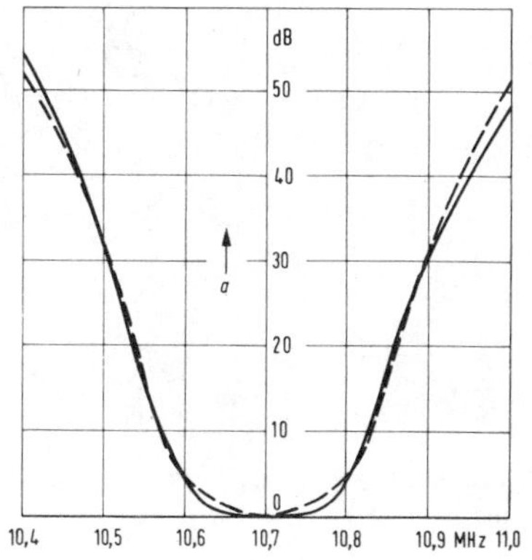

Fig. 21. Transmission characteristic of FM IF filter (Fig. 18) with two monolithic filter elements (solid line: measured, dotted line: calculated).

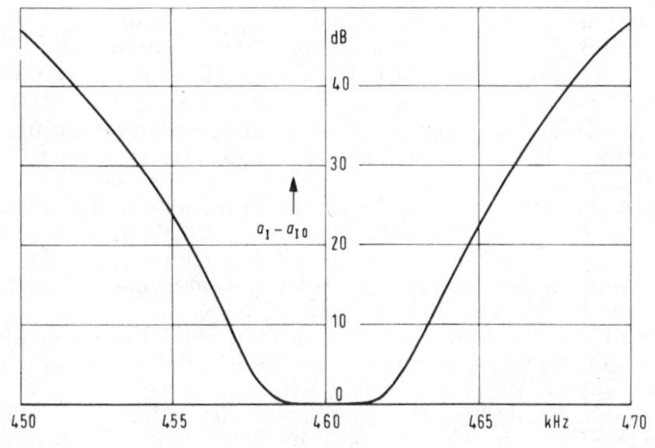

Fig. 20. Transmission characteristic of AM IF filter (Fig. 17).

VII. REALIZED FILTERS

In Fig. 2, 6–9, 13, and 14 ready realized arrangements of single resonators and monolithic filters are shown. In conclusion, three further realized filter assemblies and appropriate transmission characteristics are shown. The hybrid filters intended as AM-IF filters (two-coil circuits and two ceramic resonators) and FM-IF filters (one-coil circuit and two two-port ceramic filters) have been photographed in Fig. 19 together with a 16-pin DIP of an IF amplifier.

Cases with pressure contacts are used to hold the ceramic resonators. The entire filter is then mounted as a module on a printed circuit board. Today more modern methods can be used both for the holder as well as for mounting. For reasons of simplicity the design in Fig. 19 was chosen because the equipment manufacturers did not demand a higher packing density for standard equipment. The test ratings of the attenuation frequency responses of the AM and FM filters are shown in Figs. 20 and 21.

For the FM filter the calculated frequency response is entered for comparison due to deviations. The noticeable assymmetry in the stopband is connected with the coupling of monolithic filters being exponentially dependent

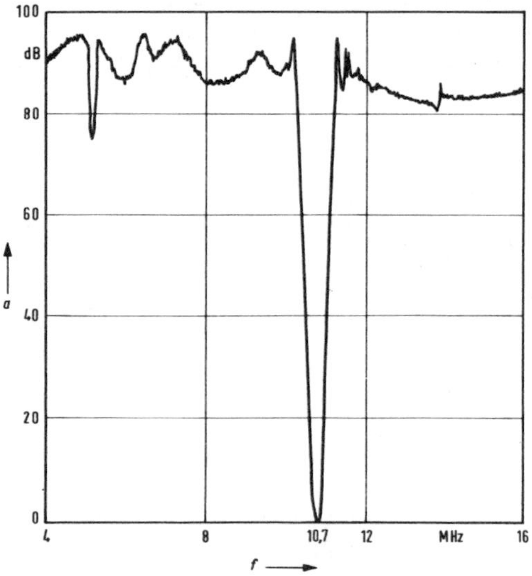

Fig. 22. Spurious responses of FM IF filter (Fig. 18).

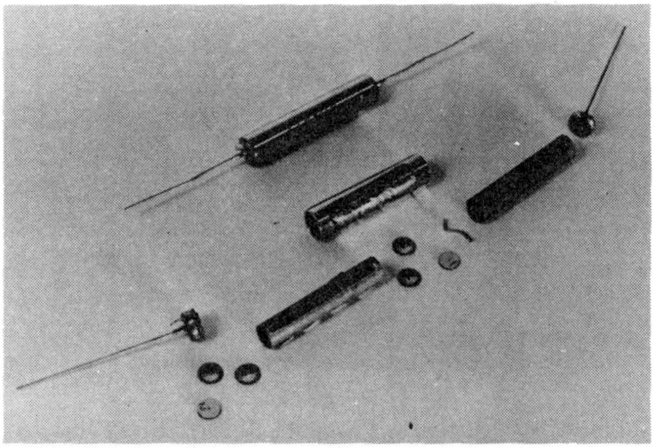

Fig. 23. Ceramic ladder filter with 17 ceramic disk (5 mm diameter) (product of Clevite Corp., Bedford, Ohio, USA).

on frequency and has also been observed with crystal filters. The measure in which the suppression of interfering spurious modes is achieved is shown in Fig. 22 for the FM filter.

Apart from these hybrid filters the ladder filter will be quoted as an example for a filter which consists solely of ceramic resonators. The filter in accordance with Fig. 23 contains 17 one-port ceramic resonators in ladder connection and features at 10% relative bandwidth a very high skirt selectivity and far-off selection of 80 dB with 50 kHz bandwidth and 500 kHz center frequency.

Papers from Japan [64] indicate the direction in which work in the field of ceramic filters will continue. However, tests must first of all show the extent to which such design may be adopted.

REFERENCES

[1] Roberts, S., Dielectric and Piezoelectric Properties of Barium Titanate, Phys. Rev. 71 (1947) pp. 890–895.
[2] Mason, W. P., Electrostrictive effect in bariumtitanate ceramics, Phys. Rev. 74 (1948) pp. 1134–1147.
[3] Jaffe, B., Roth, R. S., Marzullo, S., Piezoelectric properties of lead zironate-leaditanate solid-solution ceramics, J. appl. Phys. 25 (1954) pp. 809–810.
[4] IRE standards on piezoelectric crystals, Proc. IRE 37 (1949) p. 1378–1395; 45 (1957) p. 353–358; 46 (1958) p. 764–778 and 49 (1961 p. 1161–1169.
[5] Katz, H. W., Solid state magnetic and dielectric devices, J. Wiley Sons, New York 1959, p. 87–129, 170–232.
[6] Lungo, A., Henderson, K. W., Application of Piezoelectric Resonators to Modern Band-Pass Amplifiers, IRE National Convention Record (1958) pp. 235–242.
[7] Curran, D. R., Gerber, W. J., Piezoelectric ceramic I.F. filters, Proc. Electronic Component Conference (1959) p. 160–165.
[8] Lungo, A., Sauerland, F. A., Ceramic Band Pass Transformer and Filter Elements, IRE intern. Convention Record Vol. 9 Pt. 6 (1961) p. 189–203.
[9] Schüßler, H., Einkreisige keramische Filterelemente, Arch. elektr. Übertragung 17 (1963) p. 223–229.
[10] Schüßler, H., Filtersätze aus keramischen Schwingern, Arch. elektr. Übertragung 17 (1963) p. 519–524.
[11] Housing for ceramic ladder filter, US-Patent Nr. 2 927 285 (1960).
McSwan, A. M., Campbell, D. S., Some Use of Piezoelectric Lead Titanate Zirconate, Journal of the IEE (1962) pp. 374–378.
[12] Shockley, W., Curran, D. P., Koneval, D. J., Energy trapping and related studies of multiple electrode filter crystals, Proc. 17th Annual Symposium on Frequency Control (1963) p. 88–123
Curran, D. R., Koneval, D. J., Miniature ceramic band pass filters, Proc. National Electronic Conference (1961) p. 514–520.
[13] Onoe, M., Jumonji, H., Analysis of piezoelectric resonators vibrating in trapped-energy modes, Jour. I.E.C.E., Japan, *48* (1965) p. 1574 in English E.C.I. p. 84.
Fujishima, S., Nosaka, S., Ishiyama, H., 10 MC Ceramic Filters by Trapped Energy Modes, Report of the Acoustic Society of Japan Nov. (1966).
[14] Data sheets: Murata Cat. 761-E Feb. 1967 3K, Vernitron Piezoelectric Division 94033, Gould/Clevite Bulletin 94033, Valvo-Handbuch Piezoxide 1973.
Schumacher, H., Piezokeramische Hybridfilter für AM-und FM-Rundfunk, Stemag-Nachrichten 43 (1970) p. 1163–1170.
McDermott, J., Focus on Piezoelectric Crystals and Devices, Electronic Design 17, Aug. 16 (1973) pp. 44–54.
[15] Mitchell, R. F., Akustische Oberflächenwellenfilter, Philips techn. Rdsch. 32 (1971/72) p. 201.
[16] Bowers, K. D., Ultrasonics in Communications, Proc. 1970 Ultrasonics Symp., San Francisco p. 114–125.
[17] Land, C. E., Transistor oscillators employing piezoelectric ceramic freedback networks, IEEE International Convention Record 1965 vol. 13 pt. 7, pp. 51–68.
[18] Mason, W. P., Use of Piezoelectric Crystals and Mechanical Resonators in Filters and Oscillators, in Physical Acoustics, Mason, W. P. Ed. 1 pt. A. Academic Press, New York 1964, pp. 377–393.
[19] Nagata, T., Nakajima, Y., Sasaki, R., Pb Ti O_3 Ceramic Resonators Operating in VHF-Band, Electronics and Communications in Japan Vol. 55-C No. 7 (1972) pp. 93–98.
[20] Kohlbacher, G., Grundwellenresonatoren und Monolithische Keramikfilter im 40 MHz Gebiet mit Kalium—Natrium—Niobat, to be published.
[21] Zeverev, A. I., Handbook of filter synthesis, John Wiley and Sons, New York, London, Sydney 1967.
[22] Kohlbacher, G., Kohlhammer, B., Schüßler, H., Hybride Rundfunkfilter mit keramischen Resonatoren und Filtern, Wiss. Ber. AEG-TELEFUNKEN 43 (1970) 2, pp. 120–139.
[23] Schüßler, H., The influence of new technology on filter techniques in receivers for sound and television broadcasting, EBU Review, Part A-Technical, No. 130 (December 1971) pp. 250–260.
[24] Shaw, E. A. G., On the Resonant Vibrations of Thick Barium Titanate Disks, J. Acoust. Soc. Am 28 (1956) p. 38–50.
[25] Munk, E. C., The equivalent electrical circuit for radial modes of a piezoelectric ceramic disc with concentric electrodes, Philips Research Rep. vol. 20 (1965) pp. 170–189.
Tachibana, A., Partically Plated Thin Disk Type Piezoelectric Ceramic Vibrator, Journ. of the Institute of Electrical Comm. Eng. Japan (1965) July, pp. 17–19.
[26] Holland, R., Accurate Measurement of Coefficients in a Ferroelectric Ceramic, IEEE Trans. on Sonics and Ultrasonics Vol. Su-16 (1969) pp. 173–181.
[27] Meitzler, A. H., O'Bryan, H. M., Tiersten, H. F., Definition and Measurement of Radial Mode Coupling Factors in Piezoelectric Ceramic Materials with Large Variations in Poisson's Ratio, IEEE Trans. on Sonics and Ultrasonics Vol. SU-20 (1973) pp. 233–239.
[28] Holland, R., Contour Extensional Resonant Properties of Rectangular Piezoelectric Plates, IEEE Trans. Sonics and Ultrasonics, vol. SU-15 (1968) pp. 97–105.
[29] Data Sheet: Fa. Stettner Lauf (Murata) 32-0374-22D, Ceramic filters SFD-455D.
[30] Curran, O. R., Gerber, W. J., Low-Frequency Ceramic Band-

Pass Filters, Proc. Electronic Components Conference (1963) pp. 108–113.
[31] Johnson, R. A., Börner, M., Konno, M., Mechanical Filters—A Review of Progress, IEEE Trans. on Sonics and Ultrasonics, Vol. SU-18 (1971) pp. 155–170.
[32] De Jong, M., Thickness Vibrators of Piezoceramic Materials, IEEE Trans. Sonics and Ultrasonics, vol. SU-17 (1970) p. 69.
[33] Newell, W. E., Tuned Integrated Circuits—A State-of-the-Art Survey, Proc. IEEE, Vol. 52 (1964) pp. 1603–1608.
[34] Kohlbacher, G., Entwurf mehrkreisiger Kristall- und Keramikfilter aus monolithischen Einzelfilterelmenten mit Hilfe äquivalenter elektrischer Tiefpaßersatzschaltungen, Arch. elektr. Übertragung 25 (1971) pp. 492–501.
[35] Ortlepp, G., Racurow, B., Keramische mechanische Filter für höhere Frequenzen, Hermdorfer Techn. Mitt. (1968) 23, pp. 736–745.
[36] Holland, R., Eer. Nisse, E. P., Design of Resonant Piezoelectric Devices, Research Monograph No. 56 The M.I.T. Press, Cambridge, London, 1969.
[37] Onoe, M., Jumonji, H., Analysis of piezoelectric multipole-mode resonators vibrating in longitudinal and flexural modes, Trans. I.E.C.E., Japan, *51-A* (1968)3, p. 110 in English E.C.J., p. 35.
[38] Schweppe, H., Excitation of two-Adjacent Resonance with a Chosen Frequency Separation in a Ceramic Piezoelectric Resonator, IEEE Trans. on Sonics and Ultrasonics, vol. SU-17 (1970) 1, pp. 12–17.
[39] Jaffe, B., Cook, W. R., Jaffe, H., Piezoelectric Ceramics, Academic Press London, New York 1971.
[40] Carl, K., Härdtl, K. H., On the Origin of the Maximum in the Electromechanical Activity in Pb $(Ti_{1-x}Zr_x)$ O_3 Ceramics Near the Morphotropic Phase Boundary, phys. stat. sol. (a) 8 (1971) pp. 87–98.
[41] Zeyfang, R., Handschuh, K., Stärk, N., Temperature coefficients of resonance frequency of radially vibrating ceramic resonators as a function of concentration in binary and tenary perovskite-systems, to be published.
[42] Kulcsar, F., Electromechanical Properties of Pb (Ti Zr) O_3 Ceramics Modified with Certain Three or Five-Valent Additions, J. Am. Ceram. Soc. 42 (1959) p. 343.
[43] Kaminski, W., Aging Processes in Chemically Modified Solid Solutions of Zirconium, Titanium and Lead Oxides, Bulletin del'Academie Polonaise des Sciences, Serie des sciences techniques vol. XIII, No. 8 (1965) pp. 105–110.
[44] Thomann, H., Piezoelektrische Mechanismen in Bleizirkonat-Titanat Z. angew. Physik 20 (1966) p. 554.
[45] Murakawa, K., Niwa, K., Characteristics of Lead Zirco-Titanate Ceramics, Fujitsu sci. techn. j. 4 (1968) 1, pp. 221–224.
[46] Woollett, R. S., Le Blanc, C. L., Ferroelectric Nonlinearities in Transducer Ceramics, IEEE Trans. on Sonics and Ultrasonics, Vol. SU-20 (1973) 1, pp. 24–31.
[47] Berlincourt, D. A., Curran, D. R., Jaffe, H., Piezoelectric and Piezomagnetic Materials and their Function in Transducers, in Physical Acoustics, Mason, W. P., Ed. 1 pt. A. Academic Press, New York 1964, pp. 169–270.
[48] Jaffe, H., Berlincourt, D. A., Piezoelectric Transducer Materials, Proc. IEEE 53 (1965) p. 1372–1386.
[49] Ouchi, H. J., Piezoelectric properties and phase relations of $Pb(Mg_{1/3}Nb_{1/3})O_3 \cdot PbTiO_3 \cdot PbZrO_3$ with barium or strontium Substitutions, J. Am. Ceram Soc. 51 (1968) pp. 169–176.
[50] Handschuh, K., Investigations of tenary systems, Private Communication.
[51] Scheibner, J., Schwingquarze und piezokeramische Schwinger in Bandpässen und Bandsperren der Nachrichtentechnik, Wissenschaftliche Zeitung Elektrotechnik, Vol. 13, No. 3 (1969) p. 159–179, No. 4 (1969) p. 181–192.
[52] Sauerland, F. L., Blum, W., Ceramic IF filters for consumer products, IEEE Spectrum 5 No. 11 pp. 112–126.
Sauerland, F. L., Design of piezoelectric ladder filters, IEEE Intern. convention record part 10 (1965) pp. 111–118.
[53] Lemke, K., Zur Synthese schmalbandiger spulenfreier Brückenbandfilter, Nachrichtentechnik 19 (1969) 1, pp. 25–30.
[54] Sykes, R. A., Beaver, W. D., High frequency monolithic crystal filters with possible application to single frequency and single side band use, Proc. 20th Annual Symposium on Frequency Control 20 (1966) pp. 288–308.
[55] Sheahan, D. F., Channel Bank filtering at GTE LENKURT, Proc. IEEE Intern. Symp. on Circuits and Systems (1974) pp. 115–119.
[56] Macario, R. C., Design data for bandpass ladder filters employing ceramic resonators, Electron Eng. (1961) pp. 171–177.
[57] Waren, A. D., Schoeffler, J. D., Approximation problem for resonator ladder filters, IEEE Trans. on Circuit Theory, CT-12 (1965) pp. 215–222.
[58] Saal, R., Entwurf von Filtern mit Hilfe des Kataloges normierter Tiefpässe, TELEFUNKEN GmbH., Backnang 1963.
[59] van Bastelaer, Ph., The Design of Band-Pass-Filters with Piezoelectric Resonators, Revue H.F.-VII. 7 (1968), pp. 193–206.
[60] Kohlhammer, B. Schüßler, H., Berechnung allgemeiner mechanischer Koppelfilter mit Hilfe von äquivalenten Schaltungen aus konzentrierten elektrischen Schaltelementen, Wiss. Bericht AEG-TELEFUNKEN 41 (1968) 3, pp. 150–159.
[61] Christian, E., Circuit Transformations for Crystal Ladder Filter, IEEE Trans. on Circuit Theory, CT-14 (1967) pp. 221–224.
[62] Sauerland, F. L., Entwurfsverfahren für piezoelektrische Abzweigfilter, Thesis 1969 Rheinisch-Westfälische Technische Hochschule Aachen.
[63] Rennick, R. C., An Equivalent Circuit Approach to the Design and Analysis of Monolithic Crystal Filters, IEEE Trans. on Sonics and Ultrasonics, SU-20 (1973) pp. 347–354.
[64] Sasaki, R., Nagata, T., Matsushita, S., Piezoelectric Resonators as a Solution to Frequency Selectivity Problems in Color TV Receivers, IEEE Trans. on Broadcast and Television Receivers, BTR-17 (1971) pp. 195–201.

200 MHz Thickness Extentional Mode $LiTaO_3$ Monolithic Crystal Filter

TAKEHIKO UNO

Abstract—The thickness extentional mode energy trapping phenomenon in $LiTaO_3$ plates is studied, and it is proved that energy trapping occurs only in overtone modes. The equivalent circuit representation for the thickness extentional mode is derived in a way similar to that for the thickness shear mode. For the thickness shear mode, especially for AT-cut quartz plates, many studies have been accomplished, and various kinds of monolithic crystal filters are realized. Compared with the thickness shear mode, the thickness extentional mode in $LiTaO_3$ plates is appropriate for a higher frequency region because of its large frequency constant and high electromechanical coupling constant. A monolithic crystal filter for around 200 MHz is realized utilizing the 3rd overtone thickness extentional mode in a $LiTaO_3$ Z-cut plate.

Manuscript received September 12, 1974; revised December 18, 1974.

The author is with the Musashino Electrical Communication Laboratory, Nippon Telegraph and Telephone Public Corporation, Musashino-shi, Tokyo, Japan.

1. INTRODUCTION

PIEZOELECTRIC crystal plates have long been used in frequency selective circuits, especially in narrow width bandpass filters. Because the plate thickness of the piezoelectric resonator must be of the order of the acoustic wave length, thickness extentional mode (TE-mode) piezoelectric resonators are appropriate for very high-frequency region crystal filters above 100 MHz, compared with thickness shear mode resonators. Single-crystal $LiTaO_3$ is utilized for various kinds of filters. It is also a promising material for TE-mode resonators because of its large electromechanical coupling constant and high mechanical quality factor. The frequency constant and coupling constant of a $LiTaO_3$ Z-cut plate are 3020 kHz·mm and 0.19, respectively, and that of a 47° rotated

Reprinted from *IEEE Trans. Sonics Ultrason.*, vol. SU-22, pp. 168–174, May 1975.

Y-cut plate are 3080 kHz·mm and 0.29 [1], respectively. These are cuts for which only the extentional mode is excited by the perpendicular electric field. The high electromechanical coupling constant makes it possible to fabricate filters of relatively wide bandwidth and low terminal impedance that are difficult to fabricate in the case of quartz filters.

In order to realize TE-mode spurious-free resonators or monolithic crystal filters in $LiTaO_3$ plates, it is necessary to study the energy trapping phenomenon in such plates. This paper will be concerned with the energy trapping phenomenon of the thickness extentional mode in $LiTaO_3$ plates and monolithic crystal filters in the frequency region around 200 MHz. In Section 2 dispersion relations of thickness extentional waves is obtained, and it is proved that, in $LiTaO_3$ plates, the energy trapping phenomenon occurs only in overtone TE-modes. It is shown that measured resonant frequencies of overtone trapped energy mode resonators fit well with calculated curves. In Section 3 an equivalent circuit representation for overtone TE-mode resonators is derived in a way similar to that for thickness shear mode resonators. The result concerning trial production of a monolithic filter for around 200 MHz is described in Section 4.

2. TE-MODE ENERGY TRAPPING PHENOMENON IN $LiTaO_3$ PLATES

2-1. *Dispersion Relations*

A Z-cut $LiTaO_3$ plate is shown in Fig. 1, where H is the plate thickness and X_1, X_2, and X_3 axes correspond to crystallographical X, Y, and Z axes, respectively. In order to see the possibility of energy trapping for the thickness extentional mode, a traveling wave along the X_1 axis is treated.

Displacement components u_i $(i = 1,2,3)$ and electric potential φ accompanied with the wave are assumed as

$$\begin{aligned} u_1 &= A \cos(\xi X_3) \sin(\eta X_1 - \omega t) \\ u_2 &= 0 \\ u_3 &= B \sin(\xi X_3) \cos(\eta X_1 - \omega t) \\ \varphi &= C \sin(\xi X_3) \cos(\eta X_1 - \omega t) \end{aligned} \tag{1}$$

where it is assumed that the plate is long enough along the X_2 axis, and u_i and φ are independent of X_2.

Substituting (1) for the kinetic and Poisson's equations, we obtain equations that are identical with those for the piezoelectric ceramic plates [2]. From these equations and boundary conditions on the plate faces, dispersion curves are calculated.

Fig. 2 shows the dispersion curves for a $LiTaO_3$ Z-cut plate. The material constants used in these calculations are those of [1]. Horizontal axis and vertical axis are the normalized wave number $p(= \eta H/2)$ and normalized frequency $\Omega(= f/f_0, f_0 = v_l/2H)$, respectively, where v_l is the longitudinal acoustic wave velocity along the X_3 axis. Signs TE-n $(n = 1,3,5,\cdots)$ and TS-m $(m = 0,2,4,\cdots)$ represent the nth thickness extentional mode and the mth thickness shear mode, respectively.

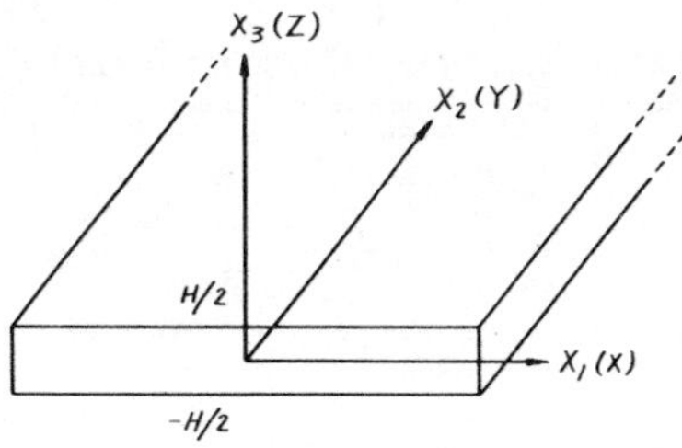

Fig. 1. Z-cut $LiTaO_3$ plate.

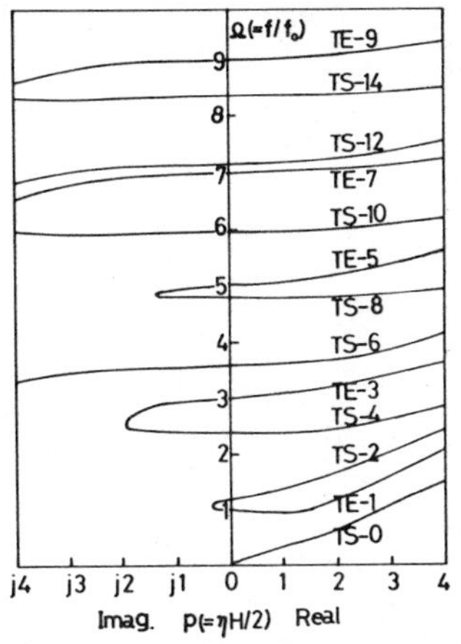

Fig. 2. Dispersion curves for Z-cut $LiTaO_3$ plate. TE-n: nth thickness extentional mode. TS-m: mth thickness shear mode. Horizontal axis and vertical axis correspond to normalized wave number and normalized frequency, respectively.

The possibility of energy trapping depends on the sign of $\partial\Omega/\partial(p^2)$ at the cutoff frequency. In the case of

$$[\partial\Omega/\partial(p^2)]_{p=0} > 0,$$

energy trapping occurs, on the other hand, in the case of

$$[\partial\Omega/\partial(p^2)]_{p=0} < 0,$$

energy trapping does not occur. From Fig. 2, $[\partial\Omega/\partial(p^2)]_{p=0} > 0$ is obtained for overtone TE-modes. On the other hand, $[\partial\Omega/\partial(p^2)]_{p=0} < 0$ is obtained for the fundamental mode. Therefore, the energy trapping phenomenon occurs only in overtone thickness extentional modes. Near the cutoff frequencies, the dispersion relations from 3rd to 9th overtone modes can be represented approximately as

$$\eta = \frac{\pi}{\gamma_n H}[(\Omega/n)^2 - 1]^{1/2} \tag{2}$$

where n is the order of overtone, γ_n is an anisotropic constant for each mode, depending on the cut angle and propagating direction, and η is a wave number, as is well known. Analysis similar to that for a Z-cut plate can be applied for a 47° rotated Y-cut plate.

In Table I, calculated values of γ_n are listed for Z-cut and 47° rotated Y-cut plates, where the propagating direction is along the X axis. From (2), the trapped energy resonant spectrum of the thickness extentional mode is obtained in a way similar to the case of the thickness shear

TABLE I
CALCULATED ANISTROPIC CONTANTS VALUES

cut angle	γ_3	γ_5	γ_7	γ_9
Z-cut	0.324	0.225	0.115	0.085
47° rotated Y-cut	0.25	0.21	0.097	0.108

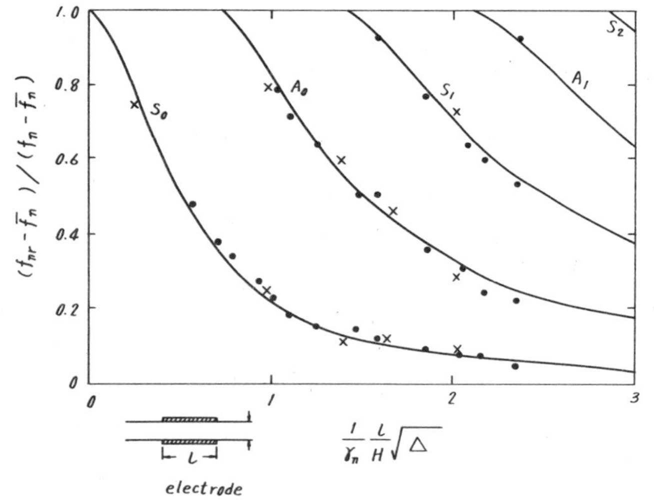

Fig. 3. Trapped energy resonators resonant frequency spectrum. s: symmetric mode and A: anti-symmetric mode. Solid lines show the calculated values, "·" and "×" correspond to measured 3rd and 5th overtone frequencies, respectively.

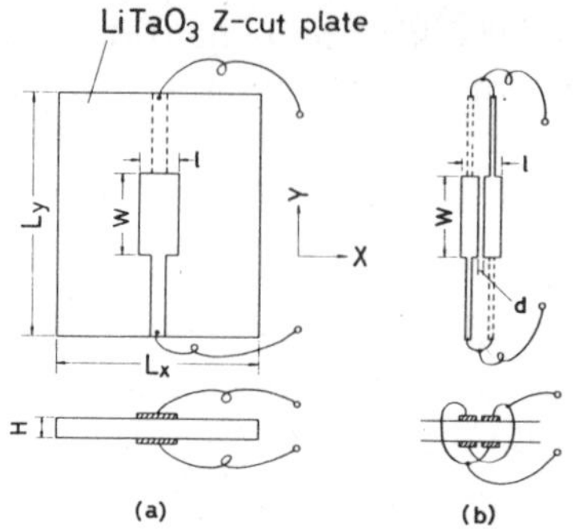

Fig. 4. Resonators electrodes construction. (a) Symmetric mode exciting method. (b) Anti-symmetric mode. Numerical values of resonators: $H = 0.3$ mm, $L_X = 8$ mm, $L_Y = 10$ mm, $w = 2.0$ mm, $l = 0.7 \sim 1.3$ mm, $d = 0.05$ mm, plate back $\Delta = 0.01 \sim 0.02$.

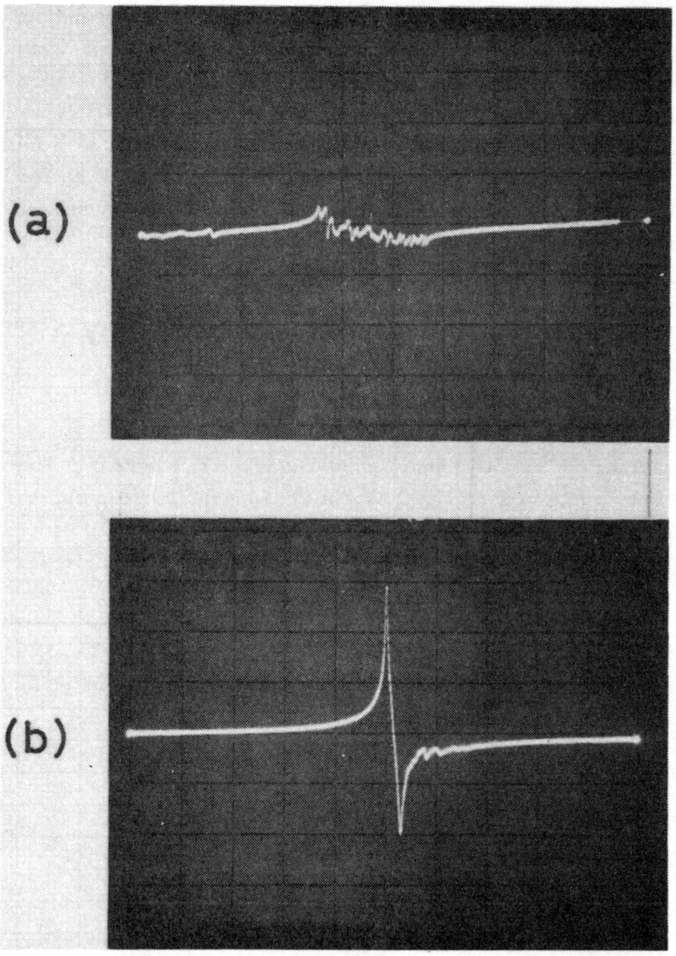

Fig. 5. Typical frequency responses of Z-cut plate resonator. (a) Fundamental mode: center frequency-9.9 MHz, horizontal-250 kHz/div, vertical-10 dB/div. (b) 3rd overtone: center frequency-30.1 MHz, horizontal-100 kHz/div, vertical-10 dB/div.

mode [3]. In Fig. 3, solid lines show the resonant spectrum, where signs S_n and A_m represent symmetric and anti-symmetric modes, respectively. The horizontal axis is degree of energy trapping $(l/\gamma_n H)\Delta^{1/2}$, where l is the electrode width and Δ is the plate back. The vertical axis is $(f_{nr} - \bar{f}_n)/(f_n - \bar{f}_n)$, where f_{nr} is the resonant frequency of the nth overtone, f_n and $\bar{f}_n[=f_n/(1+\Delta)]$ are cutoff frequencies of the nth overtone in the unelectroded and electroded regions, respectively.

2-2. Energy Trapping Experimental Results

An experiment on the energy trapping phenomenon in the Z-cut $LiTaO_3$ plate was carried out. As shown in Fig. 4(a), the symmetric mode is excited by a pair of electrodes evaporated on the top and bottom surfaces of the plate. On the other hand, to excite the anti-symmetric mode, electrodes are divided into two parts and connected crosswise, as shown in Fig. 4(b). Resonant frequencies are measured for various values of electrode size and plate back. Single crystal $LiTaO_3$ used for this experimental work is produced and poled commercially by Sumitomo Electronic Metal Division. The Curie temperature is 610°C and the poling is carried out by field cooling down through the Curie point.

Fig. 5 shows the observed resonant curves of the fundamental mode and 3rd overtone modes. Obviously, in the fundamental mode, energy trapping does not occur, but in the 3rd overtone mode, a clear resonant and an anti-resonant spectrum is observed.

In Fig. 3, the sign "·" and "×" correspond to observed 3rd and 5th overtone frequencies, respectively. The experimental points fit well with the calculated curves.

3. EQUIVALENT CIRCUIT REPRESENTATION FOR $LiTaO_3$ TE-MODE

From the experimental results for the trapped energy resonator, as mentioned before, it is easily expected that the monolithic crystal filter (MCF) [4], [5], can be realized utilizing the $LiTaO_3$ plate overtone thickness extentional mode. Such a filter may be suitable for the high-frequency region because of its large frequency constant. For example, in the case of the 3rd overtone, the Z-cut plate thickness is about 45 μm for 200 MHz. On the other hand, the AT-cut plate thickness is about 25 μm, which is very difficult to grind.

To design a monolithic crystal filter, it is necessary to know the equivalent circuit representation of the MCF. Several authors have derived [6], [7] the thickness shear or the thickness twist mode MCF. In this section, an equivalent circuit representation for the thickness extentional mode in the $LiTaO_3$ Z-cut plate will be derived in a way similar to that in [7].

Fig. 6(a) shows the dimentional configuration of a resonator. Force F and particle velocity v at the mechanical port are connected as follows:

$$F \cdot v^* = P_{X_1} = U \cdot v_g \tag{3}$$

where the asterisk represents the complex conjugate, P_{X_1} is the average time value of elastic wave power flow along the X_1 axis, U is the energy of the wave in X_1 unit length, and v_g is the group velocity.

U and v_g are given by

$$U = \frac{w}{\lambda} \int_{-H/2}^{H/2} \int_0^{\lambda} \tfrac{1}{2}(C_{ij}S_iS_j^* + \rho v v^*)\, dX_1\, dX_3 \tag{4}$$

$$v_g = \partial\omega/\partial\eta \tag{5}$$

where w is the electrode width along the X_2 axis, λ is the wave length, and other symbols are the same as in [8].

The dispersion relation (2) can be modified as

$$\omega^2 = \gamma_n'^2 v_t^2 \eta^2 + v_l^2 \xi_n^2 \tag{6}$$

where

$$\gamma_n' = n\gamma_n v_l/v_t \tag{7}$$

$$\xi_n = n\pi/H. \tag{8}$$

Here v_l and v_t are given by

$$v_l = (C_{33}^D/\rho)^{1/2} \tag{9}$$

$$v_t = (C_{55}^E/\rho)^{1/2}. \tag{10}$$

From (5) and (6), one obtains

$$v_g = \gamma_n'^2 v_t^2 \eta/\omega. \tag{11}$$

From the numerical calculation, near the cutoff frequency of the thickness extentional mode, the amplitude of displacement u_1 becomes negligible, compared with that of u_3. Therefore, energy density U can be represented as

$$U = wH\rho w^2 B^2/2 \tag{12}$$

where B is the amplitude of u_3. Equation (12) holds in a fairly good approximation for 3rd, 5th, 7th, and 9th overtone modes in the $1 \sim 2\%$ frequency deviation range from the cutoff frequency of each overtone.

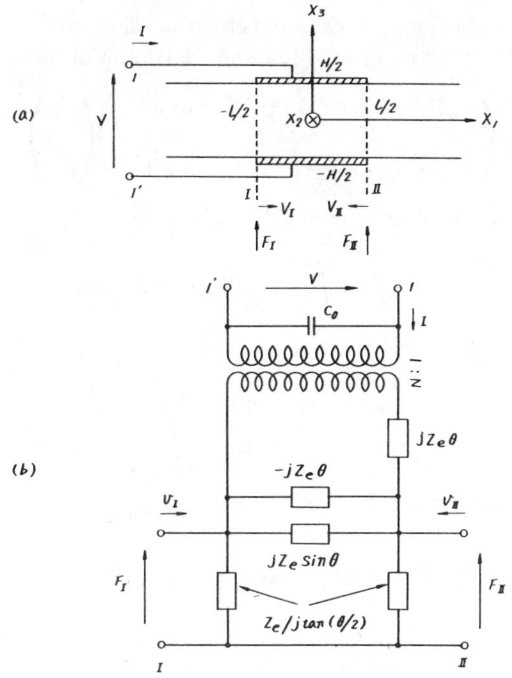

Fig. 6. Equivalent electroded region circuit representation. (a) Resonator dimensional configuration. (b) Equivalent circuit.

From (3), (11), and (12), power flow P_{X_1} is obtained as

$$P_{X_1} = wH\gamma_n'^2 C_{55}^E \omega\eta B^2/2. \tag{13}$$

It is reasonable to define velocity v by

$$v = \frac{j\omega u_3}{\sin \xi_n X_3} = j\omega B \exp\left[-j(\eta X_1 - \omega t)\right]. \tag{14}$$

Therefore, from (3), (13), and (14), force F is defined by

$$F = -\frac{wH\gamma_n'^2 C_{55}^E (\partial u_3/\partial X_1)}{2 \sin \xi_n X_3}. \tag{15}$$

In the preceding calculation, a traveling wave is considered, however, in a trapped energy mode resonator, a standing wave, which is represented as

$$u_3 = (B_1 \sin \eta X_1 + B_2 \cos \eta X_1) \sin \xi_n X_3, \tag{16}$$

must be treated. Substituting (16) for (14) and (15), one obtains

$$v = j\omega(B_1 \sin \eta X_1 + B_2 \cos \eta X_1)$$
$$F = -\omega Z_e (B_1 \cos \eta X_1 - B_2 \sin \eta X_1) \tag{17}$$

where Z_e is the mechanical impedance defined by

$$Z_e = wHC_{55}^E \gamma_n'^2 \eta/2\omega. \tag{18}$$

At mechanical ports I and II shown in Fig. 6(a), v and F are given by

$$v_{\mathrm{I}} = v(X_1 = -l/2), \qquad F_{\mathrm{I}} = F(X_1 = -l/2) \quad \text{at port I,}$$

$$v_{\mathrm{II}} = -v(X_1 = l/2), \qquad F_{\mathrm{II}} = F(X_1 = l/2) \quad \text{at port II.} \tag{19}$$

First, consider a case where the electrical applied voltage is $V = 0$. From (17) and (19), one obtains

$$\begin{pmatrix} v_{\rm I} \\ v_{\rm II} \end{pmatrix} = \begin{pmatrix} (jZ_e \tan \eta l)^{-1} & -(jZ_e \sin \eta l)^{-1} \\ -(jZ_e \sin \eta l)^{-1} & (jZ_e \tan \eta l)^{-1} \end{pmatrix} \begin{pmatrix} F_{\rm I} \\ F_{\rm II} \end{pmatrix}. \quad (20)$$

Furthermore, short current I_s is given by

$$I_s = -j\omega w \int_{-l/2}^{l/2} D_3(X_3 = H/2)\, dX_1 \quad (21)$$

where

$$D_3 = e_{3i}S_i + \varepsilon_{3j}{}^S E_j$$
$$= e_{33}S_3 + \varepsilon_{33}{}^S E_3, \quad (22)$$

and electric field E_3 can be written, approximately, as

$$E_3 = -\frac{e_{33}}{\varepsilon_{33}{}^S}(B_1 \sin \eta X_1 + B_2 \cos \eta X_1)$$
$$\cdot \left\{ \xi_n \cos (\xi_n X_3) - \frac{2 \sin (\xi_n H/2)}{H} \right\}. \quad (23)$$

Therefore, I_s is given by

$$I_s = j\frac{2we_{33} \sin (\xi_n H/2)}{\eta H Z_e}(F_{\rm I} - F_{\rm II}). \quad (24)$$

Second, treat a case where the mechanical load is free, i.e., $F_{\rm I} = F_{\rm II} = 0$. In this case, the mechanical term of the resonator is represented by a series circuit of inductance L_1 and capacitance C_1. From (17), the solution, on condition that the mechanical load is free, is given by

$$\cos (\eta l/2) = 0, \qquad B_2 = 0$$

or

$$\sin (\eta l/2) = 0, \qquad B_1 = 0.$$

As the electrically excitable mode is only in case of $\eta = 0$, the displacement is expressed as

$$u_3 = B_2 \sin \xi_n X_3, \qquad u_1 = 0. \quad (25)$$

Inductance L_1 is defined by

$$L_1 = 2E_k/|I_0|^2 \quad (26)$$

where E_k is the kinetic energy given by

$$E_k = \tfrac{1}{2}\rho\omega^2 wl \int_{-H/2}^{H/2} u_3{}^2\, dX_3 \simeq \rho\omega^2 B_2{}^2 wH/2, \quad (27)$$

and I_0 is the displacement current given by

$$I_0 = j\omega wl D_3(X_3 = H/2). \quad (28)$$

From the preceding equations, one obtains

$$L_1 = \rho H^3/\{8wle_{33}{}^2 \sin^2 (\xi_n H/2)\}. \quad (29)$$

Capacitance C_1 is given by

$$C_1 = 1/L_1\omega_r{}^2, \quad (30)$$

where ω_r is the resonant angular frequency expressed as

$$\omega_r = \xi_n v_l. \quad (31)$$

Therefore, admittance Y_M of the series resonant circuit (L_1, C_1) is obtained as

$$Y_M = \frac{8wle_{33}{}^2 \sin^2 (\xi_n H/2)}{\rho H^3} \cdot \frac{j\omega}{\omega_r{}^2 - \omega^2}. \quad (32)$$

With the help of (6), (18), and (31), total admittance Y is given by

$$Y = j\omega C_0 + \frac{4w^2 le_{33}{}^2 \sin^2 (\xi_n H/2)}{jZ_e \eta H^2} \quad (33)$$

where C_0 is the parallel capacitance.

From (24) and (33), total current I is expressed as

$$I = j\omega C_0 V + \frac{N^2}{j\eta l Z_e} V - \frac{N}{j\eta l Z_e}(F_{\rm I} - F_{\rm II}) \quad (34)$$

where N is given by

$$N = 2wle_{33} \sin (\xi_n H/2)/H. \quad (35)$$

Particle velocity v', accompanied with the second term of the right side of (34), is given by

$$v' = -\frac{N}{j\eta l Z_e} V. \quad (36)$$

From (20), (34), and (36), one obtains

$$\begin{pmatrix} I \\ v_{\rm I} \\ v_{\rm II} \end{pmatrix} = \begin{pmatrix} j\omega C_0 + N^2/jZ_e\theta & -N/jZ_e\theta & N/jZ_e\theta \\ -N/jZ_e\theta & (jZ_e \tan \theta)^{-1} & -(jZ_e \sin \theta)^{-1} \\ N/jZ_e\theta & -(jZ_e \sin \theta)^{-1} & (jZ_e \tan \theta)^{-1} \end{pmatrix} \begin{pmatrix} V \\ F_{\rm I} \\ F_{\rm II} \end{pmatrix} \quad (37)$$

where

$$\theta = k_e l$$

and

$$k_e = \frac{\pi}{\gamma_n H}[(\bar{\Omega}/n)^2 - 1]^{1/2}. \quad (38)$$

Here, $\bar{\Omega}$ is the normalized frequency in an electroded region given by

$$\bar{\Omega} = \Omega(1 + \Delta), \qquad 1 < \bar{\Omega}/n < 1 + \Delta \quad (39)$$

where Ω is the normalized frequency in an unelectroded region and Δ is the plate back.

The equivalent circuit representation of (37) is shown in Fig. 6(b). However, the configuration of Fig. 6(b) is similar to that of [7], the equation that gives each circuit component contains γ_n or γ_n', as shown in the preceding derivation.

In the unelectroded region, the electrical terms can be taken out and the angular θ is replaced with imaginary value $j\Theta$ given by

$$j\Theta = jk_s d \tag{40}$$

where

$$k_s = \frac{\pi}{\gamma_n H}\left[1 - (\Omega/n)^2\right]^{1/2} \tag{41}$$

and d is the distance of unelectroded region.

Therefore, the equivalent circuit of a monolithic filter can be shown in Fig. 7, where d_i is the distance between ith and jth $(j = i + 1)$ electrodes $\Theta_i = k_s d_i$, and it is assumed that the plate back of each electrode is equal to Δ.

Equivalent inductance L of the electroded region is given by

$$L = \frac{1}{2j}\left[\frac{\partial Z}{\partial \omega}\right]_{\omega=\omega_0;\Theta_1=\infty} \tag{42}$$

where Z is the input impedance at port $1 - 1'$ and ω_0 is the root of

$$[Z(\omega)]_{\Theta_1=\infty} = 0. \tag{43}$$

Therefore, inductance L is obtained as

$$L = \frac{\rho\pi}{32 e_{33}{}^2 \gamma_n}\frac{H^2}{w}\left\{\frac{1}{(1-\Delta)^2} - 1\right\}\frac{[F/(1-\Delta)]^2 - 1}{1 - F^2} \cdot\left\{(1 - F^2)^{-1/2} + \frac{\pi}{2\gamma_n}\frac{l}{H}\left(\frac{F}{1-\Delta}\right)^2\right\} \tag{44}$$

where

$$F = 1 - \Delta \sin^2 \phi \tag{45}$$

$$\frac{\phi}{\cos\phi} = \frac{\pi l}{\sqrt{2}\gamma_n H}. \tag{46}$$

The coupling coefficient between ith and jth $(j = i + 1)$ electrodes k_{ij} is given in [9], that is

$$k_{ij} = \frac{\Delta \sin^2 2\phi}{1 + \phi\tan\phi}\exp\left[-(2d_i\phi\tan\phi)/l\right]. \tag{47}$$

When the characteristics of a filter are given, plate thickness H, inductance L, and coupling coefficients k_{ij} are calculated. For the calculated values of H, L, and k_{ij}, the suitable relation among Δ, w, and l is given from (44)–(47). Thus a bandpass monolithic crystal filter can be designed.

The series inductance L and capacitance C of a 3rd overtone trapped energy mode resonator in a Z-cut plate, for example, are estimated as

$$L = 180\ \mu\text{H}$$

and

$$C = 3.47 \times 10^{-3}\ \text{pF},$$

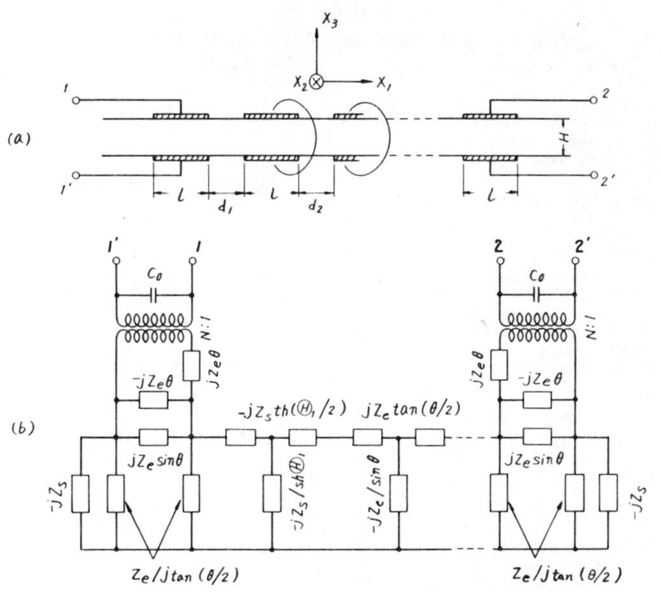

Fig. 7. MCF equivalent circuit representation. (a) MCF dimensional configuration. (b) Equivalent circuit.

for a case of $H = 45\ \mu$m, $l = 0.30$ mm, $w = 0.42$ mm, and $\Delta = 0.0035$, and the capacitances ratio is about 310, where the material constants used in calculation are those of [1].

4. COUPLED RESONATOR EXPERIMENTAL RESULTS

On the basis of the equivalent circuit given in previous section, a $LiTaO_3$ Z-cut plate monolithic crystal filter was trial produced.

The designed characteristics and obtained characteristics are listed in Table II. Fig. 8 shows the electrode configuration of the sample. The distance between 2nd and 3rd electrodes d_2 is large enough to avoid mechanical coupling. That coupling is carried out by the coupling capacitor C_M, instead of mechanical coupling (tandem connection).

Fig. 9 shows the measured frequency response of the filter in wide frequency range and narrow frequency range, where the horizontal axis and vertical axis are frequency and attenuation, respectively. The shoulder observed on the upper skirt region for narrow frequency range may be due to misadjustment of the resonant frequencies and coupling factors.

In Fig. 9, weak spurious response was observed at about 6 MHz below the pass band region. This response is due to the 5th overtone of the thickness shear mode (TS-5 mode), a similar response was observed at about 120 MHz, which corresponds to the 3rd overtone frequency of the thickness shear mode. The reason is thought to be that the shear mode may be excited by a parallel electric field component between resonators #1 and #2 and/or between #3 and #4 shown in Fig. 9. From [10], the electromechanical coupling constant of parallel field excitation of $LiTaO_3$ Z-cut plate is about 0.41. It is a problem to reduce the spurious response.

As shown in Table II, the 3 dB pass bandwidth of the

TABLE II
DESIGNED AND OBTAINED MCF CHARACTERISTICS

	designed	obtained
Center frequency(f_0)	205 MHz(about)	204.6 MHz
3 dB band width	50 kHz($2.5\times10^{-4}f_0$)	45 kHz
Amplitude characteristics	Butterworth	
Number of resonators	4(2+2 tandem)	
Terminal impedance	75 ohms	75 ohms
Overtone order	3rd	
Dead loss		5 dB

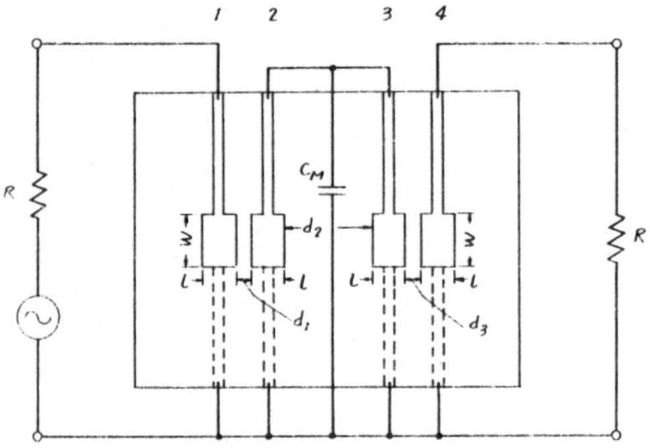

Fig. 8. MCF construction of electrodes. Electrode material: aluminum. Numerical values: $l = 0.30$ mm, $w = 0.42$ mm, $d_1 = d_3 = 0.06$ mm, $d_2 = 1.2$ mm, plate thickness $H = 45\ \mu$m, plate back $\Delta = 0.0035$, $C_M = 15$ pF, $R = 75$ ohms.

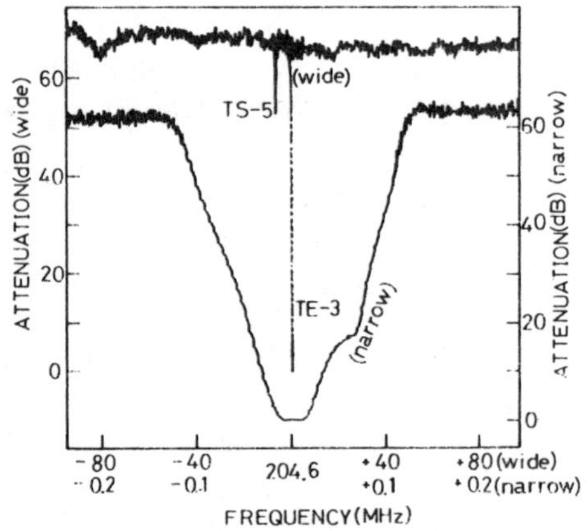

Fig. 9. MCF observed frequency response. Frequency axis upper scale and lower scale correspond to wide frequency region and narrow frequency region, respectively. Vertical scale left side corresponds to wide frequency region.

experimental sample agrees well with the designed value, and the flat loss is considerably small. Q factor of resonators is in the range of $2.2 \times 10^4 \sim 2.7 \times 10^4$, and the flat loss agrees well with the calculated value from the Q factor. The terminal impedance is low, compared with the quartz MCF (in case of AT-cut quartz, the impedance is around a few hundred ohms).

The temperature coefficient of the center frequency $(\Delta f/f_0)/\Delta T$ is about $-38 \times 10^{-6}/°C$ in the temperature range of $0° \sim 80°C$, that agrees approximately with the calculated value from [11]. Therefore, in a practical application, a simpler oven may be necessary.

5. CONCLUSION

The thickness extentional mode energy trapping phenomenon in $LiTaO_3$ plates is studied, and it is proved that the energy trapping occurs only in overtone TE-modes. The resonant frequency spectrum is calculated for the $LiTaO_3$ trapped energy mode resonator, and experimental values fit well on the theoretical curves.

An equivalent circuit representation for the thickness extentional mode monolithic crystal filter is derived, and utilizing the 3rd overtone mode, an MCF for around 200 MHz is realized. The pass bandwidth agrees well with the designed value, and the flat loss is considerably small.

By utilizing the overtone thickness extentional mode in $LiTaO_3$ plates, monolithic crystal filters, in the very high-frequency region, from 100 MHz to a few hundred MHz, may be possible to realize because the Q factor of resonators is large enough so that the maximum bandwidth of that monolithic crystal filter is almost limited by the capacitances ratio. For the 3rd overtone mode in a Z-cut plate, the maximum obtainable relative bandwidth is about 1×10^{-3}, and the terminal impedance is below 100 ohms, that is impossible in the case of the quartz MCF. In the case of a 47° rotated Y-cut plate, more wide band MCF than the Z-cut MCF may be realized.

ACKNOWLEDGMENT

The author wishes to thank Dr. R. Kaneoya and Dr. K. Sawamoto for their continual encouragement and valuable suggestions and Mr. M. Kiyomoto for polishing the sample plates.

REFERENCES

[1] A. W. Warner, M. Onoe and G. A. Coquin, "Determination of Elastic and Piezoelectric Constants for Crystals in Class (3m)," J. Acoust. Soc. Am., vol. 42, pp. 1223–1231, June 1968.
[2] H. F. Tiersten, "Wave Propagation in an Infinite Piezoelectric Plate," J. Acoust. Soc. Am., vol. 35, pp. 234–239, Feb. 1963.
[3] M. Onoe and H. Jumonji, "Analysis of Piezoelectric Resonators Vibrating in Trapped Energy Modes," J. Inst. Elect. Comm. Engrs. Japan, vol. 48, pp. 1574–1581, Sept. 1965.
[4] M. Onoe, H. Jumonji and N. Kobori, "High Frequency Crystal Filters Employing Multiple Mode Resonators Vibrating in Trapped Energy Modes," Proc. 20th Ann. Symp. on Frequency Control, pp. 266–287, April 1966.
[5] R. A. Sykes and W. D. Beaver, "High Frequency Monolithic Crystal Filters with Possible Application to Single Frequency and Single Side Band Use," Proc. 20th Ann. Symp. on Frequency Control, pp. 288–308, April 1966.
[6] W. P. Mason, "Equivalent Electromechanical Representation of Trapped Energy Transducers," Proc. IEEE vol. 57, pp. 1723–1734, Oct. 1969.
[7] K. Nakamura and H. Shimizu, "Equivalent Circuit for Thickness Modes of Elastic Waves Propagating in the Plane of the Piezoelectric Plates," Trans. Inst. Elect. Comm. Engrs. Japan, vol. 55-A, pp. 95–102, Feb. 1972.
[8] H. F. Tiersten, "Linear Piezoelectric Plate Vibrations," Plenum Press, 1969.
[9] T. Ashida, "Eigen Frequencies of Monolithic Filters," Trans. Inst. Elect. Comm. Engrs. Japan, Vol. 54-A, pp. 323–330, June 1971.
[10] T. Yamada and N. Niizeki, "Formation of Admittance for Parallel Field Excitation of Piezoelectric Plates," J. Appl. Phys., vol. 41, pp. 3604–3609, Aug. 1970.
[11] R. T. Smith and F. S. Welsh, "Temperature Dependence of the Elastic, Piezoelectric, and Dielectric Constants of Lithium Tnatalate and Lithium Niobate," J. Appl. Phys., vol. 42, pp. 2219–2230, May 1971.

Design and construction of monolithic-crystal filters using lithium tantalate

M.C. Hales, B.Sc., Ph.D., and J.W. Burgess, B.Sc., Ph.D.

Indexing terms: *Crystal filters, Lithium compounds*

Abstract

Owing to its larger electromechanical coupling coefficient, the use of lithium-tantalate substrates in place of quartz for the construction of monolithic-crystal filters provides a useful increase in available filter bandwidth. Crystal plate orientations are identified with piezoelectric properties suitable for filter applications, and filter-design criteria are derived and confirmed experimentally for a number of selected orientations in lithium tantalate. Applying these criteria to the 163° rotated Y-cut plate, third-overtone monolithic-crystal filters have been constructed operating in the frequency range 35–55 MHz with bandwidths up to 0·75%.

1 Introduction

The construction of band-pass radio-frequency filters using energy trapped resonators has been established over many years. The principles of energy trapping in which a vibrational mode is spatially confined beneath the electrode area of a piezoelectric plate resonator are now well understood, and resonators can be designed to optimise the energy-trapping effect.[1,2] Single resonators employing these principles form the basis of the high-frequency crystal filter. However, in addition to the resonators, such a filter requires inductors, capacitors and transformers to function correctly.[3] Subsequently,[4] it was realised that the filtering operation could be done on a single piezoelectric plate if a number of resonator pairs are deposited so that acoustic coupling can take place between them. The minimum number of resonator pairs is two, forming input and output transducers. A simple way to proceed is then to connect a number of these 'dual resonators' in tandem to produce the desired filter response. This requires less exacting processing technology and offers some design advantages over the true monolithic in which all electrodes are deposited on one crystal plate. Apart from possessing advantages of size, cost and reliability, monolithic-crystal filters are capable of improved performance over conventional crystal filters and are now used widely in communication systems.

Quartz is the piezoelectric material traditionally used in both resonators and filters. The most useful plate orientation for high-frequency devices (> 5 MHz) is the AT-cut which possesses a simple resonance structure with a near-zero frequency coefficient close to room temperature. The only significant limitation with quartz lies in its low electromechanical coupling coefficient (≈ 9%) which restricts the maximum available filter bandwidth to approximately 0·3% of the centre frequency for fundamental-mode operation. The desire to make wideband filters has prompted a search for alternative materials, and this paper examines the application of one such material, lithium tantalate, to monolithic-crystal filters.

Lithium tantalate is a nonhygroscopic crystalline material with a coupling coefficient of 40%. This makes possible the construction of filters with bandwidths of up to 6% at fundamental frequencies. In high-quality single-crystal form it may be expected to be as stable as quartz in respect of its aging characteristics but cannot match the quartz AT plate for temperature stability. In Fig. 1 the frequency drift is plotted for two orientations in quartz and the most stable orientation in lithium tantalate.[11]

Lithium tantalate monolithic-crystal filters have been designed and fabricated to provide a wideband filter in monolithic form in the frequency range 10–100 MHz.

Paper 7720E, first received 22nd September 1975 and in revised form 28th April 1976

Dr. Hales and Dr. Burgess are with the Allen Clark Research Centre, The Plessey Company Limited., Caswell, Towcester, Northants., England

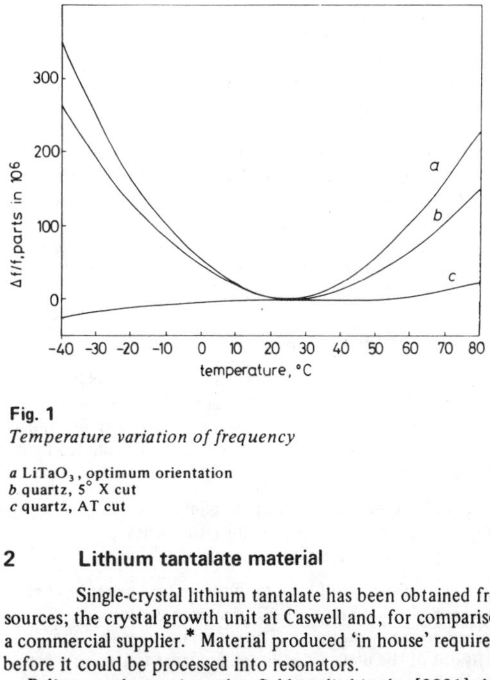

Fig. 1
Temperature variation of frequency

a $LiTaO_3$, optimum orientation
b quartz, 5° X cut
c quartz, AT cut

2 Lithium tantalate material

Single-crystal lithium tantalate has been obtained from two sources; the crystal growth unit at Caswell and, for comparison, from a commercial supplier.* Material produced 'in house' required poling before it could be processed into resonators.

Poling was done using a d.c. field applied in the [0001] direction (the direction corresponding to the piezoelectric *z*-axis[6]) as the boule was cooled from its Curie point (610°C) to room temperature. Since, for ease of growth, the majority of boules were grown on the *b*-axis (the piezoelectric *y*-axis) this involved cutting two parallel flats oriented perpendicular to the *z*-axis. Silver or platinum-paste electrodes were applied to the flats, and fields of 200 V/cm applied in 1 ms pulses with a 1% duty cycle. The current density through the boule during the pulse was limited to 1 $mAcm^{-2}$ If the current is allowed to rise, or the duty cycle increased, the crystal can discolour. Under these conditions, a yellow discolouration originating at the cathode may spread through part or all of the crystal. If the current rises still further, local heating can produce cracking. However, the presence or absence of the yellow discolouration was not found to affect either the poling or the subsequent piezoelectric behaviour.

The effectiveness of poling both house-grown and bought-in material was monitored by measurement of the pyroelectric coefficient. This was done by a charge-integration technique, and

*Crystal Technology Inc., Mountain View, California, USA

Reprinted with permission from *Proc. Inst. Elec. Eng.*, vol. 123, pp. 657–661, July 1976.

room-temperature pyroelectric coefficients between 1·9 and 2·0 × 10^{-8} C/cm^2 K were obtained for the material used.

2.1 Blank processing

The blank-processing procedure closely followed that commonly used for quartz-crystal blanks.[7-9] The crystal boules were oriented by standard X-ray techniques to an accuracy of ± 6′ of arc. The close plate-orientation tolerance required for quartz AT-cut devices is not needed in the case of lithium tantalate since the most critical properties are not rapidly-varying functions of orientation.

The crystal boules are then cut into 400 μm thick wafers on a Metals Research annular saw. The cutting speed is kept low (200 rev/min) and plenty of coolant is applied to prevent local heating since the anisotropy of the thermal-expansion coefficients[10] makes $LiTaO_3$ susceptible to breaking under thermal shock.

For producing relatively small quantities of crystal blanks, it was found convenient to shape the blanks individually by cutting from the wafers using an ultrasonic drill. The drill bit was essentially a thin-walled stainless-steel tube, 8·2 mm diameter, and incorporated a 2 mm long flat which could be aligned with a prescribed crystallographic direction.

Lapping of the blanks was carried out using a series 2000 double-sided planetary lapping machine manufactured by Lapping Services Ltd. A 5 μm abrasive (Lapmaster 1900 grade) was used suspended in Lapmaster lapping oil. Blanks lapped in batches of 15 from a thickness of 400 μm down to 130 μm showed a thickness spread of less than 2 parts in $10.^3$ In the final stages of lapping the blank thickness was monitored continuously, in terms of frequency, by a swept network analyser (Hewlett Packard 8407).

While a high degree of flatness or parallelism cannot be assessed quantitatively on lapped blanks, the subsequent electrical performance indicated that both were better than approximately $\lambda/4$.

Preliminary results on polished blanks of lithium tantalate show that considerable improvement in device Q-factor may be obtained, particularly at overtone frequencies. However, a ready supply of high-quality polished crystal blanks was not available and the results presented were obtained with blanks lapped in the manner described.

Electrode patterns were deposited by vacuum evaporation of chromium and silver through metal masks. Gold-plated copper wires were soldered to the electrode tabs using a hot-gas soldering technique. With a suitable hot plate and a low-melting-point solder, thermal shock could be avoided. Plated crystal blanks were mounted on either style-J or multi-lead transistor headers for assessment.

2.2 Choice of plate orientation

The basic properties of a crystal resonator are largely governed by the orientation of the plate, relative to the crystallographic axes in the material. For wideband-filter applications, the vibrational-mode structure of the plate should be dominated by a single shear mode with

(*i*) strong associated electromechanical coupling
(*ii*) low temperature coefficient of resonance frequency
(*iii*) freedom from competing modes.

The vibrational mode spectra of lithium-tantalate thin plates has been analysed as a general function of orientation using the full piezoelectric equations of motion.[11] The model assumes that the lateral dimensions of the plate are infinite compared to its thickness, and this allows a complete solution for wave propagation in an arbitrary direction for the lithium-tantalate crystal symmetry (point group R3c). Associated with each plate orientation are three thickness-vibrational modes and three sets of resonance-frequency constants. Because of the strong electromechanical coupling, the three sets of characteristic resonance frequencies are not in harmonic sequence; it is only the antiresonance frequencies which are truly harmonic. For most examples of plate orientation suitable for device exploitation, there is an approximate correspondence between each major vibrational mode and a particular set of resonance frequencies, and this correspondence will be assumed in the subsequent data presentations.

The analysis provides values of electromechanical coupling, directions of particle displacement, and resonance and antiresonance frequency constants as a general function of plate orientation and temperature. Several orientations have been identified for which individual properties are optimised, but there is no single orientation which gives optimum performance in all the properties (*i*) – (*iii*).

Suitable orientations for preliminary device investigation are those having a strong shear mode with good isolation from competing modes, as with the quartz AT plate. The 163° rotated Y-cut plate in $LiTaO_3$[12,13] has been selected for its near-ideal mode structure. The X-cut plate has also received some attention owing to its low temperature coefficient of resonance frequency,[14-17] but coupling between different thickness modes gives rise to some operational problems. The 163° rotated Y-cut plate is illustrated in Fig. 2 and has a plate normal rotated 163° from the Y about the x-axis. This corresponds closely to a direction of pseudothreefold symmetry which makes its identification relatively easy.

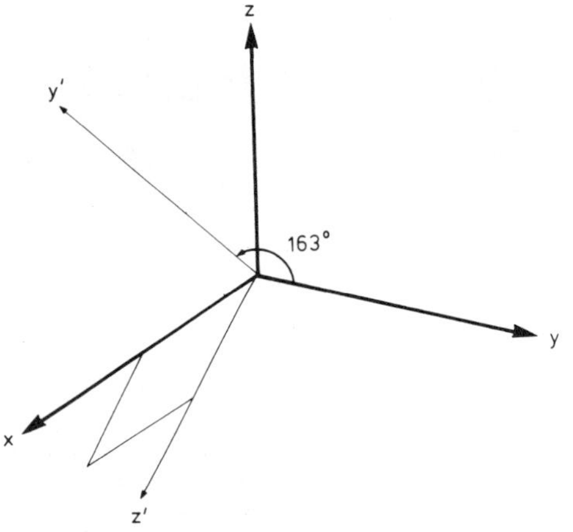

Fig. 2
The 163° rotated Y-cut plate

Thin plates cut on this orientation possess essentially a single mode with properties, summarised in Table 1, suitable for both fundamental and overtone-frequency operation.† The direction of particle displacement lies within 3° of the z′-axis so that the mode is close to a pure shear mode. A longitudinal mode with a frequency constant of 2·758 kHz mm at fundamental is also present, but its level of activity is 40 dB down from the main mode and it is sufficiently removed in frequency to be ignored. The third major mode, a further shear mode, is absent in this case.

The values quoted in Table 1 are derived from the theory but all have been checked and found to lie within 1% of corresponding experimental values.

Table 1
RESONANCE PROPERTIES OF $LiTaO_3$ 163° ROTATED Y-CUT PLATE

Parameter	Fundamental	Third overtone
Electromechanical coupling factor	0·41	
Frequency constant of resonance (MHz mm)	1·845	5·922
Frequency constant of anti-resonance (MHz mm)	1·989	5·967
Temperature coefficient of resonance frequency (parts in 10^6/°C)	− 22	− 53

3 Resonator design

3.1 Plateback and mass loading

The plateback Δ of a crystal resonator is defined as the fractional frequency lowering produced by the plating,

$$\Delta = \frac{\omega_u - \omega_e}{\omega_u}$$

where ω_u is the frequency of the lowest anharmonic symmetric mode in the nonelectroded region and ω_e is the corresponding resonance frequency in the electroded region. Plateback may be represented as the sum of contributions from the electrode mass loading and a piezoelectric effect which is due to the change in boundary conditions when an infinitesimal thickness of electrode is deposited. The mass loading parameter is

$$R = \frac{\rho' h'}{\rho h}$$

†It should be noted that the electromechanical coupling factor (k), specified in Table 1, refers to the static coupling factor defined in Reference 24, and is related directly to the elasto-piezo-dielectric properties of the material. It is therefore constant for different modes and overtone orders and is not to be confused with the motional coupling factor [defined through the capacitance ratio as $\sqrt{(C1/C_0)}$] which is a function of the mode and resonance order.

where ρ and ρ' are the densities of the crystal and electrode, respectively, and h and h' are the plate half-thickness and the thickness of one electrode.

Resonance frequencies for $LiTaO_3$ 163° rotated Y-cut plates have been calculated for a semi-infinite strip electrode, with finite dimension in the x direction.[18] Numerical results for values of R in the range (0, 0·07) are fitted by the following expressions for plateback

$$\text{fundamental: } \Delta = 0{\cdot}061 + R \qquad (1)$$

$$\text{third overtone: } \Delta = 0{\cdot}0066 + R \qquad (2)$$

The relations found experimentally have been

$$\text{fundamental: } \Delta = (0{\cdot}050 \pm 0{\cdot}002) + (1{\cdot}3 \pm 0{\cdot}3)R \qquad (3)$$

$$\text{overtone: } \Delta = (0{\cdot}0058 \pm 0{\cdot}001) + (1{\cdot}2 \pm 0{\cdot}2)R \qquad (4)$$

It is seen that for typical values of R, (0·001 – 0·01), the piezoelectric contribution dominates the plateback, contrasting the situation for AT quartz. The piezoelectric contribution to the plateback was found to be 0·052 for ω_u measured by an airgap technique and 0·048 when measured using a plate and weight method. The difference arises primarily from the small mass-loading contribution present in the second method.

3.2 Energy trapping

A careful design of electrode geometry and a consideration of plateback will control the necessary trapping of vibrational energy under the electrode area of the plate. Too large an electrode will trap unwanted anharmonic overtones just as in quartz. In terms of its vibrational-mode characteristics, the 163° rotated Y-cut lithium-tantalate plate is complementary to the quartz AT-cut plate and the conventional nomenclature of thickness shear (TS) and thickness twist (TT) directions being parallel and perpendicular to the direction of particle displacement will be used. However, in contrast to quartz, the TS direction for lithium tantalate lies along the z'-axis.

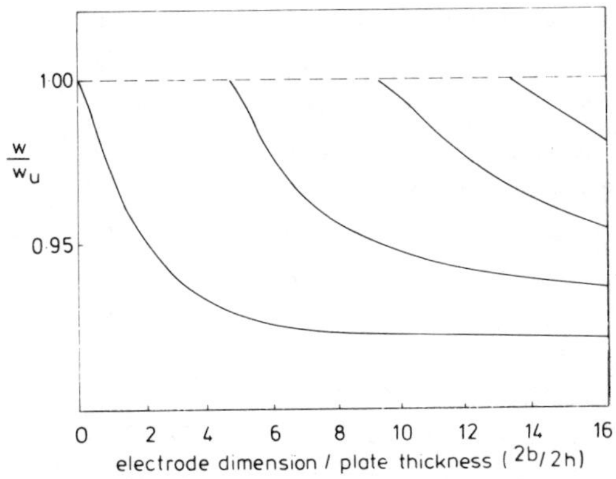

Fig. 3
Normalised resonance frequencies for thickness-twist vibrations against ratio electrode-dimension/plate-thickness (mass loading, R = 0·010)

A solution has been obtained for the resonance frequencies of the 163° lithium-tantalate plate that is partly electroded, finite in the TT direction and infinite in the TS direction. Characteristic resonance frequencies of the plate have been evaluated as a function of electrode width and plateback. In Fig. 3, resonance frequencies near the thickness twist fundamental region are plotted against the ratio of electrode width ($2b$) to plate thickness ($2h$) for plateback corresponding to a mass loading of 0·010. It may be seen that the condition for single mode trapping for typical values of mass loading is that the electrode dimension in the TT direction should satisfy

$$\frac{b}{2h} \leqslant 5 \qquad (5)$$

A similar calculation for third-overtone operation yields the conditions for a single-mode trapping at overtone to be

$$\frac{b}{2h} \leqslant 4{\cdot}5 \qquad (6)$$

The similarity of electrode dimensions for fundamental and overtone operation is a consequence of the difference of plateback expressions (eqns. 1 and 2) arising from the strong electromechanical coupling.

The corresponding analysis for a strip electrode, finite in the TS direction and infinite in the TT dimension, is complicated by a coupling between the thickness-shear mode and a flexure mode. The

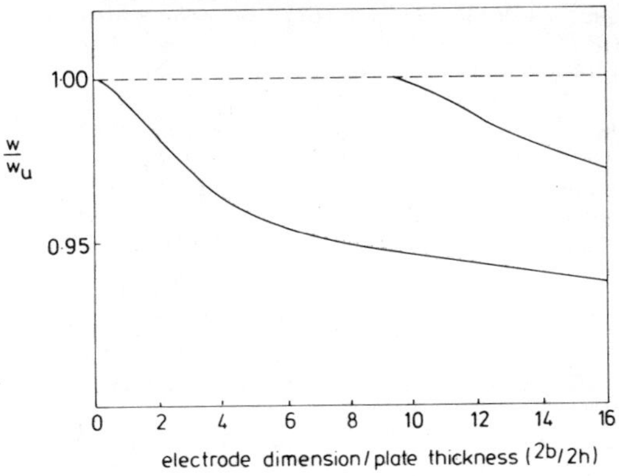

Fig. 4
Normalised resonance frequencies for thickness-shear vibrations against ratio electrode-dimension/plate-thickness (mass loading, R = 0·010)

techniques of Tiersten and Mindlin[19] have been adapted to the partly electroded 163° lithium-tantalate plate to determine the appropriate resonance frequencies,[20] which are of the characteristic form shown in Fig. 4. The resulting Bechmann conditions for the TS electrode dimension c are

$$\frac{c}{2h} \leqslant 9 \qquad (7)$$

for fundamental frequency operation, and are expected to be

$$\frac{c}{2h} \leqslant 8 \qquad (8)$$

for third overtone.

For each of the two types of electrode geometry considered, the lowest resonance occurs when the finite dimension becomes large compared to the plate thickness. Both solutions then approach the frequencies of the infinite fully-electroded plate model used to describe basic resonator properties. The effect of decreasing electrode dimensions in the TS and TT directions is to increase the resonance frequency (Fig. 4) to approach the unelectroded frequency (ω_u). Resonance frequencies for a rectangular electrode geometry can be derived from the two strip electrode solutions in an analogous manner to the quartz AT plate.[21,22] Little change in Bechmann conditions results, the upper limit of dimensions b and c erring slightly in the overtrapping direction.

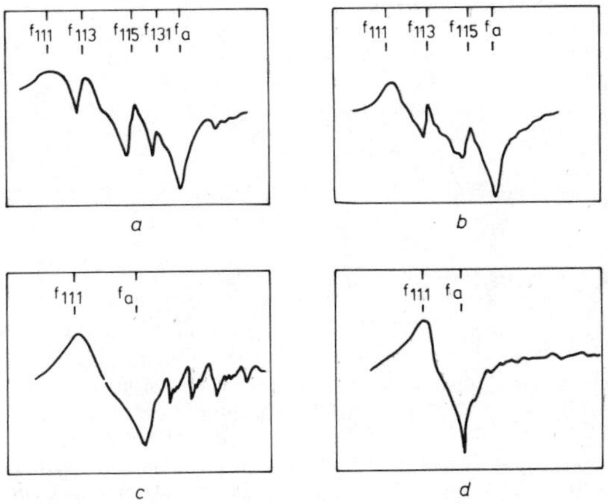

Fig. 5
Response of $LiTaO_3$ 163° rotated Y-plate resonators

Resonator plate thickness is 120 μm
Electrode dimensions are:
a 1·3 mm × 1·3 mm
b 1·0 mm × 1·0 mm
c 0·5 mm × 0·5 mm
d 0·3 mm × 0·3 mm

These predictions of trapped anharmonic resonance frequencies have been compared with experimental observation for a wide range of electrode dimensions and show good agreement. In Fig. 5, the overtone resonance characteristics are displayed for four different electrode sizes. For the thickness of plate used (120 μm) the anharmonics become trapped as the electrode dimension in the TT direction exceeds approximately 0·5 mm. At 0·3 mm square, the trapped resonance is quite clearly a single mode, just remaining so at 0·5 mm square but includes successively more trapped anharmonics as the electrode increases further in size. At 1·3 mm square, the electrode is sufficiently large to trap the first thickness-shear anharmonic overtone (f_{131}). Listed in Table 2 are the predicted modes together with their frequencies for the four electrode dimensions illustrated. Measured values are included for comparison.

Table 2
FREQUENCY RESPONSE OF ENERGY-TRAPPED $LiTaO_3$ PLATE RESONATORS

Sample	Electrode Dimension *lx*	Plate thickness *t*	Resonance frequencies		
			mode	calculated	observed
	mm	mm		MHz	MHz
1 (Fig. 5*a*)	1·36	0·117	f_{111}	16·51	16·54
			f_{113}	16·78	16·77
			f_{115}	17·30	17·34
			f_{131}	17·59	17·69
			f_a	18·00	18·07
2 (Fig. 5*b*)	1·00	0·122	f_{111}	15·95	15·88
			f_{113}	16·43	16·40
			f_{115}	17·25	17·12
			f_a	17·26	17·28
3 (Fig. 5*c*)	0·50	0·124	f_{111}	16·18	16·08
			f_a	16·98	16·87
4 (Fig. 5*d*)	0·32	0·125	f_{111}	16·41	16·29
			f_a	16·85	16·80

4 Coupled dual resonators

When two electrode pairs are deposited in close proximity on a single-crystal plate, piezoelectric coupling can take place between them. Such an arrangement is called a coupled dual resonator. A number of dual resonators may be coupled together electrically to form a filter, often called a monolithic-crystal filter but perhaps more accurately termed a polylithic.[5] Strictly, the true monolithic-crystal filter incorporates all the electrodes on one crystal plate. However, in addition to simplifying both the filter design and fabrication, the polylithic approach can yield a better filter performance since unwanted coupling between nonadjacent resonators is eliminated.

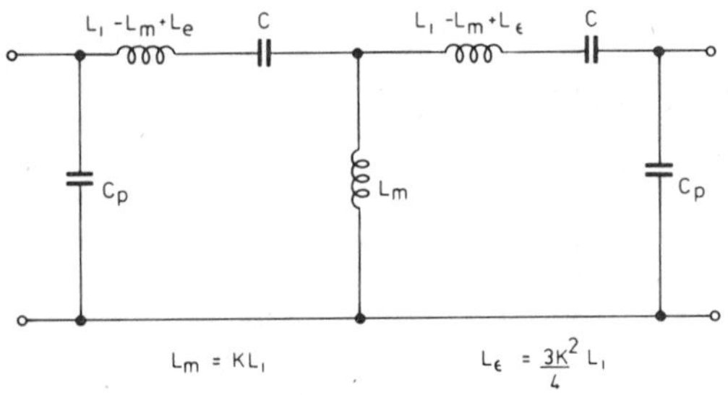

Fig. 6
Equivalent network for a high-coupling dual resonator

The electrical equivalent circuit of a symmetric dual resonator for a crystal with arbitrary piezoelectric coupling[15] is shown in Fig. 6, and is of a form compatible with that used for the quartz dual resonator.[20] Therefore, filter-network design with lithium tantalate follows a similar procedure to the standard quartz approach. The relation between the network of Fig. 6 and the physical characteristics of the dual resonator have been established for the 163° rotated Y-cut plate in lithium tantalate. In particular each single resonator may be approximately modelled by the circuit in Fig. 6, the expression for the resonator inductance being[23]

$$L = \frac{0{\cdot}154 N_m^3}{A f_m^3}\left[1 - \frac{0{\cdot}224^2_m}{N_m^2}\right] \text{henrys}$$

where N_m and f_m are the frequency constant in megahertz millimeters and the resonance frequency in megahertz of the mth order resonance, and A is the electrode area in square millimeters. Additionally, the parallel capacitance can be specified and is close to the high-frequency plate capacitance.[23]

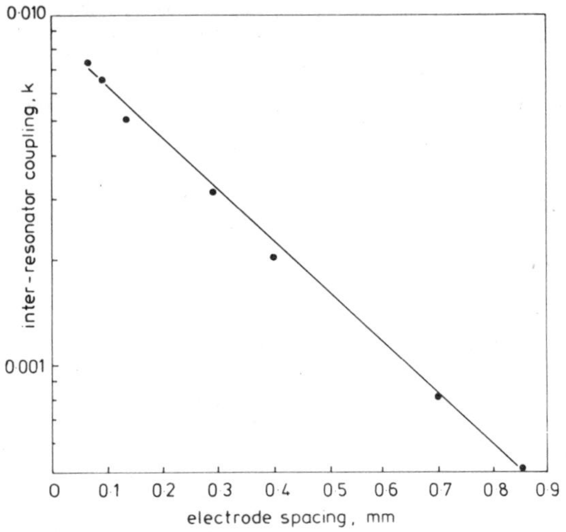

Fig. 7
Thickness-shear inter-resonator coupling against electrode spacing

For typical values of electrode area, the device inductance and impedance may be a factor of 10 lower than for an equivalent quartz device. This makes the problem of matching the filter to external circuitry much easier since the impedance is now of the order of a few hundred ohms, rather than the few thousand common to quartz devices.

Having determined the energy-trapping conditions, plateback relations and inductance formula, it remains to derive an expression for inter-resonator coupling to complete the specification of the dual resonator.

Making use of the similarity between the 163° rotated Y-cut plate and AT-cut quartz, a calculation has been developed using the techniques of Tiersten and Mindlin,[19] retaining piezoelectric effects arising from the greater coupling. This leads to an expression for inter-resonator coupling K of the form

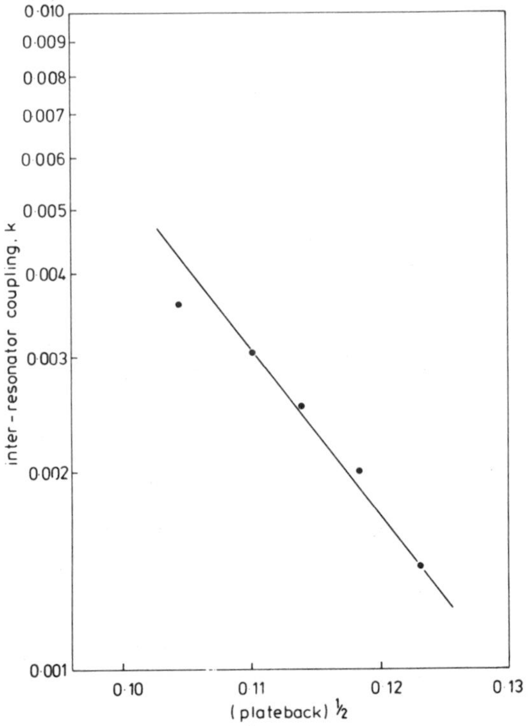

Fig. 8
Thickness-shear inter-resonator coupling against square root of plateback

$$K = A' \exp\left(-B'(1-2\Delta^{1/2})\Delta^{1/2}\frac{d}{2h}\right)$$

in which A' is a function of electrode dimensions and d is the inter-resonator spacing. Both A' and B' are functions of the coupling direction (i.e. TS or TT). The experimental coupling results from a large number of dual resonators, covering a wide range of values of coupling gap (d) and plateback (Δ), are represented in Figs. 7 and 8. Results are only presented for coupling in the TS direction which leads to the greater bandwidths. Clearly, the logarithmic form of the dependence is confirmed but detailed agreement is lacking at this state (i.e. predicted and observed values of A' and B' differ). Nevertheless, knowing the form of the relationship and possessing experimental values of coupling over a wide range of conditions permits filter design to proceed.

Prototype third-overtone monolithic-crystal filters have been fabricated using the 163° rotated Y cut in lithium tantalate. Filters covering the frequency range 35 to 55 MHz have been made with bandwidths up to 0·75%. The frequency response of a filter operating at 46·629 MHz is shown in Fig. 9, the measured specifications being:

centre frequency	46·629 MHz
insertion loss	6 dB
ripple in passband	< 0·5 dB
3 dB* bandwidth	123 kHz
6 dB* bandwidth	145 kHz
ultimate stop-band rejection	80 dB

*relate to the insertion loss at centre frequency

The inband insertion loss of 6 dB may be expected to be reduced considerably by the use of polished crystal blanks in place of the lapped ones used in this filter. The filter was composed of dual resonators individually encapsulated in style-J holders.

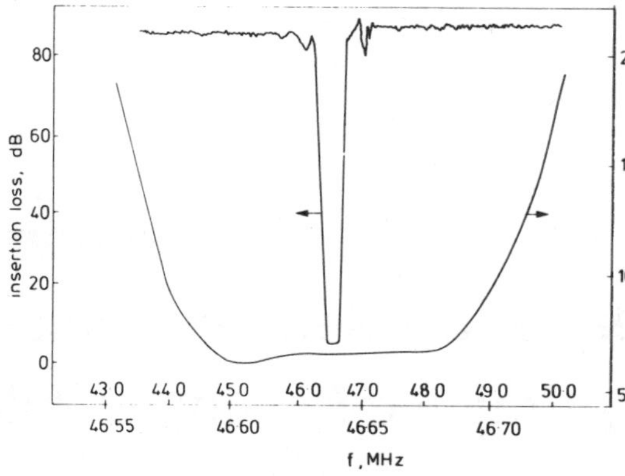

Fig. 9
Response of a $LiTaO_3$ filter comprising three dual resonators and coupling capacitors

5 Conclusions

The design criteria necessary for the construction of resonators and filters have been derived for the high-coupling material lithium tantalate. The establishment of Bechmann conditions permits single mode energy trapping to be achieved for individual resonators and the coupling of resonators on a single substrate has been shown to follow a predictable form. An equivalent circuit has been derived for high-coupling materials and individual equivalent-circuit parameters have been derived. In particular the inductance is considerably less than that obtained with similar quartz devices leading to a lower device impedance and, therefore, easier matching. The form of the equivalent circuit is such as to permit the use of established quartz-filter design programmes with only slight modification.

Using the 163° rotated Y-cut plate, wideband monolithic-crystal filters have been successfully fabricated and tested. The fabrication techniques are basically similar to those currently used for quartz-filter production but the omission of a polishing stage is responsible for the high inband insertion observed in these devices.

Third-overtone monolithic-crystal filters have been constructed with centre frequencies in the range 35 to 55 MHz and bandwidths up to 0·75%. A bandwidth of this magnitude is larger than can be achieved with a quartz monolithic operating at fundamental frequencies. However, the limits of operation of lithium tantalate (in respect of frequency or bandwidth) have not yet been reached and this material possesses considerable potential for use in advanced communications systems.

6 Acknowledgments

Grateful thanks are offered to F.W. Ainger, A.J.L. Muir and R.J. Porter for many helpful discussions, to B. Barrington for assistance in the construction of some of the devices, and to the directors of the Plessey Company for permission to publish this work. The work was carried out with the support of the Procurement executive, Ministry of Defence, sponsored by DCVD.

7 References

1 GERBER, E.A., and SYKES, R.A.: 'A quarter century of progress in the theory and development of crystals for frequency control and selection', *Proc. AFCS*, 1971, **25**, pp. 1–45

2 HAFNER, E.: 'Crystal resonators', *IEEE Trans.*, 1974, **SU-21**, pp. 220–237

3 MASON, W.P.: 'Electromechanical transducers and wave filters' (D. Van Nostrand, New Jersey, 1948)

4 SHOCKLEY, W., CURRAN, D.R., and KONEVAL, D.J.: 'Trapped energy modes in quartz filter crystals', *J. Acoust. Soc. Am.*, 1967, **41**, pp. 981–993

5 PEARMAN, G.T., and RENNICK, R.C.: 'Monolithic crystal filters', *IEEE Trans.*, 1974, **SU-21**, pp. 238–243

6 'Standards on piezoelectric crystals', *Proc. IRE*, 1949, **37**, pp. 1378–1394

7 BALLATO, A. *et al.*: 'Design and fabrication of modern filter crystals', *Proc. AFCS*, 1966, **20**, pp. 131–159

8 BYRNE, R.J.: 'Monolithic crystal filters', *Proc. AFCS*, 1970, **24**, pp. 84–91

9 MILLER, A.J.: 'Preparation of quartz crystal plates for monolithic crystal filters', *Proc. AFCS*, 1970, **24**, pp. 93–103

10 KIM, Y.S., and SMITH, R.T.: 'Thermal expansion of lithium tantalate and lithium niobate single crystals', *J. Appl. Phys.*, 1969, **40**, p. 4637

11 BURGESS, J.W., and HALES, M.C.: 'Temperature coefficients of frequency in lithium niobate and lithium tantalate plate resonators', *Proc. IEE*, 1976, **123**, (6), pp. 499–504

12 BURGESS, J.W. *et al.*: 'Single-mode resonance in $LiNbO_3$ for filter design', *Electron. Lett.*, 1973, **9**, (11), pp. 251–252

13 ONOE, M. *et al.*: US Patent 3 461 408, 1969

14 WARNER, A.W., and BALLMAN, A.A.: 'Low temperature coefficient of frequency in a lithium tantalate resonator', *Proc. IEEE*, 1967, **55**, pp. 450–451

15 ONOE, M., ASHIDA, T., and SAWAMOTO, K.: 'Zero temperature coefficient of resonant frequency in X-cut lithium tantalate at room temperature', *Proc. IEEE*, 1969, **57**, pp. 14–46

16 ASHIDA, T., SAWAMOTO, K., and NIIZEKI, N.: 'Temperature dependence of X-cut $LiTaO_3$ crystal resonator vibrating in thickness shear mode', *Rev. Electr. Commun. Lab. (Japan)*, 1970, **18**, pp. 854–861

17 SAWAMOTO, K.: 'Energy trapping in a lithium tantalate X-cut resonator', *Proc. AFCS*, 1971, **25**, pp. 246–250

18 BURGESS, J.W.: 'Thickness twist anharmonic modes in lithium tantalate and lithium niobate plate resonators', *J. Phys. D*, 1975, **8**, pp. 283–298

19 TIERSTEN, H.F., and MINDLIN, R.D.: 'Forced vibrations of piezoelectric crystal plates', *Q. Appl. Math.*, 1962, **20**, pp. 107–119

20 BURGESS, J.W., TOPOLEVSKY, R., and PORTER, R.J.: 'Ladder equivalent circuit for a monolithic crystal filter with strong electromechanical coupling'. Private communication

21 MINDLIN, R.D., and GAZIS, D.C.: 'Strong resonances of rectangular AT-cut quartz plates', *Proc. 4th US Nat. Congress Appl. Mech.*, 1962, pp. 305–310

22 SHEAHAN, D.F.: 'An improved resonance equation for AT-cut quartz crystals', *Proc. IEEE*, 1970, **58**, pp. 260–261

23 BURGESS, J.W., HALES, M.C., and PORTER, R.J.: 'Equivalent circuit parameters for lithium tantalate plate resonators', *Electron. Lett.*, 1975, **11**, (19), pp. 449–450

24 'Standards on piezoelectric crystals', *Proc. IRE*, 1958, **46**, pp. 764–778

Section II-F
Nonlinear and Tolerance Effects in Acoustically Coupled Resonator Filters

Once the manufacture of crystal filters was started, it became necessary to examine the tolerances to which the processes would have to be controlled in order to obtain economically acceptable yields. Since crystal filter manufacturing is a multistage operation that does not always allow for measurement in between each stage, a proper understanding of tolerances is very essential to the attainment of good yields.

Glowinski, in his paper, has studied the influence of blank parallelism and misalignments on resonator *Q* values, and he shows that proper electrode alignment is essential to the attainment of both good *Q* values and minimum spurious response.

Horton and Smythe discuss intermodulation problems that appeared in receivers containing crystal filters. These intermodulation problems were, in turn, traced to nonlinear effects in the crystal filters themselves. These nonlinearities are somewhat unusual in that they occur at low drive levels. Lack of manufacturing cleanliness has been found to be one cause of these effects, but the degrees of cleanliness involved here are very minute and difficult to measure, thus compounding the problem.

There is one other major nonlinearity problem, and this one is not unexpected as it occurs at high drive levels. Tiersten has performed a mathematical analysis of this phenomenon using higher order elastic constants. His analysis shows that on the basis of including these higher order terms, the resonant frequency can be expected to vary as the drive level is increased.

EFFECTS OF ASYMMETRY IN TRAPPED ENERGY PIEZOELECTRIC RESONATORS

Albert Glowinski, René Lançon* and René Lefèvre

Centre National d'Etudes des Télécommunications

92131 Issy les Moulineaux, France

Summary

One of the major achievements of energy trapping theory is unwanted modes control in thickness resonators. However some experiments seem to disagree with a concept commonly accepted, both theoretically and practically, that unwanted mode activity increases with plateback. Such observations lead to an optimum, rather than a maximum, plateback.

We give an explanation based on resonator asymmetry. Any asymmetry outside the electrodes gives rise to quasi-antisymmetric modes which have a finite electrical activity. Increasing the plateback makes the trapping more and more efficient, thus the electroded region less and less sensitive to the external cause of asymmetry : quasi-antisymmetric modes activity decreases.

We discuss three models approximately describing practical causes of asymmetry. In monolithic filters, where asymmetry is not due to defects but directly related to the coupling mechanism, the theory gives an optimum design for unwanted modes suppression.

Introduction

Energy trapping theory states that anharmonic activity of a thickness resonator appears and increases with the electrode plateback. The maximum plateback depends upon electrodes and resonator geometries.

However, if many experiments confirm this theory, a different behaviour has been observed both in single resonators and monolithic filters, where a minimum plateback is required to achieve a convenient mode suppression. Correct concept would be then of an optimum plateback.

Systematic evidence of such an effect in monolithic filters where the end resonators operate in an asymmetrical configuration, suggested to us the role of resonator asymmetry in the phenomenon.

In fact, quasi-antisymmetric modes have been already identified, since they have an electrical activity as soon as the resonator is not perfectly symmetric, but they have always been kept out of theoretical discussions.

We discuss three theoretical models, approximations of pratical cases : lack of plate parallelism, electrodes misalignment or finite defect, end resonator in monolithic filters.

The discussion is carried out in the elastic bi-dimensionnal model, where electrical considerations are introduced in a second step.

We use two approaches of thickness vibrations analysis, the determinant method and the chain matrix formalism. We show the relation between them, both being used together to compute the frequencies and the electrical activity of the resonant modes.

The main result is that the energy trapping theory predicts the opposite behaviour of quasi-symmetric and quasi-antisymmetric modes activity as a function of plateback.

Thickness resonator analysis

The bi-dimensionnal elastic model

The bi-dimensionnal elastic approximation is based on a model introduced by Shockley, Curran and Koneval[1].

The elastic plate, constant in thickness, is constituted by one – or several – region surrounded by two semi-infinite regions (fig. 1). All phenomena are assumed independent of the third coordinate.

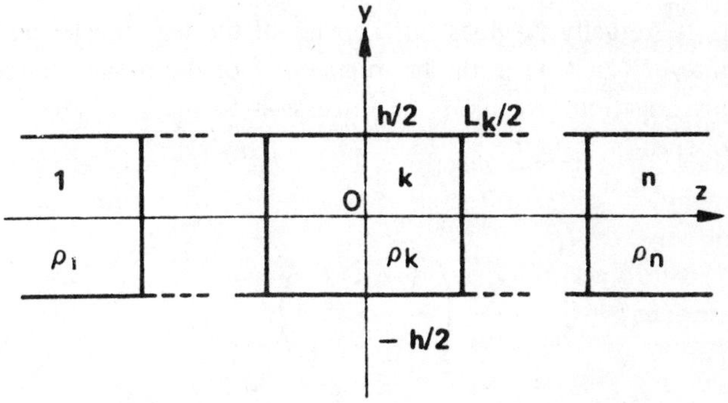

Fig. 1 - The bi-dimensionnal thickness-twist model

* Deceased.

The elastic constants of the material are the same all along the plate, while the density may vary for each region. Such a density variation was introduced as the simplest analytical way to take into account various effects occuring in an actual resonator – mass loading of the electrodes, superficial electrical conditions on a metallized face, electrical termination at the resonator input – and contributing to modify the cut-off frequency of the electroded regions.

We consider a purely elastic material with the crystal symmetry of the AT-cut of quartz, propagation taking place in the YZ plane.

We use the thickness-twist approximation of R.D. Mindlin[2], the mechanical displacement being parallel to the X-axis.

With a time dependence in exp $(i\omega t)$, omitted hereafter, the bi-dimensional equations are :

$$t_5 = \gamma_{55} \frac{\partial u}{\partial z} \tag{1}$$

$$t_6 = c_{66} \frac{\partial u}{\partial y} \tag{2}$$

$$\frac{\partial t_6}{\partial y} + \frac{\partial t_5}{\partial z} = -\rho\omega^2 u \tag{3}$$

where $\gamma_{55} = c_{55} - \frac{c_{56}^2}{c_{66}}$.

In Eq. (3) the density ρ is a step-function $\rho(z)$. In each region, numbered k, the displacement can be written :

$$u_k = (A_k \cos \alpha_k z + B_k \sin \alpha_k z) \sin \beta_k y \,. \tag{4}$$

The traction-free conditions on both faces, i.e. $t_6 = 0$ at $y = \pm h/2$,

requires that $\beta_k = (2n - 1) \frac{\pi}{h}$.

Actually β_k does not depend of the region. The odd integer $(2n - 1)$ is the harmonic rank of the modes under investigation. We limit the discussion to the fundamental operation, $\beta = \pi/h$.

Then

$$\alpha_k^2 = \frac{\pi^2}{h^2} \frac{c_{66}}{\gamma_{55}} \left(\frac{\omega^2}{\omega_{ck}^2} - 1 \right), \tag{5}$$

ω_{ck} being the cut-off frequency of the region k :

$$\omega_{ck} = \frac{\pi}{h} \sqrt{c_{66}/\rho_k} \,. \tag{6}$$

In semi-infinite regions it is more relevant to define

$$u_k = A_k \exp(i\alpha_k |z|) \sin \beta y \,, \tag{7}$$

where

$$\alpha_k = \frac{\pi}{h} \sqrt{\frac{c_{66}}{\gamma_{55}}} \sqrt{\frac{\omega^2}{\omega_{ck}^2} - 1} \quad \text{if} \quad \omega \geqslant \omega_{ck} \,,$$

$$\alpha_k = i\zeta_k = i\frac{\pi}{h} \sqrt{\frac{c_{66}}{\gamma_{55}}} \sqrt{1 - \frac{\omega^2}{\omega_{ck}^2}} \quad \text{if} \quad \omega \leqslant \omega_{ck} \,. \tag{8}$$

At each boundary between two adjacent regions, we have to express the continuity of the displacement and of the stress-component t^5.

The resonance frequencies equation can be given by a chain matrix analysis[3], or directly by the determinant of the linear homogeneous system of the boundary conditions[4].

The chain matrix method is based on a two-port representation of an homogeneous section of a plate vibrating in a thickness-twist mode, the port quantities being the stress t_5 and the particle velocity $\dot{u}$ at each boundary plane.

With notations of figure 2, we have :

$$\begin{pmatrix} t_5 \\ \dot{u} \end{pmatrix}_I = \begin{pmatrix} \cos 2\theta_k & -iZ_k \sin 2\theta_k \\ \frac{1}{iZ_k} \sin 2\theta_k & \cos 2\theta_k \end{pmatrix} \begin{pmatrix} t_5 \\ \dot{u} \end{pmatrix}_{II} \tag{9}$$

with

$$\theta_k = \alpha_k L_k/2 \,, \tag{10}$$

$$Z_k = \gamma_{55} \alpha_k/\omega \,. \tag{11}$$

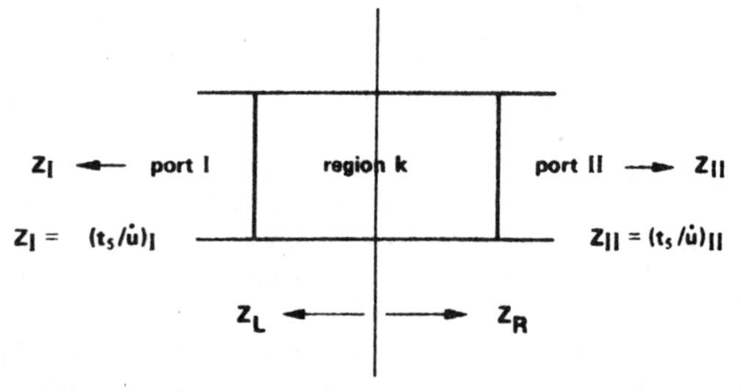

Fig. 2 - The two-port representation

(Cf. Onoe[3], with some changes in the symbols).

At any abscissa in the structure, we may define two mechanical impedances Z_R and Z_L, seen from this plane to the right and to the left :

$$t_5 = Z_R \dot{u} = - Z_L \dot{u} .$$

The resonance condition is obsviously

$$Z_R + Z_L = 0 \ , \quad (12)$$

equation which can be written anywhere in the structure.

Writing eq. (12) in the center of any region will lead to the recursive equation given by the determinant discussion[4,5].

Z_I and Z_{II} being the impedances loading the ports of region k (fig. 2), Z_R and Z_L at the center of this region can be calculated using the chain matrix of the half region, eq. (9) where $2\theta_k$ is replaced by θ_k. Eq. (12) writes then :

$$\frac{Z_I/Z_k + Z_{II}/Z_k}{1 + Z_I Z_{II}/Z_k^2} = i \tan 2\theta_k . \quad (13)$$

For a symmetric structure, in the symmetry plane $Z_R = Z_L$, which yields with eq. (12) :

$$Z_R = Z_L = 0, \text{ or } \quad 1/Z_R = 1/Z_L = 0$$

It has been shown that the first equation leads to symmetric modes (S modes) where u (z) is an even function, and the second to anti-symmetric modes (AS modes) where u (z) is an odd function[4,6].

Other forms of these equations are given by eq. (13) which separates, since $Z_I = Z_{II}$, into :

$$Z_I/Z_k = i \tan \theta_k \ , \quad (14)$$

for S modes, and

$$Z_I/Z_k = 1/i \tan \theta_k \ , \quad (15)$$

for AS modes.

The resonance frequencies are the zeroes (S modes), and the poles (AS modes) of the impedance

$$Z_R = Z_L = \frac{Z_I \cos \theta_k - iZ_k \sin \theta_k}{-iZ_I \sin \theta_k/Z_k + \cos \theta_k} . \quad (16)$$

Since Z_I and Z_{II} are the impedances Z'_L and Z'_R for a lower order structure obtained by suppression of the central region, eq. (16) is a recursive relation between the frequency equations of a sequence of cascaded structures.

Such a relation provides an iterative process of research of the resonance frequencies, zeroes and poles of Z_R. Initiated with symmetric structures it has been extended to any asymmetric structure in thickness mode[5].

On the other hand, the computation of the elastic field distribution, and of the stored energy, used in the electrical activity evaluation, is much easier in the matrix formalism.

Electrical considerations

We assume first that, in the vicinity of each resonance ω_n, the piezoelectric resonator has an impedance approximated by the lossless network of figure 3.

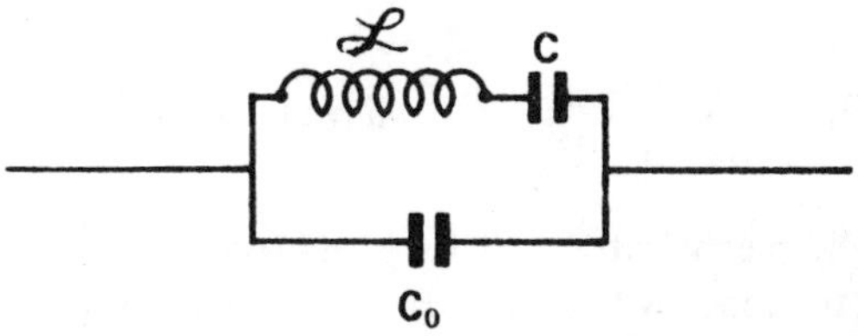

Fig. 3 - Electrical equivalent network

Then we assume that the surface charge density which appears on the electrode is proportionnal to the mechanical displacement in the corresponding region of the elastic model.

Such an approximation is of course limited to low piezoelectric coupling.

With the origin of coordinates at the center of the electroded region, numbered 2, the electrical current is proportionnal to the integral of u_2 :

$$i \sim \int_{-\frac{L_2}{2}}^{\frac{L_2}{2}} u_2(z)\, dz = \frac{2A_2}{\alpha_2} \sin \theta_2 . \quad (17)$$

Following Onoe[6], we evaluate the stored energy in the electrical and mechanical forms. Due to equipartition at resonance of time-averaged energy :

- between inductive and capacitive energies, in the lossless series arm (neglecting purely electric effects in the static capacitance) ;

- betwen kinetic and potential energies in the mechanical non dissipative model, we get

$$\mathscr{L} i^2 = W_m$$

where W_m is twice the kinetic energy :

$$W_m = \frac{h}{2} \int_{-\infty}^{\infty} \rho(z)\,|\dot{u}(z)|^2\,dz = \frac{\omega^2 n h}{2} \int_{-\infty}^{+\infty} \rho(z)\,|u(z)|^2\,dz\,.$$

Actually, the experimental evaluation of the electrical activity of the nth mode at its resonance frequency ω_n, is made by measurement of the electrical resistance of the resonator at ω_n. If we assume that the losses are independent of the mode – we neglect the influence of trapping efficiency, for instance, on the mounting loss – the activity of the mode is related to an impedance level, referred usually to the first mode of the anharmonic series. When activity goes to zero, the impedance level goes to infinity.

We shall then follow the same approach, defining the impedance level of the nth mode by a quantity M_n proportionnal to $\mathscr{L}\omega$, or to $W_m\omega/i^2$, evaluated at ω_n :

$$M_n = \frac{\alpha_2^{\,2}\,\omega_n^{\,3}}{A_2^{\,2}\sin^2\theta_2} \int_{-\infty}^{+\infty} \rho(z)\,|u(z)|^2\,dz\,. \qquad (18)$$

The piezoelectric activity of the nth mode, relative to the first mode, will be expressed in a logarithmic scale, as P_n :

$$P_n = -20 \log M_n/M_1\,. \qquad (19)$$

Obviously for AS modes where A_2 is zero, P_n goes to $-\infty$. The spectral purity criterion expresses that eq. (13) has a single root. Let us introduce the so-called *energy trapping factor* Ω,

$$\Omega = \theta_2(\omega_0) = \frac{\pi}{2}\frac{L_2}{h}\sqrt{\frac{c_{66}}{\gamma_{55}}}\sqrt{\frac{\omega_0^{\,2}}{\omega_{c2}^{\,2}} - 1}\,, \qquad (20)$$

where ω_0 is a reference frequency, usually the cut-off frequency of the surrounding plate. In ω_{c2}, various effects lowering the cut-off frequency are taken into account, as noticed before (mass loading, electrical conditions) while the elastic constants are the piezoelectrically stiffened values.

For a symmetric resonator eqs. (14, 15), the single root condition is $\Omega < \pi/2$. However, as pointed out by Horton and Smythe[7], it would lead to a strong criterion. A lesser requirement is $\Omega < \pi$, which traps the first AS mode, electrically inactive, but untraps the second S mode. It yields the well known condition

$$\frac{L_2}{h}\sqrt{\frac{\omega_0 - \omega_{c2}}{\omega_0}} < \sqrt{2}\sqrt{\frac{\gamma_{55}}{c_{66}}}\,. \qquad (21)$$

The electrical disparition of AS modes, due to charge cancellation on the electrode, is uncomplete as soon as the resonator is not perfectly symmetric.

Asymmetric resonator discussion

Existence of AS modes, electrically active in not perfectly symmetric resonators, has been identified[7,8].

For small symmetry defects, the modes are slightly different from purely S or AS modes, and will be called quasi-symmetric (QS) and quasi-antisymmetric (QAS) modes.

If the cause of asymmetry takes place out of the electroded region, it affects the behaviour of any trapped mode more and more weakly as the trapping is more and more efficient : if the plateback is increased, QS and QAS modes tend to S and AS modes. They are more and more trapped, but the electrical activity of the QAS modes decreases, since the charge cancellation is more and more complete.

We shall discuss theoretically three models, using analysis techniques defined above.

Defect of parallelism

A cause of asymmetry may be an angle between the plate faces. A rough approximation is shown on figure 4, the cut-off frequencies of the extreme regions being unequal.

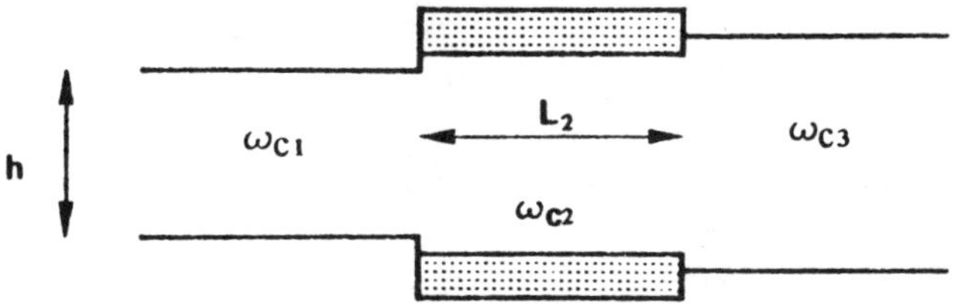

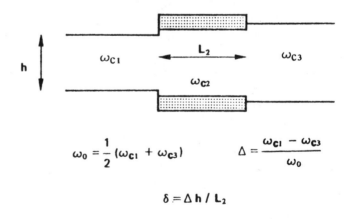

$$\omega_0 = \frac{1}{2}(\omega_{c1} + \omega_{c3}) \qquad \Delta = \frac{\omega_{c1} - \omega_{c3}}{\omega_0}$$

$$\delta = \Delta h / L_2$$

Fig. 4 - Non - parallel plate model

The frequencies equation eq. (13), yields, with :

$$Z_I = \gamma_{55}\,\alpha_1/\omega = i\gamma_{55}\,\zeta_1/\omega\,,$$

and

$$Z_{II} = \gamma_{55}\,\alpha_3/\omega = i\gamma_{55}\,\zeta_3/\omega\,.$$

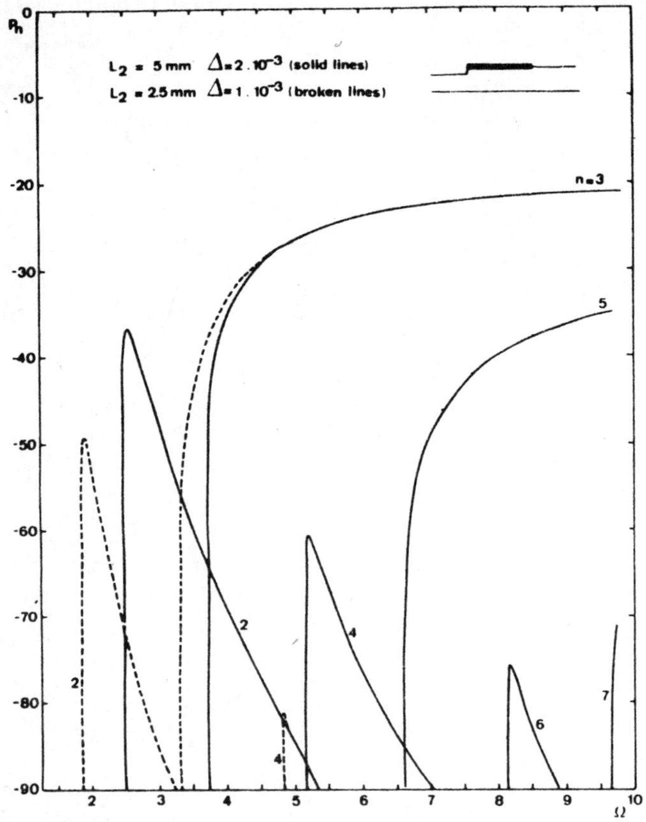

Fig. 5 - Non-parallel plate ($\delta = 6.\ 10^{-5}$). Mode activity versus trapping factor

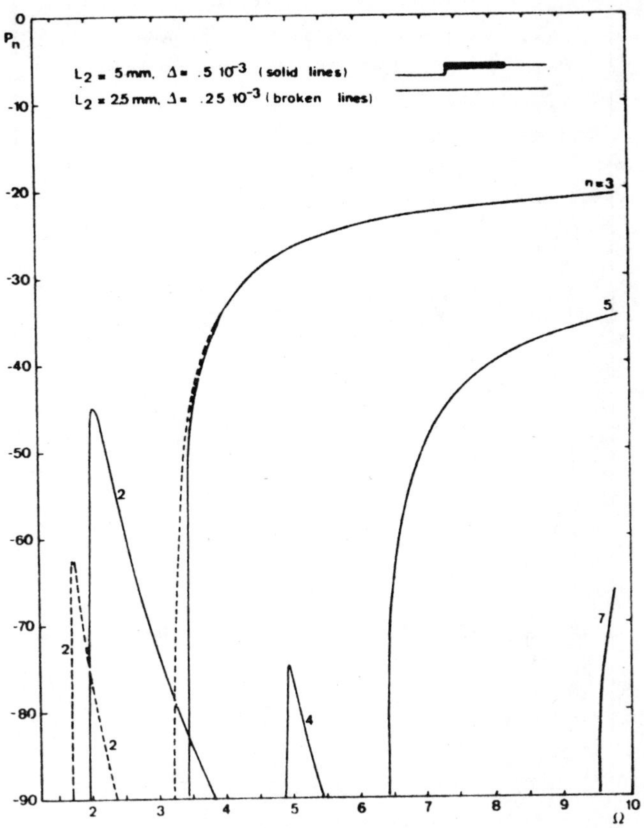

Fig. 6 - Non-parallel plate ($\delta = 1.\ 5.\ 10^{-5}$). Mode activity versus trapping factor

$$\left(\frac{\zeta_1 + \zeta_2}{2} - \alpha_2 \tan\theta_2\right)\left(\frac{\zeta_1 + \zeta_2}{2} + \alpha_2/\tan\theta_2\right) \quad (22)$$

$$= \frac{1}{4}(\zeta_1 - \zeta_3)^2 .$$

Eqs. (14) and (15) of the S and AS modes are coupled by a term of asymmetry. In the expression of the displacement u_2 (z), both cosine and sine terms exist.

Curves of figures 5-7, show the computed activity of QS and QAS modes, versus the trapping factor of region 2, Ω.

Computations have been carried out with the following data :

- plate thickness h, around 154 μm,
- cut-off frequency around 10.8 MHz,
- electrode length L_2 : 2.5 mm and 5 mm,
- relative difference between cut-off frequencies Δ
- *angle* $\delta = \Delta \dfrac{h}{L_2}$: 6.10^{-5}, $1.5.10^{-5}$, $.6.10^{-5}$.

Figure 8 shows the evolution of the ratio A_2/B_2 versus the trapping factor Ω, in the case : $L_2 = 5$ mm, $\Delta = 5.10^{-4}$.

As expected the ratio A_2/B_2 goes to zero or to infinity, according to QAS or QS nature of the modes.

Activity curves show the strong dependence of QAS electrical activity on asymmetry, while QS modes are only weakly modified.

The decrease of QAS mode activity when Ω increases shows also that the optimum value of Ω is close to the 2nd QS mode trapping criterion.

As practical design considerations, the following remarks can be made : for $L_2 = 5$ mm, $\Delta = 5.10^{-4}$ for example, the 1st QAS mode activity can be as high as – 45, for an electrode plateback around .7 %, the optimum plateback being about 1.6 %.

Concerning electrode dimension, it is seen that, for the same *angle* the small electrode gives lower QAS activity for the same trapping factor. However, for the same plateback, the trapping factor depends on L_2, and the practical results depend highly on the real defects, with no general conclusion.

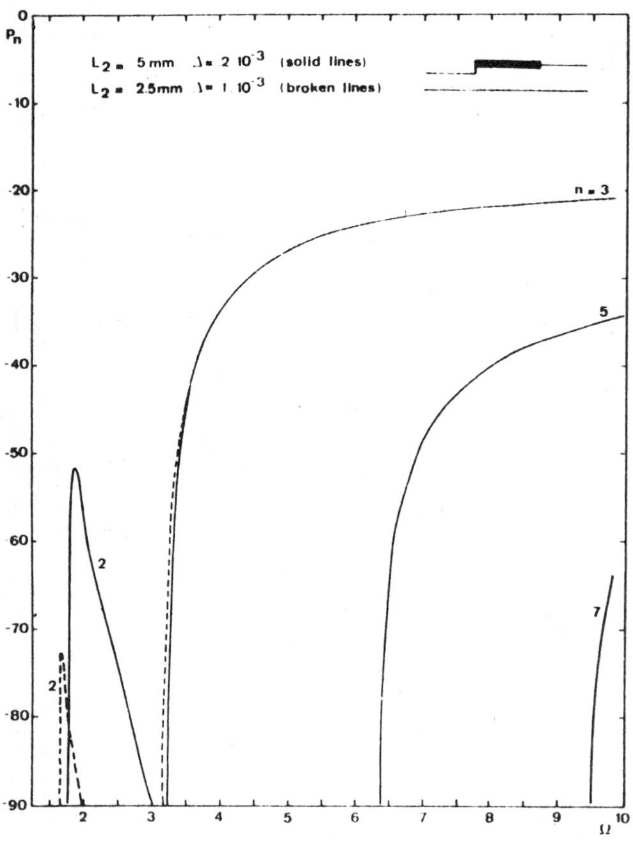

Fig. 7 - Non-parallel plate ($\delta =\ .6.\ 10^{-5}$). Mode activity versus trapping factor

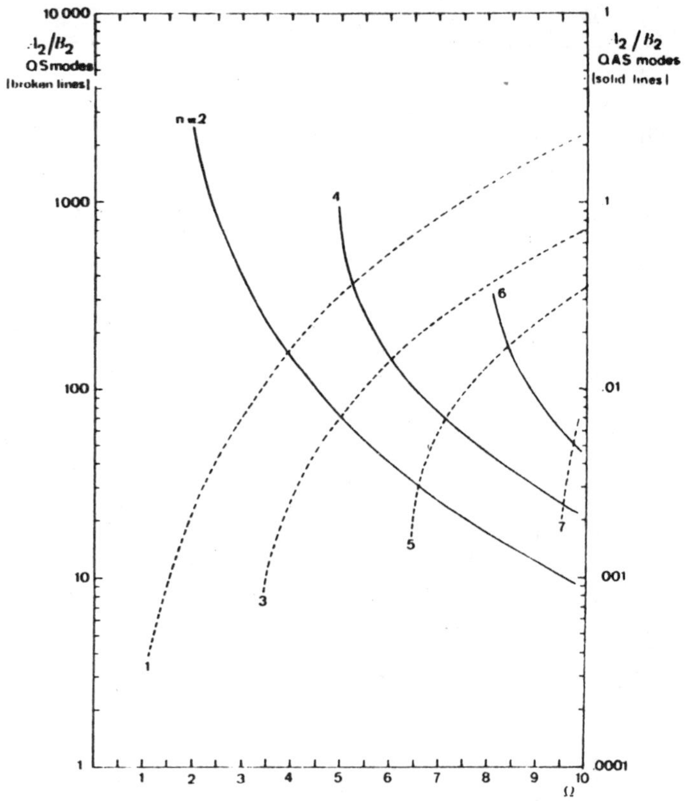

Fig. 8 - Non-parallel plate ($L_2 = 5$ mm, $\Delta = 5.\ 10^{-4}$). Mode *asymmetry* versus trapping factor

A better approximation could of course be obtained with more sophisticated step functions, but has not been yet discussed due to time limitations.

Finite defect

Such a defect is illustrated by figure 9a. Two situations can be considered :

1. the finite defect is constant,
2. the finite defect, due for example to a misalignment of electrodes, varies with the plateback on region 2.

Figures 10-12 show the evolution of mode activity versus Ω, in the second case : mass loading in region 2 is twice the mass loading in region 3, additionnal lowering due to electrical field being taken into account in region 2. Actually, if an electrode on one face is larger than on the other face, there should be also some effect on the y dependence, which should be no more in pure sine for u (y). We neglect this effect in the present discussion.

The geometrical data are :

- electrode length L_2 : 2 mm and 5 mm,

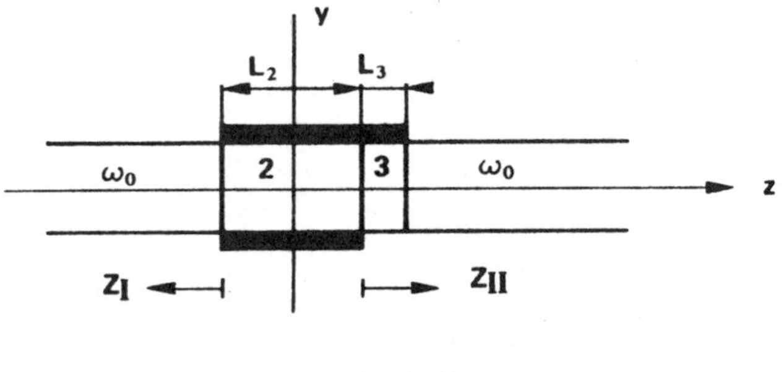

a) Finite defect : electrode misalignment

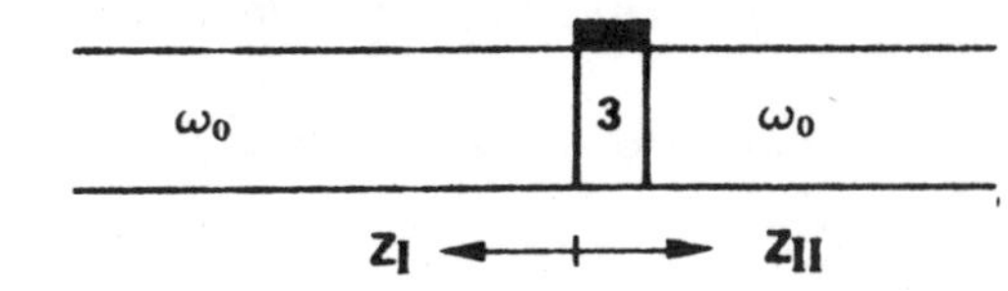

b) Lower order structure

Fig. 9 - Finite defect model

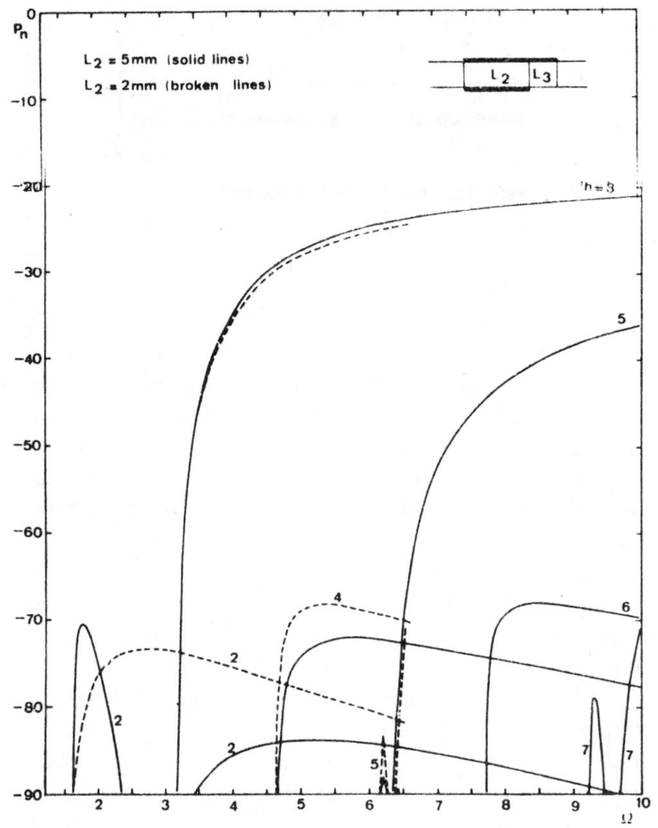

Fig. 10 - Finite defect resonator ($L_2 / L_3 = 25$). Mode activity versus trapping factor

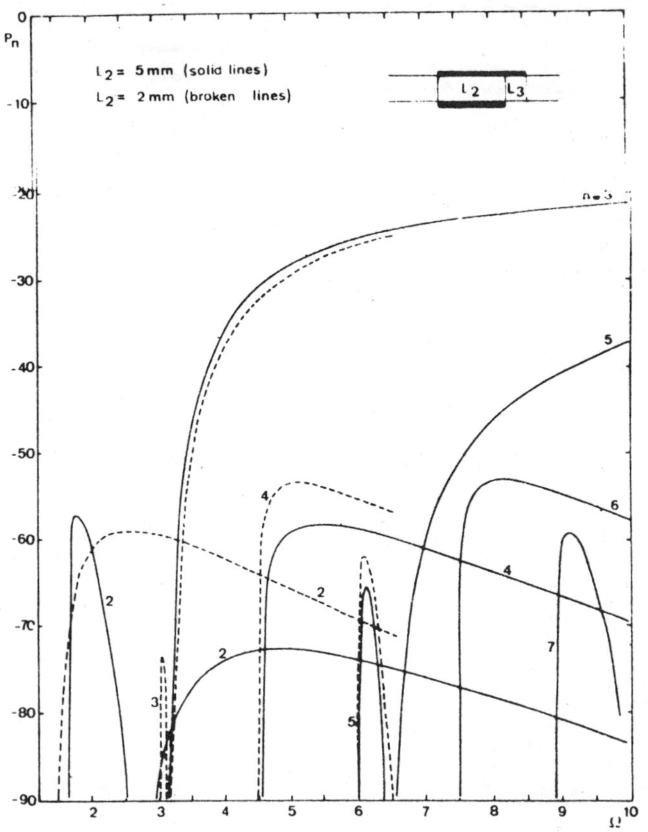

Fig. 11 - Finite defect resonator ($L_2 / L_3 = 10$). Mode activity versus trapping factor

- defect length L_3 : $L_2/25$, $L_2/10$, $L_2/5$.

In both cases 1 and 2, the finite nature of the defect gives rise to an effect shown on figures 10-12, i.e. mode disparition for convenient values of Ω. It can be explained as follows.

From the expression of the electrical current, eq. (17), it is obvious that the current vanishes if :

$$\theta_2 = k\pi \qquad (23)$$

This condition means that the elastic field distribution along z has an integer number of « wavelengths » in region 2. Eq. (23) is satisfied for a discrete sequence of frequencies, depending on Ω.

In addition, since the chain matrix of the region 2 becomes the unity matrix at these frequencies, eq. (9) requires

$$Z_I = -Z_{II} \qquad (24)$$

Eq. (24) is actually the resonance condition for the structure of figure 9b where the region 3 is now the « resonator » region, equation expressed in the left boundary plane.

If a resonance frequency of the structure 9b satisfies simultaneously eq. (23), it is also a resonance frequency of structure 9a, but without electrical activity.

In the vicinity of such a coincidence, an activity drop occurs.

Finally, as Ω varies, the roots of, at least, one of eqs. (23) and (24) vary. For discrete values of Ω eqs. (23) and (24) will be simultaneously satisfied, and the concerned mode activity vanishes.

So, in addition to the fact that QAS mode activity decreases with plateback, while QS modes increase, the activity of a mode may go through a minimum (- ∞).

This result for the bi-dimensionnal model, should be valid in real resonator, yet not experimentally demonstrated.

Monolithic filters

In both examples discussed before, the cause of asymmetry was a defect. As a result of the random nature

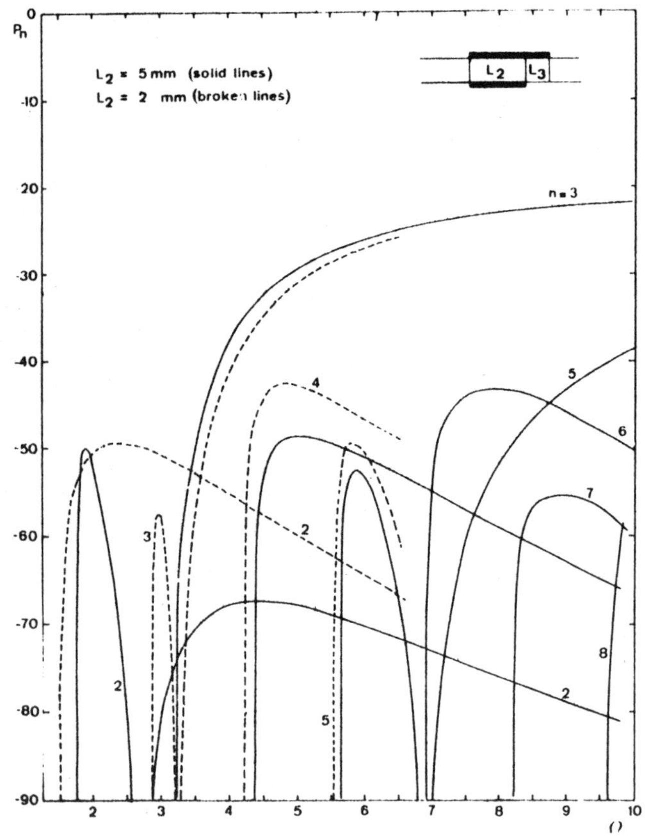

Fig. 12 - Finite defect resonator ($L_2/L_3 = 5$). Mode activity versus trapping factor

of the defects, QAS activity in crystal production has also a random distribution, as observed practically.

In one case however, the cause of asymmetry is not random, but associated with a well defined mechanism : it is in monolithic filters, where the end resonators are coupled on one side to an other resonator (fig. 13). In fact, this is the systematic evidence of unwanted mode suppression in monolithic filters by high trapping, which suggested to us the role of asymmetry.

X-ray topogram of figure 14 shows the typical QAS vibration pattern of the first unwanted mode in a two-resonators monolithic filter. Of course, QAS nature of the mode must be understood from the resonator, not from the filter, point of view.

Theoretical investigation of monolithic filter, using the same method, in the bi-dimensionnal model, leads to results essentially analog to the foregoing sections.

However, since in monolithic filters, the cause of asymmetry can be associated with the coupling mechanism, if we assume of course a defect-free plate, electrode, etc., in

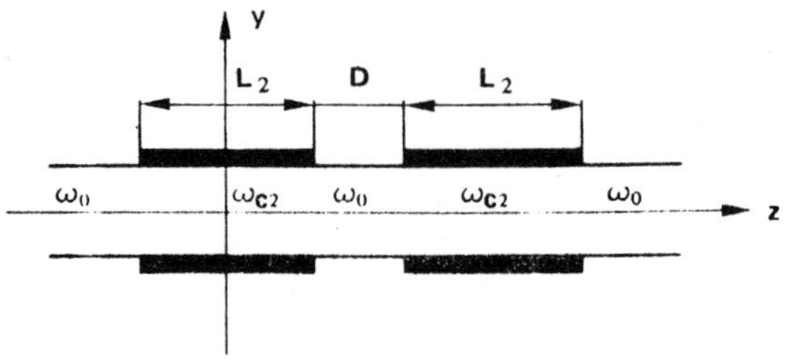

Fig. 13 - Monolithic filter model

order to avoid the superposition of several effects, it is possible to derive optimum design considerations.

For a given electrode length, we plot the variation of the bandwidth B against the trapping factor of the electrode, Ω, for several values of the resonators separation D (fig. 15).

As already noticed[4,9], the second main mode is trapped only for a sufficient Ω, above which it can be talked of bandwidth. Curves of bandwith are also well known.

As the trapping increases, unwanted activity appears, QAS mode referred to the end resonator center. For a two-pole filter, two QAS modes are successively trapped, the activity decreasing with an increasing trapping. Then appear two QS modes.

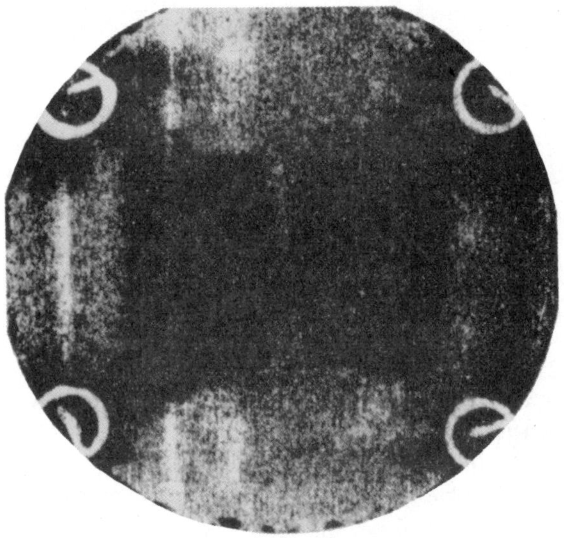

Fig.14 - Monolithic filter : X- ray topogram of the first unwanted mode

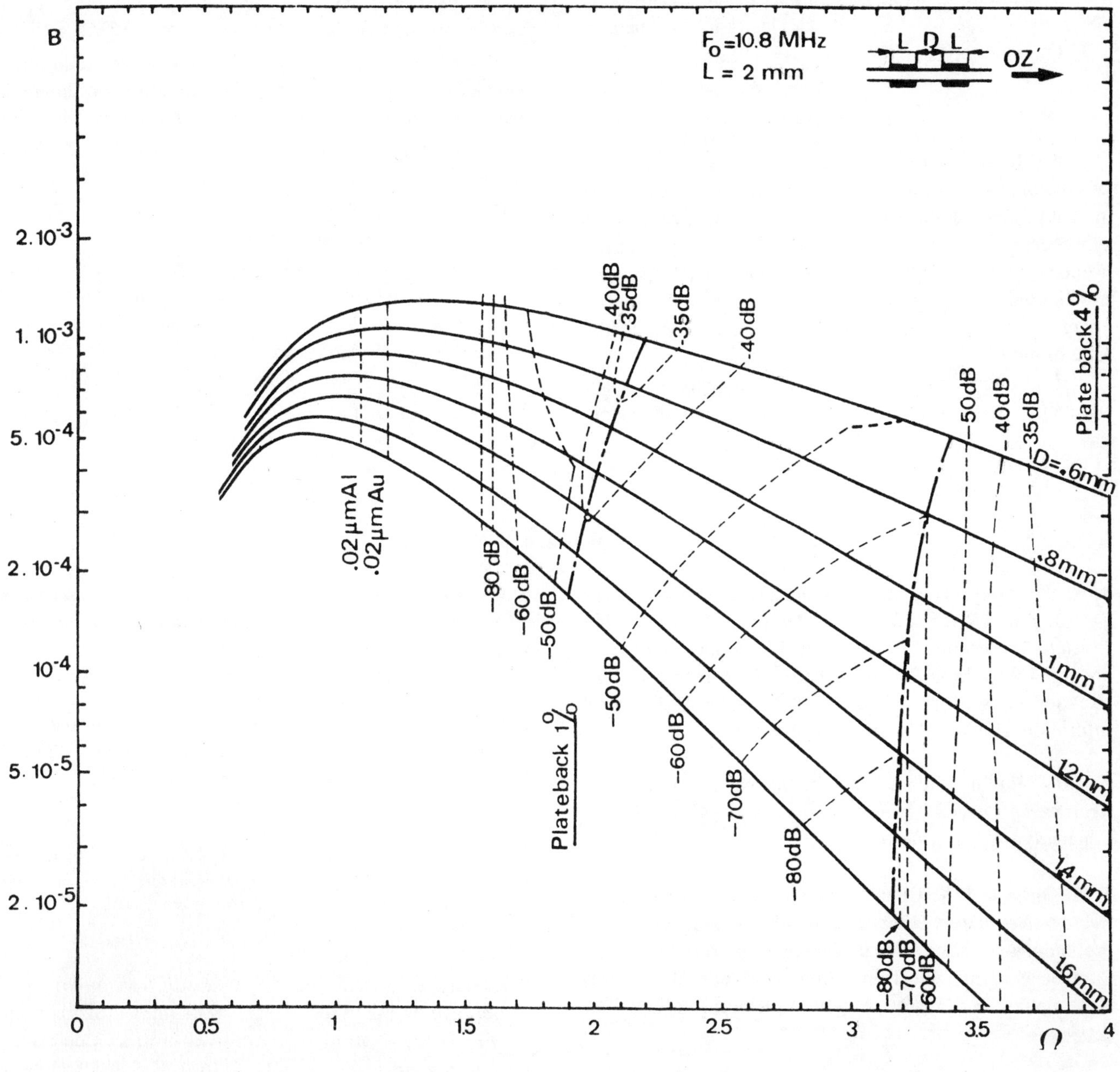

Fig. 15 - Two-resonator monolithic filter :
bandwidth and unwanted modes activity versus trapping factor and electrode separation

On figure 15, are plotted the contour lines of unwanted modes activity. This activity expresses the rejection in dB between the main modes and the unwanted modes, measured in low impedance transmission line. With the correct matching impedances, the rejection is smaller. However, these numbers give an idea of the relative magnitude of the phenomena.

The major results are the existence of a crest and of a valley in the activity surface, the crest being the maximum QAS activity, and the valley being the intersection curve of the QAS decreasing and the QS increasing surfaces.

Taking into account the electric field effect, QAS modes are trapped, if with gold electrodes, the electrode thickness on each face is greater than .05 μm ($\Omega = \pi$), always with the same geometrical data (10.8 MHz, L_2 = 2 mm).

The maximum QAS activity is found with an electrode thickness around .1 μm, on each face (plateback : 1 %).

The optimum range corresponds to the apparition of SS activity, for $\Omega \cong 2\pi$, i.e. an electrode thickness of .35 μm on each face (plateback : 2.7 %).

Conclusion

We think that asymmetry could be an important cause of unexpected behaviour of piezoelectric resonators. Actually, beyond the rough models presented here, many other defects should be taken into account, as mounting asymmetry for example. Of course, asymmetry influence is not limited to thickness vibration, although the effect of energy trapping on QAS modes activity is specific of this kind of mode.

The observed nature of random activity in crystal production is a confirmation of the role of random defects, since small variations of the asymmetry or of the trapping factor can produce large fluctuations of the QAS activity.

For monolithic filters, it shows that electrode thickness larger than for conventional single resonator, should be required to achieve convenient mode suppression.

The discussion has been limited to thickness-twist propagation. In thickness-shear vibration, along X-axis, the effects should be still more important, since the trapping is less efficient for the same plateback.

Experiments have shown a good qualitative agreement with theory. However they also demonstrated a poor accuracy in our electrode thickness measurements, what explains that no quantitative experimental results could be presented.

We thank J. Detaint for the X-ray topography work.

References

1. **W. Shockley, D.R. Curran and D.J. Koneval**, *Energy Trapping and Related Studies of Multiple Electrode Filter Crystals*, Proceedings, 17 th Annual Symposium on Frequency Control, US Army Electronics Command, Fort Monmouth, N. J., pp. 88-126, (1963). National Technical Information Service Accession Nr. AD 423381.

2. **R.D. Mindlin and D.C. Gazis**, *Strong Resonances of Rectangular AT-cut quartz plates*, Proc. 4 th US Nat. Congress of Applied Mechanics, pp. 305-310 (1962).

3. **M. Onoe and K. Okada**, *Analysis of Contoured Piezoelectric Resonators Vibrating in Thickness-Twist Modes*, Proceedings, 23rd Annual Symposium on Frequency Control, US Army Electronics Command, Fort Monmouth, N. J. pp. 26-38, (1969). National Technical Information Service Accession Nr. AD 746209.

4. **A. Glowinski and R. Lançon**, *Resonance Frequencies of Monolithic Quartz Structures* Proceedings, 23rd Annual Symposium on Frequency Control, US Army Electronics Command, Fort Monmouth, N. J. pp. 39-55, (1969). National Technical Information Service Accession Nr. AD 746209.

5. **R. Lançon**, *Résonances des structures monolithiques non symétriques*, Annales des Télécommunications (France), to be published.

6. **M. Onoe and H. Jumonji**, *Analysis of Piezoelectric Resonators Vibrating in Trapped-Energy Modes*, Electr. Communic. Eng. Japan, vol. 48, pp. 84-93 (1965).

7. **W.H. Horton and R.C. Smythe**, *On the Trapped-Wave Criterion for AT-cut Quartz Resonators with Coated Electrodes*, Proc. IEEE pp. 598-599 (april 1967).

8. **K. Haruta and W.J. Spencer**, *X-Ray Diffraction Study of Vibrational Modes*. Proceedings, 20 th Annual Symposium on Frequency Control, US Army Electronics Command, Fort Monmouth, N. J. pp. 1-16, (1966). National Technical Information Service Accession Nr. AD 800523.

9. **N.H.C. Reilly and M. Redwood**, *Wave-propagation analysis of the monolithic-crystal filter*, Proc. IEEE, vol. 116, n° 15, pp. 653-660 (may 1969).

EXPERIMENTAL INVESTIGATIONS OF INTERMODULATION IN MONOLITHIC CRYSTAL FILTERS

W. H. Horton & R. C. Smythe

Summary

Requirements on intermodulation of in-band and/or out-of-band signals are being imposed upon crystal filters in a number of advanced applications. We have observed intermodulation of in-band or out-of-band signals in large numbers of monolithic crystal filter units at frequencies from 5 MHz to 160 MHz. On the basis of experimental evidence it is now possible to postulate a surface-related mechanism for intermodulation of in-band and out-of-band signals at low strain levels. It may be further postulated that at high strain levels, other non-linear mechanisms come into play. Our understanding of the mechanisms underlying intermodulation is enhanced by regarding IM simply as one of a number of non-linear effects observable in quartz crystal devices.

Introduction

Intermodulation requirements for in-band and/or out-of-band signals are being imposed upon crystal filters in a number of advanced applications, both military and commercial. These include front-end uses and up-converter applications, as well as the more familiar IF filters. It seems likely that in the future many of the most important new applications for crystal filters will have associated intermodulation requirements.

Intermodulation of out-of-band signals has been observed and reported in both monolithic and discrete-element crystal filters.[1,2] However, the causative mechanism has not previously been established. While the existence of in-band intermodulation has been fairly widely recognized, it has not previously been studied.

During the past three years we have conducted an experimental investigation of intermodulation in monolithic crystal filters at frequencies from 5 MHz to 160 MHz. Initially, we studied the out-of-band case, in which the two generating signals (test tones) lie in the filter stopband while one or more (usually just one) intermodulation product lies in the filter passband. We next studied the in-band case in which the test tones and one or more intermodulation products lie in the passband.

This paper summarizes the results of our investigation of intermodulation for two-resonator monolithic crystal filters of one specific type. The observations reported are typical, however, of all the units we have studied. Intermodulation can be regarded simply as one of a number of non-linear effects observable in quartz crystal devices. On the basis of experimental evidence it is now possible to postulate a mechanism for intermodulation of in-band signals at low strain levels. It may be further postulated that at high strain levels, other non-linear mechanisms come into play. Moreover, it is possible to see, in broad terms, some of the relationships between in-band and out-of-band intermodulation and other non-linear phenomena.

Experimental Results

In-band intermodulation was investigated for two-resonator monolithic crystal filter units having a nominal center frequency of 11.7 MHz, 3 dB bandwidth of approximately 15 kHz and an electrode area of 4.5 mm^2. The two test tones were of equal amplitudes, with frequencies of 11,699 kHz and 11,701 kHz. The lower third order intermodulation product at 11,697 kHz was measured. Figure 1 plots the average IM ratio for 49 units versus test tone level*. In the range above approximately 10 dBm the average IMR exhibits a change of very nearly 2 dB for a 1 dB change in test tone level. Moreover, at these power levels IMR is rather consistent from unit to unit, as the histogram of figure 2 indicates. At lower levels the behavior of individual units becomes increasingly unpredictable, as shown in figure 3, while the average IMR, figure 1, fails to follow the extrapolation of the high level portion of the curve.

The latter effect is due, in part, to an upward bias of the average IMR at low levels as a result of measurement limitations. But, in addition, the mechanisms causing intermodulation at low power levels appear to be different from those at higher power levels. For the units just discussed, at test tone levels of 0 dBm and below, IMR is strongly dependent on surface condition. In particular, the effects of contamination can readily be demonstrated. This is illustrated statistically in figures 4 and 5 which show the distribution of IMR for a group of 35 units before and after ultrasonic cleaning in isopropyl alcohol. The fraction of units having an IMR in excess of 80 dB was increased from 40% to 71%.

Further evidence of the existence of surface mechanisms is obtained in the out-of-band case. Out-of-band intermodulation was investigated for the two-resonator 11.7 MHz monolithic crystal filter units described above. Test tone frequencies were 20 kHz and 40 kHz, respectively, above the nominal center frequency, so that the lower third order product fell at nominal center frequency while other IM products fell in the stopband.

Using test tone levels of 0 dBm., the principal causes of out-of-band IM were found to be surface-related. Specifically, surface contamination, either before or after base plating, was found to be a factor. This was indicated both by cleaning experiments and by deliberate introduction of surface contamination. In the latter instance units having negligible IM were coated with a mixture of isopropyl alcohol and rubber particles

* The Intermodulation Ratio (IM Ratio or IMR) is defined for purposes of this paper as the ratio of the available power, at the filter input, of either test tone to the power in a specified intermodulation product, measured at the filter load.

and dried under a gentle flow of warm air. The units were then found to have very poor and, frequently, erratic IM ratios. The use of a vehicle (the alcohol) was found to be necessary to the consistent production of intermodulation. In addition to contamination, plating defects, such as particles of plating resulting from plating scratches or the like, or nodules of plating material have been observed to produce IM. This was demonstrated in several instances by carefully removing the offending particle. With careful attention given to cleaning and handling, intermodulation ratios of 80 dB were easily realized in the out-of-band case with 90 to 100 dB being typical of production units.

Experimentally, then, it appears that at low levels of strain the non linearities causing intermodulation, whether in band or out of band, are predominantly related to surface defects of one kind or another.

The causes of intermodulation at high levels of strain are less easily determined, although our experience indicates the decreasing importance of surface conditions as the strain amplitude increases. Possible mechanisms include non-linear stress-strain relations in the electrode films or at the interface between the electrodes and the quartz, and elastic and piezoelectric non-linearities in the quartz plate. It is possible that static deformations due to mounting forces enhance these mechanisms. For the 11.7 MHz units the small unit-to-unit variation in IM ratio at high strain levels suggests a highly reproducible mechanism, most probably elastic and piezoelectric non linearities of the quartz. In our measurements of units in the 140 to 160 MHz range at similar strain levels, considerably more variability in the IM ratio has been observed, implying the presence of significant process variables. These remain to be isolated.

Other Non-Linear Effects

We now wish to discuss briefly the relationship between intermodulation and other non linear effects in quartz devices.

The increased resistance of some AT-cut resonators at low drive level is well known. Bernstein[3] has attributed its occurrence to the presence of contaminating oil films in combination with loose particles of quartz or other materials. He also studied the effects of the lapping, etching and polishing processes used in blank fabrication upon the starting resistance. More recently, Nonaka, et al[4] studied AT-cut resonator behavior at low current levels, especially the effects of plating defects, observing anomalous changes in resonator resistance and, in addition, small changes in resonance frequency which could not be attributed to the effect of resistance changes upon the frequency measurement.

The surface conditions to which Bernstein and Nonaka attribute resistance variations are similar to those to which we attribute intermodulation at low levels. It seems highly likely that both effects are attributable to the same non-linear mechanism. No attempt has been made, however, to measure resonator resistance variation in monolithic units exhibiting IM at low test tone levels.

A second well-known effect is that of non linear resonance, which has been studied in precision AT cut resonators by Warner[5] and Hammond, et al[6]. Both investigators suggest the likelihood of elastic non-linearity as the principal cause. In support of this, Warner refers to studies indicating that the frequency change is a function of the strain amplitude and is independent of frequency and Q. We believe that both non-linear resonance and intermodulation at levels above those where surface effects predominate may be caused by the same non-linearities.

For the special case of X-cut extensional mode resonators, Gagnepain and Besson[7] have related non linear resonance effects to both elastic and piezoelectric non-linearities. However, it should be noted that the sign of the frequency shift is in this case opposite to that observed for AT-cut resonators.

Measurements of the higher-order elastic, piezoelectric, and dielectric constants of quartz are needed to advance the study of intermodulation and other non-linear effects in quartz devices, particularly at high strain levels. It should then be possible to determine a lower bound on the intermodulation ratio due to material non linearities.

Acknowledgment

The authors wish to acknowledge helpful discussions with H. F. Tiersten.

References

1. Rider, Lynn S., "Final Development Report, Microelectronics Techniques Study" (N00039-68-C-2575) General Electric Co., Syracuse, New York, Feb 197[illegible]

2. Malinowski, S., & Smith, C., "Intermodulation in Crystal Filters", Proc. 26th Annual Frequency Control Symposium, pp. 180-186; 1972.

3. Bernstein, M., "Increased Crystal Resistance At Oscillator Noise Levels", Proc. 21st Annual Frequency Control Symposium, pp. 244-258; 1967.

4. Nonaka, S., Yuuki, T., & Hara K., "The Current Dependence of Crystal Unit Resistance At Low Drive Level", Proc. 25th Annual Frequency Control Symposium, pp. 139-147; 1971.

5. Warner, A.W., "Design & Performance of Ultraprecise 2.5-mc Quartz Crystal Units", B.S.T.J., v. 39, pp 1193-1217; Sept. 1960.

6. Hammond, D. L., Adams, C., & Cutler, L., "Precision Crystal Units", Proc. 17th Annual Frequency Control Symposium, pp. 215-232; 1963.

7. Gagnepain, J.J., & R. Besson, "A Study of Quartz Crystal Nonlinearities: Application to X-Cut Resonators", Phys. Lett., v. 41A, No. 5, pp 443-444; 23 Oct. 1972.

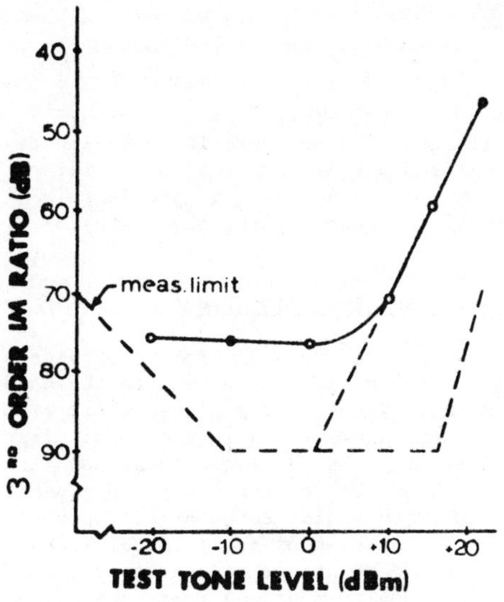

FIG. 1 Average Intermodulation Ratio vs. Test Tone Level.

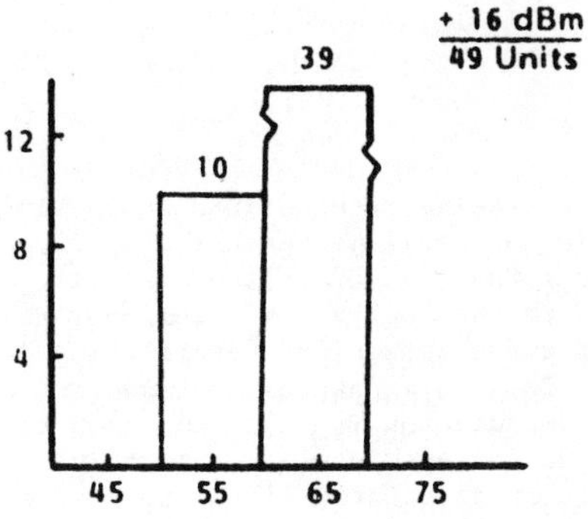

FIG. 2 Distribution of IM Ratio for Test Tone Level of + 16 dBm.

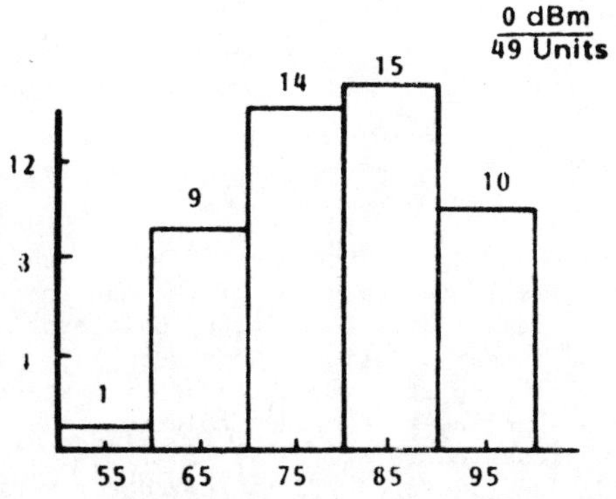

FIG 3 Distribution of IM Ratio for Test Tone Level of 0 dBm.

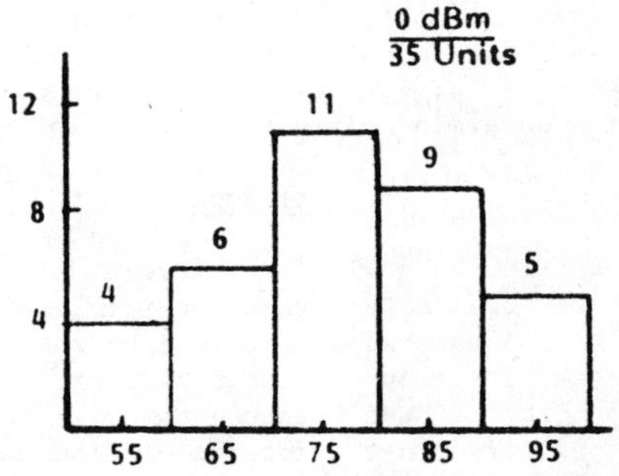

FIG 4 Distribution of IM Ratio for Test Tone Level of 0 dBm, before wash.

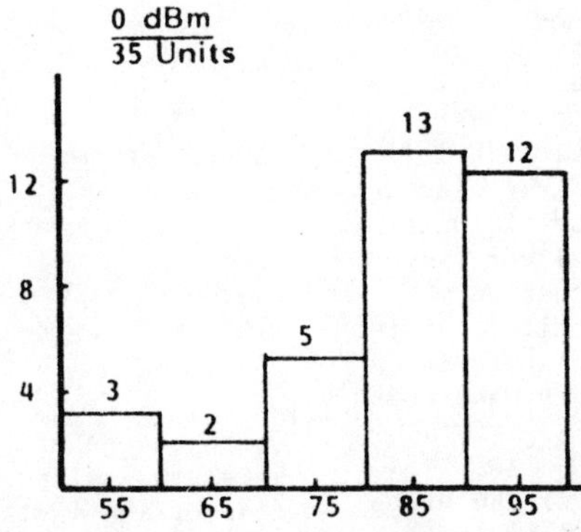

FIG 5 Distribution of IM Ratio for Test Tone Level of 0 dBm, after wash.

ANALYSIS OF NONLINEAR RESONANCE IN ROTATED Y-CUT QUARTZ THICKNESS-SHEAR RESONATORS

H.F. Tiersten
Department of Mechanical Engineering,
Aeronautical Engineering & Mechanics
Rensselaer Polytechnic Institute
Troy, New York 12181

Abstract

Electroelastic equations containing terms up to cubic in the small mechanical displacement field, but no higher than linear in the electric variables, are applied in the analysis of nonlinear resonance in rotated Y-cut quartz plates oscillating in thickness-shear. This is a natural continuation of earlier work on intermodulation in the same resonator. Since in each equation each nonlinear term is negligible compared to an associated linear term, the solution is obtained by employing an asymptotic iterative procedure and expanding in the eigensolutions of the associated linear problem and, in the vicinity of a resonance, retaining only that nonlinear term correcting the dominant eigensolution. A lumped parameter representation of the solution, which is valid in the vicinity of a resonance and relates the amplitude of the dominant thickness-shear mode and its cube to the voltage across the crystal, is presented. The expression for the current through the crystal is determined, the influence of the external circuitry is included in the analysis and, ultimately, an expression cubic in the mode amplitude and linear in the driving voltage is obtained. The analysis holds for the fundamental and odd overtone thickness-shear modes. Nonlinear resonance curves are calculated for AT-cut quartz using the nonlinear coefficient γ determined in earlier work on intermodulation.

1. Introduction

Recently, an analysis of intermodulation in rotated Y-cut quartz thickness-shear resonators was presented[1,2]. A natural and logical extension of this work is the treatment of the nonlinear resonance, i.e., the dependence of the response on drive level, of the modes considered in the intermodulation analysis. The simplifying assumptions of small piezoelectric coupling, isotropic nonlinear elastic with anisotropic linear piezoelectric behavior, which were employed in the intermodulation analysis naturally are employed here also. Since the magnitude of the mechanical displacement is very small, all nonlinear terms are negligible compared to an associated linear term. However, since in the vicinity of a resonance the linear terms in the equations of motion nearly cancel algebraically, the nonlinear terms associated with those linear terms are not negligible and must be retained.

The solution of the steady-state nonlinear resonance problem is obtained by means of an asymptotic iterative procedure and an expansion in the linear eigensolutions while retaining only that nonlinear term having its natural frequency in the vicinity of the driving frequency. A lumped parameter representation of the solution, which is valid in the vicinity of a resonance, is presented. The external circuitry is incorporated in the analysis, and an equation relating the thickness-shear displacement nonlinearly to the driving voltage and other circuit parameters is obtained. Current-frequency response curves have been obtained for a range of driving voltages.

Before proceeding with the treatment of the problem, we note that the pure thickness solution[3] is not physically realistic because all functional dependence along the plate is ignored. However, it is an important mathematical formalism that is required as the first step in the treatment of more realistic problems in which the functional dependence along the plate is included.

2. Thickness-Shear Vibrations

The crystal is referred to a Cartesian coordinate system x_1, x_2, x_3, with the x_2-axis normal to the major surfaces of the rotated Y-cut quartz plate, which are located at $x_2 = \pm h$, and the x_1-axis in a digonal direction, as shown in Fig.1. The basic nonlinear equations required for the problem of interest here are given in Sec.I of Ref.2 along with a discussion of the approximations involved. In the case of the pure thickness-modes treated here these equations reduce to (24) - (29) of Ref.2, which, for completeness, we reproduce here

$$\underset{2}{c}^{E}_{66} u_{1,22} + e_{26}\varphi_{,22} - \rho^{\circ}\ddot{u}_1 = -\gamma[(u_{1,2})^3]_{,2} - 2\beta[u_{1,2}u_{2,2}]_{,2} , \quad (2.1)$$

$$\underset{2}{c}^{E}_{22} u_{2,22} - \rho^{\circ}\ddot{u}_2 = -\beta[(u_{1,2})^2]_{,2} , \quad (2.2)$$

$$e_{26} u_{1,22} - \epsilon^{S}_{22}\varphi_{,22} = 0 , \quad (2.3)$$

$$\underset{2}{c}^{E}_{66} u_{1,2} + e_{26}\varphi_{,2} \pm \rho^{\circ} h'\ddot{u}_1 = -\gamma(u_{1,2})^3 - 2\beta u_{1,2}u_{2,2} , \quad \text{at } X_2 = \pm h , \quad (2.4)$$

$$\underset{2}{c}^{E}_{22} u_{2,2} \pm 2\rho^{\circ} h'\ddot{u}_2 = -\beta(u_{1,2})^3 , \quad \text{at } X_2 = \pm h , \quad (2.5)$$

$$\varphi = \pm \frac{1}{2} V e^{i\omega t} , \quad \text{at } X_2 = \pm h , \quad (2.6)$$

where the notation is defined in Refs.1 and 2. Since the amplitude of the largest mechanical displacement gradient is very small, the solution of the nonlinear resonance problem may readily be obtained by iteratively correcting the linear solution. Consequently, for completeness and because it forms the essential starting point for the nonlinear resonance solution, the linear solution presented in Eqs.(30), (35), (36), (41) - (43), (50) and (54) of Ref.2 is reproduced here, but in a somewhat different form

$$u_2 = 0 , \quad u_1 = u(X_2)e^{i\omega t} , \quad \varphi = \tilde{\varphi}(X_2)e^{i\omega t} , \quad (2.7)$$

$$u = \hat{u} + KX_2 , \quad \tilde{\varphi} = \hat{\varphi} + VX_2/2h , \quad (2.8)$$

$$K = -e_{26}V/2h(c_{66} - 2\rho^{\circ}{}'h'\omega^2 h) , \quad (2.9)$$

$$\hat{u} = \sum_n \hat{u}_n , \quad \hat{\varphi} = \sum_n \varphi_n , \quad (2.10)$$

where $\hat{u}_n$ and $\hat{\varphi}_n$ are the linear eigensolutions, which are given by

$$\hat{u}_n = A_n \sin \eta_n X_2 , \quad \hat{\varphi}_n = (e_{26}/\epsilon_{22})A_n \sin \eta_n X_2 + C_n X_2 , \quad (2.11)$$

$$C_n = -(e_{26}/\epsilon_{22}h)A_n \sin \eta_n h , \quad \bar{\omega}_n = (\bar{c}_{66}/\rho^{\circ})^{\frac{1}{2}}\eta_n , \quad (2.12)$$

$$\eta_n h = (n\pi/2)[1 - (4k^2_{26}/n^2\pi^2) - R] , \quad (2.13)$$

$$A_n = \frac{-4(-1)^{(n-1)/2} e_{26} V}{c_{66} n^2\pi^2 [(\bar{\omega}^2_n/\omega^2) - 1]} , \quad (2.14)$$

where the quantities appearing here and not defined here are defined in Sec.II of Ref.2 and use has been made of the fact that $R \ll 1$ and $k_{26} \ll 1$. In the vicinity of a resonance, say the Nth, one term dominates the series and the solution may be written in the form

$$u_1 = [-(e_{26}V/c_{66}2h)X_2 + A_N \sin(N\pi X_2/2h)]e^{i\omega t},$$

$$\varphi = \left[\frac{VX_2}{2h} + \frac{e_{26}}{\epsilon_{22}} A_N\left(\sin\frac{N\pi X_2}{2h} - (-1)^{(N-1)/2}\frac{X_2}{h}\right)\right]e^{i\omega t}, \tag{2.15}$$

where A_N is given by (2.14) with n = N.

For simplicity and clarity, initially we ignore the influence of the nonlinear coefficient β, which is considered later in this section. Inasmuch as we are interested in obtaining the steady-state solution of the nonlinear forced vibration problem at the frequency ω, we write the solution in the form

$$u_1 = \frac{1}{2}(u(X_2)e^{i\omega t} + u^*(X_2)e^{-i\omega t}), \tag{2.16}$$

where $u(X_2)$ is complex, the asterisk denotes complex conjugate and the phase in time is arbitrary. From (2.16) we obtain

$$u_1^3 = \frac{1}{4}\frac{1}{2}[u^3e^{i3\omega t} + u^{*3}e^{-i3\omega t} + 3(u^2u^*e^{i\omega t} + uu^{*2}e^{-i\omega t})], \tag{2.17}$$

and since we are interested in the steady-state solution at frequency ω, we ignore the terms at 3ωt when we substitute from (2.16) and (2.17) into (2.1) and (2.4) to obtain

$$c^E_{66}u_{,22} + e_{26}\tilde{\varphi}_{,22} + \rho^\circ\omega^2 u = -\frac{3}{4}\gamma[(u_{,2})^2u^*_{,2}]_{,2}, \tag{2.18}$$

$$c^E_{66}u_{,2} + e_{26}\tilde{\varphi}_{,2} \mp 2\rho^\circ{}'h'\omega^2 u = -\frac{3}{4}\gamma(u_{,2})^2u^*_{,2}$$

$$\text{at } X_2 = \pm h, \tag{2.19}$$

in which we have introduced the usual complex notation and after multiplying a function by $e^{i\omega t}$ the real part is understood and $\tilde{\varphi}$ is defined in $(2.7)_3$.

Guided by the linear solution presented in Sec.II of Ref.2 and reproduced above, we take the nonlinear solution in the form given in (2.8), where K is to be selected so that the linear inhomogeneous term that occurs in the boundary condition (2.19) is transformed into the mechanical differential equation (2.18), as in the linear solution. Substituting from (2.8) into (2.18), (2.19) and (2.6) and selecting K as in (2.9), we obtain, respectively,

$$c^E_{66}u_{,22} + e_{26}\hat{\varphi}_{,22} + \rho^\circ\omega^2\hat{u} + \rho^\circ\omega^2KX_2 = -\frac{3}{4}\gamma[(\hat{u}_{,2}+K)^2(\hat{u}^*_{,2}+K^*)]_{,2}, \tag{2.20}$$

$$c^E_{66}\hat{u}_{,2} + e_{26}\hat{\varphi}_{,2} \mp 2\rho^\circ{}'h'\omega^2\hat{u} = -\frac{3}{4}\gamma(\hat{u}_{,2}+K)^2(\hat{u}^*_{,2}+K^*),$$

$$\text{at } X_2 = \pm h, \tag{2.21}$$

$$\hat{\varphi} = 0, \quad \text{at } X_2 = \pm h. \tag{2.22}$$

Since we are interested in obtaining the steady-state solution of the nonlinear forced vibration problem for driving frequencies in the vicinity of a resonant frequency of the associated linear thickness vibration problem, we will expand the nonlinear inhomogeneous solution in the appropriate linear and nonlinearly corrected eigensolutions. To this end, we seek the nonlinearly corrected eigensolutions to the homogeneous form of (2.20) - (2.22), i.e., with V = 0(K = 0), and

$$e_{26}\hat{u}_{,22} - \epsilon^S_{22}\hat{\varphi}_{,22} = 0, \tag{2.23}$$

which is obtained by substituting from (2.7) and (2.8) into (2.3).

We now take the linear eigensolutions given in (2.11) as the zeroth order iterates of the homogeneous form of (2.20) - (2.23) and substitute from (2.11) into the nonlinear terms on the rhs of (2.20) and (2.21) to obtain

$$c_{66}\,{}_1\hat{u}_{n,22} + e_{26}\,{}_1\hat{\varphi}_{n,22} + \rho^\circ\omega_n^2\,{}_1\hat{u}_n =$$

$$\frac{3}{4}\gamma\, 3\eta_n^4A_n^2A_n^*\cos^2\eta_nX_2\sin\eta_nX_2, \tag{2.24}$$

$$c_{66}\,{}_1\hat{u}_{n,2} + e_{26}\,{}_1\hat{\varphi}_{n,2} \mp 2\rho^\circ{}'h'\omega_n^2\,{}_1\hat{u}_n = -$$

$$\frac{3}{4}\gamma\,\eta_n^3A_n^2A_n^*\cos^3\eta_nh, \quad \text{at } X_2 = \pm h, \tag{2.25}$$

where here and in the sequel we have eliminated the superscripts E and S and lower script 2 and ω_n is the amplitude dependent natural frequency of the nth mode, which is to be determined from (2.24), (2.25) along with

$$e_{26}\,{}_1\hat{u}_{n,22} - \epsilon_{22}\,{}_1\hat{\varphi}_{n,22} = 0, \tag{2.26}$$

$${}_1\hat{\varphi}_n = 0, \quad \text{at } X_2 = \pm h, \tag{2.27}$$

which are the homogeneous eigenmode equations corresponding to (2.23) and (2.22), respectively. The notation ${}_1\hat{u}_n$ and ${}_1\hat{\varphi}_n$ employed in (2.24) - (2.27) is intended to denote the first iterate eigensolution of the nonlinear problem and the aforementioned linear eigensolution is the zeroth iterate, which should be denoted ${}_0\hat{u}_n$ and ${}_0\hat{\varphi}_n$. Employing the addition theorem for products of trigonometric functions in (2.24) and (2.25) we obtain

$$c_{66}\,{}_1\hat{u}_{n,22} + e_{26}\,{}_1\hat{\varphi}_{n,22} + \rho^\circ\omega_n^2\,{}_1\hat{u}_n =$$

$$\frac{9}{16}\gamma\eta_n^4A_n^2A_n^*[\sin\eta_nX_2 + \sin 3\eta_nX_2], \tag{2.28}$$

$$c_{66}\,{}_1\hat{u}_{n,2} + e_{26}\,{}_1\hat{\varphi}_{n,2} \mp 2\rho^\circ{}'h'\omega_n^2\,{}_1\hat{u}_n = -$$

$$\frac{3}{4}\gamma\eta_n^3A_n^2A_n^*\left[\frac{3}{4}\cos\eta_nh + \frac{1}{4}\cos 3\eta_nh\right],$$

$$\text{at } X_2 = \pm h. \tag{2.29}$$

As a solution of the linear differential equations (2.26) and (2.28), we take

$${}_1\hat{u}_n = A_n\sin\eta_nX_2 + B_n\sin 3\eta_nX_2,$$

$${}_1\hat{\varphi}_n = \frac{e_{26}}{\epsilon_{22}}[A_n\sin\eta_nX_2 + B_n\sin 3\eta_nX_2] + C_nX_2, \tag{2.30}$$

the substitution of which into (2.26) and (2.28) yields

$$\omega_n^2 = (\bar{c}_{66}/\rho^\circ)\eta_n^2(1+\mu), \tag{2.31}$$

$$B_n = (9/16)\gamma\,\eta_n^4A_n^2A_n^*/[\rho^\circ\omega_n^2 - \bar{c}_{66}9\eta_n^2], \tag{2.32}$$

where

$$\bar{c}_{66} = c_{66} + e_{26}^2/\epsilon_{22}, \quad \mu = (9/16)(\gamma/\bar{c}_{66})\eta_n^2A_nA_n^*. \tag{2.33}$$

Equation (2.31) gives the dependence of the natural frequencies of thickness-shear vibration on the amplitude of the mode. Since the amplitude A_n is very small, from (2.31) - (2.33), we have

$$B_n \approx -\mu A_n/8, \tag{2.34}$$

which shows that B_n is negligible for our purposes except insofar that it might result in a change in the eigenvalue ω_n by virtue of the boundary conditions. Substituting from (2.30) into (2.27) we find

$$C_n = -\frac{e_{26}}{\epsilon_{22}h}A_n\sin\eta_nh\left[1 - \frac{\mu}{8}\frac{\sin 3\eta_nh}{\sin\eta_nh}\right]. \tag{2.35}$$

Substituting from (2.30), (2.34) and (2.35) into (2.29), we obtain

$$\tan \eta_n h = \eta_n h \frac{[1+\mu - (\mu/24)(\cos 3\eta_n h/\cos \eta_n h)]}{\left[(k_{26}^2 + R\eta_n^2 h^2)\left(1 - \frac{\mu}{8}\frac{\sin 3\eta_n h}{\sin \eta_n h}\right) + R\eta^2 h^2 \mu\right]}, \tag{2.36}$$

where

$$k_{26}^2 = e_{26}^2/\bar{c}_{66}\epsilon_{22}, \quad R = 2\rho'h'/\rho^\circ h. \tag{2.37}$$

The roots of Eq.(2.36), with (2.31), determine the amplitude dependent eigenfrequencies ω_n of thickness-shear vibration of this piezoelectric plate with shorted electrodes in the nonlinear case. In the absence of the nonlinear term μ, Eq.(2.36) reduces to Eq.(43) of Ref.2, which is the frequency equation in the linear case. Since μ is small, the roots $\eta_n h$ of (2.36) may be obtained iteratively from the roots in the linear case. Moreover, since for small piezoelectric coupling k_{26} and mass loading R, Eq.(2.36) in the absence of μ indicates that the roots $\eta_n h$ differ from $n\pi/2$ (n odd) by small quantities, say Δ_n, we may write

$$\eta_n h = (n\pi/2) - \Delta_n, \quad n \text{ odd}. \tag{2.38}$$

Substituting from (2.38) into (2.36), using the addition theorem for trigonometric functions, expanding the trigonometric functions as a power series in Δ_n and retaining terms linear in Δ_n, R, k_{26}^2 and μ, we obtain

$$\Delta_n = (2k_{26}^2/n\pi) + n\pi R/2, \tag{2.39}$$

which shows that the nonlinear boundary conditions yield the same result as in the linear case. Substituting from (2.38) and (2.39) into (2.31), we obtain

$$\bar{\omega}_n^2 = \bar{\omega}_{n0}^2(1+\mu), \tag{2.40}$$

where $\bar{\omega}_{n0}$ denotes the natural thickness-shear frequencies for the linear case and is given by

$$\bar{\omega}_{n0} = (n\pi/2h)(\bar{c}_{66}/\rho^\circ)^{\frac{1}{2}}[1 - (4k_{26}^2/n^2\pi^2) - R]. \tag{2.41}$$

Since $\mu \ll 1$, we may write

$$\bar{\omega}_n = \bar{\omega}_{n0}(1+\mu/2), \tag{2.42}$$

which, with $(2.33)_2$ and (2.41) on account of the smallness of Δ_n and μ, yields

$$\bar{\omega}_n = \frac{n\pi}{2h}\left(\frac{\bar{c}_{66}}{\rho^\circ}\right)^{\frac{1}{2}}\left[1 - \frac{4k_{26}^2}{n^2\pi^2} - R + \frac{9}{32}\frac{\gamma}{\bar{c}_{66}}\frac{n^2\pi^2}{4h^2}A_n A_n^*\right], \quad n = 1,3,5 \ldots, \tag{2.43}$$

which gives the resonant frequency of the nth odd overtone of thickness-shear vibration for small piezoelectric coupling and electrode mass loading, including the dependence on the amplitude of vibration.

The solution to the nonlinear forced vibration problem at a driving frequency ω in the vicinity of ω_N may now be obtained by means of an expansion in the eigensolutions, while retaining the nonlinear correction in the dominant Nth eigensolution only. Accordingly, we write

$$\hat{u} = A_N \sin \eta_N X_2 + B_N \sin 3\eta_N X_2 + \sum_{n \neq N} A_n \sin \eta_n X_2,$$

$$\hat{\varphi} = \frac{e_{26}}{\epsilon_{22}}[A_N \sin \eta_N X_2 + B_N \sin 3\eta_N X_2] + C_N X_2 + \sum_{n \neq N}\left[\frac{e_{26}}{\epsilon_{22}}A_n \sin \eta_n X_2 + C_n X_2\right], \tag{2.44}$$

where B_N and C_N are determined from (2.34) and (2.35), respectively, the C_n are determined from $(2.12)_1$ and all $A_n \ll A_N$. In the iterative procedure we employ we substitute the large term in the expansion in (2.44), i.e., $A_N \sin \eta_N X_2$, for $\hat{u}$ in the nonlinear terms on the r.h.s. of (2.20) and (2.21), and note, as in Sec.II of Ref.2, that since A_N effectively has a resonant denominator while K does not, K is negligible compared to $\eta_N A_N$ in the nonlinear terms on the r.h.s. of (2.20) and (2.21). Moreover, since $\eta_N \approx N\pi/2h$, we note that the boundary condition (2.21) becomes linear. Taking the foregoing into consideration, employing the addition theorem for products of trigonometric functions and substituting from (2.44) into the l.h.s. of (2.20), we obtain

$$[-\bar{c}_{66}\eta_N^2(1+\mu) + \rho^\circ\omega^2]A_N \sin \eta_N X_2 + [\bar{c}_{66}\eta_N^2 - \rho^\circ\omega^2](\mu/8)A_N \sin 3\eta_N X_2 + \sum_{n\neq N}[-\bar{c}_{66}\eta_n^2 + \rho_\circ\omega^2]A_n \sin \eta_n X_2 = -\rho^\circ\omega^2 K X_2, \tag{2.45}$$

in which we have employed (2.33) and (2.34). Multiplying Eq.(2.45) by $\sin \eta_N X_2$, integrating from -h to h, using the orthogonality of the $\sin \eta_n X_2$ and employing (2.31), we obtain

$$(\omega^2 - \omega_N^2)A_N h = -\omega^2 K 8h^2(-1)^{(N-1)/2}/N^2\pi^2, \tag{2.46}$$

and we do not bother to obtain any of the other A_n because, as already noted, we are interested in the solution only when ω is in the vicinity of ω_N and the Nth eigenmode is dominant so that all other eigenmodes are negligible. At this point we note in passing that the amplitude A_N has an influence on the amplitude A_{3N} by virtue of the term $(\mu A_N/8)$ appearing in (2.45). Substituting from (2.9), $(2.33)_2$ and (2.40) into (2.46) and recalling that $R \ll 1$, we obtain

$$N^2\pi^2 A_N[\omega^2 - \bar{\omega}_{N0}^2(1+\alpha A_N A_N^*)] = 4(-1)^{(N-1)/2}(e_{26}/c_{66})V\omega^2, \tag{2.47}$$

where $\bar{\omega}_{N0}$ is given in (2.41) and

$$\alpha = (9/16)(\gamma/\bar{c}_{66})(N^2\pi^2/4h^2). \tag{2.48}$$

Equation (2.47) gives the nonlinear relation between the amplitude of the thickness-shear displacement and the voltage across the crystal. Thus, for an ω in the vicinity of $\bar{\omega}_{N0}$, the solution can very accurately be written in the form

$$u_1 = \left[-\frac{e_{26}V}{c_{66}}\frac{X_2}{2h} + A_N \sin\frac{N\pi X_2}{2h}\right]e^{i\omega t},$$

$$\varphi = \left[\frac{VX_2}{2h} + \frac{e_{26}}{\epsilon_{22}}A_N\left(\sin\frac{N\pi X_2}{2h} - (-1)^{(N-1)/2}\frac{X_2}{h}\right)\right]e^{i\omega t}, \tag{2.49}$$

where A_N must satisfy (2.47), because the nonlinear term $\alpha A_N A_N^*$ is negligible in all expressions except (2.47), in which it is compared essentially to zero instead of one. Equations (2.47) and (2.49) constitute the solution of the steady-state nonlinear forced thickness-shear vibration problem for driving frequencies ω in the vicinity of $\bar{\omega}_N$.

As noted earlier in this section, the influence of the nonlinear coefficient β has been ignored in the foregoing treatment. However, an examination of the terms containing β in Eqs.(2.1), (2.2), (2.4) and (2.5), along with a consideration of the solution discussed in the Appendix of Ref.2, reveals that the resulting additional terms occurring in the nonlinear solution will always have a thickness dependence different from $\sin \eta_N X_2$. Since such terms contain amplitude coefficients that are determined recursively, as was B_N in the solution already obtained, the additional amplitude coefficients will be negligible for the same reason as

B_N and, similarly, will have a negligible influence on the nonlinear solution already obtained. Thus, the nonlinear coefficient β has absolutely no influence on nonlinear resonance, although it can have a significant influence on intermodulation and other nonlinear effects. As a result, the solution for nonlinear resonance is as given in (2.47) and (2.49).

3. Inclusion of Circuitry

Substituting from (2.49) into Eq.(20) of Ref.2, which we reproduce here as

$$D_2 = e_{26} u_{1,2} - \epsilon_{22} \varphi_{,2} , \qquad (3.1)$$

which is then substituted into

$$I_c = - \int_S \dot{D}_2 \, dS , \qquad (3.2)$$

where S is the area of and I_c is the current through the crystal, we obtain

$$I_c = (i\omega 4w\ell\epsilon_{22}/2h)[(1+\hat{k}_{26}^2)V - (2e_{26}/\epsilon_{22})\ddot{A}_N] , \qquad (3.3)$$

where 2ℓ and 2w are the length and width, respectively of the crystal and

$$\hat{k}_{26}^2 = k_{26}^2/(1 - k_{26}^2) , \quad \hat{A}_N = A_N(-1)^{(N-1)/2} . \qquad (3.4)$$

Application of Kirchhoffs voltage equation to the circuit shown in Fig.2 yields

$$V_g + I_c(R_g + R_L) + V = 0 , \qquad (3.5)$$

where V_g is the generator voltage, R_g and R_L are the generator and load resistance, respectively, and V is the voltage across the crystal. Substituting from (3.3) and (3.5) into (2.47), we obtain

$$(N^2\pi^2\hat{A}_N/4\omega^2)[\omega^2 - \bar{\omega}_{N0}^2(1+\alpha\hat{A}_N\hat{A}_N^*)][1 + i\omega(R_g + R_L)(2w\ell\epsilon_{22}/h)(1+\hat{k}_{26}^2)] - i\omega(R_g + R_L)(4w\ell\epsilon_{22}/h)\hat{k}_{26}^2\hat{A}_N = -(e_{26}/c_{66})V_g . \qquad (3.6)$$

Equation (3.6) is the nonlinear equation relating the amplitude of the thickness-shear displacement to the generator voltage. Since Eq.(3.6) contains $\hat{A}_N^*$ as well as $\hat{A}_N$, it is not simply a complex cubic equation. Consequently, a certain elementary manipulation facilitates the solution of (3.6) considerably. However, before we perform this elementary simplification, we note, as in Ref.2, that since a resonant mode has a finite quality factor Q_N, $\bar{\omega}_{N0}$ in Eq.(3.6) is to be replaced by $\hat{\omega}_{N0}$, where

$$\hat{\omega}_{N0} = \bar{\omega}_{N0} + i\bar{\omega}_{N0}/2Q_N , \qquad (3.7)$$

in which Q_N is the unloaded quality factor of the resonator in that mode, and serves to prevent the amplitude from becoming unbounded. If the dependence of Q_N on the amplitude of the mode is known it may be included. In order to facilitate the solution of (3.6), we multiply Eq.(3.6) by its complex conjugate and after some algebraic manipulations obtain

$$\mathcal{A}[d^2 + c^2 - 2(d+ca)\bar{\omega}_{N0}^2\alpha\mathcal{A} + (1+a^2)\bar{\omega}_{N0}^4\alpha^2\mathcal{A}^2] = e^2\mathcal{V}_g , \qquad (3.8)$$

where

$$\mathcal{A} = \hat{A}_N\hat{A}_N^* , \quad \mathcal{V}_g = V_g V_g^* , \qquad (3.9)$$

$$a = \omega(R_g + R_L)(2w\ell\epsilon_{22}/h)(1+\hat{k}_{26}^2), \quad d = \omega^2 - \bar{\omega}_{N0}^2 + \bar{\omega}_{N0}^2 a/Q_N ,$$
$$c = -(\bar{\omega}_{N0}^2/Q_N) + a(\omega^2 - \bar{\omega}_{N0}^2) - b , \quad e = e_{26}/c_{66}\nu ,$$
$$\nu = N^2\pi^2/4\omega^2, \quad b = \omega(R_g + R_L)(4w\ell\epsilon_{22}/h)\hat{k}_{26}^2/\nu . \qquad (3.10)$$

Equation (3.8) is a real cubic algebraic equation in the square of the thickness-shear vibration amplitude $\hat{A}_N\hat{A}_N^*$. For a solution $\hat{A}_N\hat{A}_N^*$ to be meaningful, it must be real and positive. Consequently, there is at least one and possibly three physically meaningful $\hat{A}_N\hat{A}_N^*$. When a physically meaningful $\hat{A}_N\hat{A}_N^*$ has been determined from (3.8), it may be substituted in (3.6), which then becomes a linear equation in $\hat{A}_N$ that may readily be solved for the phase if desired. When an $\hat{A}_N$ is known, the current I_c through the circuit may be evaluated from (3.3).

Numerical results have been obtained for the fundamental mode of an AT-cut quartz thickness-shear resonator having the dimensions 2h = .14 mm, $2h' = 3 \times 10^{-4}$ mm, 2ℓ = 3.0 mm, 2w = 3.0 mm, for the specific resistances $R_g = R_L$ = 10 ohms and a few values of driving voltage. The amplitudes of the currents obtained are plotted in Figs.3 and 4 as a function of driving frequency for the specific driving voltages considered. The value of γ given in Eq.(233) of Ref.2 was employed along with Bechmann's constants[4] for the AT-cut of quartz. The response curve shown in Fig.3 is exactly the same as the one that would be obtained from a linear analysis. This is a result of the fact that increments in frequency of 2 kHz were employed and such increments are too large to show the nonlinear effect. Increments in frequency of 10 Hz were employed in obtaining the response curves shown in Fig.4 and these curves clearly show the influence of the nonlinearity. However, it should be noted that since the thickness solution is not physically realistic for the reasons discussed at the end of Sec.1, the curves in Fig.4 overestimate the nonlinear effect somewhat for a real AT-cut quartz oscillator and, consequently, are to be regarded as qualitative.

Acknowledgement

The author wishes to thank R.C. Smythe of Piezo Technology, Inc. for many valuable discussions and B.K. Sinha and M. Jahanmir of Rensselaer Polytechnic Institute for performing the calculations and plotting the curves.

This work was supported in part by the Office of Naval Research under Contract No. N00014-67-A-0117-0007 and the National Science Foundation under Grant No. GK-38118.

References

1. H.F. Tiersten, "Analysis of Intermodulation in Rotated Y-Cut Quartz Thickness-Shear Resonators," Proceedings of the 28th Annual Symposium on Frequency Control, U.S. Army Electronics Command, Fort Monmouth, New Jersey, 1 (1974).

2. H.F. Tiersten, "Analysis of Intermodulation in Thickness-Shear and Trapped Energy Resonators," J. Acoust. Soc. Am., 57, 667 (1975).

3. The precise delineation of the range of applicability of the thickness solution for the determination of frequency has been provided by R.D. Mindlin. See R.D. Mindlin, "Thickness-Shear and Flexural Vibrations of Crystal Plates," J. Appl. Phys., 22, 316 (1951). R.D. Mindlin and D.C. Gazis, "Strong Resonances of Rectangular AT-Cut Quartz Plates," Proceedings 4th U.S. National Congress of Applied Mechanics, 305 (1962).

4. H.F. Tiersten, Linear Piezoelectric Plate Vibrations (Plenum, New York, 1969), Chap.7, Eqs.(7.36). The piezoelectric constants for AT-cut quartz appearing in Ref.4 were determined from

Bechmann's constants for left-hand quartz.
R. Bechmann, "Elastic and Piezoelectric Constants of Alpha Quartz," Phys. Rev., 110, 1060 (1958).

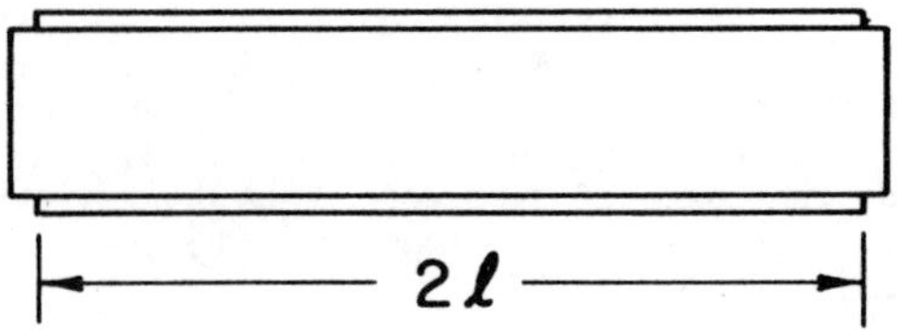

Fig. 1

Schematic Diagram of the Thickness-Shear Resonator

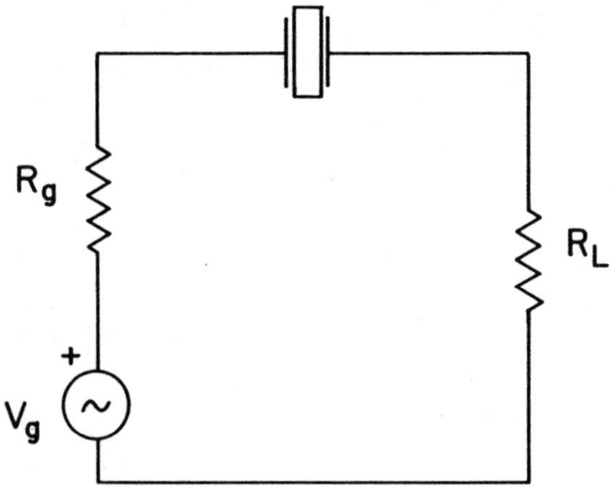

Fig. 2

Schematic Diagram of the Reduced Test Circuit

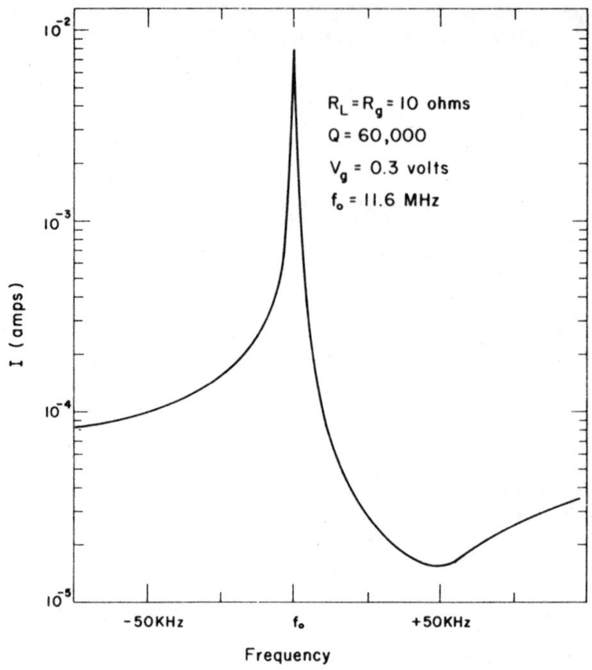

Fig. 3

Current vs Frequency Response Curve for Frequency Increment of 2kHz

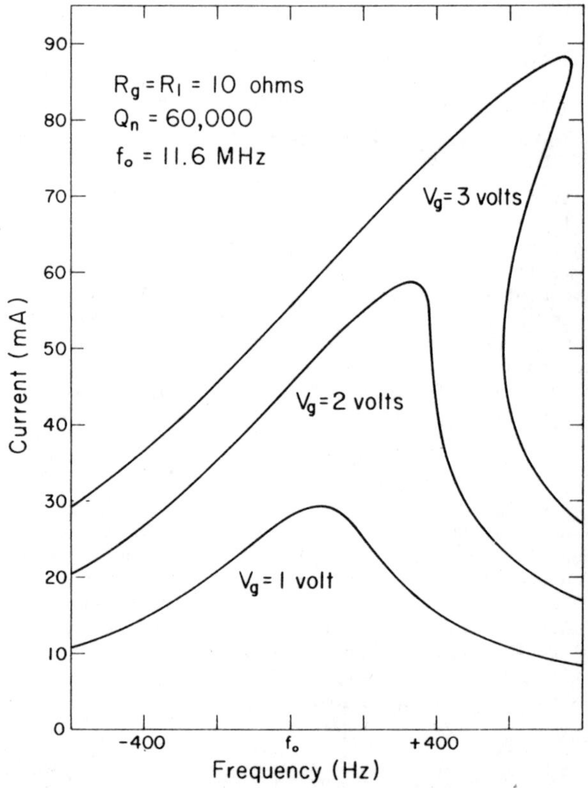

Figure 4

Current vs Frequency Response Curves for Frequency Increment of 10 Hz

Mechanical Filter Bibliography

1) R. Adler, "Compact electromechanical filters," *Electronics*, vol. 20, pp. 100–105, Apr. 1947.

2) R. Adler, "Magnetostrictively driven mechanical wave filter," U.S. Patent 2 501 488, Mar. 1950.

3) H. Albert and I. Pfeiffer, "Anomalien der Temperaturabhängigkeit des Elastizitätsmoduls von Niob-Zirkonium-Legierungen und reinem Niob," *Z. Metallkde*, vol. 58, no. 5, pp. 311–316, 1967.

4) H. Albert and I. Pfeiffer, "Über die Temperaturabhängigkeit des Elastizstätsmoduls von Niob-Titan-Legierungen," *Z. Metallkde*, vol. 63, no. 3, pp. 126–131, 1972.

5) H. Albsmeier, "Mechanische Bandpässe mit mechanisch erzeugten frei wählbaren Dämpfungspolen," *Frequenz*, vol. 17, pp. 442–448, Dec. 1963.

6) H. Albsmeier, "Über Antriebe für elektromechanische Bandpässe mit piezoelektrischen Wandlern," *Frequenz*, vol. 19, pp. 125–133, Apr. 1965.

7) H. Albsmeier and A. Günther, "Verfahren zum Abgleich eines mechanischen Filters," German Patent 2048125, Sept. 30, 1970.

8) H. Albsmeier, "A comparison of the realizability of electromechanical filters in the frequency range of 12 kHz to 10 MHz" (in German), *Frequenz*, vol. 25, no. 3, 1971.

9) H. Albsmeier, A. E. Günther, and W. Volejnik, "Some special design considerations for a mechanical filter channel bank," *IEEE Trans. Circuits Syst.*, vol. CAS-21, pp. 511–516, July 1974.

10) H. Albsmeier, M. Dombrowski, and F. Künemund, "Mechanical channel filter with bending mode resonators from THERMELAST" (in German), *Siemens-Z.*, Special Issue "Nachrichten-Übertragungstechnik," vol. 48, pp. 41–43, 1974.

11) P. Amstutz and H. Carru, "Vibrations d'un coupleur cylindrique circulaire encastré entre deux résonateurs demi-onde," Document de Travail CNET/EST/DEF/4012, Mar. 1975.

12) P. Amstutz and M. Bon, "Deviating from Darlington's equiripple pattern in filter design," in *Proc. 3rd Int. Symp. on Network Theory*, Split, Yugoslavia, Sept. 1975, pp. 641–648.

12a) T. Ashida, "Design of piezoelectric transducers for temperature-stabilized mechanical filters," *Electron. Commun. Japan*, vol. 57-A, pp. 10–17, May 1974.

13) H. Bache, "A practical electro-mechanical filter," *Marconi Rev.*, vol. 22, pp. 144–153, 3rd Quarter, 1959.

14) H. Baker and J. R. Cressey, "H-shaped resonators signal upturn in tone telemetering," *Electronics*, pp. 99–106, Oct. 1967.

15) M. Battini and F. Caviglia, "Electromechanical channel filters for FDM telephone equipment" (in Italian), *Elettron. e Telecomun.*, vol. 22, pp. 183–193, Sept.-Oct. 1973.

16) A. Bauer and B. Racurow, "Elektromechanische Filter," *Radio und Fernsehen*, vol. 15, no. 22, pp. 677–679, no. 23, pp. 727–730, 1966.

17) H. Betzl, "Ein Beitrag zur Berechnung von eingliedrigen Quarz-Brückenbandpässen mittlerer Bandbreite nach der Betriebsparametertheorie," *Frequenz*, vol. 19, pp. 206–209, June 1975.

18) B. Birn, "A modified insertion loss theory for mechanical channel filter synthesis," in *Proc. 1976 IEEE ISCAS*, Apr. 1976, pp. 754–757.

19) B. Birn, "Design and computer aided development of mechanical filters for channel modems" (in German), *Techn. Mitt. AEG-Telefunken*, Special Issue "Trägerfrequenztechnik," vol. 64, pp. 22–25, Apr. 1974.

20) M. Börner, E. Kettel, and H. Ohnsorge, "Mechanische Filter für die Nachrichtentechnik," *Telefunken-Z.*, vol. 31, pp. 105–114, June 1958.

21) M. Börner, "Biegeschwingungen in mechanischen Filtern," *Telefunken-Z.*, vol. 31, pp. 115–123, June 1958, pp. 188–196, Sept. 1958.

22) M. Börner, "Mechanische Filter für die Trägerfrequenztechnik," *Nachrichtentechnische Fachberichte*, vol. 19, pp. 34–37, 1960.

23) M. Börner, "Berechnung Mechanischer Filter," *Elektron. Rundsch*, vol. 15, pp. 11–14, Jan. 1961.

24) M. Börner, "Mechanische Filter mit Biegekopplung," *Arch. Elek. Übertr.*, vol. 15, pp. 175–180, Apr. 1961.

25) M. Börner, "Transformation der Kopplung in mechanischen Filtern mit beidseitig freischwingenden Biegeresonatoren," *Arch. Elek. Übertr.*, vol. 16, pp. 355–358, July 1962.

26) M. Börner, "Zum Nebenwellenverhalten von Mikromodulfiltern (Biegenebenwellen in biegegekoppelten Longitudinalfiltern)," *Arch. Elek. Übertr.*, vol. 16, no. 10, pp. 532–534, 1962.

27) M. Börner, "Mechanische Filter mit biegegekoppelten Biegeresonatoren und kopplungsfreie Biegefilter," *Arch. Elek. Übertr.*, vol. 16, pp. 459–464, Sept. 1962.

28) M. Börner, "Mechanische Filter mit Dämpfungspolen," *Arch. Elek. Übertr.*, vol. 17, pp. 103–107, Mar. 1963.

29) M. Börner, "Mechanical frequency filters," U.S. Patent 3 086 182, Apr. 1963.

30) M. Börner, E. Dürre, and H. Schüssler, "Mechanische Einseitenbandfilter," *Telefunken-Z.*, vol. 36, pp. 272–280, May 1963.

31) M. Börner and H. Schüssler, "Miniaturisierung mechanischer Filter," *Telefunken-Z.*, vol. 37, pp. 228–246, Fall 1964.

32) M. Börner, "Progress in electromechanical filters," *Radio Electro. Eng.*, vol. 29, pp. 173–184, Mar. 1965.

33) M. Börner and H. Schüssler, "Mechanische Miniaturfilter für Zwischenfrequenzen," International Federation of Automatic Control, Tagungsbericht d. Herbsttagung, München, pp. 717–730, 1965.

34) M. Börner, "Magnetische Werkstoffe in elektromechanischen Resonatoren und Filtern," *IEEE Trans. Magn.*, vol. MAG-2, pp. 613–620, Sept. 1966.

35) M. Börner, J. Deckert, V. Knapp, H. Schüssler, S.

Schweizerhof, and K. Schleswig, "Untersuchungen an Ferriten für Ultraschallverzögerungsleitungen," *Wiss. Ber. AEG-Telefunken 42*, vol. 42, no. 2, pp. 109-134, 1969.

36) M. Bon, R. Bosc, and P. Loyez, "New materials for transducers and resonators," in *Proc. 1976 IEEE ISCAS*, München, Apr. 1976, pp. 739-742.

37) R. Bosc, "Apercus sur la technique des filtres electromechaniques," *Cables et Transm.*, vol. 18, no. 4, pp. 296-306, 1964.

38) R. Bosc, F. Collombat, and P. Loyez, "Application de nouvelles technologies de filtrage à un système à 12 voies," *Cables et Transm.*, no. 1, pp. 103-125, Jan. 1973.

39) R. Bosc and P. Loyez, "Design of an electromechanical filter at 128 kHz in a two step modulation system," in *Proc. 1974 IEEE ISCAS*, San Francisco, Apr. 1974, pp. 111-114.

40) R. Bosc, F. Collombat, F. Duffaut, H. Pillon, and P. Loyez, "Elinvar type alloy for telecommunication mechanical filters," presented at the Int. Symp. on Materials for Electronic Components, Paris, Apr. 1975.

41) L. Brier, "Der Entwurf von HF-Bandfiltern und mechanischen Filtern mit Dämpfungspolen nach dem Betriebsparameterverfahren," in *X. Intern. Wiss. Kolloqu. TH Ilmenau (Digest of Papers)*, Summer 1965, pp. 73-83.

42) L. Bulhak, "Generator for measurement of electromechanical filters" (in Polish), *Przegl. Elektroniki*, vol. 2, pp. 415-419, 1961.

43) H. A. Burgess, "Mechanical wave filter," U.S. Patent 1 666 681, Apr. 1928.

44) C. M. van der Burgt, "Performance of ceramic ferrite resonators as transducers and filter elements," *J. Acoust. Soc. Amer.*, vol. 28, pp. 1020-1032, Nov. 1956.

45) C. M. van der Burgt, "Piezomagnetic ferrites. Applications in filters and ultrasonics," *Electron. Technology*, vol. 37, no. 9, pp. 330-341, 1960.

46) L. L. Burns and W. Van B. Roberts, "Mechanical filters for radio frequencies," *RCA Rev.*, vol. 10, pp. 348-365, Sept. 1949.

47) L. L. Burns, "A band-pass mechanical filter for 100 kilocycles," *RCA Rev.*, vol. 13, pp. 34-36, Mar. 1952.

48) J. Červený, "Properties of polarized ferrite magnets" (in Czech.), Research Report, Tesla Strašnice, 1970.

49) J. Červený, "Physical parameters of MGS ferrites" (in Czech.), Research Report, Tesla Strašnice, 1970.

50) J. Červený and V. Sobotka, "Measurements of physical properties of special materials" (in Russian), in *Proc. SSCT 71*, Short Contributions, vol. 2, Tále, Czechoslovakia, Sept. 1971.

51) J. Červený, "Magnetostrictive transducers and their application in EMF," in *Proc. SSCT 71*, Tále, Czechoslovakia, Sept. 1971, pp. 13/1-12.

52) T. Chodnikiewicz and A. Smolinski, "Experiences with electromechanical filters consisting of torsionally vibrating rods" (in Polish), *Przegl. Elektroniki*, vol. 2, pp. 382-383, 1961.

53) E. Costamagna and P. Falcicchio, "Considerations on the analytical description of mechanical resonators vibrating in flexure" (in Italian), *Alta Frequenza*, vol. 44, pp. 391-397, July 1975.

54) S. Cucchi and F. Molo, "Bridging elements in mechanical filters: Design procedure and an example of negative bridging element realizations," in *Proc. 1976 IEEE ISCAS*, Apr. 1976, pp. 746-749.

55) J. Deckert and P. Güls, "Mechanische Filter für Kanalumsetzer," *Techn. Mitt. AEG-Telefunken*, vol. 64, pp. 74-76, 1974.

56) M. L. Doelz, "Electromechanical filter," U.S. Patent 2 615 981, Oct. 1952.

57) M. L. Doelz and J. C. Hathaway, "How to use mechanical I-F filters," *Electronics*, vol. 26, pp. 138-142, Mar. 1953.

58) B. Ecotiere, "Filtres à résonateurs mécaniques—Problèmes soulevés par leur étude et leur réalisation," *Cables et Transm.*, pp. 126-136, Jan. 1973.

59) K. Ey, F. Hornung, and W. Volejnik, "Channel modem features electromechanical filters," *Siemens Rev.*, vol. 39, pp. 293-298, July 1972.

60) Z. Faktor, "Spurious vibrations in component parts of electromechanical filters and their investigation" (in Czech), *Slaboprodý Obzor*, vol. 30, pp. 444-450, 1969.

61) A. Fiok, "Electromechanical filters" (in Polish), *Postepy Telekomun.*, vol. 4, no. 2, pp. 19-39, 1959.

62) H. Fränkel, "Magnetomechanische Filter," *Radio und Fernsehen*, vol. 14, no. 19, pp. 580-583, no. 20, pp. 618-620, 1965.

63) E. M. Frymoyer, R. A. Johnson, and F. H. Schindelbeck, "Passive filters: Today's offerings and tomorrow's promises," *EDN*, vol. 18, pp. 22-30, Oct. 5, 1973.

64) S. Fujishima, R. Ohno, and S. Nozaka, "Study on a mechanical filter using radial mode vibration of a circular plate," Studies on Electronics Materials and their Applications, Division of Electronics Inst. for Chem. Res., Kyoto Univ., Kyoto, Japan, pp. 44-56, 1959, and Main Res. on Solid State Elect. studied in Kyoto Univ. and Murata Mfg. Co., pp. 31-37, Sept. 1959.

65) R. W. George, "Electromechanical filters for 100-kc carrier and sideband selection," *Proc. IRE*, vol. 44, pp. 31-35, Jan. 1956.

66) R. W. George, "Concentric-shear-mode 455-kilocycle electromechanical filter," *RCA Rev.*, vol. 18, no. 2, pp. 186-194, 1957.

67) G. L. Grisdale, "Electromechanical filters for use in telecommunication equipment," *Brit. Commun. and Electronics*, vol. 6, no. 11, pp. 768-772, 1959.

68) W. Gruszczyński, "On a design method for electromechanical filters" (in Polish), in *Proc. Symp. on Problems of Modern Telecommunications Systems*, section S 2 (Passive Systems), Warsaw, 1963.

69) A. Günther, "Verfahren zum Abgleich mechanischer Filter," German Patent 2047899, Sept. 29, 1970.

70) A. Günther, "Bemerkungen zum Entwurf breitbandiger mechanischer Filter," *Nachrichtent. Z.*, vol. 25, pp. 345-351, Aug. 1972.

71) A. E. Günther, "Electromechanical filters: satisfying additional demands," in *Proc. IEEE Int. Symp. Circuit Theory*, Apr. 1973, pp. 142-145.

72) A. E. Günther, "High-quality wide-band mechanical

filters, theory and design," *IEEE Trans. Sonics Ultrason.*, vol. SU-20, pp. 294-301, Oct. 1973.

73) P. Güls, "Mechanical filters for channel modems," (in German), Techn. Mitt. AEG-Telefunken, vol. 64, Special Issue "Trägerfrequenztechnik," pp. 17-22, April 1974.

74) W. Haas, "Channeling equipment technology using electromechanical filters," *Electrical Communication*, vol. 48, no. 1 and 2, pp. 16-20, 1973.

75) C. Hälsig, *Moderne Mechanische Frequenzselektion*. Berlin: VEB Verlag Technik, 1970.

76) C. Hälsig, "Nomogramm-Kette zur optimalen Dimensionierung magnetomechanischer Filterkörper," *Nachrichtent.*, vol. 22, no. 8, pp. 261-264, 1972.

77) C. Hälsig, "Mechanical filters of Kombinat VEB Elektronische Bauelemente Teltow and their application," in *Proc. 1976 IEEE ISCAS*, München, Apr. 1976, pp. 761-766.

78) J. Harešta, "Some properties of magnetostrictive generators and their applications" (in Russian), in *Proc. SSCT 71*, Short Contributions, vol. 2, Tále, Czechoslovakia, Sept. 1971, pp. 12-18.

79) H. C. Harrison, "Device for the transmission of mechanical vibratory energy," U.S. Patent 1 678 116, July 1928.

80) H. C. Harrison, "Mechanical transmission system," U.S. Patent 1 784 871, Dec. 1930.

81) R. V. L. Hartley, "Frequency selective transmission system," U.S. Patent 1 654 123, Dec. 1927.

82) J. C. Hathaway and D. F. Babcock, "Survey of mechanical filters and their applications," *Proc. IRE*, vol. 45, pp. 5-16, Jan. 1957.

83) D. P. Havens and P. Ysais, "Characteristics of low frequency mechanical filters, in *Proc. 1974 Ultrasonics Symp.*, Nov. 1974, pp. 599-602.

84) R. Heider, "Zur Dimensionierung mechanischer Filter mit optimaler Biegekopplung," *Hochfrequenztech. u. Elektroak.*, vol. 77, no. 3, pp. 108-112, 1968.

85) J. Hrušková and J. Trnka, "Signalling electromechanical filter" (in Czech), *Výzkumná Zpráva*, 6/6506/106/75/090, Tesla Strašnice, 1975.

86) I. Ishihara *et al.*, "Miniaturized LC channel filters" (in Japanese), in *Proc. IECE Conf. of Japan*, Tokyo, no. 109, Mar. 1973.

87) S. Jasiński, "Preliminary experimental investigations of electromechanical filters employing length-extensional mode resonators" (in Polish), *Przegl. Elektroniki*, vol. 2, pp. 383-387, 1961.

88) R. A. Johnson, "Mechanical filters for FM mobile applications," *IRE Trans. Vehic. Commun.*, vol. VC-10, pp. 32-37, Apr. 1961.

89) R. A. Johnson, "A single-sideband disk-wire type mechanical filter," *IEEE Trans. Component Parts*, vol. CP-11, pp. 3-7, Dec. 1964.

90) R. A. Johnson and R. J. Teske, "A mechanical filter having general stopband characteristics," *IEEE Trans. Sonics Ultrason.*, vol. SU-13, pp. 41-48, July 1966.

91) R. A. Johnson, "A twin tee multimode mechanical filter," *Proc. IEEE* (Corresp.), vol. 54, pp. 1961-1962, Dec. 1966.

92) R. A. Johnson, "Electrical circuit models of disk-wire mechanical filters," *IEEE Trans. Sonics Ultrason.*, vol. SU-15, pp. 41-50, Jan. 1968.

93) R. A. Johnson, "Application of electro-acoustic models to the design of a miniaturized mechanical filter," in *Rep. 6th Int. Congress on Acoustics*, vol. 4, Aug. 1968, pp. 93-96.

94) R. A. Johnson, "New single sideband mechanical filters," in *WESCON Tech. Papers*, vol. 14, sec. 10/1, Aug. 1970, pp. 1-10.

95) R. A. Johnson, M. Börner, and M. Konno, "Mechanical filters—A review of progress," *IEEE Trans. Sonics Ultrason.*, vol. SU-18, pp. 155-170, July 1971.

96) R. A. Johnson and W. D. Peterson, "Build stable compact narrow-band circuits," *Electron. Design*, pp. 60-64, Feb. 1, 1973.

97) R. A. Johnson, "Mechanical filters," in *Proc. 1973 Int. Symp. Circuit Theory*, Apr. 1973, pp. 402-405.

98) R. A. Johnson, "Mechanical bandpass filters," in *Modern Filter Theory and Design*, G. C. Temes and S. K. Mitra, Eds. New York: Wiley, 1973.

99) R. A. Johnson and W. A. Winget, "FDM equipment using mechanical filters," in *Proc. 1974 IEEE ISCAS*, San Francisco, Apr. 1974, pp. 127-131.

100) R. A. Johnson and A. E. Günther, "Mechanical filters and resonators," *IEEE Trans. Sonics Ultrason.*, vol. SU-21, pp. 244-256, Oct. 1974.

101) R. A. Johnson, "The design of mechanical filters with bridged resonators," in *Proc. 1975 IEEE ISCAS*, Boston, Apr. 1975, pp. 1-4.

102) R. A. Johnson, "The design and manufacture of mechanical filters," in *Proc. 1976 IEEE ISCAS*, München, Apr. 1976, pp. 750-753.

103) F. Jummel, "Das mechanische Einseitenbandfilter MF450 ± E-0340," *Radio und Fernsehen*, vol. 13, no. 13, pp.. 399-400, 1964.

104) H. Jumonji and M. Onoe, "Analysis of piezoelectric multiple-mode resonators vibrating in longitudinal and flexural modes," *Electron. Commun. Jap.*, vol. 51, pp. 35-42, May 1968.

105) H. Jumonji, "Analysis of longitudinal and flexural multiple-mode resonators," in *Proc. 6th Int. Congr. Acoust.*, Tokyo, Aug. 1968, pp. G105-G108.

106) J. Jungwirt, "Equivalent circuits for EMF's with uniformly spaced torsional resonators" (in Czech.), in *Proc. Tesla VÚT*, Prague, 1962.

107) J. Jungwirt, "Synthesis of EMF based on the theory of operating parameters" (in Czech.), in *Proc. Tesla VÚT*, Prague, 1963.

108) J. Jungwirt, "The design of EMF's with relatively short coupling sections" (in Czech.), Research Report, Tesla VÚT, Prague, 1967.

109) J. Jungwirt, "Analysis and synthesis of electromechanical filters" (in Russian), *Sborník konf. Elektromechanické a aklivní sel. obvody*, Liberec, Czechoslovakia, June 1973, pp. 16-27.

110) J. Jungwirt, "The design of EMF's for telecommunications," *TESLA Electronics*, vol. 7, pp. 3-9, Mar. 1974.

111) Z. Kaczkowski, "Magnetostrictive ferrites employed

in electromechanical filter" (in Polish), *Przegl. Elektroniki*, vol. 2, pp. 397-407, 1961.

112) Z. Kaczkowski, "Simplified measurement device for the determination of the mechanical resonant frequency and of the magnetomechanical coupling coefficient k of magnetostrictive materials" (in Polish), *Arch. Elektrotech.*, vol. 11, pp. 635-639, 1962.

113) Z. Kaczkowski, "Magnetostrictive equations and their coefficients" (in Polish), *Rozpr. Elektrotechn.*, vol. 7, pp. 245-275, 1971.

114) Y. Kagawa and G. M. L. Gladwell, "Finite element analysis of flexure-type vibrators with electrostrictive transducers," *IEEE Trans. Sonics Ultrason.*, vol. SU-17, pp. 41-49, Jan. 1970.

115) Y. Kagawa, "Analysis and design of electromechanical filters by finite element technique," *J. Acoust. Soc. Amer.*, vol. 49, no. 5 (part 1), pp. 1348-1356, May 1971.

116) Y. Kagawa and T. Yamabuchi, "Finite element simulation of two-dimensional electromechanical resonators," *IEEE Trans. Sonics Ultrason.*, vol. SU-21, pp. 275-283, Oct. 1974.

117) F. Kamiński, "Electromechanical filters for carrier frequency telephone transmission" (in Polish), *Postepy Telekomun.*, vol. 7, no. 1/2, pp. 6-14, 1962.

118) F. Kamiński, "Synthesis of broadband chainlike electromechanical filters having regular structure. Exact solution" (in Russian), *Bull. Acad. Polon. Sci.*, Ser. Sci. Tech., vol. 15, no. 2, pp. 1(165)-5(169), 1967.

119) F. Kamiński, "Remarks on the synthesis of electromechanical chainlike filters having regular structure" (in Polish), *Prace Inst. Tele- i Radiotech.*, vol. 11, no. 4, pp. 33-46, 1967.

120) F. Kamiński, "Methods of synthesis of electromechanical filters with the aid of distributed element equivalent circuits" (in Polish), *Arch. Elektrotech.*, vol. 17, pp. 449-467, 1968.

121) F. Kamiński, "On the synthesis of very narrowband electromechanical filters having regular structure" (in Russian), *Bull. Acad. Polon. Sci.*, Ser. Sci. Tech., vol. 17, no. 1, pp. 1(101)-5(105), 1969.

122) F. Kamiński, "Synthesis of chainlike electromechanical filters with the aid of distributed element equivalent circuits" (in Polish), *Rozpr. Elektrotech.*, vol. 15, pp. 717-749, 1969.

123) F. Kamiński, "Transfer matrix of a half-wave magnetostrictive transducer employed in an electromechanical filter" (in Polish), *Arch. Elektrotech.*, vol. 19, pp. 349-353, 1970.

124) F. Kamiński, "Relation between the algebraic method of synthesizing narrowband electromechanical filters and a synthesis employing lumped element equivalent circuits" (in Polish), *Arch. Elektrotech.*, vol. 20, pp. 825-830, 1971.

125) F. Kamiński, "Synthesis of electromechanical chainlike filters with the aid of the transfer matrix of an inhomogeneous transmission line consisting of homogeneous sections (Review of basic results)" (in Polish), *Postepy Elektroniki*, vol. 16, no. 2/3, pp. 74-88, 1971.

126) F. Kamiński, "On the synthesis of a certain group of electromechanical filters with the aid of lumped or distributed element equivalent circuits" (in Polish), *Arch. Elektrotech.*, vol. 22, pp. 517-521, 1973.

127) F. Kamiński, "Synthesis of an electromechanical filter having simple couplers" (in Polish), *Arch. Elektrotech.*, vol. 22, pp. 525-537, 1973.

128) F. Kamiński, "Sketch of a theory for synthesizing electromechanical and microwave filters having regular structures" (in Polish), *Rozpr. Elektrotech.*, vol. 20, pp. 295-312, 1974.

129) F. Kamiński, *Linear Distributed Circuit Synthesis Appropriate to the Polynomial Electromechanical Filter and Microwave Network Design, I. Stepped Transmission Line Synthesis.* Warsaw: Wanstwowe Sydawnictwo Naukowe, 1976.

130) T. G. Kinsley, "Wave filter," U.S. Patent 2 342 869, Feb. 1944.

131) J. Křiva, "Magnetostrictive materials for mechanical resonators" (in Czech.), in *Proc. Tesla VÚT*, 1965.

132) S. S. Kogan and A. S. Stepanov, "Electromechanical filters for long distance communications systems," *Vopr. Radioelektron.*, TPS, no. 7, 1967.

133) S. S. Kogan and A. S. Stepanov, "Electromechanical channel filters" (in Russian), *Elektrosviaz*, vol. 25, pp. 58-65, Nov. 1971 and in *Telecom. and Radio Eng.*, part 1, vol. 25, no. 11, pp. 44-50, 1971 (in English).

134) Y. Koh, "The mechanical filter: evolution to technical maturity," *Japan Elec. Eng.*, pp. 32-37, June 1973.

135) Y. Koh, I. Imaguchi, and T. Nagashima, "Low frequency mechanical filter and channel mechanical filter," in *Proc. Crystal and Mechanical Filter Workshop*, Rancho Santa Fe, CA, Oct. 1974, pp. 1-14.

136) B. Kohlhammer and H. Schüssler, "Berechnung allgemeiner mechanischer Koppelfilter mit Hilfe von äquivalenten Schaltungen aus konzentrierten elektrischen Schaltelementen," *Wiss. Ber. AEG-Telefunken*, vol. 41, pp. 150-159, 1968.

137) B. Kohlhammer, "Ein neuartiges Entwurfsverfahren zur Synthese von mechanischen Filtern und von Gyrator-Filtern," *Wiss. Ber. AEG-Telefunken*, vol. 43, pp. 170-177, Fall, 1970.

138) B. Kohlhammer, "Ein Beitrag zum Entwurf mechanischer Filter, " dissertation, Fakultät für Maschinenwesen und Elektrotechnik der Techn. Univ. München, 1971.

139) B. Kohlhammer and H. Schüssler, "Mechanische Torsionsfilter mit piezoelektrischen Wandlern als Kanal- und Signalfilter für ein neues Trägerfrequenzsystem," in *Proc. 7th Int. Congress on Acoustics*, Budapest, 1971, no. 20E6, pp. 341-344, Akadémiai Kiadó, Budapest, 1971.

140) B. Kohlhammer and H. Schüssler, "Bemerkungen zum Einfluss der mechanischen Schwinggüte auf die Übertragungseigenschaften von Kanalfiltern," *Frequenz*, vol. 25, no. 9, pp. 287-288, 1971.

141) B. Kohlhammer, "Vergleich verschiedener elektromechanischer Kanalfilter mit Hilfe von Empfindlichkeitskriterien," *Frequenz*, vol. 26, no. 6, pp. 169-176, 1972.

142) M. Konno, "Mounting principle of tuning-bar Onpen"

(in Japanese), *J. Acoust. Soc. Jap.*, vol. 12, pp. 75-80, Feb. 1956.

143) M. Konno, "Theoretical considerations of mechanical filters" (in Japanese), *J. Inst. Elec. Commun. Eng. Jap.*, vol. 40, pp. 44-51, 1957.

144) M. Konno, "Theory of mechanical filter with resonator disk" (in Japanese), *J. Acoust. Soc. Japan*, vol. 13, pp. 144-148, Mar. 1957.

145) M. Konno, "Electromechanical filters" (in Japanese), *J. Acoust. Soc. Jap.*, vol. 14, pp. 241-262, 1958.

146) M. Konno, H. Nakamura, and K. Sugiyama, "Driving point immittance of transverse vibrating uniform rods, General expression, concentrated element expression" (in Japanese), *J. Acoust. Soc. Japan.*, vol. 17, pp. 183-193, Mar. 1961.

147) M. Konno and H. Nakamura, "Equivalent electrical network for the transversely vibrating uniform bar," *J. Acoust. Soc. Amer.*, vol. 38, pp. 614-622, Oct. 1965.

148) M. Konno, Y. Tomikawa, and C. Kusakabe, "Some design methods for a mechanical filter with attenuation poles" (in Japanese), *Bulletin of Yamagata University*, vol. 9, pp. 287-314, Jan. 1967.

149) M. Konno, C. Kusakabe, and Y. Tomikawa, "Electromechanical filter composed of transversely vibrating resonators for low frequencies," *J. Acoust. Soc. Amer.*, pp. 953-961, Apr. 1967.

150) M. Konno and Y. Tomikawa, "An electro-mechanical filter consisting of a flexural vibrator with double resonance," *Electron. Commun. Jap.*, vol. 50, pp. 64-73, Oct. 1967.

151) M. Konno, Y. Tomikawa, and H. Izumi, "Electromechanical filter consisting of a multiplex mode vibrator," in *Rep. 6th Int. Congress on Acoustics*, vol. 4, Aug. 1968, pp. 117-120.

152) M. Konno, T. Tanno, and H. Nakamura, "A tuning fork with double resonance and its application to mechanical filters" (in Japanese), *J. Acoust. Soc. Jap.*, vol. 25, pp. 1-6, Jan. 1969.

153) M. Konno, K. Aoshima, and H. Nakamura, "H-shaped resonator and its application to electro-mechanical filters" (in Japanese), *Eng. Bull.*, Yamagata Univ., vol. 10, pp. 261-285, Mar. 1969.

154) M. Konno and Y. Tomikawa, "Electro-mechanical filters—Part 1, introduction" (in Japanese), *J. Inst. Electron. Commun. Eng. Jap.*, vol. 52, pp. 303-312, Mar. 1969.

155) M. Konno, Y. Tomikawa, T. Takano, and H. Izumi, "Electromechanical filters using degeneration modes of a disk or ring," *Electron. Commun. Jap.*, vol. 52A, pp. 19-28, May 1969.

156) H. Kopp and A. Mehr, "Channel modem with electromechanical filters," *Siemens Rev.*, vol. 38, pp. 297-300, July 1971.

157) H. Kopp, "A mechanical filter channel bank," *IEEE Trans. Commun.*, vol. COM-20, pp. 64-67, Feb. 1972.

158) F. Künemund and K. Traub, "Mechanische Filter mit Biegeschwingern," *Frequenz*, vol. 18, pp. 277-280, Sept. 1964.

159) F. Künemund, "Dimensionierung überbrückter Bandpässe mit Dämpfungspolen," *Frequenz*, vol. 24, no. 6, pp. 190-192, 1970.

160) F. Künemund, "Materials for high-grade electromechanical frequency filters," *Elec. Eng. Trans. I. E. Aust.*, EE8, no. 1, pp. 41-42, 1972.

161) F. Künemund, "Channel filters with longitudinally coupled flexural mode resonators," *Siemens Forsch. -u. Entwickl. -Ber.*, vol. 1, no. 4, pp. 325-328, 1972.

162) C. Kurth, "Eine Wellenparametertheorie für mechanische Vierpole in Kompressionsoder Torsionschwingungen," *Nachrichtent.*, vol. 9, pp. 490-503, Nov. 1959.

163) C. Kurth, "Magnetostriktive Wandler als selektive Vierpole," *Frequenz*, vol. 14, pp. 272-288, Aug. 1960.

164) C. Kurth, "Anwendung der Betriebsparametertheorie bei der Berechnung von Siebschaltungen, die sich aus in Kette geschalteten Leitungsstücken zusammensetzen," *Frequenz*, vol. 16, pp. 482-495, Dec. 1962.

165) C. Kusakabe and M. Konno, "Electro mechanical filter composed of transversely vibrating bars connected by torsionally vibrating bar," *Electron. Commun. Japan*, vol. 47, no. 12, pp. 1845-1854, Dec. 1964.

166) C. Kusakabe, M. Konno, and Y. Tomikawa, "Resonant frequencies of transversely vibrating bar excited by electrostrictive transducer," *Electron. Commun. Japan*, vol. 48, pp. 1938-1944, Nov. 1966.

167) C. Kusakabe and M. Konno, "Vibration modes of transversely vibrating bar excited by electrostrictive transducer," *Electron. Commun. Japan*, vol. 49, pp. 42-47, Dec. 1967.

168) E. Kuźniak and A. Smoliński, "Experiences with electromechanical filters consisting of length-extensionally vibrating disks" (in Polish), *Przegl. Elektroniki*, vol. 2, pp. 380-382, 1962.

169) E. Kuźniak, "Works on electromechanical filters having disk-shaped length-extensional resonators" (in Polish), *Zeszyty Nauk. Politech. Warsz.*, vol. 72 (Elektryka 29), pp. 53-72, 1963.

170) S. P. Lapin, "Electromechanical filters," in *Proc. IEEE Nat. Electron. Conf.*, vol. 9, Feb. 1954, pp. 353-362.

171) R. Lappa, "Ferrite transducers for electromechanical filters" (in Polish), *Przegl. Elektroniki*, vol. 2, pp. 391-397, 1961.

172) A. Lenk, "Die Vierpolersatzschaltbilder der elektromichanischen Wandler, Teil 2," *Acoustica*, vol. 6, pp. 303-316, 1956.

173) J. Lhoták, J. Trnka, and V. Sobotka, "The electromechanical filters of Tesla Strašnice," *Tesla Electronics*, vol. 5, no. 2, pp. 57-59, 1972.

174) A. K. Losev, "Filters with multielement coupling," *Telecommun. Radio Eng. (USSR)*, pp. 1-8, Jan. 1964.

175) A. K. Losev, "One member tuning fork filters" (in Russian), *Elektrosvjiaz*, vol. 20, pp. 31-37, Jan. 1966.

176) A. K. Losev, "Multiple member tuning fork filters" (in Russian), *Elektrosvjiaz*, vol. 20, pp. 11-16, Mar. 1966.

177) I. Lucas, "Plattenförmige elektromechanische Filter mit Dämpfungspolen," *Arch. Elek. Übertr.*, vol. 17, pp. 230-236, May 1963.

178) I. Lucas, "Berechnung der Eigenfrequenzen quaderförmiger Biegeschwinger," *Arch. Elek. Übertr.*, vol. 20, pp. 64-70, Jan. 1966.

179) D. L. Lundgren, "Electromechanical filters for single-sideband applications," *Proc. IRE*, vol. 44, no. 12, pp. 1744-1749, 1956.

180) W. P. Mason, "The motion of a bar vibrating in flexure, including the effects of rotary and lateral inertia," J. Acoust. Soc. Am., vol. 6, pp. 246-249, April 1935.

181) W. P. Mason, *Electromechanical Transducers and Wave Filters*. New York: Van Nostrand, 1942.

182) W. P. Mason, "Mechanical wave filter," U.S. Patent 2 342 813, Feb. 1944.

183) W. P. Mason, "Wave transmission network," U.S. Patent 2 345 491, Mar. 1944.

184) W. P. Mason, *Physical Acoustics and the Properties of Solids*. New York: Van Nostrand, 1958.

185) W. P. Mason and R. N. Thurston, "A compact electromechanical band-pass filter for frequencies below 20 kilocycles," *IRE Trans. Ultrason. Eng.*, vol. UE-7, pp. 59-70, June 1960.

186) W. P. Mason, Ed., *Physical Acoustics. Principles and Methods*, vol. 1A. New York: Academic, 1964.

187) J. P. Maxfield and H. C. Harrison, "Methods of high quality recording and reproducing of music and speech based on telephone research," *Bell Syst. Tech. J.*, vol. 5, pp. 493-523, July 1926.

188) C. Moeller, "Piezoelektrische und magnetostriktive Wandler in elektrischen Bauelementen," *Nachrichtent.*, vol. 14, no. 11, pp. 407-411, 1964, vol. 15, no. 3, pp. 88-94, 1965.

189) G. S. Moschytz, "Inductorless filters: A survey—Part I: Electromechanical filters," *IEEE Spectrum*, vol. 7, pp. 30-36, Aug. 1970.

190) K. Nagai and M. Konno, "Electro-mechanical vibrators and their applications" (in Japanese), Corona Company, Mar. 1974.

191) H. Nakamura, K. Sugiyama and M. Konno, "Dynamical analogy and its applications," *Bull. Yamagata Univ. (Eng.)*, vol. 2, pp. 131-160, Mar. 1962.

192) H. Nakamura, M. Konno, and T. Toda, "Experiments on lateral vibration of a hooked cantilever," *Bull. Yamagata Univ. (Eng.)*, vol. 7, pp. 713-730, Mar. 1964.

193) H. Nakamura and M. Konno, "On the electro-mechanical filter using tuning forks with couplers," *Bull. Yamagata Univ.*, vol. 9, pp. 513-535, Jan. 1968.

194) H. Nakamura, Y. Tomikawa, and M. Konno, "Finite element simulation of U type tuning fork," *Electron. Commun. Jap.*, vol. 59-A, pp. 21-27, Jan. 1976.

195) H. Nakamura, N. Matsuda, Y. Tomikawa, and M. Konno, "Finite element simulation of quartz tuning fork," Paper of Technical Group on US 73-39, IECE, Japan, Oct. 1975.

196) K. Nakamura *et al.*, "Multi-channel electromechanical filters using orthogonal modes in single mechanical system," in *Proc. 6th Int. Cong. Acoust.*, Aug. 1968, pp. G109-G112.

197) S. Nakao, "Electro-mechanical wave separating filter synthesis-channel filter application," presented at the European Conf. on Circuit Theory and Design, Genoa, Italy, Sept. 1976.

198) Y. Nakauchi, "A synthesis of parallel ladder circuits using equivalent transformation techniques," in *Proc. 1975 IEEE ISCAS*, Boston, Apr. 1975, pp. 305-308.

199) J. Näser, "Kurze Darlegung der Theorie elektromechanischer Filter mit Plattenresonatoren," *Hochfrequenztech. u. Elektroak.*, vol. 71, pp. 123-132, Oct. 1962.

200) J. Näser, "Exakte Berechnung der Biegeresonanzen rechteckiger und zylindrischer Stäbe," *Hochfrequenztech. u. Elektroak.*, vol. 74, pp. 30-36, 1965.

201) J. Näser, "Ein Beitrag zur Theorie isotroper Biegeresonatoren unter Berücksichtigung ihrer Anwendung in mechanischen Filtern," dissertation, Tech. Univ. Dresden, 1966.

202) E. L. Norton, "Wave filter," U.S. Patent 1 681 554, Aug. 1928.

203) T. Okamoto, K. Yakuwa, and S. Okuta, "Low frequency electromechanical filter," *Fujitsu Sci. Tech. J.*, vol. 2, pp. 53-86, May 1966.

204) M. Onoe and T. Yano, "Analysis of flexural vibrations of a circular disk," *Electron. Commun. Jap.*, vol. 51, pp. 33-36, Apr. 1968.

205) M. Onoe and T. Yano, "Analysis of flexural vibrations of a circular disk," *IEEE Trans. Sonics Ultrason.*, vol. SU-15, pp. 182-185, July 1968.

206) M. Onoe and T. Yano, "Electromechanical wave separating filters," in *Proc. 20th IEEE Electronic Component Conf.*, May 1970.

207) S. Oyama and M. Konno, "Equivalent networks for mechanical vibrating systems and their applications," *Bull. Yamagata Univ.*, vol. 13, pp. 31-70, July 1974.

208) M. Panetti, "On the calculation of the transversal vibrations of an excited elastic bar" (in Italian), *Atti R. Acad. delle Sc. di Torino*, vol. 36, pp. 6-26, 1901.

209) E. P. Papadakis, "Improvements in a broadband electromechanical bandpass filter in the voice band," *IEEE Trans. Sonics Ultrason.*, vol. SU-22, pp. 406-415, Nov. 1975.

210) P. Pavlik, "Computer program for research and production of EMF," in *Proc. SSCT 71*, Short Contributions, vol. 2, Tále, Czechoslovakia, Sept. 1971, pp. 79-82.

211) W. D. Peterson, "Electromechanical filter with center drive of disk," U.S. Patent 2 693 580, Nov. 1954.

212) R. Pfleiderer and P. Wollmershäuser, "Electromechanical pilotfilter with improved temperature characteristics," in *Proc. 1976 IEEE ISCAS*, München, Apr. 1976, pp. 743-745.

213) H. D. Piper, "Mechanische Filter," *Telefunken-Z.*, vol. 32, pp. 279-283, Dec. 1959.

214) W. Poschenrieder, "Physik und Technik elektromechanischer Wellenfilter," *Frequenz*, vol. 12, pp. 317-325, Aug. 1958.

215) W. Poschenrieder and F. Schöfer, "Das elektromechanische Quarzfilter - ein neues Bauelement für die Nachrichtentechnik," *Frequenz*, vol. 17, pp. 88-94, Mar. 1963.

216) W. Poschenrieder and W. Zitzmann, "Communications systems for wire and radio paths," *Siemens Rev.*, vol. 38, pp. 490-493, Nov. 1971.

217) P.-M. Prache, "Introduction to the theory of linear electromechanical networks" (in French), *Câbles et Transm.*, vol. 28, pp. 304-327, Oct. 1974.

218) P.-M. Prache and H. Ernyei, "Vibrations of homoge-

neous bars within the harmonic region" (in French), *Câbles et Transm.*, vol. 28, pp. 328-366, Oct. 1974.

219) B. Pšenička , J. Pomichálek, and K. Šlěpánek, "Graphical design method of EMF," in *Proc. SSCT 71*, Tále, Czechoslovakia, Sept. 1971, pp. 83-89.

220) B. Pšenička and J. Trnka, "Calculations of the basic pole producing mechanical filter sections" (in Russian), in *Proc. Sborník konf. Elektromechanické a aklivní sel. obvody*, Liberec, Czechoslovakia, June 1973, pp. 103-107.

221) B. Pšenička and J. Trnka, "EMF's with poles of attenuation at finite frequencies," in *Proc. 5th Collaq. on Microwave Comm.*, Budapest, MLR, June 1974.

222) W. van B. Roberts and L. L. Burns, "Mechanical filters for radio frequencies," *RCA Rev.*, vol. 10, pp. 348-365, Sept. 1949.

223) W. van B. Roberts, "Mechanical filter," U.S. Patent 2 578 452, Dec. 1951.

224) W. van B. Roberts and L. Burns, "Electromechanical filter," U.S. Patent 2 647 948, Aug. 1953.

225) H. J. Rohde, "Erfahrungen mit mechanischen Bandfiltern," *Hochfrequenztech. u. Elektroak.*, vol. 70, no. 1, pp. 31-38, 1961.

226) H. H. Rudolf, "Das mechanische Zweiseitenbandfilter MF 450-3500," *Radio und Fernsehen*, vol. 13, no. 12, pp. 367-368, 1964.

227) A. Russen, "Electromechanical filters for 50 to 500 kilohertz," *Elec. Commun.*, vol. 39, no. 3, pp. 423-434, 1964.

228) K. Sawamoto *et al.*, "Electromechanical filter using torsional mode resonators" (in Japanese), presented at the IECE Conf. of Japan, Tokyo, no. 67, Mar. 1973.

229) K. Sawamoto, E. Sasaki, S. Kondo, T. Ashida, and K. Shinozaki, "A torsional mode mechanical filter," *Trans. IECE Japan*, vol. 57-A, no. 8, pp. 575-582, 1974.

230) K. Sawamoto, S. Kondo, and E. Sasaki, "A torsional mode mechanical channel filter," *Rev. Elec. Comm. Labs (NTT)*, vol. 23, pp. 429-438, May-June 1975.

231) K. Sawamoto, S. Kondo, N. Watanabe, K. Tsukamoto, M. Kiyomoto, and O. Ibaraki, "A torsional-mode pole-type, mechanical channel filter," *IEEE Trans. Sonics Ultrason.*, vol. SU-23, pp. 148-153, May 1976.

232) H. Schüssler, "Mechanische Filter mit piezoelektrischen Wandlern," *Telefunken-Z.*, vol. 39, no. 3, pp. 429-439, 1966.

233) H. Schüssler, "Mechanische Filter in der Nachrichtentechnik," *Elektro-Anzeiger, Essen*, pp. 483-486, Dec. 1966.

234) H. Schüssler, "Darstellung elektromechanischer keramischer Wandler als Dickenscherschwinger mit piezoelektrischer und piezomagnetischer Anregung," *Arch. Elek. Übertr.*, vol. 22, pp. 399-406, Aug. 1968.

235) H. Schüssler, "Filter mit mechanischen Resonatoren," *Bull. SEV*, vol. 60, pp. 216-222, Mar. 15, 1969.

236) H. Schüssler, "Uberblick über die Entwicklung Nachrichtentechnischer Bauelemente mit elektromechanischen Schwingern, Teil 2," *Elektro-Anzeiger*, vol. 23, pp. 521-523, Dec. 9, 1970.

237) H. Schüssler and B. Kohlhammer, "Mechanical filters with piezoelectric converters as channel filters in FDM-systems" (in German), in *Proc. 7th Int. Congress on Acoustics*, Budapest, Hungary, 1971, pp. 341-344.

238) H. Schüssler, "Consideration about channel filters for a new carrier frequency system with mechanical filters," in *Proc. 25th Annual Freq. Control Symp.*, 1971, pp. 262-270.

239) S. Schweitzerhof, "Über Ferrite für magnetostriktive Schwinger in Filterkreisen," *Nachrichtent. Z.*, vol. 11, pp. 179-185, Apr. 1958.

240) T. Sekine, M. Konno and Y. Tomikawa, "Three resonator mechanical filters with two attenuation poles," *Acoust. Soc. Japan Report*, Oct. 1975.

241) R. L. Sharma, "Equivalent circuit of a resonant, finite, isotropic, elastic circular disk," *J. Acoust. Soc. Amer.*, vol. 28, pp. 1153-1158, Nov. 1956.

242) R. L. Sharma, "Dependence of frequency spectrum of a circular disk on Poisson's ratio," *J. Appl. Mech.*, vol. 24, pp. 53-54, Mar. 1957.

243) R. L. Sharma, "Magnetostriction transducers for mechanical filters," in *IRE Nat. Conv. Rec.*, vol. 6-III, pt. 6, pp. 223-234, 1958.

244) R. L. Sharma, "Theory and application of electromechanical filters," in *Proc. 3rd Int. Congress on Acoustics*, Stuttgart, vol. 2, 1959, pp. 729-731. Elsevier Publ. Co., Amsterdam, 1961.

245) D. F. Sheahan and R. A. Johnson, "Crystal and mechanical filters," *IEEE Trans. Circuits Syst.*, vol. CAS-22, pp. 69-89, Feb. 1975.

246) K. Shibayama and Y. Kikuchi, "Studies on vibration of short-column," (a) Part I, Sci. Rep. Ritu, B-(*Elec. Comm.*), vol. 8, no. 3, pp. 133-150, 1956. (b) Part II, Sci. Rep. Ritu, B-(*Elec. Comm.*), vol. 9, no. 2, pp. 113-122, 1957. (c) Part III, Sci. Rep. Ritu, B-(*Elec. Comm.*), vol. 11, no. 3-4, pp. 203-217, 1960. (d) Part IV, Sci. Rep. Ritu, B-(*Elec. Comm.*), vol. 13, no. 2, pp. 81-101, 1961.

247) K. Shibayama and Y. Kikuchi, "Magnetostrictive transducers and filters," *Electron. Commun. Jap.*, vol. 48, pp. 60-65, Nov. 1965.

248) K. Shibayama, "Electromechanical filters," *Electron. Commun. Jap.*, vol. 48, pp. 66-72, Nov. 1965.

249) K. Shibayama, "On the design of bar-type wide band mechanical filters," in *Proc. 6th Int. Congress on Acoustics*, Tokyo, Japan, Aug. 1968, pp. G89-92.

250) M. Skowrońska and R. Lappa, "Contribution to a method of designing electromechanical filters" (in Polish), *Przegl. Elektroniki*, vol. 2, pp. 370-380, 1961.

251) A. Smoliński, "Electromechanical filters" (in Polish), *Przegl. Elektroniki*, vol. 2, pp. 355-364, 1961.

252) A. Smoliński, "Electromechanical filters" (in Polish), *Rozpr. Elektrotech.*, vol. 7, pp. 439-484, 1961.

253) V. Sobotka, "Analyzing passive circuits from the viewpoint of component parts tolerance—Realizing the method, analyzing concrete filters" (in Czech.), Research Report, Tesla VÚT, Prague, 1967.

254) V. Sobotka and J. Trnka, "The use of digital computers in modelling of transfer characteristics of serially produced electromechanical filters," in *Proc. SSCT 71*, Tále, Czechoslovakia, Sept. 1971, pp. 14/1-20.

255) V. Sobotka and J. Trnka, "Die Benutzung von Digitalrechnern beim Entwurf elektromechanischer Filter für die Grosserienfertigung," *Frequenz*, vol. 26, pp. 177–182, June 1972.

256) V. Sobotka, "Die Genauigkeit der Messungen von physikalischen Parametern der Koppeldrähte für EMF," *Sborník kof. Elektromechanické a aklivní sel. obvody*, Liberec, Czechoslovakia, pp. 142–147, June 1973.

257) V. Sobotka and J. Sĕdina, "Constant elastic modulus material PY-42-Mo and Aurelast-Super for EMF resonators," Research Report, Tesla Strašnice, 1972.

258) G. Stacchiotti, P. Costa, G. F. Piacentini, and G. Muzzin, "Electromechanical quartz filter design considerations and experimental results," presented at the IEEE Crystal and Mechanical Filter Workshop, Rancho Sante Fe, CA, Oct. 1974.

259) K. Štepánek, "128 kHz electromechanical channel filter" (in Czech.), Research Report, Tesla Strašnice, 1975.

260) R. Straube, "Die Anwendung der Störungsrechnung auf freie Biegeschwingungen von Stäben bei Berücksichtigung der Schubverformung und der Drehträgheit," *Wiss. Z. Techn. Univ. Dresden*, vol. 12, pp. 1173–1176, May 1963.

261) W. Struszinski, "A theoretical analysis of the torsional electro-mechanical filters," *Marconi Rev.*, vol. 22, pp. 119–143, Fall 1959.

262) S. Sugawara, H. Yamada and M. Konno, "Analysis of U-shaped coupler in low frequency mechanical filter," Paper of Technical Group on US73-39, IECE Japan, Jan. 1974.

263) W. Taeger, "Mechanische Filter für die Nachrichtentechnik," *Frequenz*, vol. 14, no. 9, pp. 321–323, 1960.

264) Y. Tagawa and T. Hatano, *Design of Mechanical Filter and Crystal Filter Circuits* (in Japanese). Tokyo: Ohm, 1964.

265) I. Takahaski, N. Yoshida, and Y. Ishizaki, "An analysis of a torsional mode transducer for electromechanical filters," in *Proc. 1976 IEEE Ultrason. Symp.*, Sept. 1976.

266) K. Takahashi, "Low-frequency mechanical filters with U- or I-shaped couplers and resonators," *Electron. Commun. Jap.*, vol. 48, pp. 199–205, Nov. 1965.

267) M. Takahashi, F. Yamauchi, and S. Takahashi, "Stabilization of resonance frequencies in piezoelectric ceramic resonators against sudden temperature change," in *Proc. 28th Annual Symp. on Freq. Control.*, May 1974, pp. 109–116.

268) T. Tanaka, "Torsional electromechanical filter," Japan Patent S. 34-8453, Sept. 1956.

269) T. Tanaka and T. Inoguchi, "A mechanical filter," *Studies on Electronics Materials and Their Applications*, Division of Electronics Materials, Institute for Chemical Research, Kyoto Univ., Kyoto, Japan, pp. 22–25, 1959.

270) T. Tanaka and H. Kawamura, "Mechanical filter using inner coupled characteristics of resonator," *Studies on Electronics Materials and Their Applications*, Division of Electronics Materials Inst. for Chem. Res., Kyoto Univ., Kyoto, Japan, pp. 25–27, 1959.

271) T. Tanaka, T. Kawai, and S. Miura, "Mechanical filter using spherical ball shaped resonators," *Studies on Electronics Materials and Their Applications*, Division of Electronics Materials Inst. for Chem. Res., Kyoto Univ., Kyoto, Japan, pp. 28–29, 1959.

272) G. C. Temes, "Asymmetrical loss-pole mechanical filter," U.S. Patent 3 725 828, Apr. 1973.

273) H. Thomann, "Modern principles of solid state passive filters," in *Proc. 4th European Conf. Solid-State Devices*, 1974, pp. 97–108.

274) Y. Tomikawa and M. Konno, "A low-pass filter consisting of a ceramic resonator and some electrical elements" (in Japanese), *Bull. Yamagata Univ.*, vol. 9, pp. 315–325, Jan. 1967.

275) Y. Tomikawa, M. Konno, and H. Izumi, "Electromechanical filters consisting of a multi-mode plate" (in Japanese), *J. Acoust. Soc. Jap.*, vol. 25, pp. 114–121, Mar. 1969.

276) Y. Tomikawa and M. Konno, "Electromechanical filter consisting of a triple resonance vibrator of longitudinal and bending resonance modes," *Electron. Commun. Jap.*, vol. 52, pp. 58–60, Apr. 1969.

277) Y. Tomikawa *et. al.*, "Electromechanical filters consisting of longitudinal and bending resonance mode vibrators," *Electron. Commun. Jap.*, vol. 52, pp. 1–8, Nov. 1969.

278) Y. Tomikawa and M. Konno, "Electromechanical filter with attenuation poles consisting of multi-mode resonators," in *Proc. 7th Int. Congress on Acoustics*, Budapest, Hungary, 1971, pp. 637–640.

279) Y. Tomikawa, "Electro-mechanical filter with attenuation-poles consisting of multi-mode vibrators," *Electron. Commun. Jap*, vol. 54-A, Aug. 1971.

280) Y. Tomikawa, "Multi-mode mechanical filter with attenuation poles" (in Japanese), *J. Acoust. Soc. Jap.*, vol. 28, pp. 108–117, Mar. 1972.

281) Y. Tomikawa, K. Sato, and M. Konno, "Consideration on small size of low frequency vibrators" (in Japanese), *J. Acoust. Soc. Jap.*, vol. 31, May 1975.

282) Y. Tomikawa, D. P. Havens, R. A. Johnson, and S. Sugawara, "Resonances in flexure-mode mechanical filters," in *Proc. 1976 IEEE Ultrason. Symp.*, Sept. 1976.

283) K. Traub, "Mechanische statt elektrischer Schwingkreise in Trägerfrequenzfernsprech systemen," in *CONSTRONIC '72, Dig. Papers*, Apr. 1972, pp. 139–152.

284) J. Trnka, "Mechanical structures of EMF with attenuation poles—Realisation by resonators," in *Proc. SSCT 71*, Short Contributions, vol. 2, Tále, Czechoslovakia, Sept. 1971, pp. 90–101.

285) J. Trnka, "Electromechanical filters for telecommunications systems and measuring instruments and perspectives on their development at Tesla-Strašnice" (in Russian), *Sborník konf. Elektromechanické a aklivní sel. obvody*, Liberec, Czechoslovakia, June 1973, pp. 6–12.

286) J. Trnka, "EMF networks with pole producing resonators" (in Russian), *Sborník konf. Elektromechanické a aklivní sel. obvody*, Liberec, Czechoslovakia, June 1973, pp. 109–114.

287) J. Trnka and V. Sobotka, "Computer modelling of MF transfer characteristics," in *Proc. 1975 IEEE ISCAS*, Boston, 1975.

288) E. Trzeba, "Elektromechanische Vierpole ais Kopplungsfilter," *Hochfrequenztech. u. Elektroak.*, vol. 69, no. 3, pp. 108-117, 1960.

289) E. Trzeba, "Die Messung der charakteristischen Grössen von elektromechanischen Kopplungsfiltern," *Hochfrequenztech. und Elektroak.*, vol. 69, no. 4, pp. 119-123, 1960.

290) E. Trzeba, "Einfügedämpfung elektromechanischer Kopplungsfilter," *Hochfrequenztech. u. Elektroak.*, vol. 70, no. 1, pp. 17-20, 1961.

291) E. Trzeba, "Übertragungseigenschaften elektromechanischer Kopplungsfilter als Funktion des Koppelelementes," *Hochfrequenztech. u. Elektroak.*, vol. 70, no. 5, pp. 166-170, 1961.

292) E. Trzeba, "Hochfrequenzbandfilter mit Dämpfungspolen bei endlichen Frequenzen und geebnetem Betrag im Durchlassbereich," *Nachrichtentechnik*, vol. 12, no. 12, pp. 450-455, 1962, vol. 13, no. 1, pp. 36-40, 1963.

293) E. Trzeba, "Mechanische Bandfilter," *Nachrichtentechnik*, vol. 13, no. 3, pp. 82-88, 1963.

294) E. Trzeba, "Grundzüge der Synthese von Hochfrequenzbandfiltern-Mechanische Filter," *Nachrichtentechnik*, vol. 14, no. 2, pp. 59-66, 1964.

295) E. Trzeba, "Entwicklungstendenzen frequenzselektiver Bauelemente," *Hochfrequenztech. u. Elektroak.*, vol. 78, no. 1, pp. 9-16, 1969.

296) Y. Tsuzuki *et al.*, "The holographic investigation of spurious modes of mechanical filters," *Electron. Commun. Jap.*, vol. 54, pp. 31-38, May 1971.

297) J. S. Turnbull, "Considerations in i.f. mechanical filter design," in *IRE Wescon Conv. Rec.*, vol. 1, pt. 9, 1957, pp. 31-37.

298) R. P. Walters, "Balanced action toroidal transducer for electromechanical filters," *Electron. Eng.*, pp. 238-245, Mar. 1969.

299) K. Wittmann, G. Pfitzenmaier and F. Künemund, "Dimensionierung reflexionsfaktor- und laufzeitgeebneter versteilerter Filter mit Überbrückungen," *Frequenz*, vol. 24, pp. 307-312, Oct. 1970.

300) A. Wolski, "Equivalent circuits of magnetostrictive transducers employed in electromechanical filters" (in Polish), *Zeszyty Nauk Politech. Warsaw*, vol. 49 (Elektryka 21), pp. 37-50, 1960.

301) K. Yakuwa, "Characteristics of an electro-mechanical filter using piezoelectric composite bending vibrators," *Electron. Commun. Jap.*, vol. 48, pp. 206-215, Nov. 1965.

302) K. Yakuwa and S. Okuta, "Miniaturization of the mechanical vibrator used in an electro-mechanical filter," in *Rep. 6th Int. Congress on Acoustics*, vol. 4, Aug. 1968, pp. 97-100.

303) K. Yakuwa, "Electro-mechanical filters—Part 2: High frequency electro-mechanical filters" (in Japanese), *J. Inst. Electron. Commun. Eng. Jap.*, vol. 52, pp. 568-577, May 1969.

304) K. Yakuwa, S. Okuda, and T. Tanaka, "Spurious characteristics of the mechanical filter using torsionally coupled bar resonators" (in Japanese), Report of the Committee of Electro-Mechanical Functional Devices, Inst. of Elec. Eng. of Japan, July 18, 1970.

305) K. Yakuwa, S. Okuda, Y. Kasai, and Y. Katsuba, "Reliability of electromechanical filters," *Fujitsu*, vol. 24, no. 1, pp. 172-179, 1973.

306) K. Yakuwa, S. Okuda, and M. Yanagi, "Development of new channel bandpass filters," in *Proc. 1974 IEEE ISCAS*, San Francisco, Apr. 1974, pp. 100-105.

307) K. Yakuwa and S. Okuda, "Design of mechanical filters using resonators with minimized volume," in *Proc. 1976 IEEE ISCAS*, München, Apr. 1976, pp. 790-793.

308) T. Yano, T. Futami, and S. Kanazawa, "New torsional mode electromechanical channel filter," in *Proc. 1974 European Conf. on Circuit Theory and Design*, London, July 1974, pp. 121-126.

309) H. Yoda, "Quartz crystal mechanical filters," in *Proc. 13th Ann. Symp. Frequency Control*, May 1959.

310) T. Yuki, "Wire-coupled crystal mechanical filter," *IRE Trans. Comp. Parts*, vol. CP-9, pp. 89-95, Sept. 1962.

311) T. Yuki and T. Yano, "Electro-mechanical filters—Part 3: Low frequency electro-mechanical filters" (in Japanese), *J. Inst. Electron. Commun. Eng. Jap.*, vol. 52, pp. 727-732, June 1969.

312) W. Zochowski, "Development of the theory of a magnetostrictive transducer" (in Polish), *Rozpr. Elektrotech.*, vol. 9, pp. 373-394, 1963.

313) M. Wasiak, "Electromechanical channel filters for a telecommunication system with initial channel modulation" (in Polish), *Prace ITR*, vol. 54, no. 1, pp. 5-25, 1971.

314) M. Wasiak, "Electromechanical filters" (in Polish), *Postepy Elektroniki*, vol. 16, no. 2/3, pp. 65-73, 1971.

Crystal Filter Bibliography

1) R. Adler, "Compact electromechanical filter," *Electronics*, vol. 20, pp. 100-105, Apr. 1947.

2) S. W. Anderes, "A polylithic filter channel bank," *IEEE Trans. Commun.*, vol. COM-20, pp. 48-52, Feb. 1972.

3) T. Ashida, "Characteristic frequencies of monolithic filters," *Trans. IECE Japan*, vol. 54-A.6, p. 323, June 1971.

4) ——, "Eigenfrequencies of monolithic filters," *Electron. Commun. Japan*, vol. 54-A, no. 6, pp. 41-49, 1971.

5) ——, "Design and characteristic analysis of monolithic crystal filters," *Electron. Commun. Japan*, vol. 57-A, no. 5, pp. 1-9, 1974.

6) T. Ashida, N. Sawamoto, and N. Niizeki, "Temperature dependence of X-cut $LiTaO_3$ crystal resonator vibrating in thickness shear mode," *Rev. Elect. Commun. Lab. (Japan)*, vol. 18, pp. 854-861, 1970.

7) A. D. Ballato and T. Lukaszek, "A novel frequency selective device. The stacked-crystal filter," in *Proc. 27th Ann. Symp. on Frequency Control*, 1973, pp. 262-269.

8) A. Ballato *et al.*, "Design and fabrication of modern filter crystals," in *Proc. 20th Ann. Symp. on Frequency Control*, 1966, pp. 131-159.

9) A. D. Ballato and R. V. McKnight, "Frequency control development," *Phys. Today*, vol. 19, no. 8, 1966.

10) L. C. Barcus, "Nonlinear effects in the *AT*-cut quartz resonator," *IEEE Trans. Sonics Ultrason.*, vol. SU-22, no. 4, pp. 245-250, July 1975.

11) R. L. Barnes *et al.*, "Production and perfection of *r*-face quartz," in *Proc. 29th Ann. Symp. on Frequency Control*, 1975, pp. 98-105.

12) R. Bechmann, "Quartz AT-type filter crystals for the frequency range 0.7 to 60 MHz," *Proc. IRE*, vol. 49, 1961, pp. 523-524.

13) W. D. Beaver, "Theory and design of the monolithic crystal filter," in *Proc. 21st Ann. Symp. on Frequency Control*, 1967, pp. 179-199.

14) W. D. Beaver, "Theory and design principles of the monolithic crystal filter," Ph.D. dissertation, Lehigh University, Bethlehem, PA, 1967.

15) W. D. Beaver, "Analysis of elastically coupled piezoelectric resonators," *J. Acoust. Soc. Amer.*, vol. 43, pp. 972-981, May 1968.

16) E. Beck, E. Schultze, and H. Meyr, "Improved coupling formula for dual monolithic crystal filters," *Electron. Lett.*, vol. 12, pp. 494-496, Sept. 16, 1976.

17) E. Beck, E. Schultze, and H. Meyr, "An admittance approach to the dual monolithic crystal filter," in *Proc. IEEE Int. Symp. on Circuits and Systems*, Munich, Germany, 1976, pp. 316-319.

18) V. Belevitch and Y. Kamp, "Theory of monolithic crystal filters using thickness twist vibrations," *Philips Res. Rep.*, vol. 24, pp. 331-369, Aug. 1969.

19) R. E. Bennett, "Manufacturing guide for "*AT*" type units," Dep. Army, Union Thermoelectric Co., Niles, IL, PB 171839, 1960, pp. 77-97.

20) M. Bernstein, "Increased crystal resistance at oscillator noise levels," in *Proc. 21st Ann. Symp. on Frequency Control*, 1967, pp. 244-258.

21) H. Betzl, "Ein Beitrag zur berechnung von eingliedrigen quarzbrückenbandpassen mittlerer bandbreite nach der Betriebsparametertheorie," *Frequenz*, vol. 19, pp. 206-209, June 1975.

22) L. Bidart, "Semi-monolithic quartz crystal filters and monolithic filters," in *Proc. 25th Ann. Symp. on Frequency Control*, 1971, pp. 271-279.

23) G. W. Bleisch, "The *A*-6 channel bank," in *IEEE Int. Conf. Communications*, 1971, pp. 6-7-6-12.

24) G. W. Bleisch and W. P. Michaud, "The *A*-6 channel bank: Putting new technologies to work," *Bell Lab. Rec.*, vol. 49, no. 8, pp. 251-254, Sept. 1971.

25) J. L. Bleustein and H. F. Tiersten, "Forced thickness-shear vibrations of discontinuously plated piezoelectric plates," *J. Acoust. Soc. Amer.*, vol. 43, p. 1311, 1968.

26) G. Bosse and H. Matthes, "Quarzbandsperren für breite übertragungsbereiche" *Nachrichtentech. Z.*, vol. 2.17, pp. 515-519, 1964.

27) V. E. Bottom, *The Theory and Design of Quartz Crystal Units*. Abilene, TX: McMurray Press, 1968.

28) K. D. Bowers, "Ultrasonics in communications," *Bell Lab Rec.*, vol. 49, no. 5. pp. 139-145, May 1971.

29) E. Bücherl, "Bandstop filters incorporating double-tuned monolithic crystal units," *Siemens Forsch u Entwickl. Ber. Bd.*, vol. 2, no. 5, pp. 283-287, 1973.

30) T. W. Burgess, "Thickness twist anharmonic modes in lithium tantalate and lithium niobate plate resonators," *J. Phys. D.*, vol. 8, pp. 283-298, 1975.

31) J. W. Burgess *et al.*, "Single-mode resonance in LiNbO for filter design," *Electron. Lett.*, vol. 9, no. 11, pp. 251-252, 1973.

32) J. W. Burgess, M. C. Hales, and R. J. Porter, "Equivalent circuit parameters for lithium tantalate plate resonators," *Electron. Lett.*, vol. 11, no. 19, pp. 449-450, 1975.

33) ——, "Temperature coefficients of frequency in lithium niobate and lithium tantalate plate resonators," *Proc. IEE*, vol. 123, no. 6, pp. 499-504, 1976.

34) R. J. Byrne, P. Lloyd, and W. J. Spencer, "Thickness shear vibrations in rectangular *AT*-cut quartz plates with partial electrodes," *J. Acoust. Soc. Amer.*, vol. 43, pp. 232-239, 1968.

35) R. J. Byrne, "Monolithic crystal filters," in *Proc. 24th Ann. Symp. on Frequency Control*, 1970, pp. 84-92.

36) R. J. Byrne and J. L. Hokanson, "Effect of high-temperature processing on the aging behavior of precision 5 MHz quartz crystal units," *IEEE Trans. Instrum. Meas.*, vol. IM-17, pp. 76-79, Mar. 1968.

37) W. G. Cady, "Piezoelectricity, an introduction to the theory and applications of electromechanical phenomena in crystals," (two volumes), New York: Dover Publications, Inc., 1964.

38) H. F. Cawley *et al.*, "Manufacture of monolithic crystal filters for *A*-6 channel bank," in *Proc. 29th Ann. Symp. on Frequency Control*, May 1975, pp. 113-119.

39) E. Christian and E. Eisenmann, "Consideration for the design of crystal filters," in *Proc. 3rd Allerton Conf. Circuit System Theory*, 1965, pp. 806-816.

40) E. Christian and G. C. Temes, "On the Szentirmai transformation," *IEEE Trans. Circuit Theory*, vol. CT-13, pp. 450-452, Dec. 1966.

41) P. Costa, G. F. Piacentini, and G. Stacchioti, "*LC* conventional, electromechanical quartz and polylithic filters," *Telettra Tech. Inform. Bull.*, no. 27, pp. 3-22, Nov. 1975.

42) D. R. Curran and D. J. Koneval, "Energy trapping and the design of single- and multi-electrode filter crystals," in *Proc. 18th Ann. Symp. Frequency Control*, May 1964, pp. 93-119.

43) C. R. Dillon and L. F. Lind, "Cascade synthesis of polylithic crystal filters containing double-resonator monolithic crystal filter (MCF) elements," *IEEE Trans. Circuits and Systems*, vol. CAS-23, no. 3, pp. 146-154, Mar. 1976.

44) M. Dishal, "Design of dissipative bandpass filters producing desired exact amplitude-frequency characteristics," *Proc. IRE*, pp. 1050-1069, Sept. 1949.

45) H. Ekstein, "High frequency vibrations of thin crystal plates," *Phys. Rev.*, vol. 68, pp. 11-23, July 1945.

46) H. Ekstein, "Forced vibrations of piezoelectric crystals," *Phys. Rev.*, vol. 70, pp. 76-84, July 1946.

47) E. M. Frymoyer, R. A. Johnson, and F. H. Schindelbeck, "Passive filters: Today's offerings and tomorrow's promises," *EDN*, vol. 18, pp. 22-30, Oct. 5, 1973.

48) J. J. Gagnepain and R. Besson, "A study of quartz crystal nonlinearities: Application to *X*-cut resonators," *Phys. Lett.*, vol. 41A, pp. 443-444, Oct. 23, 1972.

49) J. L. Garrison, A. N. Georgiades, and H. A. Simpson, "The application of monolithic crystal filters to frequency selective networks," in *Digest of Tech. Papers 1970 Int. Symp. on Circuit Theory*, 1970, pp. 177-178.

50) E. A. Gerber and R. A. Sykes, "A quarter century of progress in the theory and development of crystals for frequency control and selection," in *Proc. 25th Ann. Symp. on Frequency Control*, 1971, pp. 1-45.

51) A. Glowinski and R. Lancon, "Resonance frequencies of monolithic quartz structures," in *Proc. 23rd Ann. Symp. on Frequency Control*, 1969, pp. 39-55.

52) A. Glowinski, R. Lancon, and R. Lefévre, "Effects of symmetry in trapped energy piezoelectric resonators," in *Proc. 27th Ann. Symp. on Frequency Control*, 1973, pp. 233-242.

53) R. P. Grenier, "A technique for automatic monolithic crystal filter frequency adjustment," in *Proc. 24th Ann. Symp. on Frequency Control*, 1970, pp. 104-110.

54) E. Hafner, "Crystal resonators," *IEEE Trans. Sonics Ultrason.*, vol. SU-21, pp. 220-237, 1974.

55) M. C. Hales and J. W. Burgess, "Design and construction of monolithic crystal filters using lithium tantalate," *Proc. IEE*, vol. 123, no. 7, pp. 657-661, July 1976.

56) D. L. Hammond, C. Adams, and L. Cutler, "Precision crystal units," in *Proc. 17th Ann. Symp. on Frequency Control*, 1963, pp. 215-232.

57) K. Haruta and W. J. Spencer, "X-ray diffraction study of vibrational modes," in *Proc. 20th Ann. Symp. on Frequency Control*, 1966, pp. 1-16.

58) R. A. Heising, *Quartz Crystals for Electrical Circuits*. New York: Van Nostrand, 1946.

59) W. Herzog, *Siebschaltungen mit Schwingkristallen*. Braunschweig, Germany: F. Vieweg, 1962.

60) J. L. Hokanson, "Laser machining thin film electrode arrays on quartz crystal substrates," in *Proc. 23rd Ann. Symp. on Frequency Control*, 1969.

61) J. L. Hokanson, "The monolithic crystal filter: The device, its operation and choice of piezoelectric materials," in *1969 IEEE Proc. 6th Annu. Integrated Circuits Seminar*, Hoboken, NJ, Apr. 1969, pp. 32-43.

62) R. Holland and E. P. Eer Nisse, *Design of Resonant Piezoelectric Devices*. Cambridge, MA: MIT Press, 1969.

63) W. H. Horton and R. C. Smythe, "On the trapped-wave criterion for *AT*-cut quartz resonators with coated electrodes," *Proc. IEEE*, pp. 598-599, Apr. 1967.

64) W. H. Horton and R. C. Smythe, "The work of Mortley and the energy-trapping theory for thickness-shear piezoelectric vibrators," *Proc. IEEE*, vol. 55, p. 222, Feb. 1967.

65) W. H. Horton and R. C. Smythe, "Experimental investigations of intermodulation in monolithic crystal filters," in *Proc. 27th Ann. Symp. on Frequency Control*, June 1973, pp. 243-245.

66) D. S. Humphreys, *The Analysis Design and Synthesis of Electrical Filters*. Englewood Cliffs, NJ: Prentice-Hall, 1970.

67) D. Indjoudjian and P. Andrieux, *Les Filtres a Cristaux Piezoelectriques*. Paris, France: Gauthier-Villars, 1953.

68) "IRE standards on piezoelectric crystals," in *Proc. Inst. Radio Eng.*, vol. 37, pp. 1378-1395, 1949.

69) H. Jumonji, N. Watanabe, and K. Tsukamoto, "Design of high performance monolithic crystal filters," in *Review of the Electrical Communication Laboratories. NTT Japan*, vol. 23, no. 5-6, pp. 439-452, May-June 1975.

70) H. Jumonji and M. Onoe, "Analysis of a trapped-energy filter with multiple modes by means of equivalent electrical transmission line theory," in *Joint Con. Rec. Electrical and Electronics Engrs. of Japan*, vol. 1233, 1969.

71) Y. S. Kim and R. T. Smith, "Thermal expansion of lithium tantalate and lithium niobate single crystals," *J. Appl. Phys.*, vol. 40, p. 4637, 1969.

72) J. E. Knowles, "On the origin of the second level of drive effect in quartz oscillators," in *Proc. 29th Ann. Symp. on Frequency Control*, pp. 230-237, 1975.

73) G. R. Kohlbacher, "The design of compact monolithic crystal filters for portable telecommunications equipment," in *Proc. 26th Ann. Symp. on Frequency Control*, pp. 187-193, 1972.

74) G. Kohlbacher, "Entwurf mehrkeisiger kristall und keramikfilter aus monolitischen einzelfilterelemten mit hilfe equivalenter elektrischer tiefpassersatzschaltungen," *A. E. U. Band 25 Heft-11*, pp. 492-501, 1971.

75) D. I. Kosowski, "Synthesis and realization of crystal filters," *MIT Res. Lab. of Electron.*, Cambridge, MA, Tech. Rep. 298, June 1955.

76) C. F. Kurth, "Analog and digital filtering in multiplex

communication systems," *IEEE Trans. Circuit Theory*, vol. CT-20, no. 4, pp. 408-415, July 1973.

77) J. Lang and C. E. Schmidt, "Crystal filter transformations," *IEEE Trans. Circuit Theory*, vol. CT-12, pp. 454-457, Sept. 1965.

78) M. S. Lee, "Equivalent network for bridged crystal filters," *Electron. Lett.*, vol. 10, no. 24, pp. 507-508, Nov. 1974.

79) P. C. Y. Lee and W. J. Spencer, "Shear-flexure-twist vibrations in rectangular *AT*-cut quartz plates with partial electrodes," *J. Acoust. Soc. Amer.*, vol. 45, no. 3, pp. 637-645, 1969.

80) M. S. Lee, "Polylithic crystal filters with attenuation peaks at finite frequencies," in *Proc. IEEE Int. Symp. on Circuits and Systems* (*Special Session S1*), Newton, MA, April 1975, pp. 297-300.

81) J. A. Lewis, "Effect of driving electrode shape on the electrical properties of piezoelectric crystals," *Bell Syst. Tech. J.*, vol. 40, pp. 1259-1280, Sept. 1961.

82) N. C. Lias and D. W. Rudd, "Growth of synthetic quartz for use in high frequency monolithic crystal filters," *Western Elec. Eng.*, vol. 13, p. 23, 1969.

83) P. Lloyd, "Equations governing the electric behaviour of an arbitrary piezoelectric resonator having *N* electrodes," *Bell Syst. Tech. J.*, vol. 46, pp. 1881-1900, 1967.

84) P. Lloyd, "Monolithic crystal filters for frequency division multiplex," in *Proc. 25th Ann. Symp. on Frequency Control*, 1971, pp. 280-286.

85) P. Lloyd and M. Redwood, "Finite-difference method for the investigation of the vibrations of solids and the evaluation of the equivalent-circuit characteristics of piezoelectric resonators," *J. Acoust. Soc. Amer.*, vol. 39, pp. 346-360, 1966.

86) T. J. Lukaszek, "Improvements of quartz filter crystals," in *Proc. 19th Ann. Symp. on Frequency Control*, 1965, pp. 269-296.

87) H. Mailer and D. R. Beuerle, "Incorporation of multiresonator crystals into filters for quantity production," in *Proc. 20th Ann. Symp. on Frequency Control*, 1966, pp. 309-342.

88) S. Malinowski and C. Smith, "Intermodulation in crystal filters," in *Proc. 26th Ann. Symp. on Frequency Control*, 1972, pp. 180-186.

89) Y. Masuda, I. Kawakami and M. Kobayashi, "Monolithic crystal filter with attenuation poles utilizing 2-dimensional arrangement of electrodes," in *Proc. 27th Ann. Symp. on Frequency Control*, 1973, pp. 227-232.

90) W. P. Mason, Ed., *Physical Acoustics*. New York: Academic Press, 1972.

91) W. P. Mason, "Equivalent electromechanical representation of trapped energy transducers," *Proc. IEEE*, vol. 57, pp. 1723-1734, Oct. 1969.

92) W. P. Mason, *Piezoelectric Crystals and Their Application to Ultrasonics*. New York: D. Van Nostrand, 1950.

93) W. P. Mason and R. A. Sykes, "Electric wave filters employing crystals with normal and divided electrodes," *Bell Syst. Tech. J.*, vol. 19, pp. 221-248, Apr. 1940.

94) W. P. Mason, *Electromechanical Transducers and Wave Filters*, 2nd ed., Princeton, NJ: D. Van Nostrand, 1948.

95) H. J. McSkimin, "Theoretical analysis of modes of vibration for isotropic rectangular plates having surfaces free," *Bell. Syst. Tech. J.*, vol. 23, p. 151, 1944.

96) A. J. Miller, "Preparation of quartz crystal plates for monolithic crystal filters," in *Proc. 24th Ann. Symp. on Frequency Control*, 1970, pp. 93-103.

97) R. D. Mindlin, "Forced thickness-shear and flexural vibrations of piezoelectric crystal plates," *J. Appl. Phys.*, vol. 23, pp. 83-88, 1952.

98) R. D. Mindlin, "High frequency vibrations of crystal plates," *Quart. Appl. Math.*, vol. 19, no. 51, 1961.

99) R. D. Mindlin and D. C. Gazis, "Strong resonances of rectangular *AT*-cut quartz plates," in *Proc. 4th U.S. Nat. Cong. Appl. Mech.*, 1962, pp. 305-310.

100) R. D. Mindlin, "Thickness-twist vibrations of an infinite, monoclinic, crystal plate," Columbia Univ., New York, NY, ONR Contract NONR 266(09) Tech. Rep. 52; USAERDL Contract DA-36-039-AMC00065 (E) Interim Tech. Rep. 1; June 1964.

101) R. D. Mindlin, "Bechmann's number for harmonic overtones of thickness-twist vibrations of rotated *Y*-cut quartz plates," *J. Acoust. Soc. Amer.*, vol. 41, p. 969, 1967.

102) R. D. Mindlin and M. A. Medick, "Extensional vibrations of elastic plate," *J. Appl. Mech.*, vol. 26, p. 561, 1959.

103) R. D. Mindlin and P. C. Y. Lee, "Thickness-shear and flexural vibrations of partially plated crystal plates," *Int. J. Solids Structures*, vol. 2, pp. 125-139, 1966.

104) R. D. Mindlin and W. J. Spencer, "Anharmonic, thickness/twist overtones of thickness/shear and flexural vibrations of rectangular, *AT*-cut quartz plates," *J. Acoust. Soc. Amer.*, vol. 42, pp. 1268-1277, 1967.

105) W. C. Morse and R. C. Rennick, "Adjusting frequency of monolithic crystal filters with an automatic vapor plater," *J. Vac. Sci. Technol.*, vol. 9, no. 1, p. 28, Jan. 1972.

106) W. S. Mortley, "Frequency-modulated quartz oscillators for broadcasting equipment," *Proc. IEE* (London), vol. 104B, pp. 239-253, Dec. 1957.

107) W. S. Mortley, "Priority in energy trapping," *Phys. Today*, vol. 19, no. 12, pp. 11-12, Dec. 1966.

108) W. S. Mortley, "Energy trapping," *Marconi Review*, xxx (165) 2nd quarter, 53, 1967.

109) T. Nagata, Y. Nakajima, and R. Sasaki, "$PbTiO_3$ ceramic resonators operating in VHF band," *Electron. and Commun. Japan*, vol. 55-C, no. 7, pp. 93-98, 1972.

110) K. Nakamura and H. Shimizu, "Equivalent circuit for thickness modes of elastic waves propagating in the plane of the piezoelectric plates," *Trans. IECE Japan*, vol. 55-A, pp. 95-102, Feb. 1972.

111) Y. Nakauchi, "A synthesis of parallel ladder circuits using equivalent transformation techniques," in *Proc. IEEE Int. Symp. on Circuits and Systems*, Newton, MA, 1975, pp. 305-308.

112) Y. Nakazawa, "High frequency crystal electromechanical filters," in *Proc. 16th Ann. Symp. on Frequency Control*, 1962, pp. 373-390.

113) S. Nonaka, T. Yuuki, and K. Hara, "The current dependence of crystal unit resistance at low drive level," in

Proc. 25th Ann. Symp. on Frequency Control, 1971, pp. 139-147.

114) R. J. Nunamaker, "Frequency control devices for mobile communications," in *Proc. 25th Ann. Symp. on Frequency Control*, 1971, pp. 74.

115) J. F. Nye, *Physical Properties of Crystals*. New York: Oxford, 1957.

116) S. H. Olster *et al.*, "A6 monolithic crystal filter, design for manufacture and device quality," in *Proc. 29th Ann. Symp. on Frequency Control*, 1975, pp. 105-112.

117) M. Onoe, Paper X111-70-437, The Papers of Barium Titanate Appl. Res. Comm., IECE Japan, pp. 113-114, July 1964.

118) ——, "Trapped energy mode filters having two-dimensional electrode pattern," 1972 Rep. Spring Meeting, Acoust. Soc. Japan.

119) M. Onoe and H. Jumonji, "Analysis of piezoelectric resonators vibrating in trapped-energy modes," *J. Electron. Commun. Eng.*, (Japan) vol. 48, pp. 84-93, 1965.

120) M. Onoe and N. Kobori, "The theory of coupling between laterally spaced energy-trapped resonators," Paper X111-70-438, The Papers of Barium Titanate Appl. Res. Comm., IECE Japan, pp. 115-116, July 1964.

121) ——, Paper X111-71-450, The Papers of Barium Titanate Appl. Res. Comm., IECE Japan, pp. 195-196, July 1964.

122) M. Onoe, H. Jumonji, and K. Kobori, "High frequency crystal filters employing multiple mode resonators vibrating in trapped energy modes," in *Proc. 20th Ann. Symp. on Frequency Control*, 1966, pp. 266-287.

123) M. Onoe and H. Jumonji, "Useful formulae for piezoelectric ceramic resonators and their application to measurement of parameters," *J. Acoust. Soc. Amer.*, vol. 41, pp. 974-980, 1967.

124) M. Onoe, T. Ashida, and K. Sawamoto, "Zero temperature coefficient of resonant frequency in *X*-cut lithium tantalate at room temperature," *Proc. IEEE*, vol. 57, pp. 14-46, 1969.

125) M. Onoe and K. Okada, "Analysis of contoured piezoelectric resonators vibrating in thickness-twist modes," in *Proc. 23rd Ann. Symp. on Frequency Control*, pp. 26-38, 1969.

126) H. J. Orchard and G. C. Temes, "Filter design using transformed variables," *IEEE Trans. Circuit Theory*, vol. CT-15, no. 4, pp. 385-408, Dec. 1968.

127) G. T. Pearman and R. C. Rennick, "Monolithic crystal filters," *IEEE Trans. Sonics Ultrason.*, vol. SU-21, no. 4, pp. 238-243, Oct. 1974.

128) W. Poschenrieder and F. Schöfer, "Das elektromechanische quarzfilter-ein neues bauelement fur die nachrichtentechnik," *Frequenz*, vol. 17, pp. 88-94, Mar. 1963.

129) Y. Rainsard, "Sur la suppression due transformateur differentiel utilise dans les structures de jaumann," *Cables and Transmission*, vol. 21, no. 4, pp. 226-237, 1967.

130) A. Reading and R. J. Martin, "A new modulation concept in channel translating equipment," in *IEEE Int. Conf. on Communications*, Montreal, Que., Canada, June 1971, pp. 6.13-6.18.

131) M. Redwood and N. H. C. Reilly, "New method of providing coupling between resonators in an electromechanical filter," *Electron. Lett.*, vol. 2, pp. 220-222, June, 1966.

132) M. Redwood and N. H. C. Reilly, "Theoretical analysis of the monolithic crystal filter," *Electron. Lett.*, vol. 3, pp. 257-258, 1967.

133) N. H. C. Reilly and M. Redwood, "Wave propagation analysis of the monolithic crystal filter," *Proc. IEE*, vol. 116, no. 5, pp. 653-659, May 1969.

134) R. C. Rennick, "An equivalent circuit approach to the design and analysis of monolithic crystal filters," *IEEE Trans. Sonics Ultrason.*, vol. SU-20, no. 4, pp. 347-354, Oct. 1973.

135) R. C. Rennick, "Modeling and tuning methods for monolithic crystal filters," in *Proc. 1975 IEEE Int. Symp. on Circuits and Systems*, 1975, pp. 309-312.

136) D. A. Roberts, "*Cds*-quartz monolithic filters for the 100-500 MHz frequency range," in *Proc. 25th Ann. Symp. on Frequency Control*, 1971, pp. 251-261.

137) G. E. Roberts, "The design of coupled resonator *AT*-cut quartz crystals for operation on the third thickness-shear overtone," *Proc. IEEE*, pp. 1527-1528, Oct. 1975.

138) Sasaki, Nagata and Matsushita, "Piezoelectric resonators as a solution to frequency selectivity in color TV receivers," *IEEE Trans. Broadcast Telev. Receivers*, vol. BTR-17, no. 3, p. 195, 1971.

139) G. Sauerbrey, "Amplituden verteilung und elektrische ersatzdaten von schwing quarz platten," (*AT-Schnitt*), *A. E. U.*, vol. 18, no. 10, pp. 624-628, 1964.

140) K. Sawamoto, "Energy trapping in a lithium tantalate *X*-cut resonator," in *Proc. 25th Ann. Symp. on Frequency Control*, 1971, pp. 246-250.

141) P. Schnabel, "Frequency equations for *n* mechanically coupled piezoelectric resonators," *Acustica*, vol. 21, p. 351, 1969.

142) H. Schumacher, "Piezokeramische hybridfilter fur AM und FM rundfunk," *Stemag-Nachrechten H.*, vol. 43, pp. 1163-1170, 1970.

143) H. Schüssler, "Ceramic filters and resonators," *IEEE Trans. Sonics Ultrason.*, vol. SU-21, no. 4, pp. 257-268, Oct. 1974.

144) H. Schüssler, "Filters for channel bank filtering," in *Proc. 1974 IEEE Int. Symp. Circuits Systems*, Apr. 1974, pp. 106-110.

145) A. Seed, "High *Q*, *BT*-cut quartz resonator units," *Brit. J. Appl. Phys.*, vol. 16, no. 9, pp. 1341-1346, 1965.

146) "Session on channel bank filtering," in *Proc. 1974 IEEE Int. Symp. Circuits Systems*, Apr. 1974.

147) E. A. G. Shaw, "On the resonant vibrations of thick barium titanate disks," *J. Acoust. Soc. Amer.*, vol. 28, no. 1, p. 38, 1956.

148) D. F. Sheahan, "An improved resonance equation for *AT*-cut quartz crystals," *Proc. IEEE*, vol. 58, pp. 260-261, Feb. 1970.

149) D. F. Sheahan, "Single sideband filters for short haul systems," in *Proc. Int. IEEE Conv. on Systems, Networks, and Computers*, Oaxtepec, MX, Jan. 1971, pp. 744-748.

150) D. F. Sheahan, "Channel bank filtering at GTE Lenkurt," in *Proc. 1974 IEEE Int. Symp. Circuits Systems*, April

1974, pp. 115-120.

151) D. F. Sheahan, "Polylithic crystal filters," in *Proc. 29th Ann. Symp. on Frequency Control*, 1975, pp. 120-123.

152) D. F. Sheahan and R. A. Johnson, "Crystal and mechanical filters," *IEEE Trans. Circuits and Systems*, vol. CAS-22, no. 2, pp. 69-89, Feb. 1975.

153) D. F. Sheahan and C. E. Schmidt, "Coupled resonator quartz crystal filters," *WESCON*, (Session 8), Aug. 1971.

154) W. Schockley, D. R. Curran, and D. J. Koneval, "Energy trapping and related studies of multiple electrode filter crystals," in *Proc. 17th Ann. Symp. on Frequency Control*, 1963, pp. 88-126.

155) W. Shockley, D. R. Curran, and D. J. Koneval, "Trapped-energy modes in quartz-filter crystals," *J. Acoust. Soc. Amer.*, vol. 41, pp. 981-993, 1966.

156) H. A. Simpson, E. D. Finch, and R. K. Weeman, "Composite filter structure incorporating monolithic crystal filters and *LC* networks," in *Proc. 25th Ann. Symp. on Frequency Control*, 1971, pp. 287-296.

157) W. L. Smith, "The application of piezoelectric coupled resonator devices to communications systems," in *Proc. 22nd Ann. Symp. on Frequency Control*, 1968, pp. 206-225.

158) R. T. Smith and F. S. Welsh, "Temperature dependence of the electric, piezoelectric and dielectric constants of lithium tantalate and lithium niobate," *J. Appl. Phys.*, vol. 42, pp. 2219-2230, May 1971.

159) R. C. Smythe, "HF and VHF inductorless filters for microelectronic systems," in *Proc. IEEE/EIA Electron. Components Conf.*, 1969, pp. 115-119.

160) R. C. Smythe, "Intermodulation in thickness shear resonators," in *Proc. 28th Ann. Symp. on Frequency Control*, 1974, pp. 5-7.

161) W. J. Spencer, "Modes in circular quartz plates," *IEEE Trans. Sonics Ultrason.*, vol. SU-12, no. 1, pp. 1-5, 1965.

162) W. J. Spencer, "Transverse thickness modes in *BT*-cut quartz plates," *J. Acoust. Soc. Amer.*, vol. 41, pp. 994-1001, Apr. 1967.

163) W. J. Spencer, "Monolithic crystal filters," presented at the 3rd Asilomar Conf., Dec. 1969.

164) W. J. Spencer, "Monolithic crystal filters," in *Physical Acoustics*, vol. IX, W. P. Mason and R. N. Thurston, Eds. New York: Academic Press, 1972.

165) W. J. Spencer and R. M. Hunt, "Coupled thickness shear and flexure displacements in rectangular *AT* quartz plates," *J. Acoust. Soc. Amer.* vol. 39, no. 5, part 1, pp. 929-935, 1966.

166) "Standards on Piezoelectric Crystals," *Proc. IRE*, vol. 37, pp. 1378-1394, 1949.

167) J. K. Stevenson and M. Redwood, "The motional reactance of a piezoelectric resonator—a more accurate and simple representation for use in filter design," *IEEE Trans. Circuit Theory*, vol. 16, pp. 568-572, 1969.

168) J. A. Stewart, "Systems considerations for light route multiplex using crystal filters," in *IEEE Int. Conf. Commun.*, Montreal, Que., Canada, June 1971, pp. 6.25-6.29.

169) W. Stoddard," Design equation for plano convex *AT* filter crystals," in *Proc. 17th Ann. Symp. on Frequency Control*, 1963, pp. 272-282.

170) R. A. Sykes, "Modes of motion in quartz crystals, the effects of coupling and methods of design," *Bell Syst. Tech. J.*, vol. 23, no. 52 1944.

171) R. A. Sykes, "New approach to the design of high frequency crystal filters," *IRE Nat. Con. Rec.*, vol. 6, part 2, pp. 18-29, 1958.

172) R. A. Sykes and W. D. Beaver, "High-frequency monolithic crystal filters with possible application to single frequency and single sideband use," in *Proc. 20th Ann. Symp. Frequency Control*, Apr. 1966, pp. 288-308.

173) R. A. Sykes, W. L. Smith, and W. J. Spencer, "Monolithic crystal filters," in *IEEE Int. Con. Rec.*, vol. 15, part II, pp. 78-93, 1967.

174) R. A. Sykes and W. L. Smith, "A monolithic crystal filter," *Bell Lab. Rec.*, pp. 52-54, Feb. 1968.

175) G. Szentirmai, "Crystal and ceramic filters," in *Modern Filter Theory and Design*, G. C. Temes and S. K. Mitra, Eds. New York: Wiley, 1973.

176) Tanaka and Shimizu, "On the energy-trapping condition of the thickness modes along a piezoelectric ceramic plate," 1970 Rep. Autumn Meeting, Acoust. Soc. of Japan, vol. 1-2-5, p. 37, 1970.

177) Tanaka and Shimizu, "Thickness vibrations traveling along a piezoelectric ceramic plate," *Papers of Technical Group on Ultrasonics, IECE Japan*, vol. US 71-3, 1971.

178) H. F. Tiersten, "Wave propagation in an infinite piezoelectric plate," *J. Acoust. Soc. Amer.*, vol. 35, pp. 234-239, Feb. 1963.

179) H. F. Tiersten, *Linear Piezoelectric Plate Vibrations.* New York: Plenum Press, 1969.

180) H. F. Tiersten, "Analysis of intermodulation in rotated *Y*-cut quartz thickness-shear resonators," in *Proc. 28th Ann. Symp. on Frequency Control*, 1974, pp. 1-4.

181) H. F. Tiersten, "Analysis of trapped energy resonators operating in overtones of thickness shear," in *Proc. 28th Ann. Symp. on Frequency Control*, 1974, pp. 44-48.

182) H. F. Tiersten, "Analysis of intermodulation in thickness-shear and trapped energy resonators," *J. Acoust. Soc. Amer.*, vol. 57, pp. 667-681, Mar. 1975.

183) H. F. Tiersten, "Analysis of nonlinear resonance in rotated *Y*-cut quartz thickness-shear resonators," in *Proc. 29th Ann. Symp. on Frequency Control*, 1975, pp. 49-53.

184) H. F. Tiersten, "Analysis of trapped energy resonators operating in overtones of coupled thickness shear and thickness twist," in *Proc. 29th Ann. Symp. on Frequency Control*, 1975, pp. 71-75.

185) H. F. Tiersten and R. D. Mindlin, "Forced vibrations of piezoelectric crystal plates, *Quart. Appl. Math.*, vol. 20, pp. 107-119, 1962.

186) H. Topolevsky and J. W. Burgess, "Interresonator coupling for overtone monolithic quartz filters," *Electron. Lett.*, vol. 11, pp. 476-477, Nov. 14, 1974.

187) R. B. Topolevsky and M. Redwood, "A general perturbation theory for elastic resonators and its application to the monolithic crystal filter," *IEEE Trans. Sonics Ultrason.*, vol. SU-22, no. 3, pp. 152-161, May 1975.

188) R. B. Topolevsky and M. Redwood, "The electrical

characteristics of symmetric and asymmetric monolithic crystal filters: A new analytical approach," *IEEE Trans. Sonics Ultrason.*, vol. SU-22, no. 3 pp. 162–167, May 1975.

189) T. Uno, "200 MHz thickness extensional mode $LiTaO_3$ monolithic crystal filter," *IEEE Trans. Sonics Ultrason.*, vol. SU-22, no. 3, pp. 168–174, May 1975.

190) C. M. van der Burgt, "Performance of ceramic ferrite resonators as transducers and filter elements," *J. Acoust. Soc. Amer.*, vol. 28, pp. 1020–1032, Nov. 1956.

191) A. W. Warner, "Design and performance of ultraprecise 2.5-Mc quartz crystal units," *Bell Syst. Tech. J.*, vol. 39, pp. 1193-1217, Sept. 1960.

192) ——, "New piezoelectric materials," in *Proc. 19th Ann. Symp. on Frequency Control*, 1965, pp. 5–21.

193) A. W. Warner and A. A. Ballman, "Low temperature coefficient of frequency in a lithium tantalate resonator," *Proc. IEEE*, vol. 55, pp. 450–451, 1967.

194) A. W. Warner, D. B. Fraser, and C. D. Stockbridge, "Fundamental studies of aging in quartz resonators," *IEEE Trans. Sonics Ultrason.*, vol. 12, pp. 52–59, June 1965.

195) A. W. Warner, M. Onoe, and G. A. Coquin, "Determination of elastic and piezoelectric constants for crystals in class (3m)," *J. Acoust. Soc. Amer.*, vol. 42, pp. 1223–1231, June 1968.

196) N. Watanabe and K. Tsukamoto, "High performance monolithic crystal filters with stripe electrodes," *Electron. Commun. Japan*, vol. 57-A, no. 4, pp. 53–60, 1975.

197) J. F. Werner, "The bilithic quartz-crystal filter," *J. Sci. Technol.*, vol. 38, no. 2, pp. 74-82, 1971.

198) J. F. Werner, A. J. Dyer, and J. Birch, "The development of high performance filters using acoustically coupled resonators on AT-cut quartz crystals," in *Proc. 23rd Ann. Symp. on Frequency Control*, May 1969, pp. 65–75.

199) T. Yamada and N. Niizeki, "Formation of admittance for parallel field excitation of piezoelectric plates," *J. Appl. Phys.*, vol. 41, pp. 3604-3609, Aug. 1970.

200) H. K. H. Yee, "Finite pole frequencies in monolithic crystal filters," *IEE Proc.*, vol. 59, pp. 88-89, Jan. 1971.

201) H. Yoda, "Quartz crystal mechanical filters," in *Proc. 13th Ann. Symp. on Frequency Control*, May, 1959.

202) H. Yoda, Y. Nakazawa, S. Okano, and N. Kobori, "High frequency crystal mechanical filters," in *Proc. 22nd Ann. Symp. on Frequency Control*, 1968, pp. 188-205.

203) H. Yoda, Y. Nakazawa, and N. Kobori, "High frequency crystal monolithic (HCM) filters," in *Proc. 23rd Ann. Symp. on Frequency Control*, 1969, pp. 76–92.

204) ——, "High frequency crystal monolithic (HCM) filters," in *Proc. 26th Ann. Symp. on Frequency Control*, 1972, pp. 76–93.

205) J. Zelenka, "The influence of material constants on the parameters of high frequency monolithic crystal filters," in *Proceedings Summer School on Circuit Theory*. Partizan Tăle hotel Czechoslovakia, 1971, pp. 12-1—12-11.

206) A. I. Zverev, *Handbook of Filter Synthesis*. New York: Wiley, 1967.

Author Index

Subject Index

688811

Editors' Biographies

Desmond F. Sheahan (M'61) was born in Boyle, Ireland, in 1937. He received the B.E. degree from the National University of Ireland, Dublin, in 1958, and the M.S. and Ph.D. degrees in electrical engineering from Stanford University, Stanford, CA, in 1963 and 1968, respectively.

Between 1958 and 1961 he was employed by Standard Telephones and Cables Ltd., London, England. In 1961 he joined GTE Lenkurt, San Carlos, CA, where he was involved in circuit design for microwave radio systems. From 1966 to 1969 he worked on techniques for building active telephone channel filters. In 1969 he started work on crystal filters, which culminated in the development of the polylithic crystal filter technique and the manufacturing processes required to produce them. He is currently Chief Engineer for Components and Processes at GTE Lenkurt. He holds a number of patents on active filters and on crystal filters.

Dr. Sheahan served as a member of the Administrative Committee of the IEEE Circuits and Systems Society from 1975 to 1977, and he was also Associate Editor for Filter Design of the IEEE TRANSACTIONS ON CIRCUITS AND SYSTEMS. Because of his invention of the polylithic crystal filter concept, he was the recipient of the GTE Warner Award in 1976. He is a member of Sigma Xi.

Robert A. Johnson (M'61) was born in Chicago, IL, on September 27, 1932. He received the B.S. and M.S. degrees in engineering from the University of California, Los Angeles, in 1955 and 1963, respectively.

Since joining the Collins Radio Group of Rockwell International in 1957, his work has been centered about the design and development of electromechanical filters. This work has included the design of single and multiple resonant element mechanical filters, transducers, computer design aids, as well as conventional crystal and monolithic filters. He is presently Technology Coordinator for the Mechanical Filter Design and Development Group.